Indian Herbal Remedies

Springer
Berlin
Heidelberg
New York
Hong Kong
London
Milan
Paris
Tokyo

C. P. Khare (Ed.)

Indian Herbal Remedies

Rational Western Therapy, Ayurvedic and Other Traditional Usage, Botany

With 255 Figures

Springer

C.P. KHARE
Founder President
Society for New Age Herbals
B-1/211, Jank Puri
New Delhi-110058
India
E-mail: chandrma_@yahoo.com

ISBN 3-540-01026-2 Springer-Verlag Berlin Heidelberg New York

Cataloging-in-Publication Data applied for

A catalog record for this book is available from the Library of Congress

Bibliographic information published by Die Deutsche Bibliothek
Die Deutsche Bibliothek lists this publication in the Deutsche Nationalbibliografie; detailed bibliographic data is available in the Internet at http://dnb.ddb.de

Springer-Verlag Berlin Heidelberg New York
a member of BertelsmannSpringer Science+Business Media GmbH

http://www.springer.de

This title is also available as *Encyclopedia of Indian Medicinal Plants* for sale in South Asia.

Production Editor and Typesetting: Frank Krabbes, Heidelberg; wiskom e.K., Friedrichshafen
Cover design: Erich Kirchner, Heidelberg

Dedicated to the pioneers of
scientific herbal healthcare.

Foreword

The guiding force behind this project is Col. Sir Ram Nath Chopra, the father of Modern Indian Herbal, who gave us the first scientific glossary of Indian medicinal plants 45 years ago. We, at the Society of New Age Herbals, got trapped in the labyrinth of the vast and scattered data, during a long-drawn-out phase of literary research, but ultimately succeeded in unveiling and highlighting the potential of Indian herbs on a scientific platform for our own scientists and for the West, where phytotherapy is emerging as a movement to be reckoned with.

Each monograph carries the family of the plant, scientific name, followed by authentic Ayurvedic, Unani, Siddha and common English nomenclature. A few plants, which are not of Indian origin but are used in Unani medicine, have also been incorporated.

Classical uses and therapeutic coverage of the plant drugs have been documented on the basis of their long-term application from 1000 BC to the 18th century, weeding out the bane of indication pluralism. (For classical treatises cited in the text see Appendix I.).

Classical jargon of *Vata*, *Pitta* and *Kapha* (*Tridosha*) and vague assumptions of a drug's properties on the weighing scale of *Rasa* (taste), *Guna* (physicial properties; cold, hot, dry, sharp, etc., divided into twenty segments), *Veerya* (potency or therapeutic efficacy) and *Vipaaka* (taste after ingestion), have been discarded.

The relationship between *Rasa* and *Tridosha* was made a basis of the medicinal property of herbs in the Database series on medicinal plants used in Ayurveda, published by The Central Council for Research in Ayurveda & Siddha (CCRAS). It is surprising that the Indian Council of Medical Research also attributed medicinal properties to the plant drugs on the basis of *Rasa, Guna, Veerya, Vipaaka*. Even scientists like Asima Chatterjee and Satyesh Chandra Pakrashi described the plant drug action on the same basis. In addition, as done by the CCRAS, they quoted Sanskrit verses (*slokas*) from the texts of the medieval period, to add a cosmetic classical touch to the scientific text of the Treatise of Indian Medicinal Plants. But they (and CCRAS) faltered when they cited Sanskrit verses from *Dravyaguna* and *Priyanighantu*, composed recently by a former professor of Banaras Hindu University, Dr Priyavrata Sharma.

The Indian National Science Academy (INSA) was the only exception where scientists rose above the traditional approach, discarded the age-old, predetermined format, and presented literary research in a scientific environment. The basic approach of INSA has been followed while documenting classical uses.

In my earlier book, "Indian Herbal Therapies—Application of Research Findings," processing techniques of herbs were discussed at length in the chapter "Harnessing the Herbal Medicine through Microcosm." Suggestions given therein for converting the herb into a potent medicine (by triturating it in its own decoction) need to be considered globally.

In India, we are still living in the age of *Bhaishajya Ratnaavali*, while, globally, an era of single standardized herbs has been ushered in. Sooner or later, Indian classical compound formulations will have to be phased out. Exponents of Indian medicine can no longer undermine the importance of active principles of the herb in the name of "pure herbal legacy." Active principles of a herb give a clear picture of the herb's therapeutic properties and toxicity, and we find ourselves in a better position to judge the balancing potential of the total herb. Also, it will be easy for scientists to select or reject herbs which have been used in various compounds for ages. (Col. Ram Nath Chopra recommended that after systematic investigations, drugs of questionable and doubtful value must be excluded.)

In India, there is a paucity of literature dealing with the biological activities of chemical constituents extracted from the plant. In the West, scientists like V. E. Tyler, Michael T. Murray, Simon Y. Mills and Donald J. Brown have done a commendable job in this field. In India, we had to depend mostly on the Wealth of India, monographs published by CCRAS, to some extent on the Compendium of Indian Medicinal Plants series of Central Drug Research Institute (CDRI), and material available at Central Institute of Medicinal and Aromatic Plants (CIMAP).

Scientific findings on the interaction of a herb, when given simultaneously with other drugs of modern medicine, could hardly be provided in the monographs.

In this context, a few examples of commonly used herbs will elucidate the importance of this specific area:

Ananas comosus (pineapple), the source of the enzyme bromelain, may not behave as expected when taken with an allopathic drug. An increased tendency for bleeding may occur in the case of simultaneous administration of anticoagulants. The levels of tetracyclenes in plasma and urine are increased by simultaneous intake of bromelain.

Garlic consumption substantially increases anticoagulant effects of warfarin. Blood-clotting times have been reported to double in patients taking warfarin and garlic supplements.

In the case of liquorice, potassium loss due to other drugs, e.g., thiazide diuretics, can be increased. With potassium loss, sensitivity to digitalis glycosides increases.

Senna leaf is usually prescribed in occasional constipation. In case of chronic use or abuse, loss of potassium may potentiate cardiac glycosides and have an effect on antiarrhythmic medications. Potassium deficiency may be exacerbated by simultaneous administration of thiazide diuretics, corticoadrenal steroids, or liquorice root.

These findings have been quoted from Herbal Medicine—Expanded Commission E Monographs. An independent chapter is devoted to this subject in PDR (Physicians' Desk Reference).

This area should be taken up seriously for providing authentic information to the practitioners of herbal as well as modern medicine.

While discussing the use of Indian medical plants in Western herbal, (German) Commission E and World Health Organization (WHO) monographs have been quoted.

In "Indian Herbal Therapies" I cited references within the text as it was addressed to the researchers, while in the present work I have followed the basic style of The Complete German Commission E Monographs. It is an attempt to make the text simple, straightforward and easy to follow.

The sources of material have been enlisted at the end as Annexure II. I recommend that herbal researchers should consult these reference books and journals for specific references and further studies. At many places within the text, important leads have been provided.

Finally, I would like to acknowledge the services rendered by my colleagues at the Society for New Age Herbals. Dr V. K. Agarwal, Senior Scientist, National Institute of Science Communication and the Wealth of India; Dr Prem Kishore, former Director, Central Council for Research in Ayurveda and Siddha; and Prof (Hakeem) Anwar Ahmad, an authority on Unani medicine, took time out of their busy schedules and reviewed the monographs.

My research assistant, Sarita Joshi, remained associated with the project since its inception. Her devotion, perseverance and patience went a long way in completing this project.

I am trying to move forward with the hope that this exercise will strike the right balance between the so-called holistic and scriptural Indian medicine and scientific phytomedicine, and will facilitate the entry of Indian herbs into the arena of modern herbal.

New Delhi. C. P. KHARE

A

Pinaceae

ABIES

Figure 1 Abies smithiana (Pinaceae) [WOI]

Abies spectabilis (D. Don) Spach.
Abies webbiana Lindl.
Pinus webbiana Wall.

Habitat
Himalayas from Kashmir to Assam at altitudes of 1600–4500 m.

Classical & common names
Ayurvedic: Taalisa, Taalisapatra, Dhaatriparni, Shukodara. A related variety: Granthikaapatra, Patraaddhya, Tulasichhada. (Ayurveda Saukhyam.)
Unani: Taalispattar, Zarnab (Arabic), Taalisfar (Arabic).
Siddha: Taalispatri.
English: Himalayan Silver Fir.

Parts used
Leaves (needles), bark, oil.

Dose
Powder 2–5 g.

Classical use
The powdered leaves (needles) were used by Charaka internally for cough, anaemia, digestive disorders, colic pain. The powdered leaves were given along with the juice of Adhatoda vasica or honey in cough, bronchial asthma, hoarseness of voice and intrinsic haemorrhage. (Vrindamaadhava, Haarita Samhitaa.)

The powdered leaves 1 part, Piper nigrum (Maricha) 2 parts, dried Zingiber officinale (Shunthi) 3 parts, Piper longum (Pippali) 4 parts, Bamboo-manna (Vansalochana) 5 parts, Greater cardamom (Sthula-elaa) 1/2 part, Cinnamomum zeylanicum (Tvak) 1/2 part and purified sugar 32 parts—this compound forms the reputed Taalisaadya Churna of Shaarangadhara Samhitaa, which is still being prescribed for cough and bronchitis, asthma and fever (due to throat or chest infection).

Taalisapatra is also an ingredient in Vyoshaadi Vati (Shaarangadhara Samhitaa), indicated in cough, cold, bronchitis and indigestion.

In Siddha medicine, Taalisaadi Choornam (Agaththiyar Eraththina Churukkam) is prescribed for gastritis, colic, distention of stomach, indigestion and cough. The compound consists of 24 herbs (while in Ayurvedic compound there are only 8 herbs).

Thippli Rasaayanam (Agaththiyar Paripoornam) of Siddha medicine, is prescribed in cough, bronchitis, asthma, also contains Taalisapatri.

In folk medicine, the juice of fresh leaves is given to infants suffering from chest infections and fever; also during dentation.

In West Bengal, the juice of fresh leaves is given as a tonic after parturition.

The resin, mixed with oil of roses, is applied externally in neuralgia.

Active principles and pharmacology

The leaves contain a bioflavonoid, abiestin, and two glycosides, methylbetuloside and betuloside; n-triacontanol and beta-sitosterol. They also contain an essential oil (0.5 %). The major constituents of the oil are: alpha-pinene, beta-pinene, camphane, limonene, bornyl acetate and carvone.

Bioflavonoid contents maintain normal conditions in the wall of small blood vessels and act as a stabilizing and calming factor in the peripheral circulation. Synergistic action exhibits anti-inflammatory, antiseptic, antispasmodic and diuretic properties. Thus, the leaves are effectively used as a carminative, astringent, antipyretic in asthma and bronchitis; also in catarrh of bladder.

Traditionally claimed antifertility activity of the leaves has not yet been scientifically established. According to the CCRAS, both benzene and alcoholic extract of leaf powder produced 51 % anti-implantation activity experimentally. The extracts had no effect on post-implantation period, oestrus cycle and teratogenicity (Jamnagar unit, 1996).

Use in Western herbal

The genus Pinus comprises about 95 species in the West. Used as a traditional remedy in coughs and colds, as expectorant and demulcent and also as a valuable remedy in bladder, kidney and rheumatic affections.

The pine oil, distilled from needles of pine spp. is included in a number of decongestant and expectorant mixtures and lozenges for cough and cold, available over the counter.

The oil is topically antiseptic and rubefacient, and is used in a variety of external rubs for muscular stiffness and rheumatism. Pine oil is added to steam inhalations for colds and catarrh.

The important addition is a derivative from the pine bark, known as pycnogenol, which is being marketed in the US since 1992 as an antioxidant. An average of 2 to 4 capsules (or 60 to 120 mg in divided doses) a day is being recommended for a number of ailments: sinusitis, bronchitis, elevated cholesterol, fatigue, fever, free radical activity, variose veins, to name a few—wherever antioxidants are indicated. (Pycnogenol is a registered trademark in the United States for a bioflavonoid, proanthocyanidin, extracted from the bark of the French maritime pine tree.)

As a folk medicine in the West, the ashes of burnt pine boughs are sprinkled over the skin of young children afflicted with measles, mumps and chickenpox. It is claimed that the itching soon ceased and the rash cleared up of its own accord.

Caution

Taxus baccata Linn. (Taxaceae) should not be accepted as a substitute for Abies webbiana, as advised in The Ayurvedic Formulary of India, Part I. Its constituents are a mixture alkaloids known as taxine and also diterpenes (incl. taxol in some varieties), lignans, tannin and resin. Not in use in herbal medicine and is currently under research as an anticancer drug of modern medicine. Extreme toxicity makes it unsafe as a herbal medicine.

Zarnab (Arabic), Budul, Bhrangi, Chilaa, Chiliraadh are doubtful synonyms of Taalispatra. Zarnab has been equated with Salix aegyptiaca Sprengel. National Formulary of Unani Medicine has equated Telispattar with Zarnab and Zarnab with Flacourtia cataphracta Roxb., also with Cinnamomum tamala Nees. Taalisfar (Arabic), known as Taalispatra in Punjab and Nepal, has not yet been equated scientifically with Abies spp. Another classical name Sthauneyaka of Sushruta Samhitaa has now been identified as Clerodendrum infortunatum Linn. by INSA scientists. Birmi, Thuno, Thuni, Thuner have been equated with Taxus baccata.

Abies webbiana is not to be confused with Cinnamomum tamala; known as Taalishpattri (Tamil) and Taalispatri (Telugu) or with Flacourtia cataphracta/Flacourtia jangomas, known as Taalisa and Taalispatri.

In Punjab, Rhododendron anthopogon D. Don. or Rhododendron campanulatum D. Don. is used as Taalispatra and is locally known as Taalisfar. But the leaves possess absolutely different properties. They are administered as errhine to produce sneezing.

In classical texts, Swarnataali was mentioned as a substitute for Taalisa. But Swarnataali could not yet be identified.

Related species

Abies alba Mill., syn. Abies pectinata DC. is also used in catarrh, cough and asthma. The yellowish needles showed high concentration of flavanols, whereas browning needles contain very low amounts of flavanols.

The needles contain an essential oil (0.55 %–0.82 %). The main components of the oil are limonene and phellandrene. The needles of a sample from Netherlands contained glycoside bound volatile compounds.

Two lectins (glycoproteins) AAA1 and AAA2 have been isolated from the bark. They showed agglutinating activity on human blood group A, B and 0. The fatty acid in the bark are palmitic, linoleic, linolenic, oleic, and 14-methylhexadecanoic acid. The mixture of diterpenic acids contained mainly abietic, dehydroabietic, neoabietic, livopimaric, palustric, isopimaric, sandaracopimaric and pimaric acid.

The resin yields 19 mono- and 24 sesquiterpenoids.

In the West, the essential oil is obtained from the fresh needles and twig tops of branches of Abies alba, Abies sachalinensis, Abies sibirica or Picea abies (syn. Picea excelsa).

Abies alba contains bornyl acetate 2–10 %, Picea abies 5–25 %; limonene 25–55 %, and 10–30 %; camphene 9–20 % and 10–25 %; alpha-pinene 6–35 % and 10–25 % respectively.

German Commission E monograph recognized the Scotch Pine's efficacy in the following areas:

- Common cold
- Cough/bronchitis
- Fevers and cold
- Inflammation of the pharynx
- Muscular and nerve pains.

The oil, as well as fresh fir shoots, are used in medicine.

The needles and young buds of Pinus sylvestris (Scotch Pine), Pinus pinaster, Pinus pinea and Pinus nigra are used in cases of bronchitis, sinusitis and upper respiratory catarrh; mostly, 1–2 ml of tincture 3 times a day is prescribed.

According to researchers, various species of Pines possess common medicinal properties (M. Grieve).

Papilionaceae

ABRUS

Abrus precatorius Linn.

Figure 1 Abrus precatorius [CCRAS]

Habitat

Found throughout India, ascending to an altitude of about 1050 m in the outer Himalayas. Red seed variety is similar to the white seed one anatomically and phytochemically.

Figure 2 Abrus precatorius [CCRAS]

Classical & common names

Ayurvedic: Gunjaa, Chirihintikaa (Sushruta), Gunjaka, Raktikaa, Kakanti, Kaakchinchi, Kamboji, Kubjaka, Chirmiti.
Unani: Ghungchi, Ghumchi, A'in-ud-deek, Chashm-e-khuroos, Ratti.
Siddha: Kundrimari.
English: Jequirity. (Other synonyms—Indian root or wild liquorice root are misnomers.)

Parts used

Leaves, roots, detoxified seeds.

Dose

Seed powder 60mg–150 mg.

Classical use

The herb has been classified as a sub-poison. Seeds were used only after they have been boiled in cow's milk for 3 hours.

The herb was used for sciatica, stiffness of shoulder joint (the part was incised with fine razor and the paste of seeds was applied thereon), for baldness, dandruff and other hair diseases (as a paste or medicated oil), for erysipelas and obstinate skin diseases (as a paste with butter). Sushruta applied the herb for cautery, for skin eruptions, goitre and scrofula, baldness, glandular swellings and also in urinary calculi.

By the 16th century, it was widely used for treating skin diseases, incl. leprosy and ulcer. On the basis of long-term use, it was clinically established that the herb was an efficacious hair vitaliser, cured catarrhal affections, nervous disorders, ophthalmic complications, helminthic infestation and possessed tonic and aphrodisiac properties. The root of the herb, pounded with goat's urine, has been recommended for treating defects of vision and blindness in Gadanigraha.

If we are to deviate from the medicinal aspect, we will find that the herb had enjoyed a history due to its seed, which was known as Ratti and was considered equal to 1 carat. The weight of the famous Koh-i-noor diamond was ascertained by means of these seeds.

In the folk medicine, seeds are ground into paste and made into needles, which are inserted under the skin of the animal. Thus the animal will be poisoned for obtaining the skin. Similar needles have also been used to produce criminal abortion.

The seeds are an ingredient in many herbal formulations in the form of hair oils, which are being prescribed for treating alopecia. Neelibhrangaadi Tailam is the most popular one.

The leaves steeped in warm mustard oil are applied over painful swellings or they are warmed over the fire and applied after smearing them with warm castor oil. Juice of fresh leaves mixed with some bland oil is also prescribed.

Decoction of the leaves is used for cough, cold and hoarseness; also being prescribed in biliousness and colic.

Active principles and pharmacology

The herb is very poisonous because the seed contains the toxic lectin abrin and isolectins. Severe poisoning among adults, following the intake of one half to two seeds, as well as cases of death among children after consuming two seeds, have been recorded. The seeds were used for criminal poisoning in India against cattle and other livestocks; also against human beings. Doses of about 1/1000 mg to 1/2000 mg of abrin per kg body weight injected subcutaneously are poisonous.

An infusion of bruised seeds, when applied to the conjunctiva, may cause fatal poisoning due to absorption of the toxic abrin through the conjunctiva. Abrin produces irritation, oedema and ecchymosis at the sight of inoculation. Water extract of seeds is, somehow, used for treating inflamed eyelids and for inducing conjunctivitis for treating the corneal opacity. This practice was also prevalent in China, but was discontinued long ago due to dangerous consequences.

Use in Western herbal

An infusion and a paste of the seeds are included in the British Pharmacopoeia; also its use in certain ophthalmic diseases. The root is considered irritant and poisonous.

According to R.C. Wren, the herb may be cautiously used in eye diseases and not until other means have been exhausted, as it produces a violent conjunctival inflammation and is likely to destroy the corneal structures.

Caution

Due to toxicity the total herb should no more be used as a herbal medicine. Its constituents should be screened and assessed for their use in modern medicine—the root for anti-estrogenic activity, and as a diuretic; leaves for leucoderma and other skin diseases; seeds for anti-tumour properties.

The herb has already fallen into disuse in China, as better alternatives have been identified for ailments covered by the herb.

Abrus precatorius has been mentioned as Indian or wild liquorice. This nomenclature is misleading. The name wild liquorice has also been given to Aralia nudicaulis Linn., indigenous to Canada and the United States. The root of Cephalanthus occidentalis Linn., a member of Madder family, and Ononis arvensis Linn. (Rest-harrow) are also called wild liquorice. None of these match with the phytotherapeutic profile of liquorice.

Malvaceae

ABUTILON

Abutilon indicum (Linn.) Sw.

Sida rhombifolia Linn.

Habitat

Abundantly found as a weed in the sub-Himalayan tract and other hills up to 1200 m, and in the hotter parts of India. Common throughout India.

Figure 1 Abutilon indicum [ADPS]

Classical & common names

Ayurvedic: Atibalaa, Kankatikaa, Rishyaproktaa; Vaatyaayani, Vaatyapushpi (Charaka); Valikaa, Bhaardwai, Uraksha gandhini, Naagbala, Vishvadevaa, Gavedhuka (16th century). (Naagbala has now been identified as Grewia tenax, syn. G. populifolia.)

Unani: Kanghi, Kangahi, Kakahiya, Kakahi; Musht-ul-ghoul (Arabic), Darakht-e-shaan.

Siddha: Thuthi.

English: Country Mallow, Flowering Maples, Chinese Bell-flowers.

The plant is variable and is divided into subspp. Abutilon populifolium (Lam.) Wright & Arn. ex Mast.; Abutilon asiaticum Wright & Arn. non (Linn.) Sweet; Abutilon graveolens Britten non (Roxb. ex Hornem.) Wight & Arn. ex Wight. (WOI, Rev Vol. 1, 21.)

All the varieties of the plant or its species possess medicinal properties of mallow or marshmallow. (Moodeen Sheriff, PV Sharma.) Sida rhombifolia has been used as synonym of Atibalaa by INSA in its Scientific Synopsis of Charaka Samhitaa and Sushruta Samhitaa. Though the properties of Abutilon indicum and Sida rhombifolia are the same, Sida rhombifolia is known as Mahaabalaa. This nomenclature has been accepted by CCRAS.

In some parts of Kerala, Urena spp. are also used as Atibalaa.
Marshmallow of the West (Althaea officinalis) is used medicinally in much the same way as Abutilon indicum. (Andrew Chevallier.)

Parts used
Root, seeds.

Dose
Powder 3–6 g, fresh juice 10–20 ml.

Classical use
During 1000 BC, the extract of the entire plant was used in prescriptions for loss of vitality, fever and rheumatic conditions (Charaka). During Sushruta's period thick aqueous decoction, as an ingredient of a massage cream, was used for promoting growth of normal healthy tissue after surgery. Charaka and Sushruta both used the drug as a rasayana (aphrodisiac, revitalising nervine tonic). The powdered herb was used during the 16th century for meno-metrorrhagia (Bhaavaprakaasha).

The herb is being used in both Ayurvedic and Unani systems as a styptic, diuretic and anti-inflammatory agent. The leaves contain mucilage. They are cooked and eaten for bleeding piles. The extract is demulcent and useful as an emollient fomentation. Along with clarified butter, it is considered a remedy for diarrhoea. A decoction of the leaves is used as a mouthwash in toothache and gum ailments; also as a wash for wounds and ulcers.

Powdered flowers are used in cough. Leaves, due to mucilage, are used as pectoral.

The root is a nervine tonic and antipyretic. Its decoction alleviates all types of dysuria. It was used in the same way during the 16th century (Bhaavaprakaasha). In Unani medicine, the herb is given in strangury polyuria and haematuria. In the South, the root is given for neurological disorders like hemiplegia, facial paralysis, sciatica and debility.

The seeds, known as Beejband*surkh* and*siyah*, are used in Unani medicine for treating impotency and spermatorrhoea.

Active principles and pharmacology
The plant contains gallic acid. Showed analgesic action in albino rats.

The leave contains 11.5 % asparagine, a diuretic principle.

Mucilage of the herb exerts wound-healing properties, reduces inflammation, pain and irritation and protects the mucous membrane (during bronchitis) and urinary system. The demulcent action of the herb acts on gastro-intestinal inflammations, lesions and ulcers, reduces excessive stomach acid and gives relief to acid dyspepsia.

Due to tannins, the herb used externally, proves haemostatic, checks haemorrhaging and subdues exposed inflammations. Internally it not only controls diarrhoea but also reduces the inflammation of the small intestine.

The plant is reported to contain fructose, galactose, glucose n-alkaline mixture, an alkanol fraction, beta-sitosterol, vanillic acid, p-coumaric acid, p-hydroxybenzoic acid, caffeic acid, fumaric acid, p-beta-D-glycosyloxybenzoic acid, leucine, histidine, threonine, serine, glutamic acid, aspartic acid and galacturonic acid. (WOI, Rev Vol. 1.) The plant also contains two (new) sesquiterpene lactones, alantolactone and isoalantolactone. Presence of alkaloids, leucoanthocyanins, flavonoids, sterols, triterpenoids, saponins and cardiac glucosides is also reported.

The seeds contain protein (31.0 %). The amino acid composition of protein is: asparagine 11.5; threonine 3.9; serine 6.9, glutamine 17.6; proline 2.5; glycine 8.6; alanine 7.0; cysteine 5.9; methionine 0.6; isoleucine 3.5; valine 5.9, leucine 5.3; tyrosine 0.4; phenylalanine 2.4; histidine 2.7; lysine 3.0; and arginine 6.5 g/100 g. The seeds act as an aphrodisiac.

An alcoholic extract of flowers and fruits has shown activity against Micrococcus pyogenes var. aureus and E. coli.

The synergistic effect of herb's constituents substantiate its various clinical applications, as discussed above, and also reaffirms its immunological value, which was pointed out by Pliny in 77AD (quoted below).

Use in Western herbal
Abutilon indicum and its related spp. Abutilon trisulcatum, native to Central America, is used to protect respiratory and urinary systems. A decoction of the root is given for bronchitis and other chest affections; mucilage is used for the skin; infusion, poultice or paste made from the pow-

dered root or bark is applied to wounds, boils and ulcers; the seeds are used as laxative and for killing threadworms; the plant exhibits an antiseptic effect within the urinary tract. (Andrew Chevallier.)

Althaea officinalis (Malvaceae), known as marshmallow, is native to Europe, naturalized in Americas. The leaves, flowers, root contain about 37 % starch, 11 % mucilage, 11 % pectin, flavonoids, phenolic acids, sucrose and asparagine.

The herb protects and soothes mucus membranes, counters excessive stomach acid, and is used in peptic ulceration and gastritis.

The herb is a mild laxative and is prescribed for many intestinal problems, incl. regional ileitis, colitis, diverticulitis, irritable bowel syndrome.

A warm infusion of the leaves is given to treat cystitis and frequent urination.

The herb's demulcent qualities bring relief to dry coughs, bronchial asthma, bronchial catarrh and pleurisy.

Flowers crushed fresh, or in warm infusion, are applied to help soothe inflamed skin.

The root is used in an ointment for boils and abscesses and in a mouthwash for inflammation.

Other species

Althaea rosea (Hollyhock) and Malva sylvestris (common mallow).

All members of the mallow family possess similar properties.

In 77 AD, Pliny wrote: "...whoever swallows daily half a cyathus of the juice of any of the mallows will be immune to all diseases."

Caution

The plant is considered as abortifacent.

Mimosaceae

ACACIA

Acacia arabica Willd. var. indica Benth.

Habitat

Found throughout the drier parts of India, ascending to an altitude of 900 m.

Figure 1 Acacia arabica [ZANDU]

Classical & common names

Ayurvedic: Babbuula, Aabhaa, Shuulikaa, Shitaka, Babbuuri, Kinkiraata, Yugmakantka, Sukshmapatra, Pitapushpaka. Sthula baboolikaa (wild spp.)
Unani: Babool, Kikar, Samur (Arabic), Mughilaan (Pers.).
Siddha: Karuval.
English: Babul, Akakia.

Parts used

Bark, pods, leaves, gum.

Dose

Fruit powder 3–5 g, decoction 50–100 ml (herb 6 12 g), leaves 2–4 g, gum 3–6 g, paste 5–10 g.

Classical use

During the 16th century Babul's bark was used for skin diseases and as an anthelmintic (Bhaavaprakaasha). The gum was used as a styptic and for urinary and vaginal discharges. Pods were used for their astringent properties (Nighantu Ratnaakara).

In folk medicines pods are used in impotency and in urinogenital disorders. The flowers, pods and gum-resin are used in diarrhoea and dysentery. Powdered leaves, flowers, bark and gum-resin (in equal quantity), if taken as first thing in the morning (2–11 g) with water, proves efficacious in premature ejaculation, seminal deficiency and spermatorrhea. This combination is also given to females for leucorrhoea.

The gum stops bleeding and urinary and vaginal discharges, is given in skin diseases, bleeding piles, sore throat; as a demulcent in asthma; also as an anthelmintic.

Active principles and pharmacology

The tannin content of the bark varies considerably. Sometimes, it is as high as 20 %; at the other times, as low as 12 %. It decreases somewhat on storage. The bark from branches yields 7–12 % tannin.

Several polyphenolic compounds identified are: (+)catechin, (-)epicatechin, (+)dicatechin, quercetin, gallic acid. The polyphenols of the bark are fungitoxic. The bark also contains sucrose.

All these constituents make the bark on effective medicinal astringent and demulcent. A decoction of the bark is used in asthma and bronchitis; also for gargling. The decoction exhibited hypoglycaemic property in alloxan diabetic rats, but hypogycaemic property of the drug could not be substantiated, as the seeds were hypoglycaemic in normal rats but not in the diabetic ones.

The pods contain 12–19 % tannin in the whole pod and 18–27 % after the removal of seeds. The polyphenols present in the pods are gallic acid, m-digallic acid and (+)catechin, chlorogenic acid, gallolyated flavan 3, 4-diol and robidandiol.

Use in Western herbal

Medicinally employed, in the West, as a substitute for oak bark. It has special use in diarrhoea. The British Pharmacopoea has recommended a decoction of 6 parts of the bark in 100, to be given half to 2 fl oz. (15 cc to 60 cc). The decoction is also used as an astringent gargle, lotion or injection.

The flowers, leaves and pods were used as medicine in ancient Egypt to expel worms.

Acacia arabica, as a strong astringent, is used to contract and strengthen mucus membranes throughout the body in much the same way as Hamamelis virginiana (witch hazel) or Quercus robur (oak bark).

Related species

Acacia nilotica Delile, subspp. indica (Benth.) Brenan; Acacia senegal (Willd.). Gum Senegal is a substitute for gum Acacia.

Caution

Not to be confused with the Australian species Acacia decurrens, commercially available in the West under the same name (Acacia).

Large doses, taken internally, are constipative. Do not take for more than 2–3 weeks.

Acacia nilotica subspp. indica may cause pollen allergy.

Acacia species are subject to legal restrictions in some countries.

Acacia catechu (L. f.) Willd.

Figure 2 Acacia catechu [CCRAS]

Figure 3 Acacia catechu [CCRAS]

Habitat

Occurring throughout India in dry types of mixed forests. Common from Punjab to Assam to an altitude of 1200 m. Common in peninsular region.

Classical & common names

Ayurvedic: Khadira, Raktasaara, Dantdhaavan, Kantaki, Baalpatra, Gaayatri (Sushruta), Somavalka (Charaka). (Khadir, Khadirkaa are also synonyms of Mimosa pudica in certain Ayurvedic classics).
The white variety was known as Shweta Saara, Kaarmuka, Kubja Kantaka.
Unani: Kattha, Khair, Kaat (Arabic, Persian).
Siddha: Kalippakku.
English: Cutch, Catechu.

Parts used

Heartwood (concentrated extract).

Dose

Powder 3–6 g, decoction 50–100 ml.

Classical use

Charaka advocated Acacia catechu heartwood decoction as the best medicine for all skin diseases, internally as well as externally. He prescribed the flowers or the fermented wine from the resinous exudate in prescriptions for internal use in haemothermia, urinary disorders and blood poisoning.

Sushruta prescribed the drug internally in obesity, urethral discharges (also externally) and jaundice. He used the drug as principal ingredient in many prescriptions for internal use in chronic and suppurating skin diseases.

By the 16th century, Acacia catechu was established as a potent drug for obstinate skin diseases, incl. leprosy, erysipelas, obstinate urinary disorders, polyurea, diabetes, diseases of the mouth, cough, hoarseness of voice and intrinsic haemorrhage. Even now, Khadiraarishta is the most efficacious medicine for chronic skin diseases and Kadiraadivati for sore throat, hoarseness of voice, excessive cough and as an adjuvant in asthma.

In Chinese medicine, cutch is used for poorly healing ulcers, weeping skin diseases, oral ulcer with bleeding and traumatic injuries.

Active principles and pharmacology

The chief constituents of the heartwood extract, catechin and catechutannic acid, makes it a potent remedy for chronic skin diseases, cough, relaxed conditions of throat, mouth and gums; also for diarrhoea. 25–60 % tannins, 20–30 % mucilage, flavonoids and resins make it a strong astringent and clotting agent.

Use in Western herbal

Pale catechu is an extract made from the leaves and young shoots of Uncaria gambier (Roxb.), a Rubiaceae. The source of catechu dark or black is Catechu nigrum/Acacia catechu (Willd.). Both the substances are sold under the name of catechu.

Both the dark and pale catechu are employed in Western herbal; the former is more astringent, the latter, being sweeter, is less disagreeable.

Catechu is employed as an astringent to overcome relaxation of mucus membrane. An infusion is employed to stop nose-bleeding and is also employed as an injection for uterine haemorrhage and leucorrhoea. Externally it is applied in the form of powder to boils, ulcers and cutaneous eruptions. Internal dose of powdered catechu is 300 mg to 900 mg.

Caution

Do not take for more than 2–3 weeks, or if suffering from kidney inflammation.

Cutch is subject to legal restrictions in some countries.

Acacia leucophloea Willd.

Figure 4 Acacia leucophloea [CCRAS]

Acacia alba Willd.

Habitat

Found in the plains of Punjab and in dry forest tracts throughout India.

Classical & common names

Ayurvedic: Arimeda, Arimedaka, Irimeda, Vitakhadir, Kalskandh, Godhaaskandha, Raamaka, Arimanja (Sushruta).

Figure 5 Acacia leucophloea [CCRAS]

Unani: Vilaayati Babool.
Siddha: Valval.
English: White Babool.

Parts used

Bark.

Dose

Decoction 50–100 ml.

Classical use

Charaka used the bark or pitch in decoctions externally for dental and oral diseases. Arimeda is one of the ingredients in Charaka Samhitaa's Khadiraadi Gutikaa and oil.

Vaagabhatta prescribed semi-solid extract of the bark. Chakrapaanidatta made it the chief drug in Irimedadyaadi oil, which is still available over the counter and is said to be useful in sores of the mouth, swelling of gums, dental caries, spongy and bleeding gums, pyorrhoea and weak roots of teeth. The oil is to be applied with a cotton swab to the affected part.

The bark is light yellowish grey to nearly white outside and light red inside. It is bitter, astringent and cooling. Its decoction (1 in 20) is used in bronchitis. The decoction together with ginger (Zingiber officinale) is an astringent wash for the teeth and is said to be useful in bleeding gums.

The powdered bark and gum are used for boils and ulcers.

Active principles and pharmacology

Stem bark constituents—n-hexacosanol, beta-amyrin and beta-sitosterol, and heartwood constituents n-octacosanol and beta-sitosterol make it an antiseptic, anti-inflammatory and antibacterial agent for medicinal use.

Caution

In Ayurvedic reference books, Irimeda has been equated with Acacia farnesiana Willd., known as Durgandh Khair, Guyaa Babul. This tree is extensively cultivated in France for extracting the cassie perfume. The bark is astringent and is used in malaria; the leaves in eye complaints; inflorescence in venereal diseases and root for antifertility (the root of Acacia leucophloea Willd. is abortifacient).

Other medicinal uses of Acacia farnesiana are in cholera, diarrhoea; convulsions, delirium, epilepsy, insanity; sterility in women. It is also used as an antiseptic agent for curing carbuncle, sores, gums and loose teeth.

CCRAS has identified Acacia leucophloea Willd. for Arimeda. It should be incorporated in Ayurvedic reference books.

Acacia polycantha Willd.
Acacia suma Buch.-Ham.

Habitat

Found in moist localities in West Bengal, Bihar and peninsular India.

Classical & common names

Ayurvedic: Shweta khadera, Kadara, Somavalkala, Shweta Saara.
Unani: Khor, Safed Khair.
Safed Khair has been equated with Acacia ferruginea DC., Acacia leucophloea Willd., Acacia senegal Willd.. In some Ayurvedic and Unani reference books, Shweta Khadira has also been equated with Uncaria gambier Roxb.

Parts used

Heartwood.

Dose

Decoction 50–100 ml.

Classical use

The heartwood was mainly used for treating diabetes. Sushruta prescribed the decoction of Kadara and Puuga (Areca catechu Linn.) nut. Ashtaanga Sangraha added both Kadara (Acacia suma) and Khadira (Acacia catechu) in the same compound with the gum of Commiphora mukul for diabetes. Vrindamaadhava added Puuga (Areca catechu) with both Kadara and Khadira.

It was obvious that Kadara and Khadira were treated as two different drugs. Sushruta included Khadira and Shweta Khadera in Saalsaaraadi group of herbs (identified for treating obstinate skin diseases, obstinate urinary disorders, incl. diabetes; anemia and obesity).

In Unani medicine, decoction of the bark of Safed Khair is used for mouthwash, cough, stomatitis, gut infections, diarrhoea, ulcers and cutaneous affections. The decoction of leaves is prescribed as a styptic and for treating old wounds. It is considered a good tonic for liver and an efficacious treatment for urinogenital problems.

Active principles and pharmacology

The presence of quercetin and hyper-3, 7-dimethylether is reported in the heartwood. Cutch is prepared from the heartwood.

The seeds are reported to have marked hypoglycaemic effect on normal albino rats. Proteins, Essential amino acids, incl. methionine, are present in the seeds.

It could not yet be proved phamacologically that Khadira and Kadara of Indian medicine are two different drugs.

Use in Western herbal

As mentioned earlier, Uncaria spp. (Rubiaceae) and its related species Uncaria gambier have been equated with Shweta Khadira in some Ayurvedic and Unani reference books. These species are known as Gou Teng in China. Stem and thorns are used. The drug contains alkaloids (incl. rhyncophylline and corynoxeine, isorhyncophylline, and hirsutine) and nicotinic acid. The herb is a sedative and antispasmodic and is mainly used to ease symptoms such as tremors, seizure, spasm, headache and dizziness. It is also prescribed for reducing high blood pressure. It contains catechin that protects the liver.

This analysis proves that Uncaria spp. does not match with Acacia species found in India.

In the West only Acacia catechu or black catechu is used as an astringent. White catechu is not in use in Western herbal.

Caution
Same as for Acacia catechu.

Euphorbiaceae

ACALYPHA

Acalypha indica Linn.
Acalypha ciliata Frosk.

Habitat
Common in plains as a weed in gardens; also in wastelands, especially in Bangalore and Pachmarhi, up to 210 m.

Classical & common names
Ayurvedic: Harita-manjari, Mukta-varchaa (non-classical).
Vernacular names: Khokali, Kuppi (Hindi); Kuppi gidda (Kannada); Kuppimeni (Tamil); Kuppaichettu (Telugu); Kuppamani (Malyalam).
English: Indian Acalypha.

Parts used
Leaves, root, stalks.

Dose
As decided by the physician.

Classical use
There is no mention of this herb in the texts of Charaka and Sushruta. It was mentioned for the first time in Vaidyamanoramaa for toothache (with Zingiber officinale), internally as well as externally. In Punarnavaadi Churna, a compound formulation of Vaidyamanoramaa, it has been included as an appetiser and stomachic.

The herb is said to possess carminative, expectorant, diuretic and emetic properties in folk medicine.

Leaves are credited with laxative properties. Used in the form of powder or decoction. Mixed with common salt, the leaves are applied to scabies and other skin diseases. Their juice, mixed with oil, is massaged in rheumatic arthritis. Expressed juice in smaller doses is an expectorant and is said to be useful in chronic bronchitis, asthma and consumption. Powder of dry leaves is used in bed-sores. Paste of leaves is applied to burns.

A decoction of the herb is used for toothache and earache by old-timers, and aqueous extract for loss of appetite.

Active principles and pharmacology
Alcoholic extracts of tender shoots, leaves and roots showed antimicrobial activity. The juice is found to possess antitumour properties.

The plant contains a cyanogenetic glucoside and two alkaloids, acalyphine and triacetonamine, possibly a degradation product of glucoside. The other constituents are n-octacosanol, beta-sitosterol, kaempferol, quebrachitol, tannin, resin and an essential oil. In addition to hydrocyanic acid, the herb contains other substances which caused intense, dark chocolate brown discolouration of blood and gastro-intestinal irritation in rabbits.

Use in Western herbal
Tincture of the fresh plant from East Indies is used in homoeopathy. John Henery Clarke (1900 AD) equated Acalypha indica with Cupameni and identified the chief employment of the tincture in violent dry cough followed by bloody expectoration; expectoration of pure blood in the morning and dark clotted blood in the evening; cough most violent in the night. The drug is also prescribed for burning in intestines, flatulence, diarrhoea and griping pain in abdomen.

Caution
The raw herb is poisonous and emetic. Causes intestinal irritation. The shoots should not be eaten uncooked. The pollen may cause allergy. Should be used only as a homoeopathic drug under medical supervision.

Asteracea

ACHILLEA

Achillea millefolium Linn.

Habitat

Commonly distributed in the Himalayas from Kashmir to Kumaun at the altitude of 1050–3600 m.; also in Mumbai and Belgaum areas. Native to Europe and Western Asia, growing wild in temperate regions.

Classical & common names

Unani: Biranjaasif, Al-Biranjaasf (Arabic), Biranjaasp, Palangasp, Buye-maaderaan (Persian).
English: Yarrow, Thousand Leaf, Milfoil.

Parts used

Flower heads, all aerial parts.

Dose

Herb 3–5 g, flowers 3 g.

Classical use

In India, used only in Unani medicine. Decoction of flower heads or of the whole plant is prescribed as a diaphoretic during fever; as a diuretic and anti-inflammatory agent. The herb acts as a liver tonic, helps relax the muscles of the digestive tract, stops internal bleeding, regulates menstruation and acts as a blood purifier. It helps recovery from colds, flu and fever. Helpful for circulatory disorders and acts as a deobstruent.

Active principles and pharmacology

The chief constituents of the whole herb are: volatile oil containing cineol, azulene, eugnol, pinene, thujone, camphor, camphene, sabinene, achillin; bitter constituents, such as ivain; cyanogenic glycosides; salicylates; asparagine; flavonoids incl. luteolin, apigenin, kaempferol, quercitrin; glycoalkaloid (achilleine); tannins; aconitic and isovalerianic acids; fluorescent substance, hydroxycoumarins; resins.

Cineol has antiseptic and expectorant properties. Helps the body control fevers, flu and the common cold.

Azulenes are effective anti-inflammatory and antispasmodic agents reducing histamine-induced tissue reactions and calming the nervous system. Azulene also stimulates the formation of granulation tissue and helps in wound healing. This property confirms the herb's traditional first aid use as emergency styptic to stop bleeding. In Scotland, a traditional wound ointment was made from the herb.

Achillin has experimentally reduced clotting time without toxic side-effects (the decoction of the whole herb has the same effect in vitro and in vivo). The herb helps arrest internal and external bleeding.

All these constituents confirm the key actions of the herb, as discussed earlier.

Flowers are rich in active principles. When converted by steam into anti-allergenic compounds, can be used for various allergic catarrhal problems, incl. hay fever. The leaves encourage clotting, can be used fresh for nose-bleeds (leaves not to be inserted into nostril). Paste should be used.

Aerial parts are found useful in catarrhal conditions, as a bitter digestive tonic and as a diuretic. They can be used for high blood pressure; also in menstrual disorders. Aerial parts exhibit diaphoretic activity and bring down fever.

Use in Western herbal

John Gerard (1545–1612) recommended yarrow for "swellings of the privie parts."

John Parkinson (17th century) advised: "If it be put into the nose, assuredly it will stop the bleeding of it."

Nicholas Culpeper (1616–1654) wrote about yarrow: "An ointment of them cures wounds and is more fit for such as have inflammations. Being boiled in the white wine and the decoction drunk, it stops terms in women, as also the bloody flux. The ointment is not only good for green wounds but also for ulcers and fistulas, especially such as around the moisture. It stays shedding of hair, the head being bathed with the decoction of it. Inwardly taken, it helps the retentive faculty of the stomach. It helps the whites in women and helps such as cannot hold their water. The leaves chewed in the mouth eases the toothache. The herb is drawing and binding. A very profitable herb in cramps."

In classical times in Europe, it was known as "herbal militaris", since it was being used to treat war wounds.

Flowers, leaves, aerial parts and essential oil are being used by Western herbalists.

German Commission E monograph recognized the herb's efficacy in the following areas:

- Loss of appetite
- Dyspeptic complaints
- Liver complaints
- Gallbladder complaints.

The herb is recommended in Europe as a cholagogue, antibacterial, astringent and antispasmodic.

In the United States, the yarrow capsules are being prescribed for restoring hormonal balance in females. In endometritis the herb flushes out excess estrogen. As yerrow helps to balance hormones, it is often prescribed to treat heavy menstrual bleeding, fibroid tumours and menopausal hot flushes. Yarrow capsules are also prescribed to ease uterine and abdominal muscles and to relax muscles that cramp or contract.

Caution

Prolonged use can increase the skin's photosensitivity. In rare cases, can cause allergic skin rashes.

To be avoided during pregnancy, as the herb is a uterine stimulant.

High doses of yarrow may turn urine dark brown.

Amaranthaceae

ACHYRANTHES

Achyranthes aspera Linn.

Habitat

Distributed throughout the tropical and subtropical regions.

Classical & common names

Ayurvedic: Apaamaarga, Chirchitaa, Shikhari, Shaikharika, Adahshalya, Mayura, Mayuraka, Kharamanjari, Kharapushpaa, Pratyakpushpaa, Aaghaat, Vashira, Kihini.
Unani: Atkum (Arabic), Chirchitaa, Latjeeraa.
Siddha: Nayuruvi.
English: Prickly Chaff Flower.

Figure 1 Achyranthes aspera

Figure 2 Achyranthes aspera

Parts used

Root, leaves, all aerial parts.

Dose

20–50 g of the drug for decoction, juice 10–20 ml, alkaline ash 500 mg to 2 g.

Classical use

Charaka prescribed dried fruits alone or in prescriptions internally in parasitic infections, hemicrania and as errhine. Sushruta used alkaline ash of the plant for promoting growth of normal tissues after surgery; thick aqueous extract as ingredient of a massage cream for ulcers; internally for skin diseases; as an ingredient of a medicated salt in rheumatism, internal tumour, cough, piles, intestinal parasites. According to the 16th century texts, the herb proves very potent for stimulating digestion, cures anorexia. It overcomes the troubles due to worms and pathogenic organisms. Arrests bleeding and dysentery; also can be used for treating ascites.

Siddha physicians use the herb effectively against bronchial asthma.

Unani physicians prescribe the alkaline ash of the herb with the milk of female camel for ascites. They also use it as a lithotriptic.

Active principles and pharmacology

The whole plant contains the alkaloids achyranthine and betaine. Achyranthine, a water soluble alkaloid, is reported to dilate blood vessels, increases the amplitude of respiration and lowers blood pressure. It also showed diuretic and purgative action. The presence of ecdysterone is also reported in the stem and leaves. The ashes of the plant yield large quantities of potash. This justifies its use in traditional medicine for dropsy, asthma and cough.

Alcoholic and aqueous extracts of the leaves showed antibiotic action against Micrococcus pyogenes var. aureus and E. coli. The extract is used for leprosy as an antibacterial agent.

Roots are astringent and are applied to wounds as haemostatic. The decoction of the root is used for stomach troubles and an equeous extract for stones in the bladder.

The flowers, ground and mixed with curd and sugar, are given for menorrhagia.

All these properties have been confirmed by various pharmacological studies.

Use in Chinese herbal

Achyranthes of the Amaranth family was named by Linnaeus in 1753 and Chinese Achyranthes bidentata (white var.) was first mentioned in a 1825 publication by Carl Ludwig von Blume. Called Niu-xi, this spp. is also in use in the United States.

Triterpenoid saponins are the main constituents of the plant.

In Chinese medicine, Niu-xi is believed to invigorate the blood flow. It is used to stimulate menstruation when a period is delayed or scanty. It relieves pain due to a kidney stone. The herb is also taken for mouth ulcers, toothache, bleeding gums and nosebleeds.

The root is considered an anodyne, analgesic, antirheumatic, diuretic, emmenagogue, anti-inflammatory, antispasmodic, hypotensive and tonic.

Caution

Contraindicated during pregnancy.

Ranunculaceae

ACONITUM

Aconitum bisma (Ham.) Rap.
Aconitum palmatum D. Don
Aconitum heterophyllum Wall. ex Royle

Habitat

Occurring in the alpine Himalayas of Sikkim, Nepal and adjoining parts of southern Tibet; also from Kashmir to the Kumaun hills at altitudes of 2500–3900 m.

Classical & common names

Aconitum bisma and Aconitum palmatum: Prativishaa, Shyamakandaa, Patis, Bikhmaa.
Aconitum heterophylum: Ativishaa, Shuklakandaa, Bhanguraa, Ghunavallabhaa, Kaashmiraa, Shishubhaishajyaa.
Unani: Atees.
Siddha: Athividyam (white variety).

Parts used

Root.

Dose

Powder 500 mg–1 g, decoction 3–5 g.

In Ayurvedic classical texts (Dhanvantari Nighantu), two varieties of the root have been mentioned. Ativishaa, the white (shukla) and Prativishaa, the blackish (shyama) root. In Charaka Samhitaa Ativishaa and Prativishaa were considered as one herb.

Classical use

Charaka used the root and tuber in internal prescriptions for fevers, rheumatic conditions and for loss of vitality.

Figure 1 Aconitum heterophyllum—flowering portion [WOI]

Charaka recommended a liquid gruel processed with Ativishaa, sours and Shunthi (Zingiber officinale) for treating diarrhoea. Many compounds, with Ativishaa as the principal herb, are found in Bhaavaprakaasha, Ashtaanga Hridaya, Vrindamaadhava and other classical texts for chronic diarrhoea, malabsorption syndrome and colitis.

Sushruta prescribed the herb internally for deranged metabolic conditions, internal tumours, colic, gastralgia, rheumatic, uterine and vaginal disorders, and for purifying breastmilk.

Ativishaa was known as Shishubhaishajyaa due to its use in pediatrics. Many prescriptions are recommended in Bhaavaprakaasha, Ashtaanga Hridaya and Vrindamaadhava for treating children suffering from cough, fever and vomiting.

Ativishaa was given as a single drug with honey or was combined with Karkatashringi (Pistacia intergerrima) and Pippali (Piper longum.)

It seems, during the 16th century, Ativishaa was not easily available and in its place Mustaa (Cyperus rotundus) was used in prescriptions. Both the herbs were known for alleviating cough and digestive disorders.

A compound formulation of Shaarangadhara Samhitaa, Sudarshana Churna, with Ativishaa as an ingredient, is still available over the counter. It is prescribed for intermittent and malarial fevers, and is considered an antiperiodic, digestive, and blood purifier for treating skin diseases.

Active principles and pharmacology

Aconitum bisma root, light brownish in colour, contains diterpenoid alkaloids-15-deacetyl vakognavine, palmadine, palmasine, 6-acetylheteratisine heteratisine, isoatisine and hetidine.

The root is extremely bitter and is used in combination with Piper longum for pain in bowels, for diarrhoea, and vomiting. Externally its paste is applied for rheumatism; also on cuts and wounds as an anti-inflammatory and antiseptic agent.

Aconitum heterophyllum roots yield 0.79 % of total alkaloids. Atisine (yield 0.4 %), heteratisine (yield 0.03 %), histisine, heterophyllisine, heterophylline, heterophyllidine, atidine, hetidine, benzolheteratisine, F-dihydroatisine and hetisinone have been isolated. The root also exhibited antiviral activity against spinach mosaic virus.

As atisine is much less toxic than aconitine and pseudaconitine, the species is often regarded as non-poisonous. Although the alkaloid atisine produces hypotention, the whole aqueous extract of the root induces marked hypertension through an action on the sympathetic nervous system.

The roots are astringent and give good results in diarrhoea, especially in infants, and in dyspepsia, vomiting, cough, irritability of the stomach and colic.

The plant, in a study, was found to possess potent immunostimulant property.

Use in Western herbal

Both the species are not used in Western herbal, not even in Germany. Only passing remarks are found in the West reconfirming the non-poisonous nature of the roots of Aconitum heterophyllum and Aconitum palmatum and their tonic and anti-periodic principles. Both the species were also not included in homoeopathic materia medica, while other Aconite spp. were used extensively.

Aconitum ferox Wall. ex Springs
Aconitum napellus Linn.

Habitat

Wild in the alpine Himalayas of Kashmir and Nepal at an altitude of about 3600 m.

Classical & common names

Ayurvedic: Vatsanaabha, Shringi visha, Visha, Amrit, Shrangikavisha. (Vishaa has been equated with Ativishaa, Visha with Aconitum ferox and Visa with lotus stalk.)
Unani: Bish (Arabic), Bishnaag (Persian), Bachhanaag, Telia bis.
Siddha: Nabi, Vashnabi.
English: Wolfsbane, Monkshood.

Parts used

Root.

Dose

Powder 10–15 mg.

For detoxification, soak the drug pieces in cow's urine for three days, changing the urine every day. Dry them in sun. Put all pieces in a cloth bag, suspend in cow's urine and boil for six hours. Wash the pieces in water and dry.

Classical use

In Chinese medicine, Aconite is used for ague, fevers, rheumatism, nervous disorders, neuralgias, dropsy, dysentery, as a stimulant, diaphoretic, diuretic, alterative, sedative and deobstruent. The same application is found in various classical texts in India.

Classical Ayurvedic compounds (like Aanand Bhairava Rasa) are highly toxic. Aconite is no more an indispensable herb for fever, diarrhoea, cough or asthma. Many formulations, incorporated in the Ayurvedic Formulary of India, Part 1, Vol. II, are not being used even by experienced physicians.

External preparations of the drug can still be used for rheumatism or for neurological disorders. Vishgarbha oil is one such over-the-counter drug. Again, this is to be used under medical supervision. The massage of this oil is efficacious in diseases associated with pain and chronic swellings. This oil, inspite of 47 other ingredients, is highly toxic and poisonous, and hands must be washed with soap after every massage.

Unani physician use the detoxified root internally to treat fever associated with acute internal inflammations, especially that of lungs or pleura and externally in sciatica, muscular rheumatism, arthritis, myalgia and joint affections.

Active principles and pharmacology

The drug Aconitum ferox of Indian commerce is in fact a mixture mainly of Aconitum atrox and Aconitum deinorrhizum with Aconitum chasmanthum, Aconitum laciniatum and Aconitum spicatum added sometimes.

The total alkaloid content in commercial Aconitum ferox varies from 0.63 to 4.7 %.

Pure roots contain the alkaloids pseudaconitinine, chasmaconitine, indaconitine and bikhaconitine. The roots, after mitigation i.e. soaking them in cow's urine or milk for 2–3 days, become soft and their active principles lose their depressant action on the heart and instead become stimulants having mild cardio-tonic property.

Aconitum napellus, known as Wolfs bane in European classical literature, is a potent poison. The efficacy of the drug is based on the di-ester alkaloids–aconitin, mesaconitin and hypaconitin. Aconitin raises membrane permeability for sodium ions, and retard polarisation. Aconitin is initially stimulating and then causes paralysis in the motor and sensitive nerve ends, and in the CNS. Other di-ester alkaloids play the similar role. Hypaconitin works more intensively. Aconitin applied in small doses trigger bradycardia and hypotension; in higher doses it shows initially a positive inotropic effect, followed by tachycardia, and finally cardiac arrest.

Di-ester alkaloids were shown to be analgesic in animal experiments. In fever the drug causes outbreak of sweats and exerts an anti-febrile effect. When applied topically, the drug is initially stimulating and then anaesthetizing. (PDR.)

Use in Western herbal

Aconite entered into British herbal during the tenth century. It was called Wolfsbane. In the Middle Ages, it became Monkshood. In 1524 and 1526, two criminals, to whom the root was given as an experiment, died instantly.

John Gerard (1545–1612) described the herb as venomous and deadly.

After the chemical constituents of the drug came into light, Aconite entered into herbal medicine during the 19th century as an anodyne, diuretic and diaphoretic. Now both tincture and liniment of Aconite, also ointment, are in general use in Europe.

Official tincture, taken internally, diminishes the rate and force of pulse in the early stage of fevers and slight local inflammations, such as feverish cold, laryngitis, first stage of pneumonia and erysipelas. It relieves the pain of neuralgia, pleurisy and aneurism.

In acute tonsillitis children have been treated well by a dose of 1 to 2 minims (for 5–10 year old), and adults by 2 to 5 minims, 3 times a day.

Tincture of the root is used in Germany in facial and trigeminal neuralgia.

Aconite is preferred as a homoeopathic medicine, rather than a typical herbal preparation all over Europe. In the UK and the US Aconitine, an alkaloid obtained from the roots and stems of Aconitum napellus, tincture of the root of Aconitum cammarum and Aconite ferox are available in homoeopathic pharmacies. Tincture of the fresh plant of Aconitum lycoctonum, when coming into bloom, is the only aconite which does not contain aconitine. It acts well on the swelling of the cervical, axillary and mammary glands, and is used for scrofulous glands.

Caution

The drug is highly toxic. While taking traditional herbal preparations, signs of poisoning can appear even with the administration of therapeutic doses. Better to switch over to homoeopathic preparations.

Araceae

ACORUS

Acorus calamus Linn.

Figure 1 Acorus calamus [CCRAS]

Figure 2 Acorus calamus [CCRAS]

Habitat

Growing wild and also cultivated throughout India, ascending to an altitude of 2200 m in the Himalayas. Thrives best in marshy and moist places. Cultivated in Koratagere taluk in Karnataka.

Three more varieties of Vachaa have been mentioned in Ayurvedic and Unani texts: 1. Baalvach, also called Paarseek vachaa, Khuraasaani vach, Dudh vach—Shweta vachaa; Shukla, Haimvati of Ayurvedic texts. This has been equated with Acorus gramineus. Also with Iris germanica Linn. 2. Malaya vachaa or Mahaabhari vachaa has been equated with Alpinia galanga Willd. and Aplinia officinarum Hance. 3. Dweepaantar vachaa has been identified as Smilax china Linn.

Classical & common names

Ayurvedic: Vachaa, Ugragandhaa, Jatilaa, Vijayaa, Bhadraa, Ikshupatrikaa (Charaka); Shad-granthaa, Ugraa (Sushruta); Shatapatrikaa, Shataparvaa, Tikshnagandhaa, Golomi, Kshudra patra, Lomashaa, Maangalyaa.
(Ayurvedic physicians use the red variety, Unani physicians the white one: Haimvati.)
Unani: Bachh, Waj-e-turki, Baj, Uud-ul-waj, Al baj (Arabic); Agar turki, Kaarunak (Persian).
Siddha: Vasambu.
English: Sweet Flag.

Parts used

Dried rhizome.

Dose

Powder 250–500 mg.

Classical use

Charaka used the leaves, stalks in powder form or decoction in internal prescriptions as purgative, appetizer, in haemorrhage, fever, urinary and skin diseases.

Sushruta used the juice of the plant with milk as a part of daily diet for promoting intellect and longevity. The medicated clarified butter was prescribed internally in epilepsy, insanity, asthma, consumption, haemorrhage, menorrhagia, pastular eruptions, impotency and also as a cardiac tonic.

Rhizomes were given internally for intestinal catarrh, indigestion, colic pain, dysentery; in vaginal discharges and other urinogenital diseases; also for purifying breast milk.

Sushruta used Vachaa as an ingredient of a medicated powder for cleansing and sterilising wounds.

By the 16th century, it was established that Vachaa promotes longevity, memory and intellect. A classical preparation, Saaraswata Churna, is still available over the counter. Its principal ingredient is Vachaa. (Ayurvedic Pharmacopoeia of India, Part I, Vol. II, included Saaraswataarishta in its monograph on Vachaa. In Saaraswataarishta Bacopa monnieri (Braahmi) is the principal ingredient, Vachaa is in minimal quantity.) Saaraswata Churna (3–6 mg) is being used for treating mental weakness and insomnia; also as an adjuvant in insanity.

Vachaa is an ingredient in many herbal formulations due to its classically confirmed carminative, antispasmodic, anthelmintic, as well as sedative, hypotensive, and tranquillizing properties. In Unani medicine, the drug is used for epilepsy, hysteria, neuralgia; also for dyspepsia, colitis and as an antispasmodic in asthma.

Active principles and pharmacology

The important constituents of the Indian calamus oil are asarone (up to 82 %) and its beta-isomer. Other constituents are: calamenol 5.0, calamene 0.4, calamenone 1.0, eugenol 0.3, methyl eugenol, 1.0, alpha-pinene and camphene 0.2 %. Other fractions are in small quantities. Due to synergistic action of these active principles calamus oil is credited with carminative, antispasmodic and antibacterial properties.

The essential oil-free alcoholic extract of the rhizome was found to possess sedative and analgesic properties and to cause a moderate depression of blood pressure and respiration.

The oil, when administered intraperitoneally to experimental animals, in a dose of more than 25 mg/kg, produced generalized depression of the nervous system. Lethal, if the dose is 0.22 g/kg body weight.

Elimination of phenolic and aldehydic fractions from the oil resulted in the increase in toxicity and sedative potentiating activity.

It has been pharmacologically established that asarone is a mild sedative, a potent tranquillizer, and like chlorpromazine (Largactil, an anti-psychotic drug of modern medicine) a mild hypotensive and hypothermic substance. When administered intraperitoneally to experimental animal (3 mg/kg body weight), it exhibits CNS depressant activity, producing no side-effects. (A dose of 0.0275 ml/100 g of asarone proved lethal to guinea-pigs.)

Asarone and beta-asarone both enhance anaesthic activity of pentobarbitone, hexabarbitone. Hypnotic potentiating as also tranquillising activity of beta-asarone is significantly higher than that of asarone. They cause a fall in rectal temperature, show anti-convulsant action, decrease sociability score and an anti-cholinergic action.

Choline (0.26 %) and acoric acid have been isolated from the roots. The plant also yields a flavone diglycoside luteolin-6, 8-C-diglucoside. Tannins, mucilage and calcium oxalate are also reported.

Asarone also inhibited the growth of Mycobacterium tuberculosis and showed antibacterial activity similar to Alpinia galanga.

Freshly prepared steam-distilled oil exhibited marked antibacterial activity against P. solanaceum, C. diptheriae, S. typhi, S. abrus. Both aqueous and alcoholic extracts from the root exhibited antibiotic activity against Staphylococcus aureus. The essential oil from rhizome produced bactericidal effect on streptococcal spore, Tubercle bacillus.

In a study, aqueous extract of the rhizome was administered in the dose of 400 mg/kg to experimental animals. The sluggishness started within 1 hour of the drug administration and reached the peak level at 3 hours, after which it became steady. Righting reflex, though somewhat delayed, remained intact. Pinna reflex was also intact but the duration of reflex was progressively increased.

A dose of 400 mg/kg of the aqueous extract was found to be as potent as 25 mg/kg of chlordizeproxide. This study was carried out to justify the use of Waj-e-turki (Acorus calamus) in Unani medicine for various neurological disorders.

Use in Western herbal

In 1653, Nicholas Culpeper wrote about the Sweet Flag: "The spicy bitterness of the root bespeaks it as a strengthener of the stomach and head."

In 1692, William Saloman, in The Complete English Physician, gave no less than sixteen preparations of the calamus. He wrote: "It is a good stimulant and carminative."

In 1822, John Augustine Walter, in The New British Domestic Herbal, wrote: "It (calamus) is of great service in all nervous complaints, vertigoes, headaches and hypocholdriacal affections." He also recommended the drug in dysentery and chronic catarrh.

A tonic medicine called Stockton Bitters, was once in much esteem in some parts of England. It was made from the roots of calamus and that of Gentiana campestris.

During the 19th century, calamus was subject to controversy due to common names like Sweet Flag, Yellow Flag, Blue Flag, Bastard Flag, which involved plants of species quite different from each other, like Calamus draco (Willd.), Iris versicolor (Linn.), Iris pseudacorus (Linn.).

R. C. Wren in Potter's Cyclopedia, during 1900, equated the Sweet Flag with Calamus aromaticus, though it was called Sweet Sedge and ancient literature placed it with Gentian family. His nomenclature of Sweet Flag as Acorus calamus (Araceae) found a place in all future encyclopedias. Wren advised the use of the root (not rhizome) for flatulence, wind, colic, dyspepsia, ague. According to him, it was an old remedy for ague. (In India it was given as a supporting medicine in fever, but not in malarial fever.)

Now on the basis of chemical constituents and their biological effect, following four varieties of Acorus calamus have finally been recognized as Sweet Flag: Type 1 is found in North America, contains no beta-asarone and exhibited considerable spasmolytic properties. Type 2 is from Eastern Europe, containing less than 10 % beta-asarone. The spasmolytic activity is lesser than type 1 in similar doses. Types 3 and 4 contain up to 96 % beta-asarone and exhibit no spasmolytic properties at similar dose levels. Type 1 is derived from diploid plants, type 2 from triploid plants and types 3 and 4 from tetraploid plants.

Experimental studies have shown that beta-asarone was responsible for malignant tumours produced in the duodenal region of rats by long-term administration of beta-asarone rich calamus oils.

Beta-asarone also showed a chromosome-damaging effect on human lymphocytes, mutagenic properties in bacteria.

On the basis of these findings, calamus as a food or food additive has been banned in the US. There is also restriction on the use of calamus in Australia, New Zealand and UK.

As calamus combines demulcent effects of the mucilage with the carminative and anti-spasmodic effect of the volatile oil and stimulating effect of the bitters, it has been accepted in the West as a tonic for gastro-intestinal tract and as a remedy for gastritis, nervous and acid dyspepsia, hyperacidity, peptic ulceration, intestinal colic, flatulence, anorexia and as an appetite stimulant.

It is not being used for neurological disorders and as a rejuvenator for the brain and nervous system, though this is the preferred application of the drug in India.

Only Acorus calamus var. americanus (Raf.), which is a diploid and its oil is iso-asarone-free, is used in the West. Calamus of European origin (triploid strain upto 15 % beta-asarone in volatile oil) is also used but its long-term use is avoided. Indian calamus is not used anywhere in the world, as it contains over 80 % beta-asarone in volatile oil.

Caution
Take only under medical supervision.

Acorus gramineus Sol. ex Ait.
Iris versicolor Linn.

Habitat
Cultivated in Kashmir; also found run wild on graves.

Classical & common names
Ayurvedi: Acorus gramineus has been equated with Haimavati, Paarseek Vachaa, Shweta Vachaa, Baalvacha.
Iris versicolor has been equated with Haimvati by SG Joshi.
Unani: Iris versicolor is used as Sosan (white-flowered); and as Irsaa (blue-flow ered).
English: Orris Root (used in Unani medicine).

Classical use
Sushruta prescribed Haimvati internally in uterine and vaginal disorders, for purification of breast milk and as a digestive aid.

A paste of Haimvati vachaa was prescribed externally in psoriasis (Bangasena).

Haimvati was prescribed mainly for abdominal diseases, colic and tympanitis (Bhaavaprakaasha).

In Unani medicine, Sosan and Irsaa are used in cough and bronchitis as an expectorant; as a deobstruent in colic. Mixed with vinegar, the root is prescribed in the inflammation of liver and spleen. Externally, its paste was applied over hard swellings and skin affections.

Irsaa is an important ingredient in Zimad-e-Khanazeer (Qarabadeen-e-Azam), prescribed for external use in scrofula (cervical adenitis).

Active principles and pharmacology
Acorus gramineus: the air-dried plant yield 0.5–0.9 % of an essential oil having asarone as a major constituent. The rhizomes show stimulant, tonic, antispasmodic properties and are used for the same medicinal purpose as Acorus calamus.

Iris versicolor Linn. bears blue flowers and is known as Blue Flag. It contains triterpenoids, salicylic and isophthalic acids, a very small amount of volatile oil, starch, resin, an oleo-resin and tannins.

Active principles of Blue Flag detoxify the body, increase urination and bile production and exert a mild laxative effect. This combination of cleansing action makes Blue Flag a useful herb for chronic skin diseases, gallbladder affections, constipation, biliousness and glandular problems.

Polypodiaceae

ACTINIOPTERIS

Actiniopteris radiata (Sw.) Link.
Adiantum incisum Frosk.
Adiantum caudatum L.

Actiniopteris radiata is equated with Mayurashikhaa by CCRAS. Adiantum incisum/Adiantum caudatum is also equated with Mayurashikhaa, Neelakantha shikhaa of Bhaavaprakaasha and Barhi shikhaa of Ashtaanga Hridaya and Gadanigraha. Largely used as a substitute for Adiantum capillus-veneris Linn., the Maidenhair Fern of the West.

Habitat
Found throughout India, below an altitude of 1200 m. Common in the Kumaun hills and in the Nilgiris.

Classical & common names
Ayurvedic: Mayurashikhaa, Sahastrahi, Saaharsra, Neelkantha shikhaa, Barhi shikhaa, Varhishikhaa.
English: Peacock's Tail.

Parts used
Root.

Dose
Root powder 3–5 g, decoction 10–20 ml.

Classical use
Mayurashikhaa is said to be one of the fertility-promoting herbs of classical texts. According to NAA, Neelkantha shikhaa may be used as a substitute for Lakshmanaa.

At many places in Ayurvedic texts Mayurashikhaa has been recommended to women for conceiving "male" child. In the absence of proper identification, peacock's ornamental portion of the feather entered into folk medicine in place of the actual herb.

In Gadnigraha, the root of Mayurashikhaa, pounded with milk, was prescribed to the woman in season for achieving conception.

According to Ayurvedic texts, Mayurashikhaa is cooling, astringent, anti-inflammatory, haemostatic, tonic to genitourinary tract, alleviates vitiated blood; indicated in cough, bronchitis, asthma, diarrhoea, dysentery, dysuria; used internally as well as externally for infected wounds, ulcers, erysipelas (Kaiyadeva Nighantu, Bhaavaprakaasha). Ashtaanga Hridaya recommends the root with rice water for gravels and calculus.

The herb was not used in Unani medicine, though its references were available during the 16th century.

Active principles and pharmacology
Actiniopteris spp.: The fern is triploid but the occurrence of diploid apogamous specimen have also been collected.

The stem and leaves contain rutin, hentriacontane, hentriacontanol, beta-sitosterol, beta-sitosterol palmitate, beta-sitosterol-D-(+)-glucoside, an unidentified glucoside, glucose and fructose.

These constituents exert stabilizing and calming effect on peripheral circulation, check haemorrhage. The herb's active principles are anti-inflammatory, diuretic and tonic to capillaries. The herb is reported to possess antibiotic properties.

Adiantum incisum, syn. Adiantum caudatum: The fern is aromatic, astringent, tonic and febrifuge. It is also used in hemicrania. The fonds are externally used in skin diseases and their juice for diabetes.

Dried plant yields adiantone, isoadiantone, fernene, hentriacontane, hentriacontanone-16, beta-sitosterol.

Use in Western herbal
In India, fonds of Adiantum incisum syn. Adiantum caudatum are largely used as a substitute for Adiantum capillus-veneris, which is equated with Hamsapadi of Ayurvedic texts and Maidenhair fern of Western herbal. Properties will be discussed under Adiantum capillus-veneris Linn.

After reviewing the literature of the West, it can be said that no fern has so far been used for promoting conception in women. This is a typical Indian application, based on indirect inference from the genitourinary healing properties of ferns.

Caution
Emetic in large doses.

Mimosaceae

ADENANTHERA

Adenanthera pavonina Linn.

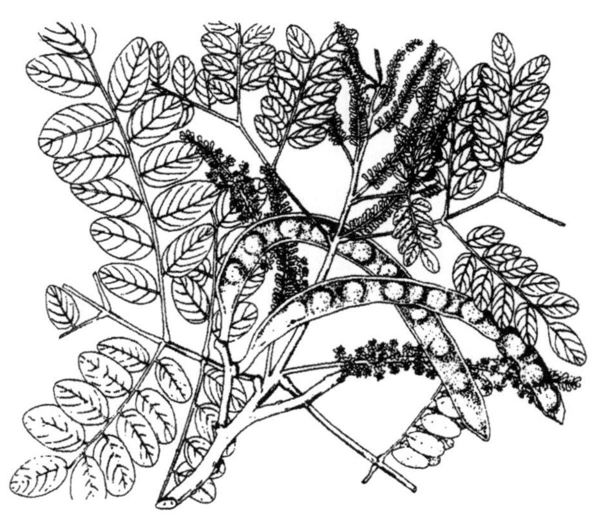

Figure 1 Adenanthera pavonina—flowering and fruiting branch [WOI]

Habitat
Found in sub-Himalayan tract, ascending up to an altitude of 1200 m in Sikkim, West Bengal, Assam,

Meghalaya, Gujarat, Maharashtra, South India and in the Andamans.

Classical & common names

Ayurvedi: Rakta Kambal, Rakta Kanchana; Kuchandana, Anaigundumani (Tamil).
English: Coral Wood.

Parts used

Leaves, bark.

Dose

Decoction 50–100 ml.

Classical use

The heartwood of the plant is red. It was used as a dye and as a substitute for red sandalwood (Pterocarpus santalinus). That is the reason it was called Kuchandana. Sushruta used it internally and externally in haemoptysis, bilious fevers, persistent dysentery, non-healing ulcers, urethral discharges, skin diseases, jaundice, and as an appetizer.

Dalhan in his commentary of drugs of Sushruta equates Kuchandana with Rakta chandana (red sandalwood) and Malayaadri chandana (Coscinium fenestratum). But Ayurvedic reference books treat Adenanthera pavonia as an adulterant to Pterocarpus santalinus.

According to Bhaavaprakaasha, four types of chandanas were used in Ayurvedic medicine: Chandana (Santalum album), Rakta chandana (P. santalinus), Kaaliyaka, Kalambaka or Pita chandana and Kuchandana (Santalum flavum—according to the scientists of INSA, P. Ray et al, Kaaliyaka and Kuchandana are synonyms). Whichever variety was available locally, was used in prescriptions as Chandana. Caesalpinia sappan Linn. was also called Kuchandana.

Active principles and pharmacology

The leaves contain octacosanol, dulcitol, glucosides of beta-sitosterol and stigmasterol. The seeds contain stigmasterol and its glucosides, dulcitol and polysaccharide. The bark yields a saponin having oleanolic and echinocystic acids as sapogenins. The wood contains robinetin, chalcone, butein, amelopsin and dihydrorobinetin.

These constituents give a better picture of the drug's coverage, which is indicated in diarrhoea, haemorrhage from bowels, haematuria, skin eruptions; also in rheumatism and gout.

Seeds, leaves and bark exhibited anti-inflammatory activity in pharmacological studies.

Caution

The roots are reported to be emetic.

Acanthaceae

ADHATODA

Adhatoda vasica Nees

Figure 1 Adhatoda vasica [ZANDU]

Adhatoda zeylanica Medic.
Justicia adhatoda Linn.

Habitat

Distributed throughout India up to an altitude of 1300 m.

Classical &common names

Ayurvedic: Vaasaa, Vaasaka, Vaasikaa, Simhaasya, Vaajidanta, Vrasha, Atarushaka, Simhaanana.
Unani: Arusaa, Baansaa, Basauntaa; Hashees-tussuaal (Arabic); Khwaazaa, Baansah (Persian).
Siddha: Adathodai.
English: Malabar Nut, Vasica.

Parts used

Flowers, whole plant.

Dose

Fresh juice 5–10 ml, decoction 50–100 ml, flowers 500 mg–1 g, root bark powder 250 mg–500 mg.

Classical use

Charaka used fresh leaves and flowers, or their decoction, alone or in prescriptions, internally in pulmonary consumption and fever. Sushruta prescribed flowers, raw or cooked, in cough and pulmonary diseases. As cooked potherb, the plant was given in dyspepsia, anorexia, haemoptysis, cough and fever.

Vaasaa Ghrita (the extract of Vaasaa in clarified butter) was prescribed by Charaka and Sushruta in consumption and intrinsic haemorrhage. The formulation has been incorporated in Bhaishajya Ratnaavali, but is not available over the counter.

Sushruta prescribed clarified butter cooked with Vaasaa, Balaa (Sida cordifolia) or Guduuchi (Tinospora cordifolia) in chronic fever, oedema and anaemia.

Vaasaa Ghrita of Bhela Samhitaa has been recommended for malarial fever.

A classical compound of Bhaavaprakaasha, Vaasaavleha, is still available over the counter and is prescribed for consumption, bronchitis, asthma, haemorrhage and respiratory diseases.

Vaasaasava, a compound formulation of Yoga Ratnaakara, is another over-the-counter medicine, being prescribed in asthma and cough.

According to Shaarangadhara Samhitaa, the cold infusion of Vaasaa checks cough, intrinsic haemorrhage and fever. Vrindamaadhava recommends the juice with honey.

In classical texts, for treating cough, asthma and fever, Curcuma longa (Haridraa), Tinospora cordifolia, (Guduchi), Solanum xanthocarpum (Kantakaari) or Abies webbiana (Taalisapatra) were added to Vaasaa in prescriptions. (Shaarangadhara Samhitaa, Bhaavaprakaasha, Vrindamaadhava.)

Charaka added three myrobalans (Triphalaa)—Terminalia chebula (Haritaki), Terminalia bellirica (Bibhitaka) and Emblica officinalis (Aamlaki)—to Vaasaa for internal as well as external use in obstinate skin diseases. Vaasaa was also added to other blood-purifying, antiseptic herbs.

Sushruta was of the view that decoction of the roots should be taken regularly for longevity.

Active principles and pharmacology

As long ago as 1888, Hooper carried out the first chemical analysis of the drug and the main alkaloid vasicine was isolated.

Clinically, an extract made from fresh and dry Adhatoda leaves was given an extensive trial in Carmichael Hospital for Tropical Diseases, Kolkata, during 1925. The drug showed a definite expectorant action. In acute bronchitis, especially where the sputum was thick and tenacious, the cough was relieved and sputum was liquefied. The depression of the vagal terminations further relieved irritation and spasm of the bronchioles.

During the same period the extract was also tried in a number of cases of bronchial asthma, but relief afforded by the drug was not marked. Even synergistic action of atropine and vasicine was not very satisfactory.

The classical claim of the application of Adhatoda in tuberculosis was also examined. The drug was not found effective in preventing or curing the disease in experimental animals or human beings, though it relieved the irritable cough by its soothing action. Thus, the classical use of the drug in consumption could not be substantiated scientifically. The Indigenous Drugs Committee recorded these conclusions seventy five years ago.

Now a near-complete picture of active principles of the drug is available. The leaves, root and flowers contain vasicine, vasicinine, vasicinone, tritriacontane, beta-sitosterol, which impart a marked therapeutic property to the drug.

Recent investigations on vasicine showed bronchodilatory activity, comparable to theophylline, both in vitro and in vivo. Vasicine also exhibited strong respiratory stimulant activity, moderate hypotensive activity and cardiac depressant effect. Vasicinone was devoid of this activity. The cardiac depressant effect was significantly reduced when a mixture of vasicine and vasicinone was used.

Vasicinone showed bronchodilatory activity in vitro but bronchoconstrictory activity in vivo. Vasicine and vasicinone in combination (1:1) showed pronounced bronchodilatory activity in vitro and in vivo.

The utility of vasicine and vasicinone in asthma is not yet fully known. The bronchodilator action of vasicone was found to be 4000 to 5000 times less than isoprenaline, thus its inclusion as a drug in modern treatment of bronchial asthma seems difficult. The anti-anaphylactic activity of vasicinone compared well with disodium chromoglycate, an antiasthmatic drug. But it could not show any similarity with known bronchodilators, adrenaline, theophylline and aminophylline.

In addition to vasicine and vasicinone, a number of other alkaloids have been isolated from plant samples of other countries. The main application of the herb, in the form of fresh juice, decoction, infusion, powder, alcoholic or liquid extract, is for cold, cough, allergic cough, whooping cough, chronic bronchitis and, to some extent, in asthma, but not in consumption and haemoptysis.

Both vasicine and vasicinone were found to be devoid of antibacterial and antifungal activity, but the alcoholic and other extracts of the leaves and root showed antibacterial activity against Micrococcus pyogenes var. aureus and E. coli. The aqueous extract showed activity only against the former organism. An extract of the bark showed antiviral activity against potato-virus X.

In Bhaavaprakaasha, a claim was made that Adhatoda cured chronic skin diseases. This might be true to a great extent. An extract of the leaves showed significant antifungal activity against ringworm fungi. In traditional medicine, flowers and leaves form a good application to scabies and dermatosis. Some alkaloids present in the plant showed protection against allergin-induced symptoms. The leaf powder is used as a counter-irritant and to treat urticaria.

Cinna aatatotai (Tamil) has been equated with Adhatoda beddomei C.B. Clarke. This plant also contains bioactive alkaloids of Adhatoda vasica viz., vasicine and vasicinone, and is used in medicinal preparations in Kerala as antiemetic, antibechic and haemostatic.

Haemostatic properties of Adhatoda spp. have not yet been proved clinically. On the basis of classical references, the plant is used in traditional medicine for controlling postpartum haemorrhage.

Use in Western herbal

European practitioners have used it with success in intermittent and typhus fever, and also in diptheria as an adjuvant (Wren).

Adiantaceae

ADIANTUM

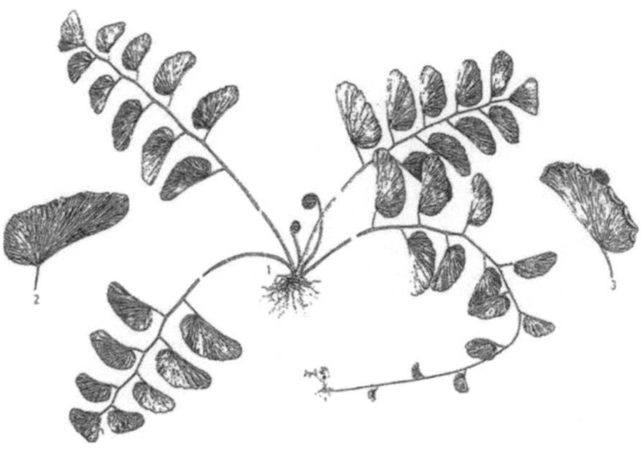

Figure 1 Adiantum philippense [ADPS]

Adiantum capillus-veneris Linn.

Habitat

Found mainly in Western Himalayas, ascending to an altitude of 2400 m and extending into Manipur. Common in Punjab, Bihar, Maharashtra and South India.

Classical & common names

Ayurvedic: Hamsapadi, Hamsavati, Hamsapaadikaa, Raktapaadi, Tripaadikaa, Prahlaadini, Kitamaari, Kitanaamaa, Madhusravaa, Godhaapaadika, Vikrantaa (Charaka). (Hamsapadi has also been equated with Adiantum philippense Linn., syn. Adiantum lunulatum Burm. INSA scientists, P. Ray et al, have equated Godhaapaadi of Charaka with Vitis pedata, syn. Cayratia pedata. In Kerala Desmodium triflorum (Linn.) DC. is used as Hamsapadi.
Unani: Parsiaoshan (Persian); Hansraaj.
English: Maidenhair Fern.

Parts used

Whole herb.

Dose

Powder 2–6 g, juice 10–20 ml.

Classical use

Charaka prescribed the herb for affections due to vitiated blood. The seeds were prescribed externally in suppurations due to poisonous bites and rheumatic conditions.

Sushruta used the herb internally in respiratory troubles, wasting diseases, internal tumours; also in swellings due to glandular enlargement.

The nasal drops, prepared by boiling the root in oil, were instilled in nose as a decongestant in hoarseness of voice (Ashtaanga Hridaya).

During the 16th century, Hamsapaadi was the drug of choice for alleviating vitiated blood, cough, asthma, dysurea, diarrhoea, hard swellings, inflammatory diseases and septic conditions. It was used as a demulcent, emollient, stimulating pectoral, emmenagogue, diuretic, blood purifier, febrifuge, styptic, cooling and antiseptic. Many prescriptions for foul ulcers, infected wounds, spider-poison, erysipelas contained this herb.

Active principles and pharmacology

Adiantum capillus-veneris: contains flavonoids (incl. rutin and isoquercitin), terpenoids (incl. adiantone), tannins and mucilage. These active principles provide the therapeutic coverage already known to Ayurvedic and Unani physicians in India.

Adiantum philippense/ Adiantum lunulatum, known as Walking Maidenhair in the West, and equated with Hamsapaadi, contains higher carotenoids. The fonds, made into a plaster, are applied to chronic, gouty and other swellings, incl. tumours. Most of the properties match with those of Adiantum capillus-veneris.

Use in Western herbal

In the West, the fern is known for its use as a demulcent, emollient, expectorant and emmenagogue. The fern is boiled in wine and drunk in cases of affections of spleen, liver and other viscera. It is also found useful in cold inposthumes (purulent swellings or abscess) of the uterus; also in hard swellings and sores.

John Gerard (1545–1612) wrote of it: "It consumeth and wasteth away King's Evil and other hard swellings and it maketh the haire of the head or beard to grow that is fallen and pulled off." (Swollen lymph glands caused by tuberculosis were called King's Evil.)

Nicholas Culpeper (1616–1654) advocated the virtues of the herb in no uncertain terms: "The decoction of the Maidenhair being drunk helps those that are troubled with cough, shortness of breath, the yellow jaundice, diseases of the spleen, stopping of the urine, and helps exceedingly to break the stone in the kidneys. It provokes women's courses, and stays both bleedings and fluxes of the stomach and belly, especially when the herb is dry; for being green, it loosens the belly and voids choler and phlegum, cleanses the lungs and rectifies the blood. The herb, boiled in oil of camomile, dissolves knots, allays swellings and dries up moist ulcers."

The herb has entered into many compositions in the West. It is employed as an emmenagogue under the names of polytrichi, polytrichon or kalliphylon, administered as a sweetened infusion of 1 oz (30 cc) to a pint (568 cc) of boiling water. A syrup is made in France from the herb's fronds and rhizomes, Sirop de Capillaire, and is given in pulmonary catarrh. The herb brings up phlegum, provides relief in severe cough, bronchitis and whooping cough.

The herb is also used in jaundice, gravel and other affections of the kidneys as a mild diuretic.

Caution

Emetic in large doses.

Rubiaceae

ADINA

Adina cordifolia Benth & Hook. f.

Habitat

Found scattered in deciduous forests throughout the greater part of India, ascending to an altitude of 900 m in sub-Himalayan tract.

Classical & common names

Ayurvedic: Haridru, Pitadaaru, Kadambaka (not to be confused with Kadamba). Known as Haldu in the trade.

Figure 1 Adina cordifolia—branch with inflorescences [WOI]

Siddha: Majakadambu.

Parts used
Bark.

Dose
Decoction 50–100 ml.

Classical use
According to Raaja Nighantu, the bark is cooling, antiseptic to wounds; cures skin diseases and improves complexion.

The decoction is prescribed as a tonic in anaemia and after fevers. Used internally and externally in skin diseases.

The bark is acrid and bitter, and is used in biliousness; also as a febrifuge. The roots are used as an astringent in dysentery.

Active principles and pharmacology
The heartwood contains a flavone, minor alkaloids and beta-sitosterol. The bark also contains 7–10 % tannins. These basic constituents match with the traditional use of the herb.

Chopra and Nadkarni described the herb as an antifertility agent, but when the plant was pharmacologically examined, it was found to be devoid of any such activity.

Caution
Not to be confused with Haridraa (Curcuma longa).

Rutaceae

AEGLE

Aegle marmelos (Linn.) Correa ex Roxb.

Habitat
Growing wild throughout the deciduous forests in India, ascending to an altitude of 1200 m in the western Himalayas.

Classical & common names
Ayurvedic: Bilva, Shaandilya, Shailusha; Shriphala Shailpatra, Tripatra, Mahaaphala, Mahaakapittha (Charaka); Maalura, Rudrajaataa, Rudranirmaalya, Shivajataakhya, Shalaatu.
Unani: Bel, Safarjal-e-Hindi (Arabic), Beh-Hindi, Shull (Persian). Pulp—Maghaz-e-Bel.
Siddha: Vilvam.
English: Bengal Quince, Bael Tree.

Parts used
Leaves, fruits, rootbark.

Dose
Fresh juice 10–20 ml, decoction 50–100 m, powder 3–5 g.

Classical use
Charaka prescribed tender fruits of Bilva with buttermilk in diarrhoea. In case of diarrhoea with blood, Sushruta prescribed tender fruits mixed with liquid jaggery, honey and oil.

Sushruta used unripe fruits internally in migraine, internal abscesses and for reducing obesity. He used ripe fruits internally as an appetizer and laxative, leaves for producing heat in the body and for toning up the digestive system.

According to Vrindamaadhava, decoction of Bilva and mango seed, mixed with honey and sugar, checks vomiting and diarrhoea. For colitis steamed tender fruits of Bilva with honey, followed by intake of buttermilk added with Chitraka (Plumbago zeylanica) powder, were advised. Paste of the tender fruit of Bilva, mixed with dry ginger powder and jaggery, was also said to alleviate severe colitis. The patient was kept on buttermilk diet during the course of treatment.

Bilvaadi Churna, a compound formulation of Chakradatta, and Bilvaadi Gutika, a compound formulation of Ashtaanga Hridaya are still available over the counter, and are indicated in diarrhoea, dysentery, sprue.

Bilvaadi Lehya, based on Sahasrayoga, is also a drug of choice in the South for diarrhoea, dysentery, dyspepsia and vomiting.

Bilva's dry fruit pulp is an ingredient in Shaarangadhara Samhitaa's Gangaadhara Churna and Ashtaanga Hridaya's Dadimaashtaka Churna. Both the compounds are in use extensively in diarrhoea and dysentery. Buttermilk is invariably given to the patient.

The rootbark of Bilva is one of the ingredients in Dashamuulaarishta, a renowned restorative tonic for women.

Active principles and pharmacology

Roots and fruits contain coumarins—scoparone, scopoletin, umbelliferone, marmesin and skimmin. Fruits, in addition, contain xanthotoxol, imperatorin and alloimperatorin alkaloids—aegeline and marmeline. Beta-sitosterol and its glycosides are also present.

The fruit and rind yield 7–9 % and 18–20 % tannin respectively.

Leaves yield essential oil containing alpha- and beta-phellandrene, rutin and marmesinin.

The essential oil from the rind of fruit showed broad spectrum and anti-fungal activity comparable to that of 0.5 % hamycin.

The unripe or half-ripe fruit is regarded as astringent, digestive and stomachic, used in chronic diarrhoea and dysentery. In after-treatment of bacillary dysentery, the fruit is a useful adjuvant, as it helps to remove constipation which hinders the healing of ulcerated surfaces of the intestines. The leaf extract reduced a period of convalescence in patients with choleraic diarrhoea.

Clinical trials of unripe fruits showed antiviral activity against Ranikhet disease virus and significant results against intestinal parasites, viz. Ascaris lumbricoides, Entamoeba histolytica and Giardia spp.

The Aegle is also used in diabetes. The juice of leaves with black pepper (Piper nigrum) is prescribed by herbal practitioners. The active principle in aqueous leaf extract shows activity similar to insulin. (Suppl. WOI, 2000) Clinical and pharmacological trials of unripe fruit and root exhibit promising results.

Aqueous and alcoholic extract of the leaves are reported to possess cardiotonic effect on amphibian and mammalian hearts. An alkaloid, aegeline, showed sympathomimetic property, and also produced stimulation of respiration and contraction in anaesthetised cats. The root as well as the bark is used in the form of a decoction as a remedy in melancholia and palpitation of the heart. Aurapten, the isolated constituent of the rootbark, is comparable with verapamil (a calcium channel blocker).

Marmin, a coumarin isolated from the roots, shows anti-inflammatory effect against carrageenin-induced inflammation in rats.

As in case of fruits, the roots also exhibited anti-amoebic property.

The most significant pharmacologically proved properties of Aegle are anthelmintic, antiviral, antidiarrhoeal, hypoglycaemic and cardiac stimulant. (CCRAS.)

Use in Western herbal

Aegle marmelos active principles—tannins, saccharides, starch, fatty oil and furocoumarins, furoquinolin alkaloids—have now entered into Western pharmacology and Aegle has been credited with digestive and astringent properties. Astringent leaves are being tried for treating peptic ulcer, as the drug is an established one for treating diarrhoea and dysentery and is capable of reducing irritation in the digestive tract.

Aegle is available in the West as a liquid extract.

Caution
Safe in designated therapies. Constipation is possible due to tannins, if taken in large quantities.

Amaranthaceae

AERVA

Aerva lanata (L.) Juss. ex Schult.

Habitat
Found throughout tropical India as a common weed in field and waste places.

Classical & common names
Ayurvedic: Gorakhagaanjaa, Aadaanpaaki, Shatkabhedi, Bhadraa. Accepted by The Ayurvedic Formulary of India as a substitute for Paashaanabheda.
Siddha: Serupeelai.

Parts used
Root.

Dose
Decoction 50–100 ml.

Classical use
Aerva lanata is cooling, diuretic, lithotriptic and used in haemetemesis, diabetes, lithiasis. It helps arrest haemorrhage associated with pregnancy. The roots are demulcent and are used in the treatment of headache and strangury. The whole plant is medicinal, though roots are often preferred.

Active principles and pharmacology
The plant is reported to contain alpha-amyrin, campesterol, beta-sitosterol, its palmitate, chrysin and flavonoid glucosides from heartwood.

These constituents justify the traditional use of the herb in the South; also its equation with Paashaanabheda. Pharmacological studies have confirmed that the roots possess diuretic, anti-inflammatory, anthelmintic, antibacterial and mild analgesic effects.

Aerva lanata is used in Tamil Nadu and Andhra Pradesh as Paashaanabheda, while in Kerala the

Figure 1 Aerva lanata [ADPS]

two plants, widely used as the sources of Paashaanabheda are Rotula aquatica Lour. (Boraginaceae) and Homonoia riparia Lour. (Euphorbiaceae).

Ayurvedic reference books, incl. the Ayurvedic Formulary of India, have equated Paashaanabheda with Bergenia ligulata (Wall.) Engl.

In addition to Aerva lanata six other herbs are also used as substitues for Paashaanabheda:Bridelia retusa Spreng. (Euphorbiaceae); Coleus amboinicus Benth, syn. Coleus aromaticus Benth. (Labiatae); Iris pseudacorus Linn. (Iridaceae); Kalanchoe pinnata Pers., syn. Bryophyllum calycinum Salisb. (Crassulaceae); Ocimum basilicum Linn. (Labiatae) and Rotula aquatica Lour. (Boraginaceae), syn. Rhabdia lycioides Mart.

Diuretic and lithotriptic properties are attributed to all these herbs.

After reviewing the literature, it is clear that Paashaanabheda is a medicinal property and not a specific herb in itself.

Use in Western herbal

None of the lithotriptic Indian herbs, known as Paashaanabheda, are being used in the West. The herbs of choice are: Crateva nurvula (Capparaceae—Varuna of Indian medicine); Ammi visnaga (Umbelliferae); Hydrangea arborescens (Hydrangeaceae); Ononis spinosa (Leguminosae) and Parietaria officinalis (Urticaceae).

Hippocastanaceae

AESCULUS

Aesculus hippocastanum Linn.

Habitat

Endemic to the mountains of Balkan Peninsula and Western Asia. Introduced in India and is occasionally grown as an ornamental tree. Indian spp. is equated with Aesculus indica.

Classical & common names

Vernacular: Pu (Punjab), Eka-paangaaraa (Almora region); Mandaar (non-classical).

Figure 1 Aesculus indica [WOI]

Unani: Baloot (National Formulary of Unani Medicine).
English: Horse Chestnut, Spanish Chestnut, Buckeye. (Entirely different from Sweet Chestnut.)

Parts used

Seeds, leaves, bark.

Potentially toxic, if ingested. Tincture of ripe kernel, trituration of dry kernel, tincture of fruit with capsule are used in homoeopathy.

Dose

Tincture and trituration up to third potency can be used under medical supervision. Lotion, ointment or gel can be applied externally to unbroken skin.

Classical use

The Horse Chestnut was first documented as a medicinal plant in 1565 in Pierandrea Matthioli's translation of Dioscorides Materia Medica. In the West, the bark was traditionally used as a tonic and febrifuge and in intermittent fevers. The fruits were employed in the treatment of rheumatism, neuralgia and haemorrhoids.

After 1956 more than 60 research papers brought into limelight many clinical uses of the herb which were hitherto not known to the medical world.

Active principles and pharmacology

Biologically active compounds of the leaf are triterpene, hydroxycoumarins (chief component is aesculin in addition to fraxin and scopolin), flavonoids incl. rutin, quercetin, isoquercetin, tannins.

The leaf is used in deep varicose veins, leg pains, phlebitis, thrombophlebitis, spastic pains before and during menstruation.

Active compounds of the seed are: Triterpene saponins (3–8 %, saponin mixture is known as aescin), beta-aescin, alpha-aescin; flavonoides (particularly biosides and triosides of quercetins); oligosaccharides; polysaccharides (starch 50 %); oligomeric proanthocyanidins, condensed tannins (only in the seed coat).

Use in Western herbal

German Commission E monograph recognized Horse Chestnut seed's efficacy in the following areas:

- Venous conditions
- Chronic venous insufficiency
- Nocturnal cramps in the calves
- Pruritus and swellings of the legs.

Horse Chestnut seeds are also used for postoperative soft tissue swelling, sprains, oedema and haemorrhoids.

Horse Chestnut reduces fluid retention by increasing the permeability of the capillaries and allowing the reabsorption of excess fluid back into the circulatory system. It prevents build up of fluids. The Horse Chestnut also improves the tone of blood vessels; veins become more elastic. With the boost in elasticity, they can contract more strongly and relax better. It is reported to combine well with other herbs that improve peripheral circulation.

In France, an oil extracted from the seeds has been used as an external treatment for rheumatism.

Standardized extract of Horse Chestnut seed, which provides a daily dose of 50 mg of aescin, "Vein Tea" or "Haemorrhoid Tea" as well as pharmaceutical preparations for the treatment of venous symptoms are available in the West.

Caution

Susceptible patients may exhibit some side-effects, such as itching, nausea, stomach discomfort. Contraindicated during pregnancy, Horse Chestnut may also interfere with the action of other drugs, especially warfarin (Coumadin, Marevan—a coumarin anticoagulant drug of modern medicine).

Agaricaceae

AGARICUS

***Agaricus albus* Linn.**

Habitat

Punjab, Asia Minor.

Classical &common names

Unani: Ghaariqoon.
English: White Agaric, Touchwood.

Parts used

Fungus of the Larch, Quercus and Fagus species.

Dose

In prescriptions, as specified; liquid Extract–3 to 20 minims., tincture (1 in 10)–20–60 minims.

Classical use

Used in Unani medicine to abort cough and cold, as an expectorant. Also prescribed in asthma.

Habb-e-Ghaariqoon, a compound drug, is given in a dose of 5 to 10 g daily.

Habb-e-Iyarij, a compound Unani drug with Ghaariqoon as one of the ingredients, is prescribed (250 to 500 mg daily; 7 to 1 g as a purgative) in facial paralysis, epilepsy, melancholia and constipation. The compound is a Unani brain tonic, diuretic and purgative.

Iyarij-e-Loghaziya, another Unani compound drug, is given (5–10 g daily) in hemiplegia, facial paralysis, cephalalgia, sciatica, gout, rheumatism, scrofule and skin diseases. Saffof-e-Lajjward (5–10 g daily) is prescribed in melancholia.

(All these compounds have been incorporated in National Formulary of Unani Medicine, Part I.)

In folk medicine, Ghaariqoon is given with honey in eruptive fevers to promote the rising of eruptions.

Active principles and pharmacology

Principal constituents of the drug are: resin, bitter extractive matter, gum, vegetable albumen and wax. Active principles are agaric and fungic or larcic acid; also phosphoric acid, potash, lime, ammonia, sulphur.

Agaricin, the resin, contains 97 % of agaric acid and 3 % of agaricol. Agaric acid occurs in minute crystals, soluble in alcohol, chloroform and ether. Boiled with water it forms a gelatinous solution. The drug exhibits astringent, expectorant, diuretic, and cathartic properties. In prescriptions it exerts a potent synergistic action in neurological disorders and inflammatory affections.

Use in Western herbal

Agaricaceae order of gall-bearing fungi comprises about 4600 species. Some members are poisonous, as Amanitas (Fly or Deadly Agaric), whereas others, as Agaricus, Cantharellus, are among the edible varieties.

The name Agaricum (as in Pliny) was applied by Dioscorides to a particular drug supplied by the Polyporus of the Larch, which was obtained principally from Agraria, a region in Sarmatia. Other Polypori were often substituted for that of the Larch and the name Agaricus became to a certain extent generic for Polyporus.

Polyporus officinalis was once a celebrated drug, known as white Agaric or Larch Agaric. It is not in use any more, though still available in herbal shops.

Caution

The drug paralyses the nerves of the sweat glands. Large doses act as an irritant to the stomach and intestines

Agaricus campestris Linn.
Psalliota campestris (Linn.) Fries

Habitat

The fungi is distributed in many parts of India, particularly on the hills and plains of northern and eastern India. Grows during the rainy weather on dead organic matter, e.g. rotting leaves and manure.

Classical & common names

Ayurvedic: Chhatraka, Chhatrakam, Samsvedana, Bhumi-chhatra, Khumbi.
Unani: Fitr (non-edible), Kamaat (edible).
Siddha: Venkodiveli.
English: Field mushroom.

Parts used

Whole fungi.

Dose

Powder 3–6 g.

Classical uses

Charaka considered it cool, sweet, heavy and used it for sinusitis and to abort cough and cold.

During the 16th century it was used in Ayurvedic as well as in Unani medicine as an aphrodisiac. Boiled in milk, it was given in debility and consumption. It was considered more nutritious than meat and was used like any other vegetable. (If taken in excessive quantity, it is irritating to the digestive system.)

Non-edible variety of the fungi was known as Sarpkshatri.

Active principles and pharmacology

Agaricus compestris contains 95.2 % moisture, 2.74 % protein, 1.6 % carbohydrates, 0.37 % fat, and 0.15 % ash. The mineral composition is as follows: calcium 0.002 %, phosphorus 0.15 %, and potassium 0.50 %; total iron 19.5 ppm.; available iron 5.95 ppm.; copper 1.35 ppm., and iodine 130–230 mcg/kg. Potassium salts and phosphates are the main constituents of the ash; alkaline in reaction.

Field mushroom is an excellent source of vitamins of the B-complex. The vitamins are well retained during cooking and processing. It also contains vitamin C (8.6 mg/100 g), vitamin K and vitamin D (0.63 i.u./g).

Field mushroom contains amylase, maltase, glycogenase, protease, catalase, tyrosinase, two phosphomonoesterases, two polyphosphatases and polyphenoloxidase. Dehydropeptidases I and II are also present.

Use in Western herbal

Agaricus is among the best edible variety of fungi in the West.

A highly toxic variety, Agaricus muscarius, is used only in homoeopathy for neurological problems characterised by involuntary movements, that too under strict medical supervision.

Agaricus muscarius was one of the most ancient hallucinogens with muscarine, muscimol, ibotenic acid as active constituents. It affects the central nervous system, causing twitching, vomiting, comatose states and hallucinations. Not in use any more.

Caution

Use only edible variety of field mushroom.

Agavaceae

AGAVE

Agave americana Linn.

Habitat

Distributed from southern United States to tropical South America. Introduced into India through Portugal as an ornamental plant.

Classical & common names

Folk medicine: Bana Kewaraa, Ghaayapaat (NAA); Kantala.
English: Century Plant, American Aloe.
(Incorporated in Homoepathic Formulary of India.)

Parts used

Sap.

Classical use

At the time of the Spanish conquest of America, the Aztec and Maya people were skilled in wound healing. They used Agave sap (often with egg-white) to bind powder or gums in pastes or poultices to be applied to wounds. A treatise of 1552 describes an Aztec treatment for diarrhoea and dysentery with Agave juice, combined with freshly ground Zea mays (maize) and extract of Utricularia spp. (bladderwort), was given as an enema.

Mexican alcoholic drinks, both Tequila and Mescal, are distilled from the fermented sap or juice of Agaves (alcohol content 3–4 %). These were used by Mexican people to treat nervous conditions.

Active principles and pharmacology

The leaves contain ten steriodal saponins designated as agavasaponins, six of these are spirostanolic and four furostanolic. The saponins, on hydrolysis, yield an aglycone, hecogenin, in each case. The sugars present are glucose, galactose, rhamnose and xylose. Leaves also contain piscidic acid.

Flowers gave chlorogenin, kempferol-3-glucoside and kempferol-3-rutinoside. Seeds gave neotigogenin, hecogenin and kammogenin.

The root and leaves contain following minerals (in ppm): Ca 1170/1260, Cr 1.1/1.4, Co 0.4/0.9, Fe 38.5/120, Mg 250/1120, K 170/1000 and Zn 3.4/9.5 respectively.

A proteolytic enzyme agavain (SH II), having strong hydrolytic activity on casein, has been isolated from the leaf juice.

From air-dried aerial parts of the plant a flavanone, agamanone has been isolated. A pentasaccharide glycoside has been isolated from the plant.

The saponins extracted from the plant are found to be lethal to Culex quinquefasciatus.

The sap contains estrogen-like isoflavonoids, alkaloids, coumarins and vitamins pro-A, B1, B2, C, D and K.

The root possesses diuretic properties, sap laxative, diuretic, emmenagogue and antiscorbutic.

Fresh juice of leaves is used externally for bruises and contusions and for its antiseptic and anti-inflammatory properties.

Related spp., Agave cantala Roxb. equated with Kantala, is naturalised in many parts of India.

Agave sisalana, cultivated in sub-tropical America and Kenya, is a source of hecogenin, used in the manufacturing of corticosteroids.

Use in Western herbal

Agave is considered demulcent, laxative and antiseptic. The sap is a soothing and restorative remedy for many digestive ailments. It is used to treat ulcers and inflammatory conditions affecting the stomach and intestines and protects them from infection and irritation, encourages healing. Agave is also employed to treat liver diseases.

In American folk medicine, Agave is used for spasm, coughs, accumulated phlegm, poor urination kidney inflammation and pain, urinary tract infection, wounds.

Tincture of Agave americana is used in homoeopathy and is indicated in constipation, poor appetite, scurvy, stranguary, swollen and bleeding gums; legs covered with dark, purple blotches, swollen, painful and hard.

Caution

Not to be used during pregnancy. Used in excess, may cause digestive irritation and even liver damage. Homoeopathic doses are safer.

Simaroubaceae

AILANTHUS

Ailanthus excelsa Roxb.

Habitat

Indigenous to central and southern India and found throughout Madhya Pradesh, in the Broach and Panchmahal districts of Gujarat, some coastal districts of Andhra Pradesh and of Ganjam and Puri districts of Orissa.

Classical & common names

Ayurvedic: Aralu, Katvanga, Dirghavranta, Putivrksha.
Siddha: Peruppi.
English: Tree of Heaven.

Parts used

Bark, leaves.

Dose

Decoction 50–100 ml.

Classical use

Charaka gave decoction of the leaves of Katvanga in prescriptions internally in pectoral leisons, diarrhoea, menstrual disorders.

Charaka considered Aralu and Shyonaka as two drugs but, later on, by the 16th century, Aralu entered into synonyms of Shyonaka. INSA scientists (P. Ray et al) equated Dirghavranta and Katvanga with Shyonaka, and Shyonaka with Bigonia indica Linn., Colosanthes indica Bl.; while Shyonaka has been equated with Oroxylum indicum Vent. (Bignoniaceae) by CCRAS. Aralu and Shyonaka should be treated as two different drugs.

Dirghvranta, Katvanga of Sushruta, was considered as an antiseptic, antitoxic, astringent and styptic and was used for non-healing ulcers, female disorders, persistent diarrhoea and dysentery. The leaves were used in affections of the mouth.

Mahaanimba has wrongly been used as a synonym of Aralu by CCRAS, it has been correctly equated with Melia azedarach Linn. (Bakaayan).

Active principles and pharmacology

Quassinoids, isolated from the stem and rootbark, have shown substantial anti-tumour and cytotoxic activities against the P388 lymphocytic leukemia and KB test systems respectively. Four alkaloids—canthin-6-one; laevo-methoxycanthin-6-one; 5-methoxycanthin-6-one and 8-hydroxycanthin-6-one—have also been isolated from the rootbark.

The bark also contains ailantic acid; 2, 6-dimethoxybenzoquinone; beta-sitosterol, malanthin, triacontane and hexatriacontane.

The bark is a good substitute for Holarrhena antidysenterica (Kurchi).

The leaves are reported to be used as an adulterant of Adhatoda vasica (Vaasaa). They contain ailantic acid, beta-sitosterol and vitaxin. The green and yellow summer leaves contain 1.18 and 0.3 mg/g carotenoids respectively.

Due to bitter, astringent, febrifuge, anthelmintic, antispasmodic, antiseptic and expectorant properties, the bark is used for asthma, bronchitis, dysentery, dyspepsia and in earache.

A decoction of the leaves is used for washing wounds, swellings and skin eruptions.

Use in Western herbal

In the West, Ailanthus altissima, syn. Ailanthus glandulosa is known as the Tree of Heaven or Chun Pi in Chinese medicine.

The bark contains quassinoids (such as ailanthone and quassin), indole alkaloids, flavonols and tannins like related species found in India.

In Chinese herbal medicine, as well as in the West, the Tree of Heaven is used to treat dysmenorrhea, diarrhoea and dysentery, especially if there is blood in the stool. Excellent results have been achieved.

The bark of the tree has been used in Australian medicine to counter worms, excessive vaginal discharge, gonorrhoea and malaria, and it has been given for asthma due to its marked antispasmodic properties. In Africa, the Tree of Heaven is used for cramps, asthma, fast heart rate, gonorrhoea, epilepsy and tapeworm infestation. The anticancer properties of quassinoids are being extensively investigated in the West.

Caution

To be used only under medical supervision.

Alangiaceae

ALANGIUM

Alangium salvifolium (Linn.f.) Wang.

Alangium lamarkii Thwaites

Habitat

Widely distributed over the plains and foothills throughout the greater part of India.

Classical & common names

Ayurvedic: Ankola, Nikochaka (Charaka); Ankota (Charaka, Sushruta); Ankokalaka, Dirghakeelaka, Tamraphala, Guptasneha.

Figure 1 Alangium salvifolium [ADPS]

Siddha: Azhinjil.
English: Sage-leaved Alangium.

Parts used

Rootbark, fruits, leaves, seeds.

Dose

Powder 1–2 g.

Classical use

Charaka prescribed the fruits extracted in clarified butter in chronic poisoning, and in epilepsy as a tonic and sedative. Sushruta used the herb externally as a cleansing and antiseptic agent.

Powdered seeds and fruits were applied externally in sinusitis. Flowers were prescribed internally in migraine, internal abscesses, calculi, urinary diseases and for reducing obesity.

In the form of a medicated salt, the drug was given by Sushruta in abdominal swellings, dyspepsia, intestinal parasites, piles and in rheumatism.

According to Sushruta, the seeds, cooked into an edible preparation, relieve bronchial asthma.

The rootbark, taken with buttermilk, is an efficacious remedy for diarrhoea (Bangasena). Chakradatta prescribed 5 g rootbark, pounded with rice-water, for treating diarrhoea and dysentery; also with Berberis aristata (Daaruharidraa) and Cissampelos pareira (Paathaa) for balancing the purgative action of Ankola.

Active principles and pharmacology

Rootbark is a rich source of alkaloids structurally related to ipecac alkaloids, e.g. emetine. From the various parts of the plant benzoquinolizine and benzopyridoquinolizine alkaloids and a protoberberine alkaloid have been isolated. Roots, leaves and fruits yield loganic acid and alangiside, a monoterpenoid alkaloidal glucoside.

Rootbark is purgative. In Ayurvedic medicine, it is invariably given with other intestinal antiseptics and soothing drugs for treating diarrhoea, dysentery, biliousness and colic. The extract of the rootbark shows potent anthelmintic effect.

The fruits, astringent and acidic, exhibit antiphlegmatic, laxative, tonic and refrigerant properties and are found useful in emaciation and haemorrhages, strangury and consumption.

The leaves are applied as poultice in rheumatism. The alkaloidal extract of the leaves showed mild adrenolytic, non-specific antispasmodic, hypotensive and anti-cholinesterase activity.

The seeds exhibit antiseptic, soothing and anti-inflammatory properties in the treatment of haemorrhage as well as skin diseases. In Ayurvedic medicine, seeds have been in use for bronchial asthma as a soothing agent.

Caution

To be used internally with supporting herbs: bufallo's buttermilk, rice-water, Berberis aristata, Cissampelos pareira; Aconitum heterophyllum (1 part to 3 parts of Ankola in colitis).

Mimosaceae

ALBIZIA

Albizia lebbeck (L.) Willd.

Habitat

Throughout India, up to 900 m in the sub-Himalayan tract and also in the Andamans.

Classical & common names

Ayurvedic: Shirish, Kapitana, Mridupushpa, Shukapriya, Bhandila, Bhandi.
Unani: Siras, Darakht-e-Zakaria (Arabic).
Siddha: Vagai.
English: Siris Tree.

Parts used

Bark, seeds, leaves, flowers.

Dose

Decoction 50–100 ml.

Classical use

Charaka prescribed the juice of flowers with Piper longum (Pippli) and honey in asthma. Sushruta gave flowers with the flowers of Musa paradisiaca (Kadali) with Piper longum (Pippli), followed by the rice water, for asthma.

For skin diseases, the bark with flowers of Gossypium herbaceum (Kaarpaasa), leaves of Cassia fistula (Aaragavadha) and the whole plant of Solanum nigrum (Kaakamaachi) was prescribed internally as well as externally (Charaka Samhitaa, Sushruta Samhitaa, Ashtaanga Sangraha, Ashtaanga Hridaya).

For eruptive boils the bark of Shirish along with the barks of Ficus glomerata (Udumbara) and Syzygium cuminii (Jambu) was prescribed in the form of a paste for external application (Vrindamaadhava).

For erysipelas, the flowers of Shirish, mixed with little clarified butter, were to be applied (Charaka Samhitaa).

Sushruta used seeds as an ingredient of an antiseptic dusting powder, and of an ointment for a virulent form of ringworm.

Both Charaka and Sushruta used powdered seeds, bark and leaves in toxic conditions, hemicrania, urinary anomalies, oedema, urethral discharges, in prescriptions.

No classical compound of the herb is available over the counter. General practitioners still use powdered rootbark for strengthening gums and teeth, and the juice of leaves in night blindness. Unani physicians use the seeds and flowers as a spermatic tonic. Flowers are given as a sex drug for achieving retention and to retard premature ejaculation. Seeds are given with milk for promoting the quality of semen. The powder of seeds is also prescribed in impotency. All these traditional uses are to be treated as a part of folk medicine with few classical references.

In China, the uses of Albizia julibrissin Durazz (known as Laal Shirish in India) are said to be similar to the Indian species, Albizia lebbeck. But medicinal applications are quite different.

In China, dried bark is used in decoction for depression, restlessness and insomnia. Flowers are also used for poor memory, insomnia and as a tonic for weak nerves.

In India, when the bark and flowers are used in asthma, their tranquillizing properties must be playing a significant role in allaying restlessness and violent bouts of cough.

Active principles and pharmacology

The bark yields tannins (7–11 %) of condensed type, and a new leucoanthocyanidin, lebbecacidin. It also gives friedelin and beta-sitosterol. Extract of the bark possesses anthelmintic activity and expectorant action.

A decoction of the bark and flowers was found efficacious in bronchial asthma. It protected the guinea-pig against histamine as well as acetylocholine-induced bronchospasm. Prolonged treatment with bark decoction protected sensitized guinea-pigs against antigen challenge.

Tropical pulmonary eosinophilia patients treated with the flower powder (200 mg twice daily for six weeks) showed excellent response.

The leaves are reported to contain caffeic acid, alkaloids, kaempferol and quercetin.

The seeds contain saponin, lebbekanin A, and echinocystic acid, and are astringent. Give encouraging results in piles and diarrhoea.

Taking into account the tranquillizing properties identified in Chinese medicine and antihistaminic properties pharmacologically proved in India, Albizia lebbek can safely be used in urticaria, allergic dermatitis, bronchial asthma and topical eosinophilia.

Pods of the plant exhibited antiamoebic property against Entamoeba histolytica. The root was found to possess antifungal property against Helminthosporium sativum. The bark possessed mild antibacterial property against E. coli. Thus, the plant can be used for its antiamoebic, antidiarrhoeal, antifungal and antibacterial properties. Anticancer activity of the pods is being investigated.

Papilionaceae

ALHAGI

Alhagi camelorum Fisch. ex DC.
Alhagi pseudalhagi (Bieb.) Desv.

Habitat

Widely spread in the Ganges valley and in the acrid tracts in Punjab, Rajasthan, Gujarat and Maharashtra, ascending to an altitude of 900 m.

Classical & common names

Ayurvedic: Yavaasa, Yaas, Yavaas, Duhsparshaa, Dhanyavaas, Dhanvayaasa, Kunaashak. Yavaasaka (Charaka); Dhamaasaa; Yaas-sharkaraa (Manna).
Unani: Jawaasaa; Haaj, Al-ghaul, Aaqool (Arabic); Khaar-e-shatur (Persian); Turanjbin (Manna).
English: Persian Manna Plant, Camel Thorn.

Parts used

Whole plant, Manna.

Dose

Decoction 50–100 ml.

Classical uses

Charaka prescribed the decoction of the plant in prescriptions internally for fever. Also used the herb in prescriptions for intrinsic haemorrhage. Sushruta used it in cough in the form of an avaleha

(Agastyaavaleha) and in chronic fever in the form of a medicated clarified butter. The compound was known as Kalashyaadi Ghrita.

By the 16th century, Yaas was the drug of choice as a laxative for cleansing the digestive system. It was prescribed in fever, vomiting, cough, biliousness and in diseases of the skin.

According to Bhaavaprakaasha, Yavaasa or Yaas and Duraalabhaa possess similar properties. Yavaasa has now been equated with Alhagi pseudalhagi, syn. Alhagi maurorum, Alhagi camelorum; and Duraalabhaa or Dhanvayaasa with Fagonia cretica. Ayurvedic Formulary of India also equated Dhanvayaasa with Fagonia cretica and Alhagi pesudalhagi as its substitute.

In Unani medicine, Turanjbin, a sweet sugary excretion of the plant, is used as an expectorant, anti-emetic and laxative. The secretion is not found on the Indian plants and is collected from the plants of Turkey, Iraq and Iran.

Infusion of the plant is used for its diaphoretic action. A decoction of the twigs is given for cough. The medicated oil prepared from the leaves is used as an external application for rheumatism, swellings and abscesses. The plant, due to its cooling, bitter, anti-bilious, antiseptic, laxative, diuretic, diaphoretic, expectorant properties, has been included in compound formulations of Duraalabhaarishta, Dashmuulaarishta and Raasanndi Kashaaya (all from Sahasrayoga).

Active principles and pharmacology

The plant contains tannins, flavonoids, coumarin derivatives, ascorbic acid and essential oil. Rutin and quercetin are the main flavonoids.

The root and fruits show the highest level of tannins. The highest concentration of flavonoids (mainly rutin) is found in the leaves.

Quercetin, gum, resin and a colouring matter are reported in the root. The alkaloids found in the twigs have also been extracted from the root, but the yield is poor.

A neutral proteinase, alhagain, has been isolated from the dried shrub.

Turanjbin of Unani medicine contains: water 4.6, sucrose 41.5, melizitose 25.0, reducing sugars 9.0, ash 3.4 and impurities 14.0 %.

An alkaline extract of the shoot showed antibiotic activity against Micrococcus pyogenes var. aureus.

Use in Western herbal

The term Manna, in Western herbal, is applied to the saccharine exudence of a number of plants: Alhagi maurorum (Alhagi Manna), Quercus vallones and persica (Oak Manna), Tamarix gallica var. mannifera (Tamarisk Manna) and Larix europoea (Briancon Manna).

Manna is used as a children's laxative and is official in the United States Pharmacopoea.

Alliaceae

ALLIUM

Allium ascalonicum Linn.

Habitat

Cultivated in all countries of the East.

Classical & common names

Ayurvdic: Granjan, Grnja, Utt (Sushruta); Ek-potia lehsun.
Unani: Gandanaa (Persian), Gandhavaa, Karaas (Arabic); Ek kanda-lasun.
English: Shallot, Leek, Porret.

Parts used

Whole plant, bulb, seeds.

Classical use

Sushruta prescribed Grnja, in prescriptions, internally in indigestion, anorexia, colic pain and in severe diarrhoea.

Gandanaa is used mainly in Unani medicine..

The bulbs are used in the same way as garlic for culinary purpose. Leaves are not used as vegetable.

The whole plant is applied externally for inflammations and for allaying irritations. The seeds are diuretic and expectorant.

Compound formulations for bleeding piles are processed in the juice of Gandanaa leaves. The seed is also used in prescriptions for bleeding piles.

Two varieties of Gandanaa have been mentioned in Unani texts—Shaami and Nafti. The seeds of both the varieties are used in aphrodisiac formulations, as a tonic to the reproductive system. Shaami variety is specific for dysentery. Roasted seeds are prescribed.

Active principles and pharmacology

The aqueous extract of the bulb contains mainly fructose (50 %), besides glucose, arabinose, galactose, mannose, rhamnose, glucosamine and uronic acid. It also contains lysine, glutamic acid and aspartic acid.

In experiments on rabbits, the alcoholic extract of the bulbs showed significant anti-coagulant, fibrinolytic and hypo-cholesteremic properties.

The brown skin of the bulbs contains quercetol. Fresh leaves contain flavonoids.

Caution

Constipative.

Allium cepa Linn.

Habitat

A native to Palestine, cultivated in all the countries of the East.

Classical & common names

Ayurvedic: Palaandu, Durgandh, Mukha-doshaka, Raktakanda (Charaka).
Unani: Piyaaz (Persian), Basl (Arabic).
Siddha: Vellai Vengayam.
English: Onion.

Parts used

Bulb, seeds.

Dose

Juice 10–20 ml, seeds 1–3 g.

Classical use

Charaka gave the tubers and the stalks in vegetable salad and soups; also in prescriptions, as appetizer and aphrodisiac. Cough, scanty stools and haemorrhage were treated with the red variety of Palaandu.

According to Charaka, Palaandu alone or mixed with meat-soup or gruel checks bleeding. For bleeding piles, Charaka prescribed vegetable of Palaandu soured with Badari fruits (Ziziphus jujuba) and taken with buttermilk.

According to Siddha-bheshaja-manimaalaa, Palaandu juice should be taken in a dose of 80 ml for gastroenteritis. As a home remedy, paste of bulbs, mixed with curd and salt, is given in diarrhoea and dysentery.

In Unani medicine, seeds are used in sex-tonics. Majoon-e-Piyaaz is a reputed aphrodisiac with retentive properties and is prescribed in sexual debility, spermatorrhoea and premature ejaculation. This classical compound has been incorporated in National Formulary of Unani Medicine. Seeds pounded with honey are applied externally for removing freckles.

In Unani medicine, Piyaaz is considered to be diuretic, anti-inflammatory, analgesic, antirheumatic and an antibiotic for cough, cold and flue.

Ayurvedic physicians give more importance to Allium sativum (Rasona).

Active principles and pharmacology

Onion contains a volatile oil with many organic sulphur compounds, such as allicin and alliin, flavonoids, phenolic acids and sterols, its esters and glycosides.

Skin of the bulb yields phenolic acids. Skin also yields flavonoids—quercetin and its glycosides. Roots yield gibberillin.

Onion juice also contains monosaccharides and oligosaccharides.

Pharmacological studies have added hypocholesteramic, hypoglycaemic and antibacterial properties to the onion.

CCRAS has reported a clinical trial on ischaemic heart disease patients. 60 mg of onion was given per patient and effect on serum lipid and blood coagulation was investigated. Onion decreased the value of euglobin lysis time; counteracted the raised level of serum cholesterol; serum lipids; triglycerides; clotting time; prothrombin time.

Onion produced a sustained hypocholesteraemic and fibrinolytic enhancing effect on a long-term basis in experimentally induced hypercholesteraemia in rabbits. In induced alimentary lipaemia in man, onion taken raw or boiled prevented the rise in serum cholesterol. The effect was evident on taking onion after 2 hours of taking fatty meal.

Hypoglycaemic activity of onion was observed when regular use of onion (50 g/day) reduced the insulin requirement of a diabetic patient from 40 to 20 units a day.

Onion extract also reduced blood sugar during intravenous glucose tolerance test and adrenaline-induced hyperglycaemia in man.

Graded doses of aqueous extract of the bulb lowered the blood sugar as well as serum cholesterol and serum urea level in alloxan diabetic albino rats.

Bactericidal activity of the fresh juice of onion, when tested, exhibited promising results. It inhibited the growth of Bacillus subtilis, Micrococcus pyogenes var. aureus, E. coli, Pseudomonas pyocyaneus and Salmonella typhi.

Aphrodisiac potential of onion was also screened. Studies on genotoxicity of onion on estradiol-treated mice revealed a significant increase in the weight of testes and epididymes and the sperm count was also significantly higher. This, in a way, validates the use of onion in Unani medicine as a sex-tonic.

Use in Western herbal

German Commission E monograph recognized the herb's efficacy in the following areas:

- Loss of appetite
- Dyspeptic complaints
- Common cold
- Cough/bronchitis
- Fevers and colds
- Inflammation of the mouth and pharynx
- Hypertension
- Atherosclerosis.

Antiasthmatic and antiallergic effect of onion was studied by Wagner et al during 1988 in Germany. The oral administration of onion juice protected the guinea-pigs (sensitized using ovalbumin) from asthma attack. Administration of an ethanol onion extract significantly reduced allergy-induced bronchial constriction in asthma patients.

In homoeopathy, the tincture of onion is used in third potency for coryza with acrid nasal discharge and laryngeal symptoms, influenza, whooping cough, chronic neuritis, neuralgic pains and catarrhal headache. Clarke wrote in 1900 that a raw onion eaten just before going to bed is a remedy for sleeplessness. The red variety is used in homoeopathy.

Allium sativum Linn.

Habitat

Native to Central Asia and cultivated all over India.

Classical & common names

Ayurvedic: Rasona, Lashuna, Urgragandha, Yavaneshta, Mahaushadh, Arishta.
Unani: Seer (Persian); Soom, Foom (Arabic).
Siddha: Poondu.
English: Garlic.

Parts used

Bulb.

Dose

Paste 3–5 g.

Classical use

According to Ayurvedic texts, Charaka Samhitaa, Sushruta Samhitaa, Raaja Nighantu and Bhaavaprakaasha, Rasona is a nourishing tonic for the whole body, age-sustainer, rejuvenator, aphrodisiac; boosts up immune system, promotes strength, complexion and eyesight. It is prescribed in cardiac diseases, chronic fevers, tumours; also for throat infections, cough, asthma, oedema. It helps in the healing of fractures. Cures constipation, colic, anorexia, deranged digestion, intestinal parasitic infections, piles. It was used in prescriptions as a carminative and laxative.

Rasona was used internally as well as externally for treating obstinate skin diseases incl. leprosy. The paste was applied to wounds for killing organisms. It was used as a potent antiseptic, disinfectant and healing agent in a host of degenerative diseases.

In a prescription of Charaka, dehusked and dried 160 g Rasona is boiled in four times of milk and four times of water till water is evaporated. Intake of this milk was prescribed in cardiac disorders, fevers, tumours, abscesses, oedema, sciatica, arthritis, neuralgia.

According to Kaashyapa Samhitaa, the herb is one of the best fertility-promoting drugs. Ashtaanga Sangraha also incorporated this property in its text while describing it as an aphrodisiac and rejuvenating agent.

Among Ayurvedic compounds, Lashunaadi Kashaaya (Sahasrayoga) is used for digestive ailments; Lashunaadi Bati for loss of appetite, indigestion, flatulence and anorexia; Lashunaadi Ghritam (Sahasrayoga) is prescribed in colic pain, internal tumours, piles, anemia, splenic disorders, parasitic infestation, fever and diseases of the female genital tract; Rasonaadi Kwaath (Bhaishajya Ratnaavali) and Rasona Tailam (Chakradatta), the extract of the herb in oil, are given internally for rheumatic affections.

Among Unani medicine, two compounds are still in use—Majoon-e-Seer Alvikhani, a nervine tonic for hemiplegia and debility; and Raughan-e-Seer, a medicated oil, indicated in arthritis, rheumatism and hemiplegia for external application.

Garlic oil as pearls was in use for quite some time, but the consumer did not accept it as a herbal supplement.

Medicines, based on garlic, are facing oblivion in India.

Active principles and pharmacology

Garlic's antibiotic constituent remained a mystery until the 1920s, when researchers at Sandoz Pharmaceuticals in Switzerland isolated alliin from the herb.

Louis Pasteur, who developed the germ theory of disease, first demonstrated back in 1858 that allicin is a strong antibacterial. Before the advent of modern antibiotics in the 1930s, cuts and abrasions were treated by expressing garlic juice into the wound. During World War II the Soviet army relied on garlic to treat the wounded when penicillin and sulfa drugs were not available.

Researchers have found that one medium-size garlic clove packs the antibacterial punch of about 100, 000 units of penicillin. Depending on the type of infection, oral penicillin doses range from 600, 000 to 1.2 million units. The equivalent in garlic range from 6 to 12 cloves (3 cloves at a time 2 to 4 times a day). Garlic must be chewed or crushed to transform its medicinally inert alliin into antibiotic allicin.

While treating infectious diseases, garlic should be included in the sensitivity test in which several antibiotics are tested against the germ. (The one that is most effective is prescribed by the physician). Otherwise, garlic should be considered as a defence against bacterial infections.

The chemistry of garlic has been extensively investigated. Its bulbs contain an odourless sulfur-containing amino acid derivative known as alliin (S-allylcysteine sulfoxide). When the bulbs are ground, alliin comes into contact with the enzyme alliinase, which converts alliin into allicin (allyl-2-propenthio-sulfinate), a potent antibacterial agent, which then forms the odorous volatile diallyl disulphide.

Diallyl disulphide is a constituent that is antibacterial even when diluted to 1:125, 000.

Due to these active principles, the garlic extract showed antibacterial activity inhibiting the growth of E. coli, Salmonella typhosa, Shigella dysentriae and Micrococcus pyogenes var. aureus.

Scientists were also able to kill Helicobacter pylori by using garlic extract. (H. pylori is considered to be involved in the development of peptic ulcer and stomach cancer.) (Gale Maleskey et al.)

The garlic extract also showed an antifungal effect on yeast and moulds. (In China, garlic is used for vaginitis, Candida and other fungal infections.)

The ability of garlic to provide some protection against atherosclerosis, coronary thrombosis and stroke is believed to be related, at least in part, to its ability to inhibit aggregation of the blood platelets. This property, initially attributed to allicin, is now known to be due to an ajoene, a self-condensation product of allicin. As an antithrombotic agent, ajoene is at least as potent as aspirin and its activity is enhanced by other breakdown products of the garlic.

Allicin is found to lower serum cholesterol by blocking its biosynthesis. Fresh garlic, garlic juice, aged garlic extracts or the volatile oil—all lowered cholesterol and plasma lipids, lipid metabolism, and atherogensis both in vitro and in vivo.

Methyl-allyl trisulfide of the garlic helps expand constricted blood vessels, thereby preventing high blood pressure.

The synergic action of active principles of garlic (volatile oil; alliin, allicin; scordinins; selenium; vitamin A, B, C and E) exhibits antioxidant effect. Experimentally it has been established that diallyltetra-, -penta-, -hexa- and -heptasulfides are potential antioxidants. Thus, garlic can also help prevent secondary infection in people who have lowered immunity.

Garlic is one herb that is recommended for everything from tuberculosis to cancer, but Tyler has rightly advised herbalists to utilize the herb for its hypocholesterolaemic, hypotensive and vasodilatory properties, in addition to its traditional application in cough, cold, flu, dyspepsia and digestive disorders.

Use in Western herbal

German Commission E monograph recognized the herb's efficacy in the following areas:

- Atherosclerosis
- Elevated levels of lipids in blood
- Common cold
- Cough/bronchitis.

The drug is also recommended as a preventive measure for age-related vascular changes.

For abdominal distress, blenching, flatulence and nausea, a dried garlic preparation is given (1000 mg/day).

For controlling hypertension 600–900 mg daily (equivalent to 1.8–2.7 g/daily fresh garlic) is given as a supplement (clinical efficacy has not yet been established).

For hyperlipidaemia, 600–900 mg of dried garlic powder or 10 g of raw garlic or 18 mg of garlic oil or aged garlic extract (dosage as per product literature) is being recommended; for patients suffering from atherosclerosis 600–1200 mg of dry garlic powder. A daily dose of 900 mg of garlic powder is indicated in fibrinolysis and platelet aggregation.

Hypoglycaemic effect of garlic has not yet been substantiated clinically, though both onion and garlic are given in natural form (50 g daily) for reducing the dose of insulin and for keeping diabetes under control.

In health-food stores liquid or capsules of Kyolic aged garlic extract, Kyolic EPA (a combination of aged garlic extract and fish oil) are available in the United States.

Oil-based garlic preparations are questionable, since allicin is unstable in oil.

According to Tyler, dried garlic preparations are most effective if they are enteric coated. This will allow them to pass intact through the stomach to the small intestine where the enzyme can act under the alkaline conditions present there. Such enteric coated capsules or tablets of freeze-dried garlic will be more effective than even fresh garlic in which alliinase is destroyed in the stomach. Fresh garlic releases its active constituents mainly in the mouth during chewing, not later in the stomach.

In homoeopathy, tincture of fresh garlic bulbs is given in periodical asthma, bronchitis, catarrh, cough, constipation, colic, dyspepsia, diarrhoea, worms, rheumatism, menorrhagia and other menstrual disorders, skin infections. Acts directly on intestinal mucus membranes. It increases peristalsis.

Araceae

ALOCASIA

Alocasia indica (Roxb.) Schott.

Habitat

A genus of topical Asia, Malaysia and the Pacific. Found wild and cultivated all over India.

Classical & common names

Ayurvedic: Maanaka, Maankanda, Maankandaka, Maan, Mahaapatra.
Tamil: Merukankilangu.
English: Giant Taro.

Parts used

Tubers, leaves.

Dose

Powder of old tuber 3–10 g.

Classical use

Bhaavaprakaasha attributed anti-inflammatory properties to the tubers, which were prescribed in abdominal inflammations, internal tumours, diseases of the spleen, anasarca, colitis. Chakradatta's confection, milk boiled with the tubers, was known as Maanaka-paayasa. Another tonic of Chakradatta, Maan-manda, was prepared by boiling 80 g old tubers, 160 g rice flour, 240 ml milk and 240 ml water, till water evaporated. Powder alone was also prescribed. In Bhaishajya Ratnaavali, only one compound formulation of the herb,

Maankaadi-gudikaa, has been incorporated. This compound is specific for diseases of the spleen, abdominal swellings, piles and colitis.

None of these classical compounds are available over the counter and are used sparingly.

Active principles and pharmacology

Tubers contain sterols—campesterol, cholesterol and beta-sitosterol, and trypsin/chymotrypsin inhibitor. Biological activity is diuretic, mild laxative, cooling. The rootstock is found efficacious in inflammations, diseases of the abdomen and spleen, and in piles.

The leaves are styptic, astringent, used against tumours. Leaves greased with coconut oil, are heated and applied hot to sprains, bruises, dislocated joints and swellings.

Young leaves contain up to 0.018 % hydrogen cyanide and a mixture of triglochinin and iso-triglochinin. All parts of the plant, except rhizome, contain cyanogenic principle. The presence of a flavonoid and a saponin has been reported from the plant.

Classical use of tubers in anasarca, spleen enlargements and abdominal tumours is still pharmacologically valid.

Liliaceae

ALOE

Aloe barbadensis Mill.
Aloe vera (L.) Burm. f.

Figure 1 Aloe vera [ZANDU]

Habitat

Cultivated throughout India, wild on coasts of Maharashtra, Gujarat and South India.

Classical & common names

Ayurvedic: Kumaari, Grihkanyaa, Ghritkumaarika, Kanyaa, Kanyakaa, Mandalaa, Maataa, Gheekuaar. Dry concentrate: Kumaari Saara.
Unani: Sibr (Persian). Dry concentrate: Musabbar (Arabic), Ailwaa.
Siddha: Kattrazhai.
English: Barbados Aloe, Curacao Aloe, Indian Aloe, Jafarabad Aloe.

Dose

Fresh pulp 10–20 ml, dry concentrate 120 mg.

Classical use

Aloe barbadensis entered into Ayurvedic medicine after the 12th century. Bhavaaprakaasha attributed cooling and cathartic properties to its dry concentrate. It was prescribed in the diseases of liver, spleen, internal tumours, chronic cough and fever, as well as in vitiated blood, skin diseases and toxic conditions. Raaja Nighntu and texts of the 17th century included the herb's dry concentrate in abortifacient and emmenagogue compound formulations. It also entered into anthelmintic prescriptions.

Rajhpravartini Vati (Bhaishajya Ratnaavali) is even now the largest-selling emmenagogue compound being prescribed in amenorrhoea and dysmenorrhoea (dose: 250 mg to 750 mg in divided doses/day under medical supervision, as it also contains clax of borax , purified asafoetida and purified ferrous sulphate).

Kumaari Saara-based Kumaarikaa Vati (Bhaishajya Ratnaavali) is specific for female genital tract ailments.

Kumaaryaasava (Shaarangadhara Samhitaa) is also available over the counter and is indicated in the treatment of enlargement of liver and spleen, anaemia and chronic constipation. In this compound the principal ingredient is Aloe leaf juice.

In Unani medicine, Habb-e-Mudir, is based on aloe dry concentrate (Musabbar). Other ingredients are saffron and green vitreol. This is prescribed in amenorrhoea, seven days earlier to the date of period.

Habb-e-Iyarij and Iyarij-e-Faiqra are purgative and diuretic Unani compounds, which are still used for cleansing stomach, intestines and kidneys, while giving treatment for rheumatism, paralysis and other nervine disorders.

Active principles and pharmacology

Leaves contain cathartic anthraquinone glycosides as active principles, most of them being C-10-glucosides, mainly barbaloin, a glucoside of aloe-emodin. Other constituents of the plant are aloesin and aloesone.

Indian aloes contain aloinosides as major constituents with only traces of aloin.

Three forms of products are prepared from leaves: Aloe gel, Aloe powder, and Barbados aloe.

Aloe vera gel is obtained from the parenchymatous cells in the leaves. It is a colourless mucilaginous gel. To avoid unwanted laxative effect, the gel is kept free from aloe latex.

The mucilage of leaves contain glucose, galactose, mannose and glacturonic acid, in addition to an unidentified aldopentose and a protein (0.013 %) with 18 amino acids.

The gel or mucilage is a wound-healing agent in its fresh state. Scientific evidence of aloe's wound-healing properties was first documented in 1935 when an American medical journal reported the case of a woman whose X-ray burns were successfully treated with aloe gel scooped straight from leaves.

Several studies have shown antibacterial, antifungal and anti-inflammatory properties in aloe gel.

Fresh juice inhibits the growth of Micrococcus pyogenes var. aureus, Streptococcus pyogenes, Salmonella paratyphi and Corynebacterium xerose. Ethanolic stem-extract shows antibacterial activity against E. coli. The leaf extract inhibits the growth of Mycobacterium tuberculosis.

Aloe stimulates fibroblast and epithelial cell growth, induces lectin-like responses in human cells and promotes neuron-like cell growth.

Aloe powder is prepared by spray-drying Aloe gel. Intragastric administration of Aloe vera powder (60 mg/kg) improved fertility rate in rabbits. An Ayurvedic Aloe compound was tried to improve fertility in women. It exhibited improvement of fertility in 85 % of the 250 cases of sterility and improved menstrual function in 44.6 % of the cases. (CCRAS unit, Jamnagar.) Aloe powder was not used in this area in Western herbal.

Solidified juice originating in the cells of the pericycle and adjacent leaf parenchyma, when dried, forms dark chocolate-brown, usually opaque, masses. This is called Curacao or Barbados Aloe, which contains not less than 28 % hydroxyanthracene derivatives, expressed as barbaloin.

The laxative effects of Aloe are primarily due to the 1,8-dihydroxyanthracene glycosides, aloin A and B (formerly designated as barbaloin).

As Aloe contains anthraquinone glycosides, it is not advised to use its prescriptions continuously for longer than 1–2 weeks.

Abdominal spasm and pain may occur after even a single dose. The correct individual dose is the smallest amount required to produce a soft-formed stool (0.04–0.11 g). The major symptoms of overdose are griping with severe diarrhoea, watery stools and consequent losses of fluid and electrolytes.

Use in Western herbal

German Commission E monograph recognized the herb's efficacy in constipation.

In the West, a number of Aloe gel preparations are available over the counter. These are used for burns, scalds, scrapes, stretch-marks, sunburns, frostbites and wounds. The gel has been credited with a property which increases epidermal cell proliferation and helps in skin regeneration. It is also said to reverse the degenerative skin changes due to regeneration. This is being exploited on a very big commercial scale.

Ointments, creams, lotions and shampoos, with Aloe as an ingredient, are in vogue in the West. A toothpaste containing the gel has been prepared for preventing gingivitis, controlling plaque and reducing hazards of bacterial contamination.

Caution

Aloe should not be administered to patients with inflammatory intestinal diseases. It is also contraindicated during pregnancy and lactation.

Aloe may reduce absorption of orally administered drugs.

Chronic abuse may lead to hepatitis.

Zingiberaceae

ALPINIA

Alpinia galanga Willd.

Habitat

Sub-Himalayan region of West Bengal; also found in Bihar and Assam. Cultivated throughout India.

Classical & common names

Ayurvedic: Kulanjana, Malaya vachaa, Sugandha, Ugradandhaa, Mahabharivachaa.
Unani: Khulanjaan (Arabic).
Siddha: Kanda.
English: Greater Galangal, Java Galangal.

Parts used

Rhizome.

Dose

Powder 1–3 g; decoction 50–100 ml.

Figure 1 Alpinia galanga [ZANDU]

Classical uses

The drug entered into Indian medicine during the 12th or 13th century. First reference was found in Shodhal Nighantu and the first compound was Kulinjanaavaleha, incorporated in Gadanigraha.

According to Bhaavaprakaasha, the rhizomes are expectorant and anticatarrhal, antiseptic and disinfectant to the mouth and pharynx. Rhizomes were used for freshening the breath, for toning up mucus membrane for improving the tone of the voice. In Raaja Nighantu, stomachic and digestive properties were also attributed to the herb. Nighatu Ratnaakar extended the herb's antiseptic and anti-inflammatory coverage to rheumatic and skin diseases.

In Unani medicine, the herb is an ingredient in Javarish-e-Jalinoos, the reputed tonic for cold, catarrhal conditions of naso-pharynx and adipose conditions of the body; also in a digestive tonic, Javarish-e-ood-e-Sheereen.

All these compounds are still available over the counter. Kulanjanaavaleha is no more in use.

Arabs gave Khulanjaan to horses to make them "fiery". This prompted the Unani physician to use the drug in sex tonics. They prescribed 3 g powder of rhizomes with milk for impotence. The patient was asked to keep the rhizome in the mouth and suck it slowly for achieving orgasm. With this in view, the herb was included in a number of sex tonics. Habb-e-Ambar, Laboob Sagheer and Majoon-e-Saleb are a few such Unani tonics.

In the South and in Siddha medicines, Alpinia galanga has been equated with Raasanaa. All Raasanaa compounds of Sahasrayoga are based on Alpinia galanga. While in the North, according to Ayurvedic formulary of India, Pluchea lanceolata is used as Raasanaa.

Active principles and pharmacology

Several volatile constituents have been identified in the essential oil obtained from rhizome (yield 0.3 %). The major constituents were myrcene 94.51, (Z)-beta-ocimene 2.05 and alpha-pinene 1.6 %. The essential oil from the Thai sample of the plant contained methyl eugenol and caryophyllene.

The rhizomes are used for their stomachic, carminative, diuretic, expectorant, anticatarrhal and antirheumatic properties. It is a popular drug for clearing the voice. The rhizomes exert an antispasmodic effect which alleviates asthma; in moderate doses inhibit excessive peristaltic movement of intestines.

The use of rhizomes in Unani medicine in impotency and as an aphrodisiac was examined during 1992. The plant caused a significant gain in the weight of sexual organs and increased sperm motility and sperm count in male mice. The ethanolic extract of the rhizomes did not show any spermatotoxic activity in mice under acute toxic doses of 0.5, 1.0, 3.0 g/kg body weight (for 24 hours) and chronic dose of 100 mg/kg/day for 90 days. The powdered drug mixed with other spermatic and aphrodisiac drugs—Orchis latifolia (Saleb misri) and Mucuna pruriens (Konch seeds)—on oral administration to male albino rats in a dose of 300 mg/100 g body weight for 5 days showed significant improvement in male sexual function.

Use in Western herbal

Alpinia officinarum, known as lesser galangal, is used in the Western herbal. It yields a volatile oil (about 1 %) containing alpha-pinene, cinole, linalool; and sesquiterpene lactones (galangol, galangin).

Galangal was introduced into Europe by Arabian physicians well over a thousand years ago. In line with the Chinese and Indian herbal traditions, it is mainly used as a stomachic, anti-inflammatory, expectorant and a nervine tonic.

In Chinese research trials, a decoction of galangal had an antibacterial action against a number of pathogens, incl. anthrax.

Research during 1988 indicated that galangal is distinctively effective against Candida albicans.

The herb's antispasmodic, antiphlogistic as well as antibacterial role has been accepted in the West.

German Commission E monograph recognized the herb's efficacy in the following areas:

- Loss of appetite
- Dyspeptic complaints
- Liver and gallbladder complaints
- Common cold
- Fevers and colds
- Cough and bronchitis

Inflammation of the mouth and pharynx.

Caution

The herb can be an irritant in higher doses.

Apocynaceae

ALSTONIA

Alstonia scholaris R. Br.

Habitat

Found almost throughout moist regions of India, up to an altitude of 600 m, especially in West Bengal and west coast forests.

Classical & common names

Ayurvedic: Saptaparna, Vishaal-tvak, Vishamchhada, Shaarad, Deva-vriksha, Madagandhaa (Charaka). Saptachhada (Sushruta); Saptaahva.

Figure 1 Alstonia scholaris [ZANDU]

Unani: Chhativan, Sataunaa, Chhatni.
Siddha: Pala, Ezhilaippali, Mukkam palai.
English: Devil's Tree, Dita-bark Tree.

Parts used

Bark, flowers.

Dose

Decoction: 50–100 ml.

Classical use

Charaka used paste of the bark in ointments for chronic skin diseases and in prescriptions for oedema and urinary diseases.

Sushruta prescribed the drug internally as well as externally in urinary diseases, poisoning, fever, malignant ulcers, leprosy and other virulent skin diseases, anal fistula. Sushruta used decoction of Saptaparna as a specific remedy for polyurea.

Saptaparna was one of the ingredients in Sushruta's Triphalaadya Ghrita, prescribed in fevers.

The juice of flowers mixed with Piper longum (Pippli) and honey was prescribed in asthma and chronic cough. (Charaka Samhitaa, Sushruta Samhitaa, Ashtaanga Sangraha, Ashtaanga Hridaya.)

Sushruta used a suspension of the milky exudate as a purgative. Vrindamaadhava prescribed it externally for infected wounds.

Active principles and pharmacology

The bark contains alkaloids, losbanine and picraline deacetyl, and non-alkaloidal compounds along with lupeol acetate and amyrin acetate.

The alcoholic extract of the bark was found active against Streptomyces aureus, Salmonella paratyphi-B-and Aspergillus niger.

Out of the alkaloids isolated from flowers, picrinine (0.001 % dry weight basis) is the major alkaloid and acts as a depressant on the central nervous system. Stem bark is also hypotensive. Echitamine induced an initial fall in blood pressure in experimental animals. Larger doses caused irregularities in rhythm and heart-block. Found toxic to mice in doses of 0.3–0.5 mg/20 g body weight.

The plant is a constituent of an Ayurvedic preparation developed by CCRAS, Ayush 64, which is said to be effective in microfilaraemia. The total alkaloids and tincture made from the bark showed little or no demonstrable action in malaria; also no synergistic action with quinine.

Echitamine chloride, a promising anticancer drug isolated from the plant, did not show any toxic effect on experimental rats.

The bark can be used for its astringent, styptic, antiseptic, anthelmintic and bitter properties. The bark, leaves, milky juice of latex possess therapeutic properties for treating ulcers, sores, tumours and rheumatic pain. Decoction of leaves is found efficacious in congestion of liver; also in dropsy.

Use in Western herbal

Alstonia scholaris is used as a bitter tonic for its astringent properties. An infusion of the bark, 5 parts to 100 parts of water, is prescribed in a dose of one fl oz. The dose of powdered bark is 120–140 mg/day.

The bark is used in homoeopathy (tincture to third potency) in weak digestion, anaemia, low fever often with diarrhoea, dysentery, and as a tonic after exhausting fever. Applied locally for ulcers and rheumatic pains.

Alstonia constricta, a native to Australia, contains indole alkaloids as well as reserpine. Though known as Fever Bark and Australian quinine, its efficacy against malaria is not yet scientifically established. The bark is antispasmodic, lowers blood pressure as it contains reserpene. Bitter constituents prove effective in diarrhoea.

Caution
Toxic in large doses. Alstonia spp. are subject to legal restrictions in some countries.

Amaranthaceae

ALTERNANTHERA

Alternanthera sessilis (L.) R. Br.

Figure 1 Alternanthera sessilis [ADPS]

Alternanthera triandra Lam.

Habitat
A weed found throughout the hotter parts of India, ascending to an altitude of 1200 m.

Classical & common names
Ayurvedic: Matsyaakshaka, Matsyaakshi, Vaahlikaa, Matsyagandhaa, Matsyadini; Gudari-saaga, Machheti.
Unani: Machhechhi, Paanachooni.
Siddha: Ponnangaani.
Alternanthera sessilis is equated with Lonikaa (Koluppa) in Kerala.

Parts used
Whole plant.

Dose
Juice 10–20 ml.

Classical use
Charaka used the entire plant in prescriptions for promoting memory and intelligence, and externally for complexion. Charaka incorporated Matsyaakshaka in Aindra Rasaayana, a rejuvenating tonic.

Sushruta's intellect-promoting compound formulation consisted of Matsyaakshaka, Shankhapushpi (Convolvulus pluricaulis) and Svarna Bhasma (calcined gold). It was prescribed to children for promoting their physical development, strength and intellect.

Bhaavaprakaasha attributed blood-purifying properties to the herb. During the 16th century, it was used in diseases due to vitiated blood, skin diseases and ulcers. Its active principles, extracted in oil, were used to treat infected wounds. The herb also proved styptic in colitis.

Active principles and pharmacology
The plant contains a saturated hydrocarbon, aliphatic ester. Also stigmasterol and beta-sitosterol.

The leaves contain uronic acid, oleanolic acid, a flavone glycoside and a saponin. The total lipid content of the leaves is 0.54 % (dry weight basis). The leaves also contain good amounts of alpha- and beta-tocopherols.

The leaves suppress carcinogenesis and act as a protective agent against cancer.

The seed extract showed antibacterial activity.

The plant is used in Indian medicine due to its nutritive value. The edible portion of the plant gave the following values: protein 5.0, fat 0.7, fibre 2.8, carbohydrates 11.6, minerals 2.5 g/100 g; calcium 510.0, phosphorus 60.0, iron 16.7, riboflavin 0.14, niacin 1.2 and vitamin C 17.0 mg/100 g; carotene 1926 mcg/100 g and energy 73 kcal.

All these constituents make the herb a potent tonic with a wide coverage for children; for male a good adjuvant with sex tonics and for females a natural galactagogue.

The plant is used in folk medicine for post-natal complaints. The leaves are used in soup for indigestion and biliousness.

Use in Western herbal

Alternanthera Forsk is basically native to the sub-tropical regions of Australia and South America.

Alternanthera philoxeroides (Mart.) Griseb. is used in the treatment of certain viral diseases such as hepatitis, epidemic parotitis, hemorrhagic fever and influenza along with other specified herbs. The levels of essential amino acids in the leaf protein compare favourably with FAO reference pattern.

This plant is also found in India, but not used in herbal medicine

Malvaceae

ALTHAEA

Althaea officinalis Linn.

Habitat

Found in the Himalayan region from Kashmir to Punjab.

Classical & common names

Unani: Khatmi (Persian), Khairu (Arabic). Tukm-e-Khatmi (Seeds). Gul-khairu (flower).
English: Marshmallow.

Parts used

Leaves, flowers, carpels, root.

Dose

Powder 3–6 g.

Classical use

The herb is an ingredient in a number of Unani compounds as a demulcent and emollient, due to the presence of mucilage.

The root is given in inflammation and irritation of alimentary canal and of the respiratory and urinary organs. In the form of a decoction, the root is found helpful where the natural mucous membrane is abraded, and in healing bruises. It is used in the form of an emollient enema in irritability of the urinogenital and alimentary passages and other nephritic and stone complaints.

The decoction of the root is used as an expectorant in cough, hooping cough and hoarseness of voice. The flowers are an ingredient of various Unani cough mixtures.

The mucilage content is highest in the root. The leaves also contain mucilage and are used in the same way as roots.

Leaves are used as poultice. Mixed with oil, leaves and flowers are applied to burns and insect bites.

The seeds are an ingredient in Bayaz-e-Kabir's compound, Dayakhuza, prescribed for removing phlegm from the lungs; in Lavookh-e-Sual (Qarabadeen-e-Jadeed), prescribed for treating dry cough and other inflammatory conditions of the air passage.

The seeds have also been incorporated in Marham-e-Dakhyleyun (Qarabadeen-e-jadeed). The ointment is used in anal fissures, chronic ulcers in and around anus, in vulvo-vaginitis.

Active principles and pharmacology

Marshmallow root contains about 37 % starch, 11 % mucilage, 11 % pectin, flavonoids, phenolic acids, sucrose and asparagine.

Mucillage is a mixture of collodially soluble polysaccharides, particularly glacturonic rhamnans, arabinoglactans, arabans and glucans.

Active principles of the drug alleviate local irritation, inhibit mucociliary activity, stimulate phagocytosis. The drug functions as an anti-inflammatory, soothing and healing agent in catarrh of respiratory tract, gastro intestinal tract and urinary tract.

Use in Western herbal

German commission E monograph recognized the herb's efficacy in cough/bronchitis.

Herbalists in the West also use the root for countering excess stomach acid, peptic ulceration and gastritis. Warm infusion of leaves is given to treat cystitis and frequent urination. The root, as in Unani medicine, is used in an ointment for boils and abscesses, and in a mouthwash.

Seeds are not used in Western herbal. Only root, leaves and flowers are used.

Amaranthaceae

AMARANTHUS

Amaranthus hypochondriacus L.

Habitat
Found in gardens throughout the world. Native to warm, temperate climate of Asia, the Americas, and Africa.

Classical & common names
Ayurvedic: Shitivaaraka. (Not yet correctly identified.)
Unani: Suriyaali, Survaali. Tukhm-e-Survaali (seeds).
English: Cockscomb var., best known as Celosia cristata L. and Celosia argentea L.
Chinese: Ji-guan-hua and Qing-xiang-zi.

Dose
Seed powder, flowers 3–5 g, seeds in decoction 3–15 g.

Classical use
In Unani medicine, the seeds (Tukhm-e-Survaali) are considered as a spermatogenetic drug and are used in spermatorrhea; also in tonics for sexual debility. Seeds are also given as a single drug with milk.

A decoction is given in heavy menstrual bleeding, bleeding piles, in excessive and frequent urination.

Leaves are used as a vegetable in biliousness.

Flowers are astringent and are used for diarrhoea, dysentery, cough and haemorrhages.

In Chinese medicine, dried flowers are used in prescriptions to treat leucorrhea, dysentery, haemorrhoids, haemoptysis, haematemesis, epistaxis, haematuria, urinary tract infections and uterine bleeding. Powdered red cockscomb flowers, dried in the sun, are given with alcohol before meals in profuse menstruation.

The seeds are used for similar purposes and for treatment of cloudy urine.

Active principles and pharmacology
Amaranth contains saponins, tannins, incl. a red pigment. Proven properties are astringent and styptic.

Use in Western herbal
Same as in Chinese medicine. Used primarily as an astringent to reduce blood loss and as a remedy for heavy menstrual bleeding, excessive vaginal discharge, diarrhoea and dysentery.

Nicholas Culpeper (1616–1654) wrote of it: "Amaranthus flowers stop all fluxes of blood, whether in man or woman, bleeding either in the nose or wound. The white flower stops the whites in women."

Caution
Nauseating in large doses.

Amaranthus spinosus Linn.

Figure 1 Amaranthus spinosus [ADPS]

Amaranthus polygamus Willd.

Habitat

Native to tropical America and found throughout India as a weed in cultivated as well as fallow lands.

Classical & common names

Ayurvedic: Tanduliya, Meghnaad, Alpamaarish, Kaandru, Tandulerak, Tanduliyaka, Bahu-viryaa, Kanderaa, Vishaghna (Charaka).
Unani: Chaurai, Chaulai, Baqaltul-arabiya, Baqal-tul-yamania (Arabic).
English: Prickly Amaranth.

Parts used

Whole plant.

Dose

Powder 400–800 mg, juice 10–20 ml.

Classical use

The entire plant was cooked as vegetable. Used in toxic conditions, intoxication, internal haemorrhage. Sushruta also recommended that the whole plant should be cooked as potherb and used regularly in diet for its diuretic, laxative and antitoxic properties. The herb was specific in curing drowsiness due to poison, alcoholism or vitiated blood.

According to Bangasena, the root of Tanduliya, mixed with honey and sugar and taken with rice water, checks bleeding.

Ashtaanga Hridaya recommends the clarified butter with Tanduliya root and shoot along with milk for all sorts of poison.

The primary application of the fresh juice of leaves was in poisoning, from snake bite, rat bite, insect bite to chemical poisoning and in toxic conditions.

Active principles and pharmacology

Leaves and stems contain alpha-spinasterol and hentriacontane. Root alpha-spinasterol and saponins. Leaves also contain amino acids with high contents of lysine.

The plant is spasmolytic and antagonises amphetamine hyperactivity. It exerts a harmonizing effect on the stimulated central nervous system, especially in the mental diseases. Diuretic, purgative, styptic, mucilaginous and cooling properties of the herb eliminate toxins and provide required tonic support to the body.

The plant is also used as an emmenagogue and galactagogue due to synergistic action of amino acids, carbohydrates, calcium, phosphorus, iron, nicotinic acid, ascorbic acid, and saponins.

A decoction of the plant improves digestion, prevents vomiting. As a diuretic, it allays kidney affections. As a styptic, it is found efficacious in piles, internal bleeding. The root is given in menorrhagia. As an emollient, it is applied to bruises, abscesses and inflammations.

Use in Western herbal

Amaranthus spinosus is taken to reduce heavy menstrual bleeding and excessive vaginal discharge, and to arrest the coughing up of blood. Also used as a diuretic.

Related species

Amaranthus blitum Linn. var. oleracea Duthie, Amaranthus gangeticus Linn., Amaranthus mangostanus Linn., Amaranthus tricolor Linn: cultivated as annuals all over India as a vegetable. Classical name: Marisha (Charaka). Common names: Laal Saag, Chaulai, Chinese Spinach.

Cooked as a potherb, it was used in Ayurvedic medicine as an aid to elimination and to assist the excretory process of the body. The vegetable is mucilaginous and a wholesome diet, rich in amino acids, calcium and magnesium.

The whole plant is considered a blood purifier, astringent, diuretic, demulcent and cooling.

The seeds contain fatty acids—palmitic 20.8, stearic 2.2, oleic 43.7 and linoleic 23.7 % (South Indian variety). Given to women in leucorrhoea and as a galactagogue. To men in sexual debility.

The root, leaves and stems are eaten in bilious disorders.

Sterculiaceae

AMBROMA

Ambroma augusta Linn. f.
Abroma augusta Linn. f.

Habitat

Found, both wild and cultivated, throughout the hot and moister parts of India from Punjab and Uttar Pradesh and eastward to Arunachal Pradesh, Assam, Meghalaya and Tripura and southward in peninsular India.

Classical & common names

Ayurvedic: Pishaacha kaarpaasa, Pivari, Karnikaara (non-classical).
Unani: Ulat-kambal.
Siddha: Sivapputtutti.
English: Devil's Cotton.

Parts used

Fresh or dried root, rootbark, leaves.

Dose

Root powder 1–3 g, fresh root 4–8 g, root juice 5–10 ml, leaf juice 10–20 ml.

Classical use

Not mentioned in Ayurvedic or Unani classical texts. It has wrongly been equated with Bhaaradwaaji, which is actually wild cotton.

Dr W. Roxburgh, author of Flora Indica and Plants of the Coromandel Coast, was first to highlight the therapeutic property of the herb and its use for dysmenorrhoea in 1801 AD. On the basis of this lead, the herb was used as an emmenagogue and for amenorrhoea by herbal physicians of Bengal. In 1872, Dr Bhuvan Mohan Sarkar discussed these properties of the herb in Indian Medical Gazette and recommended a dose of 180 mg of the fresh root juice. In 1896, G. Watt covered the herb in "A Dictionary of Economic Products of India" (1889–1893).

The herb entered into Western herbal as a uterine tonic during 1873 when a document appeared in "American Medical Science."

In folk medicine, 4 g of rootbark powder mixed with the powder of Piper nigrum (black pepper 21 number; normal recommended dose of the powder is 500 mg–1 g) is given during menstruation for 7 days. Recommended diet is rice and milk. To be repeated during two to four periods. This treatment is said to regulate menses and cures many uterine afflictions. In cases where the pain precedes the menstruation, the compound is given two days in advance.

Fresh viscid juice (sap) abounding in the thick, easily separable bark of the root is found to be equally efficacious and is given in 2 g doses a day in dysmenorrhoea from the first day of the menses to 7 days successively. A single dose during the menses is said to regulate the menstrual flow and act as a uterine tonic.

Active principles and pharmacology

The root contains abromine, choline, betaine, beta-sitosterol, stigmasterol, friedeline, a basic compound (pictrate, mp 223–270) and a fixed oil.

The root regulated the menstrual flow. Showed contractile action on the uterus and is considered to be a uterine tonic and an emmenagogue; used for the treatment of dysmenorrhoea, amenorrhoea, sterility and other menstrual disorders.

The leaves contain taraxerol and its acetate, beta-sitosterol acetate, an aliphatic alcohol, octacosanol and probably a mixture of long chain fatty oils.

The stem bark contains beta-sitosterol and friedelin. The aqueous extract of the stem contains: glucose, mannose, myo-inositol glycerol, erythritol, threitol, fucitol, glucitol, mannitol and sucrose.

The alcoholic extract of the root exhibited 40 % anti-implantation effect in rats. Oxytocic effect was noted on isolated uteri of guinea-pig, rat and dog uterus in situ. The roots exhibited uterotonic activity on isolated rat, guinea-pig, rabbit and dog uterus in situ.

Pet. ether, benzene, chloroform and alcoholic extracts of the root administered in a dose of 50 mg/kg, p.o. on D1 to D6 on pregnancy in albino mice exhibited 95 %, 30 % and 40 % abortifacient activity respectively.

Both leaf juice (250 mg/100 g p.o.) and root powder (200 mg/100 g p.o. × 5 days) on immature female albino rats of 20–30 days old exhibited best estrogenic activity. Cold water extract of the root was found to cause increase in growth of mammary gland and significant galactotropic effect in albino rats.

The seed is not used in Indian medicine, though its oil is rich in linoleic acid (71.5 %) and can play an important dietary role in the control of arteriosclerosis.

No obvious toxicity has been reported during the screening of the herb at CCRAS Pharmacological Research Unit, Varanasi.

Zingiberaceae

AMOMUM

Amomum subulatum Roxb.

Habitat
Found in the eastern Himalayas and cultivated in Nepal, northern West Bengal, Sikkim and Assam.

Classical & common names
Ayurvedic: Brihad-elaa, Sthula, Bahulaa, Elaa-mahati, Bhadrailaa, Prithvikaa, Triputaa, Nishkuti.
Unani: Heel Kalan (Persian); Qaqul-e-Zakar, Heel Zakar (Arabic).
Siddha: Periyadan.
English: Greater cardamom, Nepal cardamom.

Parts used
Seeds.

Dose
Powder 1–3 g.

Classical use
Greater cardamom and Lesser cardamom are two different drugs with different properties. According to Ayurvedic texts, Amomum subulatum, equated with Elaa or Bhrihad-elaa cures morbid thirst, vomiting, nausea, cough and skin diseases. Elettaria cardamomum, equated with Sukshma-ilaa or Heel Khurd, is used in asthma, bronchitis, piles and dysuria.

In classical Unani formulations, both the varieties are included due to their specific properties. Heel Kalan and Heel Khurd both are used in Unani liver tonics Jawarish-e-Bisbaasaa, Jawarish-e-Narmushk.

In Indian medicine, Amomum subulatum is not treated as a substitute for Elettaria cardamomum.

In classical medicine, the seeds of Amomum subulatum are used as a tonic to the liver and heart, as astringent to the bowels, as a hypnotic and as an appetiser.

The paste of paricarp (without seeds), green leaves of coriander (Dhaanyaka) and flowers of Pterospermum acerifolium (Muchukunda) is applied over the forehead in migraine. Only paricarp is given in stomatitis.

The decoction of seeds is used as a gargle in affections of teeth and gums. In combination with the seeds of melon Amomum subulatum is used as a diuretic in cases of gravel of the kidneys. In certain disorders of the digestive system, marked by scanty and viscid secretion from intestines, the seeds promote elimination of bile and are found useful in congestion of liver. Unani liver tonics invariably use seeds of Amomum subulatum in compound preparations.

The seeds have also been used in neuralgia in large doses (1.8 g) in conjunction with quinine.

Active principles and pharmacology

The seeds, contain a chalcone—cardamonin; a flavanone, alpinetin; the glycosides-petunidin-3, 5-diglucoside, leucocyanidin-3-O-beta-D-glucopyranoside and subulin. The seeds, on steam distillation, yield an essential oil containing cineol as a principal constituent.

Biological activity of all these constituents is found to be stomachic, astringent, antibilious, antiemetic; provides tonic properties to heart, liver and spleen.

Use in Western herbal

The official Cardamoms in the United States are those produced in Malabar and Mysore, while in Britain those imported from Srilanka. Amomum subulatum is called West Bengal cardamoms, Morung elaichi or Buro elaichi. Both the varieties (Amomum subulatum and Elettaria cardamomum) are called cardamom and are used in indigestion, flatulence for their carminative, stimulant and aromatic properties.

In the United States and Great Britain, the seeds are employed to a small extent as an ingredient of curry powder. In Russia, Sweden, Norway and parts of Germany the seeds are largely used for flavouring cakes. In Egypt they are ground and put in coffee. In France and America the oil is used in perfumery.

Medicinal use of both the varieties of cardamoms has not yet entered into Western herbal, possibly due to lack of scientific data from India.

Araceae

AMORPHOPHALLUS

Amorphophallus campanulatus (Roxb.) Blume ex Decne

Habitat

Wild plant found all over the plains of North India, extending to West Bengal and Assam, and in Konkan (Maharashtra). Widely cultivated throughout the upper Gangetic plains and in peninsular India.

Figure 1 Amorphophallus paeonifolius [ADPS]

Classical & common names

Ayurvedic: Soorana, Kandula, Arshoghna, Kandayak, Gudaamaya-hara.
Unani: Soorana, Zaminkand (Persian).
Siddha: Karnsa.
English: Elephant-foot Yam.

Parts used

Tuber (Corm).

Dose

Powder 3–5 mg.

Classical use

The tuber was prescribed by Sushruta as a part of diet for treating piles, rectal polyp formations and condyloma. The tuber, firmly closed inside the well-kneaded clay lump, is cooked by closed heating. The clay is removed after the baking is over and the tuber is recovered and a medicinal ash is prepared. The ash is prescribed with oil and salt in haemorrhoids (Ashtaanga Hridaya, Vrindamaadhava, Shaarangadhar Samhitaa).

The ash mixed with clarified butter and jaggery is applied externally on tumours (Haarita Samhitaa). The paste of the tuber mixed with honey and clarified butter is applied to alleviate filaria (Haarita Samhitaa).

In Ayurvedic medicine, the Panchaagni or Panch-hutaashna group of herbs, when taken together, cures piles, deranged digestion and diseases of liver. The group contains Amorphophallus campanulatus, Marsdenia tenacissima (Muurvaa), Emblica officinalis (Aamalaka), Plumbago zeylanica (Chitraka, syn. Dahana) and sea-salt (Saindhava). When taken in the form of linctus, this group also cures gout.

In Unani medicine, Sooran is given as a vegetable in sluggish liver, deranged digestive system and for astringent properties in piles and haemorrhoids, also prescribed as expectorant in asthma and bronchitis.

Active principles and pharmacology

The corm contains an active diastatic enzyme—amylase, betulinic acid, beta-sitosterol, stigmasterol, beta-sitosterol palmitate, lupeol, triacontane and amino acids.

The corms are irritant due to presence of calcium oxalate. They are to be consumed after they are washed well and boiled in tamarind water or buttermilk. If not properly boiled, they produce an irritating sensation in the mouth.

Analysis of corms gave the following values: protein, 1.2, fat 0.1, fibre 0.8, carbohydrate, 18.4, starch 17.7, oxalic acid 1.3 and minerals 0.8 %; calcium, 50.0, phosphorus 34.0, iron 0.6, thiamine 0.06, riboflavin, 0.07, niacin 0.7 mg/100 g, carotene 260 mcg/100 g, vitamin A 434 i.u., and 79 kcal/100 g.

The corms are found efficacious as antiasthmatic, antiemetic, carminative, antidysenteric and are used in piles, enlargement of spleen, diseases due to vitiated blood, elephantiasis.

The corms and seeds are applied externally as an irritant to treat acute rheumatism. The paste is prepared with clarified butter or honey for external application.

Related species

Amorphophallus dubius Blume (Purple-stalked Dragon) is allied to Amorphophallus campanulatus, found occurring in Kerala. The tuber is administered in piles and apoplexy. It is also eaten after boiling in tamarind juice.

Amorphophallus sylvaticus (Roxb.) Kunth, syn. Synantherias sylvatica Schott (Kuttukaranai in Tamil, Vanakanda in Telugu) is a tuberous herb, suspected to be poisonous to human beings and livestock. The fruits and seeds are crushed into a paste and are used externally for bruises and glandular enlargements.

Asteraceae

ANACYCLUS

Anacyclus pyrethrum DC.

Habitat

Native to the Mediterranean region. Cultivated in Algeria.

Classical & common names

Ayurvedic: Aakaarkarabha, Aakallaka, Aakulakrit, Arkakarbha.
Unani: Aaqarqarha (Arabic).
Siddha: Akkirakaram.
English: Pellitory.

Parts used

Root, essential oil.

Dose

Powder 500 mg–1 g.

Classical use

The herb was not known to the Ayurvedic practitioners during the period of Samhitaas. Bhaavaprakaasha, during the 16th century, incorporated the herb in compounds which were adopted from Unani medicine. It was used in facial paralysis, epilepsy, neurasthenia, insanity; also as an aphrodisiac and sex tonic. Aaqarqarha was already an ingredient in many Unani nervine and sex tonics. Majoon-e-Baladur, a brain tonic prescribed in neurasthenia; Majoon Arqaruya-e-Kabir, a nervine tonic for hemiplegia, facial paralysis and epilespy; Laboob-Sagheer and Majoon-e-Salab, inspissant to semen, spermatogenic, and aphrodisiac; these are a few compounds which are still available over the counter. The classical Ayurvedic compound, Aakaarkabhaadi Churna, is not in use any more.

Active principles and pharmacology

The root contains anacyclin, pellitorine, enetriyne alcohol, hydrocarolin, inulin, traces of volatile oil and sesamin. They also contain tyramine amides corresponding to isobutylamides. The presence of polyacetylenic compounds is also reported in the plant.

Pellitorine, the active and intensely pungent constituent of the root, is a mixture of isobutylamides, and produces the sialagogue effect; also tonic properties for the nervous system.

A decoction of the root is useful as a gargle in carious teeth, toothache, sore throat and tonsillitis.

Use in Western herbal

Nicholas Culpeper (1616–1654) wrote of Pellitory. "One of the best purgers of the brain. Either the herb or root, dried and chewed in the mouth, purges the brain of phlegmatic humours; thereby not only easing pains in the head and teeth, but also hinders the distilling of the brain upon the lungs and eyes, thereby preventing coughs, phthisicks and consumption, the apoplexy and falling sickness.

"The powder of the herb or root being snuffed up the nostrils, procures sneezing, and eases the headache.

"Being made into and ointment with lard, it takes away black or blue spots occasioned by blows or falls and helps both the gout and sciatica.

"It is an excellent approved remedy in the lethargy."

Patients seeking relief from rheumatic or neuralgic affections of the head or face or for palsy of the tongue have been advised to chew the root daily for several months.

The British Pharmacopoea approves its use as a masticatory, and in the form of lozenges for its reflex action on the salivary glands in dryness of the mouth and throat.

The diluted essential oil is used in mouthwash and to treat toothache.

Caution

In larger doses the powdered root is an irritant to the mucus membrane of the intestines, causing bloody stools, tetanus-like spasms and profound stupor.

Bromeliaceae

ANANAS

Ananas comosus (L.) Merr.
Ananas sativus Schult. f.

Habitat

Native to America. Cultivated in Assam, West Bengal and along the west coast of Kerala.

Classical & common names

Ayurvedic: Anaanaasa, Bahunetra.
Unani: Anannaasa, Ainunnaas (Arabic).
Siddha: Annasi.
English: Pineapple.

Parts used

Fruit.

Dose

Fruit juice 25–30 ml.

Classical use

Anaanaasa was better known as Bahunetra in Ayurvedic texts. Raajavallabha Nighantu described its tonic properties in detail: energizing, refrigerant, anthelminithic, purgative, diuretic, stomachic, dyspeptic, anti-rheumatic.

In folk medicine, fresh juice of the leaves is given with sugar to relieve hiccup. It is also used as a purgative. The juice of unripe fruit is given as an emmenagogue and as a styptic.

In Unani medicine, the ripe fruit is used as a murabbaa (boiled fruit soaked in sugar syrup or honey). Murabbaa-e-Anannaasa is given as a cardiac tonic and in palpitation. Sharbat Anannaasa is prescribed as a diuretic.

Active principles and pharmacology

The leaves contain starch (3 %), ergosterol peroxide, 5-stigmastene-3 beta-7-alpha-diol, besides beta-sitosterol, campesterol, stigmastanol, campestanol, myricyl alcohol, myristic acid and two minor aliphatic esters. The presence of pipecolinic acid, an oil wax and ascorbic acid is also reported.

The fresh juice of the leaves is a blood purifier and is considered to be useful in encouraging the onset of menstrual periods and easing painful ones. The juice of leaves is also used as a purgative and anthelmintic. The aqueous and alcoholic extracts of leaves have shown anthelmintic activity in animal studies.

The plant, incl. the fruit, contains several enzymes: the major one being bromelain, a protein-splitting enzyme that aids digestion.

The stem contains two antioxidants, traces of citric, malic and ascorbic acids.

The fresh fruit is a good source of vitamin A, B and C. It also contains iron and phosphorus. It has low sugar and fat content, and is good for obese people.

Analysis of the edible portion (60 %) of the fruit gave following values: moisture 87.8, protein 0.4, fat 0.1, fibre 0.5, carbohydrates 9.0; and minerals 0.4 g/100 g; calcium 20.0, phosphorus 9.0, iron 1.2, thiamine 0.2, riboflavin 0.12, niacin 0.1, vitamin C 39.0 and oxalic acid 5.0 mg/10 g; carotene 18 µg and cal val. 46 cal/100 g. Other minerals present are magnesium 20.0, sodium 34.7, potassium 37.0, copper 0.36, sulphur 20.0 and chlorine 13 mg/100 g.

The sour, unripe fruit improves digestion, increases appetite and relieves dyspepsia. It also acts as a uterine tonic.

The ripe fruit cools and soothes, reduces excessive gastric acid. It is both a digestive and diuretic tonic. Used as an adjuvant in gastric irritability, and as a cholagogue in jaundice.

Use in Western herbal

Bromelain, the protein-digesting enzyme complex of pineapple was introduced as a medicinal agent in the West in 1957. Since then, over four hundred scientific papers on its therapeutic application have appeared in medical literature.

Chewable Bromelain with Papin (an enzyme found in the papaya fruit) is available in health food stores. It is taken with meals as a digestive aid: for speeding wound-healing and reducing inflammation on an empty stomach. (By breaking down fibrin, Bromelain reduces swelling.)

Bromelain is also being combined with antibiotic therapy. The drug is not considered safe if taken with aspirin or anticoagulants.

The juice of pineapple is used topically in France to tonify skin. Weleda Larch Resin Lotion contains Bromelain.

(Bromelain is found in both fruit and stem of the pineapple, but the enzyme in supplements comes from the stem.)

Caution

Leaves, unripe fruit are contraindicated during pregnancy.

Acanthaceae

ANDROGRAPHIS

Andrographis paniculata (Burm. f.) Wall. ex Nees

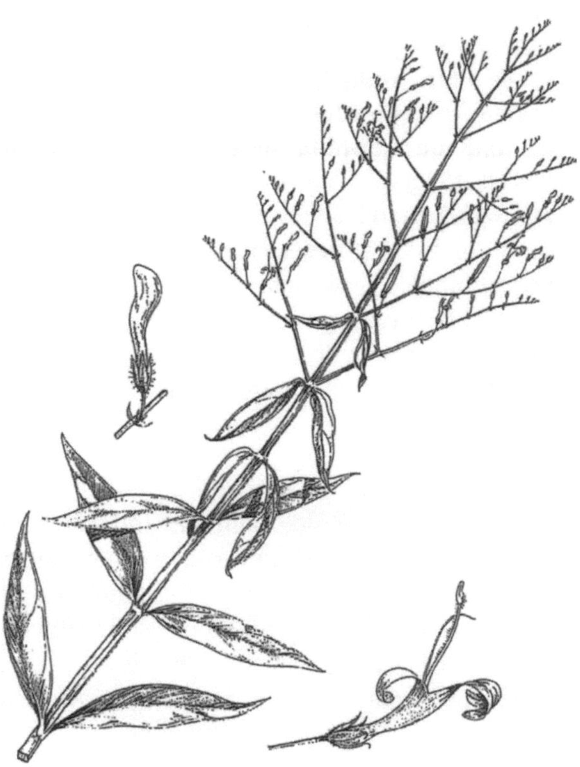

Figure 1 Andrographis paniculata [ADPS]

Habitat

Found in plains throughout India from the Himachal Pradesh to Assam and Mizoram and all over South India.

Classical & common names

Ayurvedic: Kaalmegha, Bhuunimba, Bhuuminimba, Vishvambharaa..
Unani: Desi-chiraayataa.
Siddha: Amgaravalli.
English: Creat.

Parts used

Whole plant.

Dose

Fresh juice 5–10 ml, powder 1–3 g, decoction 50–100 ml.

Kiraatikta and Bhuunimiba of Sushruta Samhitaa and Kiraatatika, Kiraatatiktaka, Katutikta, Trnanimba of Charaka Samhitaa have been equated with Swertia chirata by P. Ray et al, the scientists of INSA, and also by PV Sharma in "Dalhan and his Comments on Drugs". CCRAS has equated Bhuunimba with Andrographis paniculata. This may create confusion while applying the drug in practice, as Bhuunimba is also equated with S. chirata.

CCRAS has also equated Andrographis paniculata with Yavatiktaa. Yavatiktaa was a bitter plant found in yava (Hordeum vulgare) fields, while Andrographis paniculata grows abundantly in moist, shady waste grounds and sometimes in dry forests. It is also cultivated. To avoid further confusion, Andrographis paniculata should not be called green chiretta.

Classical use

In folk medicine, the decoction or infusion of the leaves is used in sluggish liver; dyspepsia; griping, irregular bowels; loss of appetite, flatulence and diarrhoea, especially of children; also in convalescence after fevers and general debility.

The herb, due to its bitter properties, was tried in malaria, but was not found effective.

The herb is often combined with certain spices—cumin, aniseed, capsules of greater cardamoms, cloves, cinnamon, for their carminative properties.

There are many over the counter liver tonics in India, with Andrographis paniculata as an ingredient. To name a few: Adliv (Albert David), Hepatinic (American Remedies), Hepatogard (Themis Pharma), Stimuliv (Franco-Indian), Tefroli (TTK).

Active principles and pharmacology

The plant contains the diterpenoids andrographolide, andrographiside, and a non-bitter constituent neoandrographolide. The roots gave flavones, andrographin, panicolin and alpha-sitosterol.

Biological activity of the plant has been found to be hyptoprotective, cholinergic, stomachic, antityphoid, antibacterial, antithrombogenic, antidiarrhoeal, anthelmintic and immunostimulant.

The principal use of the herb is in torpid liver and jaundice.

The plant inhibited triglyceride accumulation in hepatic cells in rats. The plant also showed reversal of paracetamol-induced hepatopathy in sheep. The crude extract of the plant showed in vitro inactivation property against hepatitis-B surface antigen.

The antihepatotoxic activity of the drug has been attributed to the diterpenoids, andrographolide, andrographiside and neoandrographolide present in the plant. Andrographolide produces a significant dose-dependent cholerotic effect, as evidenced by increase in bile flow, bile salts and bile acids in normal rats and anaesthetised guinea-pigs. The paracetamol-induced decreases in volume and contents of bile are prevented significantly by andrographolide pre-treatment. It is found to be more potent than silymarin, a clinically used hepatoprotective agent.

The total flavones isolated from the root, on intravenous administration in dogs, stimulated the synthesis of cyclic AMP in platelets, impeded aggregation of platelets and prevented the formation of thrombi as well as the development of myocardial infarction. The crude extract of the plant, as proved experimentally, possesses a promising antithrombogenic property and can be used in preventing and treating arterial thrombotic diseases.

When screened for diarrhoea, in animal models, andrographolide was found very effective against E. coli (heat stable) enterotoxin, the most common cause of epidemics of neonatal diarrhoea. The plant extract also exhibited antityphoid activity against Salomonella typhi (Schoter) and antifungal activity against Helminthosporium sativum. The shoot extract in saline and ether showed antibiotic activity against Micrococcus pyogenes var. aureus.

The ethanolic extract of the plant and the diterpenoids induced significant stimulation of antibody in mice. The total extract of the plant can be used as an immunostimulant tonic.

Caution

The drug should be administered to children only under medical supervision.

Ranunculaceae

ANEMONE

***Anemone obtusiloba* D. Don**
***Anemone pulsatilla* Linn.**

Habitat

Native to Europe, thrives in dry grassland in central and northern parts of the continent. Anemone obtusiloba D. Don occurs in the temperate and alpine Himalayas from Kashmir to Sikkim between altitudes of 2100 and 4200 m and in the Nilgiri hills, above an altitude of 1800 m. Both, Anemone pulsatilla and Anemone obtusiloba, have been equated with Sha-q-un-Noman of Unani medicine.

Classical & common names

Unani: Shaqaaq-un-Noman, Shaqiq (Arabic).
Vernacular: Gul-e-Laalaa (not to be confused with red poppy—Papaver rhoeas flower, which is also known as Gul-e-Laalaa); Ageli (Garwal area).
English: Wood Anemone, Wind Flower, Pasque Flower, Pulsatilla.

Parts used

Dried aerial parts.

Dose

Powder 1–3 g.

Classical use

Hakeem Jaalinoos prescribed the herb as an emmenagogue, galactogogue and diuretic. Same properties have been attributed to the herb in Mufeet Aazam. In Unani medicine, decoction of the herb and the husk of Hordeum vulgare (Yavaa) is used as a potent diuretic; decoction of the herb and Zataria multiflora (Saatar) as a good expectorant in asthma, whooping cough, bronchitis.

The leaves are styptic and antiseptic, but acrid and irritating, may cause blisters when applied to the skin. Converted into ointments, are applied externally in inflammations and fungal infections of the skin and eruptions due to vitiated blood.

Active principles and pharmacology

Pasque flowers contain the lactone protoanemonin (which on drying forms anemonin), triterpenoid saponins, tannins and volatile oil.

Biological activity of the herb is antispasmodic, sedative, styptic and diuretic. It is found efficacious in nervous exhaustion and emotional distress.

Use in Western herbal

Nicholas Culpeper (1616–1654) described the following virtues of the herb: "The leaves provoke the terms mightily, being boiled and decoction drunk. The body being bathed with the decoction of them, cures leprosy. The leaves being stamped and the juice snuffed up in the nose, purges the head mightily; so does the root being chewed in the mouth, for it procures much spitting, and brings away many watery and phlegmatic humours. Being made into an ointment, it is excellently good to cleanse malignant and corroding ulcers."

Modern herbalists, however, hesitate to recommend the chewing of the root on account of the acrid, irritant properties present in it.

The older herbalists recommended application of various parts of the plant for headache, tertian agues and rheumatic gout.

By the end of the 18th century, the herb had fallen out of use, because it was found strongly irritant.

When the herb entered into homeopathy, its use was revived in herbal medicine and it was accepted as a remedy for spasmodic pain of the reproductive system, in both male and female, premenstrual tension, period pain and emotional distress.

In France, the herb remained in use traditionally for treating cough and insomnia.

In homoeopathy, Pulsatilla is an effective female remedy. It is prescribed in amenorrhoea, suppressed menses due to nervous debility, tardy menses, delayed menses, scanty, thick, dark, clotted and intermittent; also in leucorrhoea, acrid, burning, creamy; diarrhoea during or after menses. Jonosia asoca (Saraca indica), introduced by ND Ray of Kolkata in homoeopathy, acts powerfully with Pulsatilla in amenorrhoea and menorrhagia.

Among related species, Anemone pratensis is used interchangeably with pasque flowers. Fresh plant is not used in any form, only dried flowers or leaves are used.

Caution
Pulsatilla is contraindicated during pregnancy.

Apiaceae

ANETHUM

Anethum graveolens Linn.

See figure 1.

Anethum sowa Kurz.
Peucedanum graveolens Linn. (in part)

Habitat
Cultivated throughout India, chiefly in Punjab, Uttar Pradesh, Gujarat, Maharashtra, Assam and West Bengal.

Classical & common names
Ayurvedic: Shatpushpaa, Madhuraa, Misi; Karavi, Atilambaa, Sitacchatraa, Samhit-cchatrikaa; Shataahva (Sushruta). Foeniculum vulgare Mill. has also been equated with Shatpushpaa, Madhura and Misi, Saunf of commerce.
Unani: Shibt, Soyaa.
Siddha: Sadakuppai.
English: Indian Dill, Sowa.

Figure 1 Anethum graveolens—flowering portion [WOI]

Parts used
Fruits, leaves.

Dose
Powder 1–3 mg.

Classical use
The common use in Ayurvedic medicine is in abdominal discomfort, colic, wind, and for promoting digestion. Specific uses are in fever, jaundice, intestinal and genito-urinary tract infections; also in rheumatic affections as an anti-inflammatory and analgesic agent.

Charaka prescribed the paste of Linseed (Linum usitatissimum), castor seeds (Ricinus communis) and Shatapushpaa pounded with milk for external application in rheumatic and other swellings of the joints.

Kaashyapa Samhitaa attributed tonic, rejuvenating and intellect-promoting properties to the herb. According to its text, the herb alleviates the disorders of the female genital tract and provides healthy progeny.

Unani physician use the herb with

Habb-ul-Qutn (Gossypium herbaceum seed kernel) and Jeera (Cuminum cyminum) as a galatogogue. Ark-soyaa is used in Unani medicine in colic and digestive problem; also in gripe-water.

Active principles and pharmacology

Several compounds have been isolated from the seeds: myristicin; coumarins; and umbelliferone; flavonoids; phenolic acids; steroids and probably phytofluene; xanthones and triterpenes. The essential oil of seeds contains carvone (30–63 %). A sample of dill oil from Haldwani, Uttaranchal, contained carvone 34.5; dihydrocarvone, 12.0; D-limonene, 10.0; D-phellandrene 6.0; alpha-terpinene 6.0; carveol 4.0; dihydrocarveol, 3.5; iso-eugenol 2.3; and dill-apiol 3.0 % v/v. The oil and its emulsion in water is stomachic, diuretic, anthelmintic and antiflatulent. It is an important ingredient of gripe-water preparations in India.

Research supports dill's 3000 years of use as a digestive aid. The herb helps relax the smooth muscles of the digestive tract and acts as an anti-flatulent. Dill seed oil is found to inhibit the growth of several bacteria that attack the intestinal tract. It may, in prescriptions, help prevent infectious diarrhoea. Dill inhibited E. coli and can be used as an adjuvant in recurrent urinary tract infections due to E.coli.

Anethum graveolens Linn. is Europian Dill, while Anethum sowa is an Indian species. Though there is a variation in chemical constituents, both are considered as one herb in Indian medicine.

Use in Western herbal

Records found in 3000 year-old Egyptian tombs show that ancient physicians used dill as a digestive aid and intestinal gas remedy. The first century Greek physician Dioscorides prescribed dill so frequently that it was known for centuries as the herb of Dioscorides.

Nicholas Culpeper (1616–1654) wrote of it: "The dill being boiled and drunk, is good to ease swellings and pain. It also stays the belly and stomach from casting. The decoction therefore helps women that are troubled with the pains and windiness. It stays the hiccough. The seed is of more use than leaves and is used in medicines that serve to expel wind. The seed, being roasted or fried, and used in oils and plasters, dissolves the imposthumes in the fundament, and dries up all moist ulcers, especially in the fundament; an oil made of dill is effectual to warm or dissolve humours and imposthumes and the pain, and to procure rest. The decoction of dill, be it herb or seed, in white wine, being drunk is the provoker of the terms."

In Western herbal dill is used with other antispasmodics to relieve period pain.

When taken regularly by nursing mothers, dill increases milk production and helps prevent colic in their babies.

German Commission E monograph recognized the herb's efficacy in the following areas:

- Loss of appetite
- Liver and gallbladder complaints
- Inflammation of pharynx
- Common cold, cough
- Bronchitis
- Fevers and colds.

Antispasmodic and bacteriostatic properties of the seed and its use in dyspepsia have been accepted in the West.

Caution

Take essential oil internally only under medical supervision. In sensitive individuals dill might cause skin rash.

Apiaceae

ANGELICA

Angelica archangelica Linn.
Angelica sinensis (Oliv.) Diels.
Angelica polymorpha Maxim. var. sinensis Oliv.

Habitat

Cultivated in China in provinces of Gansu, Ningxia, Sichuan, Yunnan, Hubei, Shaanxi and Guizhou. Occurs in Kashmir at 1000–3900 m.

Classical & common names

Ayurvedic: Chandaa, Chandaa-shuka. Choraka variety.

Chinese: Dang-gui, Pinyin, Wade-Giles, Tang-kuei.
English: Chinese angelica, Dang-quai.
See also Angelica glauca.

Parts used
Root.

Dose
Root 4.5–12 g in decoction.

Classical use
Angelica archangelica: for bronchial catarrh, coughs, digestive disorders and as a liver stimulant, diuretic, diaphoretic, uterine stimulant and anti-inflammatory.

Angelica sinensis: for anaemia, menstrual irregularities, period pain, and as a circulatory stimulant, laxative, tonic after childbirth, tonic against hepatotoxins.

The Chinese angelica is prescribed for anemia and in blood loss. It regulates the menstrual cycle, relieves period pain and cramps. Chinese angelica is given in prescription to women with heavy menstruation, along with other tonic herbs. The herb is also given for improving blood circulation in the body, for treating digestive disorders, skin diseases, abscesses, boils, Herpes zoster.

Active principles and pharmacology
Key constituents are: coumarins; volatile oil (butylidine, phthalide, ligustilide, sesquiterpenes, carvacrol); vitamin B12, beta-sitosterol.

Research in China from the 1970s has shown that the herb helps regulate uterine contractions and gives relief in period pain. Research also shows that the whole plant, incl. the rhizome, strengthens liver function.

Chinese angelica helps the liver to utilize more oxygen. Preparations containing the herb have been tried clinically to treat hepatitis and liver cirrhosis.

Animal studies have confirmed the influence of Chinese angelica on heart and cardiovascular system, incl. decrease in contractions while strengthening ventricular contractility, a definite effect on arrhythmia and an experimental ability to dilate coronary vessels, increase coronary flow, and reduce arterial blood pressure.

The volatile constituents act as an experimental brain sedative. The decoction of the root exhibited diuretic, strongly antibacterial, anti-inflammatory, analgesic and antispasmodic activity in various experiments.

Use in Western herbal
The whole root, powdered root, sliced root, alcoholic extracts, capsulated products are commonly available in American market.

Its use is mostly confined to menstrual disorders and for relieving cramps. A practitioner usually recommends it from 14 days after the period begins until the start of the next period; the drug is not given while menstruating, as it can increase blood loss. A typical dose might be one to two 550 mg capsules twice a day, or as per dose directions on the package.

Chinese angelica is also used in PMS with dysmenorrhea; also in hot flashes and during menopause. Powdered root tea 1–2 g, tincture (1:5) 4 ml or 1 tsp, fluid extract 1 ml or tsp three time a day is given, depending upon the individual requirement.

Chinese angelica has not yet entered into Western herbal as a circulation-enhancing, pain-relieving, tranquillizing, liver-protecting and haematogenic tonic.

Angelica glauca Edgew.

Habitat
Native to north-temperate regions and New Zealand. Found near the water-channels in Kashmir, at altitudes of 1000–3900 m; also reported from Sikkim at 3000–3300 m.

Classical & common names
Ayurvedic: Choraka, Parnachoraka, Phalachoraka, Kshemaka (Charaka); Tilaparnikaa (Sushruta); Nishaachar, Taskara; Chandaa (Angelica archangelica Linn.).
Siddha: Chengan (Angelica archangelica).
Related species: Angelica archangelica (European); Angelica sinensis (Chinese). In Kerala, Kaempferia galanga Linn. is used as Choraka.

Parts used
Root.

Dose
Powder 1–3 g.

Classical use

Charaka prescribed the root in prescriptions for headache, insanity, epilepsy; also in coryza, hiccough and bronchial asthma.

Sushruta used leaves of Choraka internally in skin eruptions, swelling and fracture of bones. Sushruta included it in the group of herbs that destroys poison. Vaagabhatta included it in Elaadi group of herbs which alleviates allergic conditions, itching, urticaria, asthmatic attacks and coughs.

In the South, Chandaa has been equated with Costus speciosus (Koenig) Smith and the same has been incorporated in Elaadi group of herbs and in Manjishthaadi Tailam, Manjishthaadi Churnam.

In Manjishthaadi Tailam both Choraka and Chandaa have been included. This is prescribed for massaging the head in headache and neurological affections.

Mahaapaishaachika Grita, a classical compound of Sahsrayoga, is considered an intellect-promoting tonic for children.

Active principles and pharmacology

Angelica root contains a volatile oil, consisting mainly alpha- and beta-phellandrenes, alpha-pinene, macrocyclic lactones, incl. penta- and heptadecanolide; furocoumarins incl. bergaptene, xanthotoxin, scopoletin, umbelliferone, caffeic acid derivatives incl. chlorogenic acid, and flavonoids.

Phellandrene has a stimulating action on the nervous system. The bitterness of furocoumarins accounts for digestive stimulant effects.

So far there is little pharmacological information on the circulatory stimulating effect of the herb, though it is found efficacious in cases of poor circulation and it improves blood flow to the peripheral parts of the body. It is considered a specific treatment for Buerger's disease, a condition that narrows the arteries of the hands and feet.

By improving blood flow and stimulating the coughing up of phlegm, the herb's warm, tonic properties bring relief from bronchitis and debilating conditions affecting the chest.

The coumarins prove valuable in reducing high-protein oedemas, especially swellings of lymph nodes, as well as in treating psoriasis often accompanying arthritis. The constituent bergapten is antipsoriatic, linalool and borneol antibacterial and antifungal.

Use in Western herbal

Parkinson, in his "Paradise in Sole" (1629), puts Angelica in the forefront of all medicinal herbs, and said of it: "Angelica resisteth poison by defending the heart, the blood and the spirits and giveth heate and comfort to them."

During the 16th and 17th centuries, the juice from the crushed roots was combined with other herbs to make "Carmelite Water", a medieval drink said to cure headache, promote relaxation, protect against poison. During the 17th century, Angelica became a popular treatment for colds and other respiratory ailments.

During the 19th century, American Eclectic physicians used Angelica for heartburn, indigestion, bronchitis and fevers.

German researchers have discovered that Angelica archangelica relaxes the windpipe as well as the intestines. Japanese researchers have reported that the herb has anti-inflammatory effect. According to reports from China, Angelica increases red blood cell counts. Its use in anemia is suggested.

German Commission E monograph recognized European angelica's efficacy in

- Loss of appetite
- Dyspeptic complaints
- Mild spasm of gastrointestinal tract.

Undisputed properties of Angelica, in the West, are: antispasmodic, cholagogue, stimulatory for secretion of gastric juices.

Caution

Photodermatosis is possible following intake of large quantities of the drug, particularly ethanolic extracts, due to photosensitizing character of furocoumarin. Contra-indicated during pregnancy.

Combretaceae

ANOGEISSUS

Anogeissus latifolia Wall.

Habitat

Found in the forests of sub-Himalayan tract and in the Shiwalik hills, as well as in the hills throughout India ascending to 1200 m.

Classical & common names

Ayurvedic: Dhava, Madhuravalka, Vakavrksha (Charaka); Veerataru (Sushruta); Gaur, Dhurandhar, Nanditaru, Shakataakhya. Not to be confused with Dhaataki (Dhaaya flower).
Unani: Dhaavaa; Samagh-e-Hindi (gum).
English: Crane Tree, Button Tree, Axle Wood.

Parts used

Heartwood, gum-resin.

Dose

Decoction 50–100 ml, gum-resin 1–2 g.

Classical use

According to Ayurvedic texts, the herb is cooling, cures polyurea, piles, anemia and inflammations. Charaka and Sushruta used the resin as an ingredient of medicinal oils for external application in chronic skin diseases; in prescriptions for dysuria and senility. Sushruta used the herb as an ingredient of a dusting powder for quick healing of wounds. The herb was prescribed in urinary calculi, seminal weakness, urethral discharges, jaundice and in obesity.

The herb has been included in Saalasaaraadi group of herbs. Drugs of these groups cure obstinate skin diseases, urinary disorders, piles, anemia and reduce adiposity; also prescribed in the vitiation of semen.

The paste of Khadira (Acacia catechu) and Dhava is used in Ayurvedic medicine for curing skin diseases and erysipelas. Originally, this was recommended by Charaka.

Ashtaanga Hridaya recommends the heartwood of Dhava as an aphrodisiac and rejuvenating tonic.

In Unani medicine, the decoction of the flowers is prescribed in spermatorrhoea. Water extract of 20 g flowers mixed with purified sugar is given as a styptic in bleeding piles. The ash of flowers, mixed with oil, is applied externally on burns. The gum, fried and processed in boiled sugar, is prescribed in leucorrhoea. In Unani medicine, the gum is known as "Gond Zanaanaa, " as it is specific for females. (Acacia catechu gum is prescribed for males and not given to females in Unani medicine.)

Active principles and pharmacology

While the leaves contain purely hydrolysable tannins and related compounds, the bark and wood extractives contain both flavonoid tannins and compounds related to hydrolysable and flavonoid tannins. The presence of alanine and phenylalanine has also been reported in the bark. The sap wood contains ellagic acid (0.01 %). The heartwood contains quercetin 0.4, myricetin 0.07, and trimethylellagic acid 0.03 %.

The gum is principally the calcium salt of a complex high molecular weight polysaccharide acid, ghattic acid, containing mainly laevo-arabinose, D-galactose, D-mannose, D-xylose and D-galacturonic acid.

According to pharmacological findings, the stem-bark is hypothermic, CNS active, antagonistic to amphetamine hyperactivity. Root stomachic, bark and fruit astringent. Bark's use in anemia, urinary discharges, piles, skin diseases and erysipelas has been validated on the basis of synergistic action of active principles.

Asteraceae

ANTHEMIS

Anthemis nobilis Linn.
Matricaria chamomilla L.
Matricaria recutita L.

Habitat

Anthemis nobilis: Cultivated in the temperate regions of the Himalayas, as a border plant.

Roman chamomile is cultivated in France, England, Belgium, Italy, Czechoslovakia, Egypt.

Matricaria chamomilla: Found chiefly in the Mediterranean region and Asia. Also found in Punjab, Himachal Pradesh and upper Gangetic plains.

Classical & common names

Anthemis nobilis :
Unani: Baabunaj-roomi/tuffaahi.
English: Roman chamomile, English chamomile, Double chamomile.
Matricaria chamomilla:
Unani: Baabunaa, Baabunaj (Arabic), Baabunah (Persian).
English: German chamomile, Single chamomile flowers.
Although botanically distinct, many constituents are common and justify the use of both the herbs for the same ailments.

Parts used

Flowers, seeds.

Classical use

The oil, known as Roghan-e-Baabunaa, and flower heads are used in Unani medicine externally and internally as an antirheumatic and antihysterical medicine. Also prescribed as a tonic in debility, jaundice, inflammatory diseases and convalescence. Decoction of flowers is given at night in insomnia, anxiety and stress. The oil is used against hard swellings.

Roughan-e-Baabunaa Saadaa and Qawi are prescribed for external use in rheumatism, pneumonia, pleuritis. An ointment, Zimaad-e-Mohallil, with Baabunaa as one of the ingredients, is massaged externally on swellings and rheumatic joints. In orchitis (inflammation of the testes) Zimad-e-Waram-e-Unsayain Muzmin, with Baabunaa as one of the ingredients, is applied externally.

The flowers of Baabunaa have been included in a nervine tonic, Majoon-e-Seer Alvikhani; the seeds are an ingredient in Majoon-e-Falasifa, a reputed Unani brain tonic for dementia, amensia and a nervine tonic for sexual debility, dysuria, polyuria and inflammations.

Active principles and pharmacology

Anthemis noblis: volatile oil contains proazulenes, bisabolol, farnesine, terpenes such as pinene, anthemal, angelic and tiglic acids; sesquiterpene lactone (nobilin); flavonoids (incl. paluletin, quercitrin, luteolin, apigenin); acetylenic salicylic derivatives, cyanogenic glycoside; bitter glycoside (anthemic acid); coumarins (incl. scopolin); valerianic acid; tannins.

Matricaria recutita: volatile oil (similar constituents as in Anthemis noblis, spiroether is the only additional constituent; flavonoids (incl.anthemidin and luteolin); bitter glycoside (anthemic acid); coumarin, malic acid; tannins.

The azulenes and bisabolol are anti-inflammatory and antispasmodic, reducing histamine-induced reactions such as anaplylaxis and hay fever, allergic asthma and eczema; they also are shown to speed up the healing of peptic ulcers.

The flavonoids (especially anthemidin) are antispasmodic. Spiroether is more strongly antispasmodic than papaverine. Valerianic acid is sedative, as are cyanogenic glycosides.

Heteroglycan polysaccharides have been shown to have significant immunostimulating properties.

It has been shown that the herb kills Candida albicans, the cause of vaginal infections, as well as staphylococcus.

Use in Western herbal

Pliny (AD 23-79) recommended chamomile as a cure for headaches and all illnesses attacking liver, kidneys and bladder.

Nicholas Culpeper (1616–1654) recommended chamomile for fevers, digestive problems, aches, pains, jaundice, kidney stones, dropsy, and for promoting menstruation. He wrote of it: "Chamomile takes away weariness. Flowers comfort the brain."

Today chamomile is one of the best-selling herbs in the United States. As a tea, chamomile is used as a calmative, soothing nerve tonic and in insomnia. It is also given in migraine. For liver problems, on an average 3–4 cups of chamomile tea per day is recommended. It is used in dyspeptic conditions with nervousness. It relaxes the gut wall, regulating peristaltic movements, relieving colic, nervous or irritable hyperactivity, diarrhoea and spastic colon; the volatile oil reduces flatulence.

Chamomile is recommended as an anti-inflammatory agent, as it is soothing to the gastrointestinal tract. Suggested dose is 400–500 mg capsules 3 times a day.

As antihistaminic properties have been confirmed in the herb, a chamomile throat spray (under Camocare lebel) has been launched in the United States for asthmatic patients.

Chamomile is applied externally to sore skin and eczema. A cream made from German chamomile was tested in 1987 for healing wounds and produced very positive results. Chamomile-containing preparations in the form of moisturisers, vanishing creams, shampoos, sun-tan lotions, deodorants are available in the West.

Kamillosan is available in Germany as well as in India over the counter and is being prescribed in inflammatory skin disorders, as follow-up treatment after topical corticoid therapy in contact dermatitis and eczema.

Caution

Avoid oil completely in pregnancy, as it is a uterine stimulant.

Rubiaceae

ANTHOCEPHALUS

Anthocephalus cadamba Miq.
Anthocephalus indicus A. Rich
Anthocephalus chinensis (Lam.) A. Rich. ex Walp.

Habitat

Found in sub-Himalayan tract from Nepal eastwards to Burma, southwards to Andhra Pradesh and Western Ghats; common in deciduous forests.

Classical & common names

Ayurvedic: Kadamba, Sidhupushpa, Bhringavallabha, Bhumikadamba (Charaka); Priyak, Vrttapushpa, Gandhmat-pushpa, Praavrshenya, Gaurakadambaka, Girikadambaka.
Related varieties: Dhuli kadamba, Raaja kadamba, Nipa. (Nipa has been equated with Adina cordifolia Benth and Hook. f.)
Siddha: Venkadambu.

Parts used

Bark, fruits, flowers.

Dose

Bark decoction 50–100 ml.

Classical use

In Ayurvedic medicine, the plant has been included in Rodhraadi and Kadambaadi group of herbs. These groups are adipogenous, retentive and inspissant to semen, aphrodisiac and cure affections of female genital tract, urinary disorders, anemia and skin diseases.

Charaka and Sushruta used the resin or the bark as a sedative, antitoxic, antiseptic, astringent, and in haemoptysis.

In folk medicine, the fruit juice, along with cumin (Jeera) and sugar is given to children in gastric irritability. The juice is also given in fever with persistent thirst, as it possesses cooling properties. The fruit is also used as a galactagogue, blood purifier, and as an expectorant.

The decoction of leaves is used for gargling in aphthous stomatitis due to their astringent properties, whereas the decoction of stem-bark is given in coughs, also as an astringent, febrifuge and anti-diuretic.

The decoction of the bark and Shunthi (Zingiber officinale) is given in diarrhoea and dysentery with bleeding. The decoction is also prescribed in leucorrhoea and spermatorrhoea.

Active principles and pharmacology

The stem bark contains saponins A-D. Saponins A and B gave cadambagenic acid as aglycone; C and D gave quinovic acid. Beta-sitosterol is also present in the stem bark. The ether-soluble alkaloid of the bark showed antibacterial activity.

The dried bark has been reported to contain steroids, alkaloids, fats and reducing sugars. The bark also contained tannins.

The leaves contain glycoalkaloids and nonglycosidic isomeric indole alkaloids, cadamine and iso-cadamine. The heartwood contains glycoside indole alkaloids—cadambine.

The essential oil from flowers contain linalool, geraniol, geranyl acetate, linalyl acetate, alpha-selinene, 2-nonanol and beta-phellandrene as major compounds.

Pharmacologically, the bark is tonic, sex-stimulant, galactogenic, febrifuge, astringent, vulnerary, alexeteric, antibilious. The bark is found efficacious in uterine complaints and blood diseases, the leaf in stomatitis, the fruits as aphrodisiac.

Meliaceae

APHANAMIXIS

Aphanamixis polystachya (Wall.) Parker
Amoora rohituka (Roxb.) W. & A.

Habitat

Distributed in sub-Himalayan tract from Gonda (Uttar Pradesh) eastwards to Bengal, Sikkim and Assam west, in Western ghats and the Andamans.

Classical & common names

Ayurvedic: Rohitaka, Shalmalikaa, Plihaari, Raktghna, Sadaapushpa. (CCRAS has equated Rohitaka with Tecomella undulata Seem.(Bignoniaceae), while the scientists of INSA, P. Ray et al, and the Wealth of India, CSIR, with Amoora rohituka.)
English: White Cedar, Amoora.

Parts used

Bark, flowers.

Dose

Decoction 50–100 ml, powder 1–3 g.

Classical use

According to Ayurvedic classical texts, Rohitaka is specific for diseases of liver and spleen, internal tumours and diseases of the abdomen incl. ascitis. Charaka prescribed powdered bark, its decoction and extract in clarified butter in jaundice, enlarged spleen, anaemia, intestinal worms, urinary disorders. The paste of root was given internally in leucorrhoea.

Rohitakaarishta (based on Bhaishajya Ratnaavali) is the only classical compound available over the counter and is being prescribed in liver and spleen diseases, oedema and anaemia. Other classical compounds, Rohitakaadya Churna, Rohitaka Ghrita, Rohitaka-lauha, are no more available.

Active principles and pharmacology

Leaves contain a diterpene, aphanamixol and beta-sitosterol. Stem-bark gave beta-sitosterol, stigmasterol and a number of glycosides.

Bark gave a tetranorterpenoid, alphanamixinin. Seeds gave a limonoid, rohitukin, an essential oil and a fixed oil. The fatty acids of the oil comprise mainly linoleic, oleic, stearic, palmitic and linolenic acids. Fruit-shell gave a triterpene—aphanamixin. Aphanamixin lactone and aphanamixolide are also present in the plant.

The bark is strongly astringent and its use in diseases of liver and spleen, for tumours and abdominal affections is being studied. An aqueous extract of the bark, when injected intraperitoneally in normal guinea-pigs, showed reduction in the absolute lymphocyte count and an increase in spleen weight. Total leucocyte count and total body weight were significantly reduced. The bark appears to be an immunosuppressive drug similar to prednisolone.

A 50 % ethanolic extract of the stem was reported to possess anticancer activity.

CCRAS has quoted a few studies to validate the classical use of Rohitaka. (As mentioned earlier, the Council has equated Rohitaka with Tecomella undulata.)

The alcoholic extract of Rohitaka, dissolved in propylene glycol and administered in a dose of 25 mg/kg i.p. exhibited decrease collagen content of liver and increase in body weight in chlopromazine damage (0.25 mg/g i.p.) in rats.

The alcoholic extract of Rohitaka in a dose of 300 mg produced relaxation in dog's spleen. The extract also produced decrease in alkaline phosphatase value in rats against CCl-4-induced liver cirrhosis.

Apiaceae

APIUM

Apium graveolens Linn.

Habitat

Wild in the foothills of north-western Himalayas and the outlaying hills of Punjab, Himachal

Figure 1 Apium graveolens—flowering tops [WOI]

Pradesh and Uttar Pradesh. Largely cultivated in Amritsar and adjoining parts of Punjab, Haryana and some areas of Uttar Pradesh, for its seeds.

Classical & common names

Unani: Karafs, Bazr-ul-karafs (Arabic), Vilayaati Ajmodaa.
English: Celery.

Parts used

Root, seeds.

Dose

Root 5–7 g, seeds 3–5 g.

Classical use

Used in Unani medicine. Considered as a carminative, antispasmodic, liver stimulant and nervine sedative. Prescribed in anasarca, rheumatic complaints, gout, gall and kidney stones and as a potent diuretic, urinary antiseptic, and deobstruent for clearing toxins from the system. Also given to women as an emmenagogue (contraindicated during pregnancy), and in cystitis for disinfecting the bladder and urinary tubules (the volatile oil is contraindicated in kidney infections.)

Active principles and pharmacology

The seeds yield a volatile oil (1.3–3 %) containing limonene (60–70 %), phthalides and beta-selinene. They also contain coumarins; furanocoumarins (bergapten), flavonoids (apiin).

Research in Germany and China during the 1970s and 1980s has shown that the essential oil has a calming effect on the central nervous system. Some of its constituents have antispasmodic, sedative and anticonvulsant actions. Studies in China have confirmed the oil's usefulness in treating high blood pressure.

Use in Western herbal

John Gerard (1545–1612) and Nicholas Culpeper (1616–1654) used celery for "provoking" urine, as an aid to weight loss and for expelling "phlegm out of the head". Later, herbalists prescribed it for nervousness, insomnia, as a menstruation promoter and as an "aphrodisiac".

Contemporary herbalists recommend celery as a diuretic, tranquilliser, sedative, menstruation promoter and as a treatment for gout, arthritis, lack of appetite and obesity (water weight lost using diuretics invariably returns).

The seeds are prescribed for helping kidneys dispose off urates and other unwanted waste products, as well as for reducing acidity, detoxifying the body and for improving blood circulation to muscles and joints.

As a cleansing drink, one cup of organic carrot and celery juice is given daily. Fresh stalks are given to stimulate milk flow after childbirth.

In homoeopathy, the tincture of seeds is given in retention of urine, headache, heartburn, rheumatic pain in muscles of the neck, also in sacrum and in dysmenorrhoea. Standard dose is first to thirtieth potency.

Caution

Bergapten in the seeds could increase photosensitivity. Do not apply essential oil externally in sunlight. The oil of the seeds can act as a uterine stimulant. Do not use during pregnancy.

Do not take celery if suffering from kidney disorder.

Thymelaeaceae

AQUILARIA

Aquilaria agallocha Roxb.

Habitat

Found on the hills of Assam, Meghalaya, Nagaland, Manipur and Tripura.

The wood infected by certain fungi, develops large and irregular patches of dark streaks charged with an oleoresin and becomes odoriferous. Thus, Agar is the pathological product of a fungal disease contracted by the tree through wounds on the trunk. Species of Aspergillus, Fusarium, Penicillium and also some fungi imperfecti are reported to be associated with Agar. Agar is frequently found in trees that are more than twenty years old.

Agar is graded into four categories: Black or true Agar, Brown Agar or Bantang, Yellowish brown Agar or Bhuta and Yellow Agar or Dhum. Dhum is the lowest grade which is chiefly distilled for oil. The oil of Bhuta is reported to be superior to that of Dhum.

Classical & common names

Ayurvedic: Aguru, Alaktaka, Jongaka, Kaalaaguru, Vasuka (Charaka); Krimij, Krimijagdha, Loha, Asitaka, Krishnaaguru.
Unani: Ood Hindi; Ood-ul-Hindi (Arabic).
Siddha: Agil.
English: Agarwood, Aloewood, Eaglewood, Malacca Eaglewood.

Parts used

Aromatic resinous wood.

Dose

Powder 1–3 g.

Classical use

Charaka gave the powdered root, alone or in prescriptions, in fevers, toxicosis, dermatosis, rheumatism, hiccough, asthma. The powder of Aguru mixed with honey was prescribed for checking hiccough, the paste of Chandaa (Angelica archangelica) and Aguru for external application in oedema.

Sushruta used the medicinal oil of Aguru as an external application to ulcers, wounds, ringworm, chronic skin diseases; the decoction of Paathaa (Cissampelos pareira) and Aguru in polyurea.

According to Ashtaanga Hridaya, Aguru powder mixed with honey relieves cough; incense of Aguru provides relief in hiccough and asthma; the herb possesses tonic properties like those of an aphrodisiac.

Lavangaadi Churna (Chakradatta), with Aguru as an ingredient, is the only classical product available at present, indicated in anorexia, indigestion, diarrhoea, polyurea, internal tumours and asthma.

In Unani medicine, Ood Hindi is a tonic to intestines; deobstruent, demulcent, sedative. Classical compounds, Jawarish-e-Ood shireen and jawarish-e-Ood tursh are antiemetic, antibilious, constipative, stomachic and are being prescribed in indigestion, anorexia, nausea, vomiting, diarrhoea.

Dawa-ul-Misk Motadil Jawahirwali is a reputed Unani tonic for liver, brain, heart, melancholia and psychosis. Ood Hindi is an ingredient in this and many other tonics—Majoon-e-Fanjnosh (digestive tonic), Majoon-e-Dabeed-ul-Ward (liver tonic), Majoon-e-Antaki (brain tonic), Majoon-e-Azaraqi (nervine tonic).

Unani physicians give one or two drops of the oil with betel-leaf (Piper betle) in asthmatic attacks.

The betal-leaf smeared with the oil of Ood Hindi is given like any other pouch of the betal-leaf, for retaining inside the mouth, to achieve retentive and aphrodisiac benefits. For its aromatic and sedative properties Ood Hindi is also used as a mood elevator.

Aguru is not easily available even at 80, 000 Rupees for one kg and the oil or attar is beyond the reach of physicians, at least in India.

Active principles and pharmacology

The main odoriferous constituent of the Agar oil, distilled from the Agarwood, is agarol, which is related to castol, an odoriferous constituent of costus root (Saussurea lappa). The other constituents present are agarospirol, hydroxyagariphilone, alpha-agarofuran, beta-agarofuran, dihydroagarofuran, nor-ketoagarofuran, 4-hydroxy-dihydroagarofuran, 3-4-dihydroxy-dihydroagarofuran and agarotetrol.

Stem-wood yields sesquiterpenes, melofuran and agarol. It also yields a coumarinolignan, aquillochin. The heartwood yielded liriodenine, a 7-oxoaporphine. Five alkaloids were found in the roots; one of them was identified as liriodenine.

Agarwood powder is no more used in Indian medicine and is mainly used in the manufacture of fumigators, pastilles and agarbatties. Agar oil, one of the highly prized essential oils, is being exported mainly to Middle East countries.

The spent wood, recovered after the distillation of Agar oil that too adulterated with cheap aromatic plant parts, is available in the herb market. Its medicinal value is not worth its cost

Arecaceae

ARECA

Areca catechu Linn.

Habitat
Distributed in coastal region from Maharashtra to Kerala and Tamil Nadu and in the Deccan plateau, in Assam, Meghalaya, West Bengal and Andman and Nicobar Islands. Also cultivated for its nuts.

Classical & common names
Ayurvedic: Puuga, Kramuk, Guwaak, Ghorant, Puugi.
Unani: Fufal (Arabic), Chhalia, Supaari.
Siddha: Kamugu
English: Arecanut Palm.

Parts used
Nut, bark.

Dose
Nut powder 1–3 g, decoction 50–100 ml.

Classical use
According to classical Ayurvedic texts, Puuga is heavy, cooling and astringent. It is intoxicating and appetiser. It disinfects the mouth. Wet Puuga, in excess, is heavy and obstructs the channels of circulation, suppresses the power of digestion and affects the vision. When boiled, it is devoid of its toxic properties.

Laghu Supaari Paak (Yoga Chintaamani) is a reputed Ayurvedic tonic for females. It is prescribed in leucorrhoea, menorrhagia and other uterine complaints. Brihat Supaari Paak is based on Unani Siddhayoga Sangraha, and is prescribed as a uterine tonic after delivery; also in leucorrhoea.

Majoon-e-Supaari Paak is a Unani compound, prescribed in leucorrhoea, sterility; also in premature ejaculation.

A paste of dry arecanut powder is used as dentifrice.

Powdered nuts are prescribed in diarrhoea, unripe nuts as laxative. Applied externally to ulcers and skin disorders.

The bark is used as a deobstruent in flatulence; also for choleric action.

Active principles and pharmacology
Betel nut contains pyridine alkaloids: arecoline, guvacoline (ester alkaloids), besides arecaidine, guvacine; and catechin type tannins.

Arecoline has an effect similar to pilocarpine. It is reported to be cholinergic, exerting a sialagogue and diaphoretic action in normal dosage.

Arecoline is converted in the central nervous system to the stimulant arecaidine through chewing. Due to arecoline content, the drug appears parasympathomimetic. It leads to increased salivation, in high doses to bradycardia, tremor, reflex excitability, spasm and eventual paralysis. Long-term use of the drug as a stimulant can result in malignant tumours of the oral cavity through formation of nitrosamines.

The toxic dose for humans is 8 to 10 g of the drug.

An aqueous extract of fresh nuts has been reported to inhibit the growth in vitro of Micrococcus pyogenes var. aureus and Trichophyton rubrum. The alcoholic extract inhibits the growth of E. coli and Candida albicans. The non-alkaloid fraction of arecanuts is reported to possess antimicrobial properties.

Use in Western herbal

In the past, arecanut was used in gum diseases and as a vermifuge. Now it is only used as a vermifuge in veterinary medicine for tape worms, for intestinal colic in horses, but is not considered safe.

Papaveraceae

ARGEMONE

Argemone mexicana Linn.

Figure 1 Argemone mexicana—flowering and fruiting branch [WOI]

Habitat

Native to America. Naturalized throughout India up to an altitude of 1500 m.

Classical & common names

Ayurvedic: Svarnakshiri, Katuparni, Pitadugdhaa (CCRAS), Hemaahvaa, Kanakakshiri, Hemakshiri, Choka (Charaka, Sushruta, Bhaavaprakaasha). Acccording to NAA, Hemaahvaa is equated with Svarnkshiri and Hemaahvah with Dhattura.

Unani: Satyaanaashi, bharbhaanda.

English: Mexican Poppy, Prickly Poppy, Yellow Thistle.

Parts used

Latex, seed oil (externally).

Classical use

According to Siddha-bheshaja-manimaalaa, the plant is used externally in skin and venereal diseases. It acts as an antiseptic, antibacterial and antifungal agent.

Hemaahvaa of Bhaavaprakaasha was used as a purgative and cured intoxication, parasitic infection, itching, vitiated cough, constipation, poisoning and obstinate skin diseases. It cannot be equated with Argemone mexicana.

Kanakpushpi of Charaka Samhitaa has been equated with Euphorbia thomsoniana Boiss and Kanakakshiri, Hemakshiri and Svarnakshiri of Sushruta Samhitaa with Cleome felina Linn., syn. Polanisia felina DC. by INSA Scientists (P. Ray et al).

Flowers of Kanakapushpi were used in an ointment which was applied over leucoderma. Kanakakshiri, Hemakshiri, or Svarnakshiri were prescribed by Sushruta internally in acute constipation, paralysis and abdominal swellings; roots as purgative; the plant, as an ingredient of an ointment, in virulent forms of ringworm.

Active principles and pharmacology

The seeds are emetic, narcotic and poisonous. They yield an oil which is nauseous and non-edible. It contains two alkaloids, sanguinarine and dihydrosanguinarine. The oil inhibits the growth of both Gram-positive and Gram-negative organisms.

The latex of the plant contains the alkaloids, berberine (0.74 %) and protopine (0.36 %).

Flowers are considered narcotic. They contain isorhamnetin, isorhamnetin-3-glucoside and isorhamnetin-3, 7-diglucoside. The roots and stem contain protopine 0.084, berberine 0.041, tannin 1.1 and resin 1.75 %.

Use in Western herbal

Fresh juice is applied to ulcers and warts.

Used in homoeopathy to treat painful neuromuscular conditions.

See Euphorbia thomsoniana.

Convolvulaceae

ARGYREIA

Argyreia speciosa Sweet
Argyreia nervosa (Burm. f.) Bojer.

Figure 1 Argyreia nervosa [CCRAS]

Habitat

Found all over India, ascending to 300 m.

Classical & common names

Ayurvedic: Vrddhadaaruka, Vrddhadaaru, Vriddhadaara, Atarunadaaru, Sthavira, Sthaviradaaru, Bastaantri, Ghaavapattaa, Vidhaara. Ipomoea petaloidea Choisy has been equated with Vrdhadaaruka, Vrdhadaaraka, Vrdhadaara of

Figure 2 Argyreia nervosa [CCRAS]

Ayurvedic texts and Argyreia speciosa as its substitute. (CCRAS.)

Unani: Bidhaara.

Siddha: Nilapoosani.

English: Elephant Creeper.

Parts used

Root, leaves. Seeds, known as Samudra-shokh (sosh), are also used.

Dose

Powder 3.6 g, fresh juice 5–10 ml.

Classical use

The plant entered into Indian medicine after the 14th century. Bhaavaprakaashaa promoted it as a rejuvenating tonic—age-sustainer, spermatogenic, aphrodisiac. To convert the root into a nervine tonic, the powder is to be soaked seven times during seven days in the juice of the tubers of Asparagus racemosus (Shataavari) and dried (Ashtaanga Samgraha, Vrindamaadhava). According to Bangasena, clarified butter, cooked with the root and taken with milk, is an excellent aphrodisiac, and should be used by those desiring progeny.

The powder of the root with sour gruel was prescribed for filaria (Vrindamaadhava, Bangasena). It was given with milk in venereal diseases. The under surface of the leaves, as an irritant, was applied to hasten maturation and suppuration of boils. Upper surface of the leaf was credited with cooling properties.

Active principles and pharmacology

The whole plant, roots, leaves, and different extracts possessed alterative and tonic properties in nervous disorders. None of the extracts exhibited anti-inflammatory effect during acute and sub-acute models of inflammations. They did not show diuretic activity and showed no effect on blood pressure and respiration in anaesthetised dogs (CCRAS).

In a clinical trial, Fortage, an indigenous product of Alarsin Pharmaceuticals, Mumbai, with Aagyreia speciosa as one of the ingredients, was administered for three months after post-vasectomy recanalisation operation. An appreciable increase in sperm count, sperm motility and sperm morphology matched with the classical use of Vrddhadaaruka as a spermatogenic agent and its inclusion in various sex tonics.

The seeds contain ergoline alkaloids; also contain caffeic acid and ethyl caffeate. A few of the ergoline alkaloids are reported to be hallucinogenic. The seeds possess significant hypotensive and spasmolytic activity.

The seeds are mostly used in Unani medicine, as an ingredient of sex tonics in spermatorrhoea.

(The seeds of Salvia plebeia R. Br. are being used as a substitute for Aagyreia speciosa seeds. In the market samples the seeds known as Samudra-sokh were found to be the seeds of S. plebeia.)

The plant showed antiviral, antibacterial and antifungal properties.

Methanol, benzene, chloroform and petrolium-ether extracts of the plant were found non-toxic (CCRAS).

Aristolochiaceae

ARISTOLOCHIA

Aristolochia bracteata Retz.
Aristolochia bracteolata Lamk.

Habitat
Found in upper Gangetic plain, West Bengal, Gujarat and peninsular India.

Classical & common names
Ayurvedic: Kitamaari, Dhumrapatra.
Siddha: Kattusuragam.
English: Bracieated Birthwort.

Figure 1 Aristolochia bracteolata ham [ADPS]

Parts used
Root.

Dose
Decoction 50–100 ml, powder 1–3 g.

Classical use
The plant is used in indigenous medicine for its bitter, purgative and anthelmintic properties. The root is reported to be used as a substitute for Aristolochia indica Linn. (Ishwari).

The root decoction is employed as an emmenagogue and is administered as a dry powder or infusion to increase uterine contractions.

The bruised leaves mixed with castor oil are applied externally in eczema.

The juice of the whole plant is applied to ulcers.

Active principles and pharmacology
Leaves and fruits yield ceryl alcohol, beta-sitosterol and aristolochic acid. The root also contains aristolochic acid. The seeds gave an alkaloid magnoflorine, aristolochic acid, and a fatty oil.

Biological activity of the plant is oxytocic. Active principles confirm its anthelmintic, cathartic, emmenagogue properties

Aristolochia indica Linn.

Figure 2 Aristolochia indica [CCRAS]

Figure 3 Aristolochia indica [CCRAS]

Habitat
Found throughout low hills and plains of India.

Classical & common names
Ayurvedic: Ishwari, Naakuli, Arkmuula
Siddha: Adagam.
English: Indian Birthwort.

Parts used
Root.

Dose
Root powder 3–6 g.

Classical use
The dried roots and rhizomes are used as a gastric stimulant and bitter tonic. Fresh juice of the leaves and also of the bark is used in the bowel complaints of children, diarrhoea and intermittent fevers. As a blood purifier, the root is used in skin diseases. It heals wounds and destroys the toxic effect of all poisons. In the olden days, it was used against snake bites in the South.

Active principles and pharmacology
The root contains nitrophenanthrene compounds. Aristolochic acid is the chief active principle of the drug, though aristolic and p-coumaric acids also appear to contribute to the activity of the drug.

Extracts or isolates of Aristolochia indica containing aristolochic acid possess anticancer activity. Kidney damage has been observed in rats treated with aristolochic acid. Aristolic acid has been found to possess anti-estrogenic activity and to prevent implantation in the early stage of pregnancy in mice. Methyl ester of aristolic acid, however, showed damage to liver and kidney and increase in uterine weight. The chloroform extract of the root has exhibited an anti-spermatogenic effect on mice and the petroleum ether, chloroform and alcoholic extracts have exhibited interceptive activity in mice and hamsters.

The plant possesses emmenagogue, abortifacient, antiarthritic, anti-inflammatory, antiperiodic, diuretic and antibilious properties. Anticancer, anti-implantation, interceptive and antifertility activities of the plant have not yet been recognized in herbal medicine, nor these properties ever had a traditional base.

Caution
In large doses the herb acts as a local irritant and provokes nausea, griping pains in the bowels and sometimes vomiting and tenesmus.

Asteraceae

ARTEMISIA

Artemisia absinthium Linn.
Absinthium vulgare Gaertin.

Habitat
Found in Kashmir at altitude of 1500–2100 m.

Classical & common names
Unani: Afsanteen (Arabic).
English: Wormwood.

Parts used
Dried herb, leaves, flowering tops.

Dose
2–5 g.

Classical use
In Unani medicine, the herb is used in prescriptions for treating inflammatory diseases of the liver and spleen. Unani practitioners prescribe Arq-e-Afsanteen in hepatitis and hepatic obstruction. (Dose: 50–100 ml); Dava-ul-Luk (Qarabadeen-e-Azam) in the enlargement of liver or spleen, in renal calculus; Qurs-e-Gul (Qarabadeen-e-Jadeed) in chronic fever and as a liver tonic.

The decoction is prescribed in dysmenorrhoea. The herb is also given in hysteria, epilepsy, nervous irritability, gastric and nervous depression, mental exhaustion, intermittent fevers, in prescriptions with other supporting herbs. As an enema, its infusion is used as an anthelmintic. A strong decoction is also given under medical supervision.

The herb, steeped in hot vinegar, is applied for fomenting rheumatic inflammations, sprains or bruises. The expressed juice of the herb is applied to the forehead to prevent convulsions.

Active principles and pharmacology
The plant contains essential oil (incl. thujone, thujol); bitter sesquiterpenes, bitter sesquiterpene lactones, guaianolides or azulenes (artabsin, absinthin, anabsinthin, arborescin, globicin) and santanolides; terpenoids; flavonoids; hydroxycoumarins; polyacetylenes; tannins; resin; silica. (Vilaayati or imported Afsanteen is used in Unani formulations.)

Most of the sesquiterpenes, flavonoids and essential oil constituents have bitter effects, which increase gastric acid and pepsin secretion, pancreatic digestive secretion, hepatic bile flow, muscle tone of stomach and small intestines. The germacranolide group of sesquiterpene lactones show significantly antitumour, the azulenes anti-inflammatory, santonin a strong vermifuge, absinthin and anabsinthin insecticidal effects. The essential oil is carminative, but thujone and thujol in isolation are stimulating to smooth muscle and also antiseptic. The tannins and resins exert an astringent action, promote connective tissue repair.

In patients with liver disorders, a suspension of 20 mg extract in 10 ml water was administered with a stomach probe after which a significant effect on alpha-amylase, lipase, bilirubin and cholesterol was observed after 70 to 100 minutes.

In rabbits, fever induced through yeast injection could be reduced with the application of diverse fractions of the drug using an esophegeal probe.

The drug stimulates the bitter receptions in the taste buds of the tongue and triggers a reflexive increase of stomach secretion with higher acid secretion.

A watery extract of the whole drug is supposed to retard the growth of Plasmodium falciparum. The essential oil of the drug may possess an antimicrobial effect.

The oil extracted from the leaves is reported to have antibacterial and antifungal activity in 1:1000 dilution. In large doses the oil is a narcotic poison.

Use in Western herbal
Dr John Hill (1772) recommended Wormwood flower tops: "The leaves have been commonly used but the flower tops are the right part. These, made into a light infusion, strengthen digestion, correct acidities and supply the place of gall, where, as in many constitutions, that is deficient."

The Wormwood is used as a tonic, stomachic, anthelmintic and febrifuge.

As a nervine tonic, it is particularly helpful against epilepsy. It is given for relieving melancholia.

The juice of larger leaves, which grow from the root, is used as a remedy for jaundice.

The drug is rarely used in nervous diseases such as neuroasthenia, though it is said to stimulate the cerebral hemispheres and is a direct stimulant of the cortex cerebri. When taken in excess, it produces giddiness and attacks of epileptiform convulsions.

German Commission E monograph recognized the herb's efficacy in the following areas:

- Loss of appetite
- Dyspeptic complaints
- Liver and gallbladder complaints.

Western herbalists generally give a weak infusion 5–10 g herb to 500 ml water.

The Wormwood is also used in various liqueurs. One such liqueur "Wermuth" was considered a nervine and mental restorative. In France, the essential oil of the Wormwood, absinthe, was used in a liqueur for its narcotic and hallucinogenic properties (Van Gogh's choice). It did not get the legal status, as it was discovered to severely damage the central nervous system.

In homoeopathic medicine, tincture of fresh, young leaves and flowers of Artemisia absinthium (not of Artemisia vulgaris) is used as Absinthium for dyspepsia, chlorosis, congestion of brain, epilepsy, nervousness, restlessness, sleeplessness.

Caution

Contraindicated during pregnancy. Due to the drug's thujone content, the administration of large doses can lead to vomiting, stomach and intestinal cramps, headache and dizziness.

Artemisia maritima Linn.

Habitat

Found in the Western Himalayas from Kashmir to Nepal at altitudes of 2100–2700 m.

Classical & common names

Ayurvedic: Chauhaara, Kitaamari (non-classical) Yavaani.
Unani: Kirmaalaa, Kirmaani Ajwaayan (not related to Ajwaayan or Khurasaani Ajwaayan).
English: Sea Wormwood.

Parts used

Whole plant.

Dose

Powder 2–6 g.

Classical use

According to Raaja Nighantu, Chauhaara possessed anthelmintic properties. In traditional medicine, the wormseed was principally used against ascarides (not tapeworms); also in therapeutic doses against neuropathy, and as an emmenagogue and deobstruent. In prescriptions, it is found useful in urinary infections, intermittent and remittent fevers.

Active principles and pharmacology

Artemisia maritima var. thomsoniana is an important santonin-bearing species occurring in India. Besides santonin, the plant contains two crystalline compounds—beta-santonin and pseudosantonin (less anthelmintic and devoid of anthelmintic property respectively).

Santonin is found very effective against round worms, less so against threadworms and ineffective on taenia. Santonin is highly toxic; should be administered in very small quantities with supporting herbs.

Studies on hormonal and anti-hormonal effects of santonin in albino mice revealed that it showed no androgenic, estrogenic, anti-estrogenic, progestational and anti-progestational effects.

Use in Western herbal

Nicholas Culpeper (1616–1654) gave the following uses of the herb: "Boiling water poured upon it produces an excellent stomachic infusion, but the best way is taking it in a tincture made with brandy. Hysteric complaints have been completely cured by the constant use of this tincture. In the scurvy and in the hypochondrical disorders of studious, sedentary men, few things have a greater effect: for these it is best in strong infusion. The whole blood and all the juices of the body are effected by taking the herb. It is often used in medicine instead of Roman Wormwood (Artemesia Pontica), though it falls short of it in virtue."

Absinthium occurs in British Pharmacopoeia in the form of extract, infusion, tincture, and is directed to be extracted from Artemisia maritima (the Sea Wormwood) which possesses same virtues as those of Artemisia absinthium, but in lesser degree.

Caution

In large doses the drug causes yellow vision, sometimes violet vision, induces headache, nausea, vomiting and convulsions. Cases of fatal poisoning have been reported.

Artemisia nilagirica (C. B. Clarke) Pamp.
Artemisia vulgaris Linn. var. nilagirica C. B. Clarke

Habitat

Found throughout the hilly regions of India, ascending to an altitude of 3600 m in the Western Himalayas and to 1500–2400 m in Sikkim and Khasi hills; also found in Mount Abu in Rajasthan, in the Western Ghats and from Konkan southward to Kerala.

Classical & common names

Ayurvedic: Naagdamani, Damanaka, Daunaa.
English: Mugwort, Indian Worm Weed.

Parts used

Leaves, oil of the seed.

Classical uses

According to Bhaavaprakaasha and Dhanvantari Nighantu, Damanaka is an antidote to poisons, cures dysentery, chronic skin diseases, nervous and spasmodic affections, inflammations, vitiated blood. The herb provides resistance to diseases and falls in the category of age-sustainers, cardiac and brain tonics.

In traditional medicine, Ayurvedic and Unani, the herb is used as an emmenagogue, anthelmintic, stomachic, febrifuge. Externally it is used in skin diseases.

Active principles and pharmacology

The chief constituents of essential oil, obtained in North India, are camphor and borneol; alpha-thujone from wild plants of the Nilgiri hills. The plant also contains vulgarol, polyacetylenes, farnesol, stigmasterol, beta-sitosterol, alpha-amyrin and its acetate and vulgarin.

The essential oil is larvicidal, insecticidal, antimicrobial, antifungal.

Active principles give emmenagogue, anthelmintic, stomachic, febrifuge, antilithic and blood-purifying properties to the herb. The leaf is found to be haemostatic, the leaf and flowering tops antispasmodic and the root antiseptic.

Use in Western herbal

The herb is considered of value as a nervine and emmenagogue, having also diuretic and diaphoretic action.

As a nervine, Mugwort is valued in palsy, fits, epileptic and similar affections, especially in persons of a feeble constitution.

John Gerard (1545–1612) says: "Mugwort cureth the shakings of the joynts to the Palsie."

Nicholas Culpeper (1616–1654) says: "The juice of the large leaves is the best against dropsy and jaundice. The infusion (of the herb) drank morning and evening for some time helps hysterics, obstructions of the spleen and weakness of the stomach. Its oil, taken on sugar and drank after, kills worms, resists poison, and is good for liver and jaundice. The root should be accounted among the best stomachics. The oil of the seed cures quotidians and quartans. Boiled in lard and laid to swellings of the tonsils and quinsy is serviceable. It is admirable against surfeits...Wormwood and vinegar are an antidote to the mischief of mushrooms...mixed with honey it takes away blackness after falls, bruises."

Herbalists prescribe infusion of aerial parts for menopausal syndrome and period problems; also as a bitter to cool the digestive tract in fever management. Decoction of 5 g of the aerial parts and equal amount of dry ginger (Zingiber officinale) is given as a warming tea for period pain.

In homoeopathy, tincture of the fresh root of Artemisia vulgaris is used for epileptic conditions, petit mal epilepsy without aura, convulsions, hysteria.

Caution

Contraindicated during pregnancy.

Moraceae

ARTOCARPUS

***Artocarpus lacucha* Buch.-Ham.**
***Artocarpus lakoocha* Roxb.**

Habitat

Distributed in evergreen, semi-evergreen and moist deciduous forests up to 1800m in eastern and northern India.

Classical & common names

Ayurvedic: Lakucha, Granthiphal, Pitanaasha.
Siddha: Ilangu.
English: Monkey Jack.

Parts used

Fruit.

Dose

Juice 5–10 ml.

Classical use

Sushruta discarded the fruit from prescriptions, as it retarded spermatogenesis and disturbed body functions.

According to Bhaavaprakaasha and Kaiyadeva Nighantu, Lakucha is heavy, difficult to digest, wind-forming in the stomach, disturbs metabolism and vitiates semen.

Vaidyamanoramaa advises intake of 40 ml Lakucha fruit juice mixed with 40 ml goat milk in the morning for checking dysentery with blood, mucus and tenesmus. The fruits are contraindicated during summer.

In traditional medicine, the fruit juice is no more given internally. Used only externally in combination with other antiseptic herbs for treating arthritic swellings, obstinate skin diseases, for cleansing wounds.

Bark in powder form is applied to sores to draw out purulent matter.

Active principles and pharmacology

Bark yields beta-sitosterol, cycloartenol, cycloartenone, alpha-amyrin acetate and lupeol acetate. It is a good source of soluble tannins.

The heartwood contains artocarpin, norartocarpin, norcycloartocarpin, cycloartocarpin, resorcinol and oxyresveratrol. Oxyresveratrol is present in the stem also, and is found effective as a vermifuge in the treatment of tape-worm.

A lectin, artocarpin, isolated from the seeds precipitates several galactomannans. Besides human and animal erythrocytes, it agglutinates rat lymphocytes and mouse ascites cells.

Poaceae

ARUNDO

***Arundo donax* Linn.**

Although Phragmites karka Trin. ex Steud. is the accepted source of Nala in Ayurvedic medicine, Arundo donax is also recognized as its bigger variety by CCRAS.

Habitat

Native to the Mediterranean region, found wild and cultivated in the lower Himalayas from Kashmir to eastern India, up to 2400 m, extending Southwards into peninsular India.

Classical & common names

Ayurvedic: Nala, Potgaal, Shunyamadhyaa, Dhamana.
Bigger variety: Mahaanala, Devanala (Dhanavantari Nighantu, Raaja Nighantu).
Siddha: Moongil.
English: Great Reed.

Parts used

Root.

Dose

Decoction 50–100 ml.

Classical use

Charaka gave the decoction of the reeds, in prescriptions, internally as well as externally, in haemothermia, herpes, erysiplas, fever.

Sushruta prescribed the herb internally in urinary diseases and calculi. It was given internally with cow's milk in haemoptysis.

Sushruta used the root as an ingredient of a medicated paste for inflammations and traumatic swellings.

In Ayurvedic medicine, Nala belongs to Trna Panch Muula group of herbs which is used for treating burning syndrome and for cleansing the urinary bladder.

Active principles and pharmacology

The leaves contain indole bases, incl. gramine (donaxine) and donoxarine. Donaxine in small doses raises the blood pressure of dogs, but in high doses, it causes lowering of the blood pressure.

The rhizomes are reported to possess sudorific, diuretic and antilactant properties. The indole bases reported in the leaves are also present in rhizomes. Besides, the latter also contain bufotenidine and dehydrobufotenine. The ethanolic effect of the rhizomes showed hypotensive and antispasmodic effect against spasms induced by histamine, serotonin and acetylcholine. The flowers also contained gramine.

Rhizomes show sudorific, emollient, diuretic, antilactant and antidropsical properties; also stimulate menstrual discharge.

Galactogogue properties attributed to the herb could not be confirmed pharmacologically.

The herb qualifies for its use as a potent diuretic.

Use in Western herbal

Tincture of the root sprouts of Arundo mauritanica, an Italian grass, is used in homoeopathy for catarrh, coryza, diarrhoea and as a diuretic in urine retention. Also given in hay fever.

Caution

The herb disturbs blood pressure.

Aristolochiaceae

ASARUM

Asarum europaeum Linn.

Habitat

Indigenous to the northern parts of southern Europe, central and east-central Europe. Cultivated in the US.

Classical & common names

Unani: Asaaroon (Arabic), Sumbul-e-barri, Naardeen-barri (Persian).

English: Asarum, Wild Nard, Hazel Wort, Asarbacca.

Parts used

Root.

Dose

2–5 g.

Classical use

In Unani medicine, Asaaroon is used as a deobstruent, diuretic, anti-inflammatory, promoter of mental faculties, emmenagogue and tonic to the central nervous system. Its sedative and tranquillizing properties are applied in cases of epilepsy, partial paralysis, dementia and sexual inadequacies. It is given in obstructive conditions of liver and spleen; also in retention of urine.

Active principles and pharmacology

Chief constituent of volatile oil is asarone. The root also contains caffeic acid derivatives incl. chlorogenic acid, isochlorogenic acid; flavonoids; sucrose; camphor; tannic acid.

Asarum is an expectorant, bronchial spasmolytic, superficial relaxant and local anesthetic. (110 g *trans*-isoasaron taken orally caused severe vomiting.)

The surface-tension reducing effect of trans-isoasaron and trans-isomethyleugenol of Asarum europaeum was studied in vitro, using stalamometry. Both substances showed a concentration-dependent surface activity, which surpassed the effect of the control substance tyloxapol in a normal treatment concentration.

Local anesthetic effect of *trans*-isoasaron and of isomethyleugenol was tested in a clinical trial in order to compare it with benzocaine. The results showed a dose-related action for both drugs.

Use in Western herbal

Nicholas Culpeper (1616–1654) says of the European species of Asarum: "The herb, being drunk, not only provoketh vomiting, but purgeth both choler and phlegm...doth much help pains in the hips and other parts. Being boiled in whey, they wonderfully help the obstructions of liver and spleen, and are therefore profitable for the dropsy and jaundice. Being steeped in wine and drank it helps those continual agues that come by plenty of stubborn tumours.... The leaves and root being boiled in lye, and the head often washed therewith while it is warm, comforteth the head and brain that is ill-affected by taking cold, and it helpth the memory."

In the past, the drug was used as an emetic. Now, in folk medicine, it is used as a cough remedy, sneezing powder for chronic rhinitis, for pneumonia, angina pectoris, migraines, liver diseases, jaundice; as an emmenagogue; though all these properties have not been pharmacologically proven.

Tincture of the root and whole plant is used in homoeopathy. Before Hahnemann's time, the herb was little known except as an emetic. His provings brought to light many remarkable characteristics of the herb. It was found efficacious in nervous irritability and exaltation of the senses, vertigo, headache, hysteria, catarrhs, cholerine, diarrhoea, dysmenorrhoea, leucorrhoea, restless sleep.

Caution

Toxic in higher doses.

Asparagaceae

ASPARAGUS

Asparagus adscendens Roxb.

Habitat

Occurring in the Western Himalayas, the Himachal Pradesh and Kumaun up to altitude of 1500 m.

Classical & common names

Ayurvedic: Mahaashataavari, Shweta Mushali.
Unani: Shaqaqul-e-Hindi (Arabic, Persian). Safed Musali.

Parts used

Tuber.

Dose

Root powder 3–6 g.

Classical use

The powder of adscendens root, in prescription, especially Mushaliyaadi Churna, was a reputed aphrodisiac and was used to arouse sex urge. (Shaarangadhara Samhitaa, Bhaavaprakaasha, Yoga Ratnaakara.)

The tubers boiled in milk and sugar were prescribed in spermatorrhoea, general debility, chronic leucorrhoea, and were credited with galactagogue, demulcent, nutritive and aphrodisiac properties. The powder of tubers, due to its cooling and antacid properties, was prescribed in diarrhoea, dysentery, dyspepsia and hyperacidity.

Mushali Paak (Yoga Chintaamani) is available over the counter and is prescribed in sexual weakness and as an aphrodisiac.

Active principles and pharmacology

The dried, debarked tubeous roots contain carbohydrates, proteins and a steroidal saponin, which inhibits the growth of several pathogenic organisms in vitro. The plant contains, sapogenin. Sapogenins A and B, isolated from the root, were identified as stigmasterol and sarsasapogenin.

The roots have wrongly been identified with those of Chlorophytum arundinaceum; also with Shataavari.

Asparagus officinalis Linn.

Habitat
Native to temperate Europe and West Asia.

Classical & common names
Unani: Halyun; Isf-e-raaz (Arabic); Maar-chob (Persian).
English: Asparagus, Sperage, Sparrow Grass.

Parts used
Tuber.

Dose
Powder 3–5 g.

Classical use
Prescribed in Unani medicine as a deobstruent, emmenagogue, laxative, diuretic; in flatulency, calculous affections, dropsy, rheumatism, gout. Also given as an adjuvant to cardiac tonics.

Active principles and pharmacology
The tuber contains steriod saponins (incl. asparagosides) and aminoacids (incl. sulphur-containing aspartic acid).

The tuber is found efficacious in flushing-out therapy for inflammatory diseases of the urinary tract and for prevention of renal gravel.

Use in Western herbal
German Commission E monograph recognized the efficacy of the herb in:
- Infections of the urinary tract
- Kidney and bladder stones.

In homoeopathic medicine, tincture of young sprouts is given in urinary disorders, cardiac depression, dropsy, rheumatic pains.

Caution
Because of the irritating effect of saponin, the drug should not be administered in the presence of kidney diseases. In the case of reduced cardiac and/or kidney function diuretic therapy should not be attempted.

Asparagus racemosus Willd.

Figure 1 Asparagus racemosus [CCRAS]

Habitat
Found growing wild in tropical and subtropical parts of India, incl. the Andamans and ascending in the Himalayas up to an altitude of 1500 m.

Classical & common names
Ayurvedic: Shataavari, Shatmooli, Atirasa, Bahusutaa, Shatpadi, Shatviryaa, Bhiru, Indivari, Vari. substitute for Medaa, Mahaamedaa.
Mahaashataavari has been equated with Asparagus sarmentosus Linn.
Unani: Satawar.
Siddha: Seemaithannervittan.
English: Indian Asparagus.

Parts used
Tuber.

Dose
Powder 3–6 g, fresh juice 10–20 ml.

Classical use

Charaka and Sushruta used Shataavari as a styptic and ulcer-healing agent in intrinsic haemorrhage; disorders of female genitourinary tract; as an intestinal disinfectant and astringent in diarrhoea; as a nervine tonic in epilepsy, migraine; as a blood purifier in skin diseases. Sushruta gave the drug in internal abscesses, oedema, consumption, respiratory troubles, urethral discharges and seminal disorders. Prescribed the decoction boiled with clarified butter internally in urinary calculi.

By the 16th century, the drug emerged as an efficacious female tonic, galactagogue and entered into most of the aphrodisiac tonics.

Shataavari Ghrita (Charaka Samhitaa) is a reputed rejuvenating and aphrodisiac tonic. Shataavari Ghrita (Sahasrayoga) is given in disorders of the urinary system, urethritis, dysuria and cystitis (5–10 g twice daily with milk). Phalaghrita (Bhaishajya Ratnaawali) is another classical compound, prescribed for treating disorders of female genitals, leucorrhoea and infertility.

Shataavaryaadi Churna (Yoga Ratnaakara) is given (3–6 g) in sexual debility; Shataavari Lehya (Sahasrayoga) as a restorative tonic, haematinic, choalagogue and vitaliser in emaciating diseases, menorrhagia, haemoptysis and jaundice, in urinary disorders and as a diuretic (2–5 g once daily).

Active principles and pharmacology

The tuber contains quercetin, rutin and hyperoside. Sitosterol saponins A4, A5, A6, A7 and A8 were isolated from dried tubers. Structures of Saponin A4 were assigned as sarsasapogenin, having two molecules of rhamnose and one molecule of glucose (CCRAS).

The roots and leaves and their different extracts were found to possess many pharmacological properties: refrigerant, demulcent, antiseptic, diuretic, antidysenteric, antidiarrhoeal, antiamoebic, enzymatic, gastric sedative, anticoagulant, galactagogue, alterative and aphrodisiac. (CCRAS.)

The root powder, 4 tablets (500 mg each) 6 times a day to patients, suffering from diarrhoea, dysentery and colic, was found to be a potent antidiarrhoeal and gastric sedative. The alcoholic extract of roots was found to possess antiamoebic effect against E. histolytica and antiviral effect against Ranikhet disease virus. The aqueous extract of both fresh and dried roots possessed amylase and lipase like activity.

The crude alcoholic extract of the root exhibited anti-oxytocic activity, gain in weight of mammary glands in post-partum and estrogen-primed rats and gain in weight of uterous in estrogen-primed group of animals. It increased the weight of adrenal glands and reduced the quantity of ascorbic acid, suggesting release of pituitary ACTH.

The plant saponin produced a specific blockade of syntocinon (oxytocin)-induced contraction of rat, guinea-pig and rabbit uteri in vitro and in situ. It also blocked the uterine spontaneous motility.

The root, when administered to buffaloes, produced galactagogue effect.

The root extracts exhibited antiallergic activity by inhibiting passive cutaneous anaphylaxis in mice and rats.

The roots, when fed orally, acted as immunomodulator against induced sepsis and peritonitis in rats and mice.

Liliaceae

ASPHODELUS

Asphodelus fistulosus Linn.
Asphodelus tenuifolius Cav.

Habitat

Native to the Mediterranean region, Asia and Mascarene islands.

Classical & common names

Unani: Piyazi; Khunsaa (Arabic and Persian). Asaraash (Persian).
English: Asphodel.

Parts used

Root, seed, whole plant.

Dose

Whole plant 10 g, seed 3–7 g, ash 1–4 g.

Classical use

Seeds are used for their diuretic and deobstruent properties in prescriptions. They are considered hot, dry, and intoxicating. The ash of the root, mixed with oil, is used externally in baldness. Mixed with vinegar used on ringworm. It is also applied to skin blotches and pigmentation, exposing to sun-rays. Internally, the ash is diuretic, anti-inflammatory and emmenagogue. Mostly the herb is used externally, internally it disturbs liver functions.

Active principles and pharmacology

The seed contains an ester, 1-O-17-methylstearylmyoinositol. The seed oil gave myristic, palmitic, oleic, linoleic, linolenic acids. The plant contains a triterpene—lupeol and quercetin.

The seed oil, due to its high linoleic acid (60–62 %), is pharmacologically of therapeutic value in preventing atherosclerosis.

Use in Western herbal

White asphodel (Asphodelus ramosus) is considered acrid, heating and diuretic. Said to be useful in menstrual obstructions and as an antispasmodic. The bruised root is recommended for dissolving scrofulous swellings.

Yellow asphodel (Asphodelus luteus) is a native of Sicily; the onion-leaved asphodel (Asphodelus fistulosus) of France and Crete; False asphodel is an American name for Tofieldia.

According to classical herbalists, Hippocrates, Dioscorides and Pliny, the roots were cooked in ashes and eaten. Basically it was a cattle fodder.

Acanthaceae

ASTERACANTHA

Asteracantha longifolia (Linn.) Nees
Hygrophila spinosa T. Anders.

Habitat

Common in moist places, paddy-fields, throughout India and Srilanka.

Classical & common names

Ayurvedic: Ikshuraka (Sushruta); Kokilaaksha (Ashtaanga Hridaya, Raaja Nighantu); Kshuraka, Kaakekshu, Bhikshu..
Unani: Taalmakhaanaa (Seeds). National Formulary of Unani Medicine has wrongly equated Taalmakhaanaa with Euryale ferox salisb. (Fox nut).
Siddha: Neermulli.
English: Hygrophila.

Parts used

Root, seeds, ash of whole plant.

Dose

Decoction 5–10 ml, seed powder 3–6 g, ash 1–3 g.

Classical use

Charaka identified it as a semen depurant. According to Sushruta, the herb powdered with the seeds of Mucuna prurita and sugar, taken with warm milk, infuses renewed vitality in the body, acts as an aphrodisiac and replenishes the depleted energy.

The decoction, in Ayurvedic texts, was prescribed for alleviating rheumatism and gout (Ashtaanga Hridaya, Chakradatta). The ash of the plant was prescribed in oedema; decoction of the root as a diuretic in diseases of the genitourinary tract, in hepatic obstruction and anasarca.

Haarita Samhitaa attributes sedative property to the drug.

In Unani medicine, the seeds are used mainly as semen depurant. In compounds prescribed in spermaturia, spermatorrhoea and vitiation of semen, the seeds are invariably incorporated. Unani physicians prefer to combine the seeds with Aspargus adscendens and Tribulus terrestris.

Majoon-e-Bandkushad (Quarabadeen-e-Azam-o-Akmal) is a reputed Unani tonic for sexual debility. The seeds are an ingredient in another popular tonic, Majoon-e-Piyaz, prescribed in sexual debility, spermatorrhoea and premature ejaculation.

Active principles and pharmacology

Lupeol, stigmasterol, triacontane, an isoflavone glycoside and an alkaloid have been isolated from the whole plant.

From seeds asterol, asteracanthine and asteracanthicine have been reported. Amino acids—histidine, lysine and phenylalanine have been detected in the seeds. Diuretic properties of seeds are due to large amounts of mucilage and potassium salts.

Fresh flowers contain apigenin glucuronide.

Alcoholic extract of whole plant exhibited antifungal activity against Trichophyton mentagrophytes, Trichophyton rubrum, Microsporum gypseum, Epidermophyton floccosum and Candida albicans; essential oil from the root exhibited antibacterial effect against Gram-positive and Gram-negative bacteria (S. aureus, S. lutea, B. cereus, B. subtillis; and Sh. boydi, E. coli, S. typhi, S. dysenteriae).

Use in Western herbal

Medicinal use is based on the demulcent and diuretic properties of the herb. Used for catarrh of the urinary organs in the form of decoction. (50 g of the root to 3 pints of water boiled down to 1 pint. Dose: 15 ml–60 ml.)

Papilionaceae

ASTRAGALUS

Astragalus hamosus Linn.

Habitat

Found in the plains of Punjab.

Classical & common names

Unani: Iklil-ul-Malik (Arabic); Naakhunaa (Persian).
English: Tonkin bean, Melilot, King's Crown, King's Clover.
National Formulary of Unani Medicine equated Iklil-ul-Malik with three leguminosae species: Melilotus alba Desv., Trigonella uncata Boiss, and Astragalus hamosus Linn.

Parts used

Beans, whole plant.

Dose

2–4 g.

Classical use

In Unani medicine, the decoction of beans is given internally in nervous disorders; liver, kidney and spleen affections; also as galactagogue, emmenagogue and diuretic. The paste is massaged and applied externally on inflammations. The paste of the plant with vinegar or rose water is applied externally for headache.

Active principles and pharmacology

The plant Melilotus officinalis pharmacologically represents Iklil-ul-Malik. Its constituents are flavonoids, coumarins (incl. hydroxycoumarin and hydrocoumarin), resin, tannins and volatile oil.

Leaves yield 3-nitropropionic acid.

The pant possesses tonic, wound healing, nervine, emollient, astringent and antirheumatic, diuretic and anti-inflammatory properties, which match with the overall coverage of the Unani drug.

Astragalus sarcocola Dymock

Habitat

Occurs in the mountainous regions from Asia Minor to Iraq and Iran.

Classical & common names

Unani: Anzaroot (Arabic, Persian); Kohal Farasi; Kohal Kirmaani.

Parts used

Gum.

Classical use

The paste of the gum is used by Unani bone-setters on fractures and weak joints. The gum is used in ointments which are applied on infected wounds and ulcers. It is an ingredient in Marham-e-Sabz, which is applied on fistula and protruding flesh.

Active principles and pharmacology

Various species of Astragalus yield a gum, which is known as Gum Tragacanth. Water-soluble polysaccharides of the Tragacanth Gum contains approximately 40 % tragacanthine, divisible into tragacanthic acid, and an arabino-galactane-protein-complex. Non-water soluble polysaccharides contain starch, cellulose, and nitrogenous substances.

In the West, it is considered superior to gum arabic. The mucilage of Tragacanth Gum is used as an application to burns.

Solanaceae

ATROPA

Atropa acuminata Royle ex Lindl.

Figure 1 Atropa acuminata—flowering and fruiting branch [WOI]

Atropa belladonna Linn.

Habitat

Atropa acuminata is endemic to India and Atropa belladonna Linn. is the European species. Indian Belladonna is found in the Western Himalayan ranges, extending from Kashmir at the altitude of 1800–3600 m to the adjoining hills of the Himachal Pradesh up to 2500 m. Atropa belladonna has been introduced in Kashmir and the Himachal Pradesh.

Figure 2 Atropa belladonna—branch with flowers and fruit [WOI]

Classical & common names

Ayurvedic: Suuchi (non-classical).
Unani: Luffah (Arabic).
English: Atropa acuminata is equated with Indian Belladonna; Atropa belladonna with Deadly Nightshade.

Parts used

Leaves, root.

Dose

Powder 30–60 mg.

Use in classical medicine

The oldest use recorded was in Bulgaria, where the root preparation was administered in the treatment of Parkinsonism, encephalitis and paralysis agitans.

The natural drug is no more prescribed in Indian medicine.

Active principles and pharmacology

The leaf and the root contain tropane alkaloids, chief alkaloid, (-)-hyoscyamine, on drying is transformed to some extent into atropine, furthermore apoatropine, scopolamine, tropine; flavonoids, hydroxycumarins (incl. scopoline, scopoletine), tannins; also volatile bases (nicotine).

Tropane alkaloids inhibit the parasympathetic nervous system, which controls involuntary body activities. This reduces saliva, gastric, intestinal and bronchial secretions; as well as the activity of the urinary tubules, bladder and intestines. Tropane alkaloids also increase heart rate and dilate pupils.

The herb exhibits pharmacological properties for treating the symptoms of Parkinson's disease, for improving speech and mobility and for reducing tremors and rigidity.

Use in Western herbal

Belladonna leaf: the comminuted drug is used for decoctions and the powdered drug is used internally for galenic preparations.

When using Belladonna powder (total alkaloid 0.28 % to 0.32 %) the average single dose is 0.05 to 0.10 g. The maximum single dose is 0.20 g which is equivalent to 60 mg total alkaloids, calculated as hyoscyamine. Under medical supervision the maximum daily dose is 0.60 g, which is equivalent to 1.8 mg total alkaloids, calculated as hyoscyamine.

Belladonna root: the average daily dose as a comminuted drug for infusions and dried extracts and as a powdered drug for other galenic preparations for internal use, is 0.3 g, which is equivalent to 1.5 mg total alkaloids, calculated as hyoscyamine. Single doses range from 0.05 g to a maximum of 0.1 g.

For Belladonna extract the total alkaloid range from 1.3 % to 1.45 %. Single doses of the extract range from 0.01g–0.05 g. The maximum daily dose is 0.15 g which is equivalent to 2.2 mg total alkaloids, calculated as hyoscyamine.

German Commission E monograph recognized the herb's efficacy in the following areas:

- Arrhythmia
- Cardiac insufficiency (NYHA I and II)
- Nervous heart complaints
- Liver and gallbladder complaints.

In folk medicine, processed leaf is used in gastrointestinal pain, for asthma, bronchitis and muscular pain.

In homoeopathy, tincture of the whole plant (when beginning to flower) is used in diseases of the nervous system, convulsions and pain. The drug exhibits a marked action on the vascular system. It is regarded as the headache medicine par excellence.

Meliaceae

AZADIRACHTA

Azadirachta indica A. Juss.

Figure 1 Azadirachta indica [ADPS]

Melia azadirachta Linn.

Habitat

Commonly found throughout the greater part of India and often cultivated.

Classical & common names

Ayurvedic: Nimba, Nimbaka, Sutiktak, Arishtaphala, Pichumarda (Charaka and Sushruta); Arishta, Arishtaka, Pichumanda, Pichumandaka, Tiktak, Paaribhadra.

Mahaanimbaa, known as Bakaayan, has been equated with Melia azedarach Linn.

Unani: Aazaad-darakht-ul-hindi (Arabic, Persian); Neem.

Siddha: Vembu.

English: Margosa Tree, Neem Tree.

Parts used

Whole plant.

Dose
Leaf-juice 10–20 ml, powder 2–4 g, bark decoction 50–100 ml, oil 5–10 drops.

Classical use
Charaka and Sushruta used decoction of all parts of the plant internally in virulent skin diseases, erysipelas, leprosy, urinary discharges, chronic fevers, poisoning, ascites, internal parasites, malignant ulcers, internal tumours, piles, oedema, jaundice.

Sushruta used expressed oil as an ingredient of a medicated oil for cleansing and sterilizing ulcers. Paste in the form of ointments was used for ringworm, skin diseases, ulcers and leprosy.

According to classical texts, the stem bark was given internally as a blood purifier and antiperiodic; stem bark, rootbark and young fruits as bitter tonic, antiperiodic and alterative; leaf and seed oil as a strong antiseptic and insecticidal in boils, ulcers, eczema, leprosy; flowers as stomachic; berries as purgative, emollient and anthelmintic. Seed oil, paste of the leaves and bark was used externally on infected surfaces and on rheumatic swellings. All the five parts (root, bark, leaf, flower and fruit), mixed with clarified butter, were used for fumigation to disinfect the atmosphere. (Charaka Samhitaa, Sushruta Samhitaa, Kaashypa Samhitaa, Haarita Samhitaa, Shaarangadhara Samhitaa, Bhaavaprakaasha, Ashtaanga Hridaya, Bangasena, Gadanigraha, Siddha-bheshaja-manimaalaa.)

Considered as a pharmacy in its own right, every part of the plant was used medicinally since Charaka-Sushruta times.

Over the counter, Nimabaadi Kwaath Churna (Sahasrayoga) is available for preparing the compound's decoction (60 g of Churna to 80 ml water). This is prescribed as a blood purifier, internal antiseptic, antiperiodic and anthelmintic. Seven formulations of Nimba-based Kashaaya (for preparing decoctions for internal use) have been given in Sahasrayoga, and are prescribed in infectious diseases and chronic skin affections.

Nimbaadi Churnam (Bhaishajaya Ratnaavali) is given in ringworm, eczema, rheumatic swellings and diseases due to vitiated blood. Nimbaadi Taila (IMPCOPS) is applied externally as an antimicrobial agent to wounds, ulcers, skin diseases; also to lepromatous ulcers. Naaraayana Taila (Shaarangadhara Samhitaa) is a classical massage oil, indicated in arthritis, hemiplegia, paralysis and other nervous disorders.

Azadirachta indica, Celastrus paniculatus seeds (Maalkanguni) and Corallocarpus epigaeous (Paataalgaruda) tubers are ingredients in Siddha medicine's Sivanaar Vembu Kuzhi Thailam (Bogar Vaidyam), prescribed for local application in weeping eczema, leprosy and chronic ulcers.

Active principles and pharmacology
The plant contains triterpenes and tetranortriterpenoids (limonoids and protolimonoids of the gedunin group): nimbin, nimbidin, nimbinine, nimbosterol, nimbolide, salanin, azadirone; along with polyphenolic compounds, tannins, various acids and sulphur.

Alcoholic extract from leaves possessed potentiality on scabies, ringworm and acute eczema.

On clinical trial, the extract of leaves was found very effective on helminthiasis.

An unsaponifiable matter from oil of seeds exhibited antibacterial property against S. aureus, Xanth. citri, Proteus sp., E. coli and B. subtilis, in particular.

Neem oil, nimbin and nimbidin were found to possess anti-inflammatory property. Nimibidin was also found to possess analgesic and antipyretic effect.

Nimbidin exhibited significant effect on gastric and peptic ulcers in experimental animals; exhibited anticholinergic, antihistaminic and antinicotinic effect in isolated muscle preparations.

The decoction of tender leaves, oil from seeds, nimbin from oil of seeds produced hypoglycaemic effect ranging from 24 % to 26 %.

Within 7 to 10 days of intake of Neem oil, hyperglycaemic patients reported relief in itching, dyspepsia, tiredness and muscular pain and wound healing started fast after therapy. The results suggest that the oil is effective in secondary diabetes. It was found suitable for mild and moderate diabetes and 50 % effective in severe diabetes.

Antifertility activity of the seed oil was tested on adult healthy male albino rats of proven fertility. The seed oil administered at 300–350 mg/kg level induced an inhibitory effect on spermiogenesis. The production of spermatids was reduced by 52 %. A reduction of 21 % was noticed in the population of primary spermatocyte, and spermatogonia did not change.

Oral feeding of the Neem oil (350 mg/kg–500 mg/kg) induced 100 % negative fertility. *In vitro* study in human spermatozoa, sodium nimbidinate showed a weak spermicidal action. In another study, the spermatozoa (human and monkey) became immobile within 30 seconds of contact with the Neem oil.

Polysaccharide fraction of the bark showed antitumour activity against sarcoma cells.

Use in Western herbal

The oil obtained from the bark is used medicinally as anthelmintic and emetic. It is applied externally for rheumatism. The decoction is considered to be cathartic and febrifuge. Also used in hysteria.

An ointment to destroy lice is made from the pulp and is also used for scaldhead and other skin diseases.

The oil from the nuts is used for cramps and obstinate ulcers and as an antifungal, antibacterial and anti-inflammatory agent.

The neem oil is showing its potential as a contraceptive for both men and women. Its spermicidal activity has already been confirmed.

In homoeopathy, the tincture of the neem bark is used in intermittent fever, afternoon fever, rheumatic pains, constipation, diarrhoea, congested spleen.

Scrophulariaceae

BACOPA

Bacopa monnieri (Linn.) Pennell.

Figure 1 Bacopa monnieri [CCRAS]

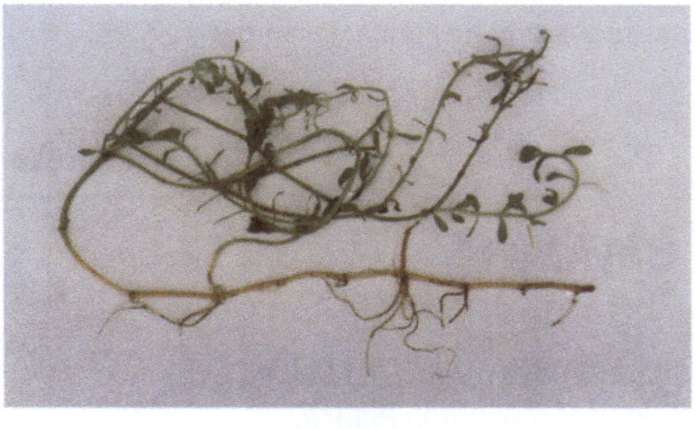

Figure 2 Bacopa monnieri [CCRAS]

Herpestis monnieria (Linn.) H.B.& K.

Habitat

Commonly growing in marshy places throughout India, ascending to an altitude of 1320 m.

Classical & common names

Ayurvedic: Braahmi, Vyasthaa, Surasaa, Somavallari (Charaka); Suvarchalaa (Charaka and Sushruta); Kapotavankaa, Somavalka (Sushruta). Jalaneem. Bhaarati, Darduradalaa; Aindri (non-classical: PV Sharma). NAA has equated Aindri with Indravaaruni (Citrullus colocynthis).
Unani: Brahmi.
Siddha: Neer Brahmi, Brahmi vazhukkai.
English: Thyme-leaved Gratiola; Indian Pennywort; Water Hyssop.
Centella asiatica was also known as Braahmi. The confusion was created during the 16th century when Bhaavaprakaasha equated Braahmi with Manduukaparni and Maanduuki. Even in the Ayurvedic Pharmacopoeia of India Part I, Vol. II, the Hindi synonym of Braahmi was documented as Manduukaparni. Braahmi and Manduukaparni are two different drugs and have been equated with Bacopa monnieri and Centella asiatica respectively. Aindri has been equated with Indravaaruni (Citrullus colocynthis Schrad) in Bhaavaprakaasha.
Charaka, Sushruta and Vagbhatta treated Braahmi and Manduuakparni as two different drugs. A critical study of comparative phytochemistry, pharmacology and therapeutic properties of these two drugs have proved that they are distinct. Braahmi was used specifically in mental diseases like insanity and epilepsy, while Manduukaparni as a general brain tonic. Braahmi promotes fertility (Charaka used it as an impregnating herb) and sustains implantation; while Manduukaparni is abortifacient. Both are used for skin diseases but their therapeutic effects are not the same.

Parts used

Whole plant.

Dose

Fresh juice 10–20 ml, whole plant powder 3–5 g, root powder 500 mg–1.5 g

Classical use

Charaka used the dried and powdered plant internally in loss of memory, senile decay and as a preventive of abortion.

Sushruta gave the plant as a potherb in fevers, haemoptysis, hiccup, cough, dysponea, urinary diseases, calculi, skin diseases; and, externally, for washing ulcers. Shushruta prescribed the juice with milk as a part of daily diet for promoting intellect and for preventing aging.

Fresh juice of Bacopa monnieri, Acorus calamus and Alpinia speciosa (zerumbet) are main ingredients in Pirami Ney (medicinal Ghee), prescribed in convulsions, nervous debility, delirium and poor memory.

Braahmi was the drug of choice in epilepsy and insanity, and was an important ingredient in a number of compound formulations of Charaka Samhitaa, Ashtaanga Hridaya, Vrindamaadhava. Kaashyapa Samhitaa included the herb in prescriptions for promoting intellect in children.

In Unani medicine, Centella asiatica is being used as Braahmi. The herb's extract in oil is used in skin diseases and itching. Internally, it is used in prescriptions for treating nervine disorders; also as a brain tonic.

Zarnab (Arabic), which was earlier taken to be Braahmi, has now been equated with Salix aegyptiaca sprengel (known as Saro Turkistaani).

In National Formulary of Unani Medicine, Zarnab has been equated with Telispattar and the use of Flacourtia cataphracta Roxb. or Cinnamomum Tamala Nees. as Zarnab has been recommended.

Over the counter, Saaraswataarishta (Bhaishajya Ratnaavali) is available for nervous debility, convulsions, and for improving memory; Braahmi Bati (Ras Tantra Saar) for mental weakness, deficient memory; Braahmi Ghrita (Charaka Samhitaa) for epilepsy, insanity and for promoting memory.

Active principles and pharmacology

The plant contains steroidal saponins: bacosides A and B; sapogenins: bacogenins A1, A4; betulic acid, stigmasterol, stigmastanol, and beta-sitosterol. The leaves gave a number of alkaloids including herpestine. Saponin, monnierin has also been isolated from the plant.

In animal studies, the extract was found to have cardiotonic, vasoconstrictor, neuromuscular blocking and tranquillizing actions. Active principles exhibited barbiturate hypnosis potentiation and anti-anxiety effect. Pharmacologically, the plant's use as nervine tonic in insanity and epilepsy and also in improving mental functions is fairly indicated, as the herb has been shown to cause a prolonged elevated level of cerebral glutamic acid and transient increase in GABA level.

Use in Western herbal

English equivalents given in Indian reference books do not represent the properties of Bacopa monnieri.

Gratiola officinale (Linn., Scrophulariaceae) is used for its diuretic, cathartic and emetic properties, and in chronic affections of liver and spleen.

Hydrocotyle asiatica (Linn., Umbelliferae) is known as Indian Pennywort in the Western herbal. The herb is used for its diuretic properties and in bowel complaints.

Hyssopus officinalis (Linn., Labiatae) is used for its expectorant, diaphoretic, stimulant and carminative properties.

The use of B. monnieri is based on research carried out in India, especially in Central Drug Research Institute and Central Institute of Medicinal and Aromatic Plants, Lucknow.

Balanitaceae

BALANITES

Balanites aegyptiaca (L.) Delile.
Balanites roxburghii Planch.

Habitat

Drier parts of Western Rajasthan and south-east Punjab to West Bengal and Sikkim.

Figure 1 Balanites roxburghii—flowering and fruiting branch [WOI]

Classical & common names

Ayurvedic: Ingudi (Charaka, Sushruta); Taapasadrum, Angaar Vrksha; Hingot (Hindi).
Siddha: Nanjunda (Tamil).
English: Zachum Oil Plant, Desert Date.

Parts used

Bark, fruit, seed, leaf, seed oil.

Dose

Decoction 50–100 ml, seed oil 5–10 drops.

Classical use

Charaka used the alkali from the plant ash in prescriptions in dermatosis and urinary diseases; the oil externally in skin diseases.

Sushruta prescribed the herb internally for intrinsic haemorrhage and deranged metabolic functions as antitoxic and aseptic; oil of the seeds as a laxative and in intestinal parasites, infected ulcers and urinary affections.

The herb was also used as a cosmetic. The fruit kernel, pounded with cold water, was used as a paste on the face for freckles (Gadanigraha). The fruit pulp, pounded with water, was applied as a paste on face (Vaidyajivana).

In folk medicine, the fruit is considered useful in whooping cough. The pulp, mixed with goat's milk, is rubbed on the chest to treat pneumonia in children. The seeds are also given in cough and colic. The paste of the root is applied locally for curing guineaworm disease.

Active principles and pharmacology

Seeds, roots and stem bark contain diogenin and yamogenin; in addition, the seeds contain higher hydrocarbons. The plant contains steroidal saponins designated as balanitisin A to G. The stem bark gave nitogenin glucoside. Tocopherols, 25-alpha- and 25-beta-spirostadiene are reported from fruits.

The oil exhibits antibacterial and antifungal properties in skin diseases, burns, excoriations and freckles; the unripe fruits, leaves, seeds and bark show anthelmintic and purgative principles.

Euphorbiaceae

BALIOSPERMUM

Baliospermum montanum Muell.-Arg

Habitat

Throughout India from Kashmir eastwards to Arunachal Pradesh, up to an elevation of 1000 m and southwards in the hills of Kerala up to an altitude of 1800 m.

Classical & common names

Ayurvedic: Dantin, Udumbaraparni, Nikumbha, Makulaka, Chitraa , Danti (Charaka, Sushruta); Erandphalaa, Shighraa, Laghudanti.
Siddha: Neeradimuthu.
English: Wild Croton.

Parts used

Root.

Dose

Powder 3–5 g.

Figure 1 Baliospermum montanum [ADPS]

Classical use

Charaka used sun-dried roots in prescriptions for internal use in oedema, jaundice, anaemia, constipation, diseases of the abdomen, piles, internal abscesses. Sushruta prescribed internally in acute constipation as a purgative, and in abdominal dropsy; externally on non-suppurating boils.

In folk medicine, powdered seeds are used as a drastic purgative, one seed being the dose for an adult. In large doses they are poisonous. The seeds are externally used as a stimulant and rubefacient.

The oil is externally applied in rheumatism. It resembles croton oil and is a powerful hydragogue.

The dried roots are considered anthelmintic, diuretic, and useful in enlarged spleen, abdominal tumours, piles, wounds and skin diseases. The root is also administered in dropsy and jaundice.

Over the counter, Dashmuul-Panchkolaadi Kashaaya (Sahasrayoga) is available and is prescribed for ascites in the South.

Active principles and pharmacology

Roots contain phorbal derivatives—montanin, 12-deoxyphorbol—13-palmitate, baliospermin, 12-deoxy-5-beta-hydroxyphorbol-13-myristate and 12-deoxy-16-hydroxyphorbol-13-palmitate.

The seeds possess purgative properties, but are narcotic poison in large doses. Oil from the seed was found to be counter-irritant in rheumatism. Latex gave relief when applied in bodyache and joint pains.

Ethanolic extract of the root and plant's ester derivatives exhibited, in vitro, activity against P-388 lymphocytic leukaemia. The plant is also being investigated for its use in abdominal tumours.

Poaceae

BAMBUSA

Bambusa arundinacea Willd.
Bambusa bambos Druce.

Habitat

Wild throughout India, especially in the hill forests of Western and Southern India, Orissa, West Bengal, Assam, ascending up to 1000 m. Also grown in other parts of India.

Classical & common names

Ayurvedic: Vamsha, Venu, Trinadhwaja, Duraruhaa, (Charaka), Shatparvaa. Vansalochana, Vansarochana (Bamboo-manna).
Unani: Qasab (Arabic), Nii (Persian); Tabaasheer (Bamboo-manna).
Siddha: Moongil.
English: Thorny Bamboo.

Parts used

Root, leaves, fruits, Bamboo-manna.

Dose

Decoction 50–100 ml, Bamboo-manna 1–3 g.

Classical use

Charaka prescribed the paste or decoction of the leaves or seeds in external and internal prescriptions for acute suppurated conditions, poisonous bites, intestinal worms, excessive fat. In polyuria, Charaka gave barley-shaped fruits of the plant in a dietary preparation.

Bamboo nodes are included in a compound formulation of Siddha-bheshaja-manimaala for amenorrhoea.

Bamboo-manna (Vansarochana) was always the drug of choice for treating bronchitis, while the whole plant was used in prescriptions for treating vitiated blood, jaundice, oedema, ulcers, skin diseases (Bhaavaprakaasha).

Even during the 16th century, genuine Bamboo-manna was difficult to procure. Curcuma angustifolia Roxb. was in use as a substitute. Its original name, Tavakshiri, was equated in due course with the Tabasheer of Unani medicine.

In folk medicine, the root is considered astringent and cooling, and is used to treat joint pain and general debility. The leaves are used to stimulate menstruation and, as an antispasmodic, to help relieve period pain. The leaves are also used as a tonic to stomach, and to expel worms.

The juice is given for strengthening the cartilage in osteoarthritis and osteoporosis.

Bamboo-manna is an important ingredient in Shaarangadhara Samhitaa's reputed compounds for cough, cold and asthma—Sitoplaadi Churna and Taalisaadi Churna. Both are available over the counter.

In Unani medicine, Qurs-e-Tabaasheer Kafoori Lului is given in hectic fever, phthisis, nervousness; Qurs-e-Tabaasheer Mulaiyin in constipation, pyrexia, sore throat, phthisis; Qurs-e-Tabaasheer Qabiz in upset stomach, diarrhoea, fever; Safoof-e-Tabaasheer in indigestion and flatulence; Jawarish-e-Tabaasheer in bilious vomiting, dyspepsia, diarrhoea.

Active principles and pharmacology

The plant gave cyanogenic glucoside and taxiphyllin. The juice contains a high level of silica.

The de-alcoholized extract of leaves has shown antibacterial activity against Bacillus subtilis, Micrococcus pyogenes var. aureus, Aerobactor aerogenes, Salmonella typhi and Salmonella paratyphi.

Bamboo-manna contains silica 90.5, potash 1.1, alumina 0.4 and iron peroxide 0.9 %.

Curcuma spp., used as a substitute for Bamboo-manna, is an entirely distinct drug. Its proximate chemical analysis showed: sugar 95.63, water 2.66, glucose 0.75 and ash 0.96 %. Bamboo-manna is mainly composed of silicic acid, which acts as a potent therapeutic agent in cough, cold, asthma, paralytic disorders and in debilitating diseases.

Use in Western herbal

Arundinaria japonica is used in Western herbal for asthma, cough and disorders of the gallbladder. It contains soluble mono-, oligo- and polysaccharides and silicic acid (to some extent water soluble).

The juice from the young shoots is hardened as bamboo sugar for medicinal use.

Bamboo-manna is available in the U.S. in the form of 270 mg (Arkocaps Phytosilica) capsules as a health supplement for bone and joint disorders and for preventing bone decay after the menopause. The capsules are also used in feverish cold, asthma and debilitating conditions.

Acanthaceae

BARLERIA

Barleria prionitis Linn.

Habitat

Found throughout the hotter parts of India. Also commonly grown as a hedge plant in gardens.

Classical & common names

Ayurvedic: Saireya (Charaka and Sushruta); Saireyaka (Sushruta); Kuranta, Kurantaka (Bhaavaprakaasha); Katsaraiyaa, Sahchara.
Unani: Katsaraiyaa, Piyabaasaa.
Siddha: Chemmulli.
English: Yellow-nail-dye Plant.

Parts used

Leaf, bark, root.

Dose

Juice 10–20 ml, decoction 50–100 ml.

Figure 1 Barleria prionitis—flowering portion [WOI]

Classical use

Charaka used the paste of the leaves as an ingredient of hot poultices and steam bath for treating stiffness of limbs, enlargement of scrotum, sciatica and erysipelas. Sushruta prescribed flowers internally in migraine, internal abscesses, oedema, haemoptysis, urethral discharges, seminal disorders and also to reduce obesity.

The herb's extract in oil was prescribed for massage in arthritis and gout (Ashtaanga Hridaya).

Saireyaka has been included in the Karanjaadi group of herbs, which are specific for curing obstinate urinary disorders and chronic skin diseases.

In folk medicine, the paste is applied over boils and glandular swellings. Fresh juice of the bark is given in anasarca. The dried bark is given in whooping cough. The leaf juice, mixed, with honey, is given to children in fevers and catarrh. The ash obtained from the whole plant, mixed with honey, is given in bronchial asthma. The leaves and flowering tops are used as a diuretic and in urinary affections.

In the South, Sahchara is widely used against neurological disorders, such as paraplegia, sciatica; also in glandular swellings, leprosy and other skin diseases.

Available over the counter, Sahcharaadi Tailam (Sahasrayoga) is prescribed in dysurea, rheumatic affections, internal abscesses, nervine disorders, chronic sinusitis.

Active principles and pharmacology

The plant contains beta-sitosterol. Flowers contain flavonol glycoside, scutellarein-7-neohesperidoside; leaves and stems iridoids, barlerin and Acbarlerin.

The plant possesses antiseptic, wound-healing, spasmolytic, anti-inflammatory and tonic properties for the central nervous system.

Lecythidaceae

BARRINGTONIA

Barringtonia acutangula (L.) Gaertn.

Habitat

Fairly common in sub-Himalayan tracts from the Ganges eastwards to Assam and in Madhya Pradesh.

Classical & common names

Ayurvedic: Hijjal, Nichula, Dhaatriphala (Charaka, Sushruta); Vidula; Samudraphala (Nighantu Adarsha). CCRAS, in Standard Nomenclature of Ayurvedic Medicinal Plants, has wrongly equated Argyreia nervosa and Argyreia speciosa with Hijjal, Nichula and Vidula.
Siddha: Kadappasi.
English: Indian Oak.

Parts used

Leaf, fruit.

Dose

Leaf juice 10–20 ml, fruit powder 1–3 g.

Classical use

Charaka used the fruits and leaves in alkaline decoction or paste in internal prescriptions for abdominal and splenic disorders, and as a purgative. Sushruta incorporated the fruit in an antiseptic ointment for venereal sores.

The juice of the leaves, mixed with honey, alleviated diarrhoea with mucus (Vrindamaadhava).

In folk medicine, the leaf juice is given in diarrhoea. The fruit is used as anthelmintic, and is prescribed as an astringent in gingivitis; also as an emetic and expectorant. The powdered seeds are inhaled as snuff for relief in headache. The bark is used in diarrhoea and blennorrhoea as astringent. A decoction of the bark is used as a mouthwash against gum trouble. The wood is used as a haemostatic in metrorrhagia.

Active principles and pharmacology

Barringtogenic acid, stigmasterol-3-beta-O-D-glucoside, beta-sitosterol, beta-amyrin, oleanolic acid, tangulic and acutangulic acids have been isolated from the leaves.

The ethanolic extract of dry and defatted fruits gave saponins, which, on acid hydrolysis, yielded three triterpenoid sapogenins, barringtogenol B, C and D, and two triterpenoid acid sapogenins. The fruit saponins possess haemolytic properties. The bark contains tannins (16 %).

Use in Western herbal

Common oak, Quercus robur, is used in Western herbal for the bark's astringent and antiphlogistic properties.

German Commission E monograph recognized the bark's efficacy in the following areas:

- Common cold
- Cough
- Bronchitis
- Inflammation of the mouth and pharynx
- Fever
- Diarrhoea.

Externally recommended for inflammatory skin diseases.

Caesalpiniaceae

BAUHINIA

Bauhinia racemosa Lam.
Bauhinia variegata Linn.

Habitat

Found throughout India, ascending to an altitude of 1300 m in the Himalayas.

Classical & common names

Ayurvedic: Kaanchanaara, Kaanchana, Ashmantaka (Bhaavaprakaasha); Kaanchanaka, Raktapushpaka, Gandhaari, Shonapushpaka. The pink flowered variety of the plant, called Kovidaara (Bhaavaprakaasha), has been equated with Bauhinia purpurea Linn., and is known as Koilaara (colloquial name).

Figure 1 Bauhinia purpurea—flowering portion and fruit [WOI]

Siddha: Sivappumanchori.
English: Mountain Ebony, Orchid Tree, Buddhist Bauhinia.

Parts used

Bark.

Dose
Decoction 50–100 ml, powder 3–6 g.

Classical use
Fresh bark of Kaanchanaara mixed with Shunthi (dry Zingiber officinale), pounded with sour gruel, was prescribed in enlarged cervical glands (Vrindamaadhava) as well as in goitre (Shaarangadhara Samhitaa, Bhaavaprakaasha).

Charaka and Sushruta prescribed the flowers of Kovidaara internally in intrinsic haemorrhage and haemoptysis. The root powder was given with buttermilk in piles (Ashtaanga Hridaya), and the decoction of bark in pox (Bhaavaprakaasha).

In folk medicine, the root is used as a carminative in dyspepsia and flatulence. The bark is used as an anthelmintic, as an astringent and antiseptic agent in skin diseases, ulcers and leprosy. A decoction of the buds is given in diarrhoea, dysentery, cough, haematuria, piles and menorrhagia.

Over the counter Kaanchanaar Guggulu (Shaarangadhar Samhitaa) is still available and is the drug of choice for treating enlarged cervical glands, goitre and scrofulous tumours; so is Kaanchan-gudikaa (Bhaishajya Ratnaavali).

Active principles and pharmacology
Quercetol glycosides are the dominant flavonoids of the plant. Myricetol glycosides are present in the seeds. Other flavonol glycosides identified are quercitroside, isoquercitroside, rutoside and a taxifolin rhamnoside.

In an experiment, goitre was produced in rats of 100 g weight by neomercaptazole. Water soluble fraction of Bauhinia variegata was administered 200 mg/day, p.o. for 20 days to the rats. The drug was found to be highly effective in goitre.

Bauhinia variegata was found less effective than Bauhinia purpuria. The drug probably possessed some thyroid hormone-like activity in addition to iodine contents which might be responsible for neutralizing the thyroid gland both morphologically as well as functionally. (CCRAS.)

The petroleum-ether, benzene, chloroform and alcohol extracts of the herb did not show any toxicity either orally or intraperitonally.

Related species
Bauhinia acuminata; Bauhinia purpurea (already mentioned in the text); Bauhinia racemosa Lamk and Bauhinia tomentosa Linn—all the varieties exhibit similar therapeutic properties.

In practice, physicians prefer Bauhinia variegata and Bauhinia tomentosa (yellow Bauhinia) as the source plants of the drug for glandular diseases.

Cucurbitaceae

BENINCASA

Benincasa hispida (Thunb.) Cogn.

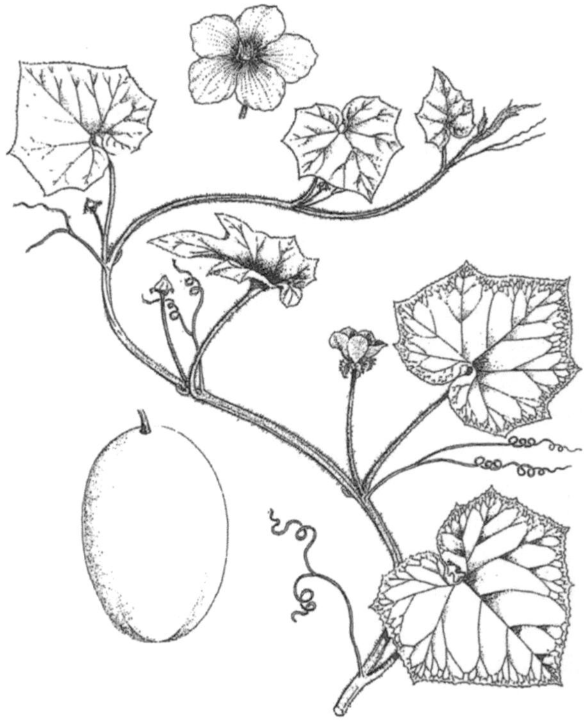

Figure 1 Benincasa hispida [ADPS]

Benincasa cerifera Savi

Habitat
Native to tropical Asia and Africa. Cultivated as a vegetable throughout India up to an altitude of 1200 m.

Classical & common names

Ayurvedic: Kuushmaanda (Charaka, Sushruta); Karakaaruka, Pushpaphala (Sushruta); Pitapushpa, Vrahitaphal, Valliphala, Somakaa, Sthirphalaa.
Unani: Mhdab (Arabic); Kaddu-e-Roomi (Persian); Kumhraa.
Siddha: Poosani.
English: White Gourd, Wax Gourd, White Pumpkin.

Parts used

Fruit, seed kernel.

Dose

Juice of the fruit 10–20 ml, powdered seed 5–7 g.

Classical use

Charaka and Sushruta prescribed it as a cooked vegetable for its laxative and diuretic properties. Sushruta gave the juice of the fruit in mental aberrations, insanity and other nervous diseases.

The juice, in prescription, was also used in haemoptysis (Bangasena) and as an antidote to intoxication (Shaarangadhara Samhitaa).

The extract of the fruit in clarified butter, mixed with the paste of Glycyrrhiza glabra, was prescribed in epilepsy and for promoting intellect (Ashtaanga Hridaya, Vrindamaadhava, Bangasena). The powder of the root, with warm water, was prescribed in asthma and cough (Bhaavaprakaasha).

In folk medicine, the juice of the leaves is rubbed on bruises as a cooling agent. The decoction of the fruit is given to cure internal haemorrhage and diseases of the respiratory tract. It is considered styptic, laxative, diuretic and nutritious. The juice is also used as an antidote for alcoholic poisoning.

Kuushmaandaka Rasaayana (Ashtaanga Hridaya and Bhaishajya Ratnaavali) is an over-the-counter tonic, used as digestive, carminative, restorative, haematinic, haemostatic, expectorant and sedative.

Kuushmaandaka Ghritam, based on Ashtaanga Hridaya, also incorporated in the text of Sahasrayoga, is still available and is prescribed in epilepsy and insanity.

Benincasa hispida juice is one of the main ingredients in Siddha medicine's Venpoosani Legiyam (Agaththyar Vaidhya, Eraththina Churukkam), prescribed in jaundice, oedema and inflammations of the urinogenital tract.

In Unani medicine, the kernel of seeds and their oil are prescribed in roundworm and tapeworm infestations. Confection of the fruit is given as a nutritious food and in dyspepsia and biliousness.

Active principles and pharmacology

See Cucurbita maxima.

The alcoholic extract of the fruit increased pentobarbitone hypnosis at 200 mg/kg i.v.; impaired muscular coordination in rats at 500 mg/kg p.o.; produced reduction of body temperature; significantly caused fall of blood pressure with simultaneous increase in respiratory rate; produced cardiac depressant effect on perfused frog's heart.

The extract could not show anti-inflammatory effect on the smooth muscles of guinea-pig ileum and uterus. It did not reduce spontaneous motor activity, lacked anti-convulsant action against MES and chemoshock, anti-nociceptive action. The extract was devoid of purgative effect (CCRAS).

The fruit juice produced tranquillizing activity and mild CNS depressant effects in mice.

The alcoholic and aqueous extract of seeds exhibited anthelmintic effect against cestodes, trematodes and nematodes.

Berberidaceae

BERBERIS

Berberis aristata DC.

Habitat

Distributed in the temperate and sub-tropical parts of Asia, Europe and America and in the Himalayas from 2000–3500 m.

Classical & common names

Ayurvedic: Daaruharidraa, Pitadaaru, Pitatvaka, Hemkaanta (Charaka); Daarvi, Daarunishaa (Sushruta, Bhaavaprakaasha); Parjanyaa, Parjani, Pita Kaaliyak, Katamkateri, Pachampachaa.

Unani: Daarhalda (Arabic, Persian). Zarishk (berry).
Siddha: Maramanjal.
English: Indian Barberry.

Parts used

Bark; berry; extract of rootbark, wood, leaf and stem (known as Rasaajana, Rasaut).

Dose

Decoction 50–100 ml, rasaajana 1–3 g.

Classical use

Charaka used the extract of the plant in prescriptions internally in haemorrhage, piles, pruritus, alopecia; Daarvi Ghrita, a classical compound, in anaemia and jaundice; Rasaanjana in diseases of the mouth, swollen gums, skin diseases and leprosy; mixed with the juice of Emblica officinalis in dysurea.

Sushruta prescribed it internally in indigestion, deficiency of breast milk, uterine and vaginal disorders; cooked with clarified butter along with other herbs in dysentery; and externally as an ingredient of ointment for wounds.

Sushruta used the milk boiled with the herb for washing eyes in eye diseases.

Daarvyaadi Kwaath was given in vaginal disorders and the decoction of the herb, mixed with honey, followed by the intake of rice-water in leucorrhoea (Vrindamaadhava, Gadanigraha).

By the 16th century, the herb was considered similar to Curcuma longa in therapeutic properties and was used in anaemia, oedema, genitourinary disorders, parasitic infections, skin diseases including ulcers, and specifically in conjunctivitis. (Ashtaanga Hridaya, Vrindamaadhava, Bhaavaprakaasha.)

In Unani medicine, the berries were used in the treatment of jaundice, enlargement of spleen, and the whole drug as a cholagogue, stomachic, laxative and antiseptic.

Among over-the-counter compounds, Daarvaadi Kashaaya is prescribed for leucorrhoea and metrorrhagia; Daarvaadi Churnam for piles and internal abscesses; Daarvibalaadi Ghritam for bleeding piles; Daarvyaadi Tailam for massage in obesity (all based on Sahasrayoga, and available in the South).

Among Unani preparations Qurs-e-Zarishk is prescribed in hepatitis and anemia; Jawarish-e-Zarishk in anorexia, indigestion, vomiting, gastroenteritis, cholera; Roghan-e-Surkh (Qarabadeen-e-Jadeed) is a reputed massage oil for rheumatoid arthritis, sciatica and gout.

Active principles and pharmacology

Berberis aristata and its related spp., Berberis asiatica, Berberis lycium, Berberis vulgaris, contain isoquinoline alkaloids (in particular berberine, berbamine, oxyacanthine and palmatine).

Berries contain isoquinoline alkaloids (traces), anthocyans, malic acid, acetic acid; are laxative and antiscorbutic; are a source of vitamin C; stimulate iron absorption and increase immune system activity.

The root of the plant possessed broad-spectrum antibiotic properties against Gram-positive bacteria.

Berberine prevented death of young rabbits suffering from cholera, given early. It acted on severe diarrhoea and cholera by reducing the increased capillary permeability. It was found to control gastroenteritis due to bacterial infections in children. Comparative studies of berberine and chloramphenicol on cholera patients showed that the former was much better than the latter as there was more reduction in mortality rate, volume and duration of diarrhoea, intake of intravenous fluid and convalescence period.

Berberine showed leishmanicial properties in in vitro-culture of Leishmani donovani. It showed potential in visceral leishmaniasis in comparison with pentamidine.

Berberine chemotherapy against intestinal amoebiasis in rats and golden hamsters showed encouraging results. It also showed promise in Giardia-infected patients.

Clinical application of berberine to chronic trachoma patients by intraconjunctival injection proved highly effective and was confirmed by the scientists of the All India Institute of Medical Sciences, New Delhi. In their view, berberine may prove a practical remedy for large-scale use in trachoma patients (CCRAS).

In Indian medicine, Berberis aristata was used for a long time as an antiperiodic and alternative in remittent type and malarial fevers. Now it has been proved that berberine is not a curative agent

Figure 1 Berberis asiatica [WOI]

in malaria. It is diaphoretic and antipyretic. It liberates the parasites into the blood circulation. Blood films taken before the administration of berberine were found negative, those taken after it were positive. Thus berberine can be used as a provocative agent for the diagnosis of latent malaria.

It has been shown that berberine first causes streptococci to lose lipoteichoic acid, then prevents the adhesion of fibronectin to streptococci and elutes already bound fibronectin.

Berberine has been shown to increase the blood supply to the spleen; it activates macrophages; improves filtration of the blood; releases compounds that potentiate immune function.

Berberine has been shown in several clinical studies to stimulate the secretion of bile and bilirubin. In patients with chronic cholecystitis, berberine caused a disappearance of clinical symptoms, a decrease in bilirubin-level, and increase in the bile volume of the gallbladder. Berberine has also been shown to correct metabolic abnormalities in patients with liver cirrhosis.

Barberry, as a whole herb, acts on the gallbladder, improves bile flow and ameliorates conditions such as gallbladder pain, gallstones, jaundice.

Barberry made history in cancer treatment when Hoxsey Cancer Formula was marketed from the 1930s to the 1950s by Harry Hoxsey (Dallas). The formula turned out to be highly controversial. Ironically, Hoxsey died of prostate cancer. He took his formula, but it did not work for him.

The Hoxsey formula is still available at the Biomedical Centre in Tijuana, Mexico. Recent studies show that 9 out of its 10 herbal ingredients—barberry, buckthorn, burdock, cascara sagarda, red clover, licorice, poke, prickly ash and bloodroot—have antitumour action. One study showed that barberry shrinks some tumours.

Berbamine has been used in China since 1972 in the treatment of depressed white blood cell counts due to chemotherapy and/or radiation.

Studies around the world show that berberine kills Staphylococci, Streptococci, Salmonella shigella, Entamoeba histolytica, Vibro cholerae, Giardia lamblia, Escherichia coli and Candida albicans. This justifies the use of berberine in wound infections, diarrhoea, dysentery, cholera, urinary tract infections and vaginal yeast infections.

The LD50-value of berberine sulphate in mice was found to be 24.3 mg/kg i.p.

Use in Western herbal

In ancient Egypt, Berberis berries were macerated with fennel seed (Foeniculum vulgare) to make a drink for fevers. After the plant was introduced from Europe to the eastern US, it was used for peptic ulcer and gastrointestinal infections.

Most present-day herbalists limit their recommendations to gargling barberry decoction for sore throat and drinking it for diarrhoea and constipation.

A few herbalists continue to recommend barberry for correcting liver functions, but the herb is not considered therapeutic for jaundice.

Jam or wine made from fresh berries are given in constipation and lack of appetite; alcoholic extract for heartburn and stomach cramps.

In Germany, a berberine preparation, Opthiole, is used to treat sensitive eyes, inflamed lids and conjunctivitis. A compress made from a herbal infusion is also used.

The decoction of the bark is prescribed during opium or morphine withdrawal.

In homoeopathy, tincture of the rootbark of Berberis vulgaris is used for hepatic and rheumatic affections, particularly with urinary, haemorrhoidal and menstrual complaints.

Caution

Berberine-containing plants are contraindicated during pregnancy. Higher doses may interfere with vitamin B metabolism.

Saxifragaceae

BERGENIA

Bergenia ligulata (Wall.) Engl.

Figure 1 Bergenia ligulata [CCRAS]

Saxifraga ligulata Wall.

Habitat

Temperate Himalayas from Kashmir to Bhutan between 2300 and 3300 m and Khasi Hills at 1300 m.

Classical & common names

Ayurvedic: Paashaanbheda, Ashmabhedaka (Bhaavaprakaasha); Ashmaghna, Ashmaribheda, Shilaabheda, Shilaabhid.
Unani: Paakhaanbheda.
Siddha: Padanabethi.

Parts used

Leaves, rhizomes.

Dose

Decoction 50–100 ml, powder 3–6 g.

Classical use

Sushruta used the root in prescription for calculus, retention of urine and dysuria.

Paashaanbhedaadya Churna and Ghrita (Chakradatta) were prescribed as antilithic, diuretic and styptic compounds.

Paashaanbheda Kashaaya (Sahasrayoga) is still available and is prescribed in retention of urine and anurea. Traikantanka Ghritam (Sahasrayoga) is given for kidney stones and urinary infections.

It is a common belief that the drug is helpful in dissolving kidney stones. In folk medicine, it is used for its lithotriptic properties in combination with other antiseptic and diuretic herbs. The rhizome is used in menorrhagia; also in urinary discharges.

Active principles and pharmacology

The rhizome contains bergenin (a C-glycoside), gallic acid, glucose, mucilage, wax, and tannins. The presence of beta-sitosterol and flavonoids is also reported. The leaves also contain flavonoids.

The acetone extract of the rhizomes is reported to be cardiotoxic in high doses and has a depressant effect on the central nervous system. The extract possesses significant anti-inflammatory activity, and is mildly diuretic; in higher doses it exhibited antidiuretic effect in experimental animals.

The alcoholic extract of Berginia ligulata showed marginal antilithic property in dissolving the formed stones. There was no preventive effect on formulation of stones (CCRAS).

The extract, on experimentally produced stone formation by introducing zinc disc in the urinary bladder of rats, could not prevent lithiasis.

Thus, classical properties of the drug could not be proved pharmacologically.

CCRAS, in addition to Bergenia spp., has also equated Didymocarpus pedicellata R. br. and Aerva spp. with Paashaanbheda.

In a patent for the herbal drug Cystone, which is commonly prescribed as an antilithic and urinary antiseptic, both the plants shortlisted by CCRAS as Paashaanbheda (Saxifraga ligulata, Didymocarpus pedicellata) have been incorporated.

Didymocarpus (Gesneriaceae), known as Shilaapushpa, occurs in the sub-tropical Himalayas from Chamba to Kumaon at an altitude of 500–2500 m. The plant contains a number of flavanones and chalcones. Used in folk medicine for stones in kidneys and bladder, urinary tract infections.

The essential oil of the plant exhibited antimicrobial activity.

In animal studies, acetone extract of leaves exhibited potentiated pentobarbital hypnosis, positive inotropic effect on perfused heart of frog, weak hypotension in anaesthetised dogs, spasmolytic effect against isolated rabbit intestine and rat uterus, produced intestinal relaxation in anaesthetised dogs (CCRAS).

Aerva lanata (Linn) Juss ex Schult, is found throughout India as a common weed in fields. The plant is reported to contain beta-sitosteryl palmitate, alpha-amyrin, beta-sitosterol and tannin. In folk medicine, the plant is used as an anthelmintic, diuretic, demulcent and in lithiasis and catarrh of bladder.

In the South, Aerva lanata, Rotula aquatica Lour. and Homonoia riparia Lour. are used as Paashaanbheda. (In Ayurvedic reference books, Ocimum basilicum Linn, Labiateae, has been suggested as a substitute for Paashaanabheda. The plant is also included in antilithic drug Cystone.)

In homoeopathy, tincture of fresh leaves of Ocimum canum (Brazillian Alfavaca)/ Ocimum incanescenes (Mart.)/ O. fluminese (Wells)/ Alfavaca (Portuguese), Labiateae, is prescribed in diseases of the kidney, bladder and urethra, uric acid diathesis, renal colic, with pronounced symptoms of renal calculus.

Betulaceae

BETULA

Betula utilis D.Don.
Betula bhojpattra Wall.

Habitat

Throughout the main Himalayan range from Bhutan onwards, ascending to an altitude of 4200 m.

Classical & common names

Ayurvedic: Bhuurja (Charaka, Sushruta); Bhuurjapatra, Chitratvak (Charaka); Charmi, Bahulavalkala, Bahuput, Lekhyapatraka; Bhojapatra..
English: Himalayan Silver Birch, Indian Paper Birch.

Parts used

Bark flakes, nodes.

Dose

Decoction 50–100 ml, powder 3–5 g.

Classical use

Charaka used nodes and bark compounded into ointment on ringworm, acute-spreading suppurations, obstinate skin diseases including leprosy, erysipelas and wounds. Nodes of the herb were included in a dusting powder for disinfecting skin eruptions.

Sushruta also prescribed the drug internally as well as externally in skin diseases, urethral discharges; in obesity and diseases of the liver. He gave unripe fruits internally in migraine, internal abscesses and for reducing obesity; ripe fruits and leaves as appetizer, laxative and antibilious.

In folk medicine, the bark is used in convulsive disorders.

Active principles and pharmacology

The bark contains betulin, lupeol, oleanolic acid, acetyloleanolic acid, leucocyanidin, betulic acid, lupenone, methylbetulonate, methylbetulate and karachic acid.

The bark possesses antiseptic, carminative, antibilious and antileprotic properties.

The Birch leaves increase the amount of urine, have a mild saluretic effect and are antipyretic.

Use in Western herbal

Nicholas Culpeper (1616–1654) wrote of Birch: "The juice of the leaves, while they are young, or the distilled water of them, or the water that comes from the tree being bored with an auger and distilled afterwards; any of these being drank for some days together, is available to break the stone in the kidneys and bladder and is good also to wash sore mouths."

A close relative of Himalayan Silver Birch, Betula pendula, syn. Betula verrucosa, has been used as a medicinal herb in northern Europe since the earlier times.

In Scotland, Silver Birch sap was drunk as an antilithic. An infusion made with the leaves is given for kidney and bladder stones, rheumatic conditions and gout. The leaves are also prescribed, in combination with diuretic herbs, to reduce fluid retention and swelling. Externally, the leaves are used for hair loss and dandruff.

The oil distilled from leaves is antiseptic and is used in preparations to treat eczema, herpes, acne and psoriasis. A decoction of the bark is used as a lotion for chronic skin problems.

In a clinical study in Russia in 1979, superficial, deep and cavity wounds were successfully treated in 108 patients with 20 % tincture of Birch buds in a 70 % alcohol solution.

In the US, betulinic acid, also found in Indian Birch, proved highly effective in the treatment of melanoma, the aggressive type of skin cancer. It is considered to be the most promising discovery among more than 2500 plant extracts so far studied for their application in the treatment of cancer. Studies have been conducted in the department of Medicinal Chemistry and Pharmacology at the University of Illinois, Chicago, since 1995.

German Commission E monograph recognized the efficacy of Birch leaves in:

- Infections of the urinary tract
- Kidney and bladder stones.

Acanthaceae

BLEPHARIS

Blepharis edulisPers.
Blepharis persica Kuntze

Habitat

Punjab and western Rajasthan.

Common name

Ayurvedic: Shikhi (CCRAS).
Unani: Uttangan (CCRUM).
In the market Uttangan seeds are sold as Anjuraa (Arabic and Persian). Anjuraa is a different drug, which has been equated with Urtica pilulifera Linn.

Parts used

Seeds.

Dose

Seed powder 3–5 g.

Classical use

In Ayurvedic medicine, according to CCRAS, the seeds are used as an aphrodisiac and sexstimulant. Physicians prescribe seeds for increasing sperm count.

In Unani medicine, the seeds are inspissating to semen, spermatogenic, retentive and are prescribed in spermatorrhoea, sexual debility and impotency. Seeds are also given in anuria and dysurea as a diuretic and urinary tract disinfectant.

The seeds of Uttangan are an ingredient in Majoon-e-Bandkushad (Quabadeen-e-Azam-o-Akmal), a reputed Unani tonic prescribed in sexual debility.

Active principles and pharmacology

The seeds gave a glucoside—blepharin—and a saponin which on hydrolysis gave lupeol.

The alcoholic extract of the seeds exhibited anti-inflammatory and diuretic activity in rats after 100 mg/kg p.o. CCRAS reported aphrodiasic property and also suggested drug's use in gout, arthritis and metabolic disorders. The seeds show deobstruent properties. This indicated their use in strangury.

The root was found beneficial in urinary discharges and dysmenorrhoea. The leaves are reported to be useful in wounds, ulcers, nasal haemorrhage, throat inflammation, disorders of liver and spleen, and ascites.

Asteraceae

BLUMEA

Blumea lacera (Burm. f.) DC.

Habitat

Throughout the plains of India, ascending to 700 m.

Classical & common names

Ayurvedic: Kukundara, Kukkuradru, Tamrachuda, Mriducchada; Kukrondaa.
Unani: Kakrondhaa.
Siddha: Narakkarandai.

Figure 1 Blumea lacera—flowering and fruiting branch [WOI]

Parts used

Leaf, root.

Dose

Extract of total herb 3–6 g, fresh juice 10 ml.

Classical use

According to Bhaavaprakaasha, the herb cures fever, bronchial affections, vitiated blood.

In folk medicine, it is used as a styptic and anti-inflammatory agent, internally and externally.

The juice of leaves mixed with black pepper (Piper nigrum) is given in bleeding piles. Also given as anthelmintic, particularly in case of threadworm. Leaves and roots are considered astringent, diuretic and febrifugal. An astringent eyewash is prepared from the leaves. Fresh juice or extract is applied to bruises and ulcers.

Active principles and pharmacology

The plant contains a diester of coniferyl alcohol, acetylenic compounds, a thiophene derivative. Aerial parts gave beta-sitosterol and campesterol. Flavonoids isolated from the leaves have also been identified.

The essential oil contains blumea camphor and possesses antibacterial and antifungal properties.

The plant and its defatted alcoholic extract exhibited anti-inflammatory and acetylcholine-like properties through a number of experiments, besides antimicrobial property and negligible toxicity. It also prevented SGPT induced inflammation (CCRAS).

Defatted alcoholic extract of the plant produced 80 % mortality at 800 mg/kg body weight oral dose and LD50 in rats was found in between 750 and 800 mg/kg body weight oral dose.

Nyctaginaceae

BOERHAVIA

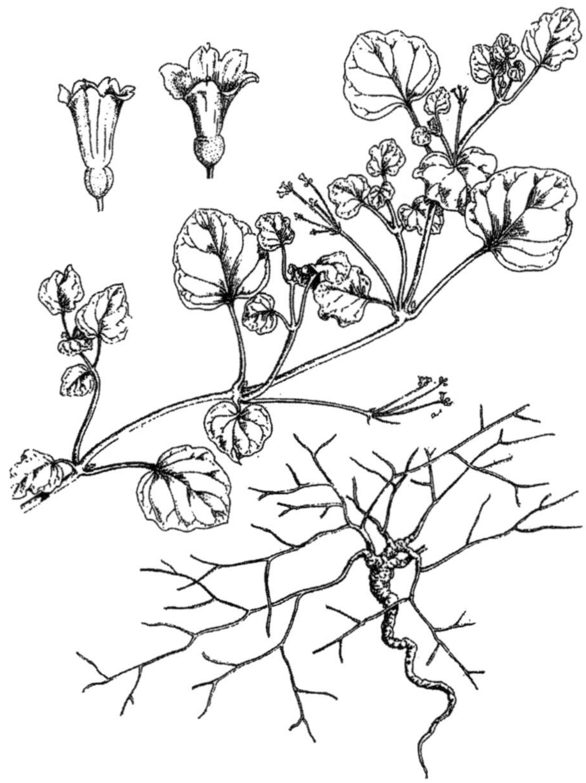

Figure 1 Boerhavia diffusa [ADPS]

Boerhavia diffusa Linn.

Boerhavia repens Linn.
Boerhavia erecta Linn.

Habitat

Abundantly occurring as a weed throughout India, up to an altitude of 2000 m in the Himalayas.

Classical & common names

Ayurvedic: Punarnavaa (Charaka, Sushruta); Shvetamoolaa, Shvetapunarnavaa, Vrishchiraa, (Charaka Sushruta); Varsaabhu, Shothaghni, Shothahrit, Vrishchira.
Unani: Biskhaparaa; Ispat (Persian).
Siddha: Mukkarattai.
English: Spreading Hogweed.

Parts used

Root.

Dose

Decoction 50–100 ml, fresh juice 10–20 ml.

Classical use

Punarnavaa (B. diffusa) has been clinically accepted as the red variety, Vrishchira (Boerhavia erecta) as the white one.

The red variety was prescribed in oedema, haemorrhage, anaemia, biliousness, diseases of the nervous system (Raaja Nighantu), in heart diseases, piles (Bhaavaprakaasha); the white one in oedema, anaemia, heart diseases, cough, intestinal colic (Dhanvantari Nighantu).

The decoction or paste of Punarnavaa; also its extract in clarified butter, was prescribed internally for alleviating oedema, internal abscesses and rheumatic affections. (Charaka Samhita, Sushruta Samhitaa, Haarita Samhitaa, Shaarangadhara Samhitaa, Vrindamaadhava, Bhaavaprakaasha.) The root was also given in enlargement of spleen, anaemia, haemoptysis, discharges, urinary diseases, skin diseases, respiratory diseases, and fever (Charaka Samhitaa, Sushruta Samhitaa).

Charaka used the decoction of the leaves and roots as an aid to rejuvenation.

According to Ashtaanga Hridaya, the paste of Punarnavaa (20 g) taken with milk daily for a few months proves an excellent revitalizing and rejuvenating tonic.

Chakradatta used Punarnavaa in the treatment of chronic alcoholism.

The practitioners of Indian medicine do not make a distinction between the red and white varieties and use both in practice.

Over the counter, Punarnavaarishta (Bhaishajya Ratnaavali) is available for the treatment of oedema, ascites, as a diuretic, alterative, haematinic, liver tonic; Punarnavaadi Taila (ibid) for massage in inflammatory conditions; Punarnavaa Mandoor (ibid) for restoring the normal functions of liver in jaundice; also for anaemia, dropsy, ascites.

Sahasrayoga has incorporated two formulations of Punarnavaadyam Ghritam; one is a tonic for alcoholics, and the other for the diseases of the abdomen and internal abscesses; seven formulations of Punarnavaa Kashaaya for preparing decoctions—all are captioned as Punarnavaadi Kashaaya and are indicated in oedema, anaemia, inflammatory disorders and fevers; Punarnavaadi Leha (confection) is prescribed as a tonic during the treatment of ailments wherever Punarnavaa is indicated.

Sukumaara Rasaayana (Ashtaanga Hridaya) is prescribed as a mild laxative and diuretic, stimulant to the urinogenital system.

In Unani medicine, water extract of total plant is used for its diuretic and anti-inflammatory properties; also prescribed in asthma, jaundice, ascites, urethritis.

Bisakhapraa, used in Unani medicine as Punarnavaa, has been equated with Trianthema portulacastrum Linn.; and Ispat Persian with Trifolium alexandricum Linn. (CCRUM).

Active principles and pharmacology

The roots contain alkaloids (0.05 %). Major constituent is a glycoside, Punarnavoside. Other constituents include rotenoids, flavones and sterols, an isofuroxanthone, boeravine, lingnans, including liridodendrin and hypoxanthine-9-L-arabinofuranoside.

The plant possesses potent antifibrinolytic and anti-inflammatory properties and is recommended for the treatment of IUD menorrhagia. Punarnavoside is a proven antifibrinolytic agent. Liridodendrin and hypoxanthine-9-L-arabinofuranoside are antihypertensive agents, the former being a Ca^{2+} channel antagonist.

The plant's diuretic and hepatoprotective properties have been confirmed in many animal studies. Punarnavaa has been reported to increase serum protein level and reduce urinary protein extraction in clinical trials in patients suffering from nephrotic syndrome.

The shoot of Punarnavaa exhibited antibacterial property against Staphylococcus aureus. The leaf extract is reported to show in vitro antifungal activity against Microsporon nanum.

Caution

The drug in large doses is emetic.

Boraginaceae

BORAGO

Borago officinalis Linn.

Habitat

Distributed in the Mediterranean region, Europe and Asia.

Classical & common names

Unani: Gaozabaan (Persian).
English: Beebread, Borage.

Parts used

Whole plant, flowers.

Dose

Whole plant 5–7 g, flowers 3–5 g.

Classical use

Araq-e-Gaozabaan is a classical preparation of Unani medicine, prescribed as a cardiac, brain and liver tonic, in palpitation, depressed mental states, polydipsia; cold, catarrh, and coryza. Sharbat-e-Sadar is given in bronchitis, asthma, catarrh and as an expectorant. Khameera-e-Gaozabaan Sada and Ambari Jawahirwala are reputed Unani compounds, prescribed as cardiac tonic, brain tonic, especially in palpitation, depression, melancholia, neurasthenia and chronic catarrh.

Active principles and pharmacology

The plant contains pyrrolizidine alkaloids, intermedine, ambiline, supinine and thesinine. Also contains ascorbic acid (38 mg/100 g), mucilage (11 %) and a cyanogenic glucoside, dhurrine. Low concentration of alkaloids may account for lack of acute toxicity in borage use.

The flowers contain choline, glucose, fructose and some amino acids, saponins, tannin (*c* 3 %) and traces of an essential oil.

The seeds contain protein 20.9 % and an oil 38.3 %. The seed oil is an important source of GLA.

The herb possesses nervine, laxative, demulcent, emollient, diuretic, diaphoretic and febrifugal properties.

Use in Western herbal

John Gerard (1545–1612) wrote in "Historie of Plants": "Those of our time do use the floures in sallads, to exhilarate and make the minde glad. There be so many things made of them, used for the comfort of the heart to drive away, sorrow and increase the joy of the mind. The leaves and flowers of Borage put into wine make men and women glad and merry, driving away all sadness, dullness and melancholy... A syrup made of flowers quieteth the phreneticke or lunaticke person". Gerard also quoted an old saying: "I, Borage, bring always courage."

With its high mucilage content, Borage is used in Western herbal as a demulcent herb that soothes the respiratory system; as an emollient (prepared either as freshly squeezed juice, in the form of a poultice or an infusion) in inflamed skin. The flowers are used to encourage sweating and leaves as diuretic.

The seed oil is considered superior to Evening Primrose oil, as it is rich in polyunsaturated fats, and is used in premenstrual and menopausal complaints, rheumatic problems; in chronic skin conditions, lupus, especially neurodermatitis. Preparations using Borage are also used for blood purification, for the prevention of chest and peritoneal inflammation, phlebitis; as a pain-relieving, sedative, cardiotonic, and performance-enhancing agent.

The flower candied and made into a conserve is given as a restorative tonic to persons weakened by long illness and to those subject to fainting.

Caution

Because of the hepatotoxic and hepato-carcinogenic pyrrolizidine alkaloid content, though in small quantity, there is some uncertainty about the use of the herb. It is subject to restrictions in some countries.

No health hazards or side effects are known in conjunction with the proper administration of designated therapeutic doses.

Restrictions and cautions do not apply to Borage seed oil, but it is contraindicated with aspirin, anticoagulant, phenothiazines, and during pregnancy or breast-feeding.

Arecaceae

BORASSUS

***Borassus flabellifer* Linn.**
***Borassus flabelliformis* Murr.**

Habitat

Throughout tropical India, in West Bengal and Bihar.

Classical & common names

Ayurvedic: Taala (Sushruta), Taadaka (Charaka); Taada. Trinraj, Mahonnat, Lekhyapatra.
Unani: Darkht-e-Taadi (Persian).
Siddha: Panai.
English: Palmyra Palm.

Parts used

Seeds, whole plant.

Dose

Decoction 50–100 ml, ash 1–3 g.

Classical use

Charak prescribed various parts fermented into beer-like liquor in insomnia, depression and anorexia.

Charaka and Sushruta prescribed the tender root, fruit and flowers, in prescriptions, in retention of urine and dysuria.

The ash of flowers mixed with jaggery was prescribed in splenomegaly (Chakradatta).

The toddy juice was given in insanity (Vrindamaadhava).

Externally, dried leaves were included in a dusting powder for skin diseases and wounds by Sushruta.

Ayurvedic scholars of the 16th century attributed spermatogenic, aphrodisiac, cooling, strength-promoting and tonic properties to the fruit. Its jaggery or sugar candy was even allowed to diabetic patients.

Over-the-counter drug, Panviralaadi Kshaara (Sahasrayoga), contains palmyra male inflorescence's ash as one of the ingredients; is prescribed in ascites, oedema, gastric and intestinal disorders, as a diuretic, digestive and for flushing out toxins.

Active principles and pharmacology

The sap is mildly acidic. It contains sucrose and starts fermenting very soon unless suitably treated with preservatives. It is a nutritious supplement to diets which are deficient in iron, ascorbic acid and vitamin B-complex. It is cooling, diuretic, laxative, antiphlegmatic and stimulant, and is also useful in inflammatory affections, ulcers, and dropsy. Slightly fermented juice is given in diabetes. The sap is given as a tonic to asthmatic and anaemic patients.

The palm-jaggery contains carbohydrates, appreciable amounts of vitamin B complex and minerals (iron is present in an easily assimilatory form). The concentration of aminoacids in palm-jaggery is much higher than in cane-jaggery. It is used as an energy food for convalescents and is given in anaemia. It exhibits a mild laxative effect. The jaggery solution is reported to be an excellent food for typhoid patients.

The low sodium and high potassium value indicates the possibility of its use in hypertension and oedema due to heart and liver diseases. It can be prescribed for diseases characterized by a marked loss of potassium, and also as a diuretic.

Kernels contain free sugars, sucrose, fructose and glucose; also polysaccharides. Amino acids are also reported to be present.

The pulp from unripe fruit is cooling, diuretic, demulcent and nutritive. The pulp from the ripe fruit is applied externally for skin diseases.

The root possesses cooling, restorative, diuretic, anti-inflammatory, and anthelminithic properties. The decoction of the young root and expressed juice from the young terminal buds and leaf stalks is given in gastritis. An ethanolic extract of roots of female palm produced hypogycaemic and hypocholesterolaemic effects in mice.

The ash of the spathe is applied externally for enlarged liver and spleen in combination with some demulcent (otherwise it will form blisters).

The leaves are styptic and arrest haemorrhage from superficial wounds.

Caution

Young shoots of the palms are reported to contain a toxic factor which is reported to be neurotoxic but not hepatotoxic to rats. Palmyra flour has been reported to produce significant alteration in the immune-competence in experimental rats.

Burseraceae

BOSWELLIA

Boswellia serrata Roxb ex colebr.

Figure 1 Boswellia serrata—flowering and fruiting portion [WOI]

Boswellia floribunda Linn.

Habitat
Commonly found in dry forests from Punjab to West Bengal.

Classical & common names
Ayurvedic: Shallaki, (Charaka, Sushruta); Sushravaa, Gajabhakshyaa, Mukundkunda, Kundaru, (oleo-gum-resin); Salai-guggul. The commercial supply is mainly from B. floribunda Linn.
Unani: Kundur (oleo-gum-resin).
Siddha: Parangisambirani.
English: Indian Olibanum Tree, Indian Frankincense Tree.

Parts used
Bark, leaf, flower, oleoresin.

Dose
Decoction 50–100 ml.

Classical use
The bark was used by Charaka and Sushruta as a corrective of deranged bowels; for infections and irritation in the digestive tract.

Sushruta prescribed the decoction of the bark in haemorrhagic diarrhoea, and as a styptic in haemoptysis and metrorrhagia. Charaka prescribed the paste or the decoction alone or processed in clarified butter. The bark was considered as antitoxic, antiseptic and astringent.

Sushruta prescribed the decoction of the bark in obesity for inhibiting accumulation of fat in the body and for exerting a tonifying effect on the digestive system and intestines.

Internally Charaka gave the oleoresin, in prescription, as a diuretic in calculus; externally in massage oils for rheumatic affections. Sushruta prescribed powder of the fruit or ash of flax for external application to wounds.

The oleoresin was used in ointments for venereal sores and other skin diseases; the bark powder for dusting.

In Unani medicine, Boswellia floribunda or Boswellia thurifera is used in compound formulations of confections and ointments.

Jawarish-e-Kundur is a renal and nervine tonic, prescribed in polyuria; Majoon-e-Kundur is prescribed in strangury, polyuria, bed-wetting, atony of the bladder, premature ejaculation; Marham-e-Ajeeb, with Kundur as one of the important ingredients, is prescribed for external application on chronic ulcers.

Active principles and pharmacology
The bark contained tannins 9.1, pentosan 18.3, lignin 28.8, holococellulose 48.7 % and beta-sitosterol. These constituents indicate its use in diarrhoea, piles, skin diseases and its external application on suppurating or bleeding wounds.

The crude gum contained uronic acid anhydride 20, nitrogen 3.2 and inorganic oxides 4.5 %. The purified gum gave pentosans 18.9, uronic acid 30.9, and methoxyl group 3.6 %. On hydrolysis, the purified gum yielded D-galactose 46.2, D-arabinose 11.7, D-xylose 6.7, D-mannose 4.5 and uronic acid anhydride 30.9 %. The gum also contained oxidizing and diastatic enzymes.

The gum is credited with diuretic, astringent and antiseptic properties and its use is indicated in diarrhoea, dysentery, piles, ulcers, tumours, skin diseases and in genitourinary affections. The oleo-gum-resin possesses anti-tumour, analgesic and sedative properties. It was found to possess marked anti-inflammatory and anti-arthritic activity. It was also shown to possess marked cholesterol and triglyceride-lowering activity. (Sushruta used the bark in obesity.)

Use in Western herbal
Frankincense or olibanum of Boswellia thurifera contains a volatile oil 5–9 %, chief components are pinene, dipentene, phellandrene; resins 60 %, components including alpha-boswellic acid, 3-acetyl-beta-boswellic acid; mucillage 12 %.

In the West, it is obsolete as a herbal drug.

Formerly, it was in great repute. Pliny (23–79AD) mentions it as an antidote to hemlock. Avicenna (10th century) recommends it for tumours, ulcers, dysentery, vomiting and fevers.

In China, the olibanum of Boswellia thurifera is used as a diuretic, astringent, carminative and alterative. It is used internally in leprosy and struma (enlargement of thyroid gland). Also recommended in spermatorrhoea, certain visical and urinary disorders, amenorrhoea, menstrual cramps and in abdominal pain.

Brassicaceae

BRASSICA

Brassica napus Linn.
Brassica campestris Hook. f. & Thoms. var. sarson Prain.
Brassica juncea Czern. & Coss.

Habitat
All varieties abundantly cultivated.

Classical & common names
Ayurvedic: Sarshapa (Charaka, Sushruta); Raajakshavaka, Kattaka, (Charaka); Katusneha, Tantubha, Siddhaartha, Siddhaarthaka, Siddharthasita, Rakshogna. Sarshapa-Gaur has been equated with B. alba Boiss. Brown Sarson, Yellow Sarson.
Unani: Sarson.
English: Mustard.

Figure 1 Brassica campestris—Brown Sarson [WOI]

Brassica juncea L. Czern. & Coss.

Classical & common names
Ayurvedic: Raajikaa (Bhaavaprakaasha, Raaja Nighantu), Aasuri Raai.
Unani: Khardal siyah (Brassica nigra Linn. Koch); Khardal safed (Brassica alba Boiss).
English: Indian Mustard, Black Mustard, White Mustard.

Parts used
Seed, oil.

Dose
Seed powder 1–3 g.

Figure 2 Brassica campestris—Yellow Sarson [WOI]

Figure 3 Brassica juncea—Indian Mustard [WOI]

Classical use

Charaka and Sushruta prescribed the powdered seeds of Sarshapa internally in intestinal catarrh, colic pain, indigestion and flatulence; Charaka in rhinitis, coryza, hemicrania and dysurea. Charaka used the oil externally in pruritus and skin diseases.

According to Sushruta, the paste of seeds matures and suppurates non-suppurating boils.

Two varieties of Sarshapa have been mentioned in Bhaavaprakaasha, Siddhaartha, the yellow variety and Rakta Sarshapa, the brown one. Siddhaartha was preferred in Ayurvedic medicine.

Sarshapa and Raajikaa were considered to possess similar properties, but Raajikaa was preferred as an appetizer.

According to Bhaavaprakaasha and Raaja Nighantu, Raajikaa is stomachic, stimulant and anthelmintic; alleviates catarrh and rheumatic affections; prescribed in the enlargement of liver, spleen and internal abscesses; applied externally in inflammations and skin diseases.

In Unani medicine, Khardal is used in ointments for external application in muscular rigidity, scrofula and fistula.

The seeds of the white variety, Brassica alba, or Khardal safed, are used like ordinary mustard. In folk medicine, the seeds are administered internally in cases of nervous diseases such as epilepsy, hysteria with Braahmi Ghrita.

Active principles and pharmacology

The seeds of Brassica alba or White Mustard are rarely used alone, they are mixed with Black Mustard in the preparations of Mustard. Black Mustard is called true Mustard.

The Black Mustard contain glucosinolates, chiefly sinigrin (allylglucosinolates 1–5 %), grinding the seed into powder and then rubbing them with warm water, as well as chewing, releases the volatile mustard oil allyl-isothiocyanate; fatty oil 30–35 %; proteins 40 %; phenyl propane derivatives including among others, sinapine (choline easter of sinapic acid 1 %).

The essential principle, or volatile oil, of Brown Mustard is allyl-isothiocyanate, while that of the White or Yellow Mustard is acrimyl-isothiocyanate. The essential principles are produced as a result of hydrolysis of their respective glycosides, sinigrin (potassium myronate) and sinalbin, by action of the enzyme myrosin, in the presence of moisture under suitable conditions.

All varieties of Indian Sarson and Rai are found efficacious in internal congestions, spasmodic, neuralgic and rheumatic affections. Their active principles are to stimulate gastric mucosa and to increase pancreatic secretions.

Use in Western herbal

Nicholas Culpeper (1616–1654) wrote about Mustard (the yellowish variety): "It is excellent for such whose blood wants clarifying ... for weak stomachs. It strengthens the heart and resists poison. The decoction of the seeds made in wine and drank provokes urine. The seed helps the spleen and pain in the sides, and gnawings in the bowels, and used as a gargle for the diseases of the throat. The outward application hereof upon the pained place of the sciatica eases the pains, as also the gout and other joint aches. The seeds bruised mixed with honey and applied or made up with wax, takes away the marks and black and blue spots of bruises, freckles or the like, the roughness or scabbiness of the skin, as also the leprosy."

At one time, Mustard seeds were a popular remedy as a laxative, especially for old people, but they were not found safe, as inflammation of the stomach and intestinal canal was reported in many cases after the ingestion of seeds in large quantities.

An infusion of the seeds, in therapeutic doses, is prescribed to relieve chronic bronchitis and rheumatism, Mustard Seed tea is given for a relaxed throat as a gargle. Black Mustard is used in the form of poultices in bronchial pneumonia, pleurisy, arthritis, neuralgia and other pains and spasms.

Mustard Linament of the British Pharmacopoea is used in chilbains, chronic rheumatism and colic. Mustard flour is considered an antiseptic and sterilizing agent, as well as an excellent deodorizer.

Tincture of the whole fresh plant of Brassica napus is used in homoeopathy for frontal headache, tension, inflamed mucous membrane of mouth and throat, spongy gums, tympanitic abdomen, torpid bowels and deficient and irritating urine.

Cucurbitaceae

BRYONOPSIS

Bryonopsis laciniosa (L.)
Bryonia laciniosa L.
Diplocyclos palmatus (Linn.) Jeffrey

Habitat

Throughout India.

Classical & common names

Ayurvedic: Lingini, Lingikaa, Ishalingi (Bhaavaprakaasha); Shivalingi.
English: Indian Bryony.

Parts used

Seeds.

Dose

45 seeds as a single dose.

Classical use

According to Bhaavapraakasha, the seeds of Shivalingi, taken with milk, promote the conception of a male child. The seeds were used for increasing sperm count; also as an aphrodisiac.

In folk medicine, 45 seeds of Shivalingi with jaggery are given to the female immediately after the menstruation on 4 consecutive days, or on alternate days for 8 days. If conception does not take place the drug is repeated after the next menstruation.

Shivalingi has been equated with Bryonopsis laciniosa by NAA.

Active principles and pharmacology
Bryonin is the bitter principle of the herb. Seed oil contains punicic acid (38.2 %). An ethanolic (50 %) extract of aerial parts is found to be spasmolytic.

Use in Western herbal
In homoeopathy, tincture of the root of white Bryony, procured before flowering, is used extensively for treating vaginal disorders—painful menstruation, dysmenorrhoea, pain in ovaries during menstrual period, inflammation of uterus, burning pain in the fundus of the uterus. It is also prescribed in threatened abortion.

Crassulaceae

BRYOPHYLLUM

Bryophyllum pinnatum Kurz
Kalanchoe pinnata Pers.

Habitat
Throughout the warm and moist parts of India, especially abundant in West Bengal.

Classical & common names
Ayurvedic: Parnabija (non-classical).
Unani: Zakhm-hayaat, Pattharchoor, Pattharchat.
Siddha: Ranakkalli.

Parts used
Leaf.

Dose
Juice 10–30 ml.

Classical use
In Unani medicine, leaves roasted over a fire or fresh bruised leaves and juice are applied as a poultice to bruises, boils and contusions to allay inflammations and prevent discolouration; as a styptic on fresh cuts, abrasions, wounds; as an antiseptic to insect bites.

Internally the juice is given with double the quantity of butter in diarrhoea and dysentery.

In folk medicine, the juice, mixed with the juice of Aegle marmelos, is given in blood and amoebic dysentery.

Leaves are also eaten to control diabetes.

Active principles and pharmacology
The leaves contain p-coumaric, ferulic, syringic, caffeic and p-hydroxybenzoic acids, flavonoids (quercetin, kaempferol); n-hentriacontane, n-tritriacontane and alpha- and beta-amyrin and sitosterol.

The leave exhibits styptic, antiseptic and astringent properties.

Anacardiaceae

BUCHANANIA

Buchanania lanzan Spreng

Figure 1 Buchanania lanzan—flowering branch [WOI]

Buchanania latifolia Roxb.

Habitat
Throughout India, ascending to 1000 m.

Classical & common names
Ayurvedic: Priyaala, Priyaalaka, Raajaaratna (Charaka); Piyaala (Sushruta); Kharskandh, Bahul valkal, Taapaseshta, Sannkadru, Dhanushpata, Chaara.
Unani: Chironji, Habb-us-Samena.
Siddha: Mudaima.

Parts used
Bark, seed kernel.

Dose
Decoction of bark 50–100 ml, seed kernel 10–20 g.

Classical use
Charaka gave the fruits internally as an invigorating tonic; Sushruta prescribed them internally in obesity, haemoptysis, menstrual disorders and vaginal discharges.

According to Bhaavaprakaasha, milk boiled with the seed kernel and Glycyrrhiza glabra checks haemorrhage.

Chakradatta gave the bark, in prescription, for treating diarrhoea with blood.

Externally, Charaka used the expressed oil of the kernel in rheumatism, glandular swellings and skin diseases. Sushruta used the pith, as an ingredient of a plaster, in goitre.

In folk medicine, kernels pounded into ointment, are applied to urticaria, prickly heat, pimples, spots and blemishes of the skin. The oil of kernels is used as a substitute for almond oil.

The leaves are used as a general and cardiotonic. Their powder is applied externally to wounds.

The gum, dissolved in cow's milk, is used internally in rheumatic pains. It is also used in diarrhoea and intercostal pains.

In Unani medicine Habb-us-Samena; which has been equated with Buchanania angustifolia Roxb., is used in tonics for sexual debility. Other uses are the same as those of Buchanania lanzan.

Active principles and pharmacology
Fatty acids are reported from the kernel oil—palmitic, stearic, oleic and linoleic. Aminoacids are also reported from the plant. The bark contains 13.4% tannin. Leaves also contain crude tannins. The presence of triterpenoids, saponins, reducing sugars and flavonoids is also reported.

Leaves contain kaempferol glucoside, quercetin.

The kernels contain moisture 3.0, protein 19.0, fat 59.1, fibre 3.8, carbohydrates 12.1 and minerals 3.0 g/100 g. Kernels yield 35.4–47.2 % oil.

Papilionaceae

BUTEA

Butea monosperma (Lam.) Taub.
Butea frondosa Koen. ex. Roxb.

Figure 1 Butea frondosa [ZANDU]

Habitat
Commonly found throughout India, except in acrid region, up to 1200 m.

Classical & common names
Ayurvedic: Palaasha, Kimshuka, (Charaka, Sushruta); Palaashaka, Raktapushpak, Kshaarashreshtha, Brahma-vraksha, Samidvar.
Unani: Darkht-e-Palah (Persian); Dhaaka, Tesu.
Siddha: Parasa.
English: Bengal Kino, Flame of the Forest.

Parts used
Flowers, seeds, bark.

Dose
Powder of flowers 3–6 g, seed or bark powder 3–6 g.

Classical use

Charaka prescribed the extract of the root, bark, leaves as ingredients of a medicinal Ghee (clarified butter) in constipation, colic, dysuria; externally in the form of an ointment in skin diseases.

Sushruta used the drug internally as antitoxic, antiseptic, styptic and astringent; prescribed in menstrual disorders, seminal weakness, obesity, piles, urinary calculi, urinary discharges, malfunctioning of the liver; in persistent dysentery; internal abscesses, abdominal glands, non-healing ulcers; in skin diseases and leprosy.

Bhaavaprakaasha recommends the seed powder for ascaris; externally, the powder pounded with lemon juice, for ringworm and herpes.

In folk medicine, the leaves, flowers and solution of the gum are used externally to disperse boils, pimples, buboes and haemorrhoids. A weak decoction of the bark is given in catarrh, cold and cough. Fresh juice is used in relaxed, congested and septic sore throat as a gargle.

In Unani medicine, tender leaves and bark is prescribed in the vitiation of semen. A confection prepared with the decoction of the bark is given in leucorrhoea, sexual debility and as an emmenagogue, diuretic and aphrodisiac. The decoction is used externally for treating the laxicity of vagina as an astringent. Other uses are the same as in Ayurvedic medicine.

Among classical Ayurvedic compounds, Palaasha Kshaar (Sahasrayoga), a medicinal ash, is prescribed for piles. Krimimudgar Rasa (Bhaishajaya Ratnaavali), a mineral-based compound with the seeds of Palaasha, is prescribed, under medical supervision, for intestinal worms.

The butea gum is an ingredient in the reputed Unani tonic, Majoon-e-Supaaripaak, specific in leucorrhoea, female sterility and premature ejaculation. The gum is also included in another popular Unani compound, Safoof-e-Habsuddam (Qarabadeen-e-Jadeed), prescribed in haematemesis, haemoptysis, bleeding piles and bacillary dysentery.

The seeds are an ingredient in Majoon-e-Mukharrij-e-Deedan (IMPCOPS), given in helminthiasis.

Active principles and pharmacology

The gum contains leucocyanidin, its tetramer, procyanidin, gallic acid and mucilaginous material. It is rich in riboflavin and also contains thiamine.

The bark contains tannins 5.82 % and non-tannins 7.98 %.

An alcoholic extract of the bark, as also of the seeds, is reported to inhibit the activity of E. coli and Micrococcus pyogenes var. aureus.

The bark, root and flowers exhibit astringent, alterative, anti-inflammatory and anthelmintic properties. The decoction is found useful in dysentery, diarrhoea, bleeding piles and various forms of haemorrhages, tumours, ulcers, and in leucorrhoea, amenorrhoea and dysmenorrhoea.

The roots contain protein, fat, carbohydrates and minerals; also glycine, a glycoside.

The major glycoside of the flower is butrin.

In a study, the seed powder internally, and in the form of a paste externally, showed much improvement in eczema.

The clinical use of seeds as an anthelmintic drug is not considered safe, as it may produce nephrotoxicity.

Use in Western herbal

Bengal or Palaasha Kino from Butea frondosa is used whenever tannin is indicated, internally in diarrhoea, dysentery and pyrosis; also in leucorrhoea; externally as a gargle for sore throat.

The seeds are mainly used to treat herpes and ringworm.

In the British Pharmacopoeia, Malabar or West Indian Kino is the only one recognized. Bengal Kino is official in the United States.

Caesalpiniaceae

CAESALPINIA

Caesalpinia bonducella (L.) Flem.
Caesalpinia bonduc Linn.
Caesalpinia crista Linn.

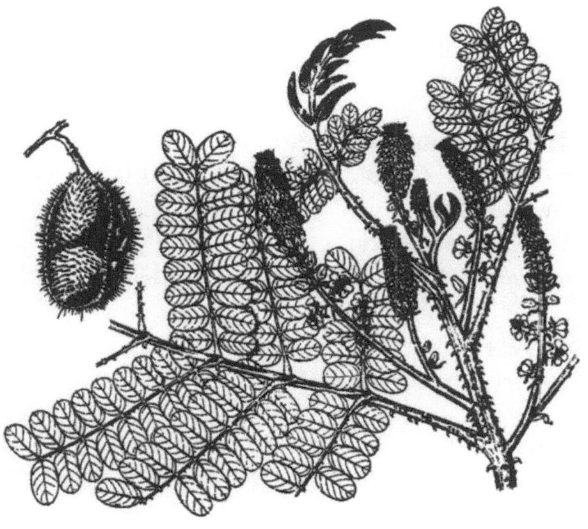

Figure 1 Caesalpinia crista [WOI]

Habitat

Wild throughout the plains of India, and up to an altitude of 1000 m in the Himalayas.

Classical & common names

Ayurvedic: Puutikaranja, Lataa-karanja, Kantaki-karanja (CCRAS). Karanji (PV Sharma). Kuber-aakshi (Seed).
Unani: Karanjwaa.
Siddha: Kazharchi.
English: Fever Nut, Bonduc Nut, Nikkar Nut.

Ayurvedic Formulary of India equated Lataa-karanja with Caesalpinia crista; and Karanja, Karanjaka, Ghrita-karanja, Naktaahvaa and Naktamaala with Pongamia pinnata (Linn.) Merr.
In Standard Nomenclature of Ayurvedic Medicinal Plants, CCRAS has created a confusion by including Ghrita-karanja as synonym of Caesalpinia bonduc; also by equataing Ghrita-karanja with Caesalpinia digyna Rottler. While documenting pharmacological investigations, CCRAS equated Karanja and Puutikaranja as synonyms.
Though The Wealth of India equated Kuberaakshi with Caesalpinia bonduc, it also quoted folk synonyms—Kanjaa, Karanju, Nata.
National Formulary of Unani Medicine equated Karanjwaa with Caesalpinia bonducella and Karanj with Pongamia pinnata.
In Shodhala Nighantu Kuberaakshi is a synonym of Kantaki-Karanja. In Raaja Nighantu, Kuberaakshi and Lataa-karanja are synonyms.

Parts used

Seed-kernel.

Dose

Powder 1–3 g.

Classical use

Sushruta prescribed Puutikaranja as a potherb for producing heat and for curing swellings; juice of the fruit in elephantiasis. According to Bhaavaprakaasha, the fruit of Karanji is hot, cures vomiting, biliousness, parasitic infections, piles, skin diseases and urinary disorders.

Siddha-bheshaja-manimaalaa recommends the seed-kernel of Karanji in fever with other supporting herbs; in colic 2 or 3 seed-kernels are to be fried and taken with salt.

Vaidya-manoramaa mentions the drug as Kuberaaksha; seed-kernel is recommended in digestive problems, dysentery, vomiting and internal abscesses.

In Unani medicine, Karanjwaa leaves, pounded with Piper nigrum are given in diseases due to vitiated blood. A prescription, Habb Karanjwaa, is given as an antiperiodic in the fourth-day fever. The seed-kernel, burnt in sesamum oil, is applied to infected wounds and cutaneous affections.

In folk medicine, the plant juice is given for 2 weeks after meals to cure intermittent fever. The seed-kernels, made into an ointment, are applied to hydrocele. As an infusion, they are prescribed in haemorrhages. Also given as an anthelmintic mixed with honey or castor oil.

Mostly, seeds ground and roasted are given internally.

Active principles and pharmacology

The seed-kernel of Caesalpinia bonducella contains diterpenes, including, among others, alpha-, beta-gamma-, eta-caesalpine; saponins; fatty oils with palmitic acid, stearic acid, oleic acid, linoleic acid; proteins; starch.

The important pharmacological properties of the plant are: antimalerial, antidiarrhoeal, antiamoebic and antiviral (CCRAS).

Caesalpinia crista nut is one of the constituents of the CCRAS compound formulation, Ayush 64, which showed significant antimalarial potentiality in Plasmodium berghei infected rats and mice. (Other constituents are: Picrorhiza kurroa, Alstonia scholaris, and Swertia chirata.) Double blind clinical trial of 58 patients suffering from P. vivax (30 cases were given 1500 mg on first day, 750 mg on subsequent 3 days; 28 cases received chloroquine 600 mg and 30 mg primaquine on first day and 15 mg primaquine on the following days) showed no difference between Ayush 64 and modern drug therapy. In field clinical trials, about two-thirds of the patients were free from parasites in the blood; in others, the parasite count was reduced significantly and reached almost negligible level. (CCRAS.)

The nuts exhibited significant antidiarrhoeal activity in mice. Oral administration of 2 g of powder 3 times a day to amoebic patients, hospitalised with symptoms of griping pain, loose motions with mucus and blood, exhibited 100 % improvement of symptoms as well as eradication of E. histolytica from stool.

Antiviral activity was observed with alcoholic extracts of the root and stems.

See Pongamia pinnata.

Use in Western herbal

Nikkar seeds are used in febrile illness. Roasted seeds are made into a kind of coffee for diabetes.

The oil extracted from the seeds is used in cosmetic preparations to soften the skin.

Asteracea

CALENDULA

Calendula officinalis L.

Habitat

Grown as an ornamental plant throughout India; wild in Punjab.

Classical & common names

Vernacular: Zergul. (Related var. Gul-jafari, Gendaa, Jhandu).
Tamil: Baantichettu.
English: Pot-Marigold; Marigold; Calendula.

Parts used

Flowers, leaves, whole plant.

Dose

Fresh juice of flower and leaf 5–10 ml.

Use in traditional medicine

The florets are used in ointments for treating wounds, herpes, ulcers, frostbite, skin damage and scars. The leaves, in infusion, are used for treating varicose veins externally.

In Indian medicine, a related variety—Tagetes erecta Linn. (Gul-jafari, Jhandu, Gendaa, French or African Marigold)—is used internally, in prescriptions for bleeding piles and gonorrhoea (Siddha-bheishaja-manimaalaa); meno-metrorrhagia; also as a blood-purifier, externally as an antiseptic and styptic.

Active principles and pharmacology

The flower contains triterpene saponins, oleanolic acid, mono- and diglycosides; flavonoids; hydroxycoumarins including umbelliferone, scopoletin; carotinoids; sesquiterpene glycosides; volatile oil; water-soluble polysaccharides, rhamnoarabinogalactans, arabinogalactans; and polyenes.

Although not containing any tannins, the herb is locally astringent, mainly due to the resin component; also probably due to other water-soluble constituents. A substance has been found to promote blood clotting and the whole plant to act against fungal, bacterial, amoebic and even viral infections, and has potent anti-inflammatory action as well; is used for swelling of glands, jaundice, wounds, eczema. The flowers stimulate the immune system, inhibit tumours, have an inhibitory effect on the CNS and are estrogenic, choleretic and haemolytic.

For the digestive system, Calendula can be used with benefit as a first-aid in unresolved infection or erosion of the upper tracts, particularly if there is evidence of bleeding into the gut.

Use in Western herbal

Externally Marigold is used for inflammations, oral and pharyngeal mucosa, as a gargle, and externally on poorly healing wounds, inflamed lymph glands, arthroma, chronic eczema, varicosis, phlebitis, thrombophlebitis, furunculosis, dry dermatosis and acne.

German Commission E monograph recognized the herb's efficacy in:

- Inflammation of the mouth and pharynx
- Wounds and burns.

An infusion for internal use is prepared by using 1–2 g of drug to a cup of water. For external use, a tincture (1:9) is prepared using 20 % alcohol/water mixture. An ointment is prepared by adding 2–5 g drug in 100 g ointment.

In the West, herbalists consider Marigold a detoxifying herb that helps treat the toxicity that underlines many fevers and infections. The herb is considered cleansing for the liver and gallbladder. It is often used in breast tenderness and sore nipples and to help reduce period pain and regulate menstrual bleeding. The infusion is used as a douche for vaginal thrush.

In homoeopathy, the tincture of leaves and flowers is used in fever, fistula, glandular swellings, jaundice, abscesses, ulcers, nipple sore, inflammation of the uterus, offensive discharge, whitlow, chilbains, burns, and inflamed eyes. It is considered one of the most remarkable healing agents, when applied locally. Locally aqueous extract and internally tincture to third potency is prescribed.

Verbenaceae

CALLICARPA

Callicarpa macrophylla Vahl

Habitat

Kashmir to Assam and Arunachal Pradesh, up to an altitude of 1800 m.

Classical & common names

Ayurvedic: Priyangu, Priyaka (Charaka, Sushruta); Phalini, Shyaamaa, Kaantaahvaa, Kanta, Gandhphali.
Unani: Habb-ul-Mihlb (Prunus mahaleb).
English: Perfumed Cherry.

Parts used

Buds, fruits, bark, root.

Dose

Powder 1–2 g.

Classical use

Charaka prescribed the fruits in fever, haemothermia, piles, dermatosis and as intestinal astringent; the seeds as a cereal.

Sushruta prescribed the fruits internally in skin eruptions, blood poisoning, haemoptysis, toxic conditions and burning sensation of the body; paste of the herb as a styptic.

Bhaavaprakaasha mentioned 2 varieties of the herb: Priyangu and Gandh-Priyangu. Priyangu is described as cooling, refrigerant, deodorant and antipyretic; tones up the digestive system, checks excessive prespiration, disinfects intestines, controls diarrhoea and allays burning sensation during fevers; Gandh-Priyangu fruits are a disinfectant and blood purifier.

In Ayurvedic skin preparations Gandh-Priyangu fruits were used to treat skin affections, freckles, blemishes.

Priyangwaadi Churna (Yoga Ratnaakara) and Priyangwaadi Tailam (Bhaishajya Ratnaavali), the classical compounds, with Priyangu as an ingredient, are prescribed in diarrhoea, dysentery and intestinal infections. Priyangu has also been included in pastes and oil extracts, which were used for promoting complexion. Kanaka Tailam (Bhaishajya Ratnaavali) is one such preparation.

Classically, Priyangu has been included in the Elaadi group (of herbs) which promotes complexion, cures itching, pimples and urticarial rashes and eliminates toxins. It also belongs to the Anjanaadi group, which is specific for curing poisoning, burning sensation inside the body and haemorrhagic conditions.

Active principles and pharmacology

The seeds and leaves contain calliterpenone and its monoacetate. Seeds also contain oleanolic acid. Besides diterpenoids, the leaves also contain fatty acids and beta-sitosterol and its beta-D-glucoside. Glycosides, terpenes, phenolic compounds, resin and saponin are the main constituents of the plant.

CCRAS has identified Priyangu with Callicarpa macrophylla, while the scientists of INSA have equated it with Aglaia roxburghiana Miq. Some Ayurvedic scholars and Unani physicians have equated Gandh-Priyangu with Prunus Mahaleb Linn. In the South, IMPCOPS is using the fruits of Aglaia roxburghiana as Priyangu.

Aglaia roxburghiana is a cooling and astringent herb, employed in inflammations, febrile complaints; its seeds are used in painful micturition.

Coumarins, salicylic acid, amygdalin and hydrocyanic acid are the main constituents of Prunus mahaleb.

In practice, Aglaia roxburghiana, Prunus mahaleb and Callicarpa macrophylla are used as Priyangu and Gandh-Priyangu.

All the 3 drugs are stomachic, diuretic, deodorant, antipyretic, blood purifier and tonic to liver.

The drug is found efficacious externally as a deodorant and antiseptic in cosmetic formulations. Internally, it may prove emetic, if not given with supporting herbs, which have been recommended in classical formulations.

Clusiaceae

CALOPHYLLUM

Calophyllum inophyllum Linn.

Habitat

Coastal regions of South India and the Andaman Islands; grown as an ornamental tree all over India.

Classical & common names

Ayurvedic: Punnaaga (Sushruta), Tunga, Sultaan Champaa, Naagchampaa, Raajchampaka.
Siddha: Punnaagam.
English: Alexandrian Laurel.

Parts used

Leaves, flowers, bark.

Dose

Decoction 50–100 ml, powder 3–5 g.

Classical use

Sushruta prescribed the herb internally in skin eruptions and diseases due to vitiated blood, in non-healing ulcers; also in presistent dysentery. It was included in a medicated liquor, which was used as a tonic.

In folk medicine, the pounded bark is applied in orchitis; its juice is taken as purgative. A decoction of root-bark is used for dressing ulcers. The leaves are given for inhalation in migraine and vertigo. The decoction of flowers is given in eczema and venereal sores. The fruits are used as purgative.

Active principles and pharmacology

The bark contains inophyllic acid, inophylloidic acid, friedelin, friedelan-3-beta-ol, beta-amyrin and sitosterol. The root-bark exhibits significant antibacterial activity towards Gram-positive organism.

The leaves contain friedelin, canophyllal, canophyllol, canophyllic acid, amentoflavone and piscicidal compounds. Quercetin, myricetin and its 7-glucoside were isolated from flowers. Petals contain only leucocyanidin.

The ripe seeds yielded calophyllolide, inophyllolide and calophyllic acid which are reputed to be useful in the treatment of leprosy.

The oil shows a beneficial influence over the mucous membrane of the genito-urinary organs and is found efficacious in the treatment of venereal diseases. It cures scabies and other cutaneous diseases. The refined oil, when injected intramuscularly, alleviates pain in leprosy.

The oil stimulates embrocation in rheumatism and gout. A paste of the kernels relives pain in inflamed joints.

The oil shows antibacterial activity against several Gram-positive bacteria. It is found useful in treating burns; mixed with camphor against ringworm.

The calophyllolide, found in defatted kernels, showed significant anticoagulant and anti-arrhythmic activity and coronary-dilatory actions in experimental animals. In addition, it raised blood pressure. It also showed significant anti-inflammatory and anti-arthritic activity in albino rats. It was found effective in reducing increased capillary permeability in mice.

Flowers and stems are often used as a substitute for Mesua ferrea Linn. (Naagakesara). Punnaga and Keshara are synonymous in classical Ayurvedic texts. Dalhana considered Naagakesara, Punnaaga and Sur-punnaga as one; actually all the three are separate herbs botanically.

The scientists of INSA (P. Ray et al) have wrongly equated Punnaaga with Ochrocarpus longifolius Benth. & Hook., which has been identified as Sur-punnaaga. Its immature buds are also sold as Mesua ferrea and are known as Red Naagakesara (the so-called red variety of yellow Mesua ferrea).

According to IMPCOPS, in Tamil Nadu and adjacent States tender fruits of Cinnamomum wighti Meissn. (Lauraceae) and Dillenia pentagyna Roxb. (Dilleniaceae), known as Sirunagappoo and Malabar Naagakesara respectively, are used as Naagakesara.

Asclepiadaceae

CALOTROPIS

Calotropis procera (Ait) R. Br.

Figure 1 Calotropis procera [CCRAS]

Figure 2 Calotropis procera [CCRAS]

Calotropis gigantea (Linn) R. Br. ex Ait.

Figure 3 Calotropis gigantea [CCRAS]

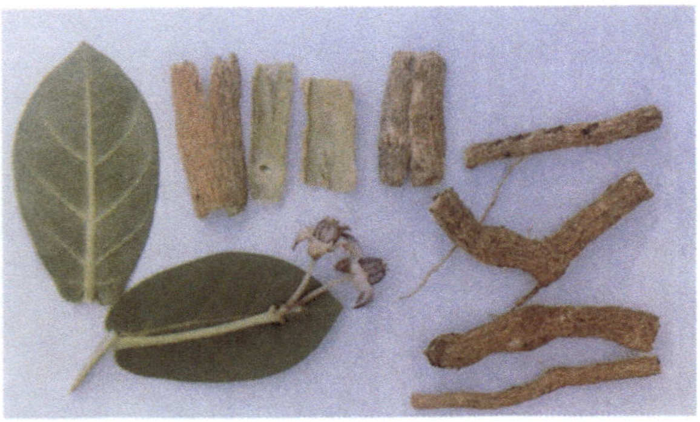

Figure 4 Calotropis gigantea [CCRAS]

Habitat

Wild throughout India, in comparatively drier and warmer areas, up to an altitude of 1050 m.

Classical & common names

Ayurvedic: Arka (Charaka, Sushruta), Suryaahvya, Kshiri, Raktaarka, Sadaapushpaa, Vikerna, Vasuka. Calotropis gigantea: Alarka, Rajaarka, Mandaar. Bhaasvanmula, Dinesha, Prabhaakara, Ravi.
Unani: Madaara.
Siddha: Erukku.
English: Swallow-wort, Milk-weed, Madar.
Calotropis procera bears purple and Calotropis gigantea white flowers.

Parts used

Latex, flowers, leaves, bark.

Dose

Latex 250 mg, flowers 1–3 g, root or bark powder 200 mg–1 g.

Classical use

Though Sushruta included both Arka and Alarka in the same prescription and both are mentioned in the Arkaadi group of herbs, both the varieties were considered to possess similar medicinal properties.

The flowers were prescribed by Charaka for their digestive, stomachic and tonic properties. Sushruta prescribed their extract in clarified butter, internally, for malignant skin diseases.

The latex was used externally in piles, boils, ulcers, scabies, eczema, enlarged glands and leprosy (Charaka Samhitaa, Gadanigraha, Vaidyamanoramaa). Charaka used the latex for emesis and purgation.

The leaves were used for covering wounds (Charaka). The powder was dusted over wounds and ulcers. Fresh or warm leaves or their poultice was applied over painful rheumatic joints, swellings, sores and wounds.

The alkali (Kshaar) of the leaves was prescribed internally in splenomegaly (Vrindamaadhava, Bhaavprakaasha), also in piles (Bangasena).

The paste of the leaves, root, rootbark was used externally on boils, scrotal enlargement, piles, elephantiasis, hydrocele and skin diseases (Chakradatta, Gadnigraha, Vrindamaadhava).

Sushruta prescribed the decoction of Arka, Alarka and Saptaparna (Alstonia scholaris) in leprosy internally.

Arka is an important ingredient in Sushruta's Samshodhana Ghrita, which was prescribed for cleansing ulcers.

During the 16th century, Arka was used primarily for leprosy and other skin diseases and abdominal diseases including ascitis. The bark, root, dried sap were specific in secondary syphilis and venereal sores. The herb is known as "vegetable mercury" in India.

In traditional medicine, the rootbark was also used as an antispasmodic and intestinal sedative for allaying spasm, tanesmus and irritation associated with dysentery and diarrhoea. Arka Vati is a compound for stimulating digestion and Ravimoolaadi Vati for dysentery and cholera (Siddhabhesaja-manimaalaa).

As flowers expel phlegm, they are used in asthma.

Calotropis gigantea white flowers, leaf juice and black pepper form Suvaasa Kudori (Siddha vaidya Thirattu) of Siddha medicine, prescribed in bronchitis, asthma with Taalisapatra leaf juice.

Unani physicians include the herb in their prescriptions for cholera; hepatic and splenic enlargement.

Active principles and pharmacology

All parts of the plant of both the species yield latex. It contains cardiac glycosides, calotropin, calotoxin, uscharin, calactin and uscharidin; alpha- and beta-amyrin and beta-sitosterol. Gigantin has been reported only in Calotropis gigantea. A bacteriolytic principle, capable of lysing Micrococcus lysodeikticus, was also found in the latex. A non-toxic proteolytic enzyme, calotropain (2–3 %) has been isolated from the latex; it is more proteolytic than papain, ficin and bromelain, coagulates milk and digests meat, gelatine and casein.

Calotropain has marked anti-blood coagulating activity which does not last long but is desirable in drugs used for treating coronary thrombosis.

The latex also contains uzarigenin, syriogenin, proceroside, sterols, calctin and voruscharin. The latex is a promising anti-inflammatory agent. It also shows anthelmintic activity. It acts as a drastic purgative and emetic.

The leaves contain calotropin and calotropagenin.

The rootbark, in small doses, is reported to give relief in diarrhoea and dysentery; also in cough and asthma; externally in leprosy and eczema. Prolonged high doses cause headache, burning in micturition and leucorrhoea.

The ethanolic (50 %) extract of the root exhibited anti-cancer activity. Calotropine showed an anti-tumour effect in vitro on human epidermoid carcinoma cells of the rhinopharynx.

The flowers possess digestive and tonic properties and are found useful in cough, cold, catarrh and asthma; latex of the flowers in eczema. The flower tops, on hydrolysis, yield holarrhenine and pyroterebic acid. They also contain cyanidin-3-rhamnoglucoside and alkaline phosphatase. Extracts of flowers showed antibacterial activity against Micrococcus pyogenes var. aureus, Treponema pallidum, Neisseria gonorrhoea, Salmonella typhi and Shigella dysenteriae. It caused widespread testicular necrosis and damage to liver when administered orally to desert-gerbil.

Tincture of the root-bark and tincture of the milky-juice are used in homoeopathy with marked success in the treatment of syphilis; also in elephantiasis, lupus and leprosy.

Caution

The drug is highly toxic. Higher doses cause vomiting, diarrhoea, bradycardia and convulsions.

Cannabinaceae

CANNABIS

Cannabis sativa L.

Figure 1 Cannabis sativa (female) [WOI]

Cannabis indica Lam.

Habitat

Cultivated all over India; found wild on the Himalaya from Kashmir to east of Assam.

Figure 2 Cannabis sativa (male) [WOI]

Classical & common names

Ayurvedic: Vijayaa, Bhangaa, Maadani, Maatulaani, Trailokya vijayaa, Tribhuvana vijayaa, Ganjaa.
Unani: Qinnab.
Siddha: Kanja.
English: Indian Hemp; Marijuana.

Parts used

Leaf.

Dose

125–250 mg of the drug in powder form.

The dried, broken flower-heads with chaff, leaves and fruiting shoots constitute Bhangaa.

Ganjaa consists of unfertilized resinous brown-green flowering heads or branches or shoots of female plant, pressed into compact masses.

Charas is cannabis resin.

Bhangaa is consumed orally, while both Ganjaa and Charas are usually smoked.

In Chinese medicine, all parts of the plant, including seeds, are used.

Classical use

The drug was not used in prescriptions by Charaka and Sushruta. It was basically used as an exhilarant.

According to Ayurvedic texts of the later period, Bhangaa entered into Indian medicine for its intoxicating, hot, digestive stimulant, diuretic, expectorant and aphrodisiac properties (Dhanvantari Nighantu). The drug was given, in prescriptions, for diarrhoea, dysentery and insomnia (Bhaavaprakaasha); also in skin diseases, internally as well as externally (Vrindamaadhava, Bangasena). Bhangaa was an ingredient in a herbal vaginal pessary which was prescribed for contracting vagina (Yoga Tantra).

In folk medicine, the juice of the leaves or their paste is applied to the head for removing dandruff and vermin. The powder of the leaves is dusted over fresh wounds for promoting granulation. A poultice of the plant is applied to local inflammations, neuralgia, haemorrhoids.

Madan Modaka and Kameshwar Modaka, sweet preparations with Bhangaa as an important ingredient, were prescribed in sexual debility; Jaatiphalaadi Churna (Bhaishajya Ratnaavali) in dysentery and colic.

Majoon-e-Falaksair, a classical Unani compound, is available over-the-counter and is prescribed in premature ejaculation and spermatorrhoea.

Active principles and pharmacology

The plant contains over 60 alkaloids of different types of cannabinoids, of which 9-tetra-hydrocannabinol (THC) is the chief active agent. Volatile oil is of a very complex composition, contains, among others, beta-caryophyllenes, humules, caryophyllene-oxide, alpha- and beta-pinenes, limonene, myrcene, beta-ocimene. The leaves contain flavonoid glycosides. Cannabis sativa is the only plant to contain THC, one of the main psychotropic constituents, which makes it a promising psychotropic drug in neurological disorders.

Antiemetic action of the drug has been reported in clinical studies involving cancer patients receiving chemotherapy.

Cannabis inhibited MES and metrazol-induced convulsions in albino rats. Anticonvulsive action of THC reduced the clinical and electrographic convulsion intensity in cats.

Cannabinoids increased heart frequency, peripheral vasodilatation caused an increase in systolic pressure in the prone position and a decrease in the supine position.

THC displayed analgesic characteristics in experiments, while, at the same time, partially increasing sensitivity to pain.

The inhalation of Marijuana smoke caused bronchial dilatation in healthy subjects.

Cannabis reduced intraoccular pressure; during the tests, intraoccular pressure dropped by 45 %.

The immune system was significantly suppressed after cannabinoid administration, in vitro, and in animal testing.

THC exhibited antibacterial effect against streptococci and staphylococci.

Most of the cannabinoids act on the CNS. The multiplicity of effects, possible interaction with cell-wall lipids or effects on prostaglandin-biosynthesis are yet to be established.

Active principles of the drug indicate its use in the following areas: painful disorders of the alimentary canal; respiratory disorders; neuralgia; migraine, mental disorders such as anxiety, neurasthenia, hysteria; urinary tract disorders; glaucoma.

The results of chronic abuse are: laryngitis, bronchitis, apathy, psychic decline and disturbances of the genital functions. Prolonged use produced sexual impotency.

Use in Western herbal

In medieval herbals, Cannabis was mostly used externally. There were recipes for balms for healing contractures and for cooling poultices for the head and joints and for podagra. In 1845, the herb's use was mentioned for internal administration for angina pectoris, choking fits and gonorrhoea. It was not until the 19th century that the hemp was described as having an euphoric effect; it was in use for insomnia, epilepsy as a sedative and mild soporific; also for neuralgia, rheumatism, acute bronchitis, painful gastrointestinal disorders.

At present, the principal use of hemp in Western herbal is for easing pain, inducing sleep and for soothing influence in nervous disorders. It is given in delirium, migraine, insanity, infantile convulsions, menstrual pain and cramps.

Hemp is used as an analgesic for cancer- and AIDS-patients undergoing chemotherapy. It reduces neurological overactivity and muscle spasm in patients suffering from multiple sclerosis, cerebral palsy and other muscular affections.

A synthetic cannabinoid (nabilone) in capsule-form is prescribed by physicians of modern medicine for the relief of nausea and vomiting in cancer patients on chemotherapy; also for glaucoma in which pressure within the eye is abnormally high.

In homoeopathy, the tincture of male and female flowering tops is prescribed in headache, vertigo, hysteria, obstinate constipation, nephritis, cystitis, painful micturition, gonorrhoea, amenorrhoea, infantile leucorrhoea, affections of the male sexual organs, and repugnance to coition.

Use in Chinese Herbal:

Cannabis was first mentioned in the pharmacopoeia of the Chinese Emperor about 3000 years ago.

Every part of the hemp plant is used in Chinese medicine the dried flowers, the achenia, the seeds, the oil, the leaves, the stalks, the root and the juice.

The flowers are recommended in menstrual disorders and in wounds.

The achenia, considered to be poisonous, is prescribed in nervous disorders, especially those marked by local anaesthesias.

The seeds or the white kernels of the achenia, are considered to be tonic, demulcent, laxative, diuretic, anthelmintic, alterative, emmenagogue. They are said to strengthen the body and prevent ageing. They are prescribed internally in fluxes, post-partum affections, constipation, obstinate vomiting, poisoning. Externally they are used for eruptions, ulcers, wounds, falling of the hair. Their oil is also used for hair loss.

The leaves are considered to be poisonous. Freshly expressed juice is used as an anthelmintic, to stop the hair from falling out and to prevent it from turning grey. They are also used for their antiperiodic properties. The stalk or its bark and the juice of the root is considered to be diuretic and is used with other drugs in gravel.

Caution

The herb is subject to restrictions in many countries.

In Ayurvedic medicine, Haritaki (Terminalia chebula Retz.) is also known as Vijayaa (first mentioned by Sushruta). It is a totally different drug.

Capparidaceae

CAPPARIS

Capparis decidua Edgew.
Capparis aphylla Roth.

Habitat
In the deserts, especially in Rajasthan, Punjab and Sindh; southward to Karnataka.

Classical & common names
Ayurvedic: Karira, Krakar, Apatra, Granthil, Marubhooruh, Karila.
Unani: Kabar-ul-Hind (Arabic), Kabar-e-Hindi (Persian); Tenti.
Siddha: Chhengan.
English: Caper Berry.

Parts used
Root-bark, fruit, stem-oil.

Dose
Powder 50–125 mg.

Classical use
The tender fruits, steamed and dried, were prescribed with fatty curd in bleeding piles; alkali of the fruit in dry piles and powdered fruit in oedema (Siddha-bheshaja-manimaalaa).

In inflammations and skin affections due to vitiated blood, the oil extracted from the fresh stem of Karira is rubbed on the affected part (ibid).

In folk medicine, the paste of the root-bark is applied to malignant tumours, boils, eruptions, rheumatic swellings and gout; infusion is prescribed internally as an antitoxic agent.

In Unani medicine, in addition to its therapeutic application in traditional medicine, the drug is used as a cardiac and nervine tonic, and for external application to indolent ulcers.

Active principles and pharmacology
A number of phenolic acids is reported in the plant.

The fruit-husk and flowers contain an alkaloid, l-stachydrine. n-pentacosane, n-triacontane. They also contain n-triacontanol, n-nonacosane, pelargonidin-3-galactoside, glucocappasalin, beta-sitosterol and phthalic acid. The flowers yield a steam-volatile sulphur compound (0.4 %), which is active against several microorganisms.

The roots contain the indole bases capparin, capparilin and capparinin. The root-bark contains l-stachydrine, capparidisine, capparisine, capparisinine, n-pentacosane, n-triacontanol and beta-sitosterol.

The protein content and mineral constituents of the fruits are comparatively much higher than common fruits like banana, grapes, guava and mango.

The bark is found alexiteric, anthelmintic and useful in inflammations, cough, asthma. A paste of the coal from burnt wood also possesses antiseptic and anti-inflammatory properties and is applied to muscular injuries.

The stem-bark is acrid, laxative and diaphoretic, efficacious in rheumatism and fevers.

Tender leaves exhibit anti-inflammatory properties and are applied on boils and swellings; decoction is used for pyorrhoea.

The alcoholic extract of flowers, fruit-husk and seeds showed antibacterial activity, particularly against Vibro cholerae Pacini, whereas the aqueous extract is anthelmintic.

Capparis sepiaria Linn.

Capparis spinosa Linn.

Habitat
Dry, rocky and stony soil of north-western India, through Punjab, Rajasthan to the South.

Classical & common names
Ayurvedic: Himsraa, Kaakdanni, Kabara.
Capparis sepiaria Linn. is also equated with Himsraa.
Ahimsraa and Himsraa are considered as synonyms; Capparis zeylanica Linn. is used as Ahimsraa.
Unani: Kabar (Persian).

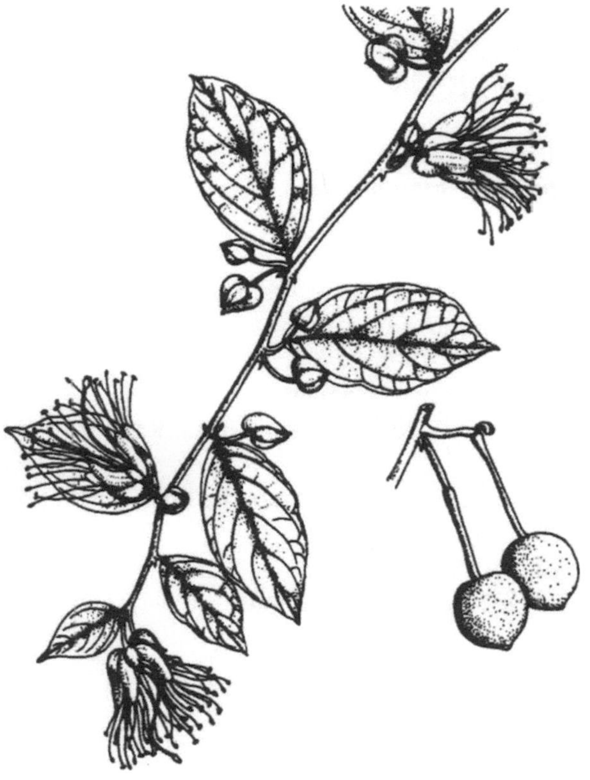

Figure 1 Capparis sepiaria [WOI]

Siddha: Karunsurai.
English: Caper Bush.

Parts used
Root.

Dose
5–10 g.

Classical use
Charaka prescribed the paste of the root in the form of vaginal pessary in disorders of the female genital tract. The paste of Capparis sepiaria root and leaves of Pluchea lanceolata (Raasanaa) was considered an efficacious remedy for swellings (Vrindamaadhava).

In Unani medicine, the decoction of the root-bark is prescribed as a deobstruent to liver and spleen, as anthelmintic, and as an anti-inflammatory agent in enlarged cervical glands, sciatica, rheumatoid arthritis, paralysis. The paste is also applied externally on swellings, skin eruptions and ringworm.

Active principles and pharmacology
The root-bark of Capparis spinosa contains stachydrine, rutic acid and a volatile substance with garlic odour. The root contains indole glucosinolates, glucobrassicin, neoglucobrassicin and 4-methoxy-glucobrassicin.

The bark is bitter, aperient, expectorant and diuretic. It is found efficacious in splenic, hepatic and renal affections.

Stachydrine, when given to dogs, rabbits and rats, quickened the coagulation of blood.

The plant is credited with anti-tubercular and hepatic-protective properties.

Brassicaceae

CAPSELLA

Capsella bursa-pastoris (Linn.) Medic.
Thlaspi bursa-pastoris

Habitat
A weed in cultivated areas and waste places throughout India, particularly in the temperate regions up to an altitude of 4200 m.

Classical & common names
Vernacular: Mumiri (Kumaun region).
English: Shepherd's Purse, St Jame's Wort.

Parts used
Whole plant, aerial parts.

Dose
Fluid extract (1:1) 0.5–2.0 ml.

Use in traditional medicine
The herb or its juice extracts are employed to check menorrhagia and haemorrhages of diverse origin, especially from renal and genito-urinary tract. The herb is found devoid of side-effects, such as headache and dyspnoea.

In atrophy of the limbs, the plant is rubbed on the affected parts with an alcoholic extract of it made with Alchemilla vulgaris (Lady's Mantle, occurs in Kashmir). An ointment is made from the herb for wounds.

The herb is also used in dysentery, diarrhoea, and as a diuretic in dropsy.

The plant is considered a substitute for Hydrastis canadensis (Golden Seal), but is a weaker styptic.

Active principles and pharmacology

The plant contains glucosinolates, sinigrin, flavonoids including rutin, luteolin-7-rutinoside; caffeic acid derivatives, including chlorogenic acid. Cardioactive steroids are found presumably only in seeds.

The herb exhibited astringent, urinary antiseptic and styptic properties, and was thoroughly screened pharmacologically and clinically.

Use in Western herbal

The herb has been used in English domestic practice from early times. Dr Ellingwood wrote of Shepherd's Purse: "This agent has been noted for its influence in haematuria...soothing in irritation of the renal and vesical organs. In cases of uncomplicated menorrhagia, it has accomplished permanent cures. The agent is also useful where uric acid or insoluble phosphates or carbonates produce irritation of the urinary tract." (Cited by M. Grieve.)

All over the West, the herb continues as one of the best remedies for preventing or arresting haemorrhage. It has long been a specific treatment for heavy uterine bleeding, and is found free from toxicity. It is used for all kinds of bleeding, from nose-bleeds to blood in urine. As an astringent it is given for disinfecting the urinary tract in cases of cystitis.

For heavy periods, cystitis and chronic diarrhoea, a strong infusion of fresh or freshly dried herb or its tincture is given.

Dried leaves 1.5 g by infusion, tincture (1:5) 4–6 ml, fluid extract (1:1) 0.5–2.0 ml and powdered solid extract (4:1) 250–500 mg is prescribed 3 times a day for the treatment of menorrhagia.

Shepherd's Purse and compound tablets with Arctostaphylos uva-ursi are available over-the-counter in the West.

Solanaceae

CAPSICUM

Capsicum annuum Linn.

Habitat

Largely cultivated for its fruit throughout the plains of India.

Classical & common names

Ayurvedic: Raktamaricha, Lankaa, Katuvira (non-classical).
Unani: Filfil-e-ahmar (Arabic), Filfil-e-surkh (Persian), Surkh-mirch.
Siddha: Milakkay.
English: Red chilli, Cayenne.

Parts used

Fruit.

Dose

Powder 30–60 mg.

Use in traditional medicine

The drug was not used during the period of Charaka and Sushruta. Its first reference is found in Aatreya Samhitaa. The drug was prescribed to revive patients with a very weak or absent pulse and in a condition of near-total collapse.

A gargle made of red chilli powder (10 g to 1 litre boiling water) is prescribed by Indian physicians in hoarseness, sore throat and in putrid throat. The drug shows powerful action on mucus membrane.

Capsicum, made into lozenges with sugar and gum tragacanth, is given for hoarseness.

Pills of Capsicum, Rheum emodi (rhubarb) and Zingiber officinale (ginger) (all in equal quantity), are prescribed by Indian physicians in hoarseness, sore throat and in putrid throat.

The whole plant, steeped in milk, is pounded and applied externally for reducing swellings and hard tumours.

The paste of Capsicum, as a single raw drug, is considered anti-poisonous and is applied to the dog-bite as a first-aid.

Mixed with Brassica (mustard) oil, the drug is applied to skin diseases. In the form of ointment, is used externally in rheumatism, lumbago, neuralgia, varicose veins.

In Unani medicine, the drug is also prescribed in impotency, spermaturia, catarrh of the prostate, and as an aphrodisiac to increase the blood flow to the genitals and for orgasm.

Active principles and pharmacology

The dried ripe fruits contain alkaloids (including capsaicin); carotenoid pigments (including capsanthin, capsorubin), flavonoids; ascorbic acid; volatile oil. Steroidal saponins (only in seeds).

Capsaicin is known to mimic the effect of some of the prostaglandins.

Capsaicin makes the fruit a most powerful and persistent heart stimulant. It increases arterial force, enlarges its calibre and slightly increases its frequency. Its influence is quite pronounced and reaches every organ through its primary influence upon the circulation.

The drug stimulates digestive secretion and increase the sterilizing capacity of the stomach against potential enteric infections.

Capsaicin produced a protective effect in rat lung and liver by strengthening of pulmonary antioxidant enzyme defence system. It "activates" sensory neurons causing release of substance P and other neuropeptides; at the same time, it causes desensitization of the respiratory tract mucosa to a variety of lung irritants.

Capsicum is used to provoke a subfebrile, catarrhal, congestive condition into a febrile one.

Skin creams containing capsaicin improve healing, reduce pain in neuralgia and diabetic neuropathy. Added to a prescription, Capsicum ensures that the ingredients are propelled vigorously into all tissues against the poorest or the worst cold conditions. It brings to cold, congested or cyanotic tissues improved arterial blood supply and removes toxins.

Applied externally in the form of ointment or plaster, Capsicum produces an appreciable counter-irritant effect.

Capsaicin stimulates and then blocks small-diameter pain fibres by depleting them of a neurotransmitter substance called substance P. Substance P is considered to be the principal chemomediator of pain impulses from the periphery. In addition, substance P has been shown to activate inflammatory mediators in joint tissues in osteoarthritis and rheumatoid arthritis. (Michael T. Murray.)

Excessive substance P levels in the skin have been linked to psoriasis. Capsaicin 0.025 per cent cream demonstrated significant relief in pruritus and reductions in scaling and redness.

Use in Western herbal

Capsicum annuum is used mostly in creams for painful muscle spasm. Prophylactically used internally for artherosclerosis, stroke and heart diseases.

It is also used as a female orgasm stimulant.

German Commission E monograph recognized the drug's efficacy in:

- Muscle tensions
- Rheumatism.

For sore throat, a pinch of chilli powder, added to 25 ml lemon juice and honey, diluted with hot water, is used as a gargle. 5–10 drops of tincture in half a tumbler of warm water is also used as a gargle in throat affections.

Half a teaspoonful of the herb is added to a cup of boiling water, then one tablespoonful of this infusion is diluted with more water to make a cupful. This is sipped for chills, cold hands and feet, shock or depression.

Tincture (5–10 drops), diluted in half a cup of hot water, is used as a circulatory stimulant.

One herbalist (Kitty Campion) recommends: "In case of heart attack, in an emergency situation while waiting for the ambulance/doctor, administer 3 teaspoonful of cayenne in warm water initially (to be drunk all at a time) and then half a teaspoon every 15 minutes."

In general, cayenne pepper reduces the likelihood of developing atherosclerosis by reducing blood cholesterol and triglyceride levels. It also reduces platelet aggregation and increases fibrinolytic activity. These properties indicate its use in cardiovascular affections.

For topical application, commercial ointments containing 0.025 or 0.075 per cent capsaicin are available over-the-counter. These are used in pain disorders, diabetic neuropathy, cluster headache, migraine, psoriasis, herpes zoster, and trigeminal neuralgia.

Capsaicin taffy is prescribed in mouth sores due to chemotherapy or radiation, as it desensitizes the neurons.

In the US, cayenne capsules (520 mg) and cayenne tincture are available in health stores. More than 50 compound tablets for indigestion, flatulence, cough, cold, influenza, arthritis are also sold. For external use, many ointments and sprays contain cayenne.

Caution
Contraindicated in hypertension, hyperacidity, peptic ulceration.

Caricaceae

CARICA

Carica papaya Linn.

Habitat
Cultivated in Uttar Pradesh, Punjab, Rajasthan, Gujarat, Maharashtra and South India.

Classical & common names
Ayurvedic: Erand-karkati (non-classical); Papitaa.
Unani: Pipitaa.
Siddha: Pappai.
English: Papaya.

Parts used
Fruit, leaf, milky juice, seed.

Dose
Leaf-decoction 40–80 ml, milky juice 3–6 g. Papain 600 mg–1.2 g, seed powder 500 mg–1 g.

Use in traditional medicine
Papaya was introduced into India in the 16th century, as such it was not used in classical Ayurvedic medicine.

Ripe fruit is eaten for correcting habitual constipation. It is advised in bleeding piles, dyspepsia, diarrhoea. Juice of the pulp is applied for removing freckles.

Milky juice of the unripe fruit, mixed with honey and followed by castor oil, is prescribed as an anthelmintic for round worms. It is also applied to ringworm.

Unripe and green fruit is made into curry and eaten by women to stimulate secretion of milk.

Dried and salted fruit, also juice of the seeds, is given in enlarged spleen and liver. Flowers are also given in jaundice.

The leaves, dipped in hot water or warmed over a fire, are applied to painful parts in nervous pains. Bruised leaves are applied as poultice to reduce elephantoid growth. The extract of seeds in oil is applied externally on diseases of the skin, and massaged in hemiplegia.

The seeds and milky juice are abortifacient, used as emmenagogue.

An infusion of flowers is used for its emmenagogue, febrifuge and pectoral properties.

Active principles and pharmacology
Green parts of the plant and seed contain an alkaloid, carpaine. Seeds also contain carpasemine. Latex contains enzymes, papain and pseudocarpaine. The fruit contains proteolytic enzymes, papain and chymopapain and traces of the alkaloid carpaine. Papain is a protein-dissolving enzyme that aids digestion.

Seeds showed abortifacient and antifertility activity. The latex showed anticoagulant activity. It has been found to increase the prothrombin and coagulation time in dogs, rabbits, rats and mice. In higher doses, it proved heart depressant and as a spasmogen on the smooth muscle of guinea-pig ileum. The alkaloid solution showed depressant action on heart, blood pressure and intestines.

The seeds exhibited anthelmintic action, due to carpasemine, against Ascaris lumbricoides.

Papain is a pharmacologically and clinically tested proteolytic enzyme, used in dyspepsia, digestive disorders, enlargement of liver and spleen, chronic infantile diarrhoea, oesophagal obstruction, andbleeding piles. It expels round worms but not tapeworms. It is also efficacious in topical application in the treatment of corns, warts, freckles, pimples, various skin lesions.

Experiments have shown that papain has an oedema reducing effect. In an animal study, papaya exerted a direct effect on stomach, helping to prevent ulcers. This finding suggests that papaya may prove of special benefit to arthritis sufferers who take high doses of aspirin, and to people with inflammatory conditions who take steroids.

Carpaine is found to possess anti-tumour activity in vitro against mouse lymphoid leukemia L 1210, lymphocytic leukemia P 388 and Ehrich ascites tumour cells.

Intraperitoneal administration of crude papain in pregnant rats during the initial days of gestation produced teratogenic and embryotoxic effects in foetus and significant changes were observed in cytoarchitecture of the placenta. The brain, liver and kidney of the rat foetus showed frequent haemorrhages and the kidneys were enlarged in treated cases.

Papain is reported to be allergenic, sometimes causing paroxysmal cough, vasomotor rhinitis, bronchial asthma, and urticaria.

Papaya latex may cause gastritis.

If chickens and pigs are fed on the leaves, it will make their flesh tender. When cooked with the leaves, the toughest meat becomes soft.

Use in Western herbal

Papaya was not used in traditional herbal medicine. But in the last 25 years or so, papaya's leaves and latex have become available through speciality herb outlets.

Papain is used as an enzyme supplement in the form of enteric-coated tablets so that they pass safely through the stomach into the intestinal tract.

After 1982, another papaya enzyme, chymopapain, is used in the treatment of herniated (slipped) vertebral disk in the back. Injected directly into the affected area, the chymopapain helps dissolve cellular debris.

Contemporary herbalists recommend the papaya fruit and infusion of leaves as digestive aids for stomach upset; to eliminate intestinal worms; for ulcers in the gastro-duodenal area; and for pancreas excretion insufficiency. The flowers, in an infusion, are used for inducing menstruation.

Over-the-counter, Papaya Plus tablets, Chewable Bromelain (Pineapple enzyme) with papain, are available in the US. A chewing-gum with papain is also sold.

Caution

Contraindicated during pregnancy. An increase in haemorrhaging tendency is not to be ruled out in coagulation disorders. Papaya extract has been shown to increase the international normalized ratio (INR) levels when used in conjunction with warfarin.

Allergic reactions, as mentioned in the text, are possible.

Carica papaya is called Papaw in the West. It should not be confused with Uvaria triloba (custard apple), which is also known as Papaw.

Asteraceae

CARTHAMUS

Carthamus tinctorius Linn.

Habitat

Tropical and sub-tropical parts of India; cultivated in Uttar Pradesh, Madhya Pradesh, Andhra Pradesh, Maharashtra and Karnataka.

Classical & common names

Ayurvedic: Kusumbha, Vahinshikha, Vastraranjaka, Kusum.
Unani: Qurtum.
Siddha: Chendurakam.
English: Safflower.

Parts used

Leaves, seeds, oil.

Dose

Powder 1–3 g, oil 2.5–5 ml.

Classical use

Charaka gave the seeds, in prescription, for calculus, gravels and dysuria.

Sushruta used the oil, in prescription, for polyuria.

In traditional medicine, the plant boiled in sesamum oil is topically massaged over rheumatic and painful joints, and paralytic limbs; also applied to ulcers and skin eruptions. Charred oil is also used for healing sores and in rheumatism.

Flowers are used as stimulant, sedative and emmenagogue. A hot infusion of dried flowers is given as a diaphoretic in nasal catarrh and muscular pain; a cold infusion as a laxative. Flowers, in prescription, are also given in jaundice.

Patients with hypertension and heart ailments use the refined oil in cooking, as it is rich in polyunsaturated fatty acids.

In Unani medicine, the seed is credited with expectorant and anticatarrhal properties, and is prescribed with honey for expelling phlegm. The seed also finds an important place in compound formulations for amenorrhoea and sexual debility. Jawarish Qurtum and Majoon Qurtum are available over-the-counter.

Active principles and pharmacology

The oil contained 13.9 to 32.5 % fatty acids; saturated 5.7–13.4, oleic 8.2–37.6, linoleic 56.7–78.4, and linolenic 1.0 %. Oleic acid type new varieties contain about 75 % of oleic acid; tocopherols; also campesterol, stigmasterol, delta-stigmastenol and beta-sitosterol.

The plant contains a propanetriol derivative which can be used for the treatment of circulatory disorders. Seedlings contain luteolin-7-glucoside. Seeds exhibit mild mutagenic activity.

The flower extract exhibited anti-inflammatory, sedative and analgesic property and inhibitory effect on spontaneous motor acitivity.

Dry flowers contained crude protein 13.3, fat 5.2, fibre 14.8 and ash 7.7 %; vitamin E 16.3 mg/100 g; and beta-carotene. The flowers also contained luteolin-7-O-glucoside, cholesterol, stigmasterol, beta-sitosterol and its beta-D-glucopyranoside; polyacetylenes, mixed esters of phytol, free fatty acids and hydrocarbons.

Green leaves (edible portion 66 %) contained moisture 91.1, protein 2.5, fat 0.6, carbohydrates 4.5 and minerals 1.3 g/100 g.

Analysis of the seeds gave moisture 5.5, protein 13.5, carbohydrates 17.9, fibre 34.9 and minerals 2.68 g/100 g.

The seed lowers blood serum cholesterol; also possesses purgative and diuretic properties.

Use in Western herbal

The use of Carthamus tinctorius, in the West, is based on its use in Chinese medicine.

In Chinese medicine, the flowers are given to stimulate menstruation and to relieve abdominal pain. The flowers are also used to cleanse and heal wounds and sores. An infusion is given to children and infants in measles, fevers and eruptive skin affections.

Chinese research indicates that the flowers and the oil can reduce coronary artery disease and lower cholesterol levels. The oil has been shown to stimulate immune function in mice.

The extract from flowers is used in China for treating blood-clotting disorders.

John Heinerman has quoted clinical experience of a naturopathic physician who prescribes Safflower capsules (4 daily), for controlling hysteria and seizures.

In the US, Safflower oil capsules (250 mg) are available in health-food stores. Efamol (Evening Primrose Oil with Linseed and Safflower) is also sold.

Caution

The flowers or seeds are contraindicated during pregnancy. Refined oil is considered safe.

Apiaceae

CARUM

Carum carvi Linn.

See figure 1.

Cuminum cyminum L.

Habitat

Carum carvi is equated with Caraway. Cultivated in the hills and plains of North India and in the hills of South India.

Cuminum cyminum is equated with Cumin. Cultivated thoughout India, except Assam and West Bengal.

Classical & common names

Ayurvedic: Caraway—Krishna Jiraka, Krishna jiraa, Sugandh, Udgaar, Shodhan.

Figure 1 Carum carvi [Woi]

Cumin—Jiraka, Karam, Ajaaji, Vilaayati jiraa, Jaaji, Dirghjiraka, Kanaa, Jarana. Safed Jiraa.
Unani: Caraway—Kamoon, Kamoon-roomi. Cumin—Kamoon-abyaj.
Siddha: Semai Seeargam and Seeragam respectively.

Parts used

Seeds.

Dose

Caraway 1–3 g, Cumin 3–6 g.

Classical use

Three varieties of cumin seeds have been mentioned in Bhaavaprakaasha Jiraka, Krishna Jiraka and Kaarvi (Prithvikaa, Upkunchikaa). Actually, Carum bulbocastanum and Carum carvi, both are being used as Krishna jiraka. Kaarvi, Prithvikaa and Upkunchikaa, though called Black cumin, have been equated with Kalaunji (Nigella sativa).

According to Bhaavaprakaasha, all the 3 varieties of cumin seeds are digestive, stomachic, carminative, astringent, uterine disinfectant; useful in diarrhoea, dyspepsia, fevers, inflammations, cough and catarrh.

Charaka prescribed Krishna Jiraka, in prescriptions, for all types of anorexia, and Jiraka as a general tonic, as an aid to virility.

Sushruta gave Jiraka seeds in indigestion, colic pain and in intestinal catarrh, as a potherb in cough and flatulence, as a medicated clarified butter internally in colic and abdominal glands.

According to Vrindamaadhava, Jiraka, taken with jaggery, alleviates irregular fever; in prescription, it is an excellent anti-emetic.

In traditional medicine, Krishna Jiraka and Jiraka, both are used similarly, as an adjuvant in diarrhoea and dysentery, as an emmenagogue, diuretic and galactagogue and in puerperal complications. Externally, both are used in cutaneous affections, inflammations, and piles.

Unani physicians use Carum carvi, but mostly Cuminum cyminum is used as a substitute.

Over-the-counter, Jirakaadyarishta (Ayurveda Saarsamgraha) is available for puerperal diseases; also for promoting quality and quantity of lactation in mothers; Jirakaadyam Churanam (Bhaishajya Ratnaavali) for diarrhoea and dysentery and deranged digestion; Jirakaadyam Ghritam (ibid) for anorexia, indigestion, acidity; Jirakaadya Tailam (ibid) for external application in skin affections.

Jawarish Kamooni, is a reputed Unani compound, for hyperacidity and flatulence.

Active principles and pharmacology

Caraway contains a volatile oil high in carvone (40–60 %), limonene; flavonoids; polysaccharides and a fixed oil.

Cumin seeds contain 2–5 % volatile oil, which consists of 25–35 % aldehydes, pinene and alpha-terpineol. The seeds also contain flavonoids, including apigenin.

The fruits and essential oils of caraway and cumin show antispasmodic, stomachic, carminative, astringent and lactagogue properties. The essential oil of both the species is antibacterial.

Use in Western herbal

German Commission E monograph recognized Carum carvi's efficacy in the following areas:

- Common cold
- Cough/bronchitis
- Fevers and colds
- Inflammation of the mouth and pharynx

- Liver and gallbladder complaints
- Loss of appetite.

Caraway is also used in mouthwashes; to improve lactation in nursing mothers; and as an emmenagogue.

Cumin is considered a close relative of caraway and is used in the same way.

Cumin has now gone out of use in European medicine and has been replaced by caraway seeds, though the older herbalists esteemed cumin superior in comforting carminative qualities to caraway.

Caesalpiniaceae

CASSIA

Cassia alata Linn.

Habitat

Introduced from the West Indies and cultivated in the gardens; also found wild almost throughout India.

Classical & common names

Vernacular: Dadrughna.
English: Ringworm Cassia.

Parts used

Leaves, fresh and dry, root, externally.

Classical Use

The fresh leaf or its juice or the decoction is used in skin diseases like herpes, blotch, eczema, mycosis (washerman's itch) and for treating syphilis and gonorrhoea. The dried leaves are used against leprosy.

The decoction of the leaves and flowers is given internally as a purgative, and in stomatitis, bronchitis, asthma. The leaves, seeds are also used internally as an anthelmintic.

The roots are used externally for treating ringworm.

Active principles and pharmacology

The leaves contain cassiaxanthone, kaempferol and its glycosides and beta-sitosterol.

The young pods contain rhein, emodin and aloe-emodin. The stem and roots contain anthraquinones.

Various extracts from leaves, aqueous, pet-ether, acetone, chloroform, menthol and ethanol, were tested for their antifungal activity in vitro against most active pathogenic fungi (dermatophytes) of man like T. ajelol, T. mentagrophytes, T. verriousum, T. rubrum and T. nanum.

Only the aqueous extracts in 50 and 100 mg/ml concentration exhibited significant activity against all types of dermatophytes, as compared to griseofulvin. (CCRAS.)

Alcoholic extract of the leaves exhibited antibiotic activity against E. coli.

The volatile oil, obtained from chloroform, ethanol extract of dry leaves, was found highly effective against Gram-positive (S. aureus and S. pyogenes) and Gram-negative (E. coli, Kl. pneumoniae and Pseudomonas) bacteria. (CCRAS.)

The stem-bark extract also showed antibacterial activity.

Cassia angustifolia Vahl
Cassia senna Linn var. senna.

Habitat

Cultivated in dry areas of Tirunelveli, Madurai and Trichirapalli districts of Tamil Nadu.

Classical & common names

Ayurvedic: Svarna-pattri, Markandikaa.
Unani: Sannaa, Sanaa-makki (Arabic); Senaai, Sonaamukhi.
Siddha: Nilavarai.
English: Senna; Indian Senna, Tinnevelly Senna; Alexandrian Senna.

Parts used

Leaves, fruits.

Dose

Powder 3–5 g.

Classical use

Senna is an Arabian name and the drug was brought into use by the Arabian physicians Serapion and Mesue. Archiarus was the first of the Greeks who used the herb, recommended not the leaves, but the fruit. Mesue also preferred the pods to the leaves, considered them more powerful, though they are less so, but do not cause griping.

The leaves and pods (shells) are administered in Ayurvedic and Unani systems of medicine as infusion for their purgative properties, combined with carminatives and aromatics.

Among Ayurvedic compounds, Panch Sakaara Churna (Siddha-bhesaja-manimaalaa) is available over-the-counter. Other compounds are Sht-shakaar Churna and Yashtyaadi Churna.

Among Unani compounds, Safoof-e-Mulaiyia and Majoon-e-Senaai are the well-known ones. All these compounds are prescribed in constipation, abdominal bloating, indigestion and colic.

Cassia angustifolia is an ingredient in Nilaavarrai Choornam of Siddha medicine, prescribed as a mild laxative in constipation, distention of the stomach and biliousness.

In India, senna extracts have entered into modern medicine and are available with chemists and druggists Laxsena (Alpine), Pursennid-IN (Novartis), Sofsena (Wander) are a few such compounds. Tablets containing sennoside A and B are also available: Glaxenna from Glaxo, Laxatin from Alembic.

Active principles and pharmacology

The leaves contain glucose, fructose, sucrose and pinnitol; mucilage consists of galactose, arabinose, rhamnose and galacturonic acid. Leaves also contain sennoside-C-(8, 8'-diglucoside of rhein-aloe-emodin-dianthrone). Pods contain sennosides A and B, glucosides of anthraquinones rhein, and chrysophanic acid. Seeds contain beta-sitosterol. The flowers are reported to contain sennosides 2.6 %, pods 3.5 %, and foliage 2.5–4.0 %.

Senna exerts a laxative effect, due to the sennosides and rheinanthrone. The effect is primarily caused by influence on the motility of the colon by inhibiting stationary and stimulating propulsive contractions. This results in accelerated intestinal passage and, because of shortened contact time, a reduction in liquid absorbed through the lumen. In addition, stimulation of active chloride secretion increases water and electrolyte content of the intestines (PDR).

The callus cultures showed a strong antimicrobial activity against Gram-positive bacteria.

Use in Western herbal

Nicholas Culpeper (1616–1656) wrote of it during the 17th century: "Senna cleanses the stomach, purges melancholy and phlegm... good in chronic agues..." (Senna does not treat fever, nor does it purge melancholy.)

Senna pods, or the dried ripe fruits, are official in British Pharmacopoeia.

German Commission E monograph recognized Senna's efficacy in constipation.

The herb is used for evacuation relief in cases of anal fissures, haemorrhoids, after recto-anal operations, and in preparation of diagnostic investigations of gastrointestinal tract.

The infusion of senna or senna tea-bag is prepared with sliced ginger, and one cup is given in the morning. United States Pharmacopoeia prefers coriander to ginger.

Many herbalists give 4 capsules of senna, 3 capsules of cinnamon with a glass of milk. Senna is often combined with psyllium.

In the West, senna capsules (300 mg) and more than 15 variants of compounds tablets are sold over-the-counter. Tincture is also available.

In homoeopathy, the tincture of Alexandrian senna is given in constipation with colic and flatulence; also to treat enlarged liver.

Caution

Contraindicated during pregnancy and while nursing. Not to be used in presence of intestinal obstruction, acute inflammatory intestinal diseases, appendicitis or to children under 12 years of age.

Long-term use may lead to losses of electrolytes, in particular potassium ions, resulting in albuminuria, haematuria and inhibition of intestinal motility. The loss of potassium may potentiate cardiac glucosides and have an effect on antiarrhythmic medication.

When used externally on wounds or burns, as is being advised by some herbalists, may cause severe dermatitis in sensitive persons.

Cassia fistula Linn.

Figure 1 Cassia fistula [ADPS]

Habitat

Found both wild and cultivated throughout India.

Classical & common names

Ayurvedic: Aaragvadha, Chaturaangula, Kritamaala, Kritamaalaka, Karnikaara, Sampaaka, Praagraha, Raajvrksha (Charaka, Sushuta); Nripapaadapa, Raajadruma, Vyaadhighaataka.
Unani: Amaltaas; Khiyaar-shambar (Arabic).
English: Purging Cassia.

Parts used

Fruit (pod) pulp.

Dose

5–10 g.

Classical use

Charaka prescribed the sap, root, bark and leaves, in prescriptions, for misperistalsis in infants, and in chronic skin affections; as diuretic and purgative. In addition, Sushruta used the drug internally in urinary troubles as antiseptic and antitoxic; medicated clarified butter in consumption, asthma, fever, and insanity; externally as an ingredient of a medicinal ointment in malignant ulcers, leprosy and other virulent skin diseases.

According to Charaka, the drug is a non-irritating and mild laxative; it cleanses stomach and intestines (Bhaavaprakaasha). In traditional medicine, the fruit (pod) pulp is given to children above 12 years of age and to patients of chronic constipation as a purgative. Also prescribed in disorders of liver and in biliousness; as analgesic and antipyretic.

Flowers are also used for their astringent, purgative, febrifugal and anti-bilious properties.

The whole seed powder is given in intestinal amoebiasis.

The pulp is an important ingredient in Aaragavadhaadi Kwaath (Bhaishajya Ratnaavali), prescribed in colic pain, acute constipation and fever. Aaragavadhaamritaadi Kashaaya (Sahasrayoga) is a compound for acidity, biliousness and skin affections.

Among Unani preparations, flowers are prescribed as Gulqand-Amaltaas in constipation. Laooq-e-khiyarshambar, with pulp as the main ingredient, is also prescribed as a purgative and for expelling phlegm.

Active principles and pharmacology

The bark contains (+)-catechin, epicatechin, kaempherol, fistacacidin, leucocyanidin, leucopelargonidin trimer, rhein glycoside, hexacosanol, lupeol and beta-sitosterol.

The pulp from the crushed ripe pods is found rich in protein (19.94 %) and carbohydrates (26.3 %). They contain sennosides A and B, rhein and its glucoside, barbaloin, aloin, formic acid, butyric acid, their ethyl esters and oxalic acids. Presence of pectin, tannin, maltose, glucose, fructose, sucrose and a small quantity of volatile oil is also reported.

Flowers gave ceryl alcohol, kaempferol, rhein and a bianthraquinone glycoside, and fistulin. Also obtained was leucopelargonidin tetramer having a free glycol unit. The leaves gave free rhein, its glycosides, sennosides A and B.

Aqueous petroleum and benzene extracts of the plant exhibited purgative property.

Acetone and alcoholic extracts of stem bark and root bark possessed marked antifungal activity.

Ether extract of fruit (pod) pulp was found to have antibacterial effect against both Gram-positive and Gram-negative bacteria (CCRAS). Maximum activity was noticed against S. aureus, S. albus, S. shigae, S. typhi para A and B. 1 g of the extract was more potent than 100 mcg of chloramphenicol in vitro.

The flower extract completely inhibited conidial germination of Bipolaris oryzae in vitro tests. Glycosides CFR I and CFR II showed antifungal activity against dermatophytes.

Pharmacologically, the plant exhibited many therapeutic properties. Bark: anti-dysenteric, decoction in jaundice, skin diseases, leprosy; decoction of stem-bark purgative, stomachic. Pods: official for laxative properties; applied heated to swellings; analgesic, antipyretic, antidysenteric. Leaves: antiperiodic, laxative, antiseptic; externally used in skin eruptions, ringworm, eczema, pruritus. Flowers: antibilious, astringent, antifungal.

Pinaceae

CEDRUS

Cedrus deodara (Roxb.) Loud
Cedrus libani Barrel.

Habitat

All over northern Himalayas, largely cultivated in India as an ornamental tree.

Classical & common names

Ayurvedic: Devadaaru (Charaka and Sushruta); Saptapatraka, Suradruma, Devadruma (Charak), Bhadradaaru, Daaru (Sushruta); Amaradaaru, Daaruka, Devaahvaa, Suraahva, Suraahvaya, Suradaaru, Surataru.
Unani: Deodaar.
Siddha: Devadhari.
English: Deodar, Himalayan Cedar.

Parts used

Heartwood, oleo-resin, oil.

Dose

Powder 3–6 g, decoction 50–100 ml.

Classical use

Charaka gave dried leaves and the extract of the wood and roots internally, in prescription for headache, intestinal worms, urinary diseases, menorrhagia, fistula; in addition, Sushruta gave for oedema, skin eruptions, blood poisoning, chronic dysentery, for purifying breast milk, and as an internal antiseptic.

Externally, Sushruta used the drug as an ingredient of a paste for filaria and swellings; of a massage oil for atrophy of the leg muscle; of a hair oil for baldness; of a medicated oil for wounds.

The oil extracted from the burnt wood was prescribed in cough (Ashtaanga Hridaya); mixed with clarified butter in skin diseases; also in chronic coryza.

Among classical compounds, Devadrumaadi Churnam (Sahasrayoga) is prescribed in diseases of the abdomen, internal swellings and in worm infestations; Mandoor-Vatakaa (ibid) tablet in anaemia, jaundice, anorexia, obstinate oedema, spleen affections, diabetes, skin diseases, piles; Devadarubalaadi Tailam (ibid) in rheumatic affections and gout.

In Unani medicine, the oil, Roghan Deodaar, is massaged over weak organs and inflammations.

The decoction of the bark is used internally as an astringent, febrifuge and antidiarrhoeal; the essential oil externally as an antiseptic in skin diseases, as an anti-inflammatory agent in catarrh and chest affections.

Active principles and pharmacology

The essential oil, extracted from Himalayan Cedrus deodara, contains alpha- and gamma-atlantone, p-methyl-3-tetrahydroacetophenone.

The oil possesses strongly antiseptic, astringent, expectorant, sedative and diuretic properties. When massaged into skin, it is found efficacious in catarrh and chest infections; topically antiseptic to wounds and ulcers.

Stem possesses spasmolytic properties; himachalol from wood is also spasmolytic and hypotensive, increases femoral blood flow. The alcoholic extract of the wood showed marked anti-inflammatory activity in mice; it also showed antibacterial activity.

The wood possesses diaphoretic, diuretic and carminative properties and is found useful in fever, and in pulmonary and urinary disorders. Its paste is applied to temples to relieve headache.

The tannin content of the bark makes it a good astringent in fevers, diarrhoea and dysentery.

Use in Western herbal

Cedar wood oil is used in aromatherapy for catarrhal conditions of the respiratory tract. Bormelin Balm, available over-the-counter, is used as a rub and for inhalation. The oil is also used on eruptions of the skin in the form of 25 % ointment with vaseline.

Celastraceae

CELASTRUS

Celastrus paniculatus Willd.

Figure 1 Celastrus paniculatus [CCRAS]

Figure 2 Celastrus paniculatus [CCRAS]

Habitat

Common all over the hilly parts of the country, up to an altitude of 2000 m.

Classical & common names

Ayurvedic: Jyotishmati, Paaraavat-padi, Kangunika, Vegaa; Maalkaanguni.
Unani: Maalkangni.
English: Staff Tree.

Parts used

Seed, oil.

Dose

1–2 g, oil 5–15 drops.

Classical use

Charaka gave the decoction of the root or seed internally, in prescriptions, as a brain tonic for headache, depression, swooning; as a laxative for cleansing digestive system.

Sushruta prescribed the oil of seeds internally in neurological disorders, urinary infections, skin affections, intestinal parasites; externally for wound-healing; leaves internally, as a purgative.

Chakradatta recommended fried leaves of Jyotishmati for inducing menstruation. The juice of leaves was also given in opium poisoning and as a deaddiction aid.

In Ayurvedic classics the intellect-promoting and memory enhancing properties were attributed to the oil of Jyotishmati. The oil, mixed with cow's ghee (clarified butter) was prescribed internally in neurological disorders, and as a brain tonic (Bhaavaprakaasha, Raaja Nighantu, Shivadatta Nighantu).

In folk medicine, the oil is used externally on sores, ulcers, wounds, scabies, eczema and in rheumatism, gout, and paralysis. The crushed root is used in pneumonia. The paste is also applied to swollen veins.

Jyotishmati Tailam (Ayurvedic) and Raughan-e-Maalkangni (Unani) medicinal oils are available over-the-counter, and are prescribed in neurasthenia, hemiplegia, Bells Palsy, lumbago, and gout.

Active principles and pharmacology

Major constituents are: fatty oil (42–45 %) with palmitic (9 %), oleic (1.9 %), linoleic (10 %) and linolenic (5 %) acids and their glycerol esters mainly alpha, alpha-1-dipalmitoylglycerol. Also, a number of sesquiterpene polyesters (mainly malkangunin) esterfied with one or more acetic, benzoic, beta-furanoic and beta-nicotinic acids; sesquiterpene alkaloids viz. celapanin, celapanigin and celapagin; quinone-methide and phenolic triterpenoids, viz. celastrol, pristimerin, zeylasterone and zeylasteral.

The alcoholic extract of the whole plant exhibited CNS effect without loss of righting reflex in albino rats and mice, potentiated pentobarbital narcosis, abolished CAR and reduced amphetamine-induced group toxicity, possessed central muscle relaxant activity and anticonvulsant activity against metrazol (a chemical convulsant); found effective against exudative and proliferative phase of inflammation (CCRAS).

The extract exhibited fall in blood pressure and decrease in rate and depth of respiration in experimental animals.

The oil obtained from seeds exhibited tranquillizing, sedative and anticonvulsant activity on rats; hastened the process of learning and memory in experimental animals.

Thus the drug justified its claim as an anti-inflammatory, tranquillizing, central muscle relaxant and anti-convulsant agent.

Pet.-ether, benzene, acetone and other extracts of the whole plant showed anti-histaminic effect on isolated guinea-pig ileum.

The alcoholic emulsion (96 %) of the oil showed antibacterial activity against Gram-positive and Gram-negative bacteria.

Caution

Emetic and toxic in large doses.

Asteraceae

CENTAUREA

Centaurea behen Linn.

Habitat

Europe, and North Africa to India and China. Imported in India from Iran.

Classical & common names

Unani: Behman safed, Behman abyaz.
English: White Rhapontic.

Parts used

Root.

Dose

5–7 g.

Classical use

In Unani medicine, both the varieties, white and red, are used in aphrodisiac tonics. The red Behman has been equated with Salvia haemotodes (Libiatae).

Both the drugs have been included in Dawa-ul-Misk Motadil Saadaa and Jawahirwaalaa, prescribed in palpitation, weakness of brain, heart and liver, hepatitis, melancholia; in Laboob Kabir and Laboob Sagheer, prescribed for sexual debility, oligospermia, attenuated seman, neurasthenia, polyuria; in Majoon-e-Piyaaz, prescribed in spermatorrhoea and premature ejaculation; in Majoon-e-Saalab, prescribed in attenuated semen; in Manjoon-e-Sangdaanaa Murgh, prescribed for hepatitis, diseases of the stomach and intestines. All these compounds are available over-the-counter and have been included in National Formulary of Unani Medicine.

Both the drugs have been found efficacious in sexual debility, fatigue, low metabolism and are considered as anabolic agents and liver tonic.

The powder of both drugs is given with milk for spermatogenesis by Unani physicians.

Active principles and pharmacology

The roots of Centaurea behen contain behenin, taraxasterol and its acetate, myristate, inulin and a glucoside which on hydrolysis yields centaurea sterol A.

The alcoholic extract of aerial parts of the plant produced hyperglycaemia in anaesthetised rats.

The aerial parts contain cynaropicrin, arguerin B, desacyl-cynaropicin, grosshemin, dehydrosolstitialin A and circimaritin.

Hydrocotylaceae

CENTELLA

Centella asiatica Linn.

Hydrocotyle asiatica L.

Habitat

Occurs in marshy places throughout the country up to 200 m.

Classical & common names

Ayurvedic: Manduukaparni, Manduukaparnikaa, Maanduuki (Charaka, Sushruta).
Siddha: Vauarai.

Figure 1 Centella asiatica [ZANDU]

English: Indian Pennywort, Gotu Kola (Chinese).

Parts used

Whole plant.

Dose

Juice 10–20 ml, powder 3.5 g, root-powder 500 mg–1.5 g.

Classical use

According to Charaka, Manduukparni is age-sustainer and a potent brain tonic. Charaka prescribed decoction of leaves internally for rejuvenation; also for pectoral lesions, ulcers and intestinal affections.

Sushruta advised prolonged use of herb's juice with milk in daily diet for intellectual vigour and longevity; as potherb in chronic skin diseases, cough, fever, urinary disorders and as an astringent. Bhaavaprakaasha recommended the herb in anaemia, dermatosis, ailments of the urinary tract and as an efficacious tonic for improving the receptive and retentive capacity of the mind.

The herb, fried in clarified butter, was given daily for a month as an intellect-promoting tonic (Ashtaaga Hridaya).

Bangasena included the herb in his prescription for insanity, as a nervine tonic.

In Chinese medicine, Gotu Kola, is indicated in premature ageing, memory and learning impairment, mental disorders including neurosis, schizophrenia, epilepsy and convulsions; chronic skin diseases, venereal diseases. The powdered herb, mixed with water or sesamum oil, is prescribed for application to eczema, psoriasis, festering sores and other chronic conditions.

The legendary Chinese herbalist, Lee Ching-yuen, who, according to a folk-lore, died in 1933 at the age of 256, recommended Gotu Kola above all other herbs for prolonging life.

Active principles and pharmacology

Major constituents are: triterpenoid saponins viz. madecassoside and asiaticoside and their aglycones viz. asiatic acid and madecassic acid. Madecassoside 0.7–5 %, asiaticoside 0.1–0.6 %, asiatic acid 0.1–0.5 %, madecassic acid 0.5–0.8 %.

The plant also contains saponins viz, asiaticoside B, brahminoside, brahmoside, centelloside, indocentelloside, thankuniside and isothankuniside; triterpenoid acids viz., brahmic acid, isobrahmic acid, betulic acid, centic acid and centoic acid; flavonoid glycosides viz., 3-glucosylquercetin and 3-glucosylkaempferol; and an alkaloid hydrocotyline.

In samples from Madagaskar, India and Sri Lanka, concentration of triterpenes can vary between 1.1 and 8 %. Most samples yield a concentration between 2.2 and 3.4 %.

The Madagaskar variety is most commonly used to produce standardized extracts internationally.

The antileprotic effects of the whole herb powder have been confirmed clinically. The glycoside (asiaticoside) has exhibited an inhibitory influence on the formation of acid-mucopolysaccharide in humans, which is essential for the carbohydrate metabolism of the leprosy bacillus and hence its growth. The inhibitory action of asiaticoside on Mycobacterium leprae as well as on M. tuberculosis has also been reported.

Centella asiatica and its three components madecassic acid, asiatic acid and asiaticoside, applied locally on wounds in rats, promoted wound-healing, but had no effect when administered orally, while the oral administration of a triterpenoid extract of Centella asiatica was found efficacious.

In a study, the triturated extract of the herb was given to patients with gastric and duodenal ulcers. The healing rate for patients of gastric ulcers was 31.6 %, 73. 7 % and 89.5 % by the end of 4, 6 and 8 weeks respectively. In cases of duodenal ulcer, the healing rate was 38.5 %, 69.2 % and 84.6 % at 4, 6 and 8 weeks respectively. The therapy combined with antacids and cimetidine showed better results.

The triterpenoid fraction of the plant improved the conditions in patients with venous hypertension and post-phlebitic syndrome.

Several investigations have been undertaken to determine Centella asiatica's effecton the brain and the central nervous system. The drug, besides exhibiting a barbiturate hypnosis potentiation, also showed anticonvulsant activity. The sedative action of the drug was attributed to the glucosidal saponins brahmosides, the mode of action was mainly on cholinergic mechanism in the central nervous system.

In a study, the whole dried herb produced significant intellectual improvement and behavioural changes in mentally retarded children. Experimentally, the alcoholic extract reduced acetylcholine content and increased histamine content and catecholamines in the brain, indicating psychotropic and antianxiety effect of the plant.

Use in Western herbal

Gotu Kola tends to be mainly regarded as a nervine tonic in the West and features in stimulating, energy-giving formulations. Capsule (520 mg) and many compounds are available over-the-counter in healthstores. The herb is also available in the form of an ointment; one such branded ointment is Emdecassol. Gotu Kola is particularly preferred as a restorative while treating drug addiction.

See also Bacopa monnieri.

Compositae

CENTRATHERUM

Centratherum anthelminticum (Willd.) Kunze
Vernonia anthelmintica (Linn.) Willd.

Habitat

Throughout India up to 2000 m in the Himalayas and Khasia Hills.

Classical & common names

Ayurvedic: Aranya-Jiriaka, Vanjiraka; Kaalijiri, Somaraaji, Karjiri.
Unani: Kamoon barri (Arabic).
English: Purple Flea-bane; Achenes.
Centratherum anthelminticum has been equated with Somaraaji, also with Baakuchi (Psoralea corylifolia Linn), which a totally different herb, though both are used in chronic skin diseases.

Classical use

The drug was used as an anthelmintic against earthworm and tapeworm, as an emetic and blood purifier; externally as a resolvent of inflammations.

According to Bhaishajya Ratnaavali, 1 g Somaraaji and 1 g black cuminum, with hot water, was prescribed in skin diseases.

Active principles and pharmacology

The seed contains a bitter principle as elemanolide lactone. Active anthelmintic principle has been isolated in 1 % yield as a bitter resin acid. Antiseptic, diuretic and stimulating effect of the achenes was found to be due to 0.02 % essential oil and resins. 60 and 90 % alcoholic extracts showed good anthelmintic action against threadworms, but were not found effective against hook-worms. Ethanolic extract of achenes exhibited good results in giardiasis patients.

Ethanolic (50 %) extract of the seed showed antiviral properties.

Ceratophyllaceae

CERATOPHYLLUM

Ceratophyllum demersum Linn.

Habitat

All over India from temperate to tropics in ponds and lakes.

Classical & common names

Ayurvedic: Shaivaala, Jalnili, Jalaj.

Figure 1 Ceratophyllum demersum [WOI]

Shaivaala has also been equated with Vallisneria spiralis Linn.
Unani: Tuhlub (Arabic); Pashm vazg (Persian).
English: Coontail, Hornwort.

Parts used
Whole plant.

Dose
Juice 10–20 ml, deocoction 50–100 ml.

Classical use
Charaka used the weed, in prescriptions, for haemothermia, giddiness, morbid thirst; externally in rheumatism, erysipelas.

Shaivaal, in Ayurvedic texts, belongs to Bhadrashriyaadi Group, which is prescribed in intrinsic haemorrhages. Shaivaala was an ingredient of a medicated clarified butter prescribed by Charaka in cardiac affections. Sushruta used the weed, in prescription, for spermaturia.

In folk medicine, Shaivaala is used as a styptic; also prescribed in leucorrhoea and venereal diseases. Its ash, mixed with sesamum oil, is applied over discoloured skin. Externally the paste is used for resolving inflammations.

Active principles and pharmacology
The herb is rich in protein (24.6 %), calcium and magnesium. It contains ferredoxin and plastocyanin. Exhibits purgative, antibilious and antimicrobial properties.

The extract of the herb has shown activity against Mycobacterium smegmatis, Candida albicans and Fusarium sambucinum syn. F. roseum.

Vallisneria spiralis L. showed stomachic, antileucorrhoeic and demulcent activities. INSA scientists equated the drug with Shaivaala of Charaka and Sushruta.

The extract of the plant yielded polysaccharides, containing D-galactose, D-xylose, L-arabinose, L-rhamnose, uronic acid and acidic xylan, and an arbino galactan.

Papilionaceae/Fabaceae

CICER

Cicer arietinum Linn.

Habitat
Cultivated in most parts of India.

Classical & common names
Ayurveda: Chanaka, Chanakaa, Harimanth, Vajimanth, Jivan, Sakalpriya.
Unani: Chanaa.
Siddha: Kodalai.
English: Gram.

Parts used
Seed.

Dose
Saktu (Sattu) with water 50–100 ml.

(Saktu is prepared by pounding the presoaked, dried and fried seeds of Cicer arietinum. Mixed with salt or sugar Saktu was prescribed during summer due to its cooling properties in the form of a thin linctus.)

Classical use
Charaka prescribed Chanaka mixed with Ushira (Vetiveria zizanioides) and Dhaanyaka (Coriandrum sativum) in vomiting caused by biliousness.

The Saktu was given in colic as a diet (Gadanigraha). In the case of excessive burning sensation and in fever, the soup of Chanaka was recommended (Bhaavaprakaasha).

Maadhava Dravyaguna attributed antibilious, blood-purifying properties to Chanaka but also considered it punsatvaghna i.e. it causes impotency. This seems to be an erroneous interpretation of the cooling properties of the herb.

Active principles and pharmacology
The foilage contains flavonoids ononin, iso-quercetrin, kaempferol-3-glucoside, astragalin, populnin, biochanin-A-7-glucoside, iso-rhamnetin, protensein and garbanzol. The seeds gave cyanogenic glycosides.

The seed possesses antibilious and diuretic properties, leaf stomachic and laxative. Pangamic acid exhibits anti-stress and antihyperlipidemic properties. The seed is considered as an adaptogenic agent.

Seeds reduced post-prandial plasma glucose.

The nutritional value of gram protein is superior to soybean, cowpea, fababean, mungabean, pea, lentil, pigeonpea. The protein content ranges from 20.9 to 25.27 %.

Low digestibility of seed protein is due to the presence of phytic acid.

Soaking, cooking of soaked seeds, sprouting and autoclaving reduce the phytic acid content and increase the digestibility. Pressure cooking and germination eliminate the alpha-amylase inhibitor activity of gram.

Asteraceae

CICHORIUM

Cichorium intybus Linn.

Habitat

Found wild in Punjab and Andhra Pradesh. Cultivated in Bihar, Punjab, Himachal Pradesh, Assam, Maharashtra, Gujarat, Tamil Nadu. Orissa, Andhra Pradesh and Kerala.

Classical & common names

Ayurvedic: Kaasani.
Unani: Kaasani (Persian).
Siddha: Kasinikkeerai.
English: Chicory; Garden Endive.

Parts used

Whole plant, seed, root.

Classical use

The root is used in Unani medicine as a carminative and bitter tonic. It is prescribed in jaundice, liver enlargement, abdominal inflammations, gout, rheumatic affections and as a diuretic, emmenagogue. Fresh leaves and seeds are also used. The leaves are bruised and applied like poultice.

Araque-Kaasani (water-extract of seeds) is a classical Unani single drug preparation, prescribed in hepatitis, jaundice, polydipsia. Fresh juice of the whole plant is also prescribed (Araque Kaasani sabz, Aab-murabbkin). Seeds have been included in a classical compound, Sikanjabeen Buzoori Motadil, prescribed in hepatic obstruction, fever, retention of urine; in Sharbat Buzoori Motadil, prescribed in amenorrhoea, retention of urine, hepatic obstruction, fever.

Active principles and pharmacology

Flowers and leaves contain escutetin and its glycosides. Roots gave flavonoids, catechol tannins, unsaturated sterols and triterpenoids. Racemose yielded coumarins cichorilin, esculin, 6, 7-dihydroxycoumarin, umbelliferone and scopoletin.

Active principles exhibit cholagogue, hepatoprotective, stomachic, and diuretic properties.

An alcoholic extract of the plant was found to be effective against chlorpromazine-induced hepatic damage in adult albino rats. The alcoholic aqueous extract showed cholagogue activity which may be attributed to polyphenols.

Recently (in 1998), researchers isolated a phenolic compound, esculetin, from the roots and confirmed hepatoprotective activity in mice against paracetamol and carbon tetrachloride-induced hepatic damage.

The extracts of roots were found to be active against several bacteria. Alcoholic extract was found antimicrobial against microorganisms causing gingival inflammation.

Chicory was not found toxic up to 60 g/day, however ingestion of 100 g/day caused slight bradycardia.

Use in Western herbal

According to Pliny (AD 23–79), chicory juice was mixed with rose oil and vinegar as a remedy for headache.

Currently, its main application is as a mild bitter tonic for liver and digestive tract, as a mild laxative and for cleansing the urinary tract.

German Commission E monograph recognized the herbs efficacy in the following areas:

- Loss of appetite
- Dyspeptic complaints
- Liver and gallbladder complaints.

Infusion of 2–4 g of comminuted drug is prescribed.

Whenever deep-fried or fatty meals are consumed, a cup or two of chicory root brew is recommended. A cup of cold root brew is given for settling an upset stomach or correcting acid indigestion and heartburn. Over-the-counter quite a few chicory products are available.

Lauraceae

CINNAMOMUM

Cinnamomum camphora Nees Eberm.

Habitat

Native to China and Japan, introduced and cultivated in India as a source of camphor, in South India, also in Dehradun, Saharanpur and Kolkata.

Classical & common names

Ayurvedic: Karpura (Sushruta); Ghanasaara, Chandra, Chandra Prabhaa, Sitaabhra, Himavalukaa, Himopala, Sitaraja, Himakara, Indu, Tushaara.
Unani: Kaafoor (Arabic).
Siddha: Indu.
English: Camphor Tree.

Parts used

Camphor crystals.

Dose

125–300 mg.

Classical use

Sushruta used camphor internally as a cleansing agent; also in halitosis (offensive odor of the breath).

The camphor, mixed with clarified butter, was applied over wounds for protecting them from infection and pus (Chakradatta, Raajmaarttanda). It was also used in retention of urine (Chakradatta, Bhaavaprakaasha) and asthma (Siddha-bhesaja-manimaalaa).

In folk medicine, camphor liniment is used for relief of pain in muscular rheumatism, sprain, fibrositis and neuralgia.

Camphor, in prescriptions, is given internally as a carminative, reflex expectorant and reflex stimulant of heart and circulation as well as respiration.

Camphor is also used as a sedative and nervous depressant in hysteria, epilepsy and convulsions.

Among over-the-counter classical compounds, Karpuraadi Churnam (Bhaishajya Ratnaavali) is prescribed to treat derangement of the digestive system also dysentery; Karpuraadyam Churnam (ibid), is prescribed in chronic cough and asthma; Karpuraasava (ibid), an alcoholic preparation, is specific for diarrhoea and dysentery. Karpura Rasa (Ayurved Saar Sangraha) is also specific for diarrhoea, dysentery and sprue.

Among Unani compounds, Qurs-e-Kaafoor is prescribed in hectic fever; Araque-e-Kaafoor in diarrhoea and sprue; Marham-e-Kaafoori is prescribed externally for wounds, chronic ulcers, ringworm, warts.

Active principles and pharmacology

The plant contains a volatile oil comprising camphor, safrole, eugenol and terpineol. It also contains lignans. Safrole is thought to be carcinogenic.

The bark of the plant is sedative, antispasmodic, diaphoretic and anthelmintic. The roots are also anthelmintic. The roots and leaves are used for hardness of spleen and liver. Ethanolic extract (50 %) of fruits showed antibacterial activity against several Gram-positive and Gram-negative bacteria. Camphor inhibited the growth of Vibrio parahaemolyticus, which is one of the causative agents of diarrhoea and dysentery. The essential oil from the plant possesses potent antifungal activity against many fungi.

The leaf samples collected from the Himachal Pradesh, during April to December, yield 0.2–0.3 per cent of an essential oil and 0.5–4.5 per cent camphor. The leaves from Cochin are reported to contain more camphor than those from other parts of India.

Camphor is locally rubefacient and anti-inflammatory; also produces a menthol-like cooling which makes it slightly anaesthetic. Inhaled, it stimulates mucus flow and acts to cleanse a congested condition.

Internally, camphor encourages the secretion of saliva and digestive juices, stimulates peristalsis and relaxes sphincters; it thus aids the whole digestive process.

It acts as a circulatory stimulant, dilating the peripheral circulation, but shows a paradoxical action on the heart, stimulating it, if failing, yet dilating the coronary circulation.

Camphor stimulates CNS-function and is an effective counter to barbiturate or morphine respiratory depression.

Camphor in large doses is toxic. Toxicity symptoms are headache, nausea, excitement, confusion and delirium; ultimately there is loss of consciousness. The ingestion of solid camphor by children is a common cause of camphor poisoning.

Use in Western herbal

Camphor is used externally as a bronchial secretolytic; internally as a bronchial antispasmodic, respiratory analeptic and circulatory tonic.

German Commission E monograph recognized camphor's efficacy in the following areas:

- Cough/bronchitis
- Rheumatism
- Hypertension
- Hypotension
- Nervous heart complaints
- Arrhythmia
- Cardiac insufficiency
- NYHA I and II.

Internal average dose is 30 to 300 mg. The lethal dosage for children is approximately 1 g and for adults 20 g. Camphor salves are not administered to infants. Toxic dose of camphor is 2 g.

Cinnamomum spp.

Habitat

Cinnamomum verum/Cinnamomum zeylanicum is native to India, in Western ghats at low levels, and Srilanka; Cinnamomum cassia to China, Indonesia, Viet Nam; Cinnamomum tamala to Himalayas from Kashmir to Bhutan and Khasi Hills at 450–2000 m.

Classical & common names

Ayurvedic: Daalchini, Tvak, Daarusitaa. Tejpatra, Tamaalpatra, Patra (Charaka, Sushruta).
Unani: Daarchini (Persian). Tejpaat, Saleekhaa (Arabic).
Siddha: Kiramboo (Daalchini). Taalishpattri (Tejpaat). Not to be confused with Taalispatra or the leaf of Abies webbiana.
The bark of Cinnamomum verum J.S.Presl./ Cinnamomum zeylanicum Brevn., also the bark of Cinnamomum cassia Blume/ Cinnamomum aromaticum Nees. are equated with Daalchini.
The leaf of Cinnamomum tamala Nees & Eberm. is equated with Tejpaat.

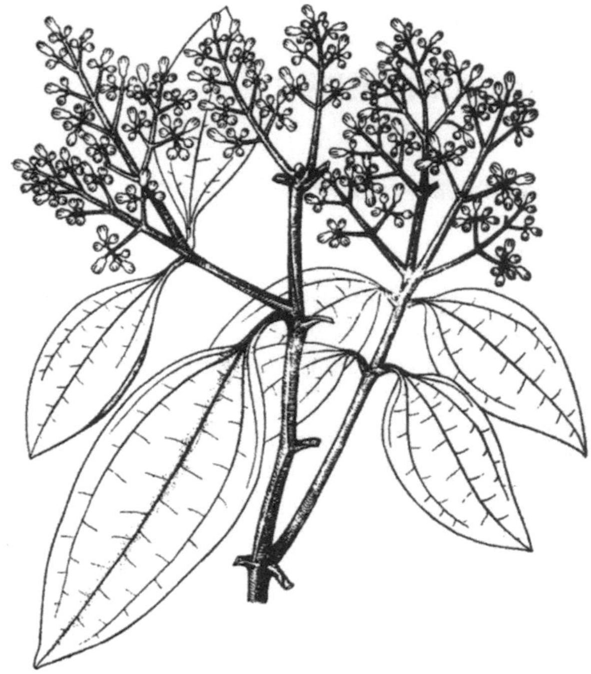

Figure 1 Cinnamomum zeylanicum [WOI]

Parts used

Daalchini: Bark; Tejpaat: leaf.

Dose

Powder 1–3 g.

Classical use

Charaka used the dried leaves and bark of Cinnamomum tamala, in prescriptions, as an aid to rejuvenation, in anaemia, cardiac disorders, fever and externally in body-odour. He included Tvak in a mouthwash.

Sushruta prescribed the leaves internally in skin eruptions and blood poisoning; also the bark of Cinnamomum zeylanicum. He used the dried bark as an antispasmodic; externally as an ingredient of a dusting powder for wounds. For erysipelas, Sushruta incorporated the bark for a medicinal plaster.

A classical compound of Ashtaanga Hridaya, Sitopalaadi Churna, incorporated Tvak, Elaa (Elettaria cardamomum), Pippli (Piper longum), Tavkshiri (Bamboo-manna) and sugar, successively double in quantity, for cough and asthma.

Among over-the-counter classical Ayurvedic drugs, Elaadi Gutikaa (Bhaishajya Ratnaavali) incorporates Tamaalpatra, and is prescribed in chronic and dry cough; Lavangaadi Churna (Shaarangadhara Samhitaa) contains Cinnamomum zeylanicum bark, and is prescribed in anorexia and flatulence.

Among Unani compounds, Iyarij-e-Faiqra contains Daarchini as well as Saleekhaa and is prescribed in rheumatism, hemiplegia, Bells Palsy, as a purgative, cathartic. Daarchini has been used as a single drug in Raughan-e-Daarchini, prescribed externally in rheumatism and headache; internally in sexual debility (2–5 drops) and neurasthenia.

Active principles and pharmacology

The major constituent in both Cinnamomum verum and Cinnamomum cassia is cinnamaldehyde at concentrations of 65–80 % and 90 % of the volatile oil, respectively.

Cinnamomum verum also contains o-methoxycinnamaldehyde. Cinnamomum verum volatile oil contains 10 % eugenol, whereas in Cinnamomum cassia only a trace quantity of the compound is found. Courmarin is present in Cinnamomum cassia (0.45 %) but not in Cinnamomum verum.

Antibacterial and antifungal activities of the essential oil have been demonstrated *in vitro.* The essential oil of Cinnamomum verum is active in vitro against Bacillus subtilis, E. coli, Staphylococcus aureus, Salmonella typhimurium and Pseudomonas aeruginosa. It was also active in vitro against fungi (Aspergillus spp., Cladosporium werneckii, Geotrichum candidum, Kloeckera apivulata, Candida lipolytica and Cinnamomum albicans). The antibacterial and fungicidal effects have been attributed to o-methoxycinnamaldehyde. (WHO.)

The essential oil of Cinnamomum verum exhibited carminative activity and decreased smooth muscle contractions in guinea-pig trachea and ileum, colon and stomach. The active antispasmodic constituent of the drug is cinnamaldehyde. A reduction in stomach motility in rats and dogs, and intestinal motility in mice and a decrease in the number of stress- and serotonin-induced ulcers in mice have been described. An ethanol extract of the drug inhibited histamine- and barium-induced contractions in guinea-pig ileum; the hot-water extract was not active. (WHO.)

Both drugs find their application in dyspeptic conditions; also in the treatment of abdominal pain with diarrhoea and pain associated with amenorrhoea and dysmenorrhoea.

C. tamala leaves yield 0.3–0.6 % essential oil. The oil from bark contains 70–85 % cinnamaldehyde as a major constituent; other constituents include alpha- and beta-pinene, limonene, beta-phellandrene, cymene, ocimene, terpinene, camphor, linalool, borneol, beta-caryophyllene, alpha-terpineol, benzyl cinnamate, benzaldehyde, eugenyl acetate, eugenol and cinnamylacetate.

The leaves are carminative and are used in colic, diarrhoea and rheumatism; with long pepper and honey in cough and cold.

In a study, two teaspoonfuls of leaf powder given to diabetic patients four times a day, accompanied by controlled diet, significantly reduced the blood sugar level and showed distinct hypogycaemic action.

The oil shows antibacterial activity.

The bark of Cinnamomum tamala is one of the common adulterants to the true cinnamon, Cinnamomum verum.

Use in Western herbal

German Commission E monograph recognized the efficacy of Cinnamomum verum in the following areas:

- Loss of appetite
- Dyspeptic complaints
- Cough/bronchitis
- Fevers and cold
- Inflammation of the mouth and pharynx.

The British Pharmacopoeia specifies cinnamon as the dried bark of Cinnamomum verum.

Caution

The drug is contraindicated in pregnancy, stomach or duodenal ulcers and in cases of fever of unknown origin.

Menispermaceae

CISSAMPELOS

Cissampelos pareira Linn.

Figure 1 Cissampelos pareira [CCRAS]

Figure 2 Cissampelos pareira [CCRAS]

Habitat

Commonly distributed throughout tropical and sub-tropical India, ascending up to an altitude 2000 m.

Classical & common names

Ayurvedic: Paathaa (Charaka, Sushruta); Ambashthaa, Varatiktaa, Aviddhakarni, Piluphalaa.
Unani: Paathaa.
Siddha: Appatta.
English: Velvet-leaf Pareira.

Parts used

Root, leaves.

Dose

Powder 1.25–3.75 g, decoction 50–100 ml.

Classical use

In classical texts 2 varieties have been mentioned: Laghu Paathaa and Raaj Paathaa. Laghu Paathaa has been equated with Cissampelos pareira and Raaj Paathaa with Stephania hernandifolia Walp.

Charaka used the roots and leaves, in prescriptions, internally in consumption, fever, urinary disorders, diarrhoea, piles. Sushruta, in addition, used the herb internally in indigestion, loss of appetite, persistent nausea, colic pain, intestinal catarrh and in vaginal discharges and other uterine affections; used decoction externally as a cleansing and antiseptic agent for wounds and ulcers.

Paathaa comes as the first drug in the Glactodepurant Group (Kaashyapa Samhitaa).

Among over-the-counter classical compounds, Vrihat Gangaadhar Churna (Shaarangadhara Samhitaa) is prescribed in amoebiasis and diarrhoea; Pushyaanuga Churna (Bhaishajya Ratnaavali) in leucorrhoea, menorrhagia and other uterine affections.

Active principles and pharmacology

The whole plant contains cissampareine; roots, leaves contain several alkaloids, the most important being hayatine.

The methiodide and methochloride derivatives of hayatine are reported to be potent neuromuscular-blocking agents. Both the derivatives produced varying degrees of fall in blood pressure. Hayatine methochloride has a direct inotropic effect on the isolated cardiac muscle.

Cissampareine showed a significant and reproducible inhibitory activity against human carcinoma cells of the naso-pharynx in cell-culture.

The roots showed more significant antibacterial activity against Gram-positive organism than against Gram-negative strains.

An ethanolic extract (50 %) of the stem and root showed CNS-depressant activity.

The poultice of leaves is found efficacious in abscesses, boils, burns, sores, pimples, scabies, itches.

The roots possess astringent, diuretic, stomachic, analgesic, antipyretic and emmenagogue properties.

Use in Western herbal

Dried root, bark of Cissampelos pareira or Pareira brava are employed as a diuretic for the relief of chronic inflammation of the bladder and urinary passages; also for calculus affections, leucorrhoea, gonorrhoea, jaundice, dropsy, rheumatism. Infusion of the root is given internally, while the bruised leaves are applied externally.

Vitaceae

CISSUS

Cissus quadrangularis (L.) Wall. ex Wt.

Habitat

Distributed throughout India, particularly in hotter parts; also cultivated in gardens.

Classical & common names

Ayurvedic: Asthisamhaara, Asthisamhrit, Asthisamyojaka, Vajravalli.
Unani: Hadjoda.
Siddha: Pirandai.
English: Square Stalked-Vine.

Parts used

Tuber.

Dose

Juice 10–20 ml, paste 10–20 g.

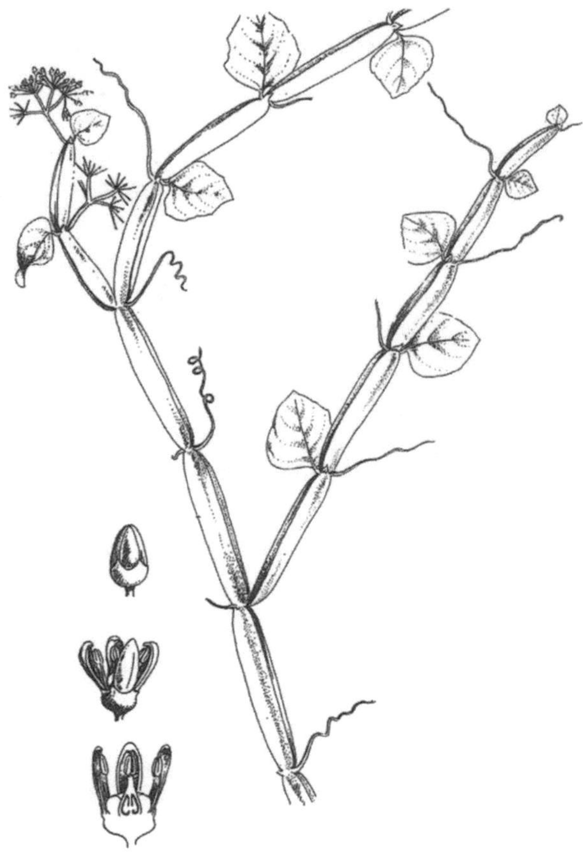

Figure 1 Cissus quadrangularis [ADPS]

Classical use

In cases of dislocation of joints and bone fractures, the herb was given internally with milk (Vrindamaadhava) and its medicated oil was massaged externally (Bhaavaprakaasha.) The herb was recommended with meat or other protein-rich foods as a preventive measure from fractures and other bone affections (Gadanigraha).

For fertility, the paste of Asthisamhaara, mixed with its juice and equal quantity of oil was prescribed to women after menstruation.

In folk medicine, the herb is prescribed in leucorrhoea and metrorrhagia.

Active principles and pharmacology

The stem contains 2 tetracyclic triterpenoids and 2 steroidal principles. The anabolic steroidal principles showed a marked influence in the rate of fracture-healing by influencing early regeneration of all connective tissues involved in the healing and quicker mineralization of the callus. The total alcoholic extract of the plant, on parenteral administration, neutralizes the anti-anabolic effect of the cortisone in healing of the fractures.

The aqueous extract of the plant reduced the total convalescent period by 33 per cent in experimental rats and dogs. It aids in recovery of the strength of the bones up to 90 per cent in 6 weeks. The drug exhibits a definite influence both on the organic and mineral phase of fracture-healing leading to quicker recovery in animals.

The drug is found useful not only in building up the chemical composition of the bone but also in bringing about its functional efficiency.

Analysis of the air-dried plant gave: moisture 13.1, protein 12.8, fat and wax 1.0, fibre 15.6, carbohydrates 36.6, mucilage and pectins 1.2, and ash 18.2 %. The ash contains mostly carbonates and to a small extent phosphates of sodium, potassium, magnesium and calcium. Presence of potassium tartarate is also reported.

The plant is rich in vitamin C.

Cucurbitaceae

CITRULLUS

Citrullus colocynthis Schrad.

Figure 1 Citrullus colocynthis [ADPS]

Habitat
Wild in the warm, arid and sandy parts throughout India, up to 1500 m.

Classical & common names
Ayurvedic: Indravaaruni (Charaka); Gavaakshi, Chitraa, Aindri, Chitraphalaa, Indraasuri; Mrigaani, Mrigairvaaru, Vishaala, Vishaalyka; Indraayana. (Aindri has also been equated with Bacopa monnieri by PV Sharma.)
Unani: Hanzal.
English: Colocynth, Bitter Apple.

Parts used
Fruits, root.

Dose
Fruit powder 125–250 mg.

Classical use
The dried pulp of the fruit is used in Indian systems of medicine, wherever strong purgative action for cleansing and detoxifying the system is required in chronic constipation, sluggish liver, jaundice, dropsy, ascitis, paralysis, internal abscesses, congestions, arthritis, and gout. Ayurvedic physicians give the drug with jaggery and/ or dry ginger, Unani physicians with gum-arabic and almond oil, to counteract the drastic action of the drug.

In folk medicine, a paste of the root is applied to cervical adenitis and various inflammations and swellings. Leaves are applied in migraine and neuralgia.

Among classical Ayurvedic compounds, Naaraayan Churna (Shaarangadhar Samhitaa) is available over-the-counter and is prescribed as a carminative and purgative; Abhayaarishta (Bhaishajya Ratnaavali) is recommended for piles, as it helps in the evacuation of accumulated faeces.

Among Unani compounds, Habb-e-Iyarij, Habb-e-Shahm-e-Hanzal are prescribed as purgatives, Habb-e-Ghariqoon in asthma for expelling phlegm.

Active principles and pharmacology
The fruit contains a glycoside, colocynthin, its aglycone-alpha-elaterin; citrullol. Plant contains colocynthetin. Roots contain elaterin and hentriacontane.

The dried pulp of mature fruits, feed from the rind and seeds, constitute the drug colocynth and exhibits hydragogue, cathartic and drastic purgative properties; is found useful in bilious derangement, chronic constipation, dropsy, fevers and cases requiring purgatives. The drug is considered emmenagogue.

The glycoside exhibits uterine depressant property and decreases the rate and amplitude of contraction; shows anti-histaminic an anti-acetylcholine activity. The glycoside also shows cardiac depressant activity.

The extract of dried pulp showed significant antibacterial activity against Salmonella paratyphi.

The ethanolic extract of flowers and leaves exhibited antibacterial activity against a number of Gram-positive and Gram-negative bacteria.

In pharmacological trials, the fruit exhibited hypoglycaemic action. Presence of a potent anticoagulant factor has been reported.

Use in Western herbal

Preparations of Citrullus colocynthis are used exclusively in fixed combinations in the treatment of acute and chronic constipation; also in the treatment of liver and gallbladder disorders. All preparations act as drastic laxatives.

Caution

The drug is severely poisonous. Vomiting, bloody diarrhoea, colic, kidney irritation follow the intake of toxic dosages (0.6–1 g) and then increased diuresis leading to anuria. 4 g (starting at 2 g) is fatal.

Rutaceae

CITRUS

Citrus limon (Linn.) burm. f.

Habitat

Cultivated all over India, common in Kumaon, northern and central India.

Classical & common names

Ayurvedic: Jambira (Sushruta); Jamiri Nimbu.
Unani: Leemu, Baraa Neebu.
Siddha: Elumishchai.
English: Lemon.

Parts used

Fruit.

Dose

Juice 5–10 ml.

Classical use

According to Sushruta Samhitaa, Jambira juice is carminative, digestive, antitoxic, cures parasitic infection, halitosis, cough, asthma, nausea, constipation, colic pain, hiccup.

The juice of Jambira alleviates acid gastritis (Chakradatta).

In folk medicine, the juice with salt is used on ringworm.

Available over-the-counter, Sikanjabeen Lemuni, a Unani compound, is prescribed in indigestion, sluggish liver, nausea, dyspepsia. Sharbat-e-Madani is another such compound.

Active principles and pharmacology

Limonene is the principal constituent of essential oil. Others are citronellal, n-nonanal, n-decanal, n-dodecanal, linalyl-acetate, geranyl acetate, citronellyl acetat, methyl anthranilate; also lipophilic flavonoids including sinensetin and furocoumarins.

Chief flavonoids are the bitter neohesperidosides naringin and neohesperidin dihydro chalcones; hesperidin, rutin.

Citroflavonoids affect vascular permeability, are anti-inflammatory and a source of vitamin C.

Use in Western herbal

Lemon juice is used as a cooling drink in fevers and for allaying thirst. It is highly recommended in acute rheumatism; also in uterine haemorrhage after delivery.

Locally, lemon juice is used as an astringent, whether as a gargle in sore throat, in pruritus of the scrotum, or as a lotion in sunburn. It is given in obstinate hiccup; also in jaundice and hysterical palpitation of the heart.

The oil is used externally as a rubefacient and internally in small doses for its stimulating and carminative properties.

Citrus medica Linn.

Figure 1 Citrus medica [ADPS]

Habitat

Cultivated all over India. Met with chiefly in Khasia Hills, the south-west of India and parts of northern India.

Classical & common names

Ayurvedic: Maatulunga (Sushruta); Bijapuura, Bijauraa.
Unani: Turanj, Baalang (Persian).
Siddha: Kadaranarathai.
English: Citron.

Parts used

Fruit, leaf, flower, root.

Dose

Juice 5–10 ml, decoction 50–100 ml.

Classical use

The pieces of Maatulunga fruit, pounded with honey and rocksalt, were to be licked for tastelessness due to fever, for nausea, biliousness, anorexia, dryness of tongue, palate, throat and trachea. Piper nigrum (Maricha) was also to be mixed in the linctus (Sushruta Samhitaa, Haarita Samhitaa, Vrindamaadhava).

The juice of the fruit, mixed with honey, Piper longum (Pippali), Piper nigrum and ginger, was prescribed in jaundice and anaemia (Sushruta Samhitaa).

For hiccup and asthma, the juice of the fruit, mixed with honey and dry ginger powder, was advised as an adjuvant (Vrindamaadhava). The heartwood cooked with milk or the root powder with clarified butter was prescribed in colic (Sushruta, Chakradatta). The paste of root and flowers, with rice water, was given for checking haemorrhage (Sushruta).

The leaves and roots, along with the roots of Mangifera indica (Aamra) and Syzygium cuminii (Jambu), cooked in closed heating and the decoction so extracted, was prescribed with honey for checking severe vomiting (Shaarangadhar Samhitaa).

Cellular covering of the seeds was given by Sushruta internally as astringent for treating nausea, abdominal glands, piles; the rind and epidermic layer for flatulence, biliousness and as vermifuge.

In Unani medicine, the peel of the fruit is the most potent medicinal part and is included in liver, cardiac and brain tonics. Among over-the-counter compounds, Jawarish-e-Ood Tursh, Jawarish-Anarain, Jawarish-e-Zarishk are prescribed as stomachic for biliousness, nausea and vomiting. Jawarish-e-Utraj, which contains Murabha-e-Turanj (confection of Turanj) as the main ingredient, is prescribed in hepatitis; also as antibilious and stomachic.

Active principles and pharmacology

The fruit contains constituents of related citrus species.

The limonoid glycosides are reported from the peel. The essential oil of the leaves contains neral, geranial and limonene. The roots contain campesterol, stigmasterol, sitosterol and cholesterol. Also xanthyletin.

Aqueous extract of the peel showed hypotensive action in dogs.

Hypocreaceae

CLAVICEPS

Claviceps purpurea (Fr.) Tul.

Figure 1 Claviceps purpurea on rye [WOI]

Habitat

A fungus parasitic on grasses and cereal crops, especially on rye (Secale cereale Linn.)

Classical & common names

Ayurveda: Annaamaya, Sraavikaa.
Unani: Agrat.
Siddha: Ergot.
English: Ergot.

Parts used

Whole fungus.

Dose

1–2 g. (More than 4 g toxic.)

Classical use

In Ayurvedic and Unani medicine, as in Western herbal, Ergot is employed to stimulate uterine contraction in the final stages of parturition. It is also employed, though rarely, for arresting uterine haemorrhage. (It has been found to raise blood pressure in pulmonary and cerebral haemorrhage.)

It is prescribed in leucorrhoea, impotency, poor erection and spermatorrhoea by Unani physicians. Also given for treating enlarged prostate.

Active principles and pharmacology

Ergot owes its activity to specific complex alkaloids, ergotoxine and ergotamine; in good Ergots the alkaloidal content may be 0.02 %. A large number of other substances have been isolated from Ergot; the most important is a fatty oil 30–35 %. A red colouring matter, sclererythrin, is extracted by alcohol and by alkalis. The drug also contains mannitol, partly combined as a glucoside, and the sugar trehalose.

Ergot is usually administered in the form of extract (Ergotin), liquid extract, infusion or ammoniated tincture.

Use in Western herbal

The earliest reference of Ergot is found in Loneer's classic in 1582, where the occurrence of Ergot on rye and its obstetric virtues are mentioned. Camerarius, at the same time, stated that it was a popular remedy for accelerating parturition. In France and Italy it was in quite general use for the same purpose for many years before it was employed by professional physicians. A Dutch physician used it for obstetrics in 1747. In 1936, it appeared in the London Pharmacopoeia as Ergot.

At present, Ergot and Ergot preparations are used in gynaecology and obstetrics e.g. haemorrhages, climacteric haemorrhages, menorrhagia and metrorrhagia; before and after miscarriage, for removal of placenta and shortening of the childbirth period and for atony of the uterus.

Caution

Ergot is contraindicated in peripheral blood flow disorders.

Verbenaceae

CLERODENDRUM

Clerodendrum serratum (L.) Moon.

Figure 1 Clerodendrum serratum [ZANDU]

Habitat

South and Eastern India, especially in West Bengal.

Classical & common names

Ayurvedic: Bhaargi (Charaka, Sushruta); Bhaarangi, Angaarvalli, Phanji, Braahmanyashtikaa, Kharshaak, Padma, Bhragubhavaa..
Unani: Bhaarangi.
English: Blue-flowered Glory Tree, Beetle-killer.

Parts used

Root.

Dose

Powder 3–6 g.

Classical use

Charaka used the root internally for cough, fevers, hard cutaneous eruptions; and as an ingredient of a medicinal oil externally. Sushruta also prescribed the root internally for catarrh, cough, asthma, as well as in dyspepsia, colic pain, parasitic worms, piles, and skin diseases. It was considered antitoxic, antiseptic, astringent, styptic.

The herb was generally prescribed with dry ginger in bronchial asthma (Vrindamaadhava); also with Brahati (Solanum indicum) (Bangasena). The paste of the root, pounded with sour gruel, was applied externally on goitre, cervical adenitis and scrotal enlargement, inguinal hernia (Chakradatta, Bangasena).

In folk medicine, a decoction of the root with Acorus calamus (Vachaa) is given in asthma.

Over-the-counter, Bhaarngyaadi Kwaatha (Sahasrayoga) is available and prescribed in coughs and bronchitis, as an expectorant and carminative. Bhaarangi Guda (Bhaishajya Ratnaavali) is also prescribed in coughs, dyspnoea.

Active principles and pharmacology

The bark contains triterpenoids serratagenic acid, oleanolic acid and queretaroic acid. Leaves contained alpha-spinasterol and flavonoids, and phenolic acids.

The extract of the root exhibited antihistaminic property on isolated guinea-pig ileum. It effectively antagonised the effect of histamine on tracheal chain of guinea-pigs. The drug pharmacologically exhibited direct inhibition of constrictor response, antihistaminic effect, hypotensive and antiasthmatic properties (CCRAS).

The root is also found efficacious in dyspepsia, anorexia, colic, flatulence.

Use in Western herbal

A related species of Clerodendrum serratum, Clerodendrum trichotomum, known as Chou Wu Tong in Chinese medicine, is used in Western herbal for joint pain, numbness and paralysis. In a Chinese trial, 171 persons with high blood pressure were given the herb, in 81 % blood pressure levels dropped significantly.

Chou Wu Tong contains clerodendrin, acacetin and mesoinositol.

(In India, Clerodendrum serratum is used mainly in respiratory conditions, though it also exhibited hypotensive effect.)

Papilionaceae

CLITORIA

Clitoria ternatea Linn.

Figure 1 Clitoria ternatea [CCRAS]

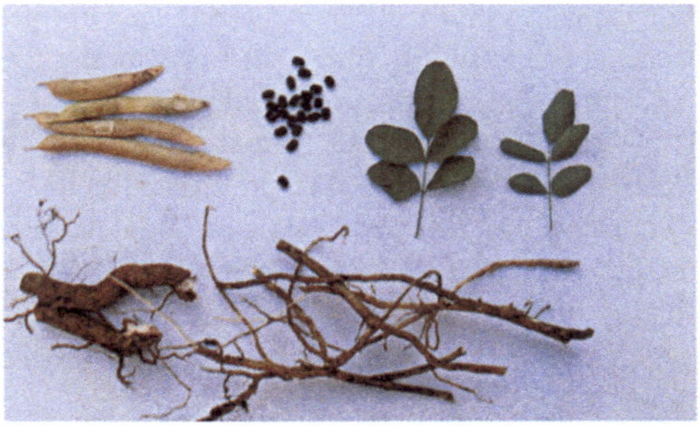

Figure 2 Clitoria ternatea [CCRAS]

Habitat

Throughout India in tropical areas; also cultivated in hedges.

Classical & common names

Ayurvedic: Girikarnikaa (Charaka, Sushruta); Aparaajitaa; Asphota, Girimallikaa, Girikarnikaa, Kokilaa, Koyal, Vishnukrantaa. (Not to be confused with Evolvulus alsinoides Linn., which is known as Vishnukraantaa, Vishnukraanti.)
Unani: Mezereon Hindi (Moodeen Sheriff).
Siddha: Kakkanam.
English: Butterfly Pea, Winged-leaved Clitoria, Mezereon.

Parts used
Whole plant.

Dose
Decoction 50–100 ml, paste 1–1.5 g.

Classical use
Charaka and Sushruta considered it as an errhine herb, which was used as a snuff for aborting accumulated secretions from the nose; and internally as an emetic.

As a snuff, the herb was used in cerebral congestions while treating mental disorders (Raajamaarttanda, Chakradatta). It was also given internally in oedema (Bangasena), goitre (Bangasena, Bhaavaprakaasha), cervical adenitis (Gadanigraha, Chakradatta); applied externally in filaria (Haarita Samhitaa), vitiligo and scrofula (Raajamaar tanda).

In folk medicine, slightly roasted seeds are given with jaggery and rocksalt as purgative and for expelling phlegm. The root and seed are prescribed internally in cough, diseases of the liver and spleen, and rheumatic affections.

According to CCRAS, Aparaajitaa is to be treated as Mezereon, while in Unani medicine, Mezereon has been equated with Daphne mezereum Linn. The leaves and latex of the plant are highly toxic. The leaves are soaked in vinegar for 48 hours (vinegar is to be changed at least three or four times), washed afterwards, dried, and pounded in almond oil before using. Detoxified leaves are prescribed as a purgative in liver and spleen affections, dropsy and in helminthic infestations; externally in skin diseases.

Active principles and pharmacology
The root contains taraxerol and taraxerone, while the seed contains cinnamic acid and an anthoxanthin glucoside. The leaf contains glycoside of kaempferol and stigmast-4-ene-3, 6-dione. The flower furnished a blue anthocyanin which contains delphinidin-3, 5-diglucoside, acylated anthocyanin based on delphinidin along with flavonols. A lactone, aparajitin, clitorin from leaves was also reported.

Pharmacologically, the seeds are aperient and purgative, the roots cathartic, purgative and diuretic. The plant also possesses significant cholinergic, antihistaminic and analgesic properties; reduces gastric acidity.

Use in Western herbal
Mezereon, equated with Daphne mezerum, was used in the past to relieve headache, joint pains and for increasing circulation in rheumatic complaints; and in anaesthetic salves for the skin.

Currently, used only in homoeopathy for skin conditions such as cradle cap, shingles, weeping eczema, weeping blisters.

Cucurbitaceae

COCCINIA

Coccinia grandis (Linn.) Voigt

Figure 1 Coccinia grandis [CCRAS]

Figure 2 Coccinia grandis [CCRAS]

Coccinia indica W.& A.
Coccinia cordifolia Cogn.
Cephalandra indica (W. & A.) Naud.

Habitat

Cultivated in Assam, West Bengal, Bihar, Orissa, Maharashtra, Andhra Pradesh, Tamil Nadu; wild in many parts of India.

Classical & common names

Ayurvedic: Bimbi, Tundi, Tundikeri; Kundaru.
Unani: Kanduri, Kunduru.
Siddha: Koovai.

Parts used

Leaf, root, fruit.

Dose

Juice 10–20 ml, decoction (10:1) 10–20 ml, powder 2 g (emetic in large doses).

Classical use

During the 16th century, the fruit was prescribed in biliousness, in vitiation of blood, bronchitis, asthma, consumption. It was included in Varunaadi group of herbs, which was prescribed for alleviating cough, catarrh, headache; also tumours and internal abscesses. It was also given as an anthelmintic (Gadanigraha). According to Kaiyadeva Nighantu, the bitter variety was used.

In Unani medicine, the juice of Kanduri is given in polyuria; root bark as a purgative; flowers in bilious affections, liver and skin diseases; fruits as blood-purifier, astringent, for direct action on kidneys and urinary disorders.

A few Unani compounds have wrongly been quoted in certain Ayurvedic reference books under the plant Kunduru. Kundur is the gum of Boswellia serrata Roxb. and is totally different from Kunduru. (Unani compound, Jawarish-e-Kundur is prescribed in polyuria; Majoon-e-Kundur in strangury, atony of bladder, polyuria, bedwetting in children, premature ejaculation.)

Active principles and pharmacology

Fruits yielded beta-amyrin and its acetate lupeol and cucurbitacin B. Aerial parts gave heptacosane, cephalandrol, tritriacontane, beta-sitosterol, alkaloids cephalandrines A and B.

The plant was screened for its hypoglycaemic activity in the Central Research Institute of Siddha Medicine, Trivendrum.

The juice, decoction of leaves and stem of Coccinia indica and decoction of fruit in 20 ml/kg showed significant hypoglycaemic response in fasting rabbits. The fruits were found to be more potent. The root did not show hypoglycaemic activity. Anyhow, the water-soluble alkaloid obtained from alcoholic extract of the root exhibited antidiabetic activity on alloxan diabetic rats.

A quarternary base isolated from the whole plant, exhibited short-lasting hypoglycaemia in guinea-pigs and showed definite activity on glucose tolerance test in guinea-pigs, and moderate hypoglycaemic activity in alloxanised rats. Alcoholic extract of the plant, as well as the whole plant and leaf extracts exhibited significant hypoglycaemic effect on somatotropin and corticotropin-induced and streptozotocin-treated hyperglycaemic rats.

Arecaceae

COCOS

Cocos nucifera Linn.

Habitat
Extensively cultivated in South India; also found in West Bengal, Costal Maharashtra, Orissa, Laccadive and the Andaman Islands.

Classical & common names
Ayurvedic: Naarikel (Sushruta), Naarikera.
Unani: Naarjeel (Arabic); Naariyal.
Siddha: Thennai.
English: Coconut Palm.

Parts used
Fresh kernel, water, oil, shells.

Classical use
Sushruta prescribed the pulp and water as a tonic for gastro-intestinal irritations, nausea and biliousness; also as a laxative and diuretic. According to Bhaavaprakaasha, it cleanses the urinary bladder.

Naarikel-khanda was prescribed in acid gastritis; in colic Naarikela-lavana (Chakradatta).

Parched grain flour mixed with equal quantity of sugar was given with coconut water in severe acidity, thirst, fainting, giddiness, palpitation of heart (Vaidyamanoramaa).

In Unani medicine, the oil extracted from the burnt shell is used in obstinate skin diseases. The ash in the form of an ointment is also applied in dermatophytosis.

Over-the-counter, Naarikel-lavana (IMPCOPS) is available and is prescribed in colic.

The kernel is an ingredient in reputed Unani compound, Majoon-e-Falasifa, prescribed in polyuria, dysuria, indigestion, anorexia, amnesia; also as a tonic in debility.

Cocos nucifera shell-oil, known as Chirattai Thailam in Siddha medicine, is prescribed for external application in eczema, ringworm, chronic skin diseases.

Active principles and pharmacology
Analysis of the coconut water gave the following values: water 95.5, protein 0.1, fat 0.1, mineral matter 0.4, carbohydrates 4.0, calcium 0.02, phosphorus 0.01, iron 0.5 mg/100 g. The percentage of arginine, alanine, cystine and serine in the protein is more than those in cow's milk.

The nut water contains sodium 105.0, potassium 312, calcium 29, magnesium 30, iron 0.10, copper 0.04, phosphorus 37, sulphur 24, chlorine 183 mg/100 g.

As a beverage, the water of unripe coconut has the advantage of being sterile. It is aperient, diuretic and anthelmintic. The fresh pulp of the tender nut is diuretic.

The coconut oil is characterised by the presence of a high percentage of lauric and myristic acids. Undecanoic and tridecanoic acids are also present. The glycerides of the coconut oil form a very intimate mixture; main glycerides being caprylolauromyristin and dilauromyristin.

The coconut shell gave the following values; moisture 8.0, ash 6.6, solvent extractives 4.2, lignin 29.4, pentosans 27.7, uronic anhydrides 3.5, cellulose 26.6, methoxyl content 5.6, and nitrogen 0.11 %. The pentosans of the shell are largely xylosans.

A polysaccharide factor isolated from coconut water is found to be immunogenic; oil from coconut shell fibre antimicrobial; leaf diuretic; endosperm aperient, anthelmintic; oil from endosperm antiseptic; water of tender fruit, root and endosperm are found efficacious in urinary affections.

The effect of alcoholic extract of coconut shell, 2 % ointment with petroleum jelly as an external medication, on 31 cases was observed at CCRAS, Trivendrum, during 1987. 23 cases were completely cured without relapse of the complaints. Symptoms flared up during the second week of treatment in 4 cases. All other patients tolerated the external application with no side-effects.

The extracts of shell fibres on pyrolysis, at 160–240 degree temperature, yielded a fraction which showed highest antimicrobial activity. On further purification with ether, it yielded an oily liquid, which was found to be highly active against six types of fungal organisms Aspergillus niger, Penicillium chrysogenum, Candida albicans, Trichophyton mentagrophytes, Rhizoctonia solani and R. betaticola. The activity was found to be due to the oily liquid and not the aqueous portion.

Colchicaceae

COLCHICUM

Colchicum autumnale Linn.
Colchicum luteum Baker

Habitat
Temperate Himalaya from Kashmir to the Himachal Pradesh at 700–2800 m.

Classical & common names
Ayurvedic: Hiranyatuttha.
Unani: Suranjaan talkh.
English: Hermodactyls, Colchicum, Meadow Saffron. Golden Collyrium (Indian substitute.)

Parts used
Corm.

Dose
Powder (sweet variety) 1–3 g. Daily dose of colchicine content must not exceed 8 mg.

Classical use
Yellow or black variety, known as Suranjaan talkh, is used in Unani medicine externally in swellings, rheumatic affections.

Internally, the white variety, known as Suranjaan shireen, equated with Merendera persica, is used as an anti-inflammatory drug. Majoon-e-Suranjaan, Sufoof-e-Suranjaan Zafrani and Habb-e-Suranjaan are given in rheumatism, gout, sciatica, arthritis; and are available over-the-counter. All these compounds contain the sweet variety of Suranjaan.

Active principles and pharmacology
Colchium contains tropolone alkaloids: colchicine, colchicoside and N-deacetyl-N-formyl-colchicine.

Colchicine analogs deacetyl thiocolchicine (DTC), deacetymethyl-colchicine (DMC) and trimethyl-colchicinic acid (TMCA) were found effective in the treatment of gout; DMC and DTC may elicit agranulocytosis.

Use in Western herbal
Colchicum autumnale, known as Meadow Saffron, is used in the West.

John Gerard (1545–1612) wrote of it: "The roots of Hermodactyls are properly given to those that have the gout. The powder of ginger, anise seed or cumine seed, and a little masticke correct the churlish working of the Hermodactyls. Those who have eaten of the common Meadow Saffron must drink the milke of a cow, or else death presently ensueth... the roots of all the sorts of Meadow Saffrons are very hurtfull to the stomacke, and being eaten they kill by choaking..."

The drug is severely poisonous. Kidney and liver damage (surprisingly, CIMAP, indicated its use in diseases of liver and spleen), hair loss, peripheral nerve inflammation, myopathias and bone-marrow damage have been observed following long-term administration. Teratogenic damage is also possible.

Despite its toxicity, Meadow Saffron is still considered one of the best remedies for acute gout pain.

Leukaemia has also been treated with Meadow Saffron and also Bechcet's syndrome, a chronic disease marked by recurring ulcers and leukaemia. Even at low doses the herb exhibited significant side-effects. Its use is possible only under strict professional supervision.

For an acute attack of gout, an initial oral dose corresponding to 1 mg colchicine, followed by 0.5 to 1.5 mg can be given by the physician. Total daily dosage must not exceed 8 mg of colchicine.

German Commission E monograph recognised the efficacy of Colchicum autumnale in the following areas:

- Gout
- Mediterranean fever.

Colchicum is also available as a homoeopathic medicine in the form of tincture.

Burseraceae

COMMIPHORA

Commiphora myrrha (Nees.) Engl.
Balsamodendron myrrha T. Nees.

Habitat
Indigenous to eastern Mediterranean countries, Somalia. Imported into India from Persia or Saudi Arabia.

Classical & common names
Ayurvedic: Bola, Gandhrasa, Hiraabola.
Unani: Mur Makki.
Siddha: Vellaibolam.
English: Myrrh.

Parts used
Gum-resin.

Dose
625 mg–1.25 g.

Classical use
According to Raaja Nighantu, Bola is bitter, hot, astringent, alleviates blood impurities, biliousness, catarrh, and cures uterine affections.

In Unani medicine, Mur Makki is used in dysmenorrhoea and amenorrhoea; in chest infections and chronic bronchitis for stimulating expectoration. Mixed with rose water or the extract of roses, it is prescribed as a gargle in ulcerated sore throat, spongy gums and dental problems. The paste is applied externally on swellings and wounds.

Active principles and pharmacology
The drug constitutes 25–40 per cent resin (containing triterpenes, alcohols and esters); around 60 per cent gum; up to 14 per cent volatile oil (containing primarily sesquiterpenes and some monoterpenes).

Antimicrobial activity has been reported in the resin *in vitro*; it is said to stimulate phagocytosis in vivo as well.

The drug is astringent, antiseptic topically and on mucosal membranes; also expectorant, antispasmodic and carminative.

Chinese researchers have identified substances that fight bacteria.

Use in Western herbal
In the West, tincture of myrrh is an important ingredient in many herbal mixtures for the treatment of infections and inflammations in the mouth and throat. It is found very effective in reducing gum inflammations, sore throats and tonsillar infections.

Myrrh is not soluble in water and is therefore normally taken as a powder or tincture, rather than as infusion, like other herbs. It is not easily absorbed by the intestines, so is generally used topically. Its astringent and antiseptic properties make it an efficacious application for acne, boils and other skin affections. In Germany, it is used externally for pressure sores.

Neat tincture of myrrh mixed with that of calendula is applied to fungal infections of the nails and skin.

Myrrh is a common ingredient in European toothpastes. It is included to help fight bacteria that cause tooth decay. Some natural food stores in America sell one such brand Merfluan.

Myrrh is not given to children under 2 years; for older children and people over 65 a lower strength is prescribed initially.

Caution
Contraindicated during pregnancy.

Commiphora mukul (Hk. ex Stocks) Engl.
Commiphora wightii (Arn.) Bhandari
Balsamodendron mukul Hk. ex stocks.

Habitat
Rajasthan, Andhra Pradesh, Assam, Madhya Pradesh, Karnataka.

Classical & common names
Ayurvedic: Guggulu (Sushruta), Devdhuupa, Kaushik, Pur, Mahishaaksha, Kumbha, Palamkasha, Uluukhal.

Figure 1 Commiphora wightii [CCRAS]

Figure 2 Commiphora wightii [CCRAS]

Unani: Muqil (Arabic), Guggul.
Siddha: Kungilyam.
English: Indian Bdellium, Gum Guggul.

Parts used
Gum-resin.

Dose
500 mg–1 g.

Classical use
Sushruta prescribed the drug internally in obesity, sluggishness of the liver and bowels, intestinal parasites, vitiated blood, sinus, oedema, internal tumours, malignant tumours, skin eruptions, itches, leucoderma, sudden paralytic seizures, and epilepsy. He also prescribed tender leaves as a rejuvenating tonic.

The gum-resin, in prescriptions, was specific for obesity, oedema and arthritis (Ashtaanga Hridaya, Ashtaanga Sangraha, Chakradatta, Vrindamaadhava, Bangasena).

The drug was also used internally as a stomachic, carminative and in hyperacidity; as an expectorant, diuretic, uterine stimulant and emmenagogue; externally in the form of a lotion for indolent ulcers and as a gargle in spongy gums, pyorrhoea, chronic tonsillitis, ulcerated throat and pharyngitis.

In folk medicine, inhalation of the fumes from burnt guggul is recommended in hay fever, acute and chronic nasal catarrh, and chronic laryngitis.

Only detoxified gum-resin is used in Indian medicine.

Among over-the-counter compounds, Kaishore Guggulu (Shaarangdhara Samhitaa) is prescribed in gout; Kaanchnaar Guggulu (ibid) in enlargement of cervical glands, scrofulous tumours and skin diseases; Triphalaa Guggulu (ibid) in piles, fistula and other inflammatory conditions; Mahaa Yogaraaj Guggulu (ibid) in rheumatic and other inflammatory conditions.

Guggulu is an important ingredient in many other Ayurvedic compound formulations: Gokshuraadi Guggulu, diuretic, demulcent and alterative, indicated in spermaturia, painful micturition, leucorrhoea, albuminuria, and enuresis; Laakshaadi Guggulu, antipyretic, antiseptic, indicated in fractures; Sinhanaad Guggulu, indicated in dyspepsia, indigestion, rheumatism, gout, paralysis, and paraplegia; Trayodashaanga Guggulu, indicated in paraplegia, and hemiplegia.

Unani compounds, Itrifal-e-Muqil Mulaiyin and Majoon-e-Muqil are prescribed in rheumatism, gout, blind piles; Majoon-e-Jograaj Guggul in hemiplegia, Bells Palsy, tremor, neurasthenia, rheumatism, arthritis, and bleeding piles.

Active principles and pharmacology

The gum-resin contains, guggulsterones Z and E, guggulsterols I–V, two diterpenoids, a terpene hydrocarbon cembrene A, and a diterpenealchohol mukulol, alpha-camphorone and cembrene; long chain aliphatic tetrol octadecan-1, 2, 3, 4-tetrol, eicosan-1, 2, 3, 4-tetrol and nonadecan-1, 2, 3, 4-tetrol.

Major components from essential oil of gum-resin are myrcene and dimyrcene.

Fractions A and B (pet.-ether soluble and neutral) when studied in experimental hyperlipidaemia and atherosclerosis in chicks and experimental myocardial infarction in rats showed significant decrease in cholesterol, phospholipids, triglyceride and total lipids. Fraction B was found less effective. Fraction A and steroidal fraction A showed effectiveness in experimental myocardial necrosis in rats produced by isoprenaline. These results suggest gum Guggul's application in hyperlipidaemic and myocardial infarction cases.

The guggulsterone fraction of gum Guggul exhibited anti-inflammatory activity in models of acute inflammation, which was comparable to approximately one-fifth that of hydrocortisone and equal to phenylbutazone and ibuprofen.

The significant pharmacological properties based on clinical trials are: hypocholestraemic, hypolipidaemic, antiarthritic, thyroid-stimulating and cardioprotective. Clinical studies with gum Guggul on psoriasis showed encouraging results.

Clinical studies with Shunthi Guggulu on 63 patients of rheumatoid arthritis showed effectiveness of the drug (CCRAS).

In India, Gugulipid was granted approval for marketing as a lipid-lowering drug in June 1986. For medicinal purposes, a standardized extract, known as Gugulipid, should contain 50 mg of guggulsterones per gram.

Gugulipid, in clinical trials, lowered total cholesterol by 24 % when compared to cholestyramine (14 %), gemfibrozil (10 %), lovastatin (34 %). Triglycerides came down by 23 % and HDL cholesterol increased by 16 %.

The soluble portion (45 %) of the gum Guggul possesses significant cholesterol-lowering and anti-inflammatory activity, the insoluble portion (55 %) is toxic.

Caution

Contraindicated during pregnancy.

Convolvulaceae

CONVOLVULUS

***Convolvulus microphyllus* Sieb. ex. Spreng.**
***Convolvulus pluricaulis* Choisy**

Habitat

All over India ascending to 2000 m in the Himalayas.

Classical & common names

Ayurvedic: Shankhapushpi, Shankhaahvaa (white-flowered).
The blue-flowered var. is equated with Convolvulus alsinoides L./Evolvulus alsinoides L., known as Vishnukraanti, Vishnu gandhi, Vishnukrandi, Vishnukraantaa.
Unani: The blue-flowered variety: Sankhaaholi.
Siddha: The white-flowered variety: Sivakrandi. The blue-flowered variety: Vishnukrandi.
Raaja Nighantu treated Vishnukraantaa as a blue-flowerd variety of Shankhapushpi, but Bhaavaprakaasha Nighantu treated Vishnukraantaa as a synonym of Aparaajitaa (Clitoria ternatea Linn.), which belongs to a totally different species (Papilionaceae). Due to white and blue flowers, Aparaajitaa was locally called Shankhapushpam in the South.
The Ayurvedic Pharmacopoeia of India, Part I, vol. II has equated Shankhapushpi with Convolvulus pluricaulis Choisy.

Parts used

Whole plant.

Dose

Powder 3–6 g.

Classical use

Charaka used Shankhapushpi (white-flowered) as a specific tonic for promoting intellect. He prescribed the clarified butter, cooked with three times juice of Shankhapushpi along with milk, as a brain tonic. He added the juice of Bacopa monnieri (Braahmi), Acorus calamus (Vachaa) and Saussurea lappa (Kushtha) for alleviating insanity and epilepsy.

According to Chakradatta, the juice of Shankhapushpi mixed with that of Saussurea lappa and honey was prescribed in insanity. According to Kaiyadeva Nighantu, Shankhapushpi is a unique brain tonic which alleviates confusion, insomnia, psycho-neurosis and neurological disorders.

In the South, Vishnukraantaa (blue-flowered) is used as a powerful brain stimulant for toning up intellectual prowess. The drug is also used in nervous debility and dysentery. The flowers are used in uterine bleeding and the roots for gastric and duodenal ulcers.

Shankhapushpi syrup (Unjha Pharmacy), and Shankhapushpi syrup with Bocopa monnieri (Baidyanath, based on Ayurveda Saar Sangraha) are available over-the-counter, and are prescribed as intellect-promoting tonics.

In Unani medicine, Sankhaaholi is used as a tranquilliser and sedative; also as a blood purifier. Its syrup prepared with Piper nigrum is prescribed in diseases due to vitiated blood, bleeding piles, and venereal diseases. The herb is also used as a laxative. The root powder is given in spermatorrhoea.

Active principles and pharmacology

Convolvulus pluricaulis plant extract contains coumarin; beta-sitosterol; fatty acids and waxy constituents; hydrocarbons; crystalline substances; sterols. Alcoholic extract of the plant yield kaempferol, its glucoside, dihydroxy-cinnamic acid and beta-sitosterol glucoside. An alkaloid, sankhpuspine has been isolated.

Hypotensive and barbiturate hypnosis potentiating activity was maximum in leaves and flowers. Water insoluble portion was most active and was found effective in anxiety neurosis as a sedative.

In Clinical trials, Convolvulus pluricaulis reduced different types of stress, including psychological, chemical and traumatic. The drug also exhibited significant antianxiety activity, tranquillising effect, also anti-thyroid property.

Petroleum extract of leaves and flowers, probably due to some flavons, was found to possess antifungal property.

Convolvulus alsinoides plant gave an alkaloid evolvine; beta-sitosterol, stearic, oleic and linoleic acids, betaine, and an uncharacterized base. Also obtained were pentatriacontane and triacontane.

In preliminary pharmacological studies, alkaloid evolvine exhibited powerful stimulation on respiration and blood pressure and possible analeptic property.

The plant and leaves were found to be intellect-promoting, and efficacious in nervine affections, epilepsy, insanity; spermatorrhoea; internal haemorrhages; bowel complaints. Aqueous extract of the flower petal was found to possess antifungal properties.

Boraginaceae

CORDIA

Cordia dichotoma Forst. f.

Habitat

Common throughout India, wild and often planted.

Classical & common names

Ayurvedic: Shleshmaataka, Shelu, Bahuvaara, Bahuvaaraka, Bhuutvrkshak, Uddaalak Shita, Picchila; Lisodaa.
Unani: Sapistaan, Lasodaa.
Siddha: Naruvili.
English: Sabestan Plum.

Parts used

Fruits.

Dose

9–15 fruits, decoction 50–100 ml.

Classical use

Sushruta prescribed fruits internally for their cooling and astringent properties in biliousness and cough, intrinsic haemorrhage. The paste of the bark was applied on erysipelas, spider-poisoning (Sushruta); eruptive boils (Vrindamaadhava); the paste of tender fruits on soft chancre (Siddha-bhesaja-manimaalaa).

In Unani medicine, the fruit is used as an adjuvant to purgatives for counteracting their irritative effect. Prescribed as demulcent and expectorant in dry cough; also in burning micturation.

A Unani compound formulation, Laooq-e-Sapistaan, available over-the-counter, is prescribed in catarrh, coryza, chronic bronchitis, influenza. Sharbat-e-Sual (IMPCOPS), with Sapistaan as an important ingredient, is indicated in dry cough, as it loosens the secretions and makes expectoration easy.

Active principles and pharmacology

The plant contains: D-glucose, D-xylose, L-rhamnose, D-glucuronic acid, D-arbinose, lactose, and L-fructose among sugars; and glycine, leucine, glutamic acid, cystine, alanine, threonine, aspartic acid and proline as free amino acids.

The fruit pulp possesses demulcent, expectorant, astringent, anthelmintic and diuretic properties. A decoction of the bark was found useful in dyspepsia and fevers; the paste of kernels in ringworm.

Apiaceae

CORIANDRUM

Coriandrum sativum Linn.

Habitat

Cultivated throughout India.

Classical & common names

Ayurvedic: Dhaanyaka (Charaka), Kustumburu (Sushruta); Dhaanyeyaka, Dhanika, Dhanak, Dhaanaa, Dhaanya Kunati, Dhenukaa, Chhatraa, Vitunnak.
Unani: Kishniz (Persian).
Siddha: Kothamalli.
English: Coriander.

Parts used

Fruit, whole plant, oil.

Dose

Powder 3–6 g.

Classical use

Charaka used the entire plant, in prescriptions, as a digestive, diaphoretic and as an aid to virility.

Sushruta prescribed dried seeds, internally, in nausea, loss of appetite, thirst, deranged metabolic functions; for allaying pain and burning sensation; and as a detoxifying agent.

The cold infusion of Dhaanyaka, mixed with sugar, was given in fevers for alleviating burning sensation; with dry ginger in indigestion and as a diuretic (Vrindamaadhava).

Dhaanyaka, mixed with sugar, was given with rice-water for treating asthma and cough of children (Bangasena).

The seeds were included in a host of prescriptions for fever, diarrhoea, vomiting, indigestion as stomachic and carminative.

In Unani medicine, Kishniz is used in the same way. Green leaf is used as a refrigerant. Fresh juice of leaves is used as a gargle in sore throat and stomatitis. Paste is applied over swellings and boils; also over cervical adenitis. The paste is prepared by pounding green leaves with barley flour. The paste of dry fruits is also applied over forehead and temples during headache.

For cooling effect on the mind and for inducing sleep, fresh juice of the leaves, mixed with sugar, is given. It is also given in biliousness, intestinal irritations, heart-burn, thirst and nausea.

Itrifal Kishnizi is a reputed Unani compound for gastric headache, flatulence in stomach, chronic catarrh; also prescribed as an adjuvant in conjunctivitis, otalgia, bleeding piles, due to its cooling properties.

Active principles and pharmacology

Coriander contains up to 1.5 % volatile oil, consisting mainly of delta-linalool (around 70 %), alpha-pinene and terpinine. It also contains flavonoids, coumarins, phthalides and phenolic acids.

The essential oil of coriander stimulates the secretion of gastric juices and is a carminative and spasmolytic.

The fruit extract inhibited mycelial growth of Pythium aphanidermatum. The essential oil exhibited strong antifungal activity at very low concentration.

Use in Western herbal

German Commission E monograph recognized the herb's efficacy in:

- Loss of appetite
- Dyspeptic complaints.

Coriander is applied as a lotion for rheumatic pain. Seeds figure in some cough remedies (Weleda Cough Drops), and as an ingredient in carminatives and antispasmodic compounds.

Caution

Allergic reactions, like contact dermatitis, are known to be associated with the use of powdered coriander and more particularly with the oil.

Rosaceae

CRATAEGUS

Crataegus oxyacantha auct. non Linn.
Crataegus crenulata Roxb.
Crataegus songarica C. Koch.

Habitat

According to the Wealth of India, Crataegus oxyacantha (English Hawthorn) is an ornamental tree, 600–900 cm high, encountered in north-western Himalayas from Indus to Ravi, at altitudes of 2000–3000 m. Crataegus crenulata Roxb., syn. Pyracantha crenulata Roem, is found in the Himalayas from Sutlaj to Bhutan, at altitudes of 800–2500 m.

According to CCRAS, Crataegus crenulata is a large shrub, found especially in the Uttarakhand and Himachal area, at altitude of 900–1500 m, in Chambaa, Kaanaataal, Ghanshtaal, Bhilanganaa valley, Jamunaa valley, Tauns. Also found in Almora, Ranikhet hilly areas.

Classical & common names

Vernacular: Ban Sengli (CIMAP), Ban Sanjli (Wealth of India), Ring, Ringo, Pingyat Ghingaaru.

English: Hawthorn.

Parts used

Berries, flowering tops.

Traditional use

The extract or juice of Crataegus crenulata, locally known as Ghingaaru, is used as a tonic in heart diseases in Chambaa, Kaanaataal, Ghanshtaal, Bhilanganaa valley, Jamunaa valley, Tauns areas of north-western Himalayas. Another variety, Crataegus songarica, found in the Kashmir region, is also used locally in heart diseases.

According to the Wealth of India, the fruits are used as a cardiac tonic. They are used as a marmalade. A beverage, similar to tea, is also prepared from the rind and pip.

Hawthorn berries have not yet entered into cardiac tonics in India.

Active principles and pharmacology

According to CIMAP, the flowers of Crataegus songarica yield phenyl-ethylamine, O-methoxyphenethylamine and tyamine. Fruits gave tannins. Several flavonoids, 2-phenylchromone derivatives and chlorogenic acid have also been reported from the plant.

Amines showed cardiotonic activity.

In experimental studies, hypolipidemic activity of the berries of Crataegus oxyancantha has been confirmed. An alcoholic extract, when administered to rats fed with hyperlipidemic diet, prevented the elevation of lipid levels. A sufficient decrease in lipid deposits in liver and arota was also observed.

Use in Western herbal

Key constituents of Crataegus species are bioflavonoids (rutin, quercitin), triterpenoids, cyanogenic glycosides, amines (trimethylamine in flowers only), polyphenols, coumarins, and tannins.

Bioflavonoids of Hawthorn relax and dilate the arteries, especially the coronary arteries. This increases the flow of blood to the heart muscles and reduces the symptoms of angina. The bioflavonoids are also strongly antioxidant, helping to prevent or reduce degeneration of blood vessels.

The cyanogenic glycosides are sedative and increase the parasympathetic (vagal) tone of the heart, so slowing it down (there is an additional anti-cholinesterase action exhibited by the whole plant that possibly contributes to this latter action); in addition, trimethylamine stimulates the pulse rate slightly, and has a peripheral vasoconstrictor effect.

The combination of these properties helps to account for the paradoxical and valuable effect of exerting a sympathetic action on the coronary circulation and parasympathetic action on myocardial muscle. (Simon Y. Mills.)

In a nutshell, Hawthorn improves the blood supply to the heart by dilating the coronary vessels; improves the metabolic processes in the heart, which result in an increase in the force of contraction of the heart muscle and elimination of certain types of rhythm disturbances; inhibits angiotensin-converting enzyme.

Following doses (3 times a day) are recommended in Europe and the United States:

- Berries or flowers (dried): 3–5 g, or as an infusion
- Tincture (1:5): 4–5 ml
- Freeze-dried berries: 1–1.5 g
- Flower extract (standardized): 100–250 mg, as advised.

Hawthorn Berry Vegicaps and tablets are also available: dose as advised by the physician.

German Commission E monograph recognized the herb's efficacy in mild cardiac insufficiency (stage II NYHA).

The drug is used in senile heart and mild forms of bradycardial arrythmia, and angina under medical supervision.

Capparaceae

CRATEVA

Crateva magna (Lour.) DC.
Crateva nurvala Buch.-Ham.

Figure 1 Crateva nurvala [ZANDU]

Habitat

All over India, wild as well as cultivated.

Classical & common names

Ayurvedic: Varuna (Charaka, Sushruta); Ashmarighna (Charaka); Kumaarak, Tiktashaaka, Varana, Setu; Barnaa.
Unani: Baranaa.
Siddha: Maranlingan.

Parts used

Bark, leaves.

Dose

Decoction 50–100 ml, powder 3–6 g.

Classical use

The name Ashmarighna, used by Charaka, indicates that the herb was specific for disintegrating and expelling gravels.

The decoction of the bark mixed with other diuretic herbs was prescribed for expelling calculus and for alleviating pain in the pelvis. The bark was also specific in urinary tract infections.

The decoction of the root, with honey, was prescribed in cervical adenitis (Vrindamaadhava, Shaarangadhara Samhitaa).

Active principles and pharmacology

The bark contains a bioactive triterpenoid, lupeol, about 0.6 %; alkaloids viz. cadabacine, cadabacine diacetate; flavonoids viz. (-)-Catechin, (-)-epicatechin-5-glucoside and (-)-epiafzelchin; an isothiocyanate glucoside, glucocapparin. The root-bark also contains lupeol acetate, spinasterol acetate, taraxasterol and 3-epilupeol.

The aqueous decoction of the stem-bark is found effective in the management of urolithiasis in clinical studies and in experimentally-induced urolithiasis.

Treatment with bark decoction lowers the levels of liver glycolate oxidase, Na^+, K^+ and ATPases of small intestines of experimental rats fed on calculi producing diet.

Lithotriptic activity of the drug is being attributed to lupeol. The drug also exhibits anti-inflammatory activity.

The alcoholic extract of the stem exhibited antibacterial property against E. coli in vitro. The root extract proved antibacterial against Gram-positive and Gram-negative bacteria.

Clinical studies indicate that the drug increases bladder tone and inhibits the formation of bladder stone, encourages the removal of stones in the urine. (CCRAS.)

Promising results have been achieved by combining the drug with antiseptic and immune-stimulating herbs in urinary tract infections including cystitis. It is found effective in bladder conditions involving poor muscle tone and urinary problems associated with prostate enlargement. (CCRAS.)

Use in Western herbal

Varuna has attracted the attention of Western herbalists after the publication of a monograph by CCRAS during 1987.

Now Varuna is used in the prevention and treatment of kidney stones. It is given to people who are prone to develop kidney stones, and for reducing the tendency to stone formation. It is also prescribed for the people who already have small stones.

Iridaceae

CROCUS

Crocus sativus Linn.

Figure 1 Crocus sativus [ZANDU]

Habitat

Cultivated in Kashmir up to 2000 m and in Chaubhattia in Uttar Pradesh.

Classical & common names

Ayurvedic: Kumkuma, Rudhira, Vadrika, Kaashmira, Kaashmiraka Agnishikhaa (Charaka, Sushruta); Ghusrrn, Rakta, Vaalhika, Kshataja; Keshara.
Unani: Zaafraan.
English: Saffron.

Parts used

Stigma of the flower.

Dose

250–500 mg. (CCRAS.)

Classical use

Charaka used the powdered tendrils as ingredient of a pill which was prescribed in cataract, night-blindness and poor vision. Sushruta prescribed Kumkuma internally in skin eruptions and blood poisoning, and flowers as a part of diet for their antitoxic, antiflatulent and antibilious properties.

According to Bhaavaprakaasha and Danvantari Nighantu, Kumkuma alleviates headaches, nausea, promotes strength, cures skin diseases, wounds and ulcers as an antiseptic and antibacterial agent, and is an excellent cosmetic aid for keeping the skin radiant and free from blemishes and eruptions.

Among over-the-counter preparations, Kumkumaadi Lepa (IMPCOPS) is prescribed in acne and allied skin affections.

Kumkumaadyam Ghitam (Bhaishajya Ratnaavali) is prescribed in consumption, cough and asthma. Kumkumaadyam Tailam (ibid) for promoting complexion and for removing blemishes and freckles.

Among Unani medicines, Dava-ul-Kurkum (Qarabadeen-e-Jadeed) is given in spleen and liver diseases. It tones up kidney and urinary bladder. It is also given in flatulence, jaundice, ascitis, and inflammations. Majoon-e-Dabeed-ul-Ward (ibid) is prescribed in hepatitis, uteritis, and anaemia. Zaafraan is also an ingredient in the reputed uterine tonic, Majoon-e-Supaari Paak (Iliaz-ul-Amraz), prescribed in leucorrhoea.

In Unani medicine, Zaafraan is used as an emmenagogue. A classical compound Habb-e-Mudir is prescribed for regulating periods in women. Also used as an aphrodisiac and sex-stimulant. It has been included in Laboob-e-Kabir, a reputed sex tonic.

Active principles and pharmacology

Saffron contains a volatile oil composed of terpenes, terpene alcohols and esters. The herb also contains bitter glycosides (including crocin), carotenoids and vitamin B1 and B2.

Small doses of saffron stimulate the secretion of the gastric juices. Large doses in situ stimulate the smooth muscle of the uterus. Overdoses cause poisoning symptoms include vomiting, uterine bleeding, intestinal colic and haemorrhages. Lethal dose is approximately 12 to 20 g.

Use in Western herbal

Nicholas Culpeper (1616–1656) wrote in 1653: "Let not above 10 grains (600 mg) be given at a time (a cordial being taken in an immoderate quantity, hurts the heart instead of helping it). It quickens the brain. It helps consumption of the lungs and difficulty of breathing. It is excellent in epidemical diseases, as pestilence, small-pox and measles. It is a notably expulsive medicine, and a notable remedy for yellow jaundice."

Despite its long history as a medicinal herb, saffron has fallen out of favour as a treatment. Cheaper and superior herbs are easily found for inducing menstruation and treating period pain and chronic uterine bleeding.

In Chinese herbal medicine, saffron stigmas are occasionally used to treat painful obstructions of the chest, to stimulate menstruation and to relieve abdominal pain.

Caution

Contraindicated during pregnancy.

Fabaceae

CROTALARIA

Crotalaria juncea Linn.

Figure 1 Crotalaria juncea [ADPS]

Habitat
Grows throughout the plains of India, especially in South India.

Classical & common names
Ayurvedic: Shana, Katutikta, Tvaksaara (Charaka); Shanapushpi.
Unani: Sunn.
English: Sunn Hemp.

Parts used
Leaves, seeds, root.

Dose
3–5 g.

Classical use
Charaka used the leaves cooked as vegetable; also a decoction of the flowers and roots, in prescriptions, for dysentery and irregular bowel movements.

According to Sushruta, one who takes fruits of Shana cooked with milk, followed by intake of milk, does not suffer from senility. The plant was also used as an emmenagogue and abortifacient (Raaja Nighantu).

Unani physicians prescribe the leaves for spermatogenesis. Cooked leaves are given as a galactagogue and for stimulating sex urge. Seeds are given to women for anti-implantation. The root is given in intestinal spasm and colic; also as a diuretic.

Active principles and pharmacology
Seed oil gave fatty acids oleic, linoleic and linolenic acids; pyrrolizidine alkaloids junceine, tricodesmine, riddelline; seneciphylline and senecionine were also obtained.

Leaves are demulcent, emetic, purgative, emmenagogue and abortifacient. The extract of leaves and seeds exhibit antifertility property in albino rats. The seed extract proved hepatotoxic in mice.

Caution
The plant causes hepatic damage. No more in use in Indian medicine.

Euphorbiaceae

CROTON

Croton tiglium Linn.

Figure 1 Croton tiglium [ADPS]

Habitat
Many parts of India, especially in Uttar Pradesh, West Bengal, Assam and South India.

Classical & common names
Ayurvedic: Jayapaala, Dravanti, Dantibija.
Unani: Jamaalgotaa.
Siddha: Neervalam.
English: Purging Croton.

Parts used
Seed, seed oil.

Dose
Seed 25–50 mg; oil ½ to 1 drop (1 ml can be fatal).

Classical use
The seeds were used in medicine after they were detoxified.

Charaka prescribed sun-dried roots of wild croton internally, in prescriptions, for oedema, abdominal abscesses, and jaundice.

Sushruta prescribed the seeds of wild croton for acute constipation, abdominal dropsy and internal abscesses; cooked with the urine of she-buffalo, in jaundice; a decoction of the root as purgative.

According to Ashtaanga Hridaya, the seed or its oil was used only in severe conditions.

Wild Croton, Danti of Ayurvedic texts, has been equated with Baliospermum montanum Muell-Arg., Dravanti and Jayapaala with Croton tiglium Linn.

Active principles and pharmacology

The seed contains diterpenes: phorbol ester, including 12-O-tridecaneolyphorbol-13-acetate (TPA, myristoylphorbol-acetate, MPA); fatty oil.

The seed oil is a drastic irritant. TPA is a carcinogen, affecting prostaglandin metabolism.

Use in Western herbal

Croton seed oil, earlier used as a drastic purgative, is now obsolete as a drug.

At present, used only in Chinese medicine, in very small doses, for gallbladder colic, obstruction of the bowels and malaria. It is used only when other drugs have failed.

Periplocaceae

CRYPTOLEPIS

Cryptolepis buchananii Roem. & Schult.

Habitat

Throughout India in deciduous forests and in hedges.

Classical & common names

Ayurvedic: Krishna saarivaa (Sushruta); Karantaa, Shyaamalataa; Shyaama, Gopi, Gopvadhu, Kaalghatika.
Siddha: Kattupala. Paal-Valli. (Ichnocarpus fruitesceus R. Br. roots are used in the South as Krishna Saarivaa.)

Figure 1 Cryptolepsis buchananii [ADPS]

English: Indian Sarsaparilla (black variety).

Hemidesmus indicus (Linn.) R. Br.

Classical & common names

Ayurvedic: Saarivaa, Shweta Saarivaa, Gopavalli, Ananatmula (Charaka, Sushruta), Gopi, Gopaa, Gopkanyaa, Krishodari, Sphotaa, Gopavalli.
Siddha: Nannari.
English: Indian Sarsaparilla (white variety).

Parts used

Root.

Dose

Powder 3–5 g.

Ayurvedic Formulary of India accepts Cryptolepis buchananii and Hemidesmus indicus as the black and white varieties of the classical drug Saarivaa, referred to as "Saarivaa dwaya" or the two Saarivaas.

Classical use

Charaka used the root of Saarivaa, in prescriptions, as a blood purifier and styptic, extensively in skin diseases, leprosy, erysipelas; chronic fevers, intrinsic haemorrhage.

Sushruta used the black variety internally in respiratory troubles and wasting diseases.

Charaka used the leaves of Saarivaa alone in decoction, internally, in dysentery, cough, loss of voice, menstrual disorders.

Sushruta also used Saarivaa in cough, asthma, consumption, in body pain and abdominal swelling, internally as well as externally. Fruits were prescribed, internally, with clarified butter in haemoptysis and urinary diseases. Clarified butter, cooked with the double quantity of the decoction of Saarivaa root, was prescribed in bronchial asthma.

According to Bhaavaprakaasha and Dhanvantari Nighantu, both the varieties of Saarivaas are indicated in obstinate skin diseases, chronic rheumatism, dyspepsia, respiratory diseases and diseases due to vitiated blood; also in dysuria, dysentery, leucorrhoea, uterine haemorrhage.

Among over-the-counter compounds, Saarivaadyaasava (Bhaishajya Ratnaavali) is prescribed for purification of blood and treatment of chronic skin lesions.

Active principles and pharmacology

Cryptolepis buchananii: stems gave alkaloids, buchananine identified as 6-O-nicotinoyl-alpha-glucopyranose and 1, 3, 6-O-trinicotinoyl-L-glucopyranose.

Hemidesmus indicus: twigs of the plant gave a pregnane ester diglycoside named desinine. Roots gave beta-sitosterol, 2-hydroxy-4-methoxy-benzaldehyde, alpha-amyrin, beta-amyrin and its acetate, hexatriacontane, lupeol octacosonoate, lupeol and its acetate.

The plant exhibits blood-purifying, anti-inflammatory, antibacterial, antifungal, and antiviral properties. (CCRAS.)

A saponin from the drug was found to possess anti-inflammatory activity in experimental conditions. The ethyl acetate fraction of the plant exhibited significant anti-inflammatory activity in both acute and subacute models of inflammation.

Fresh decoction of the root was found to possess blood-purifying property.

Pet.-ether, chloroform and alcoholic extracts of the root showed significant antibacterial activity against S. aureus, S. albus, S. typhosa, V. choleriae, E. coli, Sh. sonnei, Sh. flexneri and Sh. shigae. But the aqueous ethanolic extract of the whole plant showed no activity against bacteria. The essential oil, isolated from the plant, possessed antibacterial property against Gram-positive and Gram-negative bacteria.

The aqueous ethanolic extract of the whole plant showed antiviral activity against Ranikhet disease virus.

See Hemidesmus indicus.

Cucurbitaceae

CUCURBITA

Cucurbita maxima Duch. and related spp.

Habitat

Cultivated throughout India.

Classical & common names

Ayurvedic: Peeta Kuuhmaanda, Kuushmaandaka, Kuushmaandi, Karkaaruka; Seetaaphal.
Unani: Kaddu-e-Sheerin, Aqteen (Persian).
English: Pumpkin, Red Gourd.
Cucurbita pepo Linn. has been equated with white Gourd or Safed Kumraa, Safed Konhraa, Vilaayati Kumhadaa.
Benincasa hispida (Thumb.) Cogn., syn. Cucurbita hispida, is equated with white Pumpkin or white Gourd-melon, Ash Gourd, and is also known as Kuushmaanda.
According to NAA, smaller fruits may be called Kuushmaandaka, while Kuushmaanda and Kuushmaandi seem to be the same. Kuushmaandi has been equated with Cucurbita pepo; also with Benincasa hispada.
See entry under Benincasa.

Parts used

Seed.

Dose

3–6 g.

Classical use

Sushruta used the white gourd, cooked as a vegetable, as diuretic and in mental aberrations; also prescribed the oil of seeds, internally.

According to Ayurvedic texts, the clarified butter cooked in eighteen times Kuushmaanda juice with the paste of Glycyrrhiza glabra (Yashtimadhu) alleviates epilepsy and promotes intellect (Ashtaanga Hridaya, Vrindamaadhava, Bangasena). The powder of Kuushmaanda root, taken with warm water, alleviates asthma and cough.

Kuushmaanda was also prescribed to alleviate retention of urine, gravels, calculus, pain in pelvis, genitals and dysuria (Vrindamaadhava, Bhaavaprakaasha, Haarita Samhitaa, Vaidyamanoramaa).

In Unani medicine, Kaddu-e-Sheerin seeds are prescribed for their potent diuretic, anthelmintic and urinary tract disinfectant properties.

Active principles and pharmacology

The medicinal parts of Cucurbita pepo are the fresh and dry seeds. The drug contains steriods: delta-5-, delta-7- and delta-8-phytosterols (24-alkyl sterols), including clerosterol, isofucosterol, sitosterol, stigmasterol, cholesterol, isoavenasterol, spinasterol; fatty oil; unusual amino acids, including cucurbitin and gamma-tocopherol.

Cucurbitin has been identified as (-)-3-amino-3-carboxy-pyrrolidine; the active principle responsible for the anthelmintic effect of the seed.

The concentration of cucurbitin in different samples of Cucurbita pepo ranged from 1.66 to 6.63 %, in Cucurbita maxima from 5.29 to 19.37 % and in Cucurbita moschata 3.98 to 8.44 %.

The cucurbita seeds contain about 50 % fatty oil which is an efficient diuretic, increases urine flow and gives a feeling of symptomatic relief in prostatic hypertrophy.

Administration of unsaturated fatty acid is considered to be beneficial in the treatment of prostate affections. The seed oil contains about 25 % oleic acid and 55 % linoleic acid. Phytosterols may also play some role. Delta-7 sterols present in the seed have been shown to compete with dihydrotestosterone at androgen receptors for human fibroblasts. (Tyler.)

The anthelmintic compound, found in the seed embryo, is also able to arrest cell division at metaphase. This may also be responsible for the effectiveness of the seed in hypertrophy of the prostate.

Zinc has been shown not only to reduce the size of the prostate but also to inhibit the enzyme that converts testosterone into the potent androgen DHT that causes overproduction of prostate cells. Cucurbita pepo seeds are rich in zinc content. 2 g to 3 g shelled seeds can be taken daily as a supplement. Therapeutic dose is 1 to 2 heaping dessert spoons with liquid.

Benincasa hispida, equated with Kuushmaanda in Ayurvedic reference books, was first documented in the Tang Materia Medica, written in 659 AD. (Benincasa cerifera has been equated with Tung-Kua, Pai-Kua, Shui-chih, Ti-chih; Cucurbita maxima with Hu-lu; Cucurbita moschata with Nan-Kua.)

In Chinese medicine, the seeds are used as a diuretic, and in the treatment of gravel. The seeds are also used to treat appendicitis; in combination with Rheum palmatum for internal abscesses.

The fruit contains lupeol, alpha-sitosterol and their adenine, trigonelline and histidine. The fruit juice produced tranquillizing activity and mild CNS depressant effects in mice.

The seed contains the triterpene alcohols, euphol, tirucallol, cycloartanol, taraxeol, alpha- and beta-amyrin, butyrospermol, cycloartenol, lupeol, 24-methylene cycloartanol and multiflorenol.

Cucurbita-5, 24-dienol is isolated from the leaves, stem, pericarp and roots.

Use in Western herbal

The pumpkin seeds are taken principally as a safe de-worming agent. The seeds of Cucurbita spp., Cucurbita pepo, Cucurbita maxima, Cucurbita moschata, according to Tyler, possess similar properties and enjoy a considerable reputation as teniafuges.

In Central American herbal medicine, seeds are also used as a treatment for nephritis and other urinary problems, and as a tonic to the bladder. The seeds, high in zinc, are recommended in the early stages of the prostate problems.

The pulp is used as a decoction to relieve intestinal inflammation and is applied as a poultice for burns.

German Commission E monograph recognized Cucurbita pepo seed's efficacy in:

- Irritable bladder
- Prostate complaints.

A daily dose of 10 g of ground seeds is recommended for stages I and II of prostatic adenoma to relieve micturition difficulties, though the drug does not reduce the enlargement.

Amaryllidaceae

CURCULIGO

Curculigo orchioides Gaertn.

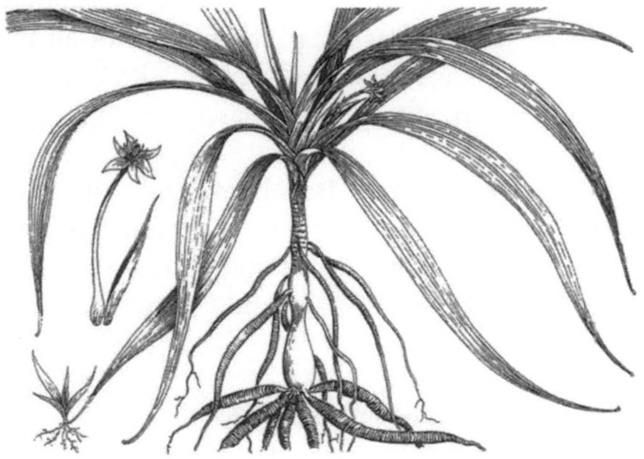

Figure 1 Curculigo orchioides [ADPS]

Habitat
Subtropical Himalaya from Kumaon eastwards; Western Ghats from Konkan southwards.

Classical & common names
Ayurvedic: Taalmuuli, Taalpatri (Charaka, Sushruta); Krishna Mushali.
Unani: Musali Siyah.
Siddha: Nilappankkizhangu.

Parts used
Tuber.

Dose
Powder 5–10 g.

Classical use
According to Ayurvedic texts, Taalmuuli is a rejuvenating and aphrodisiac herb, tonic to all vital functions of the body, including liver, lungs, cardiovascular system, spleen and kidneys. The drug is spermogenetic, age-sustaining, galactagogue, diuretic, hypotensive, antitussive. (Charaka Samhitaa, Sushruta Samhitaa, Bhaavaprakaasha, Bhel Samhitaa, Ashtaanga Hridaya, Vaidyavallabha).

The drug is also externally applied to glandular swellings (Sushruta).

In the South, Curculigo orchioides, locally known as Nilappana, is used for both, white Mushali (Asparagus adscendens) and black Mushali.

In Chyavanaprasha Lehya (Charaka Samhitaa) IMPCOPS is incorporating Krishna Mushali as a substitute for Riddhi and Vrddhi. Krishna Mushali is also an ingredient in Shataavari Lehya (Sahasrayoga), prescribed in emaciating diseases, menorrhagia, haemoptysis, jaundice, urinary disorders.

Curculigo orchioides is the main ingredient of Siddha compound, Inbooral Legiyam (Siddha Vaidya Thirattu), prescribed in asthma, cough, haematemesis, haemoptysis, tuberculosis.

Active principles and pharmacology
Rhizomes contains glycosides, 5, 7-dimethoxymyricetin-3-O-alpha-L-xylopyranosyl-4-O-beta-D-glucopyranoside, curculigoside and corhioside A; sapogenin-yuccagenin; an alkaloid lycorine; three aliphatic hydroxykentones; a long chain fatty acids.

Rhizomes exhibit tonic, aphrodisiac, demulcent, hepatoprotective, antiasthmatic, antidiarrhoeal properties; externally found efficacious in skin diseases.

The flavone glycoside of the plant showed powerful uterine stimulant property in guinea-pigs, rats and rabbits.

Zingiberaceae

CURCUMA

Curcuma domestica Valeton
Curcuma longa Linn.

Figure 1 Curcuma longa [ZANDU]

Habitat

Cultivated all over India, particularly in West Bengal, Tamil Nadu and Maharashtra.

Classical & common names

Ayurvedic: Haridraa (Charaka, Sushruta); Priyaka, Haridruma (Charaka); Kshanda, Pitaa, Gauri, Kaanchani, Krimighna, Varavarnini, Yoshitapriyaa, Hattavilaasini, Naktaahvaa, Sharvari.
Haridraa and Daaruharidraa (Berberis aristata) were known as Haridraa-dwaya in Ayurvedic texts. Rajani, Nishaa, Nishi, Raatri should be equated with Curcuma caesia Roxb, instead of C. longa.
Unani: Zard Chob.
Siddha: Manjal.
English: Turmeric.

Parts used

Rhizomes.

Dose

Powder 1–3 g.

Classical use

Charaka used the drug in liver disorders, urinary affections, dermatosis, toxicosis, piles, bronchial asthma, senility, impaired vision.

Sushruta prescribed the drug in oedema, anaemia, skin diseases, leprosy, malignant ulcers, haemoptysis, seminal and urinary disorders, urethral and vaginal discharges; also in indigestion and chronic dysentery. The drug was also given for purifying and promoting breast milk.

Sushruta included Haridraa in a medicinal oil for cleansing and sterilizing ulcers; also in an ointment for ringworm.

In folk medicine, turmeric is used as a stomachic, tonic, blood-purifier and skin-disinfectant. Externally the paste is applied to indolent ulcers, skin eruptions, inflamed joints. Mixed with calcium oxide, the paste is also applied over sprains and contusions. Mixed with warm milk, it is given for treating common cold; also as an internal antiseptic, anti-inflammatory and antispasmodic agent in bronchial asthma. The powder of all the varieties of Curcuma spp., mixed with milk, is given for expediting the healing of fractured bones.

Over-the-counter, Haridraakhand, an Ayurvedic compound formulation of Bhaishajya Ratnaavali, is available for urticaria and skin diseases. Haridraa also enters into many cosmetic formulations.

Active principles and pharmacology

Major constituents are: Curcuminoids (approx. 6 %), the yellow colouring principles, of which curcumin constitutes 50–60 %, essential oil (2–7 %) with high content of bisabolane derivatives.

Other constituents are: desmethoxycurcumin, bidemethoxycurcumin, dihydrocurcumin, common phytosterols, fatty acids and polysaccharides, A, B, C & D.

Oral administration of the drug to 116 patients with acid dyspepsia, flatulent dyspepsia or atonic dyspepsia in a randomized, double-blind study, resulted in a statistically significant response in the patients receiving the drug (500 mg powdered drug, four times daily for 7 days).

Two other clinical trials, which measured the effect of the drug on peptic ulcers, showed that oral administration promoted ulcer healing and decreased the abdominal pain involved.

Curcumin has been shown to inhibit gas formation by Clostridium perfringens in rats, given diet rich in flatulence-producing foods. In addition, sodium curcuminate has been shown to inhibit intestinal spasm. Another compound from turmeric (p-tolyl-methylcarbinol) has been shown to increase the secretion of secretin, gastrin, bicarbonate and pancreatic enzymes.

Several constituents of turmeric stimulate the flow of bile. The flavonoids in turmeric cause the contraction of the gallbladder, thereby increasing the effective emptying of the gallbladder. Turmeric has also shown synergistic action with other cholagogues. The anti-inflammatory and antibacterial properties play a major role when turmeric is employed in the treatment of gallstones, acute and chronic inflammation of the gallbladder and inflammation of the bile duct.

Curcumin, being an active choleretic, increased bile acid output by over 100 per cent. In addition to increasing biliary excretion of bile salts, cholesterol and bilirubin, curcumin also increases the solubility of bile. This suggests a benefit in the prevention of cholelithiasis. (Michael T. Murray.)

Curcumin has also exhibited hepatoprotection similar to that of glycyrrhizin and silymarin against carbon tetrachloride-and galactosamine-induced liver injury. The increase in liver enzymes, SGOT and SGPT, commonly seen in experimental models of inflammation, have been prevented by curcumin.

Turmeric exerts significant antioxidant activity. Although both water and fat-soluble extracts are effective antioxidants in various in vitro and in vivo models, curcumin is the most potent component.

The turmeric and curcumin supress the mutagenicity of several common mutagens. They also exert impressive anticancer effect against a number of chemical carcinogens in a wide range of cell types in both in vitro and in vivo studies. Curcumin has demonstrated an ability to reduce the levels of urinary mutagens. Turmeric and its derivatives not only exert antioxidant and free radical-scavenging effects, but also increase the body's natural antioxidant system, and act directly on several enzymes. Turmeric and curcumin show promise for their application as a cancer preventive and treatment adjunct.

The effects of turmeric and curcumin on the cardiovascular system include lowering of cholesterol levels and inhibition of platelet aggregation. Researchers recommend the herb as an adjuvant to the patients prone to vascular thrombosis.

In a clinical study, curcumin proved an effective anti-inflammatory drug. A 2-week double-blind cross-over study of 18 patients with rheumatoid arthritis showed that patients receiving either curcumin (1200 mg/day) or phenylbutazone (30 mg/day) got significant relief in morning stiffness, walking time and joint swelling.

In another study, the effectiveness of curcumin and phenylbutazone on postoperative inflammation was investigated in a double-blind study. Both the drugs produced a better anti-inflammatory response than a placebo, but the degree of inflammation in the patients varied greatly and was not evenly distributed among the three groups (WHO).

Clinical trial with Curcuma longa showed remarkable results in 60 % cases of asthma and cough by diminishing cough and dyspnea. (CCRAS).

Alcoholic extracts and essential oil of Curcuma longa were shown, to inhibit the growth of most organisms occurring in cholecystitis, i.e. Sarcina, Gaffkya, Corynebacterium and Clostridium. Other microorganisms that are inhibited include Staphylococcus, Streptococcus, Bacillus, and Entamoeba histolytica. The concentrations used in these studies were relatively high at 0.5–5.0 mg per ml of the alcohol extract and essential oil, and 5–100 mcg per ml of curcumin.

Essential oil, in different dilutions, also exhibited potent antifungal property on Aspergillus niger, A. flavus, Pencilium javanicum, T. viride, Curcuma oryzae, Helminthosporium and P. lapagericola. A compound preparation with Curcuma longa showed antifungal property on experimentally induced ringworm infection in calves.

Turmeric, or its alcoholic extract, administered in 2.5 g/kg and 300 mg/kg on different species of animals proved non-toxic.

Use in Western herbal

Nicholas Culpeper (1616–1654) wrote of it: "Turmerick opens obstructions, is profitable against yellow jaundice and cold distemper of the liver and spleen, half a dram being taken at night going to bed in the pulp of a roasted apple, and if you add a little saffron to it, it will be better by far." (Charaka also used the drug for jaundice. Traditional

Chinese physicians used turmeric to treat liver and gallbladder problems.)

In the West, a mixture of bromelain (a protein-dissolving enzyme from the pineapple) and curcumin is prescribed for angina. Bromelain helps reduce the formation of clots in damaged arteries, and curcumin helps to reduce the tendency for blood to clot and also reduces inflammation. (Dose: Bromelain 125–450 mg, Curcumin 400–600 mg a day.)

For Carpel Tunnel Syndrome 250–500 mg of curcumin a day is prescribed.

For rheumatoid arthritis 400–600 mg a day is recommended; in sciatica 250-500 mg three times a day.

For phlebitis, curcumin (400–600 mg a day) is prescribed with bromelain and anticoagulants. Once the clot is dissolved, the patient is asked to continue bromelain and curcumin up to three months.

German Commission E monograph recognized Curcuma domestica's efficacy in:

- Loss of appetite
- Liver and gallbladder complaints.

Curcuma amada Roxb.

Habitat

Cultivated in Gujarat, wild in parts of West Bengal, Uttar Pradesh, Karnataka and Tamil Nadu.

Classical & common names

Ayurvedic: Aamra Haridraa (Bhaavaprakaasha), Aamragandhi Haridraa, Surabhidaaru, Karpuraa.
Unani: Aambaa Haldi, Daarchoba (not to be confused with Daarhald or Berberis aristata).
Siddha: Mangaiinji.
English: Mango-ginger, Wild Turmeric.

Parts used

Rhizomes.

Dose

Juice 10–20 ml, powder 3–5 g.

Classical use

According to Bhaavaprakaasha, the rhizome is antibilious, tonic to heart and cures obstinate skin diseases.

Active principles and pharmacology

The essential oil contains alpha-pinene, alpha- and beta-curcumene, camphor, cuminyl alcohol. myristic acid and tumerone. Car-3-ene and *cis*-ocimene contribute the characteristic mango odour of the rhizome.

Rhizome exhibits carminative and stomachic properties; anti-inflammatory activity in contusion and sprains.

Curcuma angustifolia Roxb.

Habitat

Central Himalaya, Bihar, West Bengal, Maharashtra, South India.

Classical & common names

Ayurvedic: Tvakshira, Tvakshiri.
Unani: Tikhur. Tabaasheer (as a substitute for Bamboo-manna.)
Siddha: Kua.
English: East Indian Arrowroot.

Parts used

Rhizomes, starch.

Dose

Powder 5–10 g.

Classical use

Even during the 16th century, Curcuma angustifolia starch was used as a substitute for Vansalochana (Bamboo-manna) or Tavaasheer. According to Ashtaanga Sangraha, the starch is cooling and cures consumption, asthma and bronchitis (properties similar to Vansalochana or Tabaasheer). Dalhan and Chakrapaani have also attributed similar properties to the starch of Curcuma angustifolia.

In Unani medicine, the starch is used as a demulcent and is prescribed as a tonic in debility, palpitation, urinary disorders, diarrhoea, dysentery, stomatitis. Recommended for invalids and children as a milk pudding.

Active principles and pharmacology

The major constituents of the essential oil are alpha- and beta-pinene, l-arcurcumene, d-camphor, alpha-terpineol, d-borneol, zingiberol and a sesquiterpene alcohol.

The rhizome is demulcent and essential oil anthelmintic. The tubers are washed and pulped. After repeated washing and straining through cloth the starch is recovered and used in medicine after sun-drying.

Curcuma aromatica salisb.

Habitat
Wild throughout India. Cultivated chiefly in West Bengal and Kerala.

Classical & common names
Ayurvedic: Karpuraa (not to be confused with Cinnamomum camphora), Aranya-rajani Kanda, Vana Haridraa. Curcuma amada has also been equated with Karpuraa.
Unani: Daarchob.
Siddha: Kashthuri Manjal.
English: Wild Turmeric.

Parts used
Rhizomes.

Dose
1–3 g.

Classical use
According to Kaiyadeva Nighantu, Aranya-rajani Kanda is to be prescribed in diseases due to vitiation of blood, piles, inflammatory affections, cough, bronchitis, asthma and as an internal antitoxic and antiseptic agent.

In folk medicine, rhizomes are used internally as a carminative, and externally in sprains and bruises.

Active principles and pharmacology
The essential oil contains alpha- and beta-curcumene, d-camphor, d-camphene and p-methoxycinnamic acid. The colouring matter is curcumin. Several sesquiterpenoids have been identified.

Rhizomes exhibit good support in respiratory disorders.

Curcuma caesia Roxb.

Habitat
Cultivated in West Bengal.

Classical & common name
Ayurvedic: Kaali-Haldi, Nilkanth, Narkachura. (Rajani, Nishaa, Nishi, Raatri of Ayurvedic texts should be equated with C. caesia.)
Unani: Zarambaad variety (a drug of Persian Materia Medica).
Siddha: Manupasupu (Tamil).
Vernacular: Kaalihaldi (Bengal, Maharashtra, Gujarat).
English: Black Zedoary, a var. of Zingiber Zerumbet Rose. ex Smith.

Classical use
It is intriguing that in the same treatise (Bhaavaprakaasha) 32 formulations with Nishaa, 10 with Nishaa and Daarvi, and only 23 with Haridraa have been incorporated, while Haridraa was a well-identified herb since the Charaka period. In Shaarangadhara Samhitaa, 7 formulations with Nishaa, 16 with Nishaa and Daarvi, 3 with Rajani and only 9 with Haridraa have been incorporated.

In Bhaavaprakaasha, Daarvi or Daaruharidaa is incorporated in 36 and Daarunishaa only in 5 formulations. In Shaarangadhara Samhitaa, Daarvi is included in 17 and Daarunisha only in 2 formulations.

It is obvious that Nishaa, Rajani or Raatri was not used in place of Haridraa, nor the drug was its substitute. It could have been a relative or a variety of Haridraa, used for specific reasons.

According to a 16th century text (Ayurveda Saukhyam of Todaraananda), Nishaa was used for treating anaemia, oedema, parasitic infections, skin diseases including leprosy, ulcer, diseases of the eye, and cervical adenitis. The same text has quoted Haaridraka, Pitavarna, Shrimaan, Gaurdruma and Vara as synonyms of Haridraa.

Nishaa entered into Mustaadi Group of herbs, used as a carminative and galactogogue, and for treating fevers.

In folk medicine, the tubers of Kaali Haldi are used as a stimulant and carminative. Externally, the paste is used on sprains and bruises. Fresh tubers are used as a cosmetic.

Active principles and pharmacology
Steam distillation of the dried rhizomes gave 1.6 % essential oil, containing d-camphor 76.6, camphene and bornylene 8.2, sesquiterpenes 10.5 and unidentified residues 4.7 %.

Rhizomes possess carminative, stomachic and anti-inflammatory properties. Essential oil exhibits anthelmintic properties.

Curcuma zedoaria (Berg.) Rosc.

Figure 2 Curcuma zedoaria [ADPS]

Habitat

Cultivated more or less throughout India; wild in Eastern Himalaya.

Classical & common names

Ayurvedic: Karchura (Charaka); Draavida, Palaashi, Kachura Gandh-mulaka, Shati. Shati is equated with Karchura, while Shathi is equated with Gandhapalaashi (Kapurakachari/Hedychium spicatum Buch.-Ham. ex Smith.). In Kerala, Kaempfera galanga Linn. is used as Karchura and Shathi.

Unani: Zarambaad (also equated with Zingiber zerumbet Rosc. ex. Smith., known as Narkachura). Khurd or smaller variety is known as Kachura; Kalaan or bigger variety as Narakachura.
Siddha: Kichili Kizhangu.
English: Zedoary.

Parts used

Tuber.

Dose

Powder 3–6 g.

Classical use

Charaka prescribed the fruit and the tuber as appetizer; in cough, hiccup, asthma, inflammatory affections, arthritis, piles and skin diseases. Bhaavaprakaasha and Dhanvantari Nighantu attributed similar properties to the tuber.

In folk medicine, a decoction of the rhizome with long pepper, cinnamon and honey is given during colds. The decoction or powder is prescribed in flatulent colic.

Karchura Tailam (Bhaishajya Ratnaavali) is prescribed for external application in skin diseases and infected ulcers.

Safoof Chutki is a Unani compound prescribed in dyspepsia and diarrhoea as a stomachic.

Active principles and pharmacology

The essential oil from rhizomes contains alpha-pinene, d-camphor, d-camphene, cineole, sesquiterpenes and sesquiterpene alcohols, Rhizomes contain curcumene, curcumol, curdione.

Rhizhomes possess stomachic, stimulative, carminative and diuretic properties. Curcumol and curdione exhibit anti-cancer properties. The most useful properties of the plant are antibacterial and antifungal.

The essential oil was found to possess potent antibacterial property against B. subtilis, E. coli, Kl. aerogenes, S. paratyphi. S. typhi, S. aureus, Erwinia carofovora, P. solanacearum, Xantyomonas citri and X. malavacearum in nutrient agar broth. The activity was observed even in lesser dilutions, but it was less active than Curcuma angustifolia, Curcuma caesia and Curcuma longa. (CCRAS.)

Marked antifungal activity was observed with alcoholic extract of the rhizome against T. mentagophytes, M. canis, T. tonsurans, M. gypseum, E. floccsum and Candida albicans in vitro. (CCRAS.)

Use in Western herbal

Zedoary is used in much the same way as ginger (Zingiber officinale) to relieve indigestion, nausea, flatulence and to promote appetite.

The extracts of the drug are contained in numerous combinations for gastrointestinal indications and as cholagogues.

The rhizome is used in China to treat certain types of tumours. A few trials showed that zedoary reduced cervical cancer and increased the cancer-killing effects of radiotherapy and chemotherapy.

Convolvulaceae

CUSCUTA

Cuscuta epithymum L. Murr.
Cuscuta reflexa Roxb.

Habitat

Europe, Asia, South Africa.

Classical & common names

Ayurvedic: Aakaashvalli, Amarvalli. Amarvela (Cuscuta reflexa.)
Unani: Aftimoon; Kushuus .
English: Lesser Dodder, Dodder, Parasitic Climber.

Parts used

Whole plant, seeds.

Classical use

In Unani medicine, the whole plant and seeds are used for the problems of liver and gallbladder; also as stomachic and anthelmintic.

Itrifal-e-Deedaan, a compound preparation, is prescribed in intestinal worms. Sharbat-e-Deenaar (Qarabadeen-e-Jadeed) is given in sluggish liver, jaundice, constipation, pain in uterus and baldder. Itrifal-e-Aftimoon is used as a blood purifier.

Active principles and pharmacology

Cuscuta epithymum contains flavonoids (including kaempferol and quercitin), and hydroxycinnamic acid.

Cuscuta reflexa seeds contain amarbelin and kaempferol; stem gave cuscutin, cuscutalin, beta-sitosterol, luteolin, bergenin and kaempferol.

Seeds exhibit carminative, anthelmintic, alterative; stem antibilious; and whole plant purgative and febrifuge properties.

Cuscuta japonica, Dodder used in Chinese medicine, also contains the glycoside cuscutin. The drug has long been a popular ingredient in sex-tonics and longevity formulas. The drug is used in impotency, premature ejaculation, nocturnal emissions, urinary incontinence, leucorrhoea, premature senility.

Dodder appears in numerous varieties and carries the basic properties of the plant it grows upon.

Use in Western herbal

In Western herbal, Dodder is used for disorders of the urinary tract, spleen and liver; also for its laxative effect. Dodder of Thymus vulgaris is considered to be more efficacious.

Papilionaceae

CYAMOPSIS

Cyamopsis tetragonoloba (L.) Taub.

See figure 1.

Cyamopsis psoralioides (Lam.) DC.

Habitat

Cultivated in many parts of India, especially in Uttar Pradesh and Tamil Nadu.

Classical &common names

Ayurvedic: Kshudra Shimbi, Gorakshaphalini (non-classical); Guaar, Guaalin.
Tamil: Kothaveray.
English: Cluster Bean, Guar.

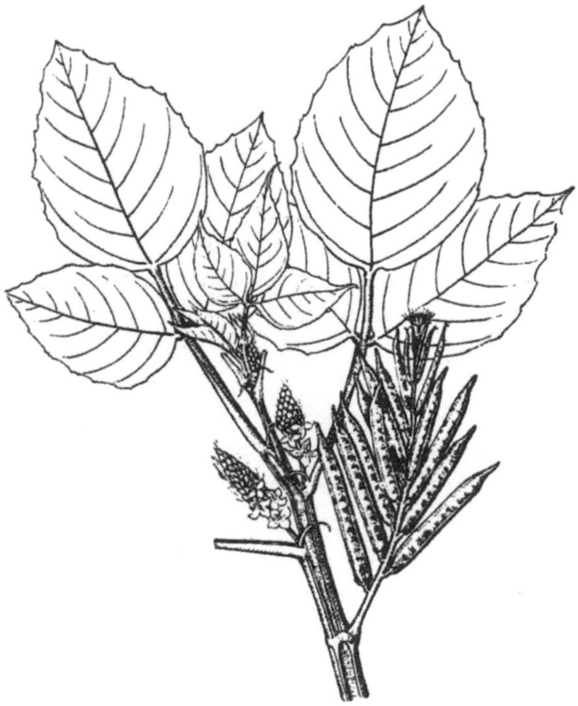

Figure 1 Cyamopsis tetragonoloba [WOI]

Parts used
Gum powder.

Traditional use
The pods are used as vegetables and seeds as a pulse. Used in biliousness. Leaves are boiled and given in night-blindness. Boiled and mashed leaves, mixed with oil, are used as a poultice in sprains and inflammations. The pods are considered constipative, though nutritive and aphrodisiac.

Active principles and pharmacology
Guarmeal contains galactomannan, 3-epikatonic acid and a saponin. The gum contains about 86 % water-soluble mucilage.

Guar gum is an effective bulk laxative, similar in action to psyllium (Plantago ovata seed hulls). It delays the emptying of the stomach and thus slows down absorption of carbohydrates. This helps stabilize blood sugar levels. The gum basically lowers cholesterol levels.

In a study at Salgren Hospital in Sweden, a number of obese subjects were put on a long-term treatment with daily intakes of Guar gum preparations, while still maintaining their normal dietary habits. In 10 weeks their hunger cravings had been significantly reduced, their blood sugar and cholesterol levels lowered and most of them had lost 10–15 lbs of weight.

A Finnish study also demonstrated the effectiveness of Guar gum in the treatment of hypercholesterolemia. 15–30 g of Guar gum per day reduced serum cholesterol 12–20 % with only expected side-effects. Laxative effects were common, as was intestinal gas. Guar gum did not inhibit nutrient metabolism in those who participated in the Finnish study.

The gum is prepared by separating the endosperm (35–42 %) from the seed coat or hull and embryo or germ, and by grinding it to a fine powder.

Major Guar producers are India, Pakistan and the US.

Use in Western herbal
Guar gum is available in health food stores in the form of capsules.

Guar gum is recommended for lowering cholesterol levels, in pre-diabetic conditions and in early stages of late-onset diabetes.

Usual dose is 2–4 capsules daily with meals.

Caution
In overdoses Guar gum can cause flatulence, abdominal distention and even intestinal obstruction.

Gramineae

CYMBOPOGON

Cymbopogon martinii (Roxb. Wats.)
Andropogon martinii Roxb.

Habitat
In drier parts of India; Maharashtra, Madhya Pradesh, Uttar Pradesh; wild as well as cultivated.

Classical & common names
Ayurvedic: Rohisha (Charaka). (varieties: motia and sofia).
English: Rosha or Rusa Grass, Palmarosa.

Parts used

Whole plant, oil.

Dose

Decoction 50–100 ml, oil 1–3 drops.

Classical use

Charaka prescribed a decoction of the grass and roots, in prescriptions, internally for abdominal diseases, splenic disorders, jaundice, fever, leucorrhoea, skin diseases. Sushruta used the herb in fever accompanied with bronchitis, cough, asthma, constriction and swelling of throat, pain in the chest, catarrh, hiccup, dyspepsia.

The extract of the herb in clarified butter is an ingredient of Dhanvantaram Ghrita (Sahasrayoga), prescribed in polyuria, jaundice, splenic affections, internal abscesses, internal and external inflammations.

Active principles and pharmacology

The major constituents of essential oil of motia variety are geraniol, geranyl acetate, citronellol and linalool. Among others are myrcene, gamma-muurolene, alpha- and beta-pinene, beta-selinene, alpha-terpinene and alpha-phellandrene.

The major constituents of sofia variety are cis-p-menthadien-1(7), 8-ol, perrillyl alcohol, limonene, iso-piperitenol and trans-p-menthadien-2, 8-ol. Geraniol is either absent or present in minor quantity.

The essential oil from the leaves is found to be antimicrobial and antifungal; also exhibits anti-inflammatory activity in rheumatism and neuralgia, when massaged externally.

Gramineae/Poaceae

CYNODON

***Cynodon dactylon* (Linn.) Pers.**

Habitat

Throughout India up to 3000 m.

Figure 1 Cynodon dactylon [ADPS]

Classical & common names

Ayurvedic: Duurvaa (Charaka, Sushruta); Bhaargavi, Shatvalli, Shatparvaa, Tiktaparvaa, Shatviryaa, Sahastravirya, Shitaa, Anantaa, Golomi.
Siddha: Arugan.
English: Couch Grass.

Parts used

Whole plant.

Dose

Juice 10–20 ml.

Classical use

Charaka and Sushruta used Duurvaa, in prescriptions, for intrinsic haemorrhage, bleeding piles, wounds and for erysipelas and skin diseases, internally as well as externally.

The paste of Duurvaa and Haridraa (Curcuma longa) was prescribed for external application in scabies, eczema, ringworm and urticaria (Vrindamaadhava). The drug was also given internally.

Duurvaa, given internally with rice water, checked vomiting (Gadanigraha). Chakradatta prescribed the paste of Duurvaa for inducing menstruation.

Duurvaadyam Ghritam (Sahasrayoga) is prescribed as a styptic, internally as well as externally. Duurvaadi Tailam (ibid) is for external application to wounds and ulcers.

Active principles and pharmacology

Cynodon dactylon contains six phenolic phytotoxins, ferulic, syringic, p-coumaric, vanillic, p-hydroxybenzoic and o-hydroxyphenyl acetic acids. Leaves contain tricin, flavone C-glycosides and a flavonoid sulphate.

The plant exhibited a number of pharmacological properties: coagulant, wound-healing, haemostatic in menorrhagia, astringent in chronic diarrhoea and dysentery, antifungal in eczema; diuretic, hypotensive, hypoglycaemic.

The paste of plant helped in immediate stoppage of blood if applied on any wound or injuries, healed the wound if kept bandaged for three consecutive days. Aqueous extract, dried and ground, mixed with paraffin applied as (1 %) ointment produced wound-healing effect in experimentally produced wound on the back of rabbits.

The plant screened against inflamed tumours, flashy excrescences on the nails produced significant antiseptic and healing activity. The plant exhibited antibiotic, antifungal and antiviral activity. The plant used in the form of pulp, mixed with turmeric (Curcuma longa) and applied externally on eczema, gave good results.

Oral administration of the juice of the plant, mixed with honey, 2–3 times a day for a few days, was found effective in menorrhagia. Local application in the form of paste upon lower abdomen reduced severe vaginal bleeding.

A decoction with sugar, used internally 2–3 times a day, was found effective in retention of urine.

Cyperaceae

CYPERUS

Cyperus rotundus Linn.

Figure 1 Cyperus rotundus [ADPS]

Habitat

Throughout India, as a weed, up to an elevation of 2000 m.

Classical & common names

Ayurvedic: Musta, Mustaa, Mustaka (Charaka, Sushruta); Abda, Ambuda, Ambhoda, Ambodhara, Bhadra, Bhadraa, Bhadramusta, Bhadramustaa, Bhadramustaka, Ghana, Jalada, Jaladhara, Meghaahvaa, Nirada, Vaarida, Vaarivaaha. (Cyperus esculentus Linn. has been equated with Naagar Mustaka.)
Unani: Saad-Kufi.
Siddha: Koraikkizhangu.
English: Nut Grass.

Parts used

Rhizomes.

Dose

Powder 3–5 g, decoction 50–100 ml.

Classical use

Charaka and Sushruta used Mustaa in diarrhoea, dysentery, indigestion; uterine and vaginal diseases.

Sushruta gave the decoction of Mustaa, mixed with honey, in diarrhoea. Mustaa rhizomes, boiled in milk, with three times water, reduced to the original quantity of milk, were prescribed in diarrhoea with mucus and pain (Sushruta Samhitaa, Ashtaanga Hridaya).

Mustaa, crushed and boiled in equal quantity of milk, reduced to one-fourth, taken with honey when cooled, was prescribed for checking diarrhoea with mucus and blood (Vaidyamanoramaa).

Over-the-counter, many compound preparations, with Mustaa as an important ingredient, are available. Gangaadhar Churna (Shaarangadhara Samhitaa), Mustakaarishta (Bhaishajya Ratnaavali) are carminative and astringent tonics used in diarrhoea, dysentery, fever.

Saad-kufi is an important ingredient in the reputed Unani compound, Jawarish-e-Jalinoos, indicated in hepatitis, flatulence of the stomach, palpitation and debility.

Active principles and pharmacology

The essential oil from rhizome contains mainly sesquiterpenoids, including alpha- and beta-rotunol, mustakone, copaene, sugeonol, sugetriol, cyperol, iso-kobusone, (+)-epoxyguaiene, (-)rotundone, 4-alpha, 5-alpha-oxidoeudesm-11-en-3-alpha-ol, cyperenone identified as isopatchoul-4(5)-en-3-one (isopatchoulenone, cyperotundone) and cyperone. Tubers have been reported to contain a triterpenoid glycoside.

Plant also contains beta-sitosterol, aureusidin and fatty oil which is constituted by linolenic, linoleic, oleic, myristic and stearic acids.

The essential oil exhibited tranquillising activity.

The alcoholic extract exhibited diuretic effect at a dose of 0.5, 1 and 1.5 g/kg p.o. in rats and this effect was biphasic.

The extract in doses of 0.5 and 1 g/kg p.o. showed protection against diarrhoea during first hour of drug administration, but after 2–3 hours there was no protection.

In a clinical trail, oral administration of the root powder (2 capsules of 500 mg each twice a day) produced significant reduction in body weight and lowered blood pressure in hypertensive obese patients. The drug had no effect on the blood pressure of normotensive obese patients.

It was found in animal studies that water and alcoholic extracts of Cyperus rotundus exhibited lipolytic action to mobilize fat from adipose tissues and helped in the reduction of obesity. Experiments showed that the drug is effective in rats and humans by releasing enhanced concentrations of biogenic amines from nerve terminals of the brain. This suppresses an appetite centre.

Cyperus rotundus has been reported to possess anti-inflammatory, antipyretic and analgesic activity in experimental models. Beta-sitosterol isolated from Cyperus rotundus possessed anti-inflammatory activity against carrageenin and cotton pellet-induced oedema in rats and was comparable to hydrocortisone and oxyphenylbutazone. A triterpenoid obtained by chromatographic separation from the petroleum ether extract was found to be 8 times more potent than hydrocortisone.

Sitosterol, in a dose of 160 mg/kg i.p. exhibited significant antipyretic effect against pyrexia by Brewer's yeast. The result was compared with acetylsalicylic acid.

In a clinical study, Cyperus rotundus 1 g thrice daily and Withania somnifera 1 g twice daily was prescribed in cases of rheumatoid arthritis. Results were quite encouraging (excellent 28 %, good 12 %, fair 45 %, poor 17 %, no response 2 %).

The extract of rhizomes exhibited effect on 14 fungal strains.

D

DAEMONOROPS

Daemonorops draco Blume

Habitat
Indo-Malayan region. The resin comes to India mostly from Sumatra and Borneo.

Classical & common names
Ayurvedic: Raktaniryaas; Khoon kharaabaa.
Unani: Damm-ul-Akhwain.
English: East Indian Dragon's Blood.

Parts used
Resin.

Dose
1–1.5 g.

Classical use
Used as styptic, astringent and antiseptic in Unani medicine for menorrhagia, bleeding piles, diarrhoea, dysentery and in dentifrices. Its powder is dusted over fresh wounds for checking bleeding.

The drug has also been equated with Dracaena cinnabari Balf. syn. Sanguis draconis; Calamus draco. (The resinous matter is collected by rubbing or shaking the fruits of D. draco and a few other related species in bags. The product is marketed in the form of rounded lumps of dull red colour, odourless and tasteless.)

In Unani compound formulations, Damm-ul-Akhwain is an ingredient in Habb-e-Bawaaseer Damiya, prescribed in bleeding piles; in Habb-e-Sil, prescribed as a styptic in phthisis; in Qurs-e-Kaknaj, prescribed in renal ulcers, vesicular ulcers, burning micturation; in Sufoof-e-Habis-ud-Dam, prescribed in polymenorrhagia, haematemesis, haemorrhagic diarrhoea.

Active principles and pharmacology
The pure resin is almost entirely soluble in alcohol, but commercial specimens contain 20 % to 40 % of insoluble substances.

The alcohol soluble resin contains 50–60 % of dracoresinotannol, mostly in the form of benzoic and benzoyl acetic esters; 13 % of a yellow resin and 2.5 % of dracoalban. Abietic acid has been isolated from the resin acids. The principal pigment is dracocarmin, an anthocyanidin. Another pigment, dracorubin has also been reported.

Use in Western herbal
Doses of 600 mg to 1.8 g were formerly given as an astringent in diarrhoea and allied ailments, but officially it is not being used internally, and is regarded as inert.

Dracaena terminalis or Chinese Colli yields Chinese Dragon's blood. It is used in dysentery and diarrhoea, and as a diaphoretic. Known as Xue jie over-the-counter.

Though the drug can be used internally for chest and abdominal problems, it is most widely used externally, often as a powder, although it is included in a number of proprietary ointments. In the US, Amber Massage Salve and Dr Shir's Liniment contain the Dragon's Blood.

Papilionaceae

DALBERGIA

Dalbergia sissoo Roxb.

Figure 1 Dalbergia sissoo [CCRAS]

Habitat

Sub-Himalayan tract, up to 1200 m from Indus to Assam and in plains throughout India.

Classical & common names

Ayurvedic: Shimshapaa (Charaka, Sushruta); Krishnashimshapaa (Charaka); Picchilaa, Krishnasaaraa.
Dalbergia latifolia Roxb. is also used as Shimshapaa. In Kerala the heartwood of Xylia xylocarpa (Roxb.) Taub is wrongly used as Shimshapaa.
Unani: Shisham (Persian).
Siddha: Sisu.
English: Sissoo, South Indian Redwood.

Parts used

Leaves, tender bark.

Dose

Decoction 50–100 ml, powder 3–6 g.

Classical use

Charaka used the expressed juice of the leaves or tender bark, in prescriptions, internally, for senility and debility.

Sushruta prescribed the drug in obesity, seminal weakness, urethral discharges, urinary calculi; in scabies and chronic skin diseases, internally as well as externally.

According to Sushruta, the decoction of the drug was a specific remedy for leucorrhoea and other vaginal discharges. Concentrated extract of the heartwood in milk was prescribed in fevers (Sushruta Samhitaa, Ashtaanga Hridaya); extract of leaves in jaundice; bark extract as an anti-inflammatory agent in piles, sciatica, and as a blood purifier (Sushruta Samhitaa, Bangasena, Bhaavaprakaasha). The oil was used externally in skin diseases and inflected ulcers (Dhanvantari Nighantu).

In Unani medicine, the drug is used for treating wounds, cutaneous affections and burning syndrome. The liquid, collected by burning the wood, is applied over ringworm. The drug is also used as an anthelmintic and emmenagogue.

Active principles and pharmacology

The stem-bark yielded dalbergenone, dalbergin, methyldalbergin and dalbergichromene.

Pods contain isoflavone glucoside, iso-caviudin, along with tectoridin, caviunin-7-O-glucoside, iso-caviunin, tectorigenin, dalbergin, biochanin A and 7-hydroxy-4-methylcoumarin.

The heartwood gave 3, 5-dihydroxy-trans-stibene, biochanin A, dalbergichromene, delbargenone and iso-dalbergin.

The wood exhibits alterative, stomachic, anthelmintic, blood-purifying, antileprotic and cooling properties. Its extract was found useful in scalding of urine.

Aerial parts showed significant spasmolytic activity.

Solanaceae

DATURA

Datura metel Linn.

Figure 1 Datura metel [CCRAS]

Figure 2 Datura metel [CCRAS]

Datura alba Nees.
Datura stramonium Linn.

Habitat

Throughout India in plains; often met with on wastelands, roadsides and gardens.

Classical & common names

Ayurvedic: Dhattuura (Sushruta); Dhuurta, Kanaka, Maatulaputraka, Maheshpriya, Pramada, Shivapriya, Unmatta, Unmattaka. (Datura metel has been equated with Raaja Dhattura of Raaja Nighantu; Datura stramonium with Krshnadhattura, the black seed variety.)

Datura stramonium: Dhuurta, Kitav, Tuuri, Maatul, Madan, Unmatta.

Unani: Dhaturaa.

English: Downy Datura, Thorn Apple.

Parts used

Seed (after detoxification).

Dose

Powder 50–100 mg.

Classical use

Sushruta included seeds in an ointment for external application in sinusitis.

Rice-milk confection cooked with the root of white Dhattura, mixed with jaggery and clarified butter, was prescribed in insanity (Chakradatta).

Externally the juice of the leaves of Dhattura, Nimba (Azadirachta indica) and Taambula (Piper betle) was applied over eczema and ringworm (Shaarangadhara Samhitaa); the juice of Manduukaparni (Centella asiatica) and the paste of Dhattura root over boils and eruptions (Gadanigraha).

The juice of Dhattura leaves was given as an anthelmintic (Bhaavaprakaasha). The juice of leaves, mixed with sugar, in a dose of 125 ml was prescribed in all types of fevers (Siddha-bheshaja-manimaalaa). Pills (30 mg) made of Dhattura seeds, pounded with the decoction of Kushtha (Saussurea lappa) for eight days, act as astringent and are prescribed in chronic coryza.

The whole plant is an ingredient in Kanakaasava (Bhaishajya Ratnaavali), prescribed in chronic bronchitis and asthma, to be taken only under medical supervision.

The seed has been incorporated in Suutshekar Rasa Vrihat (Yoga Ratnaakar), prescribed under medical supervision, for hyperacidity, acid eruction, vomiting, burning sensation in the chest.

Dhatturaadi Tailam (Sahasrayoga), the extract of Dhattura leaf and seed in oil, is prescribed, for head massage in alopecia, falling hair and cutaneous affections of the scalp.

The drug is used in Unani medicine as a hypnotic for averting premature ejaculation. Other uses are similar to those in Ayurvedic medicine.

Active principles and pharmacology

The two principal alkaloids, hyoscine and atropine/hyoscyamine were found in all parts of the plant.

Fruits contain daturaolone and daturadiol.

Roots gave several additional alkaloids, including ditigloyloxytropane derivatives, tigloidine, apohyoscine, nor-hyoscine, meteloidine, nor-hyoscyamine, cuscohygrine and tropine.

The leaf possesses anti-tumour, and anti-rheumatic properties; kills lice and other vermin. The leaf and corolla are found efficacious in swellings, flowers in asthma. The seed, root and leaf are febrifuge, antidiarrhoeal, anticatarrhal; found useful in insanity, cerebral complications. The plant and the fruit are CVS active, spasmolytic.

Use in Western herbal

Related species, Datura stramonium L., syn. Datura tatula L., known as Jimson Weed, Thorn Apple, is used in medicine for asthma, spastic or convulsive cough, pertussis during bronchitis and influenza.

The drug is used in therapy for diseases of the autonomic nervous system. It prevents muscle contractions. Gives relief in muscle spasm associated with Parkinson's disease.

The drug relaxes the muscles of the gastrointestinal, bronchial and urinary tracts and reduces digestive and mucous secretions.

John Gerard (1545–1612) recommended a salve made from the leaves for "all inflammations whatsoever".

In homoeopathy, tincture of the flowers of Datura arborea is given in brain affections and vertigo; tincture of unripe fruits of Datura ferox in delirium, mania; trituration of the seeds of Datura metel in convulsions, delirium, epilepsy, mania.

Caution

Contraindicated in glaucoma, paralytic ileus, pyloric stenosis, enlarged prostate, tachycardiac arrhythmias, acute pulmonary oedema.

Umbelliferae

DAUCUS

Daucus carota Linn. var. sativa DC. (cultivated)
Daucus carota Linn. (wild var.)

Habitat

Grown as a cold weather crop in Northern India; also in Southern and Central India.

Mostly found wild in temperate regions of Europe, Africa and Asia.

Classical & common names

Ayurvedic: Gaajara (Bhaavaprakaasha); Garjara, Granjana.
Unani: Gazar. Gazar barri, Gazar-e-dashti, Zazrul-barri.
English : Cultivated carrot. Wild carrot, Queen Anne's Lace, Bird's Nest.

Parts used

Root, seed, whole plant.

Traditional use

Cultivated carrot roots are used as a vegetable for soups, stews, curries and salads for their nutritional value. A Unani confection, Halva-e-Gazara (Qurabadeen-e-Jadeed) is recommended in debility and as a nervine tonic. Roasted roots are prescribed in palpitation, burning micturation and as an expectorant. The juice is given as a tonic.

Seeds are used for their diuretic, anthelmintic, emmenagogue and anti-colic properties. They are prescribed in anuria, dropsy, also in sexual debility (seeds are considered superior to the root). Externally, powdered seeds are dusted over wounds and ulcers.

Wild carrot is considered diuretic, deobstruent, stimulant and useful in dropsy, retention of urine and affections of the bladder.

Active principles and pharmacology

Cultivated carrot root contains sugar, pectin, carotene, vitamins, minerals and asparadine.

Wild carrot seeds contain flavonoids, and a volatile oil containing asaron and carotol, pinene, limonene, daucol, gerianol.

Cultivated carrot leaves contain significant amounts of porphyrins, which stimulate the pituitary gland and lead to the release of increased level of sex hormones.

The juice of organically grown carrots is a cleansing medicine, detoxifier, supports the liver, stimulates urine flow.

Wild carrot leaves and seeds also possess diuretic properties. They are used to counter cystitis and kidney stone formation and to diminish stones that are already formed. The seeds stimulate menstruation. Both the leaves and seeds relieve flatulence and settle the digestion.

Wild carrot root is white and inedible. While the root of Daucus carota var. sativa—the orange or red one from the cultivated variety—is edible and rich in beta-carotene, a precursor of vitamin A.

Beta-carotene is the newest antioxidant vitamin, and which protects both the insides and outsides of cells against free radical damage.

In a study 15 mg (25, 00 IU) beta-carotene supplement daily and 15 mg of beta-carotene per day from carrots were prescribed. The group that ate carrots showed more increase in white cell numbers.

Use in Western herbal

In the 1st century, Dioscorides recommended the seeds to stimulate menstruation, to relieve urinary retention and to "wake up genital virtue".

Nicholas Culpeper (1616–1654) wrote of wild carrot: "Wild carrots break wind, remove stitches in the sides, provoke urine and woman's courses, and help to break and expel the stone; the seed also showed the same effect, and is good for dropsy, and those whose bellies are swelling and rising of the mother; being taken in wine, or boiled in wine and taken, it helps conception. The leaves being applied with honey to running sores or ulcers, do cleanse them." (Seeds are not used for promoting conception. They are used as an emmenagogue and abortifacient.)

Wild carrot (Bird's Nest) is included in most of the ginseng teas, due to its diuretic, carminative and stimulating property.

Derangements of the liver are especially treated with carrots in one form or the other. Formerly, the seeds were considered as a specific in jaundice.

The seeds are used in folk medicine as a treatment for hangovers.

A poultice made of roots is used to mitigate the pain of cancerous ulcers.

Beta-carotene is recommended by beauticians as a useful supplement before sun-tanning.

The oil has shown a mild bactericidal effect, especially on Gram-positive bacteria and is used in skin creams, anti-wrinkle creams and in ointments for eczema, dermatitis, photodermatitis and pigment anomalies.

Carrot extracts, in the form of capsules, tablets, juice, oil and creams, are available over-the-counter in the West.

Ranunculaceae

DELPHINIUM

Delphinium denudatum Wall.
Delphinium pauciflorum Royle

Habitat

Common on the outer range of temperate Himalaya from Kashmir to Nepal at 1500–2700 m.

Classical & common names

Ayurvedic: Nirvishaa, Nirvishi.
Unani: Jadwaar.
English: Larkspur spp.

Parts used

Root.

Dose

Powder 500 mg–1 g.

Classical use

According to Raaja Nighantu, Nirvishi possesses anti-inflammatory, anodyne, astringent, blood-purifying properties and is indicated in catarrh, inflammations, piles, ulcers and as a detoxifying tonic.

In Ayurvedic medicine, the root is used in diseases due to vitiated blood, in dysuria, dysmenorrhoea and in convulsive disorders and diseases of the nervous system. Also in digestive disorders, sluggish liver, and abdominal affections.

In Unani medicine, Jadwaar is an ingredient in many compound formulations: Khamiraa Jadwaari; Khamiraa Gaozabaan Jadwaari (indicated in low blood pressure); Habb-e-Jadwaar (prescribed in neurasthenia, sexual debility, attenuated semen, premature ejaculation, depression, chronic fatigue, chronic catarrh); Marham Jadaawar (for ulcers, wounds, scrofula, lymphadenitis).

Active principles and pharmacology

The root contains the alkaloids denudatine, denudatidine, condelphine, talatizidine and iso-talatizidine; and the sterols campesterol, stigmasterol, sitosterol, cholesterol and avenasterol.

Aqueous extract of the root exhibited a protective effect against hepatocellular damage in rats; EtOH- (50 %) extract showed anticonvulsant activity.

Delphinium zalil Aitch. & Hemsl.

Habitat

A perennial herb with bright yellow flowers found in Persia and Afghanistan. Flowers, mixed with fragments of flowering parts and stalks, are imported and sold in India.

Classical & common names

Ayurvedic: A few Ayurvedic reference books have equated Delphinium zalil or Delphinium brunonianum Royle with Sprukkaa of Charaka and Sushruta, while INSA scientists (P. Ray et al) have equated Sprukkaa with Melilotus officinalis, syn. Trigonella corniculata, better known as Aspurka or Naakhunaa.
Unani: Zarir (Persian), Zalil, Asbarag, Gul-zalil.
English: Zalil Larkspur.

Parts used

Dried flowers, leaves, whole plant.

Classical use

Charaka prescribed the flowers of Sprukkaa in the form of a paste or as an ingredient of a medicated oil for alleviating oedema. Charaka's Balaa Taila and Amritaadya Taila were prescribed in arthritis and inflammations.

Charaka used pulverised seeds as an ingredient of pills for fever, emaciation, obstructed and difficult labour. Sushruta prescribed the drug internally in blood poisoning and skin eruptions.

In Unani medicine, Zarir is boiled with Munakkaa (Vitis vinifera) and is prescribed in the enlargement of liver and spleen. The drug, mixed with barley flour, is applied over swellings; the ash of the plant over ringworm, cutaneous affections and ulcers and wounds.

Use in Western herbal

Larkspur is Delphinium consolida of the West. Habitat: Europe, US, especially the Western US. Flowers are used in medicine as a diuretic and vermifuge, as a sedative and appetite stimulant.

The drug contains diterpene alkaloids, including delphinine. The presence of alkaloids cannot always be found.

Although the delphine pharmacologically exhibits a paralyzing effect upon periferal and motor nerve-endings and central nervous system, poisoning among humans by Delphinium consolida has never been observed (PDR).

The tincture of seeds is employed to destroy lice and nits in the hair.

The expressed juice of the leaves is considered good as an application to bleeding piles. The juice of the flowers and an infusion of the whole plant were earlier prescribed against colic.

The name consolida refers to the plant's power of consolidating wounds.

Papilionaceae

DESMODIUM

Desmodium gangeticum (L.) DC.

Habitat

Common on lower hills and plains throughout India, ascending to 1500 m in the Himalayas.

Classical & common names

Ayurvedic: Shaalaparni (Charaka, Sushruta); Vidaarigandha, Sthiraa (Sushruta); Tiparni, Amshumati, Somyaa, Guhaa. Desmodium gangeti-

Figure 1 Desmodium gangeticum [CCRAS]

Figure 2 Desmodium gangeticum [CCRAS]

cum is also used as Prishniparni; and Uraria picta Desv. as Shaalaparni.
Siddha: Pulladi.
English: Salpan.

Parts used

Whole plant.

Dose

Decoction 50–100 ml.

Classical use

Charaka prescribed Shaaliparni, boiled with milk, in cardiac pain; decoction of the leaves, in prescriptions, for constipation, intestinal infections, oedema and fever.

Sushruta prescribed the drug internally in persistent dysentery, abdominal swellings, respiratory diseases, non-healing ulcers, haemorrhagic and inflammatory affections.

In hemicrania, the juice of the plant was applied locally, it was also prescribed as a snuff (Ashtaanga Hridaya, Gadanigraha). Water decoction of Shaaliparni, Prshniparni (Uraria picta) and Puga (Areca catechu) was prescribed in all types of diarrhoeas (Bangasena).

Active principles and pharmacology

Roots afforded pterocarpanoids gangetin, gangetnin and desmodin, and several alkaloids. The aerial part gave indole-3-alkylamines.

Biological activity of the root is anti-inflammatory and analgesic. It exhibited astringent, antidiarrhoeal, carminative, and anticatarrhal properties.

Gangetin produced significant anti-inflammatory activity in exudative phases of inflammation in 50 and 100 mg/kg p.o.; produced 21.89 % of antiarthritic activity against formaldehyde-induced arthritis in rats. It also showed significant analgesic activity in acetic acid-induced writhing as well as on hot plate method. The compound was found non-toxic up to a dose of 7 g/kg. (CCRAS).

The alkaloid from seeds exhibited non-specific spasmolytic property against acetylcholine, histamine on guinea-pig ileum, and on rat uterus against 5-HT, acetylcholine and oxytocin.

The alkaloid from the root produced spasmogenic effect antagonised by pentolimium and produced anticholinesterase, smooth muscle stimulant, CNS stimulant and depressor responses, nicotine-like effect on dog's intestines in situ and carotid blood pressure.

Purified alkaloidal fraction caused marked positive inotropic effect on isolated heart preparation in doses from 0.001–0.1 mg, suggesting cardiac-stimulating activity, but the effect diminished on repetition. The effect was comparable to tyramine. The effect was enhanced after noradrenaline infusion, suggesting depletion of noradrenaline by the drug from the heart muscle (CCRAS).

Use in Western herbal

Salpan is used as a bitter and tonic for improving poor appetite and digestion and for treating dysentery, haemorrhoids, catarrhal conditions, such as bronchitis and asthma.

Desmodium spp., Desmodium adscenden, which grows in Africa, South America and other tropical areas, is used in West Africa and Europe to treat asthma and liver problems, such as hepatitis. Research in Ghana indicates that the drug has a significant protective effect on liver. (Andrew Chevallier.)

Poaceae

DESMOSTACHYA

Desmostachya bipinnata Stapf.

Habitat

Throughout the plains of India in dry and hot areas and in sandy deserts.

Classical & common names

Ayurvedic: Kusha, Suuchyagra, Kshurapatra, Yajnabhuushana.

Kusha and Darbha (Imperata cylindrica Beauv.) are separate grasses. Earlier, Darbha was supposed to be a variety of Kusha. Desmostachya bipinnata and Eragrostis cynosuroides Beauv. are synonyms, according to the Wealth of India and INSA.

Siddha: Tharubai.

English: Sacrificial Grass (smaller variety).

Parts used

Root.

Dose

Decoction 50–100 ml.

Classical use

Charaka prescribed the drug, in prescriptions, for dysuria. Charaka's Kushaadya Ghrita was specific for lithiasis.

Sushruta used the herb internally in urinary diseases and calculi; with cow's milk in haemoptysis. He used the decoction, with other herbs, for cleansing and disinfecting wounds.

The herb was used, in prescriptions, for treating bleeding piles and meno-metrorrhagia (Bangasena, Vrindamaadhava). For leucorrhoea, the root pounded with rice-water was prescribed (Vrindamaadhava).

Kusha belongs to the Trna Panch Muula group of 5 drugs (the roots of 5 grasses)—others being Kaasha (Saccharum spontaneum), Nala (Arundo donax), Darbha (Imperata cylindrica Beauv.) and Kaandekshu (Saccharum officinarum). These roots are specific in urinary affections in Ayurvedic medicine.

Pancha-trnamuul Kashaaya (Sahasrayoga), extracted in milk, is prescribed in dysuria and haematuria.

Active principles and pharmacology

A terpene was isolated from the grass. The grass contains crude protein 6.76 %, crude fibre 40.30 %, ether extract 1.61 %, N-free extract 42.22 %, and total ash 9.12 %.

Analysis of leaf culms (on a dry basis) gave water-soluble extract 8.61 %, pectose (with fat and wax) 32.13 %, lignin 10.33 %, cellulose 48.91 % and ash 3.5 %.

The drug exhibited CNS depressant, anticonvulsant and negative inotropic and chronotropic effect (CCRAS).

The whole plant extract showed diuretic, urinary tract disinfectant and blood-purifying properties and was found efficacious in menorrhagia, calculus affections and dysentery.

Scrophulariaceae

DIGITALIS

Digitalis purpurea Linn.

Habitat

Cultivated in Jammu and Kashmir at 2000–2300 m, Darjeeling district, and in the Nilgiris. (Detailed description in the Wealth of India.)

Classical & common names

Ayurvedic: Hritpatri, Tilapushpi (non-classical).
English: Common Foxglove.

Parts used

Leaf.

Dose

500 mg (in divided doses), maintenance dose 100 mg, Alcoholic extract 5–15 drops. Under medical supervision.

Classical use

Used sparingly in Indian systems of medicine as a herb.

In 1913, Dr. Gordon Sharp carried out a biological assay of digitalis grown in India. In 1920, Dr Douglas Cow assayed tincture prepared from the leaves. The drug was tried in the form of tablets, powder, tincture and injections, as a myocardial stimulant in congestive heart failure, auricular flutter and auricular fibrillation. An ointment of digitalis glycosides was used for cleansing wounds. In cases of burns, it was found more effective than tannic acid or silver nitrate in preserving cells severely injured by heat. For preparing standardized powder of the drug, Hatcher and Brody's cat method, modified by Col. RN Chopra, was adopted. It gave desired results. (Indigenous Drugs of India, 1933.)

Used as a homoeopathic medicine in India.

Active principles and pharmacology

Foxglove leaves are the leaves of Digitalis purpurea or of Digitalis lanata. The drug contains cardioactive steroid glycosides (cardenolides): including from the A-sequence (aglycone digitoxigenin) purpurea glycoside A (primary glycoside), digitoxin (secondary glycoside); B-sequence (aglycone gitoxigenin): purpurea glycoside B (primary glycoside), gitoxin (secondary glycoside), digitalinum verum; E-sequence (aglycone gitaloxigenin): glucoverodoxin, glucogitaloxin, gitaloxin; pregnane glycosides: including digipurpurin, diginin, digitalonin; steroid saponin: including desgalactotigonin, digitonine, purpureagitoside; and anthraquinones. (PDR.)

Foxglove stimulates contraction of the heart muscles, reduces the frequency of heartbeat and lowers the oxygen requirement of the heart (in relation to performance). (PDR.)

Cardiac glycosides also stimulate urine production which lowers the volume of blood, thus lessen the load on the heart.

In medical history, foxglove is best known as the discovery of William Withering, an 18th century English country doctor. Curious about the formula of a local herbalist, he explored the plant's potential medical use. His work led to the production of a life-saving medicine. Today, glycosides are extracted from Digitalis purpurea and Digitalis lanata by pharmaceutical industry to produce the heart drugs digitoxin and digoxin.

Use in Western herbal

Though the leaf has been shown to be less toxic and is required in smaller doses, only the extracted pharmaceutical drugs are used.

In homoeopathy, tincture from the second year of the plant is used in angina pectoris, dropsy and heart ailments under medical supervision.

Caution

The simultaneous administration of arrhythmogenic substances increases the risks of the appearance of cardiac arrhythmias. (PDR.)

The constant use of Digitalis may lead to hypertrophy of the heart (M. Grieve).

Dioscoreaceae

DIOSCOREA

Dioscorea bulbifera Linn.

Figure 1 Dioscorea bulbifera [WOI]

Dioscorea sativa Thumb. non L.

Habitat
Common throughout India, ascending up to 2000 m in the Himalayas.

Classical & common names
Ayurvedic: Vaaraahi, Vaaraahikanda, Grshti, Bana Aalu, Suraalu.
Unani: Baaraahikand.
English: Potato Yam, Bulb-bearing Yam.

Parts used
Tuber.

Dose
Powder 3–6 g.

Classical use
Sushruta prescribed Vaaraahi with honey and milk for rejuvenation.

According to Ayurvedic texts, Vaaraahicanda is spermatogenic, revitalizer, age-sustainer and a tonic for the whole body, easy to digest, detoxifies morbid factors (Ashtaaga Hridaya, Vrindamaadhava, Kaiyadava Nighantu, Gadanigraha).

Active principles and pharmacology
Rhizomes contain D-sorbitol, furanoid norditerpenes, diosgenin, lucetin, neoxanthin, violaxanthin, zeaxanthin, auroxanthin and cyrptoxanthin.

Analysis of tubers gave the following values (on a dry matter basis): albuminoids 7.36–13.31, fat 0.75–1.28, carbohydrates 75.11–81.39, fibre 3.28–9.64 and ash 3.31–7.08 %. The tubers are used for the preparation of starch.

Tacca integrifolia Ker-Gawl., syn. T. aspera Roxb. is also known as Vaaraahikanda (spelt wrongly in the Wealth of India). In a few Ayurvedic reference books, it has been equated with Vaaraaha. The tubers are sweet, nourishing, digestive and tonic; are used in haemorrhagic diathesis, cachexia and leprosy. The drug was known as Suuran in the South (not easily available) and was an ingredient in Suuranaadi Leham and Suuranaadi Ghritam, prescribed as restorative tonics in haemophilic conditions. Suuran and Shuuran have also been equated with Amorphophallus campanulatus Blume ex Decene.

Use in Western herbal
Related species are used in the West: Dioscorea villosa Linn, the wild yam, contains steroidal saponins based on diosgenin, including dioscin and dioscorin; starch, tannins; is used as an antispasmodic, gives relief in bilious colic and in the nausea of pregnant women, cramps, neuralgic affections, spasmodic hiccough and asthma; also in rheumatic conditions.

Dioscorea villosa of Mexico is the best source of diosgenin, a precursor of the female hormones used in contraceptive pills.

Dioscorea opposita, syn. Dioscorea batata, the Chinese Yam, is an ingredient of the reputed "Pill of Eight Ingredients", traditionally prescribed in Chinese medicine to treat hypothyroidism, nephritis and diabetes. Known as Shan Yao, it is used as a gentle tonic for tiredness, loss of weight, lack of appetite, frequent urination, chronic cough and wheezing; also in vaginal discharges and spermatorrhoea, as a source of natural hormone in menopausal syndrome.

Dioscorea hypoglauca, the Bei Xie seven-lobed Yam of China, is used mainly for cystitis and urinary tract infections as an antibacterial agent. Also used in rheumatoid arthritis due to its anti-inflammatory properties.

The extracts of Dioscorea hypoglauca are used in the synthesis of contraceptive pills.

According to Rudolf Weiss (1985), the wild yams are the only known available source of diosgenin, a precursor in the synthesis of progesterone.

Many yams are used as a starter material to produce hydrocortisones for eczema creams.

Ebenaceae

DIOSPYROS

Diospyros peregrina (Gaertn.) Gurke
Diospyros embryopteris Pers.
Diospyros malabarica Desr.

Habitat

Throughout India in shady wet places and near streams.

Classical & common names

Ayurvedic: Tinduka, Nilasaara (Charaka); Tinduki (Sushruta), Sphuurjak, Kaalaskandh, Asitkaarak.
Unani: Tendu.
English: Gaub Persimon.

Parts used

Bark, flowers, fruits.

Dose

Decoction 50–100 ml.

Classical use

Charaka used the fruits as a single drug in urticaria, phlegm, excessive bile secretion, piles.

Sushruta's Tinduki has been equated with Diospyros glutinosa Koenig, syn. of Diospyros embryopteris, by INSA scientists (P. Ray et al). Sushruta prescribed Tinduki bark internally in haemoptysis and excessive menstrual and other vaginal discharges; ripe fruits internally for catarrh and biliousness. He prescribed pulverized bark, in prescriptions, for applying over major burns and erysipelas.

The extract of the bark was prescribed in diarrhoea (Haarita Samhitaa), flowers and fruits in hiccough (Bangasena), externally the paste of the bark in burns and for repigmentation of the skin (Bhaavaprakaasha, Ashtaanga Hridaya).

In Unani medicine, the powder of unripe fruit is prescribed in vitiation of semen; spermatorrhoea and as a semen coagulant.

Active principles and pharmacology

Fruits yielded betulin, marsformosanone and lupeol. Fruit pulp and seeds furnished lupeol, betulin, gallic acid, betulinic acid, hexacosane, hexacosanol, sitosterol, a triterpene ketone.

Leaves gave beta-sitosterol, betulin and oleanolic acid. Roots yielded a flavonol glycoside.

From bark lupeol and betulinic acid were isolated.

Ethanolic extract of the bark showed potent anti-inflammatory and analgesic properties (CCRAS). The fruit, the ethanolic extracts of stem bark, exhibited antiamoebic activity against Entamoeba histolytica. Fruits possess antibiotic and antibacterial activities against E. coli, and Ranikhet disease virus. The plant extract exhibited IFM-like activity against RDV.

Alcoholic extract of the plant exhibited CNS-depressant activity in mice and rats; spasmolytic action on rabbit and guinea-pig ileum; bradycardia and hypotension in anaesthetised dogs.

Some species like Diospyros montana Roxb. are poisonous.

Visha Tinduka is a totally different herb, equated with Strychnos nux-vomica Linn.

Papilionaceae

DOLICHOS

Dolichos biflorus Linn.

Habitat
An important pulse crop, particularly in Madras, Mysore, Mumbai and Hyderabad.

Classical & common names
Ayurvedic: Kulattha, Kulittha, Kulatthikaa (Charak); Kulatthaka, Kulatthikaa. Allied species: Dolichos falcatus Linn.
Unani: Kulthi.
Siddha: Kollu.
English: Horsegram.

Parts used
Seed.

Dose
Powder 6 g, decoction 50–100 ml.

Classical use
Charaka gave the paste of seeds in misperistalsis, emaciation, cough, fever. The cooked pulse was advised as a food. He recommended the soup of Kulattha in heart disease.

Sushruta gave the soup of Kulattha in asthma and cough, prescribed the decoction of Kulattha added to milk as an anthelmintic.

Kulatthyaadi Ghrita was prescribed for litholysis (Bhaavaprakaasha), especially for the stone in the spermiduct.

The soup of seeds was prescribed in arthritis, cervical adenitis, internal tumours, piles and inflammatory affections (Chakradatta, Bangasena).

By the 16th century, Kulattha was used by Indian physicians for allergic conditions, such as urticaria, chronic rhinitis, asthma, bronchitis; and for treating flatulence and adiposity.

In Unani medicine, the concentrated water extract of Kulthi seeds and Shalgam (Brassica rapa) seeds (both in equal quantity) is given for destroying stones in the kidney. The fresh juice of Kulthi and Muuli (Raphanus sativus) leaves is also given.

The paste of the seed is prescribed for external application on piles; also as a cosmetic for blemishes and freckles.

Among classical Ayurvedic compounds, Kulatthaadyam Ghritam (Bhaishajya Ratnaavali) is prescribed for calculus and urinary tract infections; Kulatthayuusha (ibid) for colic; Kulatthaguda (ibid) for hiccough, asthma, bronchitis, and fever. In Sahasrayoga 2 Kulattha-based compounds, Kulatthaadi Kashaayas, have been incorporated; for colic and abdominal ailments; and for cough and asthma.

Active principles and pharmacology
Analysis of the seed gave the following values: moisture 11.8, crude protein 22.0, fat 0.5, mineral matter 3.1, fibre 5.3, carbohydrates 57.3, calcium 0.28, and phosphorus 0.39 %; iron 7.6 mg, nicotinic acid 1.5 mg, carotene 119 (international vitamin A units) per 100 g.

The seed is rich in urease; other chemical constituents present are: strepogenin, beta-sitosterol, a phyto-haemagglutinin, beta-N-acetylglucosaminidase, alpha- and beta-galactosidases, alpha-mannosides and beta-glucosides. 5-hydroxy-7, 3', 4'-trimethoxy-8-methylisoflavone-5-neohesperidoside has also been isolated from the seed.

During a clinical study, the aqueous extract of the seed (in a dose of 8 oz daily in divided doses for 2 weeks) on patients with splenomegaly, with cirrhosis, nutritional oedema, and chronic diarrhoea exhibited much increase in urine output and reduction in body weight.

The horsegram and green gram in the sprouted form with supplement of skimmed milk and methionine showed good growth-promoting effect on albino rats.

Alcoholic and pet.-ether extract of the seed exhibited significant hypocholesterolemic activity in rats in 250 to 500 mg/kg.

The extract produced myocardial stimulant effect on frog and rabbit hearts, but exhibited hypotensive effect in anaesthetised dogs.

Alcoholic extract of the seed showed significant diuretic effect in rats.

Both the extracts were non-toxic in rats up to a dose of 2 g/kg p.o. (CCRAS.)

Pharmacologically, the plant, seeds and extracts exhibited astringent, diuretic, antispasmodic, myocardial stimulant, hypocholesterolemic, hypotensive, haemagglutinating properties (CCRAS).

Umbelliferae

DOREMA

Dorema ammoniacum D. Don

Habitat
Persia, South-West Asia, Southern Siberia.

Classical & common names
Ayurvedic: Ushaka (Sushruta).
Unani: Ushaq (Arabic); Ushah (Persian); Kandal (Afghani).
English: Gum ammoniac.

Parts used
Oleogum resin.

Dose
500 mg–1 g.

Classical use
Sushruta refered to the drug as "Ushakaadi", indicating that it was not a single drug, but oleogum resin of Ushaka and other related species. It was prescribed in obesity, catarrh, calculus, dysuria and diabetes.

In Unani medicine, the gum resin, mixed with honey, is prescribed as an antispasmodic and expectorant in chronic cough. Zimad-e-Tehal and Zimad-e-Kibreet are recommended for external use in induration of spleen and spleenitis; Marham-e-Ushaq in muscular rigidity, scrofula and fistula.

Active principles and pharmacology
Ammoniacum contains a resin (60–70 %), gum (20 %), volatile oil (0.1–1.0 %) including ferulene and linalyl acetate, free salicylic acid and coumarins.

The drug exhibits anti-inflammatory activity on hard inflammations, especially when applied externally; and antispasmodic internally.

Use in Western herbal
The oleogum resin is listed in British Pharmacopoeia as an antispasmodic and as an expectorant that stimulates the coughing up of thick mucus. It is a specific treatment for chronic bronchitis, asthma, persistent coughs, especially in the aged when the secretion is tough and viscid.

The resin is also used for its mild diuretic action; sometimes as a diaphoretic and emmenagogue.

Externally the drug is used in the form of a plaster for swellings of the joints and for indolent tumours.

Lamiaceae

DRACOCEPHALUM

Dracocephalum moldavica Linn

Habitat
Temperate Western Himalaya in Kashmir from 2100–2400 m.

Classical & common names
Hindi: Raam Tulasi.
Unani: Faranjmushk; Tukhm-(seeds).

Parts used
Leaf, seeds.

Dose
5 g.

Classical use
In Unani medicine, the drug is used in prescriptions as a cardiac and brain tonic; also in liver and intestinal affections, considered as a deobstruent and used in headache due to cold.

Active principles and pharmacology
Citral and geranyl acetate are major constituents of the essential oil. Others include: alpha-pinene,

nerol, citronellol, linalool, geraniol, limonene and carproic acid.

The seed exhibits astringent, wound healing, febrifuge, carminative, demulcent and tonic properties. It can be used with benefit as an adjuvant in neurological disorders.

The National Formulary of Unani Medicine has equated Faranjmushk with Dracocephalum moldavica, as well as with Ocimum basilicum Linn.; it should have been equated with Ocimum gratissimum Willd. Ocimum basilicum Linn. should be equated with the Raihan of Unani medicine, Baabuii or Kaali Tulasi of Ayurvedic medicine and Barbari of Ayurvedic texts (Bhaavaprakaasha, Kaiyadeva Nighantu).

Polypodiaceae

DRYOPTERIS

Dryopteris filix-mas (L.) Schott.

Habitat
Throughout temperate regions of Americas, Europe, Asia, near damp and shady terrains.

Classical & common names
Unani: Sarakhs, Sarakhs Muzakkar.
English: Male Fern.

Parts used
Dried fonds, dried rhizomes, oleo-resin extracted from the root.

Dose
As advised by the physician.

Use in Classical medicine
In Unani medicine, the drug is used in prescriptions as an anthelmintic. The powder is dusted over wounds and ulcers; mixed with oil, applied over the scalp for killing lice. Internally the drug is irritant and is invariably prescribed with antispasmodic and soothing herbs.

Active principles and pharmacology
Male Fern contains an oleo-resin (6 %) with phloroglucinol derivatives or "filicin", in particular flavaspidic acids, filicinic acids, paraspidin, desaspidin, responsible for the de-worming action of the drug. In addition, the fern contains triterpenes, alkanes, a volatile oil and resins.

Use in Western herbal
Male Fern root or its oleo-resin is used as a specific treatment for tape-worms. It acts by paralysing the muscles of the worm, forcing it to relax its hold on the gut wall. The root is prescribed with a non-oily purgative.

Preparations of Male Fern are used externally for rheumatism, muscle pain, neuralgia, and sciatica.

Caution
Other vermifuge drugs must be tried before using Male Fern.

The drug should not be administered in the presence of anemia, cardiac, liver or kidney diseases or diabetes. Not to be prescribed to children under 4 years and to elderly persons.

Asteraceae

ECLIPTA

Eclipta alba Hask.

Figure 1 Eclipta alba [ZANDU]

Eclipta prostrata Linn. Hassk.
Wedelia chinensis Merril

Habitat
Throughout India up to 2000 m on the hills.

Classical & common names
Ayurvedic: Bhringaraaja, Bhringa, Bhringaaraka, Bhringaara (Charaka, Sushruta); Kesharaaja, Kesharanjan, Maarka, Maarkava.
Unani: Bhangraa.
Siddha: Karrisalai.
English: Trailing Eclipta Plant.

Parts used
Whole plant.

Dose
Fresh juice 5–10 ml.

Classical use
In Ayurvedic texts 3 varieties have been mentioned: white-flowered, yellow-flowered and blue-flowered. The white-flowered variety is used in Ayurvedic medicine. The yellow-flowered variety has been equated with Wedelia chinensis Merrill and is known as Kesharaaja.

Charaka used the plant juice, with honey, for asthma, cough and senility.

Bhringaraaja powder 1 part, black sesamum seeds half part, Aamalaka (Emblica officinale) half part, classically known as Bhringaraaja Churna, was prescribed as a rejuvenating and age-sustaining tonic. The juice of Bhringaraaja, followed by milk, was prescribed daily for a month for retarding old age and for tackling the problems of old age. Bhringaraaja was also used as a detoxifying, deobstruent and antiseptic herb in vitiated blood, anaemia, splenic and liver enlargements, catarrhal jaundice, hyperacidity, gastritis, dysentery (Ashtaanga Hridaya, Vrindamaadhava, Bangasena, Chakradatta).

The juice of Bhringaraaja was used for washing wounds and soft chancre (Gadanigraha). The powder of the root and Haridraa (Curcuma longa) was applied locally on skin affections (Bhaavaaprakaasha).

The oil extract of leaves was prescribed by Charaka and Sushruta for anointing the head, for hair growth and for giving natural colour to grey hair.

Over-the-counter, Neelibhringaadi Tailam (Sahasrayoga) is still available and is prescribed for promoting hair growth and for giving natural colour to grey hair.

In Unani medicine, the juice of leaves is prescribed in skin diseases, allergic urticaria, in flatulence, colic and liver affections. The seeds are used in sexual debility and as a tonic and aphrodisiac. Externally the paste of leaves is applied over swellings.

Eclipta powder is an ingredient in Siddha's Seenthil Choornam (Agaththiya Vaidya Kaviyam), prescribed with clarified butter in bronchitis, cough, rheumatism and skin diseases.

Active principles and pharmacology

Leaves gave stigmasterol, alpha-terthienymethanol, wedelolactone, desmethylwedelolactone and desmethyl-wedelolactone-7-glucoside.

Aerial parts gave beta-amyrin and luteolin-7-O-glucoside.

Roots gave hentriacontanol, heptacosanol and stigmasterol.

The plant, in a dose of 90 mg kg p.o., exhibited hepatoprotective effect on CCl_4-induced hepatic damage in albino rats. Various biochemical and histopathological studies showed significant reduction in SGOT and alkaline phosphate values and healing of liver tissues within 4 weeks.

The plant also exhibited anticatarrhal, deobstruent, spasmogenic, hypotensive properties. The juice of fresh leaves, mixed with neem oil, applied locally, promotes hair growth; the herb's extract boiled with coconut oil promotes hair growth and the black pigment of the herb makes grey hair black.

The paste of the herb applied on fracture sight produced healing effect.

The shoot extract showed antibacterial activity against Staphylococcus aureus and Eclipta coli. The alcoholic extract of the plant is reported to possess antiviral activity against Ranikhet disease virus. The water extract of the plant at absolute concentration tested on sitotroga cerebella ova recorded strong ovicidal property.

Use in Western herbal

Eclipta alba is used in the West more as a Chinese medicine. The herb was first mentioned in the Chinese Tang Medica of 659 AD.

A decoction is used to invigorate the liver, to prevent premature greying of hair, and to staunch bleeding; and for spermatorrhoea, menorrhagia.

In the Caribbean, the juice is taken for asthma and bronchitis; for enlarged glands, as well as for dizziness, vertigo and blurred vision.

The plant is employed externally for skin problems.

Elaeocarpaceae

ELAEOCARPUS

Elaeocarpus ganitrus Roxb.
Elaeocarpus sphaericus (Gaertin.) K. Schum.

Habitat

West Bengal, Madhya Pradesh, Maharastra, Orissa, Andhra Pradesh, Western Ghats.

Classical & common names

Ayurvedic: Rudraaksha.
English: Utrasum Bead Tree.

Parts used

Flesh of fruit, dry beads (stones).

Dose

Powder of dry bead 3–5 g.

Classical use

For acquiring tranquillity and relaxed mental state, the beads of Rudraaksha were worn over the body. Internally, the powder of the bead was used in neurological disorders, brain fags, psychological instability, cardiac depression, restlessness and insomnia. Externally, the paste was used for allaying burning sensation and for treating eruptions of smallpox (Raaja Nighantu, Vrindamaadhava).

In folk medicine, the flesh of the fruit is prescribed in epileptic fits. The powder of beads is used for expelling thick and sticky phlegm due to its emetic property. It is a common notion that the beads, while remaining in contact with the body, exert a balancing influence on blood pressure, nervousness and cardiac irregularities.

Active principles and pharmacology

Leaves yield 7 isomeric alkaloids: l-, D-epi- & D-si-epi-isoelaeocarpiline, D- & L-elaeocarpiline, Lalloelaeocarpiline & D-epiadloelaeocarpiline; also elaeocarpidine, DL-elaeocarpine; and a new alkaloid rudrakine. Quercetin, gallic and ellagic acids are also present.

See Planta Med. (1977) 32:197.

Ethanolic (50 %) extract of the stem-bark exhibited hypoglycaemic and CNS-active properties.

Aqueous extract of the fruit showed hypotensive, sedative, anticonvulsant, spasmolytic, choleretic, bronchodialatory and cardiostimulant properties.

Zingiberaceae

ELETTARIA

Elettaria cardamomum Maton

Figure 1 Elettaria cardamomum [WOI]

Habitat

South-Western India.

Classical & common names

Ayurvedic: Elaa, Sukshmailaa, Kshudrailaa, Bhrngaparnikaa, Tutthaa (Charaka); Draavidi, Prithvikaa, Triputaa, Truti, Upkunchikaa. Ilaayachi.
Unani: Heel Khurd.
Siddha: Ilam.
English: Lesser Cardamom.

Parts used

Seed.

Dose

1–3 g.

Classical use

Charaka and Sushruta prescribed the seed in indigestion, colic pain, intestinal mucus, gastralgia, rhinitis, catarrh, hemicrania and as an antitoxic agent.

Elaa was prescribed with the juice of Aamalaki (Emblica officinalis) fruits in suppressed urine (Sushruta Samhitaa, Ashtaanga Sangraha, Ashtaanga Hridaya, Vrindamaadhava, Gadanigraha). For dysuria Elaadi Kashaaya or Elaa with curd-water was prescribed (Gadanigraha).

Elaa was prescribed as a cardiac tonic in combination with the root of Piper longum (Bangasena).

Chakradatta prescribed Elaadi Churna in vomiting.

Elaadi Churna (Gadanigraha) is available over-the-counter and is prescribed in loss of appetite, dyspepsia, for cough and bronchitis with Vaasaarishta (Gadanigraha) and as a cardiac tonic with Arjunaarishta (Bhaishajya Ratnaavali).

Elaa is an important constituent of Sitopalaadi and Taalisaadi Churna (Shaarangadhara Samhitaa) prescribed in chronic throat infections, cold, cough and asthma. Both the compounds are available over-the-counter.

For external application in skin affections, the paste of Elaadi Churna or Elaadigana Tailam (Sahasrayoga) is prescribed. (Elaadigana group consists of 28 herbs, which are used for promoting complexion and curing itching, urticarial rashes and pimples.)

Heel Khurd is an ingredient in Sharbat-e-Madani (IMPCOPS), prescribed as an adjuvant in intestinal disorders, copious and bilious vomiting and diarrhoea.

Active principles and pharmacology

Volatile oil includes, borneol, camphor, pinene, humulene, caryophyllene, carvone, eucalyptole, terpinene, sabinene.

Research showed that the volatile oil (2–8 %) has a strong antispasmodic action, confirming the herb's effectiveness in relieving wind and treating colic, griping and dyspepsia.

Analysis of cardamom capsules gave the following values: moisture 20.0, protein 10.2, ether ext. 2.2, crude fibre 20.1, carbohydrate 42.1, mineral matter 5.4, calcium 0.13, and phosphorus 0.16 %; iron 5.0 mg/100 g.

Biological activity of cardamom is found to be antimicrobial, carminative, antiemetic and diuretic. The drug is reported to be a cholagogue and to have virustatic properties.

See also Amomum subulatum Roxb.

Use in Western herbal

German Commission E monograph recognized cardamom's efficacy in the following areas:

- Loss of appetite
- Liver and gallbladder complaints
- Common cold
- Inflammation of the mouth and pharynx
- Cough/bronchitis
- Fevers and colds.

Ground seeds as well as galenic preparations are used. Considered good for productive coughs.

Myrsinaceae

EMBELIA

Embelia ribes Burm. f.

Habitat

Throughout the hills of India, up to 1700 m.

Classical & common names

Ayurvedic: Vidanga, Bhasmaka, Ghoshaa, Chitrabijaa (Charaka); Krmishatru (Sushruta); Krimighna, Krimihaa, Krimihara, Krimihrit, Krimiripu.
Unani: Baobarang.

Figure 1 Embelia ribes—Flowering branch [ADPS]

Siddha: Vaivilangam.
English: Babreng, Embelia.

Parts used

Fruits, root.

Dose

Powder 3–5 g.

Classical use

According to Charaka and Sushruta, Vidanga is the best remedy for worms and excels as an anthelmintic. The drug is given early morning, fasting, mixed with milk or syrup, followed by a purgative.

Sushruta, in addition, prescribed the drug internally for dyspepsia, colic pain, gastralgia, catarrh, cough, asthma, and as an emetic.

Active principles and pharmacology

Berries gave quinones embelin, rapanone, homoembelin, homorapanone and vilangin.

Aqueous extract of berries in a dose of 100 mg/kg p.o. on rats from day 1 to 5 post-coitum prevented implantation in 8 out of 10 animals.

Clinical trials, with the tablets of an aqueous extract in a dose of 200 mg/day for 10 days on 56 fertile women of 25–40 years old, commencing the therapy on 5th day of menstrual cycle, exhibited no pregnancy during the period of trial and no contra-indications i.e. nausea, vomiting, headache, bleeding, weight gain and tenderness, were noticed in any of the cases under trial, contrary to the side-effects of established synthetic contraceptives. The action of the tablet under trial was due to its antiestrogenic activity (CCRAS).

50 % ethanolic extract of seed showed diuretic as well as spermicidal activity in experimental models.

In vitro studies of anthelmintic activity with disalts of embelin (disalts being remarkably active) exhibited significant results.

Alcoholic and aqueous extracts of ripe fruits produced significant anthelmintic property on motility of earthworms. The alcoholic extract was found more potent.

Alcoholic and aqueous extracts, in the form of capsules (500 mg), 2 capsules a day, cured 80 % and 50 % of the nematode-infected patients respectively. The capsules were free from toxicity and no purgative was needed during the treatment.

Embelia ribes and Butea frondosa (Palaasha) seeds had definite ascaricidal properties with doses of 6 g or more, administered either singly or in combination; favourably comparable with santonin. They also exhibit significant effect on taenia and hookworm.

Both alcoholic and aqueous extracts of the seeds of Embelia ribes showed antibacterial activity against Staph. aureus and Embelia coli.

Embelia ribes, externally, shows good results in ringworm as an antifungal and antimicrobial agent.

Euphorbiaceae

EMBLICA

Emblica officinalis Gaertn.

Figure 1 Emblica officinalis [ZANDU]

Phyllanthus emblica L.

Habitat

South Indian hills up to 1500 m and Kashmir. Cultivated also in other parts of India.

Classical & common names

Ayurvedic: Aamalaki, Aamalaka, Dhaatri (Charaka, Sushruta, Bhaavaprakaasha); Kaayasthaa, Amoghaa, Amritaphala, Aamala, Aaamalaa, Dhaatriphala, Vayasyaa, Vrishya, Shiva.
Unani: Aamlaa.
Siddha: Nellikkaina.
English: Myrobalan, Indian Gooseberry.

Parts used

Fruit.

Dose

Juice 10–20 ml, powder 3–6 g.

Classical use

In order to promote intellect and prevent senility, and for longevity, the powder of Aamalaka, mixed with honey, was prescribed (Charaka Samhitaa, Sushruta Samhitaa, Ashtaanga Hridaya, Vrindamaadhava). Sesamum seeds, Eclipta alba, Embelia ribes, Pterocarpus marsupium, calcined gold, calcined iron were used as adjuvants.

The reputed rejuvenating tonic, Chyavanpraasha was composed of Aamalaka as well as the group of 10 roots (Dashmuula), the 4 aromatics (Chaturjaatakas), the group of 8 spermogenic drugs (Ashtavarga).

The most popular supporting drugs were sesamum seeds and Eclipta alba for making Aamalaka a potent age-sustainer.

Sushruta prescribed Aamalaka, impregnated with its own juice, with honey, followed by intake of milk, as an aphrodisiac.

For treating acidity, a powder consisting of Aamalaka, Shataavari (Asparagus racemosus) and sugar in equal parts, mixed with equal quantity of honey, was prescribed with milk (Bangasena).

Aamalaka and Haridraa (Curcuma longa) were considered the most efficacious remedy for urinary disorders (Sushruta Samhitaa, Ashtaanga Hridaya, Vrindamaadhava).

To treat the cases of suppressed urine, Sushruta prescribed the juice of Aamalaka with Elaa (Elettaria cardamomum).

The juice of Aamalaka, with honey, was given in the case of dysuria, haematuria and in burning sensation in the female genitals (Bhaavaprakaasha).

The juice of grapes and Aamalaka was given as an adjuvant in the case of jaundice (Charaka Samhitaa).

Aamalaka seeds were also used in Indian medicine. For treating leucorrhoea the paste of Aamalaka seeds mixed with sugar and honey, or powder or juice of Aamalaka was prescribed (Charaka Samhitaa). Aamalaka seed powder was given with rice-water for alleviating meno-metrorrhagia.

Among over-the-counter drugs, Chyavanpraasha (Charaka Samhitaa) and Brahma Rasaayana (ibid) are prescribed extensively during winter.

Dhaatryaarishta (Bhaishajya Ratnaavali), a fermented alcoholic compound, containing Aamalaka juice, Piper longum (Pippali), honey and sugar, is an appetiser, digestive, carminative, haematinic and mildly laxative, prescribed in anaemia, jaundice, diabetes, hiccup; also in intermittent fever.

Triphala Churna (Bhaishajya Ratnaavali), containing equal parts of Chebulic myrobalans (Haritaki), Belleric myrobalans (Bibhitaki) and Aamalaka, is given for constipation with hot water, with honey in coughs and in the form of a decoction as a wash in conjunctivitis. Decoction is also used for washing and cleansing wounds and ulcers.

Dhaatri Lauha (Bhaishajya Ratnaavali) is a calcined iron-based compound, containing Aamalaka, Glycyrrhiza glabra and Tinospora cordifolia, prescribed as an appetiser, antacid and mildly laxative; also in anaemic conditions.

Among Unani compounds, Itrifal-e-Ustukhuddus (Qarabadeen-e-Jadeed) is prescribed as a tonic in cephalalgia, hemiplegia, Bells palsy, epilepsy and catarrh. Itrifal-e-Kabir (ibid) and Itrifal-e-Sagheer (ibid) are brain tonics.

All Unani itrifals invariably contain Triphala (the 3 myrobalans).

Dry Aamalaa is the main ingredient in Jawarish-e-Aaamlaa Sada (Qarabadeen-e-Jadeed), prescribed as a tonic in dyspepsia, sluggish liver, hepatitis, bilious diarrhoea, and in palpitation and nervousness.

Bhringaamalakaadi Taila (Sahasrayoga), prepared with the fresh juice of Eclipta alba and Emblica officinalis; and Nilibhringaadi Taila (Sahasrayoga), prepared with the fresh juice of Indigo leaf, Eclipta alba, Baloon vine (Cardiospermum halicacabum) and Emblica officinalis, are available over-the-counter for promoting hair growth, for retarding greying and for dyeing grey hair.

A Unani preparation, Raughan-e-Aamlaa, contains juice of fresh Aaamlaa, Barg-e-Muraad (Myrtus communis) and Post-e-Sanobar (Pinus longifolia) in Til (sesamum) oil base, is prescribed in falling of hair, as a hair tonic.

Active principles and pharmacology

The plant gave the tannins glucogallin, corilagin, cheoulagic acid, and 3, 6-digalloylglucose.

The fruit contained vitamin C, L-(+)-thioascorbic acid; cytokinin substances identified as zeatin riboside and nucleotide; suspension culture gave phyllembin.

The root yielded ellagic acid, quercetin, and beta-sitosterol.

Alcoholic extract of the plant (1 g/kg) has shown an increase in the cardiac glycogen and decrease in SGOT, SGPT and LDH in isoprotenol-treated rats, suggesting a cardio-protective action.

Aamalaka fruit is one of the richest sources of vitamin C. The fruit juice contains nearly as much vitamin C as orange juice, a single fruit is equal in antiscorbutic value to one or two oranges.

The dried fruit loses only 20 % of its vitamin in 375 days when kept in a refrigerator; loses 67 % in the same period when stored in room temperature. Vitamin C saturation is more quickly reached with Aamalaka powder than with synthetic vitamin C.

The fruit pulp contains moisture 81.2, protein 0.5, fat 0.1; mineral matter 0.7, fibre 3.4, carbohydrate 14.1, Ca 0.05, P 0.02 %; Fe 1.2 mg, nicotinic acid 0.2 mg, vitamin C 600 mg/100 g. Vitamin C content up to 720 mg/100 g of fresh pulp and 921 mg/100 c^3 of pressed juice has been recorded. The fruit is also a rich source of pactin. A tannin containing gallic acid, ellagic acid and glucose in its molecule and naturally present in the fruit, prevents or retards oxidation of the vitamin.

The antiscorbutic value can be conserved by preserving the fruits in salt solution or in the form of dry powder.

Dried fruit gives good results in haemorrhage, diarrhoea and dysentery. In combination with calcined iron, the fruit is used as a remedy for anaemia, jaundice, dyspepsia. Acute bacillary dysentery may be arrested by drinking a sherbat of Aamalaka with lemon juice.

The dried fruit possesses detergent properties and is used as shampoo for the head. A fixed oil (yield 16 %), brownish yellow in colour, exhibits hair growth-promoting property.

The fruit contains 28, twig bark 21, stem bark 8–9, and leaves 22 % tannin. The fruit contains 2 tannins, one giving on hydrolysis gallic acid, ellagic acid and glucose and the other giving ellagic acid and glucose.

Triphala finds its application as a laxative in constipation, biliousness, dyspepsia, piles, enlarged liver and ascites.

The seed, powdered, and mixed with the powdered root of Withania somnifera (Ashwagandhaa) in equal quantity is a restorative invigorator, especially during the winter.

The seed is also found useful in the treatment of asthma, bronchitis, biliousness, nausea and vomiting. Leaves are used as infusion with fenugreek seeds in chronic dysentery and as a bitter tonic.

Gentianaceae

ENICOSTEMMA

Enicostemma hyssopifolium (Willd.) Verdoorn
Enicostemma littorale Blume

Habitat

Throughout India, from Punjab and Gangetic plain to Kanyakumari up to 500 m.

Classical & common names

Ayurvedic: Naagjhvaa, Maamajjaka, Naahi, Tikshnapatraa, Chhotaa Chirayataa.
Unani: Naai, Naahi.
English: Indian Gentian.

Parts used

Whole plant.

Dose

Powder 2.5–5 g decoction 50–100 ml.

Classical use

Naagjhvaa is one of the classical antidiabetic drugs mentioned in Ashtaanga Hridaya and Shodhal Samhitaa. Shaligram Nighantu described Shukanaasaa as synonymous to Naagjhvaa. It has now been equated with Corallocarpus epigeus Benth., quite distinct from Enicostemma hyssopifolium.

Used as a substitute for Chirayataa.

In Unani medicine, a syrup prepared with the leaves of Naai, Cuminum cyminum and Allium sativum is prescribed as an emmenagogue and in retention of urine.

Also used in folk medicine as an anthelmintic and blood purifier, and in rheumatism, swellings and fevers.

Active principles and pharmacology

Various parts of the plant are extremely bitter due to glycosides and ophelic acid. Alkaloids, phytosterol, glucoside, sugars and tannins have been reported. Alkaloids include gentianine, erythrocentaurin, enicoflavine and gentiocrucine.

Flavonoids include apigenin, genkwanin, isovitexin, swertisin, saponarin, 5-O-glucosylswertisin and 5-O-glucosylisoswertisin. Glucoside swertiamarin, a triterpene betulin have also been isolated.

The alcohol-insoluble portion of the unsaponifiable matter has yielded n-hexosanol, heptacosane, nonacosane. The non-saponifiable matter of the petroleum ether extract has yielded myristic, stearic and oleic acids.

Aqueous extract of the plant on oral administration in normal as well as diabetic (alloxan-treated) rabbits produced slight lowering of blood sugar in normal rabbits in 5 hours, whereas it produced a significant hypoglycaemic activity on diabetic rabbits. (CCRAS.)

The herb qualifies for its use as a supportive agent in diabetes mellitus.

Antibacterial activity of the plant extracts against Gram-positive and Gram-negative organism has been reported.

Caution

In a toxicity study, on the 3rd day after the administration of large doses of the plant's extract in rats and pigeons, liver and kidney showed early signs of degeneration. No toxicity has been reported with the standard dose in humans.

Papilionaceae

ERYTHRINA

Erythrina indica Lam.
Erythrina variegata Linn. var. orientalis (Linn.) Merrill

Figure 1 Erythrina variegata [CCRAS]

Figure 2 Erythrina variegata [CCRAS]

Habitat

Throughout India, including the Andaman and Nicobar Islands.

Classical & common names

Ayurvedic: Paaribhadra, Paaribhadraka, Paarijaataka, Mandaara; Dadap. Kamtakipalaasha, Raktapushpa. In Ayurvedic texts, Nimba-taru is equated with Paaribhadra, Nimba and mandaara.
Siddha: Kalyanamurungu.
English: Indian Coral Tree.

Parts used

Bark, leaves.

Dose

Decoction 50–100 ml.

Classical use

According to Ayurvedic texts, the juice of fresh leaves, mixed with honey, was prescribed in parasitic infections; a decoction of leaves with Aamalaka fruits (Emblica officinalis) in acid-gastritis and biliousness. The leaves were also prescribed in retention of urine and obstinate urinary diseases including diabetes; and as a galactagogue and emmenagogue. Fresh juice of leaves was used in earache; the paste was applied externally on inflammations, glandular swellings, venereal buboes. Dry powder was used for disinfecting the skin of infants. The bark was used as astringent, anti-bilious, anthelmintic and febrifuge. (Sushruta Samhitaa, Haarita Samhitaa, Gadanigraha, Vrindamaadhava, Raaja Nighantu, Bhaavaprakaasha.)

In folk medicine, dried water-extract of the stem-bark is used as an anthelmintic and has been included in many compound formulations. The bark is also used for skin diseases including leprosy. A decoction of leaves is given for promoting digestion; also for providing relief in period pain.

Active principles and pharmacology

Plant gave docosyl alcohol and beta-sitosterol. The bark yielded erysotrine, erysodine, erythraline, erysopine, erysopitine, erysodicnone, crysonine, hypaphorine, hypaphorine methyl ester, campesterol, stigmasterol, 24-methylenelophenol, stachydrine, erysovine and isoflavones—erythrinins A, B, and C in addition to osajin and alpinumisoflavone.

Alkaloids from the bark exhibited spasmolytic, smooth muscle relaxant, CNS-depressant and anticonvulsant properties in animal studies.

Myrtaceae

EUCALYPTUS

***Eucalyptus globulus* Labill.**

Figure 1 Eucalyptus globulus—fruiting branch [WOI]

Habitat

Found in the Nilgiri and Palni Hills, Shimla and Shillong, from 1350 to 2350 m.

Classical & common names

Ayurvedic: Tailaparna; Sugandhapatra (non-classical), Haritaparna, Neelaniryaasa.
Siddha: Karpooramaram.
English: Blue Gum Tree, Australian Gum Tree.

Parts used

Leaves, oil extracted from leaves.

Classical use

Eucalyptus globulus was introduced in India in 1843. The essential oil of the leaves is not used in the Indian systems of medicine.

Indian pharmaceutical industry is using the oil largely as a mosquito and vermin repellent and as an ingredient of germicidal and disinfecting preparations.

Eucalyptus oil is used locally as an antiseptic, especially in the treatment of infections of upper respiratory tract and in certain skin diseases. Mixed with an equal amount of olive oil, it is found useful as a rubefacient for rheumatism. It is also used as a stimulating expectorant in chronic bronchitis and asthma. It is frequently given by inhalation.

Dried leaves are used in the form of tincture in asthma, phthisis and chronic bronchitis.

As a standardized product, the oil should contain at least 70 % cineole, while Indian Eucalyptus oil contains a much lesser amount of cineole.

Active principles and pharmacology

Eucalyptus oil consists of the volatile oil from various cineol-rich species—Eukalyptus globulus, Eukalyptus fructicetorum, Eukalyptus smithii.

Eukalyptus globulus oil contains over 80 % cineol, and p-cymene, alpha-pinenes, limonene, geraniol, camphene. Eukalyptus smithii oil contains 70 % cineol. Indian eucalyptus oil contains about 55 % cineole.

In vitro, eucalyptus oil exhibits antibacterial and antifungal activity. The drug inhibits prostaglandin biosynthesis, and shows a mild hyperemic, expectorant and secretolytic motor effect, when used topically.

Use in Western herbal

German Commission E monograph recognized the drug's efficacy in:

- Cough/bronchitis
- Rheumatism.

The drug is used internally and externally for catarrh of the respiratory tract and externally for rheumatic complaints.

The plant was introduced into Europe in the 19th century and is now being commercially cropped in parts of southern Europe. The essential oil of the herb is being used in rubs for muscle aches, steam inhalations for catarrh and colds; also added in a small quantity to throat pastilles. The leaf is also available over-the-counter.

Caution

Contraindicated in inflammation of gastrointestinal area and the biliary ducts; also in diseases of the liver.

Cases of eucalyptus oil poisoning have been sometimes reported. The symptoms are epigastric burning with nausea and vomiting, dizziness; muscular weakness also occurs.

Myrtaceae

EUGENIA

Eugenia aromatica Kuntze
Syzygium aromaticum (L.) Merr. & Per.

Habitat

Cultivated in South India, mainly Kerala and Coorg. Native to the Malacca islands.

Classical & common names

Ayurvedic: Lavanga (Charaka, Sushruta); Devakusum, Chandanpushpaka, Shrisangya, Shriprasuunak, Shriprasuuna, Vaarija.
Unani: Qaranful.
Siddha: Kirambu.
English: Clove.

Parts used

Dried flower buds.

Dose

Powder 1–3 g.

Classical use

Charaka prescribed dried flowers alone or in medicinal oils for cough, hiccup, vomiting, inflammatory conditions of the mouth and throat. Charaka and Sushruta used dried flowers in halitosis. Clove-water was prescribed in nausea, indigestion and colic (Vrindamaadhava, Siddha-bheshajamanimaalaa).

Lavanga is an important ingredient in Khadiraadi Gutika (Charaka, Yoga Ratnaakar), still available over-the-counter, and is prescribed in sore throat, hoarseness and excessive cough.

Avipattikar Churna (Bhaishajya Ratnaavali), an over-the-counter compound, is prescribed for hyperacidity, constipation and colic.

Lavangaadi Churna (Shaarangadhara Samhitaa) is given in anorexia and flatulence; Lavangaadi Bati (Vaiday Jiwan) in bronchitis, cold and cough.

Qaranful is an ingredient in reputed Unani tonic, Jawarish-e-Jalinoos (Qarabadeen-e-Jadeed), prescribed as a sovereign tonic. Araq-e-Ambar is prescribed in debility.

Raughan-e-Qaranful is used externally in toothache; as a local stimulant in sexual debility, and internally (1–3 drops, diluted) in flatulence and colic.

Active principles and pharmacology

Cloves contain 14–21 % of volatile oil, 10–13 % of tannin and a crystalline substance, caryophyllene. Capsaicin is also present.

The oil of clove contains phenols, sesquiterpenes and a small quantity of esters, ketones and alcohols. Medicinal oil has a phenol contents of about 82–90 %. The oils which have relatively low phenol content are mainly used in pharmacy, while the strong oils are used in the manufacture of vanillin.

In many studies Eugenia aromatica has been tested for its antibacterial activity.

The oil inhibited tuberculosis at 1:80 dilution. Ethanol extract of clove inhibited the growth of C. botulinum at 0.5 %. Eugenol and Isoeugenol also inhibited the growth of C. botulinum. Ethanol extract of Eugenia aromatica exhibited the strongest inhibition on Gram-positive and Gram-negative microorganism tested on agar media at 5000 ppm.

Eugenol exhibited marked antifungal activity against 4 mould species using agar diffusion method. It showed bactericidal activity in various dilutions. Vermicidal activity has also been shown.

A formula, 9 % clove and 22 % liquorice suspended in 0.5 % carboxymethylcellulose solution, when injected in rats with induced gastric ulcer, healed 55–72 % of the ulcerated area (Chemical Abstracts (1978) 89–65270).

Clove's use as an appetizer and digestive cordial has been substantiated pharmacologically. It was also found efficacious as an antitussive agent.

Caution

In a study, the aqueous suspension of Eugenia aromatica and its oil, used in large doses, produced oedema, congestion and necrosis. Large doses of eugenol caused pulmonary oedema in dogs when given intravenously; produced rigidity of the hind limb when injected intra-arterially.

In the doses commonly used, no adverse reactions are reported.

Eugenia jambolana Lam.

Figure 1 Eugenia jambolana [CCRAS]

Figure 2 Eugenia jambolana [CCRAS]

Syzygium cuminii (L.) Skeels.

Habitat

Throughout India, up to 1800 m.

Classical & common names

Ayurvedic: Jambu (Charaka, Sushruta), Mahaaphalaa, Phalendraa, Surabhipatra.
Unani: Jaamun.
Siddha: Naval.

Parts used

Fruit, seed, leaf, bark.

Dose

Decoction of bark 50–100 ml, powder 3–6 g.

Classical use

Charaka used seeds, leaves and the stones of the fruits in decoctions for diarrhoea, nausea, vomiting, consumption, and the bark, in prescriptions, as an astringent.

Sushruta prescribed the fruit internally in obesity, in vaginal discharges and menstrual disorders; cold infusion in intrinsic haemorrhage.

The juice of Jambu, Amra (Mangifera indica) and Aamalaka (Emblica officinalis) leaves mixed with goat's milk and honey was prescribed in diarrhoea with blood (Chakradatt, Shaarangadhara Samhitaa).

The juice of Jambu fruit or its vinegar was given in sluggish digestion (Siddha-bhashajamanimaalaa). The juice of Jambu bark, mixed with equal quantity of goat's milk, was prescribed in sprue syndrome (Chakradatta).

The bark of Jambu is an ingredient in a classical compound Ushiraasava (Bhaishajya Ratnaavali), available over-the-counter and prescribed in haematemesis, epistaxis, bleeding through rectum and urinary passage. The seed is an ingredient in Pushyaanuga Churna (ibid), indicated in leucorrhoea and other uterine affections.

In classical Unani compound, Qurs-e-Ziabetus Khaas (National Formulary of Unani Medicine) the seed of Jaamun is one of the main ingredients. It is prescribed in diabetes mellitus.

In Unani medicine, the seed of Jaamun, kernel of Mangifera indica (Aam) and flowers of Acacia arabica (Kikar) are prescribed in diarrhoea and dysentery. In folk medicine, the ash of dried bark of Rai Jaamun is given in a dose of 1.25 g with water on empty stomach or 1 hour before lunch and dinner for 40 days to diabetic patients.

Jaamun and Rai Jaamun of the North and Naval and Narva varieties of the South have been equated with Syzygium cuminii and Syzygium cerasoides (Roxb.) Chatterjee and Kanjilal f. (syn. S. nervosum DC., S. operculatum Niedenz, Eugenia operculata Roxb.) respectively. Both the varieties are used in various formulations of Indian medicine.

Active principles and pharmacology

Stems, leaves and fruits contain essential oil having alpha- and beta-pinene. limonene, *cis*-ocimene, trans-ocimene, alpha-humulene and bornyl acetate as major constituents.

Fruits contain the anthocyanins, delphinidin-3-gentiobioside, malvidin-3-laminaribioside and petunidin-3-gentiobiside. Fruits yielded citric, malic and gallic acids.

Seeds gave beta-sitosterol.

Bark exhibits astringent; bark and seed kernel hypoglycaemic; fruit stomachic, carminative, diruretic; bark and leaf antidiarrhoeal properties.

Fruit juice of Eugenia jambolana, administered orally on rabbit (fasting), produced 20–50 % rise in blood sugar in 2 hours, then there was steep fall of blood sugar to the level below the fasting blood sugar.

The seed, in a dose of 10 mg/kg p.o. on normal and alloxanised rabbits, exhibited significant hypoglycaemic effect up to 23 % and 20 % respectively (CCRAS).

In a clinical trial, 6 non-diabetics and 5 diabetics were administered 150 g of Jambu fruit (pulp approx. 115 g) after overnight fasting. Average of fasting blood sugar in non-diabetics was 75.3 (±2.63 SE), in diabetics 240 (±44.3 SE). After 1 hour, 2 hours, and 3 hours the blood sugar level in the first group was 68.1 (±3.7 SE), 67.1 (±4.6 SE), and 68.3 (± 2.11 SE), and in the second group 240 (±44.3 SE), 254 (±12.6 SE), and 272.6 (±49.67 SE). Jambu fruit (pulp) did not exhibit a significant role in the management of diabetes (CCRAS).

Powdered seeds of Eugenia jambolana, in a clinical trial, were given in the dose of 12 g daily in 3 divided doses for 3 months to 30 patients of non-insulin-dependent diabetics. Initial fasting blood sugar in mg % ± SE was 163.00 (±14.83), after 1 month 129.61 (±12.03), after 2 months 99.64 (±9.49), after 3 months 130.11 (±18.87). The reduction in blood sugar was found significant after 1 month and 2 months. It was less after 3 months (CCRAS).

Euphorbiaceae

EUPHORBIA

Euphorbia neriifolia Linn
Euphorbia ligularia Roxb.

Habitat
Common in rocky ground, often cultivated for hedges in villages. Wild in Central India, Orissa and South India. Cultivated in West Bengal.

Classical & common names
Ayurvedic: Snuhi, Samantdugdhaa, Sehunda, Singhtunda Snuk, Gudaa, Sudhaa, Vajjri, Vajjradram; Thuuhar. Euphorbia nivulia, Euphorbia antiquorum, Euphorbia trigona and Euphorbia royleana are also used as Snuhi.
Siddha: Ilaikkalli.
English: Holy Milk Hedge.

Parts used
Latex of the root.

Dose
Juice 5–10 ml, latex 125–250 ml.

Classical use
Sushruta prescribed the drug internally in acute constipation, jaundice, abdominal glands, abdominal dropsy, intestinal paralysis, urinary calculi; milky juice with water as a powerful purgative; for external application, as an ingredient of a medicated oil, for cleansing and sterilizing ulcers, dressing wounds and anal fistula. Sushruta gave Curcuma longa (Haridraa) powder mixed with the latex of Euphorbia neriifolia for external application on piles.

Steamed leaves, in the crushed form, were prescribed for topical application on piles, fistula-in-ano, itching and swellings.

According to Ayurvedic texts, the latex of Snuhi should be prescribed for patients suffering from obstinate abdominal diseases including ascitis, splenic disorders, internal tumours, obstinate skin diseases, urinary disorders including diabetes (Charaka Samhitaa, Bhaavaprakaasha).

Active principles and pharmacology
The milky latex contains euphol and a monohydroxy triterpene, nerifoliol; whereas taraxerol, beta-amyrin and glut-5-(10)-en-1-one have been isolated from leaves and stems.

Latex possesses rubefacient, diuretic, purgative, expectorant; the root and pulp of the stem antiseptic and emetic properties. Found useful in warts, scabies, unhealthy ulcers, and cutaneous eruptions, when applied topically.

Euphorbia thomsoniana Boiss.

Habitat
Kashmir above 2350 m.

Euphorbia thomsoniana is a high altitude plant seen along the Tibetan Himalayas. Classical common name Hiyaavali (Dalhan) matches well with the Kashmiri name Hirvi. The plant gives golden yellow latex.

Argemone mexicana L. (Papaveraceae), being a Mexican native, was introduced in India rather late. Ayurvedic Formulary of India considered it as a substitute for the original classical drug Svarnakshiri.

Classical & common names
Ayurvedic: Svarnakshiri, Kanchankshiri, Kanakapushpi, Pitadugdhaa, Katuparni..
Siddha: Bramadandu (CCRAS).

Parts used
Latex, root.

Dose
Latex 125–250 mg, juice 5–10 ml.

Classical use
Charaka prescribed Naarayana Churna and Kshaargutikaa, with Svarnakshiri as one of the ingredients, in abdominal disorders. The petals of Kanakapushpi were included by Charaka in an ointment for leucoderma.

Sushruta gave Kanakakshiri internally in acute constipation, abdominal swelling, intestinal paralysis and as a purgative; externally as an ingredient of an ointment for virulent form of ringworm.

In folk medicine, crushed rootstocks are employed as a detergent for washing hair, also used as purgative. (The latex is liable to cause dermatitis.) Oil from the seeds is applied externally to ulcers, herpetic eruptions and other skin diseases. The root is sometimes used as an adulterant for Saussurea lappa.

Active principles and pharmacology

Chemical constituents of Euphorbia thomsoniana have not been screened.

Argemone mexicana is found throughout India, runs wild in wastelands and along roadsides. Bears yellow flowers.

Classical medicinal properties of Svarnakshiri cannot be attributed to the plant.

Mexican poppy contains isoquinoline alkaloids similar to the opium poppy (Papaver somniferum). The fresh latex contains protein-dissolving constituents and is used for treating warts, cold sores, blemishes on the lips. The whole plant acts as a mild painkiller. An infusion of the seeds in small quantity is used as a sedative in asthma. In greater quantities, the oil in the seeds is purgative. Argemone oil has led to widespread epidemics of dropsy and glaucoma. The flowers are expectorant and are used for treating cough and other chest conditions.

In Western herbal, Argemone mexicana is used only under professional supervision.

Brahmadandi of the South has been equated with Argemone mexicana by CIMAP, creating further confusion.

Brahmadandi has already been equated with Tricholepis glaberrima DC. Tricholepis glaberrima contains betulin, spinasterol, stigmasterol and a triterpenoid. Active principles exhibit properties which justify the drug's use in Unani medicine in seminal debility, spermatorrhoea, urinary disorders, as an aphrodisiac and blood purifier.

Euphorbia thymifolia Linn.

Habitat

This smaller variety, equated with Euphorbia thymifolia, is found in tropical plains and low hills of India, ascending to 1750 m.

The bigger varieties, Euphorbia pilulifera/ Euphorbia hirta Linn. are found in warmer parts of India from Punjab eastwards, and southwards to Kanyakumari.

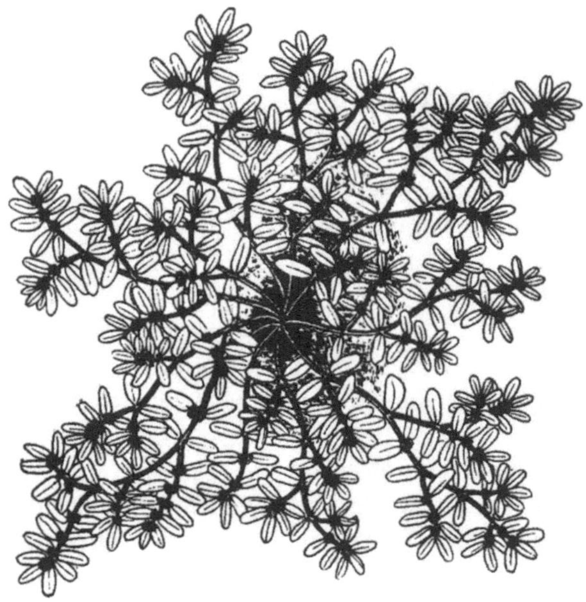

Figure 1 Euphorbia thymifolia [WOI]

Classical & common names

Ayurvedic: Chhoti Dudhi, Laghu Dugdhikaa (smaller var.), Badi Dudhi (bigger var.)
Unani: Dudhi Khurd (smaller var.)
Dudhi: Kalaan (bigger var.)
Siddha: Ammanptharisi.
English: Asthma Plant, Pill-bearing Spurge.

Classical use

Charaka prescribed Dugdhikaa as an ingredient of a vegetable soup for diarrhoea, painful bleeding piles. He gave Dugdhikaa and Solanum surattense (Kantakaari) cooked in clarified butter for alleviating bleeding and painful piles.

The latex of Dugdhikaa was applied on ringworm and eruptive boils.

According to Bhaavaprakaasha, Dugdhikaa is expectorant, cures aggravated cough, skin diseases, parasitic infection; promotes conception, possesses aphrodisiac and age-sustaining properties.

Active principles and pharmacology

Euphorbia thymifolia: Aerial parts gave epitaraxerol, n-hexacosanol, euphorbol, 24-methylenecycloartenol; 12-deoxy-4-beta-hydroxyphorbol-13-dodecanoate-20-acetate, 12-deoxy-4-beta-hydroxyphorbol-13-phenylacetate-20-acetate; 12-deoxyphorbol-13, 20-diacetate and quercetin-3-beta-galactoside.

The plant shows antispasmodic, bronchiodilatory, and antifungal activity. Pharmacologically, the root is indicated in amenorrhoea, the leaf and seed as astringent in diarrhoea, latex in ringworm and dandruff. The leaf, seed and latex possess purgative properties.

In a clinical trial, solid extract of Euphorbia thymifolia/ Euphorbia prostrata (750 mg, 16 tablets in 4 divided doses, equal to 50–60 g of crude drug), was given to 15 patients of bronchial asthma, having mild or moderate paroxysms for 2 weeks. 33.4 % were relieved of all symptoms, 60 % improved and 6.6 % remained unchanged. (CCRAS.)

Euphorbia hirta: Several terpenes, anthocyanins, alcohols and steroids have been isolated from various plant parts. Aerial parts also gave shikimic acid and choline, which are considered responsible for the antispasmodic action of the plant.

As a specific treatment for bronchial asthma, the herb relaxes bronchial tubes and eases breathing. Mildly sedative and expectorant, it is also found useful in bronchitis and other respiratory tract conditions.

Use in Western herbal

Euphorbia hirta is most often used along with other antiasthmatic herbs, notably gumplant (Grindelia camporum) and lobelia (Lobelia inflata).

Nymphaeaceae

EURYALE

Euryale ferox Salisb.

Habitat

Northern, Western and Central India in lakes and ponds. Cultivated in China for its edible seeds.

Classical & common names

Ayurvedic: Makhaanaa, Padmbijaabh.
Unani: Makhaanaa (Not to be confused with Taalmakhanna which has been equated with Hygrophila spinosa T. Anders.)
English: Gorgon Nut, Fox Nut.

Dose

Powder 5–10 g.

Classical use

The seeds, of the size of a pea or of a cherry, are eaten raw or roasted. On roasting in hot sand, the seed coat swells and bursts and can be easily peeled off. These are sold in dry fruit stores as Makhaanaa.

Makhaanaa is prescribed as a nutritious supplement in the form of various confections after parturition and as a food for invalids. Soaked overnight in milk, it is prescribed as a breakfast.

In China, the seed is used as a tonic, astringent and deobstruent. A kind of dry biscuit is prepared by the meal of the kernels and are given to children. The seeds are recommended in polyuria, spermatorrhoea, gonorrhoea and for restraining seminal gleet.

Active principles and pharmacology

Analysis of the edible part of the seeds gave the following values: moisture 12.8, protein 9.7, fat 0.1, mineral matter 0.5, carbohydrates 76.9, calcium 0.02 and phosphorus 0.09 %; iron 1.4 mg/100 g, carotene trace.

Zygophyllaceae

FAGONIA

Fagonia cretica Linn.

Habitat

North-west India and Deccan. Westwards to Afghanistan.

Classical & common names

Ayurvedic: Dhanvayaasa, Dhanvayaasaka, Dhanva, Duhsparshaa, Duraalabhaa.Yaasa has been equated with Alhagi pseudalhagi.
Unani: Dhamaasaa, Dhamaayaa.

Parts used

Whole plant.

Dose

Decoction 50–100 ml.

Classical use

Charaka prescribed Dhanvayaasa with Chandan (Santalum album), in prescriptions, in intrinsic haemorrhage; alcoholic preparation (Aasava) of the drug in dysentery; with honey as an antiemetic.

Sushruta gave the paste of the drug, processed with milk, for diarrhoea; the juice for treating the retention of urine.

The decoction of Dhanvayasaa mixed with clarified butter was prescribed for controlling vertigo (Vrindamaadhava, Chakradatta, Bangasena).

In folk medicine, the herb is credited with antibilious, antiseptic and blood-purifying properties. It is given internally in the form of a decoction in diseases due to vitiated blood; externally its paste is applied to abscesses, wounds, scrofulous glands; a decoction as a gargle in stomatitis and other diseases of the mouth. The drug is prescribed as a prophylactic against small pox and as a bitter astringent and febrifuge.

Duraalabhaadi Kashaaya (Sahasrayoga) is a classical compound, prescribed in the form of a decoction, as an antitoxic, cooling and anti-inflammatory drug.

Sahasrayoga has given separate compounds of Duraalabhaa and Dhanvayaasa. During the 16th century, in Ayurvedic texts Duraalabhaa is mentioned as a substitute for Dhanvayaasa. In practice, both the drugs are used as Dhamaasaa or Dhamaayaa.

Dhanvayaasaadi Kashaaya of Sahasrayoga is prescribed in bilious diarrhoea.

Dhamaayaa is an important ingredient in reputed Unani tonic Araq-e-Mussafi-e-Khoon Qawi (Qarabadeen-e-Jadeed), prescribed as a blood purifier in acne, pimples, boils, scabies, pruritus and other skin diseases.

Active principles and pharmacology

The plant gave diosgenin, kryptogenin, lanosterol, beta-sitosterol, harmine, fagogenin and oleanolic acid. Harman, chinovic acid obtained by hydrolysis of a glycoside, is considered to be chinovin. Betulin, campesterol, stigmasterol and triacontanol were also obtained from the plant.

Aerial parts of the plant exhibit antiviral, CVS-active, and spasmolytic activity. Active principles antagonise amphetamine hyperactivity.

A decoction of leaves and twigs qualifies for its use as a blood purifier. Stem, leaves and fruit show antimicrobial properties.

Rutaceae

FERONIA

Feronia elephantum Correa
Feronia limonia (Linn.) Swingle

Figure 1 Feronia limonia—flowering branch [WOI]

Limonia acidissima L.

Habitat

Indigenous to South India, cultivated throughout the plains of India; up to 500 m in the western Himalayas.

Classical & common names

Ayurvedic: Kapittha, Dadhiphal, Dadhith, Surabhichhad.
Unani: Kaith.
Siddha: Narivila.
English: Elephant Apple, Wood Apple.

Parts used

Bark, leaves, fruits.

Dose

Decoction 50–100 ml.

Classical use

Charaka and Sushruta included the leaves and fruits of Kapittha in prescriptions for diarrhoea, toxicosis, urinary disorders, ringworm and other chronic skin diseases.

Charaka prescribed the soup of Kapittha and Bilva (Aegle marmelos) in piles; the juice, mixed with Pippli (Piper longum) and honey, in hiccough.

Kapithaashtaka Churnam (Sahasrayoga, Shaarangadhara Samhitaa) is prescribed in diarrhoea, dysentery, internal abscesses, and piles.

Tender leaves of 5 trees—Feronia limonia, Mangifera indica (Aam), Syzygium cuminii (Jamun), Aegle marmelos (Bilva) and Citrus medica (Maatulunga)—are known as the Pancha Pallava group and are used as vaginal disinfectants due to their astringent, antimicrobial and anti-inflammatory properties.

In folk medicine, Kapittha is used as a substitute for Aegle marmelos in the treatment of diarrhoea and dysentery.

The fruit forms part of a paste applied to tone the breasts. The leaves are used for treating indigestion, flatulence, diarrhoea, dysentery and haemorrhoids.

Active principles and pharmacology

Analysis of the edible part (55–58 %) of the fruit gave the following values: moisture 69.5, protein 7.3, ether ext. 0.6, fibre 5.2, carbohydrate 15.5, mineral matter 1.9, calcium 0.13, and phosphorus 0.11 %; iron 0.6 mg/100 g, riboflavin 170 mg/100 g; and vitamin C 2.0 mg/100 g. The fruit is rich in mineral constituents, especially calcium and phosphorus. The acid content of the pulp varies from 7.6 % in unripe fruits to 2.3 % in fully ripe fruits. The fruit contains 3–5 % pectin.

Leaves contain stigmasterol, orientin, vitaxin, bergapten and saponarin, and tannins. The major constituents of essential oil from leaves are methyl chavicol, t-anethol, thymol and p-cymen-7-ol. Others include alpha- and beta-pinene, sabinene, myrcene, beta-phellandrene, p-cymene, limonene, t-ocimene, 1, 8-cineole, fenchone, borneol, eugenol, iso-eugenol, anisaldehyde and several sesquiterpene hydrocarbons.

The essential oil showed antibacterial, antifungal; fixed oil antimicrobial; the fruit astringent, stomachic, cordial, antiscorbutic; the leaf carminative, stomachic and the bark antibilious properties.

The fruit stimulates the digestive system.

Apiaceae

FERULA

Ferula assafoetida L.
Ferula narthex Boiss.

Habitat
Native to Iran, Afghanistan and Pakistan. Ferula narthex occurs in Kashmir.

Classical & common names
Ayurvedic: Hingu, Hinguka (Charaka, Sushruta); Baahlika, Jatuka, Sahasravedhi, Raamatha; Hing.
Unani: Hilteet.
Siddha: Perunkayam.
English: Asafoetida.

Parts used
Oleo-gum-resin.

Dose
125–500 mg.

Classical use
Charaka and Sushruta included Hingu in prescription for indigestion, colic pain, tympanitis, internal catarrh, and as an appetizer.

Sushruta prescribed it in hysterical convulsions and epileptic fits. Charaka also prescribed the clarified butter processed with Hingu as a nervine tonic in insanity.

During the 16th century, Hingu oleo-gum-resin was included in formulations for amenorrhoea and dysmenorrhoea.

Hingavaadi Bati (Sushruta Samhitaa, Bangasena), is still available over-the-counter and is prescribed in colic and loss of appetite.

Hingu-triguna Taila (Ashtaanga Hridaya) is prescribed internally as a laxative, digestive, appetizer, anthelmintic and anti-inflammatory drug in chronic constipation, intestinal colic. It was basically formulated for treating internal obstructions.

Rajahpravartini Vati (Bhaishajya Ratnaavali) is used as a uterine tonic in amenorrhoea, for cleansing urinogenital system and for regulating menstrual cycle.

Hingvaadyam Ghritam (Sahasrayoga) is prescribed in insanity.

According to INSA scientists, P. Ray et al, Charaka gave dried fruits of Hingu (Sutrasthan, 2, 29) as ingredients of a gruel, as a laxative, digestive and rejuvenant, and in cough, dysuria, and dermatosis. (The fruit of Ferula foetida is avate, flat, thin, flaky, reddish brown with distinct oil marks, has milky juice and a strong smell.)

Further literary research is required to establish the use of the fruit in medicine.

Habb-e-Hilteet is a classical Unani compound, used in flatulence, anorexia and indigestion.

Habb-e-Murr (IMPCOPS) is for amenorrhoea and other functional disorders of the uterus.

Active principles and pharmacology
Chemically, asafoetida contains volatile oil, resin, gum. The resinous constituents are asaresinol ferulata and free ferulic acid (Hydroxymethoxy-cinnamic acid) umbelliferone, a lactone of umbellic acid formed on boiling the resin with hydrochloric acid and filtering into ammonia, umbellic acid (dihydroxycinnamic acid) is also present. Disulphides and sugars form the other components.

The dried exudate from plant was tested for anti-implantation activity in rats. It exhibited 31 % anti-implantation and 46 % abortifacient activity. The drug showed no effect on oestrus cycle of animals and had no teratogenic effect (CCRAS).

Gum-resin was reported as an emmenagogue (CCRAS).

The essential oil from gum-resin was found to stimulate the respiratory and nervous system.

Asafoetida is bacteriostatic to E. coli, Aerobactor aerogenes, S. aureus and Shigella sonner. Another study, however, avers that Ferula assafoetida is not bacteriostatic to intestinal organism. Its role in abnormal states of the gastrointestinal tract is to exert a change in the ratio of hydrogen sulfide and carbon dioxide formed by microorganism in the digestive tract.

Ferula narthex plant gave coumarin derivatives, including umbelliferone and scopoletin. The leaf is found diaphoretic and carminative; stem tonic for brain and liver; root antipyretic; oleo-gum-resin antispasmodic, expectorant, antiasthmatic, anticholerin, anthelmintic, stimulant, nervine, and urterine tonic (CIMAP).

Ferula assafoetida fruits gave flavonoids, luteolin and luteolin-7-O-beta-D-glucopyranoside.

Ferula jaeschkeana Vatke has been equated with Hingupatri. The rhizomes gave a coumarin ferujol, a sesquiterpene ester and feruginin. Ferujol exhibited anti-implantation, essential oil antimicotic, plant abortifacient properties (CIMAP). See also Gardenia gummifera.

The dried roots and mature dry fruits of Ferula jaeschkeana yielded on distillation 1.2 % and 2.8 % essential oil respectively. The oil contained about 25 % camphene and about 60–90 % alpha-pinene. The fruit oil contained an optically active D-alpha-pinene, while the root contained inactive alpha-pinene.

On steam distillation, the fruits of Ferula alliacea yielded an oil (0.9 %) and some coumarins. A mixture of furocoumarins, byakangelicin, isopimpinellin and ferulin were isolated from the fruits. Ferulin was found to be a mixture of phellopterin and byakangelicol.

The effect of byakangelicin, isolated from the fruit of F. alliacea on human chorionic gonadotrophin (HCG), has been studied. Complete inhibition of HCG-hormone was observed.

The effect of byakangelicin has also been studied in female sex hormones and fertility in rats. The changes in gross uterine weight, histological and histochemical examination of tissues and vaginal smear indicated that byakangelicin partially blocked the uterotropic responses of exogenous estrogen. It has no effect on progesterone response of the uterus.

It seems possible that during the period of Charaka and Sushruta, plant parts other than oleo-resin were used in medicine.

Use in Western herbal

The drug is called Devil's Dung (due to its offensive odour); also Food of the Gods.

The drug is available as an extract, as well as in pill form, and is used for chronic gastritis, dyspepsia, and irritable colon.

As the volatile oil of the drug is eliminated through the lungs, it gives encouraging results when used for asthma, whooping cough, and bronchitis.

According to Penelope Ody and Andrew Chevallier, recent research has indicated that asafoetida lowers blood pressure and possesses anticoagulant activity.

The drug is used for helping in neurotic states and nervous disorders, but its effect has not yet been scientifically validated.

Tincture of the gum-resin of Narthex asafoetida is used in homoeopathy in flatulence and spasmodic contraction of stomach and oesophagus with reverse peristalsis. Also in asthma, spasmodic contraction of thorax and accumulation of stringy mucous; in hysteria of flatulence order.

Moraceae

FICUS

The Group of Four Ficus

Habitat

Throughout India.

Classical & common names

Ayurvedic: Chaturkshiri.
Siddha: Naapaalmaran.
The group of Four Ficus, all yielding latex, according to Ayurvedic texts, consists of Nyagrodha (Ficus benghalensis), Udumbara (Ficus glomerata/Ficus racemosa), Plaksha (Ficus lacor/Ficus retusa) and Ashvattha (Ficus religiosa).
The fifth one, of the Panchkshiri vrkshas (latex-bearing trees) of Ayurvedic texts, is Paarisha (Thespesia populnea). It was replaced with Vetasa (Salix caprea) in Raaja Nighantu.

Classical use

The bark and leaves of this group are used as astringent, haemostatic, anti-inflammatory, antiseptic; prescribed in diarrhoea, dysentery, and in the treatment of skin diseases, ulcers, vaginal disorders, leucorrhoea, monorrhagia, deficient lactation.

Drugs of this group promote fertility and cure debility. Act as an urinary astringent and reduce blood sugar in diabetics.

Active principles and pharmacology

Discussed under individual entries.

The analysis of the latex of the Four Ficus is as follows:

Ficus benghalensis: Coagulum: caoutchouc 2.7–15.9, resin 77.3–94.9, insolubles 0.7–20.0 %. Original latex: water and water solubles 46.0–90.6, caoutchouc 0.3–7.7 %.

Ficus glomerata/Ficus racemosa: Coagulum: caoutchouc 16.3–25.7, resin 72.3–77.6, insolubles 0.8–8.3 %. Original latex: water and water solubles 70.3–76.4, caoutchouc % 4.0–7.4 %.

Ficus lacor: Coagulum: caoutchouc 6.9, resin 90.6, insolubles 0.5. Original latex: water and water solubles 57.2, caoutchouc 3.0 %.

Ficus retusa: Coagulum: caoutchouc 2.3, resin 86.9, insolubles 10.8 %. Original latex: water and water solubles 92.4, caoutchouc 0.2 %. Ficus lacor Buch-Ham. (Urticaceae) is equated with Plaksha. In the South, Ficus retusa Linn. is commonly used as Plaksha. (NAA.)

Ficus religiosa: Coagulum: caoutchouc 5.2–7.1, resin 74.9–90.0, insolubles 4.8–18.0 %. Original latex: water and water solubles 23.5–86.4, caoutchouc 0.7–5.1 %.

Ficus benghalensis Linn.

Habitat

Found throughout India, is planted around temples and in gardens.

Classical & common names

Ayurvedic: Nyagrodha, Vata (Charaka, Sushruta), Bahupaada. In Charaka Samhitaa (Chikitsaasthaan 3-257) Nyagrodha and Vata are mentioned together. Now Vata is accepted as a synonym of Nyagrodha.
Unani: Darkht-e-Reesh, Bargad.

Figure 1 Ficus benghalensis [ZANDU]

Siddha: Ali.
English: Banyan Tree.

Parts used

Bark, latex, leaf, fruit.

Dose

Powder 3–5 g, decoction 50–100 ml, latex 5–10 drops.

Classical use

Charaka prescribed aqueous extract of leaf-buds of Nyagrodha, Udumbara (Ficus glomerata) and Ashvattha (Ficus religiosa) mixed with sugar and honey for checking diarrhoea; milk processed with the aerial roots or leaf-buds of Nyagrodha in haemorrhages and bleeding piles; the paste of Lodhra (Symplocos racemosa) with the decoction of Nyagrodha bark for leucorrhoea and other vaginal discharges.

A decoction of leaf-buds and aerial roots of Nyagrodha, mixed with honey, was given for checking vomiting and thirst; also during fevers with burning sensation (Ashtaanga Hridaya, Vrindamaadhava, Vaidyamanoramaa).

Leaf-bud of Nyagrodha was prescribed for promoting conception (Gadanigraha).

Nyagrodhaadi Churnam (Bhaishajya Ratnaavali) is prescribed in dysuria and polyuria. Nyagrodha bark is an ingredient in Saarivaadyaasava (ibid), prescribed as a blood purifier in skin diseases; also in Chandanaasava (ibid), prescribed in urinary and urinogenital disorders. The bark of the Group of Four Ficus is an important ingredient in Dineshavalyaadi Taila (Sahasrayoga), recommended for external application in skin diseases.

(Except the first one, all the compounds are available over-the-counter.)

Active principles and pharmacology

The bark yielded flavonoid compounds: compounds A and C were identified as different forms of a leucoanthocyanidin, compound B was a leucoanthocyanin. All the three were found effective as hypoglycaemic agents.

Three flavonoid compounds A, B, and C were individually found effective as hypoglycaemic agents on oral administration to normal fasting rabbits, compound B having the maximum activity. This compound was also found to be fairly effective in controlling the hyperglycaemia produced by oral administration of glucose to normal fasting rabbits (1964).

Leaves gave friedelin, beta-sitosterol, and the flavonols quercetin-3-galactoside and rutin.

Heartwood gave tiglic acid ester of si-taraxasterol.

The flavonoids relieved capillary bleeding. The latex exhibited bactericidal action.

The latex and fruits were found useful as external application to pains, bruises, sores, ulcers; to the soles of feet when cracked or inflamed; to the teeth and gums for toothache; in rheumatism and lumbago.

In a study (1992), alcoholic extract prepared from the stem bark of Ficus benghalensis was studied for its long-term feeding effects on the blood sugar of albino rats, which were made diabetic through i.v. injection of alloxan monohydrate. Different doses (25, 50, 75 mg/day/100 g body weight of the rat) over different duration showed significant hypoglycaemic potential. The blood sugar once lowered, remained unaltered when the treatment was discontinued. The blood sugar at no stage fell down below normal levels. The extract was also able to bring down the level of cholesterol and blood urea. Total protein levels remained unaffected.

During the study 18 to 23 % fall in blood sugar level was recorded after 8 days; 43 to 47 % fall after 15 days and 59 to 63 % fall after 30 days with 25 mg dose; 18 to 24 % fall after 8 days, 46 to 52 % fall after 15 days and 64 to 66 % fall after 30 days with 50 mg dose; and 31 to 43 % fall after 8 days, 66 % fall after 15 days and 63 to 73 % fall after 30 days with 75 mg dose.

In another study, stem bark of Ficus benghalensis was found to significantly lower the fasting blood glucose level and markedly improve glucose tolerance in rats. The hypoglycaemic effect produced by a therapeutic dose (20 mg/kg body weight) of the plant extract was, in fact, better than that produced by a therapeutic dose of tolbutamide. The maximum hypoglycaemic effect was observed with a dose of 40 mg/kg body weight of the plant extract.

Dimethoxy ether of leucopelargonidin-3-O-alpha-L-rhamnoside isolated from the bark was tested for antidiabetic effect during 1994. At a medium effective dose (100 mg/kg) on oral administration, the compound showed significant hypoglycaemic and serum insulin-raising action in normal, as also in moderately diabetic dogs (induced by alloxan) during a period of 2 hours. The mechanism of action of the glycoside compound was found to be similar to that of drugs which stimulate insulin.

In a recent study (1995), the treatment for one month (50 mg/kg body weight/day) with water extract of Ficus benghalensis not only controlled the blood sugar but also brought down the level of total serum cholesterol in subdiabetic and diabetic rabbits. Low density lipoprotein cholesterol and very low density lipoprotein cholesterol value came down from 34 ± 10 mg per cent and 95 ± 24 mg per cent to 16 ± 3 mg per cent and 29 ± 4 mg per cent in subdiabetic and diabetic rabbits. Triacylglycerol before treatment was 121 ± 21.6 mg per cent and 416 ± 70 mg per cent in subdiabetic and diabetic rabbits. Treatment brought it down to 45 ± 5 mg per cent and 81 ± 25.7 per cent. Glycosylated haemoglobin was brought down from 2.1 ± 0.3 per cent to 1.5 ± 0.1 per cent in subdiabetic group and from 4.28 ± 0.5 per cent to 2 ± 0.3 per cent in diabetic group. (MAPA, 9603-1487.)

Ficus carica Linn.

Habitat

Cultivated in North-western and South India.

Classical & common names

Ayurvedic: Phalgu, Manjula (Charaka, Sushruta). Phalgu is equated with C. carica. Indian fig with Ficus palmata Forsk; another species, Malayu, with Ficus cunia Buch-Ham.

Unani: Anjir vilaayati.
Siddha: Semaiatti.

Parts used
Fruit.

Dose
Juice 10–20 ml.

Classical use
Charaka gave the paste of figs in prescriptions, also as cooked vegetable, in emaciation and debility. As a diuretic in urinary stones. Sushruta included the fruit in a medicated clarified butter for internal use in fever, consumption, asthma, epilepsy and insanity.

The juice of the fruit, with honey, was prescribed for checking haemorrhage (Vrindamaadhava).

In Unani medicine, Anjeer is used as a mild laxative, expectorant, diuretic; also in the diseases of the liver and spleen as a deobstruent and anti-inflammatory agent. Ficus carica and Juglans regia (Akharot) form a good aphrodisiac tonic in Unani medicine. Anjeer as a dry fruit is also considered a good nutritional support for diabetics.

Sugar-free sweets are available in a few sweets speciality shops for diabetics. These contain Ficus carica.

The fruit pulp is heated and applied to swellings, tumours; gum to abscesses for relieving pain and inflammation.

Active principles and pharmacology
Figs contain inverted sugar (51.4 %), mucilages, flavonoids, vitamins and enzymes.

Fresh fig, on an average, consists of 84 % pulp and 16 % skin, moisture 80.8, protein 1.3, ether extr. 0.2, mineral matter 0.6, carbohydrates 17.1, calcium 0.06 and phosphorus 0.03 %; iron 1.2 mg, carotene 270 international units vitamin A, nicotinic acid 0.6 mg, riboflavin 50 mcg and ascorbic acid 2 mg/100 g.

Figs owe their value chiefly to their mineral and sugar contents. The total mineral content is 2 to 4 times that of most other fresh foods; only cheese and a few of the nuts have a higher calcium content. Figs are richer in iron and copper than nearly all fruits and vegetables and most other dried fruits. Traces of zinc are also reported to be present.

The nutritive index of fig in comparison with that of other fruits is reported to be as follows: fig 11, apple 9, raisin 8, date 6 and pear 6.

The total (average) sugar content of fresh fig is 15.5 % and that of dried figs 51.4 %. The sugar is present mostly in the form of invert sugar.

The principal acids in fresh figs are citric and acetic, small amounts of malic, boric and oxalic are reported to be present. The acid content ranges from 0.1 to 0.44 % (as citric acid).

Fresh figs also contain gum and mucilage (0.8 %) and pentosans (0.83 %). A phosphatide with a nitrogen:phosphorus ratio at 1:2 and containing palmitic and oleic acids is reported to be present.

Fig juice contains on average 20.7 % sugar.

Analysis of fig skin gave the following values: moisture 76.3, protein 1.5, fat 0.5, fibre 2.3, carbohydrates 18.7, ash 0.7, calcium 0.162, phosphorus 0.055 and potash 0.233 %; sugar 5.4 %, gum and mucilage 2.74 %.

Ficusin, bergaptene, stigmasterol, sitosterol and tyrosin are present in the leaf.

The fig tree yields a latex containing caoutchouc (2.4 %), resin, albumin, cerin, sugar, malic acid. The latex also contains rennin, proteolytic enzymes, diastase, esterase, lipase, catalase and peroxidase. The anthelmintic action of the latex has been traced to ficin, a proteolytic enzyme which has the remarkable power of digesting living helminths. Ficin is effective against both Trichurus and Ascaris.

Use in Western herbal
Nicholas Culpeper (1616–1654) wrote of it: "The milk (latex) dropped upon warts takes them away. The decoction of leaves is excellently good to wash sore heads with. It clears the face also of morphew and the body of white scruf, scabs and running sores. A syrup made of leaves dissolves congealed blood caused by bruises or falls, and helps the bloody flux. The ashes of the wood, made into an ointment, helps kibes and chilblains. A syrup made of the leaves or green fruit is excellently good for coughs and all diseases of the lungs."

Fig was one of the principal articles of sustenance among the Greeks. Athletes were fed almost entirely on figs, considering that they increased their strength and swiftness.

Figs are used for their mild and laxative action, and are employed in the preparation of laxative confections and syrups, usually with Alexandrian senna and carminatives. The three preparations of fig of the British Pharmacopoeia are: Syrup of Figs, a mild laxative suitable for children; Aromatic Syrup of Figs, Elixir of Figs or Sweet Essence of Figs for children and delicate persons. The Compound Syrup of Figs is a stronger preparation, composed of the syrup of figs, liquid extract of senna and syrup of rhubarb (Rheum palmatum) and is more suitable to adults.

The fig is also considered a demulcent to soothe irritated tissues and is taken for sore throat, coughs and bronchial problems.

Figs are available in natural form in health food stores; plant juice and syrups are also sold.

The latex is used only externally for removing warts.

Ficus hispida Linn. f.
Ficus daemona Koen. ex Vahl.
Ficus oppositifolia Roxb.

Figure 2 Ficus hispida—leaves and fruits [WOI]

Habitat

Outer Himalaya from Chenab eastwards to West Bengal Assam, Central and South India and the Andaman Islands.

Classical & common names

Ayurvedic: Kaakodumbara, Kaashtodumbera, Phalgu, Malapu,
Malayu (Charaka, Sushruta). See Ficus carica.
Unani: Anjir Dashti, Kathguular.
Siddha: Peyathi.

Parts used

Root, bark, fruit.

Dose

Decoction 50–100 ml.

Classical use

Hot decoction of the root, as a drink internally, and as an ingredient of a medicinal plaster externally, was prescribed by Charaka and Sushruta in leucoderma.

The root, pounded with rice-water, was prescribed for checking intrinsic haemorrhage (Raajmaarttanda).

The juice of Kaakodumbara fruits, mixed with honey, was prescribed in leucorrhoea (Vrindamaadhava).

The tender leaves of Kaadodumbara, pounded with and cooked in cow-milk, added with a bit of Pippali (Piper longum), were prescribed for cough and asthma (Vaidya Manoramaa).

In Unani medicine, the root-bark of Anjeer Dashti is used as a blood purifier and as a drastic purgative. It is an ingredient in Zimad-e-Kibreet, prescribed for external application in induration of spleen, as an anti-inflammatory agent. The latex is applied on ringworm; paste of the ripe fruit, pulp over goitre. Powdered fruits, mixed with vinegar, are also applied externally for treating skin affections, boils and ulcers.

Active principles and pharmacology

The plant yields 10-ketotetracosyl-arachidate, beta-sitosterol and lupeol acetale. Leaves yielded 3,6,7-trimethoxy-phenanthroindolizidine; 3,6,7-trimethoxy-14-hydroxy-phenanthroindolizidine and hispidine. Leaves also contain bergapten, psoralen and beta-amyrin.

Chloroform extract from the fruit exhibited 36 % anti-inflammatory effect against carrageenin-induced rat hind paw oedema; fall of blood pressure in anaesthetised cats which was blocked by atropine, neuromuscular blocking effect on isolated phrenic nerve diaphragm of rat, potentiated by neostigmine (CCRAS).

Different extracts of fruits, seeds and bark exhibited anti-inflammatory, antispasmodic, neuromuscular blocking, antihistaminic, hypotensive, anticolagogue and lactagogue properties. The root was found efficacious in jaundice (CCRAS).

Ficus lacor Buch.-Ham.
Ficus retusa Linn.

Habitat

Plains and lower hills of India.

Classical & common names

Ayurvedic: Plaksha (Charaka, Sushruta), Karpari, Pitana (Charaka).

NAA equated Plaksha with Ficus lacor. In the South, Ficus retusa Linn. is commonly used as Plaksha. IMPCOPS includes Ficus retusa, the shining-leaved fig bark, in Chaturkshiri or the Four Ficus Barks. Nadkarni equates Ficus benjamina Linn. with Ficus retusa. The species is found at the base of Eastern Himalayas, Khasi Hills and the Deccan Peninsula.

In Ayurvedic reference books, Ficus var. infectoria; var. lambertiana and var. wightiana are equated with Plaksha spp. (PV Sharma).

Siddha: Kurugu (Tamil), Juvvi (Telugu, Kannada), Chakkila, Itti (Malyalam).

Parts used

Bark, leaf.

Dose

Decoction 50–100 ml.

Classical use

Charaka prescribed a decoction of tender leaves in diarrhoea, haemothermia; the paste of tender leaves and bark in inflammations and erysipelas; powdered bark in the form of a lump mixed with honey, as a vaginal pessary in meno-metrorrhagia; a decoction of the bark and leaves for wound-healing.

Sushruta prescribed the bark, in prescription, for checking diarrhoea with blood; the fruit in haemoptysis.

Charaka prescribed a decoction of tender leaves in seminal debility; Sushruta gave fruits in vaginal discharges.

Tender leaves were used as a vegetable in intrinsic haemorrhage (Bhaavaprakaasha).

Active principles and pharmacology

Leaves contain hydrocarbons, alcohols and sterols. The alcohol fraction consisted of aliphatic alcohols. The sterols consisted of sitosterol, stigmasterol, campesterol, cholesterol and 28-isofucosterol. The triterpenoids—lupeol and alpha- and beta-amyrin—are also present in the leaves.

The bark exhibited antileucorrhoetic and antiulcer activity.

From the latex of Egyptian Ficus benjamina alpha-amyrin, bergapten and imperatorin have been isolated.

Ficus racemosa Linn.

Figure 3 Ficus racemosa [CCRAS]

Figure 4 Ficus racemosa [CCRAS]

Ficus glomerata Roxb.

Habitat
Throughout India. Grows wild in many forests and hills. Often found around subterranean water streams.

Classical & common names
Ayurvedic: Udumbara (Charaka, Sushruta), Audumbara, Jantuphala, Yagyaanga, Hemdugdhak.
Unani: Guular.
Siddha: Athi.
English: Cluster Fig Tree.

Parts used
Bark, fruits.

Dose
Powder 3–5 g, decoction 50–100 ml.

Classical use
Charaka prescribed tender leaves of Udumbara as an astringent and styptic in diarrhoea and haemorrhages; also the tender leaves and leaf-buds of Udumbara, Vata (Ficus benghalensis), Ashvattha (Ficus religiosa) and Plaksha (Ficus lacor/ Ficus retusa).

For treating vaginal laxicity, Charaka prescribed a vaginal pessary, prepared by the sesamum oil, impregnated 6 times with the latex of Udumbara and cooked with the decoction of the same.

Sushruta prescribed the juice of pounded fruit in intrinsic haemorrhage; the decoction of the fruit, mixed with powdered Shaali (Oryza sativa) rice, with sugar and honey for checking miscarriage; ash of the bark mixed with honey in hiccough.

In dysentery, tender fruits of Udumbara, steamed and mixed with curd, were prescribed (Siddha-bhesaja-manimaalaa). Externally the latex of Udumbara was applied on boils due to diabetes.

In Unani medicine, fruits, also the root-water, are prescribed as a tonic to diabetics.

Active principles and pharmacology
Stem-bark gave gluanol acetate, beta-sitosterol, leucocyanidin-3-O-beta-D-glucopyranoside, leucopelargonidin-3-O-alpha-L-rhamnopyranoside, lupeol, ceryl behenate, lupeol acetate and alpha-amyrin acetate, tannins are also present. Presence of psoralens is also

reported.

The stem-bark, root and fruit extract exhibited hypoglycaemic; leaf antibilious; latex antidiarrhoeal; root antidysenteric; syconium stomachic, carminative; properties. The bark and syconium proved astringent and styptic in menorrhagia and hemoptysis; latex in piles.

The alcoholic extract of the stem-bark possessed antiprotozoal activity against Entamoeba histolytica.

The extract of the bark, leaves, fruits, when used locally, is found efficacious in inflammations, lymphadenitis, in sprains and fibrositis.

Glycosides of the ethanolic extract of the leaves were found to exert hypotensive and vasodilator effect in animal studies.

The alcoholic extract of the stem-bark showed hypoglycaemic activity in albino rats.

Ficus religiosa Linn.

Figure 5 Ficus religiosa [CCRAS]

Figure 6 Ficus religiosa [CCRAS]

Habitat

Sub-Himalayan tracts. West Bengal, Central and South India; planted throughout India.

Classical & common names

Ayurvedic: Ashvattha (Charaka, Sushruta), Bodhidru, Bodhivrksha, Sebya, Gajabhaksha, Kshiradruma (Charaka); Peeppal.
Unani: Peepal.
Siddha: Arasu.
English: Bot Tree, Peepul.

Classical use

Charaka and Sushruta prescribed a decoction of the bark of Ashvattha in haemorrhages; leaves for covering wounds; the paste of tender roots or the bark for skin infections. Sushruta administered a decoction in urinary disorders and vaginal discharges. Milk cooked with the fruit, leaf-bud, bark and the root, added with sugar and honey, was prescribed as an aphrodisiac.

Powder of the dried bark was dusted over burns (Vrindamaadhava). A paste of the bark and leaves was prescribed in stomatitis (Chakradatta, Asthaanga Hridaya).

Ash-water of the dried bark was given for checking vomiting; decoction of the root-bark with salt and jaggery for alleviating severe colic (Bhaavaprakaasha, Siddha-bheshaja-manimaalaa).

In folk medicine, a decoction of the bark is administered in whooping cough and asthma. For promoting conception, powder of the fruit is given. The bark and fruits are prescribed in diabetes.

Ashvattha belongs to the group of Four Ficus and is an ingredient in classical compounds prescribed for diarrhoea, dysentery, erysipelas and other skin affections, vaginal disorders, leucorrhoea, menorrhagia, haemorrhages, sexual debility, nervous disorders, diabetes.

Active principles and pharmacology

Phytosterolin from the bark is a powerful CNS-stimulant. Beta-sitosterol and its glycoside, when isolated from the bark, showed hypoglycaemic activity, which compared favourably with that of tolbutamide.

Analysis of the dried fruits gave: moisture, 9.9, albuminoids 7.9, fatty matter 5.3, carbohydrates 34.9, ash 8.3, silica 1.85, phosphorus 0.69 and colouring matter 7.5 %.

The leaves, on an average on dry matter basis, contain crude protein 13.99, ether extract 2.71, crude fibre 22.36, N-free extract 46.02, total ash 15.06, lime 4.64, and phosphorus 0.52 %.

The bark contains 4 % tannin, stem bark vitamin K.

An aqueous extract of the bark showed antibacterial activity against Staphylococcus aureus and E. coli. The stem-bark possessed antiprotozoal activity against E. histolytica, anthelmintic activity against Ascaridia galli, antiviral activity against Ranikhet disease virus.

Apiaceae

FOENICULUM

Foeniculum vulgare Mill.

Figure 1 Foeniculum vulgare—fruiting branch [WOI]

Foeniculum capillaceum Gilib.
Foeniculum officinale All.

Habitat

Widely cultivated throughout India up to 1830 m. Also found wild.

Classical & common names

Ayurvedic: Mishreyaa, Madhurikaa, Madhuraa, Misi, Shatapushpaa, Shataahvaa. In Bhaavaprakaasha, Shatapushpaa is equated with Saunf and Shataahvaa with Soyaa. Some authors treat these as vice-versa.
Unani: Baadiyaan, Bari Saunf.
Siddha: Sogikeenai.
English: Fennel.

Parts used

Seeds, oil, root.

Dose

Seed powder 3–5 g, oil 5–10 drops, root-powder 3–5 g, extract 20–40 ml.

Classical use

According to Kaiyadeva Nighantu, Dhanvantari Nighantu and Ayurvedic texts of the 16th century, Shataahvaa cures colic pain, vomiting, flatulence, burning syndrome, biliousness, anorexia; diseases due to vitiated blood. It was used as an appetiser, carminative, spasmolytic and expectorant; also as a galactagogue.

In Unani medicine, Araq-e-Baadiyaan (Qarabadeen-e-Jadeed) is prescribed in indigestion, loss of appetite, flatulence; as a carminative.

Baadiyaan is an ingredient in Itrifal Mulaiyin (Haj-ul-Amraz), prescribed in constipation.

Baadiyaan is given as an adjuvant with purgatives due to its spasmolytic properties.

Active principles and pharmacology

Indian fennel seeds contain about 1–4 % volatile oil, the principal components of which are the phenolic ether-E-anethole (approx 60 %) and ketone fenchone (10–30 %); flavonoids, coumarins (including bergapten), tannins and stigmasterol.

Fennel seed (also fennel oil) stimulates gastrointestinal motility and exhibits antispasmodic effect in higher concentrations. Experimentally, anethole and fenchone have shown a secretolytic action on respiratory tract in frogs. The aqueous fennel extract raised the mucociliary activity of the ciliary epithelium.

Research indicates that naturally present isomers of anethole, dianethole and photoanethole are active oestrogenic agents. Anethole has structural similarity to catecholamines adrenaline, noradrenaline and dopamine. This may account for some of the sympathomimetic effects of fennel, such as ephedrine-like bronchodilator action and amphetamine-like facilitation of weight loss. The traditional lactogenic effect may be due to competitive inhibition of dopamine inhibition of prolactin secretion as much as to any direct hormonal activity. (Simon Y. Mills.)

Fennel oil showed significant antimicrobial activity.

Though fennel seeds are safe, the oil may cause skin rash in sensitive individuals. When taken internally, it may cause nausea, vomiting and possibly seizures.

Use in Western herbal

Nicholas Culpeper (1616–1654) wrote of it: "Fennel is good to break wind, to provoke urine and ease pains of the stone and helps to break it. The leaves or seeds, boiled in barley water and drank, are good for nurses, to increase their milk and make it more wholesome for the child. The leaves or rather the seeds, boiled in water, stays the hiccough, and takes away the loathings which oftentimes happen to the stomachs of sick and feverish persons, and allays the heat thereof.

"The seed and the roots much more, help to open obstructions of the liver, spleen and gall... The seed is of good use in medicines to help shortness of breath and wheezing. It helps also to bring down the courses and to cleanse the parts after delivery. The roots are of most use in broth that are taken to cleanse the blood, to open obstructions of the liver, (to) provoke urine.

"The distilled water of the whole herb, dropped into the eyes, cleanses them from mists and films that hinder the sight."

German Commission E monograph recognized the drug's efficacy in:

- Cough
- Bronchitis.

Fennel seed is used in dyspepsia, such as mild gastrointestinal ailments, fullness, flatulence; catarrh of the upper respiratory tract. Fennel Syrup and Fennel Honey are used for catarrh of the upper respiratory tract in children.

An infusion of the seeds is used as a gargle for sore throats and distilled water of the seeds as an eyewash for sore eyes and conjunctivitis.

Fennel seeds are mixed with strong laxatives for counteracting intestinal cramps.

The seeds are also used as an aid for losing weight.

Caution

Though antispasmodics soothe not only the digestive tract but also other smooth muscles such as uterus, high doses of fennel stimulate the uterus into menstruation. Fennel in medicinal amounts is contraindicated during pregnancy.

Fumariaceae

FUMARIA

Fumaria parviflora Lam. subspp. vaillantii.
Fumaria indica Pugsley.

Habitat

Asia, Europe and Africa. In India on the hills, up to 2700 m on the Himalayas.

Classical & common names

Ayurvedic: Parpata, Parpataka (Charaka, Sushruta); Renu, Sukshmapatra, Varatikta, Pittapaaparaa.
Unani: Shaahtara (Fumaria officinalis Linn.)
Siddha: Tusha, Parpataka.
English: Fumitory.

Parts used

Whole plant.

Dose

Powder 3–5 g, decoction 50–100 ml.

Classical use

Charaka and Sushruta used Parpataka as a pot herb in fevers, chronic skin diseases, urinary disorders, diarrhoea, haemoptysis. A decoction of the leaves was also prescribed.

Parpata alone, or combined with Guduuchi (Tinospora cordifolia), Aamalaka (Emblica offici-

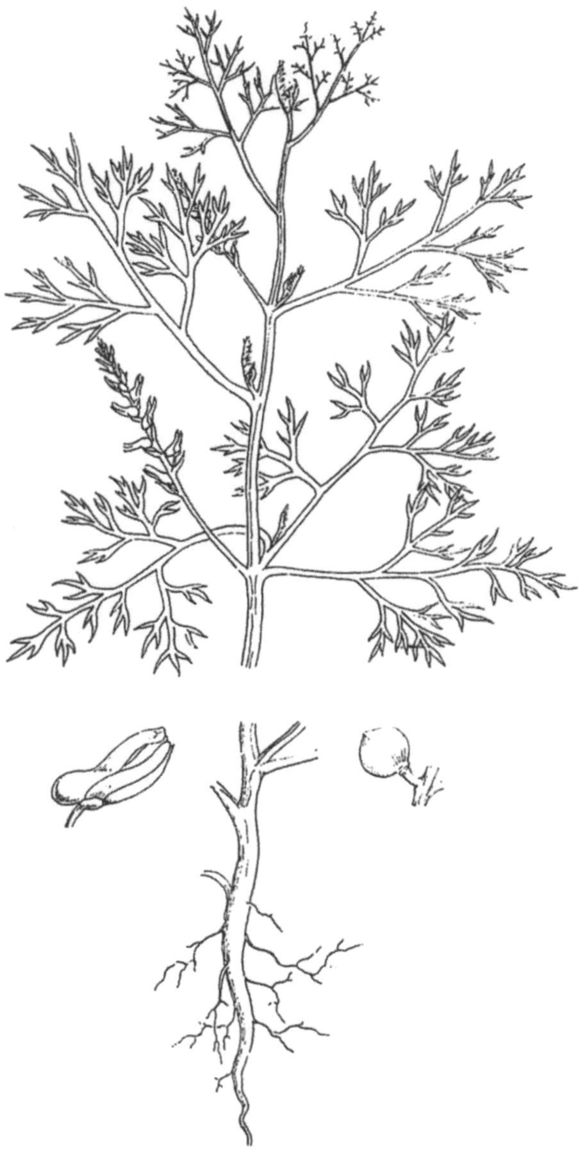

Figure 1 Fumaria indica [ADPS]

nalis), Chandana (Santalum album) or Shunthi (Zingiber officinale) was prescribed for alleviating fever (Ashtaanga Hridaya, Vrindamaadhava, Bhaavaprakaasha).

In folk medicine, the drug is used as a blood purifier, styptic and febrifuge; also in disorders of the liver.

Parpataka is an important ingredient in Amrtaarishta (Bhaishajya Ratnaavali), prescribed as an antipyretic and antiperiodic compound; Aravindaasava (ibid), prescribed as a carminative and restorative; Chandanaasava (ibid), prescribed for urinary and urinogenital disorders, Parpataadyarishtam (Sahasrayoga), prescribed in anaemia, affections of liver and spleen; Mahaatiktaka Ghrita (Ashtaanga Hridaya), prescribed as a blood purifier, anti-infective, appetiser, and restorative; Parpataka-Kashaaya (Sahasrayoga) prescribed in fevers.

In Unani medicine, the drug imported from Persia is used as Shaahtara and is an important ingredient in a number of blood-purifying compounds.

Itrifal-e-Shaahtara is prescribed in putrefaction of blood, syphilis, skin diseases.

Majoon-e-Musaffi-e-Khoon and Arq-e-Musaffi-e-Khoon are reputed blood-purifying compounds of Unani medicine, available over-the-counter.

In the North, Fumaria indica is used as Parpata. Ayurvedic Formulary of India has accepted this as the source of the drug Parpata.

IMPCOPS, Chennai, is using Mollugo cerviana Ser. (Molluginaceae) as Parpataka, while in Kerala Hedyotis (syn. Oldenlandia) spp. is used as Parpataka. In Gujarat Rungia repens Nees is used as Parpata.

All these herbs purify blood, stimulate liver, cure skin diseases, alleviate fever.

Active principles and pharmacology

Fumaria officinalis (Fumitory) contains isoquinoline alkaloids; flavonoids, including rutin; fumaric acid, hydroxycinnamic acid derivatives including caffeoylmalic acid.

Fumitory exhibits a light antispasmodic effect on bile ducts and gastrointestinal tract. Its amphicholeretic effect has also been observed.

Water soluble fraction of 90 % ethanol extract of Fumaria indica whole plant exhibited muscle relaxant, non-specific antispasmodic effect in experimental animals; the extract produced hydrocholeretic effects which were due to the presence of protopine, an alkaloid in the extract. Tertiary alkaloidal fraction also showed muscle relaxant, non-specific antispasmodic and hydrocholeretic effects. The smooth muscle relaxant property of this fraction was much less than papaverine.

The choleretic action induced by single administration, experimentally, was found to be due to the watery content of bile with solid constituents diminishing progressively. The use of the plant as alterative and in dyspepsia might be due to its effects on bile secretion (CCRAS).

The plant and its extracts were found efficacious in low fever. Also exhibited diuretic, diaphoretic, aperient, blood -purifying and vasodilatory activities (CCRAS).

Use in Western herbal

German Commission E monograph recognized the drug's efficacy in liver and gallbladder complaints.

Comminuted drug and its galenic preparations are used internally in spastic discomfort in the area of the gallbladder and bile ducts, as well as in the gastrointestinal tract. The drug is also used in skin diseases as a blood purifier and for hypoglycaemia. Is considered diuretic and mildly laxative. Dried parts, gathered during flowering season, are used.

Clusiaceae

GARCINIA

Garcinia indica Choisy.

Figure 1 Garcinia indica [CCRAS]

Garcinia purpurea Roxb.

Habitat
Western Ghats. Planted in Maharashtra.

Figure 2 Garcinia indica [CCRAS]

Classical & common names
Ayurvedic: Vrkshaamla (Charaka); Tintidika, Chukra, Amlavrkshak; Kokam, Amsula.
Siddha: Murgal.
English: Kokam Butter Tree.

Parts used
Fruit.

Dose
Juice 5–10 ml.

Classical use
Charaka used the fruits alone, or in prescriptions, internally, in flatulence, oedema, chronic alcoholism; and as a cordial. The fruit was also prescribed for stimulating digestive power, for quenching thirst and for diseases of the mouth.

During the 16th century, the fruit was used as a substitute for Daadima (Punica granatum).

Vyosaadi Gutikaa, a classical compound formulation of Shaarangadhara Samhitaa, is available over-the-counter and is prescribed in chronic catarrh, cough, asthmatic affections, bronchitis. (In the South, IMPCOPS, Chennai, is using Garcinia indica as Amlavetasa, an ingredient in the compound).

Bhaaskar Lavana Churna (Shaarangadhara Samhitaa), prescribed as a digestive, carminative, appetiser and astringent, contains Garcinia indica.

In West Bengal and Assam, acidic fruits of Garcinia pedunculata Roxb., known as Thaikal, are used as Amlavetasa. In the North, leaf stalks of Rheum spp., intertwined together, are sold in the market as Amlavetasa. According to Ayurvedic Formulary of India, Garcinia pedunculata is the source of the drug Amlavetasa and Rheum emodi can be used as a substitute.

In practice, Malabar tamarind, known as Kokam (Amsula in Maharashtra), is used as Vrikshaamla and Amlavetasa. Garcinia cambogia (Gaertn.) Desr. has also been equated with Malabar tamarind.

In folk medicine, Kokam fruit is considered antiscorbutic, cooling and cholagogue. Kokam syrup, also in the form of aerated water, is sold in Mumbai and Pune. Pieces of dried fruits are used in Kerala, Maharashtra and Gujarat for making food preparations sour.

The seeds of Kokam fruit yield an edible fat (23–26 % on the weight of seeds; approx. 44 % on the weight of kernels), known as Kokam Butter. Externally it is used for healing ulcerations, skin diseases, fissures of lips, hands, chapped skin. Also used in ointments and suppositories.

In folk medicine, Kokam Butter is also used for treating dysentery and diarrhoea with mucous.

Active principles and pharmacology

Fruit rind of Garcinia indica gave garcinol, isogarcinol, (–)-hydroxycitric acid, cyanidin-3-glucoside and cyanidin-3-sambubioside. The latex gave camboginol and cambogin.

The fruit possesses antiscorbutic, cholagogue, and antibilious; fruit and bark astringent; leaf and oil antidysenteric; and oil demulcent and emollient properties.

Kokam Butter, like other Garcinia fats, is rich in combined stearic and oleic acids. It contains about 75 % of mono-oleodisaturated glycerides and possesses a fairly low melting point. The yield of stearic acid from the fat is approx. 45.7 %.

Use in Western herbal

Commercial extract of Malabar tamarind (Garcinia cambogia) entered into weight-reducing compounds during the year 1991.

Hoffmann-La-Roche identified (–)-hydroxycitric acid (HCA) in the rind of Garcinia cambogia. In America, Sabinsa Corporation imported Malabar tamarind from India and marketed the extracted HCA to natural food and food supplement manufacturers under the trademarked name Citrin.

CitriMax, a trademark of Interhealth Co, is a popular over-the-counter Garcinia-product in the US.

CitriMax was tested on 50 overweight people. At the end of the 2-month test period, the group (25 persons) taking the Malabar tamarind-derived HCA (with chromium added) lost an average of 11.1 pounds per person. In the placebo group (25 persons), the weight loss was only 4.2 pounds.

Research showed that (–)-hydroxycitric acid curbs the appetite, thus reducing food intake. HCA is said to inhibit an enzyme (ATP-citrate lyase), responsible for converting carbohydrates into fat for storage, through competitive inhibition. The HCA binds to the ATP-citrate lyase, thus preventing it from storing fat and so surplus is instead converted to glycogen. This increase in glycogen production sends a satiety signal to the brain, suppressing the appetite.

Prescription drugs tend to act on the central nervous system to suppress food cravings, while HCA is said to act purely on digestive system's chemistry.

HCA is an ingredient in many slimming preparations—Biocare CitriMax Forte, Citri-Trim, Weight-Logic, to name a few.

Rubiaceae

GARDENIA

Gardenia gummifera Linn. f.

Habitat

Uttar Pradesh, Bihar, Andhra Pradesh, Maharashtra, Karnataka, Tamil Nadu, Kerala.

Classical & common names

Ayurvedic: Naadi-hingu, Hingupatri, Hingushivaatikaa, Dikaamaali. (Not to be confused with Hingu, Ferula narthex Boiss.) Hingupatri has now been equated with Ferula jaeschkeana Vatke (Ayurvedic Formulary of India).
Hingupatrikaa is a misleading synonym of Gardenia gummifera.
Siddha: Tikkamalli.
English: Gummy Cape Jasmine.

Parts used

Resinous exudation.

Dose

250–500 mg.

Classical use

Sushruta prescribed leaves internally as an appetizing and aromatic herb and as an ingredient in atrophy of muscles (INSA).

A resinous exudation of leaf-buds and young shoots, known as Dikaamaali, is used in folk medicine for keeping away flies from sores. It is given to children in nervous disorders and diarrhoea due to dentition. Also used for cleaning foul ulcers. A decoction is given in dyspepsia and fevers.

Active principles and pharmacology

The gum yielded flavones, including gardenin, de-Me-tangeretin and nevadensin; 3 wogonins, isoscutellarein, apigenin and de-MeO-sudachitin. Stembark gave oleanonic aldehyde, sitosterol, erythrodiol and its derivatives.

The gum possesses antispasmodic, carminative, antiseptic, expectorant, diaphoretic and anthelmintic properties.

Gentianaceae

GENTIANA

Gentiana kurroo Royle.

Habitat

Kashmir, North-Western Himalayas.

Classical & common names

Ayurvedic: Traayamaana, Traayanti; Anujaa, Balbhra, Girisaanja, Girijaa.
Unani: Ghaafis.
English: Gentian.

Parts used

Root.

Dose

Powder-1–3 g.

Classical use

Charaka prescribed Traayamaana, in prescriptions, for rigor, high body-temperature, splenic disorders, and skin diseases.

Charaka gave Traayamaana in diarrhoea for expelling impurities from the system. He prescribed milk boiled with Traayamaana as a purgative, also in chronic fever; and mixed with honey and sugar in intrinsic haemorrhage.

Sushruta prescribed the drug internally in leprosy, malignant ulcers and other virulent skin diseases, and as a vermifuge.

Traayamaana is an ingredient in Sudarshana Churna (Shaarangadhara Samhitaa), prescribed in intermittent and malarial fevers, as an antiperiodic, blood-purifying and digestive compound; also prescribed in skin diseases.

The herb is also included in Brihat-Manjishthaadi Kwaath (Shaarangadhara Samhitaa), prescribed as a blood purifier. Both the compounds are available over-the-counter.

Traayanti is an ingredient in Mahaatikta Ghrita (Shaarangadhara Samhitaa) prescribed in skin diseases, leprosy, as a blood purifier.

Traayamaana has been equated with Ghaafis of Unani medicine. Instead of Indian species, Persian gentian, Gentiana olivierii Griseb, is being used. Habb-e-Ghaafis is prescribed in inflammation of the gallbladder, hepatitis, jaundice, splenomegaly.

Ayurvedic Formulary of India has equated Traayamaana and Traayanti with Gentiana kurroo, while, earlier, the herb was also equated with Delphinium zalil Aitch & Hemsl. f., Thalictrum foliolosum DC. and Ficus heterophylla Linn.

The Wealth of India has wrongly equated the herb with a folk name, Paashaanbheda, which has now been identified with Bergenia ligulata Wall. Engl.

Active principles and pharmacology

The plant gave 6, 7-di-MeO-coumarin.

Gentiana lutea Linn., a native of Europe and Asia minor, is imported into India and is used as Gentian or yellow gentian. Constituents of yellow gentian are bitter glycosides (including gentiopicrin, gentiopicroside, amaropanin, amarogentin and amaroswerin; gentiin and gentiamarin are formed from gentiopicrin on drying); alkaloids (including gentianine); flavonoids (including gentisin).

Amarogentin is used as a bitter standard (it is able to be tested at concentrations of 1:50,000); gentianine shows anti-inflammatory activity. The drug is an archetypal bitter remedy, exhibits cooling, drying, digestive stimulant effects of the bitter principle.

Gentian stimulates the bitter taste receptors on the tongue, causing an increase in the production of saliva and gastric secretions. The herb also acts as a stimulant on gallbladder and liver, encouraging them to function more efficiently.

Use in Western herbal

Nicholas Culpeper (1616–1654) wrote of it: "Gentian strengthens the stomach exceedingly, helps digestion, comforts the heart and preserves it against faintings and swoonings. The powder of the dry roots opens obstructions of the liver and restores appetite. The herb, steeped in wine, and the wine drank, refreshes such as be over-weary with travelling...is an excellent remedy for such as bruises by falls. It provokes urine and the terms exceedingly, therefore let it not be given to women with child... It helps agues of all sorts, and the yellow jaundice."

In 1885, Moxie Nerve Food was a popular bitter brew, which carried the claim that it cured "brain and nervous exhaustion". As a beverage, for years, Moxie outsold Coca-Cola in New England. It is still available there and gentian is still one of its ingredients.

German Commission E monograph recognized Gentiana lutea's efficacy in the following areas:

- Dyspeptic complaints
- Loss of appetite.

Gentian is one of the most popular bitter digestive stimulants in Europe. It is widely used to improve appetite and digestive function and to treat a range of gastrointestinal problems including nausea, diarrhoea, indigestion, gastritis and heartburn. Gentian is taken for increasing the absorption of nutrients across the gut wall, including iron and vitamin B12; and is added to prescriptions for women with heavy menstrual bleeding and for iron deficiency. A decoction, 25 ml 3–5 times a day, is prescribed for anaemia. It is often taken as a digestive tonic in old age.

In Chinese medicine, Gentiana scabra is used for jaundice, abscesses due to toxic liver, gallbladder inflammation, diabetes, diarrhoea, rheumatic pains.

In homoeopathy, tincture of the root of Gentiana lutea is prescribed in anorexia, biliousness, dyspepsia, colic, diarrhoea, disorders of the stomach, fever, rawness in throat, and rheumatic affections.

Liliaceae

GLORIOSA

Gloriosa superba Linn.

Habitat

Throughout India, up to 2350 m on the hills.

Classical & common names

Ayurvedic: Laangaliki (Charaka), Laangalaki, Indrapushpi, Laanglaahva (Sushruta); Agnishikhaa, Anantaa, Kalihaari, Laangali; Vishalyaa, Halini, Sirikramaa, Shukrapushpikaa, Vahni-mukhi, Garbhapaatani, Garbhanut.
Siddha: Akkinichalam.
English: Tiger's Claw, Superb Lily.

Parts used

Tuberous root.

Dose

Powder 3–6 g.

Classical use

Charaka gave the fresh or dried powdered petals for inhalation during labour pain; internally in pruritus, dermatosis, and as a laxative.

Sushruta prescribed the drug in retention of the placenta; in obesity; skin diseases; as an ingredient of a medicinal oil for external application on malignant ulcers, decoction as a cleansing and antiseptic agent.

Mostly, the oil extract of the root was used externally in obstinate skin diseases.

The paste of the tuberous root or seeds was applied on haemorrhoids, boils, and scrofula (Gadanigraha, Ashtaanga Hridaya). In Ayurvedic texts this is considered as an ecbolic drug, and has also been used like Kapikachhu (Mucuna pruriens). Kushta (Saussurea lappa) has been suggested as a substitute for Laangali.

In folk medicine, the tubers are regarded as stomachic and anthelmintic when taken in small doses; they are intensely poisonous in large doses. The drug is sometimes used for promoting labour pains; also as abortifacient; externally on piles, chronic ulcers, parasitic skin diseases and as a cataplasm in neuralgic pains. The leaf juice is used for killing lice in the hair.

Over-the-counter, Kaashisaadi Taila (Bhaavaprakaasha, Shaarangadhara Samhitaa) is available, prescribed externally for application on piles, as an antiseptic. Mahaa Marichaadya Taila (Bhaavaprakaasha) is also available and is prescribed as an antiseptic and fungicide in skin diseases, scabies, pruritus, psoriasis, ulcers, and septic wounds.

Active principles and pharmacology

The flowers, leaves and tubers contain colchicine. Tubers also contain gloriosine. Leaves, in addition, contain chelidonic acid, 2-hydroxy-6-methoxybenzoic-acid and beta-sitosterol glucoside. Colchicine, demethylcolchicine and colchicoside have been isolated from seeds.

Colchicine increased basal release of growth hormone in experimental animal tissue. In primary tissue culture of malignant human gliomas, it induced mild and nonspecific reversible reduction in cell motility, and did not change adhesion.

The tuberous root was found to be oxytocic; fresh juice of the plant uterine stimulant; the root antibiotic.

Caution

The drug is a gastrointestinal irritant and may cause vomiting and purging when taken internally.

Papilionaceae

GLYCYRRHIZA

Glycyrrhiza glabra Linn.

Figure 1 Glycyrrhiza glabra [CCRAS]

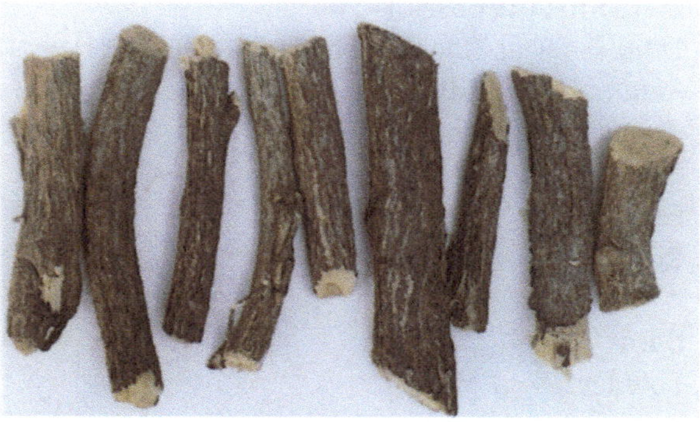

Figure 2 Glycyrrhiza glabra [CCRAS]

Habitat

Native to the Mediterranean countries and China. The drug is imported into India from Asia Minor, Iraq, Persia and other Central Asian countries.

Classical & common names

Ayurvedic: Yashtimadhu, Madhuyashtyaahvaa, Madhuli, Madhuyashtikaa, Atirasaa, Madhurasaa (Charaka), Madhuka (not to be confused with Madhuuka which is equated with Madhuka indica J.F. Gmel.); Yashtikaahva, Yashtyaahva (Sushruta); Yashti, Yashtimadhuka. Klitaka (equated with

Yashtimadhu, also with Nili/Indigofera tinctoria in Bhaavaprakaashaa).
Dalhan, in his commentary on Sushruta Samhitaa, mentioned 2 varieties of Madhuka, the wild one and the aquatic one, found in watery places. The aquatic variety, according to Bhaavaprakaashaa, was known as Klitaka.
Mulehthi nomenclature is actually Muulayashti, adopted in local dialect.
The synonym Atirasaa indicates that the drug of Ayurvedic texts was not only very "sweet" but also very "juicy", and it is possible that it was found locally; otherwise, it was not convenient to include an imported drug in more than 70 compounds. Imported drugs are always used sparingly.
The original drug, mentioned in classical texts, was replaced by the imported Unani drug Asl-us-Soos during the 16th century. Currently in practice, Ayurvedic compounds contain Glycyrrhiza glabra or Asl-us-Soos as Yashtimadhu, Atirasaa or Klitaka of Ayurvedic texts. The Egyptian variety is considered superior to Arabic and Turkish varieties. (Russian and Persian Liquorice are not recognized by the British Pharmacopoeia as suitable for medicinal purposes.)
Unani: Asl-us-Soos, Rubb-us-Soos (extract).
Siddha: Athimadhuram.
English: Liquorice, Licorice.

Parts used
Root.

Dose
Powder 3–5 g.

Classical use
According to Bhaavaprakaasha, Yashti was an ingredient in 23, Yashtimadhu in 7, Yashtaahva in 16 compounds. More than 17 medicinal oils contained the drug. Yashti was an important ingredient in medicinal oils for epilepsy, paralysis, rheumatism, haemorrhagic diseases; Yashtimadhu in compounds for dysuria, ulcers; Yashtimadhuka in prescriptions for diarrhoea, haemorrhagic diseases and skin diseases; Yashtaahva was prescribed in fevers, anuria, haemorrhagic diseases; Klitaka in high fever with delirium.

In oedema, the paste of Yashtimadhu, Tila (Sesamum indicum) and milk, mixed with butter, was prescribed. Warm, clarified butter mixed with Yashtimadhu, was applied topically on wounds, bruises, and burns. A decoction of Madhuka was applied on erysipelas. (Charaka Samhitaa, Sushruta Samhitaa, Vrindamaadhava).

Charaka prescribed 10 g Madhuka powder mixed with honey, followed by intake of milk, as an aphrodisiac and as an intellect-promoting tonic. Charaka also prescribed a paste of Maduyashti and Katuka (Picrorhiza kurroa) with sugar-water as a cardiac tonic.

A decoction of Madhuka or its powder was prescribed with honey in anaemia (Sushruta Samhitaa, Gadanigraha). Yashti mixed with cow's milk was prescribed for promoting lactation (Vaidya Manoramaa).

Sushruta prescribed the paste of Yashtimadhu, 10 g, in intrinsic haemorrhage. Chakradatta prescribed Yashtimadhu and Chandana (Santalum album), pounded with milk, in haematemesis. 10 g Madhuka powder mixed with 10 g sugar, pounded with rice-water, was prescribed in meno-metrorrhagia (Bhaavaprakaasha).

After surgery, clarified butter mixed with Yashtimadhu was prescribed for external application on haemorrhoids (Sushruta Samhitaa).

A confection of rice-milk, prepared with Yashtimadhu, was prescribed in hoarseness of voice (Sushruta Samhitaa).

Sushruta included Yashtimadhu in an eye-salve for prophylactic treatment for the eyes.

Yashtimadhu was included in a hair-oil formulation as a tonic for retarding greying of hair (Shaarangadhara Samhitaa).

Over-the-counter, Yashtyaadi Churna is available and is prescribed in coughs, bronchitis, hoarseness of voice and constipation; Yashtyaadi Kashaaya is indicated in diarrhoea with blood and in menorrhagia.

Yashti is an ingredient in Nilibhringaadi Taila (Sahasrayoga), prescribed for promoting hair growth and for preventing greying; in Duurvaadi Taila (IMPCOPS), indicated in skin diseases; in Balaashwagandhaa-laakshaadi Taila (ibid), prescribed as a reputed restorative and nervine tonic for internal as well as external use; in Asanvilvaadi Taila (Sahasrayoga), meant for external application before bath in diseases of the head and rheumatic affections.

Yashti is an important ingredient in Naarikelaanjana (IMPCOPS) eye-drops, prescribed in conjunctivitis, both acute and chronic; also in blepharitis.

In Unani medicine, Habb-e-Shaheeqa (Haj-ul-Amraz) is prescribed in migraine, dry cough, acute catarrh; Habb-e-Sil (Qarabadeen-e-Azam) in chest diseases; Habb-e-Surfa (Qarabadeen-e-Ehsani), Habb-e-Surfa Qawi (Qarabadeen-e-Azam), Laooq-e-Nazli (Bayaz-e-Kabir) in catarrh, bronchitis, coryza. Rubb-us-Soos is also an important ingredient in Laooq-e-Baadaam (Qarabadeen-e-Qadri).

Asl-us-Soos is an ingredient in Dayaqooza (Qarabadeen-e-Azam), Laooq-e-Supistan (Qarabadeen-e-Azam-o-Akmal), Laooq-e-Zeeq-un-Nafas (Qarabadeen-e-Ehsani); prescribed in catarrh, coryza, chronic bronchitis, and asthma.

Asl-us-Soos is also an ingredient in Marham-e-Jadwar (Qarabadeen-e-Azam), prescribed externally in ulcers, scrofula, and lymphadenitis.

Active principles and pharmacology

The major constituents are triterpene saponins. Glycyrrhizin (glycyrrhizic acid/glycyrrhizinic acid) is the major component (2–9 %). Other constituents are: flavonoids (isoflavones: liquiritin, isoliquiritin, formononetin), polysaccharides, sterols, coumarins, asparagin.

Research shows that on being broken down in the gut, glycyrrhizin exerts an anti-inflammatory action similar to hydrocortisone and other corticosteroid hormones. It stimulates production of hormones by adrenal glands and reduces the breakdown of steroids by the liver and kidneys. Glycyrrhizin also proved effective in the treatment of chronic hepatitis and liver cirrhosis.

Glycyrrhizia glabra, as a whole, reduces stomach secretion, produces a thick protective mucus for the lining of the stomach which protects it from inflammations, gastritis and peptic ulceration.

According to controlled clinical studies, glycyrrhizic acid and the aglycone of glycyrrhizic acid accelerate the healing of gastric ulcers. Secretolytic and expectorant effects have been confirmed in tests on rabbits. In the isolated rabbit ileum an antispasmodic action has been observed at concentration of 1:2500 to 1:5000 (PDR).

Oral administration of deglycyrrhizinated liquorice (380 mg 3 times daily) to 169 patients with chronic duodenal ulcers was as effective as antacid or cimetidine treatments. Other unidentified constituents of the herb also contribute to its anti-ulcer activity (WHO).

Use in Western herbal

The clinical application of Glycyrrhizia glabra can be divided into 3 main categories: use of oral liquorice preparations containing glycyrrhetinic acid; use of deglycyrrhizinated liquorice (DGL) and use of topical preparations containing glycyrrhetinic acid and derivatives.

Oral liquorice preparations are used in the treatment of viral infections (common cold, viral hepatitis, HIV, AIDS), premenstrual syndrome, inflammations, and as a sweetening agent.

DGL is used in ulcerative conditions of the gastrointestinal tract.

Topical preparations are used in eczema, herpes, and psoriasis.

The standard dose of powdered root is 1–2 g, of fluid extract (1:1) 2–4 ml, of solid (dry powdered) extract 250–500 mg.

The standard dose of DGL is 2–4 (380 mg) chewable tablets between or 20 minutes before meals. To heal peptic ulcers, DGL must mix with saliva. DGL may promote the release of salivary compounds that stimulate the growth and regeneration of stomach and intestinal cells. DGL in capsule form has not shown to be effective.

German Commission E monograph recognized the herb's efficacy in the following areas:

- Cough/bronchitis
- Gastritis.

Caution

The intake of higher doses (above 50 g per day) over an extended period may cause sodium retention, hypertension and cardiac complaints. The drug is contraindicated in patients with a history of hypertension, renal failure and using digitalis preparations.

Verbenaceae

GMELINA

Gmelina arborea Roxb.

Figure 1 Gmelina arborea [CCRAS]

Figure 2 Gmelina arborea [CCRAS]

Habitat

Throughout India up to 1700 m on the hills and in Andman Islands.

Classical & common names

Ayurvedic: Kaashmari, Kaashmarya, Bhadra, Mahaabhadra (Charaka, Sushruta); Gambhaari, Sadaabhadraa, Madhuparnikaa, Sriparni, Pitarohini, Hiraa, Bhadraparni. Kaashmiraka has been equated with Crocus sativus.
Siddha: Kattanam.
English: Candahar Tree, White Teak.

Parts used

Root.

Dose

Decoction 50–100 ml.

Classical use

Charaka prescribed a paste of the leaves as ingredient of a medicated clarified butter for stiffness of the back, facial paralysis; prescribed the soup of fruits in diarrhoea.

Sushruta prescribed the drug internally in bilious fever, haemoptysis, breathing trouble, asthma and for promoting adhesion of fractured bones.

Chakradatta gave ripe fruits with honey for checking haemorrhage; ripe fruits, dried and cooked with cow's milk, for urticaria.

Sesamum oil cooked with the decoction and paste of Kaashmarya was applied to underdeveloped or sagging breasts (Bangasena).

Kaashmarya belongs to the group of Five Larger Roots (Mahat Panchmuula) (four others being the roots of Aegle marmelos, Premna integrifolia, Oroxylum indicum and Stereospermum suaveolens). This group is specific for chronic fever, high fever, rheumatic affections, haemorrhages, urinary tract infections, anuria, dysuria.

In folk medicine, the drupes are used as an ingredient of a refrigerant decoction for fevers and bilious affections. A paste of leaves is applied to the head for the relief of headache in fevers. The leaf juice is used as a wash for foul ulcers. Pulverised root is applied locally for gout.

Among classical compounds, Gambhaari is an important ingredient in Shriparnyaadi Kwaath (Bhaishajya Ratnaavali), prescribed in bilious fever.

Shriparni Tailam (ibid) is prepared with the extract of Gambhaari bark, for treating sagging breasts.

Dashmuulaarishta (Bhaishajya Ratnaavali), a reputed restorative tonic, contains the root of Gambhaari.

Active principles and pharmacology

The heartwood gave lignans; leaves yielded luteolin, apigenin, quercetin, hentriacontanol and beta-sitosterol; roots gave apiosylskimmin, a coumarin characterized as umbelliferone-7-apiosylglucoside and gmelofuran.

Stem-wood bark was found to be hypoglycaemic; stem-bark antiviral; leaf demulcent; bark anticephalalgic; root galactogogue, demulcent, tonic, stomachic, laxative, antibilious, bark and root febrifuge. The leaf removes foetid discharges from ulcers. The root was found efficacious in anasarca.

Malvaceae

GOSSYPIUM

Gossypium herbaceum Linn.

Habitat
Cultivated in Punjab, Maharashtra, Assam, Karnataka, Andhra Pradesh, Tamil Nadu.

Classical & common names
Ayurvedic: Kaarpaasa, Kaarpaasi (Charaka); Kaarpaasaka, Samudraantaa, Tundikeri.
Unani: Habb-ul-Qutin, Pambadana (seed).
Siddha: Paruthi.
English: Asiatic Cotton, Levant Cotton.

Parts used
Root, seed.

Dose
Decoction 50–100 ml, seed powder 3–6 g.

Classical use
Charaka prescribed oil cooked with Kaarpaasa seeds and Horsegram for massage in rheumatic affections; flowers in leprosy; advised smoking with Kaarpaasi seeds and Ashwagandhaa (Withania somnifera) for treating cough; gave a decoction of the root in dysuria; internally and externally for treating scabies and scrofula.

Sushruta prescribed Kaarpaasa fruit-juice mixed with honey, as ear-drops for treating discharge of pus from ears; Kaarpaasi and Naagbalaa (Grewia hirsuta), pounded in equal quantity, for alcoholism.

The juice of Kaarpaasa and Plaksha (Ficus lacor), added with honey, was prescribed for diarrhoea; the root with rice-water for leucorrhoea (Vrindamaadhava).

The paste of tender fruits of Kaarpaasa was given with milk for a week in menstrual period to women desiring progeny. Cooked Shaali rice (Oryza sativa), in double quantity, with the ash of the whole plant of Kaarpaasa added, was prescribed with milk and clarified butter for treating oedema (Vaidya Manoramaa).

In folk medicine, the root-bark and seeds are used for treating poor lactation, metrorrhagia, atonic amenorrhoea, painful menstruation and climacteric complaints.

Kaarpaastyaadi Taila (Sahasrayoga) is avilable over-the-counter for massage in rheumatic affections, paralysis, hemiplegia.

Active principles and pharmacology
Cotton root-bark contains gossypol (a sesquiterpene) and flavonoids. Cotton seed contains a fixed oil, containing gossypol 2 %, and flavonoids.

Gossypol's contraceptive effect in men was first discovered in China. Studies were started in the early 1970s. In 1972, semen analysis were carried out on 5 male subjects. After administration of gossypol for 35–42 days, at a dose of 60–70 mg daily, 4 of the men were totally absent of spermatozoa in their semen, and the spermatozoa of the semen in the remaining individual was completely dead.

By 1980, more than 8000 men in the mainland of China had been treated with gossypol and two other forms of this compound. The usual dose administered was 20 mg daily for 60–70 days, followed by a maintenance dose of 60 mg on a weekly basis. In 99 % of the cases, a marked decrease in sperm count, usually 4 million per milliliter or less of semen, was noted 2–3 months after the dosing began.

Gossypol was reported to cause a degeneration of the germ cells in the seminiferous tubules of man and animal alike when given orally and to lead to absences of sperm production.

Gossypol is being explored as a possible vaginal female contraceptive.

In India, benzene and alcohol extracts of the root were studied for anti-implantation effect, post-implantation influence, effect on oestrous cycle and teratogenicity. Benzene extract in the dose of 125 mg/100 g body weight produced 82.1 % of antifertility effect in mice and 80.31 % effect in rats; with 250 mg dose, it produced 86.7 % effect in mice and 86.2 % effect on rats. Neither of the extracts had any effect on oestrous cycle. They had no post-implantation influence and any teratogenic action (CCRAS).

Hot water extract of the whole plant, seed, stem-bark and root-bark caused significant galactagogue effect in rats. The extract showed no toxicity up to 2 g/100 g body weight (CCRAS).

The root-bark stimulates uterine contractions and hastens a difficult labour. It also promotes abortion or the onset of the period and reduces menstrual flow. The root-bark encourges the blood to clot. It promotes secretion of breast-milk.

The seed oil is found efficacious in treating heavy menstrual bleeding and endometriosis.

According to John Heinerman, gossypol (from root-bark) demonstrated reasonably strong antiviral activity. Influenza virus is inactivated by treatment with gossypol, resulting in a 96–100 % protection rate. Gossypol also displayed definite action against herpes-infected mice on both oral and subcutaneous administration. A tea of the root-bark, taken several times daily, is said to help with the flu. It also controls herpes.

Use in Western herbal

In homoeopathy, tincture of fresh inner root bark of Gossypium herbaceum is used as a powerful emmenagogue.

Caution

Contraindicated during pregnancy.

Tiliaceae

GREWIA

Grewia asiatica Linn.

Habitat

Extensively cultivated in India.

Classical & common names

Ayurvedic: Parushaka, Parusha (Charaka).
In Kerala, fruits of Phoenix pusilla Gaertin. (Arecaceae) are used as Parushaka (NAA). Parushakubhe indicated towards some other variety of Parushaka.
Unani: Phaalsaa.
Siddha: Palisha.
English: Asiatic Grewia.

Parts used

Ripe fruits.

Dose

Juice 25–50 ml.

Classical use

Charaka used the fruits alone, or in prescriptions; also a decoction of the leaves, roots and bark, internally for fever, cough, diseases of the spleen, alcoholism, rheumatism, and as a purgative.

Sushruta prescribed the fruit internally as a cordial, cooling and appetizing drug; unripe fruits as astringent for treating haemoptysis. For disorders of the semen, Sushruta prescribed clarified butter cooked with Parushaka and Vata (Ficus benghalensis) fruits.

Parushaka belongs to the group of fruits which are used as a cardiac tonic.

In folk medicine, the root-bark is used internally and externally in rheumatism.

Sharbat-e-Phaalsaa, a Unani squash, available over-the-counter, is prescribed as a cardiac tonic and appetizer, and used as a refreshing summer drink.

Active principles and pharmacology

The bark gave erythrodiol, taraxasterol, beta-sitosterol and its glucoside: beta-amyrin, betulin, lupeol, lupenone and friedelin.

Fruits gave pelargonidin-3, 5-diglucoside, quercetin, quercetin-3-O-beta-D-glucoside, naringenin-7-O-beta-glucoside, and amino acids: proline, lysine, glutamic acid and phenylalanine.

Quercetin, kaempferol and their glycosides were also obtained from the leaves.

The fruit exhibited astringent, stomachic; bark demulcent; root-bark antirheumatic properties. Leaves were found efficacious in pustular eruptions.

The ether extract of leaves possesses antibacterial activity against Staphylococcus aureus and E. coli.

Asclepiadaceae

GYMNEMA

Gymnema sylvestre R. Br.

Figure 1 Gymnema sylvestre [ZANDU]

Habitat
Central and Peninsular India.

Classical & common names
Ayurvedic: Meshashrngi, Meshavishanikaa, Meshavalli (Charaka); Chhaagashringi.
Ajashrngi, a synonym of Meshashrngi, is also equated with Dolichandrone falcata Seem and Pergularia extensa N. E. Br.
Unani: Gurmaar Buti.
Siddha: Sirukurinjan.
English: Australian Cow Plant, Indian Ipecacuanha.

Parts used
Root, leaves.

Dose
Powder 3–5 g, decoction 50–100 ml.

Classical use
Charaka prescribed a decoction of leaves in fever.

Sushruta used the herb in migraine, internal abscesses, jaundice, urethral discharges, chronic skin diseases; also in gradual loss of vision. The herb was cooked with clarified butter and given internally. It was an ingredient of an eye-salve, prescribed in corneal opacity, also of a collyrium meant for improving vision.

Sushruta mentioned Meshashrngi and Ajashrngi separately. He applied the paste of Ajashrngi on lymphatic swellings and used its decoction as a cleansing and antiseptic agent. He prescribed the drug internally in migraine, internal abscesses, haemoptysis, general oedema, urethral discharges, seminal disorders, vaginal affections and as an antitoxic, astringent and styptic agent.

In Ayurvedic texts two varieties of the drug have been mentioned: Meshashrngi, Meshavalli, Ajasrngikaa and Dakshinaavarti, Vrshchikaali, Vishaanikaa.

Charaka gave a decoction of Vishaanikaa internally in constipation, assimilation disorders, as a purifier of breast milk. He prescribed Vrshchikaali, as an ingredient of a medicinal clarified butter, as an aid to memory, intellect and growth of young children.

Sushruta prescribed Vrshchikaali (syn. Vishaanikaa) internally in respiratory troubles, consumption, abdominal swelling and for parasitic worms and skin diseases. Roots were given internally as a purgative; were included in a plaster for goitre and scrofula.

During the 16th century, Meshashrngi was an ingredient in Nyagrodhaadi Churna (Bhaavaprakaasha), which was specific for polyurea and dysuria.

Active principles and pharmacology
Leaves gave several O-iso-propylidene derivatives of gymnemagenin, a hexahydroterpene, crystalline gymnemagenin, gymnestrogenin and gymnemic acid (the antisweet principle) which was a

complex mixture of at least nine closely related acidic glycosides. On hydrolysis, various genins, including gymnestrogenin and its derivatives, were obtained. Leaves also contained hentriacontane, pentatriacontane, alpha- and beta-chlorophyll, phytin, anthraquinone derivatives, inositol and D-quercitol.

Hypoglycaemic component, conduritol A, was isolated from leaves in 1992. Five new antisweet glycosides, gymnema saponins, were isolated in 1991. Leaf extract, purified by affinity chromatography, inhibited glucose absorption in the small intestines of rats. Gymnemic acid showed suppressed effect in blood glucose level after oral sucrose administration in normal rats.

Gymnema saponins showed antisweet activity; treatment of surface of rat tongue with peptide suppressed selectively neutral responses of sweet taste stimuli.

Gymnema sylvestre, should not be confused with Dolichandrone falcata or Pergularia extensa, when it is to be incorporated in an antidiabetic compound.

Capparidaceae

GYNANDROPSIS

Gynandropsis gynandra (L.) Briquet.

See figure 1.

Cleome gynandra L.

Habitat
A common weed abundant throughout warmer parts of India.

Classical & common names
Ayurvedic: Tilaparni (white-flowered variety), Ajagandhaa (yellow-flowered variety), Arkapushpi (Sushruta). (White-flowered variety has been equated with Gynandropsis pentaphylla, yellow-flowered variety with Cleome viscosa Linn., and blue-flowered with Cleome monophylla Linn.)
Unani: Hurhur, Hulhul.
Siddha: Kadugu.

Figure 1 Gynandropsis gynandra [WOI]

English: Caravella.

Parts used
Seed, leaf, root.

Dose
Seed powder 1–3 g, leaf decoction 50–100 ml.

Classical use
Charaka used the entire plant of Tilaparni and its roots, in internal prescriptions, for fever.

Cooked as a potherb, Sushruta prescribed Arkapushpi in haemoptysis, fever, cough, asthma, urinary diseases and skin diseases.

In Ayurvedic and Unani medicine, the seeds of Hurhur are considered to possess the therapeutic properties of Raai or Khardal (Brassica nigra).

Active principles and pharmacology
Seeds contain cleomin, glucocapparin and hexcosanol, beta-sitosterol and its glycosides; glucoside and kaempferol. Seeds also contain an unsaturated viscosic acid, a flavone, viscosin. The

herb and seed yield an essential oil similar to garlic or mustard oil.

Seeds of cleome viscosa contain a 7-phenoxycoumarin; leaves a diterpene lactone—cleome-olide. Seed oil contains myristic, palmitic, stearic, oleic and linoleic acids.

Seeds are carminative, antiseptic, anthelmintic; leaf sudorific; bark externally rubefacient, vesicant.

Seeds of Ajagandhaa are given for infantile convulsions. Poultice of paste of seeds is used in painful joints.

H

Orchidaceae

HABENARIA

Habenaria intermedia D. Don.

Habitat
Himalayan region.

Classical & common names
Ayurvedic: Rddhi, Vrddhi. (CCRAS.)

Parts used
Tuber.

Dose
Powder 3–6 g.

Classical use
Both the drugs belong to the group of the Eight Tonic Herbs, known as Ashtavarga. The other 6 drugs are Jivaka (Microstylis wallichi Lindl., Rishabhaka (Microstylis musifera Lindley, Ridley), Kaakoli, Kshirakaakoli (Roscoea procera Wall.) Medaa, Mahaamedaa (Polygonatum verticillatum, Linn., Allioni).

Jivaka, Rishabhaka, Kaakoli, Kshirakaakoli and Meda belong to jivaniya(vitalising) and shukrajanana (spermatogenetic) group of Charaka, and were included in rejuvenating and age-sustaining tonics. Chyavanapraasha is one such reputed Ayurvedic tonic.

These herbs, belonging to the Himalayan region, could never be identified properly and were never available in sufficient quantity. That is why, substitutes of Ashtavarga were recommended in Yogaratnaakara (1500 AD). In place of Rddhi and Vrddhi, Jivaka and Rishbhaka, Kaakoli and Kshirakaakoli, Meda and Mahaameda the drugs used are: Vaaraahikanda (Dioscorea bulbifera), Vidaarikanda (Ipomea digitata), Ashwagandhaa (Withania somnifera) and Shataavari (Asparagus racemosus) respectively. In the South, Black Mushali (Curculigo orchiodes) is used as a substitute for Rddhi and Vrddhi.

Zingiberaceae

HEDYCHIUM

Hedychium spicatum Buch-Ham.

Habitat
Central Himalayas at 1100–2500 m, East India and the hills of South India.

Classical & common names
Ayurvedic: Shathi, Gandha-shathi, Gandhapalaashi, Kapurkachari. (In Kerala, Kaempferia galanga Linn. is used as Karchura and Shathi. In other parts, Curcuma zeodoaria Rosc. is used as Karchura.) Shati and Shathi are considered synonyms in Ayurvedic Formulary of India.
Siddha: Seemaikichilikkighaga.
English: Spiked Ginger Lily.

Parts used
Rhizome, leaves.

Dose
Powder 1–3 g.

Figure 1 Hedychium spicatum [WOI]

Classical use

Charaka prescribed the powder of Shathi, Pushkaramuula (Inula racemosa) and Aamalaka (Emblica officinalis) with honey in bronchial asthma. Shathyaadi Churna of Charaka, prescribed in asthma, contained Shathi as the main drug.

For rheumatism and local inflammations a paste of Shathi, Shunthi (Zingiber officinale) mixed with the decoction of Punaranavaa (Boerhavia diffusa) was prescribed (Vrindamaadhava, Bhaavaprakaasha).

In folk medicine, Shathi is prescribed in bronchial asthma, hiccup, nausea and vomiting, halitosis, loss of appetite. The rhizomes are used as a carminative and stomachic in dyspepsia in the form of powder or decoction.

The rhizomes are considered to have insect-repelling properties and are used for preserving clothes. They are also used for disinfecting the scalp and for promoting hair growth.

Shathi is an ingredient in more than 40 classical formulations.

Among compounds available over-the-counter, Draakshaadi Churna (Vaidya Chintaamani) is prescribed in coughs, wasting diseases, and pthisis; Kantakaadyaavaleha (Shaarangadhara Samhitaa) in bronchial asthma, chronic bronchitis and painful respiration; Mahaamaashaadi Taila (Bhaishajya Ratnaavali) in nervous disorders and paralytic conditions.

Active principles and pharmacology

Rhizomes gave sitosterol and its glycoside, a furanoid diterpene-hedychenone and 7-hydroxy-hedychenone; essential oil contains cineol, gamma-terpinene, limonene, beta-phellandrene, p-cymene, linalool and beta-terpineol as major constituents.

Studies on the essential oil showed that it possesses a mild tranquillizing effect of short duration. Also possesses antimicrobial properties.

In preliminary pharmacological studies, the drug was found to have a vasodilatory effect on coronary vessels, mild hypotensive property and a non-specific antispasmodic effect on smooth muscles.

The crude ethanolic extract of rhizomes showed anti-inflammatory and analgesic activity. The analgesic activity was more prominent in benzene fraction.

Clinical trials have been conducted in tropical eosinophilia, with promising results.

Sterculiaceae

HELICTERES

Helicteres isora Linn.

Habitat

A sub-deciduous shrub found throughout Central and Western India.

Classical & common names

Ayurvedic: Aavartani, Aavartphalaa, Aavartaki.
Unani: Marorphali.
Siddha: Valamburi.
English: East Indian Screw Tree.

Parts used

Bark, fruit.

Dose

Decoction 50–100 ml, powder 3–6 g.

Classical use

The drug was used during the 16th century for its antibilious, astringent and blood-purifying properties in diarrhoea and worm infestation.

In Unani medicine, Marorphali is considered antidysenteric, anti-diarrhoeal, anti-flatulent and anthelmintic. Generally, fruit of the plant is used. Bark is also used for diarrhoea, dysentery and biliousness.

Active principles and pharmacology

Seeds gave diosgenin; roots gave cucurbitacin B and iso-cucurbitacin B; leaves yielded an ester tetratriacontanyl-tetratriacontanoate along with tetratriacontanoic acid, tetratriacontanol and sitosterol.

Aerial parts are spasmolytic; root cucurbitacin B and iso-cucurbitacin B cytotoxic. Root and stem-bark are antigalactagogue.

Fruit, root and bark are astringent in gripes; bark anti-bilious; fruit and bark anti-diarrhoeal.

In a clinical study, a group consisting of 37 patients of confirmed intestinal amoebiasis were treated with a combination of Bekh-e-Madaar (root-bark of Calotropis procera 125 mg) and Marorphali (1000 mg) 3 times daily for 26 days. The clinical efficacy was close to that of Metronidazole.

Periplocaceae

HEMIDESMUS

Hemidesmus indicus (Linn.) R. Br.

Figure 1 Hemidesmus indicus [ADPS]

Habitat

Throughout India; common in Bengal, Maharashtra and extending to Travancore and Sri Lanka.

Classical & common names

Ayurvedic: Saarivaa (white variety), Anantmuula, Gopavalli, Utpalsaarivaa, Kapuuri.
Krishna Saarivaa (black variety), Jambuupatraa-saarivaa.
Ayurvedic Formulary of India accepts Hemidesmus indicus and Cryptolepis buchananii Roem. & Schytt. as the white and black varieties of Saarivaa. But in West Bengal and Kerala, Ichnocarpus fruitescens is used as Krishna Saarivaa. In Kerala, Tamil Nadu and Karnataka, tuberous roots of Decalepis hemiltonii Wright & Arn. (Asclepiadaceae) are sold in the market as Saarivaa.
Unani: Ushbaa Hindi.

Siddha: Ninnari.
English: Indian Sarsaparilla.

Parts used

Root.

Dose

Powder 3–5 g.

Classical use

Charaka prescribed a decoction of leaves in sallow complexion, loss of voice, cough, menstrual disorders, dysentery. Charaka's Mahaa Tikta Ghrita contained both the varieties of Saarivaa, was specific for chronic skin diseases.

Sushruta gave Saarivaa in asthma, cough, consumption, abdominal swellings and aching limbs; externally as a paste for massage over inflammations. Also prescribed fruits internally with clarified butter in haemoptysis and urinary diseases.

Sushruta prescribed Krishna Saarivaa internally in respiratory affections and wasting diseases.

Among Ayurvedic compounds, available over-the-counter, Saarivadyaasava (Bhaishajya Ratnaavali) is prescribed as a blood purifier in skin diseases, syphilitic condition, and rheumatic affections. Pinda Taila (Sahasrayoga) is a massage oil, anti-inflammatory and antimicrobial, used in arthritis, inflammatory swellings, burns, wounds, fissured soles. Patraangaasava (Bhaishajya Ratnaavali) is indicated in the treatment of loss of blood, bleeding piles and anaemia. Arvindaasava (ibid) is prescribed as a carminative and restorative.

Krishna Saariva is an ingredient in Saarivaadyaasava and Patraangaasava.

Traditionally, Saarivaa was used as a blood-purifing, antitoxic, antiseptic, diaphoretic, diuretic, and anti-inflammatory tonic in fevers, chronic skin diseases, urinary tract infections and polyuria.

Active principles and pharmacology

Twigs of the plant gave a pregnane ester diglycoside named desinine.

Roots gave beta-sitosterol, 2-hydroxy-4-methoxy-benzaldehyde, alpha-amyrin, beta-amyrin and its acetate, hexatriacontane, lupeol actacosonoate, lupeol and its acetate.

Leaves, stem and root cultures gave cholesterol, campesterol, beta-sitosterol and 16-dehydropregnenolone.

Leaves and flowers also gave flavonoid glycoside: rutin, hyperoside and iso-quercitrin.

The significant utilization of the plant is as an antibacterial, antifungal, antiviral, anti-inflammatory and blood-purifying drug (CCRAS).

Pet.-ether, chloroform and alcoholic extract of the root showed significant antibacterial activity against S. aureus, S. albus, S. typhosa, V. choleriae, E. coli, Sh. sonnei, Sh. flexneri and Sh. shigae. The aqueous ethanolic extract of the whole plant showed no activity against bacteria.

The essential oil (2-hydroxy-4-methoxy-benzaldehyde) isolated from the plant possesses antibacterial property against Gram-positive and Gram-negative bacteria.

The aqueous ethanolic extract of the whole plant showed antiviral activity against Ranikhet disease virus.

A saponin from the drug was found to possess an anti-inflammatory activity in experimental animals. The ethyl acetate fraction of the plant exhibited significant anti-inflammatory activity in both acute and sub-acute models of inflammation.

The fresh decoction of roots was found to possess a blood-purifying property.

Use in Western herbal

Hemidesmus indicus has been successfully used in the cure of venereal disease, where American sarsaparilla (Aralia nudicaulis Linn.) was not found effective. The drug is used as an infusion, as boiling dissipates its volatile principle, for rheumatism, scrofula, skin diseases and thrush. Also used in nephritic complaints, and for sore mouth in children.

Smilax sarsaparilla, the drug of Western herbal, contains 1–3 % steroidal saponins, phytosterols (including beta- and alpha-sitosterol), about 50 % starch, resin, sarsapic acid and minerals.

Sarsaparilla is used as an anti-inflammatory and cleansing agent which can bring relief to skin problems such as eczema, psoriasis and general itchiness, and helps treat rheumatism, rheumatoid arthritis and gout. It possesses a tonic and specifically testosterogenic action on the body leading to increased muscle bulk and has a potential use for impotence. Sarsaparilla also has a progesterogenic action, making it beneficial in premenstrual problems and menopausal conditions, such as debility and depression.

Chinese studies indicate that sarsaparilla holds potential against syphilis.

Western use is confined to skin preparations. Sarsaparilla Root Vegicaps (520 mg), tablets, Skin Eruption Mixture are a few over-the-counter drugs.

In homoeopathy, sarsaparilla (triturations and tincture of the dried rhizome) is prescribed in scrofula and secondary syphilis and other associated conditions, such as ulcers, cutaneous eruptions, nodes, indurated glands, articular swellings, rheumatism. Blood-purifying property of sarsaparilla was confirmed by Hahnemann's proof.

Malvaceae

HIBISCUS

Hibiscus abelmoschus Linn.
Abelmoschus moschatus Medic

Habitat

Cultivated in hotter parts of India. Found wild in hilly regions of South India and in the foothills of the Himalayas.

Classical & common names

Ayurvedic: Lataakasturi, Lataakasturikaa (Sushruta), Kattaphala (Charaka), Katukaa.
Unani: Musk-daanaa, Habb-ul-mushk.
Siddha: Kattu Kasturi.
English: Muskmallow.

Parts used

Seed.

Dose

Powder 1–3 g.

Classical use

Charaka gave a decoction of seeds, in internal prescriptions, in disorders of the spleen, vomiting, pectoral lesions.

Sushruta prescribed the fruits of Lataakasturi for their cleansing, cooling, laxative and diuretic properties, and for halitosis.

During the 16th century, Lataakasturi was used for promoting eyesight and for diseases of the mouth. It was considered a good cooling herb for the nervous system and was included in antihysteric formulations.

The seeds are used in folk medicine as stomachic and carminative.

Lataakasturi is an ingredient in Mahaa-Chandanaadi Taila (Bhaavaprakaasha), still available over-the-counter as a massage oil for promoting strength, vigour and nourishment of the body.

In Unani medicine, the seed is given in spermaturia; leaves and root, with sugar, in venereal diseases, urinary discharges, painful micturition. A decoction of seeds is given in hysteria, nervous disorders.

Active principles and Pharmacology

Fatty oil of seeds contains the phospholipids alpha-cephalin, phosphatidylserine and its plasmalogen and phosphatidylcholine plasmalogen; absolute contains farnesol and ambrettolic acid lactone.

Leaves contain beta-sitosterol and its beta-D-glucoside.

Petals contain beta-sitosterol, flavonoid myricetin and its glucoside. Anthocyanins have been isolated from flowers.

Seeds contain chiefly 2-*trans*-6-*trans*-farnesyl-acetate, 2-*cis*-6-*trans*-farnesyl-acetate and ambrettolide.

A novel homologue of ambrettolide, oxacyclononadec-10-en-2-one, has also been identified.

An aromatic absolute from seeds is known as Ambrette seed oil. Seeds are used as a substitute for musk.

Pet.-ether and chloroform extracts of seeds caused dose dependent (50–100 mcg/ml) inhibition of ACh and histamine-induced responses. Methanol extract showed this effect in 200 to 400 mcg/ml.

The extracts exhibited antioxytocic effect in doses of 100 and 400 mcg/ml on uterus.

The extracts failed to exhibit excitatory or inhibitory responses on isolated tracheal chain or frog rectus abdominis muscle preparation, failed to have effect on guinea-pig auricle and blood pressure and respiration of anaesthetised cat in 10 mg/kg i.v. (CCRAS.)

Powdered seeds exhibited insecticidal properties when dusted over woolen garments.

Use in Western herbal

An emulsion made from the seeds is used as an antispasmodic. Also used as a tea or tincture for cramps, intestinal disorders, loss of appetite.

The seeds made into an emulsion with milk are used for itch and skin diseases. Seeds are also used as an insecticide.

In Egypt, the seeds are chewed as a stomachic, nervine and to sweeten the breath.

Hibiscus rosa-sinensis Linn.

Figure 1 Hibiscus rosa-sinensis [ADPS]

Habitat

Native to China, grown in gardens throughout India.

Classical & common names

Ayurvedic: Japaa (Bhaavaprakaasha), Aundrapushpa, Rudrapushpa.
Unani: Gul-e-Gurhal (flower).
Siddha: Sambarathai.
English: Shoe-flower Plant.

Parts used

Flower.

Dose

Water-extract 10–20 g.

Classical use

Japaa flowers, pounded with sour gruel, followed by jaggery, were prescribed as a contraceptive (Bhaavaprakaasha). Buds were used in Garbha Anaasthaapaka Yoga, which was given internally for contraception.

For leucorrhoea and other gynaecological disorders 10–12 buds of Japaa flower, pounded with milk, were prescribed; in amenorrhoea Japaa flowers and Jyotishmati (Celastrus paniculatus) leaves, pounded with sour gruel, were administered (Chakradatta).

A paste of Japaa flowers and Aamalaka (Emblica officinale) was prescribed externally as a hair tonic for retarding premature greying of hair (Vrindamaadhava).

In folk medicine, flowers, crushed with sugar, added to the fresh juice, are given for controlling excessive uterine bleeding. Flowers, fried in clarified butter, are also given in menorrhagia.

In Unani medicine, Sharbat-e-Gurhal (Qarabadeen-e-Jadeed) is prescribed as a refrigerant and vitaliser in palpitation, cough, fever, burning sensation in the body.

The root and flowers are given for cough and fevers. Flowers are made into a paste and applied to swellings and boils. A decoction of root is given in venereal diseases.

A black hair dye is prepared from the petals of Japaa flowers and is extensively used for blackening of hair.

Active principles and pharmacology

Flowers gave quercetin-3-diglucoside, quercetin-3, 7-diglucoside, cyaniding-3, 5-diglucoside, quercetin-3-sophorotrioside, kaempferol-3-xylosylglucoside and cyanidin-3-sophoroside-5-glucoside.

Leaves and stems gave taraxeryl acetate, beta-sitosterol and the cyclicacids sterculic and malvalic acids.

Benzene extract of the flower in 100 mg/kg exhibited post-coital antifertility effect in 80 % of treated female rats. Benzene extract was found most effective to prevent pregnancy. Total benzene extract of flowers showed anti-estrogenic activity in bilaterally ovariectomised immature albino rats. The extract played roles in reduction of weight of ovary, uterus, pituitary, as well as atrophic changes in experimental animals.

The flower extract inhibited marked depletion in germ cells and sperms of spermatogenically active testes of non-serotal bats.

The ethanolic extract of flower was devoid of antifertility effect i.e. antispermatogenic, antiovulatory, anti-implantation and abortifacient activity, in 400 mg/day p.o. (CCRAS).

In a clinical study, spirit extract of the flower was used in 750 mg dosage per day, as an oral contraceptive, for 15 days, no side-effects were recorded.

Flowers gave good results in dysfunctional uterine bleeding. The plant was found effective in suppressing estrone-induced uterine growth.

Malpighiaceae

HIPTAGE

Hiptage benghalensis Kurz

Figure 2 Hiptage benghalensis [CCRAS]

Hiptage madablota Gaertn.

Habitat
Throughout the warmer parts of Maharashtra, Konkan, Karnataka and other parts of India.

Classical & common names
Ayurvedic: Atimukta, Atimuktaka, Maadhavi
Siddha: Adigam.

Parts used
Fruit, seed, root.

Figure 1 Hiptage benghalensis [CCRAS]

Dose
Powder 3–5 g.

Classical use
Atimukta is equated with Maadhavi, also with Tinisha (Ougeinia dalbergioides Benth.) in Ayurvedic texts.

Vegetable of the leaf buds of Atimuktaka and Vata (Ficus benghalensis), processed with clarified butter, was prescribed in intrinsic haemorrhage (Sushruta).

In Bhaavaprakaasha, the kernel of Atimukta seeds is recommended for reducing abdominal girth (obesity).

Maadhavi root, with buttermilk, was also prescribed for reducing the waist girth (Chakradatta).

Sushruta prescribed Tinisha (heartwood) internally in obesity, jaundice, urethral discharges; also externally in chronic skin diseases. He prescribed the oil of seeds internally for flatulence, biliousness and for increasing slimy secretions of the body.

Active principles and pharmacology

Root-bark of Hiptage benghalensis gave a nitrogenous glucoside, hiptagin, identical with endecaphyllin and glucosyl xanthone-mangiferin. The bark also contains 8.5 % tannin. The leaves give good results in cutaneous diseases. The leaf juice shows insecticidal properties in scabies. The bark is found efficacious in rheumatoid arthritis; also gives relief in asthma.

The leaves and heartwood of O. dalbergioides contain iso-flavonoids, leaves, in addition, contain flavonoids quercetin, kaempferol and leucopelargonidin. Stem bark gave lupeol and betulin. Stem bark showed CVS- and CNS-active and spasmolytic properties.

Hiptage benghalensis or Ougeinia dalbergioides, though equated with Atimukta, Atimuktaka or Maadhavi, are not used for disorders due to deranged fat.

Apocynaceae

HOLARRHENA

Holarrhena antidysenterica Wall.

Habitat

A native of the tropical Himalayas, going up to an altitude of 1100 m. Also found throughout many forests of India in Travancore, Assam and Uttar Pradesh.

Classical & common names

Ayurvedic: Kutaja, Girimallikaa, Kaalinga, Indravrksha, Vatsaka; Indrayava (Seeds). Bitter variety.
Unani: Teewaaj, Kurchi (bark), Inderjao Talkh (seeds).
Siddha: Kudasappalai.
English: Easter Tree, Tellicherry bark.

Parts used

Seed, bark.

Figure 1 Holoarrhena antidysenterica [CCRAS]

Dose

Powder 2–4 g.

Classical use

Charaka prescribed a decoction of Kutaja bark with Shunthi (Zingiber officinale) in diarrhoea with mucus and blood; bark and seeds, in prescriptions, for fever and haemorrhage.

Sushruta prescribed seeds internally in amoebic dysentery, externally for malignant ulcers, leprosy and other virulent diseases.

Fruits were given by Sushruta as a diet in loss of appetite, persistent nausea, urinary troubles. He prescribed flowers internally in malignant skin diseases.

Kutaja seeds and bark were used in many single or compound Ayurvedic formulations—a decoction of seeds (40 g); seeds with barley-scum; linctus of Kutaja bark; a decoction of Kutaja bark, seeds and Musta (Cyperus rotundus). Kutaja was considered the best single drug for diarrhoea and dysentery (Charaka Samhitaa, Bhaavaprakaasha, Ashtaang Hridaya, Ashtaanga Sangraha, Vrindamaadhava, Bangasena).

According to Charaka Samhitaa, Kutaja is of 2 varieties: male and female. The male variety has bigger fruits, white flowers, long leaves. Its bark is extremely red and thick. The female variety gives small fruits, flowers are round in shape and greyish red in colour and bark of white colour.

The male variety, according to the 16th century texts, is astringent, constipative, cures haemorrhages, diarrhoea, menorrhagia and allied gynaecological disorders. The female variety was considered inferior in quality. A few scholars equate the male variety with Holarrhena and the female variety with Wrightia.

Available over-the-counter, Gangaadhar Churna (Shaarangadhara Samhitaa), Kutajaarishta (Bhaishajya Ratnaavali), Kutaja Ghanvati (Siddha Yogasangraha), Kutajaavaleha (Ashtaanga Hridaya) are reputed compounds for diarrhoea, dysentery and haemorrhoids.

In Unani medicine, mostly the female or white variety is used. Jawarish Teewaj, Habb-e-Teewaj are prescribed in diarrhoea, dysentery, haemorrhages and haemorrhoids. The bark powder mixed with curd is also prescribed.

Active principles and pharmacology

The bark contains about 30 alkaloids, out of which conessine, kurchine, kurchicine, holarrhimine, conarrhimine, conaine, conessimine, iso-concessimine, conimine, holacetine and conkurchine; aminoglycosteroids are important; the bark also gave nonalkaloids. The seeds gave amino acids in free state, asparatic and arginine being the major ones.

Conessine from the bark kills free living amoebae; it also kills Entamoeba histolytica in dysenteric stools of experimentally infected kittens. Conessine produces little effect on Trichomonas hominis but is lethal to the flagellate protozoon.

In a clinical study on 40 cases of amoebiasis and giardiasis, the efficacy of Kutaja in intestinal amoebiasis was 70 %. Good response was also observed in Entamoeba histolytica cyst-passers, when treated with Kutaja bark.

The seeds were found cooling, appetising and astringent to the bowels. Flowers improved appetite. The seed oil showed an inhibitory effect against pathogenic Keratinophillic fungi; maximum inhibition was noted against K. ajelloi and M. gypseum.

Various active principles of the plant exhibited following properties: conessine: amoebicidal; conkurchine hydrochloride: hypotensive, vasodilator; fruits: antiprotozoal, anticancer, hypoglycaemic; fruit and stem bark: spasmolytic; stembark: CVS-active; seed oil: antifungal.

The mother tincture prepared from the bark with 70 % alcohol exerted maximum antispasmodic activity as compared to other tinctures prepared with varying percentage of alcohol.

Wrightia tinctoria R. Br. (Wrightia antidysenterica J. Grah.) is the sweet variety of Kutaja.

Wrightia tinctoria pods contain alpha-amyrin, beta-sitosterol, ursolic and oleanolic acids; stem bark beta-amyrin, beta-sitosterol and lupeol; leaves beta-amyrin. Properties are the same as those of Holarrhena antidysenterica (CIMAP).

Ulmaceae

HOLOPTELEA

Holoptelea integrifolia Planch.

Habitat

Distributed throughout the greater part of India. Sometimes grown on the roadside.

Classical & common names

Ayurvedic: Chirabilva (Charaka, Sushruta); Karanji, Udkirya. The tree is also called Puutika, as all parts have an unpleasant smell.
The tree is one of the three species used as Karanja or Karanji, others being Pongamia pinnata Pierre and Ceasalpinia bonduc (Linn.) Roxb.
Siddha: Iya.
English: Indian Beech Tree.

Parts used

Bark.

Dose

Decoction 50–100 ml.

Classical use

Charaka gave a decoction of the sprouts in internal prescriptions for disorder of the spleen and fever.

Sushruta used Chirabilva externally as a tearing agent for inducing bursting non-suppurating wounds; internally in poisoning, fevers, jaundice, urinary disorders, migraine, internal abscesses; decoction of the fruits as an emetic. Fruits were also prescribed internally in leprosy; oil of the seeds as laxative.

Sushruta prescribed a paste of Puutika, Chitraka (Plumbago zeylanica) and Shunthi (Zingiber officinale) with the alkali of Puutika in piles.

Chirabilva was an ingredient in Karanjaadi Churna (Bhaavaprakaasha), which was prescribed with buttermilk for bleeding piles.

Active principles and pharmacology

The bark contains beta-sitosterol, friedelin and friedelan-3-beta-ol; heartwood di-OH-olean-12-en-28-oic acid; leaves hexacosanol, octacosanol, beta-sitosterol and beta-amyrin.

Stem-bark paste and seeds were found efficacious in ringworm; seeds in scabies. Stem-bark gave encouraging results when used externally and internally in rheumatism.

Poaceae

HORDEUM

***Hordeum vulgare* Linn.**
***Hordeum sativum* Pers.**

Habitat

Widely cultivated in northern India.

Classical & common names

Ayurvedic: Yava, Hayapriya.
Siddha: Baillarisi.
English: Barley.

Parts used

Whole plant.

Dose

Decoction 50–100 ml.

Classical use

Charaka prescribed barley flour, cooked in milk and with sugar and honey added, in fevers and for allaying a burning sensation; in dysentery, vomiting and debility.

For urinary disorders, barley kept overnight in the decoction of Triphalaa (The 3 myrobalans—Chebulic myrobalan, Beleric myrobalan, and Emblica officinalis) and the preparation made thereof, mixed with honey, was prescribed by Charaka. Charaka also recommended regular use of parched barley and the flour of dry, parched barley for urinary disorders.

According to Charaka, a powder of barley, mixed with Aamalaka (Emblica officinalis), was the best remedy for obesity; Vrindamaadhava recommended a diet of barley and Shyaamaaka (Echinochloea frumentacea) for alleviating obesity.

Charaka prescribed a paste of barley powder and Glycerrhiza glabra, mixed with clarified butter, for external application in erysipelas. Sushruta recommended edibles prepared with barley in skin diseases.

During the 16th century, liquid gruel of barley, mixed with honey and cooled, was prescribed in fever, colic, severe burning sensation and thirst; water boiled with dehusked barley, Vaasaa (Adhatoda vasica) and Aamalaka (Emblica officinalis), mixed with honey, for checking vomiting (Chakradatta, Bhaavaprakaasha).

More than 70 compound formulations, incorporating Yava, have been mentioned in Bhaavaprakaasha.

Yavakshaar is prepared by reducing to ashes the green spikes of the barley, dissolving the ashes in water, straining the solution through thick cloth and evaporating it over a fire. The resulting salt is chemically carbonate of potash with some impurities.

Yavakshaar (Javaasheer of Unani medicine) is used in Ayurvedic, Unani and folk medicine in urinary diseases, uric acid diathesis, uterine irritability, colic, acid eructation, dyspepsia, piles; also in diseases of liver and spleen. It enters into the composition of a number of saline medicines.

Locally, Yavakshaar solution is used in chronic skin diseases.

Active principles and pharmacology

Hordeum vulgare and other species of Indian barley on an average contain: moisture 12.5, protein 11.5, ether extr. 1.3, fibre 3.9, and carbohydrates 69.3, mineral matter 1.5, calcium 0.03 and phosphorus 0.23 %; iron 3.7 mg/100 g. Barley contains 4 classes of protein, viz. albumin, globulin, prolamin (hordein) and glutelin (hordenin); free amino acids and protein intermediates are also present.

Seeds contain ubiquinones, extract contains biflavones, proanthocyanidins, chrysoeriols; p-coumaroylagmatine from the plant shows antifungal activity. Seedlings contain glycosides of hordatines A & B, are also antifungal. Barley grits contain pangamic acid.

Seed-enriched barley meal was found slightly hypoglycaemic in human; hypocholesteremic in rats; liver protective in animals.

Hordeum distychum contains polysaccharides: starch (50 %), fructans; mono- and oligosaccharides: saccharose, raffinose, glucodifructose, glucose, fructose; protein (10 %): including, among others, prolamines, hordein-glutelins, hordenine, albumins and globulins; fatty oil (2 %): chief fatty acids are linoleic and oleic; vitamins: E, nicotinic acid, pantothenic acid, vitamin B6, B2, and folic acid.

Hydroxycoumarins, found only in stalks, include among others, umbelliferone, scopoletin, herniarin, aesculetin; whereas the sprouts contain amines: tyramine, hordenine, gramine also with certain strains (dimethylaminomethylindol). (PDR.)

Use in Western herbal

Barley is used for convalescents and in the treatment of diarrhoea, gastritis, inflammatory bowel conditions. Barley is also used to assist infants in the digestion of milk.

Made into a poultice, barley is used for reducing inflammations and swellings.

Trials undertaken in the 1990s indicate that barley may help control diabetes and that barley bran may have the effect of lowering cholesterol and preventing bowel cancer.

Chinese research suggests that barley may be used in the treatment of hepatitis.

Barley sprouts are used as a valuable aid to digestion for patients with digestive disturbances associated with Candida albicans overgrowth and hepatitis.

Barley is contraindicated for nursing mothers, as it dries mother's milk. In China, it is given to women whose children have suddenly died after birth.

Flacourtiaceae

HYDNOCARPUS

Hydnocarpus laurifolia (Dennst.) Sleumer.
Hydnocarpus wightiana Blume.

Habitat

Western ghats, up to 700 m.

Classical & common names

Ayurvedic: Tuvaraka, Katu-kapittha, Kusthavairi, Chaulmoograa.
Siddha: Maravattai.
English: Soorty Oil Tree.

Parts used

Seed, seed-oil.

Dose

Seed powder 3–5 g, oil 5–10 drops.

Classical use

Charaka prescribed a decoction of the seeds of Tuvaraka for washing vagina and the paste for topical application for allaying foul smell.

Sushruta prescribed the oil of the ripe fruits internally in diabetes, urinary diseases; in leprosy and malignant skin diseases.

According to Ashtaanga Hridaya, the seeds of Tuvaraka, Bhallaataka (Semecarpus anacardium), Baakuchi (Psoralea corylifolia), Chitraka root (Plumbago zeylanica) and Shilaajatu (a mineral pitch) were prescribed in leprosy, after processing them in their own decoction at least 7 times and converting them into a potentised tonic.

Active principles and pharmacology

Seed oil gave hydnocarpic, chaulmoogric and gorlic acids, lower homologues of hydnocarpic acid and a flavonolignan-hydnocarpin. Cyclopentene fatty acids characterise seed lipids. Pericarp gave leucopelargonidin.

Stearic acid was the main acid in the leaf lipid and linoleic acid in the lipids of trunk-wood and roots.

Hydnocarpus oil is mainly used in the treatment of leprosy and is effective in early cases, in decreasing the size of nodules, anaesthetic patches and skin lesions. It is administered internally under medical supervision, the dosage being increased gradually to prevent gastric irritation. Intramuscular injections give better results; iodised hydnocarpus oil is found to be less painful. Sodium salts of the fatty acids have also been used with benefit.

Ethyl hydnocarpate possesses the therapeutic properties of the oil and is generally preferred to the latter in the treatment of leprosy. When injected, it infiltrates over a large area and is also less irritating.

It has been shown that in culture media, various hydnocarpates have a strong action in checking the growth of acid-fast Mycobacterium tuberculosis; derivatives of the oil were more active. Sodium salts of chaulmoogric and hydnocarpic acids were reported to be bactericidal against M. tuberculosis in a dilution 1:100, 000; but definite results could not be recorded.

Related species

Hydnocarpus kurzii (King) Warb, syn. Taraktogenos kurzii King, is found in North-eastern India and is known as Chaulmoograa. The plant is also a source of Chaulmoograa oil which contains hydnocarpic, chaulmoogric, gorlic and palmitic acids and lower homologues of hydnocarpic acid.

Gynocardia odorata R. Br., syn. Chaulmoogra odorata (R. Br.) Roxb., also known and used as Chaulmoograa, is found in the Eastern Himalayas, Sikkim and Assam.

The oil from the seed is used in folk medicine for its antileprotic, antisyphilitic, antirheumatic properties in scrofula, eczema, psoriasis, and gout.

The bark yielded the triterpenes odolactone, acetylodollactone and odollactone, characterised as derivatives of friedelan 26-12-beta-lactone.

Use in Western herbal

Externally, preparations of Hydnocarpus are used in the treatment of various skin affections, such as psoriasis and treatment of leprosy.

Caution

The oil is severely irritating in local application.

Coughing, dyspnoea, laryngospasm, kidney damage, visual disorders, head and muscle pain and central paralysis are side-effects following intake of the oil.

Seeds are severely poisonous due to their glycoside content.

Acanthaceae

HYGROPHILA

Hygrophila schulli (Ham.) MR & SM Almeida.

Figure 1 Hygrophila schulli [ADPS]

Habitat

A common herb throughout India around ponds, on sides of the river and in marshy areas.

Classical & common names

Ayurvedic: Kokilaaksha, Ikshuraka (Charaka, Sushruta); Ikshura, Kokilaakshi. PV Sharma equates Ikshuraka with Astercantha longifolia (Linn.) Nees. and Kokilaaksha with H. spinosa T. Anders.
Unani: Taalamakhaanaa. (National Formulary of Unani Medicine has wrongly equated Taalamakhaanaa with Euryale ferox Salisb.)
Siddha: Neermulli.

Parts used

Seed, ash.

Dose

Seed powder 3–6 g, ash 1–3 g.

Classical use

Charaka gave a decoction of the leaves and roots, alone or in internal prescriptions, as an aid to virility; also in urinary calculus, and haemothermia.

Sushruta prescribed powdered seeds of Kokilaaksha and Kapikacchuu (Mucuna prurita), mixed with sugar, to be taken with warm milk, as an aphrodisiac.

The seeds of Kokilaaksha were prescribed in the form of a decoction in gout; ash of the seeds in oedema (Ashtaanga Hridaya, Chakradatta).

According to Haarita Samhitaa, a decoction of Kokilaaksha acts as a hypnotic and induces sleep.

Shataavari Churna (Shaarangadhara Samhitaa) contains Ikshuraka as an important ingredient; it is prescribed as an aphrodisiac and rejuvenating tonic.

In Unani medicine, the seeds of Taalmakhaanaa find an important place in many sex tonics, prescribed for spermaturia, spermatorrhoea, vitiation of semen and impotency. Majoon-e-Bandkushad (Qarabadeen-e-Azam-o-Akmal), available over-the-counter, is a reputed Unani tonic for sexual debility.

The root and ash of the plant is prescribed as a diuretic. A decoction of the root and the ash is also given as a liver tonic.

Active principles and pharmacology

Diuretic properties of the seeds are due to the large amount of mucilage and potassium salts.

Seeds contain 23 % of a yellow semi-drying oil; palmitic 7.2, stearic 0.8, oleic 11.9, and linoleic 80.1 %; acids also contain the enzymes diastase, lipase and protease. Xylose and uronic acids are present in the seed oil. Lipid extract of the plant contains lupeol, stigmasterol and straight-chain of hydrocarbons.

Alcoholic extract of the whole plant exhibited antifungal activity against Trichophyton mentagrophytes, T. rubrum, Microsporum gypseum, Epidermophyton floccosum and Candida albicans.

Essential oil from root exhibited potent antibacterial activity against both Gram-positive (S. aureus, S. lutea, B. cereus, B. subtilis) and Gram-negative (Sh. boydi, E. coli, S. typhi, S. dysenteriae) bacteria.

Pet.-ether and alcoholic extracts of root and leaves of A. longifolia in 100 mg/kg p.o. dose showed potent anti-inflammatory, diuretic and moderate antipyretic effect in rats. (CCRAS.)

Use in Western herbal

Hygrophila spinosa, though native to India, is now widely distributed throughout tropical regions. It is gathered when in flower.

The seeds are used for their aphrodisiac properties, aerial parts and ash of the plant as a strong diuretic, the root as demulcent for alleviating the inflammation produced by urinary tract infections.

The herb is also thought to support the liver in jaundice, hepatitis, hepatic obstruction.

Solanaceae

HYOSCYAMUS

Hyoscyamus niger Linn.

Habitat

Grows wild throughout the Himalayan range at an altitude of 2500–3500 m; also in Kashmir. Native to Scandinavia, Europe, Great Britain. Cultivated in the USA.

Classical & common names

Ayurvedic: Paarsika-Yavaani, Madakaarini; Turushka (Charaka, Sushruta); Yaavani, Khuraashaanikaa.
Unani: Khuraasaani Ajawaayan.
Siddha: Kurosani omam.
English: Henbane seeds.

Parts used

Seed.

Dose

Powder 500 mg–1 g.

Classical use

Charaka prescribed the liquid extract of Turushka internally for cough, consumption, pectoral lesions, debility.

Sushruta prescribed Turushka internally in skin eruptions and blood poisoning.

Turushka has been equated with Hyoscyamus niger by CCRAS. Earlier, Turushka and Sihlika were considered as synonyms (Bhaavaprakaasha) and were equated with Altingia excelsa Naronha and balsam obtained from the tree was used in medicine. INSA scientists (P. Ray et al) equated Turushka with Altingia excelsa or Liquidambar altingia Bl. Classical uses of Turushka cannot be attributed to Paarsika-ajvaayan. Turushka, in fact, indicated that the drug originated from an Islamic country.

Paarsika-Yavaani, mixed with jaggery, was administered with stale water for expelling intestinal worms (Vrindamaadhava).

In Unani medicine, Khuraasaani Ajawaayan (not to be confused with Ajawaayin or Nankhwah of Unani medicine) was used as a sedative in mental and maniacal excitement, epileptic mania, chronic dementia with insomnia, convulsions, neuralgia; also as a styptic.

The drug was also included in antispasmodic, carminative, antiasthmatic compound formulations.

Oil extract is used externally in rheumatism, gout.

In Indian herbal medicine, seeds are used. In modern medicine, leaves and flowering tops are used for preparing tincture Hyoscyamus.

Active principles and pharmacology

Young plants contain more hyoscine and less hyoscyamine; in mature plants leaves contain hyoscyamine, hyoscine, apoatropine, tropine, cuscohygrine, (–)-6-beta-hydroxyhyoscyamine, skimmianine, apohyoscine, alpha- and beta-belladonine, hyoscyaminic and hyoscine-N-oxides. Leaves also gave rutin.

Seeds contain hyoscyamine, fatty oil 25 %.

Due to its sedative effect the drug gave good results in bronchial asthma, spasmodic and hooping cough, irritable affections of lungs, bowel and genitourinary organs.

Use in Western herbal

Hyoscyamus niger (Henbane) is used extensively as a sedative and painkiller, specific for pain affecting the urinary tract, especially due to kidney stones. Used for its sedative and antispasmodic effect in Parkinson's disease for relieving tremor and rigidity during early stages of the illness; also in asthma and bronchitis.

Research shows that Henbane preparations produce a parasympatholytic or anticholinergic effect by competitive inhibition of acetylcholine. This inhibition affects the muscarinic action of acetylcholine but not its nicotine-like effects on ganglia and motor end-plates.

Henbane preparations exert peripheral actions on the autonomic nervous system and on smooth muscle, as well as the central nervous system. Because of their parasympatholytic properties, they cause relaxation of organs containing smooth muscle, particularly in the region of the gastrointestinal tract. Furthermore, they relieve muscle tremors of central nervous origin. (PDR.)

The drug is applied externally as an oil in neuralgia, sciatica, and rheumatism.

Caution

Overdosage is toxic, producing delirium, coma and even death. Cases of paralysis associated with taking hyoscine internally were observed.

Hypericaceae

HYPERICUM

Hypericum perforatum Linn.

Habitat

Temperate North-Western Himalaya from Kashmir to Shimla at 2000–3000 m.

Classical & common names

Unani: Heufariqoon (National Formulary of Unani Medicine); Bassant, Balsaan; Al-heu-fariqoon (Arabic).
English: Common St. John's Wort, Klamath Weed.

Parts used

Whole plant.

Dose

Powder 3–6 g.

Classical use

Used in Unani medicine as an emmenagogue and diuretic. Seeds are given in the fourth-day fever, mixed with Sudaab (Ruta graveolens) seeds. Fruits are also used.

Powder of dry leaves is dusted over infected wounds. A paste of leaves is also applied over wounds, sores, ulcers, and burns; and is massaged over neurological affections, rheumatic inflammations, lumbago. Leaves are also prescribed internally. The oil, prepared by infusing the fresh flowers in olive oil, is also used externally.

The fruit is used as a purgative.

Extract of the whole plant is used in urinary tract infections.

In folk medicine, an ointment, prepared from the aqueous extract of the plant, is used as a hair restorer.

In Indian systems of medicine, Hypericum perforatum is not in use as an anti-depressant drug.

Active principles and pharmacology

The herb contains: anthracene derivatives: in particular hypericin, pseudohypericin; flavonoids: in particular hyperoside, quercitrin, rutin, isoquercitrin; also biflavonoids, including, among others, amentoflavone; xanthones: 1, 3, 6, 7-tetrahydroxyxanthone; acylphloroglucinols: hyperforin with small quantities of adhhyperforin; oligomeric procyanidines; catechin tannins; caffic acid derivatives. (PDR.) The essential oil contains caryophyllene, pinene, limonene and myrcene; hypericins, flavonoids, resin.

The major compounds of medicinal value in leaves and flowers are hypericin and pseudohypericin, found in very low concentrations ranging from 0.0095 to 0.466 % in the leaves, 0.24 % in flowers. Flowers contain 16 %, leaves 12 % and whole herb 9 % flavonoids. The whole herb contains 0.13 % essential oils.

In 1988, hypericin and pseudohypericin drew attention when a report by researchers at the New York University Medical Centre and the Weizmann Institute of Science (Rehovot, Israel) showed that these constituents could inhibit retroviruses, and focus has shifted to hypericin's effect on the viral cause of AIDS.

In vitro, studies have shown that hypericin and pseudohypericin exhibit strong antiviral activity against herpes simplex virus types 1 and 2 as well as influenza types A and B and vesicular stomatitis against Epstein-Barr virus. (This is the virus that causes mononucleosis.) Hypericin also shows weak antiviral activity against a number of hepatitis B family, duck hepatitis B virus.

The clinical evaluation of St. John's Wort extract began in 1984 with an initial clinical study of 6 depressed women, aged 55–65 years. Since this initial study, more than 26 double-blind controlled studies with the standardized St. John's Wort extract (0.3 % hypericin) were carried out. The extract at a dosage of 300 mg 3 times a day was found to be as effective in relieving symptoms of depression as standard antidepressants (1992–1994).

Research indicates that other constituents, such as xanthones and flavonoids also contribute to the medicinal actions of St. John's Wort.

Studies have demonstrated that the antidepressive effect may be due to the presence of a monoamine oxidase inhibiting function in the active agents. The effect may be largely due to the ability of the herb to inhibit the reuptake of serotonin.

St. John's Wort extracts showed broad-spectrum antimicrobial activity against both Gram-positive and Gram-negative bacteria. The organisms studied included S. aureus, S. mutans, E. coli, Pseudomonas aeruginosa. Oil-based preparations of St. John's Wort demonstrate anti-inflammatory action due to their high flavonoid content and antibacterial properties, when applied topically.

Use in Western herbal

In Western herbal, standardized extract of St. John's Wort (with 0.3 % hypericin) is used as an antidepressant in dosage of 300 mg 3 times daily with meals. St. John's Wort extract standardized to 5 % hyperforin gave much better results in patients with severe depression. (Expanded Commission E Monographs, 2000). The drug is not recommended with other antidepressants.

German Commission E monograph on St. John's Wort lists psychovegetative disturbances, depressive states, fear, and nervous disturbances as clinical indications for the extract. It recognized the drug's efficacy in the following areas:

- Anxiety
- Depressed moods
- Inflammation of the skin
- Blunt injuries
- Wounds and burns (first degree).

The herbalists of Europe have long used the herb as a tonic for anxiety, tension, insomnia, depression; for menopause hot flushes, night sweats, depression, irritability, lack of concentration, and fatigue. The herb is also being used for intestinal colic, irritable bowel, dysmenorrhoea and as a antispasmodic.

The red oil of St. John's Wort is used externally as an antiseptic and tissue healer for wounds and burns and to relieve cramp and nerve pain.

The oil is also given, under medical supervision, internally for gastric inflammation and peptic ulcers, and oesophagitis.

For external use creams, ointments and sprays are sold over-the-counter. Standardized extract in the form of capsules are also sold. Even plant juice is available.

In homoeopathy, tincture of the whole fresh plant is prescribed for concussion of brain, hypersensitiveness, asthma, whooping cough, rheumatism, neuralgia, sciatica, ulceration, haemorrhoids, and wounds.

Caution

Hypericin makes the skin more sensitive to sunlight and other sources of ultraviolet light.

Lamiaceae

HYSSOPUS

Hyssopus officinalis Linn.

Habitat

Native of Mediterranean region and temperate Asia, naturalised in parts of Europe and America. One species occurs in the Himalayas, from Kashmir to Kumaon at 2700–3700 m.

Classical & common names

Unani: Zuufaa.
English: Hyssop.

Parts used

Whole plant.

Dose

3–9 g.

Classical use

An imported Persian species is used in Unani medicine. One species, Hissopus parviflora Benth., is found in the Himalayas, Kashmir and Punjab. In Sindh, Nepeta ciliaris Benth. is known as Zuufaa Yaabis.

Araq-e-Zuufaa (liquid extract) or Sharbat-e-Zuufaa (squash) is prescribed when phlegm is thick and sticky and chest is congested.

Active principles and pharmacology

Aerial parts gave an essential oil containing laevo-pinocamphone, camphor, thusone, alpha- and beta-pinene, alpha-terpinene, laevo-pinocampheol and cadinene. The plant also contains ursolic acid.

The Mediterranean species contains terpenes (including marubiin, a diterpene), a volatile oil (mainly camphor, pinocamphone and beta-pinene), flavonoids, hyssopin, tannins and resin.

Marubiin is a strong expectorant. Pinocamphone is toxic, and the volatile oil can cause epileptic seizures.

Marubiin and camphor-like constituents loosen phlegm and it can be coughed up more easily.

Preparations of Hyssop are also found efficacious in intestinal catarrhs.

Use in Western herbal

The Greek physician Dioscorides prescribed the herb in tea for cough, wheezing, and shortness of breath, in plasters and chest rubs, and as an aromatic nasal and chest decongestant.

Nicholas Culpeper (1616–1654) wrote of it: "It helps to expectorate tough phlegm, and is effectual in all cold griefs or diseases of the chests or lungs being taken either in syrup or licking medicine... boiled with figs, it makes an excellent gargle for quinsey (tonsillitis)."

Hyssop, sometimes, irritates the mucous membranes; it is best given after the infection has peaked. It is used against asthma, especially when the condition is exacerbated by mucus congestion.

German Commission E monograph recognized the herbs efficacy in the following areas:

- Fevers and colds
- Liver and gallbladder complaints.

Though in the West the prime use of Hyssop is in cough remedies, it is also used as a bitter digestive tonic for wind and colic. The herb is an ingredient in over-the-counter catarrh mixtures; also in chest-rubs.

I

Poaceae

IMPERATA

Imperata cylindrica (L.) Raeus.

Figure 1 Imperata cylindrica—flowering panicles [WOI]

Habitat

The grass found throughout the hotter parts of India, both in plains and hills, ascending up to 2300 m in the Himalayas.

Classical & common names

Ayurvedic: Darbha, Suuchiyagra, Yagnika. (Not to be confused with Kusha, equated with Desmostachya bipinnata Stapf.), Yagyabhuushana, Bahir.
English: Thatch Grass.

Parts used

Root, grass.

Dose

Decoction 50–100 ml.

Classical use

Charaka gave flowers of Darbha, in prescriptions, in calculus and gravels.

Darbha was one of the ingredients of Stanyajanana Mahaakashaaya of Charaka, which was prescribed as a galactagogue.

The herb was also an ingredient in Charaka's Brahma Rasaayana (also of Ashtaaga Hridaya), a reputed Ayurvedic tonic, still available over-the-counter, which invigorates the body and mind.

Sushruta prescribed Darbha internally in migraine and internal abscesses, with cow's milk in retention of urine, dysuria, calculi, and in haemoptysis. The herb was also prescribed for reducing obesity.

Darbha root is an ingredient in Sukumaar Ghrita (Sahasrayoga), still available over-the-counter, prescribed as a restorative, laxative and used in the treatment of diseases of the reproductive system.

Trinapanchamuulaadi Kashaaya (ibid) is prescribed as a diuretic in urinary affections, Varunaadi Kashaaya (ibid) is given for reducing fat or obesity and for harmonizing a deranged digestive system.

Active principles and pharmacology

Analysis of tender grass gave the following values (dry matter basis): crude protein 6.56, ether extract 3.33, crude fibre 34.6, N-free extract 47.6, ash 7.92, calcium 0.39, nitrogen 1.05 and phosphorus 0.22 %; starch equivalent 10.7, digestible protein 0.8 lb/100 lb. The grass is reported to be a good source of vitamin A and C.

Rhizomes gave the following values (dry basis): total sugar 22.05, reducing sugar 9.20 and invert sugar 12.45 %.

Feeding experiments on animals (young and adults) show that the grass at any stage of growth produces acidic urine. Adult animals fed on mature or partially mature grass are likely to suffer from acidosis.

The rhizomes of Imperata cylindrica exhibit restorative, tonic, astringent and antiviral properties. A decoction of rootstocks is found efficacious in diarrhoea, dysentery and gonorrhoea. Fruiting spikes possess sedative properties and seeds are styptic.

The root is used in fevers and as a diuretic, but antipyretic and diuretic properties remain unconfirmed.

Thai plant gave triterpenoids, arundoin and cylindrin. Culms and blades gave 7 triterpenoids, Me-esters. Plant contained serotonin.

Papilionaceae

INDIGOFERA

Indigofera tinctoria Linn.

Habitat

Cultivated in many parts of India.

Classical & common names

Ayurvedic: Nilikaa, Nilaa (Charaka, Sushruta); Nili, Nilini (Bhaavaprakaasha); Nilapushpa, Ranjani, Shaaradi, Tutthaa.
Unani: Nila.

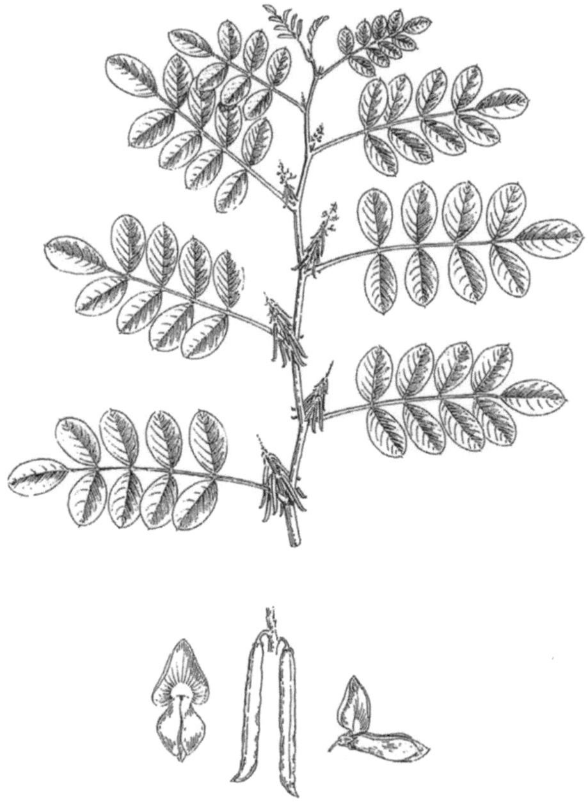

Figure 1 Indigofera tinctoria [ADPS]

Siddha: Neeli.
English: Indian Indigo.

Parts used

Root, leaves, fruits, seeds.

Dose

Decoction 50–100 ml.

Classical use

Charaka gave leaves, in prescriptions, in retention of urine, flatus, acute constipation. Sushruta prescribed the fruit mixed with clarified butter as a purgative.

Nilini was the main ingredient in Charaka's Nilinyaadi Churna, specific for abdominal diseases, ascites, enlargement of spleen and liver, intestinal obstruction.

A medicinal oil, processed with Nili, was used by Sushruta on malignant ulers; Nili's extract in clarified butter externally and internally for treating tumours.

Sushruta used pounded leaves as an ingredient of a hair oil, meant for giving natural colour to grey hair.

The root, pounded with goat's milk, was prescribed in dysuria.

For treating skin diseases, the root powder was prescribed with milk in the morning and the paste was applied locally (Vaidya Manoramaa).

According to Shaarangadhara Samhitaa and Bhaavaprakaasha, Nili was the drug of choice in polyuria, chronic diarrhoea, piles, dyspnoea, bronchitis, and premature greying of hair. Also in diseases of the liver and of spleen. In Kaiyadeva Nighantu, Nili was recommended in neuralgic and nervous symptoms, and epilepsy.

In Unani medicine, a paste of seeds is applied topically on blemishes, freckles, pimples and other skin affections. Leaves are used in hair dyes. Extract of the root is prescribed internally for piles; a paste for applying locally.

Among over-the-counter compounds, leaves are an ingredient in Nilibhringaadi Taila (Sahasrayoga), prescribed as a hair oil for promoting hair growth and for preventing greying. To a certain extent, it gives black colour to the hair. Japaapatryaadi Taila (IMPCOPS) is for external use in skin diseases. Triphalaadi Taila (Sahasrayoga) contains the root of Indigo plant, is applied externally in diseases of the scalp, arrests balding and greying.

Active principles and pharmacology

Apigenin, kaempferol, luteolin and quercetin have been isolated from various parts of plants (roots, stems, leaves, pods) and tissue culture; production of flavonoids in tissue culture was greater than in plant parts.

Alcoholic extract of aerial parts exhibited antihepatotoxic activity against carbon tetrachloride-induced hepatic injury in animals. EtOH (50 %) extract showed hypoglycaemic activity.

The plant is found efficacious in enlargement of liver and spleen, nervous disorders, epilepsy, piles, ulcers, burns, skin diseases, leucoderma; leaf possesses anti-inflammatory; root and stem antiasthmatic, expectorant, laxative, anthelmintic, antitumour, febrifuge, diuretic, internally and externally, antiseptic properties. Root, leaves, or the entire plant impart antibacterial properties to medicinal hair oils and skin oils.

Asteraceae

INULA

Inula racemosa Hook. f.

Habitat

Temperate and Alpine Himalayas from Chitral to Nepal at 1500–4200 m.

Classical & common names

Ayurvedic: Pushkaramuula, Pushkara, Paushkara, Padmapatra, Kaashmira.
Unani: Raasan (not to be confused with Raasanaa, which has been equated with Pluchea lanceolata).
English: Elecampane, Scabwort.

Parts used

Root.

Dose

Powder 1–3 g.

Classical use

According to Charaka, Pushkaramuula is a specific remedy for cough, asthma, hiccough, chest pain.

The powder of Pushkaramuula was prescribed with honey for alleviating cough, hiccough, dyspnoea, chest pain; a decoction of Dashmuula (the Group of Ten Roots), added with the powder of Pushkarmuula, was also prescribed (Vrindamaadhava).

In folk medicine, the herb is used topically in skin diseases.

Agastya Haritakyaavleha (Shaarangadhara Samhitaa) and Pushkaramuula Churnam (Sahasrayoga), prescribed in cough, asthma, and chronic bronchitis, are available over-the-counter.

Active principles and pharmacology

Essential oil and extracts from roots contained inulin, alantolactone, iso-alantolactone, and dihydroisoalantolactone.

Preliminary studies with the ethanolic extract of roots of Inula racemosa exhibited antiallergic and antiasthmatic properties, the later being more pronounced.

Specific studies for bronchodilator properties on isolated trachea were performed and found it a potent bronchodilator.

The extract also protected guinea-pigs against various experimental asthma, plant pollen etc. It possessed antihistaminic as well as anti-5-HT activity, suggesting its use in bronchial asthma.

The clinical trial in patients suffering from bronchial asthma and chronic spasmodic bronchitis showed significant improvement of pulmonary functions and reduction or absence of dyspnoeic attacks. (CCRAS.)

Inula racemosa, 5 g/day/rabbit, administered orally in the form of a decoction, produced significant fall of blood sugar level (52.2 %) during second hour after administration in fasting rabbits; against glucose (5 g/kg p.o.)-induced hyperglycaemia, the rise of blood sugar was insignificant (15.68 %).

Ethanol 70 % extract of sun-dried roots of Inula racemosa, in doses of 100, 150 and 200 mg/kg i.p., produced dose-dependent anti-inflammatory effect experimentally.

The essential oil of Inula racemosa was tested for antibacterial and antifungal activity, and was found moderately effective against S. aureus, Ps. aeruginosa, B. anthracis. The essential oil of the root exhibited anthelmintic activity against earthworms and tapeworms, but it was less potent than piperazine citrate.

Alantolactone and isoalantolactone exhibited antidermatophytic activity.

Use in Western herbal

Inula helenium Linn., Elecampane, contains alantolactone, isoalantolactone, 11,13-dihydroisoalantolactone, inulin (up to 44 %), triterpene saponins, sterols, and polyacetylenes.

Preparations of the rhizome are used to treat bronchitis, whooping cough and bronchial catarrh.

In folk medicine, Elecampane is used as a carminative, stomachic, diuretic, cholagogue, as well as in menstrual complaints.

Caution

Larger administration of the drug lead to vomiting, diarrhoea, spasm.

Convolvulaceae

IPOMOEA

Ipomoea digitata Linn.

Figure 1 Ipomoea digitata (Kshira-vidaari) [ZANDU]

Habitat

Throughout tropical India in moist regions.

Classical & common names

Ayurvedic: Kshira-vidaari, Kshirvalli (Charaka), Payasvini, Swaadukandaa, Ikshukandaa, Gajavaajipriyaa, Kandapalaasha, Bhuumikuushmaanda. Vidaari.
English: Milky Yam.

Parts used

Tuber.

Figure 2 Pueraria tuberosa (Vidaari Kanda) [ZANDU]

Dose

Powder 3–6 g.

Classical use

Two types of Vidaari tubers are mentioned in Samhitaas, but in Nighantu part, mostly, Kshira-vidaari is mentioned. Kshira-vidaari has been equated with Ipomoea digitata, syn. Ipomoea paniculata R. Br., Vidaari with Pueraria tuberosa DC. Ipomoea digitata is used for both (NAA).

In Ayurvedic texts, Vidaari has been suggested as a substitute for Jivaka and Rishabhaka.

Charaka used a decoction of the leaves in milk, and as an ingredient of medicinal clarified butter, as a restorative tonic.

Sushruta used the tuber as an aphrodisiac tonic in seminal disorders, urethral discharges, consumption, respiratory troubles, debility; and in oedema, haemoptysis. The tuber was also cooked and used as a vegetable.

The juice of Kshira-vidaari or its dry powder was prescribed as a tonic in spermatorrhoea, dysuria, venereal diseases, also in cough, bronchitis, internal abscesses, diseases of liver and spleen. For increasing lactation the powder of the tuber was given with milk (Bhaavaprakaasha, Shaarangadhara Samhitaa).

Compounds, available over-the-counter are: Shataavari Ghrita (Sahasrayoga), indicated in diseases and disorders of the urinary system, urethritis, dysuria, cystitis; Vidaaryaadi Ghrita, a restorative tonic; Ashvagandhaadi Lehya (IMPCOPS), prescribed as a tonic in wasting diseases and impotency; Shataavari Lehya (Sahasrayoga), used in menorrhagia and urinary disorders. Vidaari is also one of the ingredients in the reputed restorative tonic, Chyavanapraasha Avaleha (Charaka Samhitaa).

Kaamdeva Ghrita (Shaarangadhara Samhitaa) was used as a sperm and fertility-promoting tonic and aphrodisiac. Though the compound has been incorporated in the text of Bhaishajya Ratnaavali, it is not available now.

Active principles and pharmacology

Rhizome contains taraxerol and sitosterol, and resin similar to jalap resin. Fixed oil from tubers contained palmitic, stearic, oleic, linoleic and linolenic acids.

Fraction D (ether insoluble, alcohol soluble) raises blood pressure and stimulates respiration. The suppressor action has a peripheral component, and a small central component. The extract also stimulates rabbit gut muscle and guinea-pig uterus and relaxes the bronchus.

Pueraria tuberosa, equated with Vidaarikanda, contains modified iso-flavonoids-L-alpha-hydroxytuberosone, tuberosin, anhydrotuberosin and 3-O-methylanhydrotuberosin and a coumestan-tuberostan. Roots gave beta-sitosterol, stigmasterol, daidzin, puerarin and 4', 6"-di-O-acetylpuerarin. The tuberous root showed tonic, galactagogue, antirheumatic, hypoglycaemic properties.

Aphrodisiac activity, as reported in classical Ayurvedic literature, could not be confirmed experimentally, but when the tuber of P. tuberosa was excluded from a sex tonic, the compound was not found effective against impotency, premature ejaculation and nocturnal emissions. (CCRAS.)

Ipomoea nil (Linn.) Roth

Figure 3 Ipomoea nil—flowering branch [WOI]

Ipomoea hederacea auct. non (L.) Jacq.

Habitat

Throughout India up to 1800 m in the Himalayas; also occurs as a weed in sugar cane and wheat fields.

Classical & common names

Ayurvedic: Krishnabija, Kaalaadaanaa.
Unani: Habb-ul-Neel, Kaalaadaanaa.

Parts used

Seed.

Dose

Powder 3–6 g.

Classical use

The seeds, fried with purified butter and powdered, mixed with sugar, were prescribed as a purgative (Siddha-bheshaja-manimaalaa).

In Unani medicine, seeds are roasted and prescribed with Terminalia chebula or rose flowers as a purgative. Externally a paste of seeds is applied over blemishes and freckles.

Active principles and pharmacology

Seeds contain the alkaloids lysergol, chanoclavine, penniclavine, isopeniclavine and elymoclavine. They also contain resin 14.2 % and glucoside.

Seed oil contained palmitic, stearic, arachidic, oleic, linoleic and linolenic acids.

The seed possesses purgative, anthelmintic and blood-purifying properties. The drug is a substitute for jalap (Convolvulus jalapa Linn., Ipomoea jalapa Linn.).

Oleaceae

JASMINUM

Jasminum grandiflorum Linn.

Figure 1 Jasminum grandiflorum [ADPS]

Jasminum officinale L. var. grandiflorum (L.) Kobuski

Habitat

Subtropical north-western Himalayas; cultivated in gardens throughout India.

Classical and common names

Ayurvedic: Pravaaljaati (Charaka), Jaati, Jaatika, Saumanasyaayani, Sumanaa, Chetikaa, Hriddgandhaa, Maalati, Chameli. Yellow-flowered variety: Svarnajaati, Hemapushpikaa (equated with Jasminum humile Linn.). Gandhamaalati or Madhumaalati is a different plant, equated with Aganosma dichotoma (Apocynaceae).
Unani: Yaasmin.
Siddha: Pichippu.
English: Spanish Jasmine, Common Jasmine.

Parts used

Leaves, flowers, root.

Dose

Fresh Juice 10–20 mg, decoction 50–100 mg, powder 1–3 g.

Classical use

Charaka used the sprouts or dried flowers, in prescriptions; externally in coryza, nasal haemorrhage, and dermatosis.

Sushruta used Maalati as an ingredient of a medicated clarified butter for external application on infected wounds, for cleansing and sterilizing the interior of ulcers; as an ingredient of a hair oil for baldness and alopecia and as an ingredient of an eye-salve for loss of vision. Sushruta prescribed Maalati internally and externally, in leprosy, malignant ulcers and other virulent skin diseases.

The root of Jaati, cooked in goat's milk and mixed with sugar, was prescribed for giving relief in pain due to retention of urine and for expelling calculus (Raaja Maarttanda).

The oil, cooked with juice of Jaati leaves, was prescribed in purulent discharge from the ear

(Vrindamaadhava, Bangasena). Fresh juice was also instilled into the ear (Gadanigraha).

The leaf was chewed in stomatitis, toothache and allied ailments. A decoction of the leaf was also used as a gargle. (Bhaavaprakaasha, Vrindamaadhava.)

Paste of Jaati leaf, also of its root, was used externally for improving complexion and for removing freckles, dark shades. Paste of the root was applied over temples and forehead during headache. In neurological disorders the paste of the root was applied externally and its extract in oil was massaged over affected parts and head.

Over-the-counter, Jaatyaadi Ghrita (Ashtaanga Hridaya) is available for external application on chronic and septic ulcers. Jaatyaadi Tailam is also prescribed externally. Maalthyaadi Tailam is a bathing oil for alopecia and baldness.

Active principles and pharmacology

The jasmine concrete contains benzyl alcohol, benzyl benzoate, benzaldehyde and eugenol. Other constituents include linalool, linalyl acetate, methyl anthranilate, geraniol, farnesol, nerolidol, p-cresol, indole, cis-jasmone, alpha-terpineol, *cis*-3-hexenyl benzoate, nerol, 5-hydroxyjasmonic acid lactone, *cis*-pent-2-enyl-pentanolide, methyl heptenone, methyl-N-acetylanthranilate, methyl jasmonate, methylphytol, iso-phytol, tetramethyl hexadecanol, geranyl-linalool and phytyl acetate. Pyridine and nicotinate derivatives have also been identified in the jasmine absolute.

On topical application, the extract of Jasminum officinale in clarified butter was found to accelerate the healing time on a second degree burn in rats by six days, whereas in the unmedicated clarified butter group the acceleration was by three days.

The juice of leaves in the form of jelly on local application was found to promote wound healing, as assessed by histological, biochemical and contraction rat studies. The fresh juice of leaves showed increased and early gain of tensile strength in the treated linear wounds in rats. Studies on musculoperitoneal wounds on the abdomen of rats, with 2.5 per cent Jasminum auriculatum leaves extract injected intramuscularly confirmed the wound-healing activity of the leaves. (ICMR.)

Use in Western herbal

Jasmine was introduced in Europe in the 16th century and is mainly used as perfume. Flowers make a calming and sedative infusion, which is taken during stress and tension.

An essential oil is distilled in Tunis and Algeria. The East Indian oil of Jasmine is a compound and is found largely contaminated with sandalwood oil.

Syrup of Jasmine is made by placing in a jar alternate layers of flowers and sugar, covering the whole with wet cloth and standing in a cool place. Green, in "Universal Herbal" (1932), recommended a syrup prepared from the flowers "as an excellent medicine in coughs, hoarsenesses and other disorders of the breast." (M. Grieve.)

Jasminum angustifolium (L.) Willd.

Figure 2 Jasminum angustifolium [ADPS]

Habitat
South India.

Classical and common names
Ayurvedic: Kaanan-mallikaa, Aasphotaa.
Siddha: Kattu Malligai.
English: Wild Jasmine.

Traditional use
The bitter root is used in external application for ringworm. The juice of leaves is given as an emetic in cases of poisoning.

Active principles and pharmacology
Fresh flowers gave indole.

Jasminum auriculatum Vahl

Habitat
South India. Cultivated on commercial scale in Uttar Pradesh, Bihar, West Bengal.

Classical and common names
Ayurvedic: Yuuthikaa, Yuuthi, Mugdhee, Juhi.
Siddha: Usimalligai.

Classical use
Charaka prescribed sprouts cooked as vegetable and leaves in decoction, internally, for diarrhoea, colic, pain, and jaundice. Sushruta prescribed Yuuthikaa leaves processed with clarified butter in intrinsic haemorrhage.

Flowers are used for the production of perfumed hair oils and otto.

Jasminum sambac (L.) Ait.

Habitat.
Found througout India, mostly under cultivation.

Classical and common names
Ayurvedic: Mallikaa, Madyantikaa, Nava-Mallikaa.
Siddha: Malligai.
Unani: Mograa.
English: Arabian Jasmine, Tuscan Jasmine.

Classical use
Charaka included the herb for the treatment of insanity and epilepsy. It is an ingredient in Mahaa-panchgavya Ghrita, still available over-the-counter.

In traditional medicine, the root is used as an emmenagogue. Dried leaves, soaked in water and made into a polutice, are used on indolent ulcers.

Active principles and pharmacology
The plant contains pyridine and nicotinate derivatives. Alcoholic extract of aerial part showed CNS-depressant and hypotensive properties.

Leaves contain jasminin, quercitrin, iso-quercitrin, rutin, quercitrin-3-dirhamnoglycoside, kaempherol-3-rhamnoglycoside, mannitol, alpha-amyrin, beta-sitosterol and an iridoid glycoside named sambacin. The absolute contains several pyridine and nicotinate derivatives including methyl-5-ethyl-4-methyl nicotinate as the major one.

Euphorbiaceae

JATROPHA

***Jatropha curcas* L.**

Figure 1 Jatropha curcas [ADPS]

Habitat
Native to tropical America, grown in various parts of India as a field barrier.

Classical and common names
Ayurvedic: Sthula-eranda, Vyaaghrairanda, Kaanan-eranda.
English: Physic Nut, Purging Nut.

Parts used
Root bark, fresh viscid latex.

Dose
Used only externally.

Classical use
Fresh viscid latex, flowing from stem, is employed to arrest bleeding or haemorrhage from wounds; also for treating scabies, eczema, ringworm. Extract of the root bark, in combination with bland oil, is applied externally in rheumatism. Juice of the leaves is applied externally over piles. Stem juice is haemostatic and styptic.

The drug is not used internally in Indian medicine.

The seeds possess poisonous and purgative properties, but are rarely used as purgative. Three to five seeds, slightly roasted and decorticated, are sufficient for active catharsis; they seldom produce nausea and vomiting, but cause burning sensation in the stomach.

Seed oil is used as a cleansing agent for wounds, sores, and ulcers; also employed for treating herpes and eczema.

Active principles and pharmacology
Leaves contain apigenin, vitexin and isovitexin, alpha-amyrin, stigmasterol, stigmastenes. Seeds contain phorbol derivatives. The plant also contains curcosones and lathyrane diterpenes.

Alcoholic extract of aerial parts is found to be CNS-depressant and diuretic; showed activity against P-388 lymphocytic leukemia.

Caution
Seeds are poisonous to human beings and cattle.

Juglandaceae

Juglans

***Juglans regia* L.**

Habitat
Temperate Himalayas at 1000–3000 m; wild as well as cultivated in Khasi Hills. Native to Iran.

Classical and common names
Ayurvedic: Akshota, Akshodaka, Sailbhav, Pilu, Karparaal, Vrantphala, Kandaraal, Prthuchhada.
Unani: Akharot.
Siddha: Akrottu.
English: Walnut.

Classical use

Charaka and Sushruta used the edible nut-kernel in prescriptions, internally, for anaemia, consumption, debility, senility, and as a vitalizer and tonic. Sushruta prescribed oil of the seeds as a potent and digestive tonic.

Akshota is an ingredient in Amritpraasha Ghrita of Charaka, still available over-the-counter. It is prescribed in cough, consumption, emaciation, and sexual debility.

In Unani medicine, kernel of Akharot is an ingredient in many revitalizing and rejuvenating sex tonics—Halwa-e-Gazar, Laboob Kabir, Laboob Sagheer, Majoon-e-Falaksair, to mention but a few. Anqarua-e-Kabir, prescribed in nervous debility, is processed in the oil of Akharot.

Akharot, in combination with dry Ficus carica (Anjeer) and dry Vitis vinifera (Munakkaa), is prescribed by Unani physicians as a brain tonic. Roasted kernel is prescribed during cold and cough. Kernel mixed with purified sugar candy is given as a tonic to rheumatic patients.

Paste of fresh kernel is used for removing blemishes and freckles.

Seed oil is used for tape worms, shells of unripe fruits as vermifuge.

The nut is used in Chinese medicine as a kidney tonic.

Active principles and pharmacology

Cysteine, tryptophan, thiamin, riboflavin, nicotinic acid, pantothenic acid, folic acid, vitamin B6, biotin, vitamin A, ascorbic acid are found in the leaves and fruits. Kernels yield 60–70 % of a drying oil, known as Walnut Oil, Fresh leaves and unripe fruits are rich in ascorbic acid. Leaves yield an essential oil. Juglone, bernerine, cyclotrisjuglone, beta-sitosterol from bark and root; oxalic acid from fruit have been isolated.

Leaves exhibit astringent, bark and leaves anthelmintic, laxative and detergent properties, and give encouraging results in herpes, eczema, scrofula, skin eruptions, sores.

Use in Western herbal

White Walnut (Juglans cinerea Linn.), known as Butternut, is used in the West. The leaves possess much the same properties as the Black Walnut.

Butternut is a mild cathartic and encourages regular bowel. It does not constipate and is often used as a habitual laxative. It also acts as a vermifuge. The expressed oil of the fruit expels tapeworm.

Butternut also lowers cholesterol levels and promotes the clearance of waste products by the liver.

Butternut is often used as a cleansing remedy for skin problems due to accumulation of toxins.

Being antimicrobial and astringent, it is prescribed as a treatment for dysentery.

Juglans cinerea contains naphthaquinones (including juglone, juglandin and juglandic acid), a fixed and a volatile oil, and tannins.

Juglone exhibited purgative, antimicrobial, antiparasitic and cancer-inhibiting properties.

In Europe, Juglans regia leaves are a popular home remedy for eczema and blepharitis (eyelid inflammation). Recent research suggests antifungal properties, as well as antiseptic action. The leaves are also used for intestinal worms. Walnut Bach Flower remedy is used during menopause. Unrefined walnut oil (2 tsp.) is recommended as a dietary supplement for menstrual dysfunction. (Penelope Ody.)

The drug is available over-the-counter in combination with carminative herbs. It has also been included in slimming tablets, due to its laxative action and cleansing properties.

Cupressaceae

Juniperus

Juniperus communis Linn.
Juniperus macropoda Boiss.

Habitat

Western Himalayas from Kumaon westwards to Srinagar at 1700–4200 m.

Classical and common names

Ayurvedic: Hapushaa, Havushaa; Haaubera.
Unani: Haauber, Hubb-ul-arar, Abahal.
English: Common Juniper.

Parts used

Fruits.

Figure 1 Juniperus macropoda—branches with scale-like and acicular leaves [WOI]

Dose

3–5 g.

Classical use

Charaka prescribed the seeds of Hapushaa in decoction; also as an ingredient of a medicinal clarified butter in female diseases and dysuria. Sushruta used leaves as an ingredient of a bolus for treating paralysis.

Hapushaa belongs to the Shaalparnyaadi group of herbs which act as appetiser and digestive. According to the 16th century classical texts, Hapushaa cures obstinate abdominal diseases including ascites, internal tumours, piles, colic pain and sprue syndrome.

For abdominal affections, Hapushyaadi Churna (Charaka); for internal tumours, Hapushyaadi Ghrita (Charaka) were the compounds of choice. For cough, Kalyaanaka Guda (Sushruta) was prescribed. These compounds are no longer available.

Hapushaa is an important ingredient in Hingwaadi Gutikaa (Sushruta), still available over-the-counter. It is being prescribed for dyspepsia, loss of appetite and other digestive problems. The drug is also an ingredient in Naaraayan Churna (Charaka), available over-the-counter and used as a laxative, carminative and stomachic.

Laooq-e-Abahal is an antispasmodic expectorant of Unani medicine, being prescribed for cardiac asthma.

Active principles and pharmacology

Juniper contains 1–2 % volatile oil, consisting of over 60 compounds, which include alpha- and beta-pinene, sabinene, alpha-terpinene; small quantities or traces of several sesquiterpene hydrocarbons (with strong diuretic properties). Geijerone, characterised as *trans*-isopropenyl-4-methyl-4-vinylcyclohexanone has also been isolated from the essential oil. Diterpenes, catechin tannins, potassium salts, inverted sugar (20–30 %), flavonoids are also present in the berry. Juniper camphor is also present, its melting point being 1.65–1.66° C.

Active principles exhibit diuretic, carminative, stimulant, antidropsical, antileucorrhoeic, and abortifacient properties of the berry and its volatile oil. Animal studies have shown an increase in urine excretion as well as direct effect on smooth muscle contraction (PDR).

Terpinen-4-ol of the volatile oil increases the fluid-filtering rate of the kidneys. (1.85 mg/kg orally was found toxic in mice.)

Use in Western herbal

Nicholas Culpeper (1616–1654) wrote of it: "The berries provoke urine exceedingly, and therefore are very available to dysuries and stranguaries. It is so powerful a remedy against the dropsy, that the very lye made of the ashes of the herb being drank cures the disease. It provokes the terms, helps the fits of the mother, strengthens the stomach exceedingly, and expels the wind. The berries are admirably good for cough, shortness of breath and consumption... cramps and convulsions. They give safe and speedy delivery to woman with child."

According to herbalists, juniper is a strong diuretic and antiseptic for the urinary tract. It is a valuable remedy for cystitis and helps relieve fluid retention, but not advised in cases of kidney disease. In the digestive system, juniper is warming and settling, easing colic and supporting the function of the stomach. Juniper stimulates menstruation and tends to increase menstrual bleeding.

Taken internally or applied externally, juniper is helpful in the treatment of chronic arthritis, gout and rheumatic conditions. (Andrew Chevallier.)

German Commission E recognized the efficacy of juniper in the following areas:

- Loss of appetite
- Infection of urinary tract
- Kidney and bladder stones.

Juniper is an ingredient in over-the-counter diuretic drug Ordrinil. Juniper oil, for external use, is also available for muscular pains and arthritis.

Caution

High doses of juniper cause kidney irritation, and possibly kidney damage. It should not be taken internally for more than six weeks without a break.

Contraindicated for internal administration during pregnancy. It is a uterine stimulant and may induce heavy menstrual bleeding.

Juniper imparts violent odour to the urine.

Essential oil should not be taken internally except under medical supervision.

K

Zingiberaceae

KAEMPFERIA

Kaempferia galanga L.

Figure 1 Kaempferia galanga [ADPS]

Habitat

China (Hainan Island). Distributed in the tropics and sub-tropics of Asia and Africa. Found throughout India.

Classical and common names

Ayurvedic: Chandramuula, Chandramuulikaa (The Wealth of India). In Kerala used as Karchura and Shathi.
Siddha: Kachoram.
English: Galanga.

Classical use

In Ayurvedic medicine, Curcuma zedoaria, Kaempferia galanga and Hedychium spicatum are used as Karchura, Palaashi, Shathi and Shati. All these drugs are credited with stimulant, carminative, diuretic and expectorant properties.

Active principles and pharmacology

The essential oil from rhizomes contains n-pentadecane, ethyl-p-methoxycinnamate, ethyl cinnamate, carene, camphene, borneol, p-methoxystyrene, p-methoxycinnamate, p-methoxy-*trans*-cinnamic acid and cinnamaldehyde.

The tuberous rhizomes possess a camphoraceous odour with a somewhat bitter aromatic taste, resembling that of rhizomes of Hedychium spicatum; administered with honey give relief in coughs and pectoral affections. Rhizomes also exhibited anti-inflammatory properties. In the form of powder or ointment, when applied to wounds and bruises, reduced swellings, as a gargle provided relief in sore throat.

In the Philippines, a decoction of the rhizomes is used for dyspepsia. Boiled in oil, the rhizomes are applied externally for treating nasal congestion. Roasted rhizomes are applied hot in rheumatism.

Biological activity of the rhizomes is found to be cytotoxic (CIMAP).

Use in Western herbal

Galanga has been known in Europe for seven centuries, longer than its botanical origin. Lesser Galangal is the native of China, the Greater Galangal of Java. The drug has been used in Europe as a spice for over a thousand years, but it has now largely gone out of use, except in Russia.

The drug is being recommended for sea-sickness. The powder is used as a snuff for catarrh.

Nicholas Culpeper (1616–1654) wrote of Galanga, the Greater and Lesser: "Galanga strengthens the stomach exceedingly, and takes away the pains thereof coming of cold or wind; the smell of it strengthens the brain; it relieves faint hearts; takes away windiness of the womb; heats the reins."

Crassulaceae

KALANCHOE

***Kalanchoe pinnata* (Lam.) Pers.**
***Bryophyllum pinnatum* (Lam.) Oken**

Habitat

Throughout the warm and moist parts of India, especially abundant in West Bengal.

Classical and common names

Ayurvedic: Parnabija (non-classical), Pattharchuur, Patharchat, Hemsaagar.
Unani: Zakhm-hayaat.

Parts used

Leaves.

Dose

Fresh juice 10–20 ml.

Traditional use

Fresh juice of leaves is given in ulcerative colitis, haemorrhagic dysentery, bleeding piles, metrorrhagia. The leaves are considered styptic, astringent and antiseptic. In folk medicine, roasted or crushed leaves are applied as poultices to wounds, cuts, abrasions, ulcers, bites of venomous insects. Leaves are eaten to control diabetes.

In Malaya, leaves are used as poultice for coughs and colds; they are used in lotions for small pox.

Active principles and pharmacology

Plant contains n-alkane, n-alkanol, alpha- and beta-amyrin and sitosterol. Leaves contain malic, isocitric and citric acids. Glycosides of quercetin and kaempferol, fumaric acid, phenolic components.

A potent cytotoxic bufadienolide—bryophyllin B—has been isolated.

The 50 % ethanolic extract of Kalanchoe integra (whole plant) showed anticancer activity against human epidermoid carcinoma of nasopharynx in tissue culture.

Fresh leaves of Kalanchoe integra showed encouraging results in inflammatory wound healing in a clinical trial.

L

Asteraceae

LACTUCA

Lactuca sativa L. (garden var.)
Lactuca virosa L. (wild var.)

Habitat

Garden lettuce grows as a cold weather crop in the plains. Wild lettuce, distributed in North-western Himalayas, is found between 1800 and 3300 m.

Classical and common names

Unani: Garden var.: Kaahu, Salaad, Khass. Wild var.: Kaahu Sahraai, Khass barri.
English: Garden Lettuce, Wild Lettuce.

Parts used

Leaves, seeds, dried latex (Lettuce Opium).

Dose

Leaves for docoction 20 g, lettuce opium 180 mg–500 mg, seed powder 500 mg–1.25 g.

Classical use

In Unani medicine, garden lettuce is prescribed as a vegetable for treating bilious eructation, disturbed blood pressure, loss of appetite and insomnia. It is also prescribed as a galactagogue to nursing mothers; with vinegar as a tonic for intestines and digestive system. Raw lettuce or its decoction is given for inducing sleep.

The seeds (Tukhm-e-Kaahu) are prescribed for treating headache, insomnia, biliousness, and for suppressing sex urge. Paste of the seeds is applied over forehead during headache.

The oil of seeds (Raughan Tukhm-e-Kaahu) is massaged over the head for inducing sleep; also used as a nasal drop.

Dried latex of lettuce (Afyun-Kaahu) is used as a sedative, antispasmodic and anodyne; preferred over opium, as it does not disturb the digestive system and liver functions.

Raughan-e-Kaahu (Qarabadeen-e-Azam-o-Akmal), the over-the-counter massage oil, imparts sedative, humectant and hypnotic properties of Kaahu and is prescribed for external use in insomnia, melancholia, and epilepsy.

Active principles and pharmacology

Lactuca sativa: leaves contain calcium, phosphorus, iron, thiamine, riboflavin, niacin, vitamin C, carotine, iodine and fluorine.

Lactuca virosa: The latex contains sesquiterpene lactones, including lactucopicrin and lactucerin. The leaves also contain flavonoids and coumarins.

The sesquiterpene lactones exert sedative effect.

All lettuces possess the narcotic juice lactucarium; Lactuca virosa having the most and others in the following order: Lactuca scariola, Lactuca altissima, Lactuca candensis and Lactuca sativa. Cultivation has lessened the narcotic properties of Lactuca sativa.

Use in western herbal

John Gerard (1545–1612) wrote: "Lettuce cooleth the heat of stomacke, called the heart-burning; and helpth it when troubled with choler; it quencheth thirst, and causeth sleepe."

Nicolas Culpeper (1616–1654) highlighed the following virtues: "The juice of Lettuce, mixed or boiled with oil of roses, applied to the forehead and temples, procures sleep, and eases the headache proceeding of an hot cause. Being eaten boiled, it helps to loosen the belly. It halps digestion, quenches thirst, increases milk in nurses, eases griping pains in the stomach or bowels, that come of choler."

Water distilled from lettuce, eau de laitre, is used in France as a mild sedative.

Caution

Lactuca is contraindicated in glaucoma and prostate enlargement. (Michael McGuffin.)

Cucurbitaceae

LAGENARIA

Lagenaria siceraria (Mol.) Standl. (bitter var.)

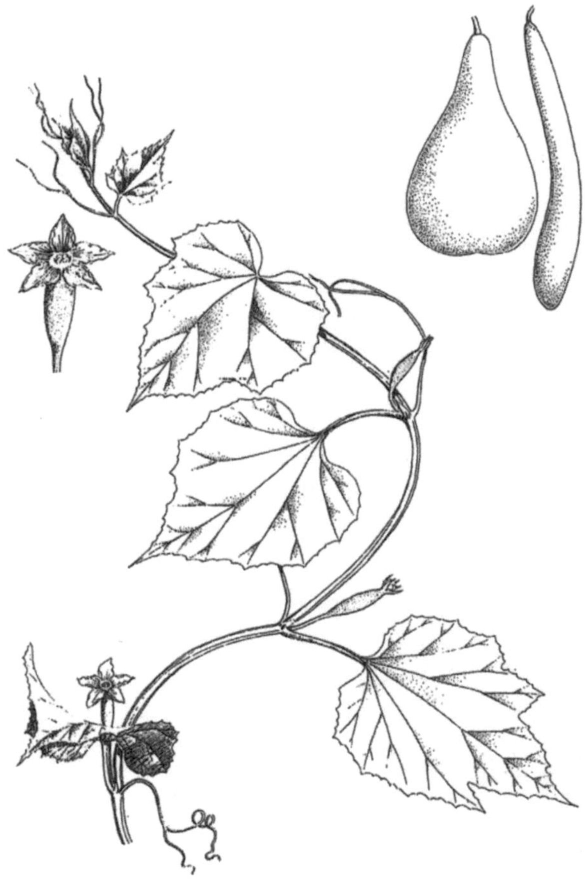

Figure 1 Lagenaria siceraria [ADPS]

Lagenaria vulgaris Ser.
Cucurbita lagenaria Linn. (sweet var.)

Habitat

Throughout India.

Classical and common names

Ayurvedic: Ikshavaaku, Katutumbi, Tiktaalaabu, Katukaalaabu, Pindaphalaa (bitter var.) Alaabu (sweet var.)
Unani: Kar-ul-murr (bitter var.), Kaddu-e-daraaz (sweet var.).

English: Bitter Bottle Gourd (bitter var.), Bottle Gourd (sweet var.).

Parts used

Fruit, root, leaves.

Dose

Bitter variety juice 10–20 ml, sweet variety fruit pulp 10–20 g, seed powder 3–5 g.

Classical use

Sushruta prescribed Alaabu in the form of cooked vegetable as a diuretic and stool-forming drug; Ikshavaaku fruits internally as emetic.

In retention of placenta after delivery, the dried leaves and tendrils of Ikshavaaku were burnt for giving local fumigation. (Sushruta Samhitaa, Bhaavaprakaasha.)

The bitter variety (Ikshavaaku) was prescribed internally in adenitis; for vaginal discharges powdered fruit of the sweet variety (Alaabu) was recommended (Bhaavaprakaasha).

The juice of mature Ikshavaaku, added with Yavakshara (impure carbonate of potash, wheat and barley straw ash) and sugar, was prescribed for breaking calculus (Chakradatta).

The bitter variety was also prescribed as a cooling and cardiac tonic, as an antidote to poisoning and for alleviating bronchitis, cough, asthma, and biligenic affections.

In folk medicine, decoction of leaves of the bitter variety is given in jaundice. The fruit pulp is applied externally in delirium. The seed oil is massaged in headaches. Seeds are given internally in dropsy and as diuretic and anthelmintic.

Active principles and pharmacology

Bitter fruits yield 0.013 % of a solid foam containing cucurbitacins B, D, G and H, mainly cucurbitacin B. These bitter principles are present in the fruit as aglycones. Leaves contain cucurbitacin B, and roots cucurbitacin B, D and traces of E. The fruit juice contains beta-glycosidase (elaterase).

Plants which yield non-bitter fruits contain no bitter principles or elaterase. The roots of such plants are not bitter.

Botanically, both sweet and bitter varieties belong to the same species, but in medicine the bitter variety, on the basis of its bitter principles and therapeutic properties, is used.

Lauraceae

LAURUS

Laurus nobilis Linn.

Habitat

Distributed chiefly in Mediterranean region. Sometimes grown in Indian gardens, but does not thrive well. Dried berries are imported into India for medicinal use.

Classical and common names

Unani: Habb-ul-Ghaar, Daphni.
English: Sweet Bay, True Laurel, Bay Laurel.

Classical use

Traditionally, both leaves and berries were formerly employed for hysteria, amenorrhoea and flatulent colic. Unani physicians consider it as a brain tonic and prescribe it in a dose of 9 g for neurological affections. For asthma, the drug (9 g) is given mixed with honey. For colic, it is administered mixed with Plantago ovata husk.

Oil of Laurel berry is sometimes applied as a stimulant in sprains.

The leaves are used as a condiment and flavouring agent in food and confectionary.

Active principles and pharmacology

Bay Laurel contains up to 3 % volatile oil (including 30–50 % cineole, linalool, alpha-pinene, alpha-terpineol acetate, mucilage, tannin, resin).

Cis- and *trans-*thuj-2-en-ol from laurel volatile oil have been identified.

D-gluco-L-glycero-3-octulose from fresh buds and young leaves has been isolated.

1, 8-cineole have been identified as major component in leaf oil.

A new germacrane derivative has been isolated from fruits.

The plant also contains sesquiterpene lactones—dehydrocostuslactone, costunolide, eremanthin, laurenbiolide; isoquinoline alkaloids, including reticulin.

Use in Western herbal

The essential oil, well diluted in a carrier oil, is used as a friction rub in rheumatic conditions.

Laurel leaves, when used as an ingredient in cooking, promote digestion. The herb is settling to the stomach, and stimulates appetite and gastric juices.

Lamiaceae

LAVANDULA

Lavandula angustifolia Mill.
Lavandula stoechas Linn.

Habitat

Lavandula angustifolia (Common lavender) is a native of southern Europe, extensively cultivated in southern France. In India, cultivation has been undertaken on an experimental basis in Kashmir at altitudes of 1500 m; also in some parts of Nilgiris.

Lavandula stoechas (French lavender) is a native of the Mediterranean region. Grown in some gardens in western India. Dried plant and flowers are imported into Mumbai from the Persian Gulf.

Classical and common names

Unani: Ustukhuddus, Dhaaru, Alfaazan, Ustaaduus, Ustikuus. Brunella vulgaris Linn., bearing blue-white flowers, is known as Kashmiri lavender. Lavandula burmanni Benth. and Lavandula vipinnate O. Kuntz are known as wild lavender. Lavandula stoechas (Arabian or French lavender) is used in Unani medicine.
English: Common lavender, French lavender.

Parts used

Leaves, flowers.

Dose

3–6 g.

Classical use

In Unani medicine, Ustukhuddus is used as a nervine tonic. Itrifal-e-Ustukhuddus (Qarabadeen-e-Azam), Khamira-e-Ustukhuddus (Q.A.), Majoon-e-Azaraqi (Bayaz-e-Kabir), Majoon-e-Zabeeb (B.K.), Majoon-e-Khadar (B.K.) are reputed Unani tonics, prescribed in cephalalgia, epilepsy, hemiplegia, Bells Palsy, neurasthenia, melancholia, debility.

Ustukhuddus has been referred to as the "broom of the brain" (a tonic to brain that clarifies the intellect) by Hakeem Gulam Gilani in Makhzan-ul-Advia (1915), and is used as a single drug for treating confused states of mind associated with behavioural disturbances, as well as memory disturbances. The drug relieves mental fatigue, controls anxiety, irritability and depression.

Hakeem Anwar Ahmad has included Ustukhuddus in his compound formulations for migraine and depression.

Ustukhuddus also finds an important place in Unani compounds for chronic catarrh and sinusitis; stomach irritations and nervous intestinal discomforts. Itrifal Mulaiyin (Haj-ul-Amraz) is prescribed for a chronic headache associated with constipation.

Active principles and pharmacology

Lavender flower contains 1.5–3 % volatile oil, of which 25–55 % is linalyl acetate, 20–38 % linalool, 4–10 % *cis*-linalyl acetate, 2–6 % *trans*-beta-ocimene, 2–6 % 1-terpinen-4-ol, 2 % 3-octanone, 0.3–1.5 % 1, 8-cineole, 0.3–1 % alpha-terpineol, 0.2–0.5 % camphor, and 0.1–0.5 % limonene; tannins (5–10 %), hydroxycoumarins, including among others, umbelliferone, herniarin; flavonoids (luteolin); phytosterols, and triterpenes. Caffeic acid derivatives include rosmaric acid.

In a series of tests, the inhalation of lavender oil vapor caused a dose-dependent suppression of convulsions induced in mice by pentylenetetrazole, nicotine and electric shock. 4 geriatric patients with sleep disorders were subjected to aromatherapy with lavender oil. Sleep time increased significantly and reached a level comparable to that earlier achieved with benzodiazepines and neuroleptics. (Rational Phytotherapy.)

In a study at A. K. Tibbya College, Aligarh, India, alcoholic extract, aqueous extract and crude powder of the drug was used in 40 cases of chronic sinusitis. Best response was produced by alcoholic extract (41.66 %), followed by crude powder (33.3 %) and aqueous extract (16.66 %). In another study Lavandula stoechas, in the form of 750 mg tablets, two tablets three times a day for a period of three weeks, was administered to 198 patients of chronic sinusitis. Promising results were recorded.

Use in Western herbal

Nicholas Culpeper (1616–1654) wrote: "Lavender is of special good use for all the grief and pains of the head and brain that proceed of a cold cause, as the apolexy, falling-sickness, the dropsy, or sluggish malady, cramps, convulsions, palsies and often fainting. It strenghthens the stomach, and frees the liver and spleen from obstructions, provokes women's courses... The flowers of Lavender, steeped in wine, help them to make water that are stopped... Two teaspoonfuls of distilled water of the flowers taken, helps them that have lost their voice, as also in tremblings and passions of the heart, and faintings and swoonings..."

In Germany, lavender is licensed as a standard medicinal tea for sleep disorders and nervous stomach. Lavender flower and extract are used as sedative and cholagogue medical preparations. In Germany and the United States, the aqueous infusion is used in balneotherapy and the essential oil in aromatherapy.

German Commission E monograph recognized the efficacy of Lavandula angustifolia in:

- Loss of appetite
- Nervousness and insomnia.

1–2 teaspoons (3–5 g) of dried herb per cup of tea or 1–4 drops of lavender oil with sugar cube have been recommended.

Caution

An infusion of lavender taken too frequently may cause griping and colic. It is advised that excessive internal use should be avoided in early pregnancy.

Lavender oil in large doses is a narcotic poison.

Lythraceae

LAWSONIA

Lawsonia inermis Linn.
Lawsonia alba Lam.

Habitat

Throughout India, chiefly in drier parts. Commercially cultivated in Punjab, Madhya Pradesh, Rajasthan, Gujarat and Tamil Nadu.

Classical and common names

Ayurvedic: Madayanti, Madayantikaa (Charaka, Sushruta), Mendika, Ranjaka. Madayantikaa is not to be confused with Mallikaa.
Unani: Hinaa.
English: Henna.

Parts used

Leaves, seeds.

Dose

Fresh juice 10–20 ml, seed powder 3–5 g, decoction 50–100 ml.

Classical use

Charaka used the entire plant of Madayantikaa as an ingredient of a medicinal Ghee (clarified butter) and prescribed it for epilepsy and hysteria. Externally, it was prescribed for promoting hair growth and for dyeing grey hair.

Sushruta incorporated Madayanti in a medicinal oil, which was used externally on malignant ulcers. Mahaanila Ghrita was prescribed for external application in obstinate skin diseases including leprosy.

Madayanti, mixed with honey and cold water, was prescribed in haematuria (Ashtaanga Sangraha). Cooled decoction of the drug, added with sugar and honey was given in intrinsic haemorrhage (Gada Nigraha). In Ashtaanga Hridaya, Madayanti was included in compounds for haemothermia.

Mahaapanchgavya Ghritam (Sahasrayoga), available over-the-counter, is prescribed in epilepsy, hysteria; also in anaemia and jaundice.

Madayanti Kwaath is administered in haemothermia.

In Unani medicine, leaves are an ingredient in blood-purifying compounds—Araq-e-Juzam and Araq-e-Musaffi. Oil extract of Neem (Azadirachta indica) and Hinaa is prescribed externally in leprosy and chronic skin diseases. Internally, an extract prepared from flowers and leaves of Hinaa is administered.

A decoction of seeds, also of the bark, is given in hepatitis, jaundice, enlarged spleen, calculous affections.

Juice of the plant, mixed with oil, is massaged over the head during headaches.

In folk medicine, a paste of leaves is rubbed over soles of feet for allaying burning sensation.

A decoction of leaves is used as a gargle in relaxed throat.

Active principles and pharmacology

The herb contains coumarins, napthaquinones (including lawsone), flavonoids, sterols and tannins. Lawsone (2-hydroxy-1, 4-napthaquinone), xanthones, laxanthones I, II, III, beta-ionone of the essential oil impart specific medicinal properties to the herb.

Various extracts of the herb exhibited marked analgesic effect in 250 mg/kg p.o. in experimental animals. Dye-stuff, extracted from leaves, produced anti-inflammatory activity against acute, sub-acute and chronic models of inflammation. Decoction of bark 2 g/kg p.o. for 26 days showed marked protection against carbon tetrachloride-induced liver damage in rats. (CCRAS.)

Leaf extract was found effective against bacterial strains of Mic. pyogenes var. aureus, Strep. pyogenes, Dip. pnemonae, B. subtitles, E. coli, Sal. typhi, Vib. comma and Sh. dysenteriae.

The ethanolic extract of the whole plant exhibited antifungal activity against Candida albicans, Cryptococcus neoformans, Trichophyton mentagrophytes, Microsporum canis and Aspergillus. The leaf extract showed antifungal activity against C. albicans, M. gypseum, T. mentagrophytes, Helminthosporium spp., Ustilago tritici, U. hordei, Diplodia natalensis.

Use in Western herbal

The drug is used externally for ulcers, eczemas, scabies, fungal infections. It is also used for amoebic dysentery and gastrointestinal ulcers. The drug is also contained in hair lotions and is used to treat dandruff. (PDR.)

Vitaceae

LEEA

Leea aequata Linn.
Leea hirta Roxb. ex Hornem.

Habitat

North-eastern part of India, Orissa, Andaman Islands.

Classical and common names

Ayurvedic: Kaakajanghaa (The Wealth of India), Nadikaantaa, Sulomaasha, Paaraavatpaadi. Peristrophe bicalyculata Nees., Acanthaceae, has also been equated with Kaakajanghaa.

Parts used

Root.

Dose

Decoction 50–100 ml.

Classical use

The root of Kaakajanghaa, pounded with milk, was prescribed in consumption (Raaja Maarttanda, Chakradatta); with rice-water in leucorrhoea (Vrindamaadhava). Externally the paste of Kaakajanghaa root was applied on wounds (Chakradatta).

Active principles and pharmacology

On steam distillation, the plant yields 0.15 % of an essential oil which inhibits the growth of Mycobacterium tuberculosis in a concentration of 10 mcg/cc. The oil also inhibits the growth of Micrococcus pyogenes var. aureus and Pasteurella pestis in concentrations of 100 mcg/cc and 50 mcg/cc respectively.

Root tubers and stem exhibit mucilaginous and astringent, leaves and twigs antiseptic and wound-healing properties. (The Wealth of India.)

Peristrophe bicalyculata: A yellowish brown, essential oil, obtained by steam-distillation of the plant, shows tuberculostatic activity *in vitro*. It inhibits growth of various strains of Mycobacterium tuberculosis in concentrations from 15 to 20 mcg/cc.

This plant has been equated with Aatarilaal in the Wealth of India, while reference books of Unani medicine consider Anthriscus cerefolium Hoffin as Aatarilaal, and Leea hirta as Kaakajanghaa.

Anthriscus cerefolium, known as Chervil in English, is native to Europe, Asia Minor, Iran and the Caucasus. Juice from the fresh plant is applied to various skin conditions, including wounds, eczema, abscesses, and can be equated with Aatarilaal of Unani medicine.

Leea macrophylla Roxb.

Habitat
Throughout hotter parts of India.

Classical and common names
Ayurvedic: Hastikarna, Hastikarna Palaasha (Bhaavaprakaasha).

Parts used
Root, oil of seeds.

Dose
Powder 3–5 g.

Classical use
Sushruta prescribed the oil of seeds, internally, as laxative, in urinary diseases, intestinal parasites and skin diseases.

The root of Hastikarna, pounded with rice-water, was applied locally for alleviating enlargement of thyroid gland (Vrindamaadhava, Bangasena). The powder of the root, with clarified butter, was prescribed in the morning as an age-sustainer.

In folk medicine, mucilaginous root is applied as a paste or poultice to wounds and sores, and used for guineaworm and ringworm. It is also employed externally to allay pain and to stop bleeding from wounds.

The leaves of the plant are eaten as vegetable. Fruits are also edible.

Though the root was used in morbid enlargement of thyroid gland, its constituents have not yet been investigated.

Papilionaceae

LENS

Lens culinaris Medic.

Figure 1 Lens culinaris—fruiting branch [WOI]

Lens esculenta Moench.
Ervum lens Linn.

Habitat
Cultivated in India for its dehusked grain (pulse).

Classical and common names
Ayurvedic: Masura (Bhaavaprakaasha), Masurikaa, Mangalyaa, Mangalyak.

Unani: Masoor.
Siddha: Misurpurpu.
English: Lentil.

Classical use

Masura Ghrita (extract of the lentil in clarified butter) was prescribed for diarrhoea and malabsorption syndrome. For chronic diarrhoea Masura soup was given with the paste of Aegle marmelos (Bilva) fruit and Zingiber officinale (Shunthi).

The parched flour of Masura, mixed with honey and the juice of pomegranate, was administered in acute vomiting. (Shaarangadhara Samhitaa.)

Masura grains, fried in purified butter, pounded with milk and mixed with honey, were applied on face topically for treating freckles. Masura grains, pounded with milk and mixed with clarified butter, were also prescribed as a cosmetic. (Ashtaanga Hridaya, Bhaavaprakaasha, Vrindamaadhava.)

Active principles and pharmacology

Lentils are rich in proteins (as much as 30 %) and a good source of minerals and vitamins of B group; tricetin, luteolin, a diglycosyldelphinidin and two proanthocyanidins from seed coat; kaempferol glycoside from cotyledons have been isolated. Seeds afforded glucose, fructose, raffinose, sucrose, stachyose, verbascose, kaempferol-3-O-alpha-L-rhamnosido-7-O-glucopyranoside and kaempferol-7-O-rutinoside.

Lentil contains amylase, proteolytic enzymes, phosphatase and phytase.

Lentils possess cooling, astringent, diuretic, mild laxative and tonic properties; improve appetite, provide relief in biliousness, strangury, diarrhoea and dysentery.

Lamiaceae

LEONURUS

Leonurus cardiaca Linn.

Habitat

Occurring in the Himalayas from Kashmir to Kumaon at altitude of 1800–3000 m. (Botanical Survey of India, 1940.)

Classical and common names

English: Common Motherwort.

Parts used

Dried leaves and flowering tops.

Dose

Infusion or decoction 50–100 ml.

Classical use

Traditionally, the herb is considered antispasmodic, analgesic, diuretic, emmenagogue and prescribed to women suffering from hysteria, heart palpitation, and amenorrhoea.

Active principles and pharmacology

The dried plant contains alkaloids (0.35 %), saponins, tannins (2.14 %), bitter principles, essential oil (0.05 %), carbohydrates (2.9 %) and traces of vitamins A and C. Stachydrine has been identified in the alkaloidal fraction; also an iridoid (leonurine). Stachydrine and leonurine are uterotropic constituents.

The drug tones up generative organs, uterine membrane, stimulates the muscles of the uterus and is found efficacious in delayed periods, period pain and premenstrual syndrome. The drug allays nervous irritability and affords relief from pain in stomach and gall bladder. (The Wealth of India.)

Use in Western herbal

Nicholas Culpeper (1616–1654) wrote of Motherwort: "There is no better herb to take melancholy vapours from the heart, to strengthen it, and make it a merry, cheerful, blithe soul than this herb. It may be kept in a syrup or conserve; therefore the Latins called it Cardiaca. Besides, it makes women joyful mothers of children and settles their wombs as they should be, therefore we call it Motherwort. It is held to be of much use for trembling of the heart, and faintings and swoonings... It also provokes urine and women's courses, cleanses the chest of cold phlegm, oppressing it; kills worms in the belly."

In the West, the herb is popular for treating menopausal upsets and is included in some remedies for premenstrual syndrome, although its main application in over-the-counter products tends to be that of a general sedative. Well-known brands are: Frank Roberts Alchemilla Compound, BP Tables, Motherwort Compound Formula, Gerard House Motherwort Compound, Lanes Quiet Life Tablets, Muir's Pick-Me-Up, Potter's Prementaid, Wellwoman Tablets.

Caution

The drug is contraindicated during pregnancy; also if menstrual bleeding is heavy.

A dose in excess of 3 g of a powdered extract may cause diarrhoea, uterine bleeding and stomach irritation (ref. cited by Michael McGuffin).

Leonurus sibiricus Linn.

Figure 1 Leonurus sibiricus—flowering branch [WOI]

Habitat

Found in Assam, Bengal and Bihar, extending southwards to Mysore.

Classical and common names

Vernacular: Guma.
English: Siberian Motherwort.

Classical use

Traditionally, the dried herb is used for its emmenagogue properties, and is employed in puerperal and menstrual disorders. Leaves and roots are used as febrifuge.

Active principles and pharmacology

The plant contains an alkaloid, leonurine, fatty oil (0.5 %), a resin (0.37 %) and resinic acid (0.83 %). Lauric, oleic, linoleic, and linolenic acids have been isolated from the leaves.

The alkaloid exerts a curare-like effect on motor-endings. It produced marked diuresis when intravenously injected into rabbit; in concentrations up to 1:1000 it displays haemolytic action. Administered to cats in small doses, it acted as a respiratory stimulant, in large doses caused respiratory paralysis. (The Wealth of India.)

The experimental results indicate that the decoction of Chinese drugs Carthamus tinctorius, Angelica sinensis and Leonurus sibiricus has stimulating action on the uterus of mouse in vitro. The stimulating action of Carthamus tinctorius and Leonurus sibiricus has been found related to stimulating H1-receptor and alpha-adrenergic receptor of uterus; but the action of Angelica sinensis is confined only to H1-receptor of uterus.

Caution

Contraindicated during pregnancy due to emmenagogue/uterine stimulant properties.

Brassicaceae

LEPIDIUM

Lepidium sativum Linn.

Habitat
Cultivated as a salad plant throughout India.

Classical and common names
Ayurvedic: Chandrashuura, Chandrikaa, Vaaspushpaa, Pashumehankaarikaa.
Unani: Habb-ul-Rashaad, Tukhme-e-Tartezak, Haalim.
Siddha: Alivirai.
English: Garden Cress, Water Cress.

Parts used
Seed.

Dose
1–3 g.

Classical use
According to Bhaavaprakaasha, Chandrashuura is indicated in asthma, cough with expectoration; also in rheumatic affections and as a restorative tonic.

In folk medicine, seeds of Chandrashuura are also used in bleeding piles; root in secondary syphilis. Mucilage of seeds is used as aperient. A paste of seeds or poultice is used externally in rheumatic affections.

Roasted seeds of Haalim, known as Muqliyaasaa in Syrian language, are an ingredient of Unani compound Suffof-e-Muqliyaasaa, being prescribed for their anti-inflammatory properties.

Active principles and pharmacology
Garden Cress contains carotene and glucosinolates; vitamin C and vitamins of B group; cucurbitacins, cardiac steroids (pharmacological studies on Lepidium sativum suggested presence of a cardioactive substance). Seedlings gave benzyl thiocyanate; seeds sinapic acid and sinapin.

Seeds exhibited spasmolytic properties. The extract showed hypotensive effect with transient respiratory stimulation.

The antibacterial activity of Garden Cress has been demonstrated in various tests. An antiviral effect against the encephalitis virus Columbia SH was also demonstrated in tests on mice (PDR).

An aqueous extract of Lepidium latifolium, given orally and intraperitoneally, considerably enhanced urinary excretion (UV) in rats with respect to control groups. A slight increase in ion excretion was also observed. (J Ethnopharmacol, Jan. 1994.)

Use in Western herbal
Lepidium sativum is used occasionally in the form of a salad as a "blood cleanser." The root of Lepidium virginicum (Verginia Peppergrass, a related spp.) is used for treating excess catarrh within the respiratory tract.

Asclepiadaceae

LEPTADENIA

Leptadenia reticulata (Retz.) W. & A.

Figure 1 Leptadenia reticulata [ZANDU]

Habitat
Sub-Himalayan tract of Punjab, Uttar Pradesh and throughout Peninsular India, up to 900 m.

Classical and common names
Ayurvedic: Jivanti, Jivaniya, Jivapushpa, Hemavati (Charaka); Jivana (Sushruta); Shaakashreshtha (the best among vegetables), Payaswini, Maangalya, Madhusravaa. Dodi-shaaka.

Siddha: Palakudai, Paalai Keerai.

Parts used
Root.

Dose
Powder 3–6 g.

Classical use
Charaka treats Jivanti as an important tonic for promoting health and longevity. Sushruta used it for spermatogenesis and as a galactagogue. Vision-promoting properties were attributed to the herb, and it was known as Chakshushya (Sushruta Samhitaa, Bhaavaprakaasha, Kaiyadeva Nighantu, Dhanvantari Nighantu). Vagabhatta included Jivanti in Jivaneeya Gana, the vitalising group of herbs.

Jivanti is mentioned by Charaka and Bhaavamishra (in Bhaavaprakaasha) as the best among leafy vegetables, but generally the fruits of the climber were used in Ayurvedic medicine. Charaka used the fruits alone or as ingredient of unguent or decoction in chest congestion, cough, and for rejuvenation and longevity. In West Bengal, fruits of Desmotrichum fimbriatum Blume or Dendrobium macraei Landl. are used as Swarna Jivanti. In the South, the root of Holostemma adakodien Schutt. is used as Jivanti.

Jivanthyaadi Ghritam is available in the South over-the-counter and is prescribed in eye diseases of nervous origin. Jivanti is an ingredient in Anu Tailam (Ashtaanga Hridaya), prescribed as nasal drops in diseases of eyes, nose, throat; also in headache and cold.

Active principles and pharmacology
Leaves and twigs contain hentriacontanol, alpha- and beta-amyrin, stigmasterol, gamma-sitosterol and flavonoids—diometin and lutcolin; pericarp of follicles gave quercetin, iso-quercitrin, rutin and hyperoside; seeds contain meso-inositol and its monomethyl ether. Alkaloids have been detected in the thin stems and leaves of the plant. A new trisaccharide—leptaculatin—has been isolated.

The potassium and calcium contents in Leptadenia reticulata were found to be 1.69 and 0.99 per cent respectively. The aerial parts on ethanolic extraction yielded the two triterpenoids taraxerol and fernenol, and beta-sitosterol.

A considerable amount of sterols is present in the free condition in the stems and roots of the plant. Stigmasterol is the major component; sitosterol is found in small quantities. A fructosan (7–8 hexose units) of the inulin type has been separated from the tubers.

The aqueous extract of the stem showed highly potent and prolonged hypotensive action in anaesthetised dogs. Hypotensive activity was also reported in dogs with Leptadenia reticulata powder.

The aqueous extract of the stem when given to rats i.p. in a dose of 4 ml/kg for a period of 6 weeks did not produce any obvious signs of local or systemic toxic reactions. Intraperitoneally, the drug (up to 6 ml/kg) did not lead to any immediate or delayed alteration in the activity of animals.

The plant is found useful in the treatment of habitual abortion in women.

The lactogenic effect of Leptadenia reticulata has been clinically assessed by many investigators. A few studies were also undertaken on the effect of the drug on uterine disorders. A compound herbal preparation, Leptaden, comprising extracts of Leptadenia reticulata and Bryonia patens, a spasmogenic herb, has been used in most of the studies.

Leaves and roots give good results in skin affections. Alcoholic (50 %) extract of roots and leaves show antibacterial activity against Gram-positive and Gram-negative bacteria, including Micrococcus pyogenes var. aureus, var. albus, and var. citreus, Bacillus megatherium, E. coli, Salmonella typhi and Proteus vulgaris; also against Trycophyton rubrum.

Lamiaceae

LEUCAS

Leucas aspera (Willd.) Link

Figure 1 Leucas aspera [ADPS]

Leucas cephalotes spreng.

Habitat
Throughout India as a weed in cultivated fields, wastelands, roadsides.

Classical and common names
Ayurvedic: Leucas cephalotes has been equated with Dronapushpi; Leucas aspera, its related spp., with Phalepushpaa, Kutumbaka.
Siddha: Thumbai (Leucas cephalotes).
English: White Dead Nettle (Leucas aspera).

Parts used
Leaves.

Dose
Fresh juice 10–20 ml.

Classical use
Dronapushpi was mostly used for eye diseases and jaundice; also as a blood purifier in psoriasis, scabies, chronic skin eruptions. During malarial fever, juice of leaves was prescribed. (Bhaavaprakaasha, Shaarangadhara Samhitaa, Raaja Nighanu, Vrindamaadhava, Gadanigraha.) The herb is an ingredient in Mahaanaarayana Taila (Bhaavaprakaasha), prescribed for external use in rheumatic affections.

In folk medicine, the plant is used as an antipyretic in villages in South India. The juice of leaves is used externally for skin eruptions and swellings. In Bengal, flowers are given with honey for coughs and colds to children. The plant is also used as a pot-herb.

Active principles and pharmacology
Leucas aspera: the plant gave oleanolic acid, ursolic acid and beta-sitosterol. Seeds contain palmitic, stearic, oleic, linoleic and linolenic acids; unsaponifiable parts gave beta-sitosterol and ceryl alcohol.

An alcoholic extract of leaves show anti-bacterial activity against Micrococcus pyogenes var. aureus and E. coli.

The bark extract showed antibacterial activity against E. coli, Sal. typhose, Vib. comma and Sh. dysenteriae.

The oil from the plant was found to exert antifungal activity against Trichophyton mentagrophytes, T. rubrum, Epidermophyton floccossum and Candida albicans. The oil completely inhibited the growth of E. floccossum even at a concentration of 1:250.

The chloroform extract (0.5 %) of the plant was found to bring about disappearance of clinical lesions in T. verrucosum, T. mentagrophytes, T. rubrum and Microsporum gypseum dermatomycosis in mice, and against T. mentagrophytes in calves.

The leaf extract showed some antiviral activity against bean common mosaic virus.

The aqueous and alcoholic extracts of the plant showed good anti-inflammatory activity in chronic anti-inflammatory model. In acute model, however, the aqueous extract was more potent.

The 50 % ethanolic extract of the whole plant, however, did not show any antibacterial, antifungal, antiviral, anti-inflammatory, or hypoglycaemic and CVS-activity. The extract revealed diuretic effect in rats, CNS-depressant and hypothermic effects in mice.

Leucas cephalotes: The plant contains beta-sitosterol glycoside and traces of an alkaloid.

The ethyl acetate extract failed to protect carbon tetrachloride-induced hepatotoxicity in mice and rats when used in a dose of 300 mg/kg, which was confirmed histologically. The extract also failed to reduce CCl_4-induced mortality in 48 hours.

Plant extracts showed only antispasmodic action.

Though freshly prepared juice from the plant is applied externally in scabies and skin eruptions, the extract of the plant showed no pharmacological action except darkening of skin on local application (CCRAS).

Linaceae

LINUM

Linum usitatissimum L.

Habitat

Cultivated throughout India, up to 2000 m, sometimes runs wild.

Classical and common names

Ayurvedic: Atasi, Umaa, Masrnaa (Charaka); Nilapushpi, Kshumaa.
Unani: Alasi, Kataan.
English: Linseed, Flax.

Parts used

Seeds, flowers, oil.

Dose

Powder 3–6 g, flowers 3–6 g, oil 5–10 ml.

Classical use

The decoction of seeds and leaves was prescribed internally, in prescriptions, by Charaka in anaemia, consumption, urinary diseases. Sushruta prescribed the oil of the seeds internally as a laxative for urinary diseases, intestinal parasites. He included seeds in medicinal plaster for goitre. Charaka included Atasi in the group of herbs used for tearing the abscess. Poultice or plaster of the seeds was externally applied on gout and inflammations. A paste made of fried linseed and sesamum seed, soaked with milk and pounded with the same, was applied on wounds. (Charaka Samhitaa, Ashtaanga Hridaya.)

According to Bhaavaprakaasha, Atasi affects eyesight and spermatogenesis adversely, though it alleviates rheumatic affections.

Laooq-e-Kataan (Haj-ul-Amraz) is a single drug preparation for pneumonia, cardiac asthma and bronchitis, given to liquify phlegm.

In folk medicine, for removing a foreign body from the eye, a linseed is moistened and placed under the eyelid.

Active principles and pharmacology

The plant contains mucilages (3–10 % in the epidermis, high swelling capacity); cyanogenic glycosides (0.05–0.1 %), linustatin and neolinustatin; fatty oil (30–45 %) with linolenic acid (40–70 %), linoleic acid (10–25 %), oleic acid (13–30 %); lignans (secoisolariciresinol-diglucoside), phenylpropane derivatives (including among others, linusitamarine); proteins (20–27 %); ballast (25 %). (PDR.)

Toxic constituents are: linataine (glutamic acid derivative) from seeds; linamarin and lotaustralin (cyanogenic glycosides) from leaves, stems, flowers, roots and 0.1–1.5 % linustatin and neolinustatin (cyanogenic glycoside) in seeds.

The cyanogenic glucoside is responsible for deaths among liverstock grazing on linseed plant.

Linseed (veterinary) meal is reported to possess goitrogenic effect on sheep and mice. Ewes fed on a diet containing linseed meal, under controlled conditions, have been observed to produce lamb with acute goitre. The goitrogenic principle has been identified as a thiocyanate.

Use in Western herbal

Linseed tea (25 g of ground or entire seeds to 30 cc of boiling water), added with little honey and lemon juice, is taken for colds, coughs, irritation of the urinary organs, gastritis and enteritis.

Seeds are taken internally for irritation in the digestive tract. (Seeds absorb fluid and swell, forming a jelly-like mass that acts as an effective bulk laxative. The use of large quantity of seeds with too little fluid intake can lead to intestinal obstruction.)

Caution

Avoid internal use of seed in early pregnancy. May cause intestinal obstruction, esophageal or gastrointestinal stricture, acute gastroenteritis or esophagitis. (Francis Brinker.)

Verbenaceae

LIPPIA

Lippia nodiflora Rich.
Phyla nodiflora (Linn.) Greene

Habitat

Common in hedges and waste areas throughout India.

Classical and common names

Ayurvedic: Jalapippali, Shaaradi, Shakulaadani, Matsyagandhaa; Laangali (now equated with Gloriosa superba L.).
Siddha: Paduthalai.
English: Wild Sage.

Parts used

Whole plant.

Dose

Juice 10–20 ml.

Classical use

According to Bhaavaprakaasha, Jalapippali is a cardiac tonic, promotes eyesight, spermatogenesis, cures haemothermia, piles, and fevers.

Active principles and pharmacology

Two glucosides, nodiflorin-A and B from dried plant have been identified. Plant also contains non-glucoside bitter substance, an essential oil, resin and a large amount of potassium nitrate. Free lactose, maltose, glucose, fructose and xylose have been detected in the aqueous extract of the plant. The diuretic action of the plant is attributed to the presence of potassium nitrate.

An alcoholic extract of leaves shows antibacterial activity against E. coli.

Use in Western herbal

Lippia spp. are credited with calming, digestive and antispasmodic properties. The drug is also used as a stimulating expectorant.

Recently, essential oils of Lippia spp. have been tested for their antibacterial and antifungal properties against some micro-organisms living on the skin of feet and armpits. The essential oils, also their main components, thymol and carvacrol, show strong antagonistic activities. Corynebacterium xerosis developing axillary odour is specially inhibited. No specific activity has been observed upon the feet microflora (Ann Pharm Fr (1996) 54(5):228–30).

In another study, preliminary *in vitro* results provide a scientific basis for the use of Lippia alba against bacterial respiratory infections (J Ethnopharm (1991) 33:57–62).

Lippia dulcis is used in Mexico for the treatment of cough and bronchitis.

Extracts of Lippia dulcis, *in vitro*, demonstrated activity against enteropathogenic E. coli, Salmonella enteritidis and Sh. flexneri. This substantiated the use of the herb in gastrointestinal disorders.

Lippia dulcis did not show *in vitro* effect against V. cholerae.

Hamamelidaceae

LIQUIDAMBAR

Liquidambar orientalis Miller

Habitat

Native to Asia Minor. Large forests of the plant are found in south-western parts of India.

Classical and common names

Ayurvedic: Turushka (Charaka, Sushruta); Silhak, Kapitaila. In Bhaavaprakaash, Turushka and Silhak are synonyms, while Shodhal Nighantu treats them as two different drugs. In Dhanvantari Nighantu, Turushka is equated with Pindaka, which indicates that it was solid storax; Silhak was taken to be a different drug, as it was liquid storax. Storax is also obtained from Altingia excelsa Nar-

onha, found in Assam and Bhutan and is equated with Shilaarasa.
English: Liquid Ambar, Asiatic Storax, Oriental Sweet Gum.

Parts used
Exudate.

Dose
1–2 g.

Classical use
Liquid extract of Turushka, as an ingredient of a medicinal oil, was given by Charaka internally in cough, consumption, pectoral lesions, debility. Sushruta used the drug internally in skin eruptions, blood poisoning and neurological disorders.

Active principles and pharmacology
Purified storax is composed principally of an alcoholic resin, storesin (33–50 %), occurring free and as cinnamic ester; it also contains free cinnamic acid (5–15 %), Cinnamyl cinnamate or styracine (5–10 %), phenylpropyl cinnamate (about 10 %) and small amounts of ethyl cinnamate, benzyl cinnamate, styrene (phenylethylene), a pleasant-smelling liquid, probably styrocamphane, and traces of vanillin.

Storax exhibits stimulating expectorant and antiseptic properties and is found efficacious in ointments for scabies and other parasitic diseases.

Altingia excelsa: Antistress, anti-inflammatory, hepatoprotective, antiulcerogenic and antipyretic properties of Altingia excelsa were examined by CCRAS at its pharmacological research units.

The ethyl acetate extract of the root of Altingia excelsa increased the physical endurance in swimming performance in mice. It protected rats and mice against stress-induced ulcers and CCl_4-induced hepatotoxicity. The extract also exhibited anti-inflammatory, antipyretic and immunomodulator activities. The plant established its efficacy as a potent antistress plant and compares well with Panax ginseng. The time taken for drowning in control mice was c 310 minutes, whereas in treated group it was c 450 minutes. (CCRAS.)

Lauraceae

LITSEA

Litsea glutinosa (Lour.) C. B. Robinson

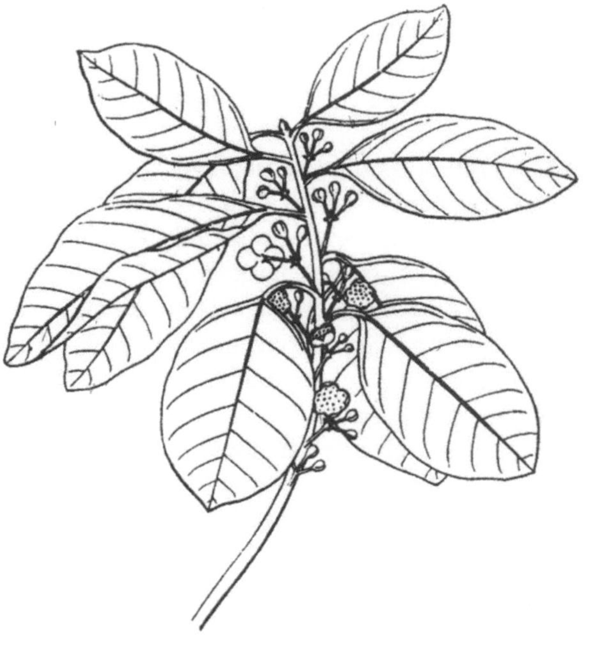

Figure 1 Litsea glutinosa—flowering branch [WOI]

Litsea chinensis Lam.
Litsea sebifera (Willd.) Pers.
Tetranthera apetala Roxb.

Habitat
Throughout India, ascending up to 1350 m in the outer Himalayas.

Classical and common names
Ayurvedic: Medaasaka (not mentioned in Ayurvedic classics); Maidaa-lakdi.
Unani: Maidaa-lakdi, Maghaase-Hindi.
English: Common Tallow Lowrel.

Parts used
Bark.

Dose
Powder 3–5 g.

Classical use

Traditionally, paste of the bark is used externally as an emollient application for sprains, bruises and rheumatic and gouty joints. Also used as a styptic dressing for wounds. Leaves and flower buds are also used as poultice for bruises and wounds.

Powdered bark, with honey, is prescribed internally in sprains, fractures, rheumatic affections, sciatica; also as a nervine tonic. Bark in infusion or decoction is given in diarrhoea and dysentery.

Active principles and pharmacology

Plant contains a polysaccharide; leaves and stem contain the aporphine alkaloids boldine, laurotetanine, N-methyllaurotetanine, actinodaphnine, nor-boldine, and methylactinodaphnine; the trunk-bark gave sebiferine and litseferine.

The whole plant, excluding roots, was found to be devoid of tannins.

The bark on extraction with menthol and hexane has been reported to yield actinodaphnine and beta-sitosterol respectively. Aqueous extract of the leaves yielded kaempferol-3-glucoside, quercetin-3-rhamnoside, quercetin, kaempferol-7-glucoside, pelargonidin-5-glucoside and a new flavonoid, glycoside naringerin, identified as naringenin-7-monorhamnoside.

The seeds yielded an essential oil comprising ocimene, alpha-pinene, D-limonene, terpinolene, alpha-terpinene, carvone, caryophyllene, beta-amyrin acetate, fenchyl alcohol, alpha-thujone and linalool oxide.

The essential oil showed antibacterial activity against E. coli, B. subtilis, Vib. cholerae, Sh. dysenteriae, C. diphtheriae, Sal. typhi, Strep. faecalis, Strep. pyogenes and Ps. solanacearum. The oil from the bark and the root extract also showed antibacterial activity.

The bark extract was found active against Fusarium solani, Fusarium moniliforme, Helminthosporium turcicum, Helminthosporium oryzae, Pythium vexans, Rhizoctonia solani, Rhizoctonia bataticola, Colletotrichum capsici, Pyricularia setariae and Alternaria helianthi; being considerably active against the last three.

The essential oil showed weak anthelmintic activity (less effective than piperazine citrate) against earthworms and tapeworms.

The 50 % ethanolic extract of the whole plant (excluding roots) did not show antibacterial, antifungal, antiprotozoal effects. It showed antispasmodic action on isolated guinea-pig ileum.

The essential oil obtained from berries exerted a prolonged hypotensive effect in anaesthetised dogs in a dose of 125 mg/kg. The oil also blocked the contractions produced by acetylcholine, histamine and barium on guinea-pig ileum.

The oil from berries caused a reduction in the spontaneous motor activity without any concomitant muscle weakness in rats. Analgesic activity of the oil was also observed.

Campanulaceae

LOBELIA

***Lobelia inflata* Linn.**
***Lobelia nicotianaefolia* Heyne**
***Lobelia excelsa* Lesch.**

Habitat

Lobelia inflata is indigenous to America and Canada. Indian substitutes, Lobelia nicotianaefolia and Lobelia excelsa, are found in western Ghats, from Mumbai to Travancore, Nilgiris, Malabar and Karnataka.

Classical use

Not used in Ayurvedic and Unani medicine.

Active principles and pharmacology

In the West, the whole herb of Lobelia inflata is used as a strong antispasmodic and respiratory stimulant in asthma and chronic bronchitis. Piperidine alkaloids (principally lobeline) afford these therapeutic properties to the drug.

The lobeline content of Lobelia nicotianaefolia revealed higher concentrations (1–1.8 %) of lobeline as compared to 0.3–0.4 % content in Lobelia inflata (as per B.P.). Norlobelanine has also been isolated from the plant. But ethereal tincture of Lobelia nicotianaefolia (with 0.3 % lobeline) showed no effect in patients of bronchial asthma. It was found of therapeutic value when administered in combination with other parasympathetic depressants (potassium iodide and tincture belladonna).

Examination of commercial samples of Lobelia nicotianaefolia revealed an adulterant, Verbascum thapsus, which does not contain lobeline alkaloids. Moreover, the alkaloid content of the plant itself showed seasonal variation.

Lobelia excelsa gave 0.3–0.4 % lobeline content. It showed diuretic and antispasmodic activity on isolated guinea-pig ileum. Lobelia nicotianaefolia whole plant showed abortifacient action in female rats.

Lobelia inflata, known as Indian Tobacco, should not be confused with Nicotiana tabacum Linn., cultivated throughout India. The plant contains alkaloids including nicotine, nor-nicotine, anabasine, nicotyrine, anatabine and several acids. The leaf is used in the manufacture of cigarettes, cigars, bidi, hookah tobacco, and for the purpose of chewing. Medicinally, it is included in tooth-powders, especially the red ones. In Unani medicine, Sanoon-e-Tambaku (Bayaz-e-Kabir) is a reputed product, prescribed in gingivitis, toothache, oral sepsis.

Use in Western herbal

Tablets, containing lobelia in combination with other herbs, are used to treat bronchial asthma. In the Anglo-American herbal tradition, lobelia has mostly been combined with Capsicum frutescens (cayenne).

Though lobelia exerts a stimulating effect on the respiratory center, it breaks down too quickly in the body to be used as a respiratory analeptic. (PDR.)

0.6–1 g of the leaves are said to be toxic, 4 g fatal.

The drug (Lobelia inflata) is used in homoeopathy as an asthma treatment and also as an aid in curing addiction to smoking.

Palmac, Arecaceae

LODOICEA

Lodoicea maldivica (Poir.) Pers.
Lodoicea seychellarum Labill.

Habitat

Native of Seychelles Islands, grown in Indian gardens.

Classical and common names

Ayurvedic: Samudra-naarikela, Dariyaayee Naariyal.
Unani: Naarjeel-e-Bahari, Naarjeel-e-Dariyaayee.
Siddha: Kadathengai.
English: Sea Coconut Palm, Double Coconut Palm.

Parts used

Kernel.

Dose

Powder 5–10 g.

Classical use

Traditionally, dried kernel of Samudra-naarikela is used in paediatric disorders. Rubbed with Terminalia chebula (Haritaki) in water and given cold, it alleviates impurities of blood, boils, skin eruptions and also acts as an anthelmintic. (Siddhabhesaja-manimaalaa.) Ayurvedic practitioners also use the kernel as an antidiabetic aid.

The water of the green fruit and its soft kernel are prescribed for their antibilious and antacid properties.

Active principles and pharmacology

Decoction of nut kernel in doses of 2 g/kg p.o. and 1.5 g/kg p.o. in fasting and alloxan diabetic rabbits reduced blood sugar level to 37 % and 29 % respectively (CCRAS).

The kernel also exhibited anti-diarrhoeal, antiemetic and antacid properties; gave good results in haemorrhoids. It is an ingredient in Arshina, a commercial compound for haemorrhoids, available over-the-counter.

Cucurbitaceae

LUFFA

Luffa echinata Roxb.

Habitat
Wild in Uttar Pradesh, Bihar, Bengal and Gujarat.

Classical and common names
Ayurvedic: Devadaali, Jimuuta, Bandaal.
English: Bristly Luffa.

Parts used
Fruits.

Dose
Powder 1–3 g.

Traditional use
The fruit is used for its purgative properties in dropsy, nephritis, chronic bronchitis, chest congestions, biliary and intestinal colic. The fruit is also used in enlarged liver and spleen.

Active principles and pharmacology
Fruits contain chrysoeriol and its glycosides as principal flavonoids; apigenin and luteolin glycosides are present in minor amounts. Also present are elaterin, its 2-glucoside, iso-cucurbitacin B, beta-sitosterol glycoside and gypsogenin glycoside.

Seeds contain cucurbitacin B, triterpene alcohols—echinatols A and B, and a saponin with oleanolic acid as sapogenin. Seed oil consists of palmitic, stearic, oleic and linoleic acids.

Both alcoholic and aqueous extracts of the fruit exhibited hepatoprotective effect in carbon tetrachloride-induced liver damage in rats. Better result was obtained with the alcoholic extract.

The aqueous extract of fruit lowered serum bilirubin level in chlorpromazine-induced jaundice in rats. The 50 % ethanolic extract of the whole plant, excluding roots, showed hypoglycaemic action in rats.

Hypotensive, cardiac depressant, antispasmodic and diuretic properties of the fruit were also confirmed experimentally. (CCRAS.)

Juice from seedlings exhibited mild antifungal property against Cephalosporium sacchar.

Lamiaceae

LYCOPUS

Lycopus europaeus Linn.

Habitat
Western Himalaya, in Jammu & Kashmir and Himachal Pradesh.

Classical and common names
Vernacular: Gandam-gundu, Jalnim.
English: Gipsywort.

Classical use
Traditionally, the herb is used as a sedative and in pulmonic and other haemorrhages. The leaf is used as poultice to clean foul wounds.

Active principles and pharmacology
Leaves contain lithospermic acid. Plant contains luteolin-7-glucoside, ursolic acid, caffeic acid, chlorogenic acid, sinapic acid and ellagic acid.

In the West, extracts of the plant are used in the treatment of hyperthyreosis and similar diseases. They inhibit the action of thyrotropic hormone and the thyroxine output of thyroid (The Wealth of India.)

Diverse effects of Lycopus extracts on pituitary thyroidal system as well as on the pituitary gonadal system were studied in various experimental models after parenteral application.

An ethanolic extract of Lycopus europaeus was applied orally to rats, diverse endocrine parameters were measured between 3 and 24 h later and the effects compared to an i.p.-treated group. The plant extract given p.o. caused a long-lasting (for a period of more than 24 h) decrease of T3-levels, presumably as a consequence of a reduced peripheral T4-deiodination. A pronounced reduction of T4- and thyroid stimulating hormone (TSH)-concentrations was observed 24 h after application of the test solution by gavage. The luteinizing hormone (LH)-decrease as well as the TSH-decrease, which was pronounced in spite of reduced T4- and T3-levels, indicate a central point of attack of the plant extract. (1994;http://www.ibiblio.org)

Plant is narcotic.

Use in Western herbal

A close relative of European Lycopus is American Lycopus virginicus, used for nervousness, insomnia, premenstrual syndrome and mild thyroid hyperfunction associated with disturbances of autonomie nervous system. Also used for mastodynia.

Caution

The herb should not be administered simultaneously with thyroid medication.

M

Sapotaceae

MADHUCA

Madhuca indica J. F. Gmel.
Madhuca latifolia (Roxb.) Macb.
Madhuca longifolia (Koen.) Macb.
Bassia latifolia Roxb.

Habitat

Throughout India, up to an altitude of 1200 m.

Classical and common names

Ayurvedic: Madhuuka (Sushruta), Gudapushpa, Madhupushpa, Madhusrav; Mahuaa.
Unani: Mahuaa.
Siddha: Kattuiluppu.
English: Mowra, Butter Tree, Mahua.

Parts used

Flowers, fruits, bark.

Dose

Juice of flowers 10–20 ml, decoction of bark 50–100 ml.

Classical use

Charaka and Sushruta included Madhuuka in prescriptions for intrinsic haemorrhage, hiccough, diarrhoea, dysentery, deficient lactation, and debility. Charaka prescribed Madhuukaasava (fermented flowers) as an alcoholic elixir. Sushruta gave Madhuuka in angina pectoris, syncope and fevers; also as a sex tonic for spermatogenesis.

Flowers of Madhuuka, devoid of stamens and pollens, cooked with sugar and clarified butter and spiced with Cuminum cyminum (Jiraka), were prescribed as a restorative tonic.

Flowers (25 g), boiled in 250 ml milk, were prescribed in sexual debility.

A decoction of flowers was given in cough, colds, bronchitis, constipation. Flowers, fried in clarified butter, were prescribed for treating haemorrhoids.

In folk medicine, the seed oil is applied externally in skin diseases, rheumatic affections; internally administered as an emetic, and in habitual constipation.

A decoction of the bark is used as an astringent and emollient for itches, ulcers, bleeding and spungy gums, tonsillitis, arthritis; and as a hypoglycemic agent in diabetes mellitus.

The leaf-ash mixed with clarified butter, is applied over burns and wounds.

Madhuukaasava is available in the South, prescribed in chronic dyspepsia, indigestion, polyuria, debility.

Active principles and pharmacology

Mahua flowers (corollas) are a rich source of sugar and contain appreciable amount of vitamins and calcium. One analysis gives following values: moisture 18.6, protein 4.4, fat 0.5; total sugars 72.9; fibre 1.7 and ash 2.7 %; phosphorus 140, calcium 140 and iron 15 mg/100 g; magnesium and copper were also present. Sugars were identified as sucrose, maltose, glucose, fructose, arabinose and rhamnose.

The vitamins present in corollas are: carotene (as vitamin A) 39 I.U., ascorbic acid 7 mg, thiamine 32 mcg, riboflavin 878 mcg, and niacin 5.2 mg/100 g; folic acid, pantothenic acid, biotin and inositol are also present.

The corollas also contain an essential oil with an unpleasant odour, anthocyanins, betaine and salts of malic and succinic acids. Catalase, oxidase, invertase, maltase, amylase and emulsin are among enzymes identified.

Mahua flowers exhibit cooling, tonic and demulcent properties when used in coughs, colds and bronchitis.

Flowers show antibacterial property against E. coli. The methanolic extract of flowers, stem and stem bark have been reported to possess antibacterial activity against B. anthracis, B. pumilus, B. subtilis, Sal. paratyphi, Vib. cholerae, Xanth. campestris and Xanth. malvacearum.

Fruit pulp yielded a number of triterpenoids, including alpha- and beta-amyrin acetate, 3-beta-monocaprylic ester of erythrodiol, 3-beta-capryloxy oleanolic acid and an acetate. In addition, n-hexacosanol, beta-D-glucoside of beta-sitosterol and free beta-sitosterol were obtained.

Trunk-bark contained lupeol acetate, beta-amyrin acetate, alpha-spinasterol, erythrodiol monocaprylate, betulinic acid and oleanolic acid caprylates.

Leaves gave beta-carotene, n-octacosanol, sitosterol, its glucoside, stigmasterol, erythrodiol, mono-palmitate, palmic acid, oleanolic acid, quercetin.

Mixture of saponins from seeds exhibited spermicidal activity in human semen at a concentration of 2 %. The saponin having bassic acid as the major sapogenin (in a dilution of 0.06 %) showed spermicidal activity in human semen.

The 50 % alcoholic extract of the stem bark showed hypotensive activity in cat and dog.

Mi-saponins A and B extracted from South Indian Mahua exhibited anti-inflammatory activity.

Euphorbiaceae

MALLOTUS

Mallotus philippensis (Lam.) Muell. Arg.

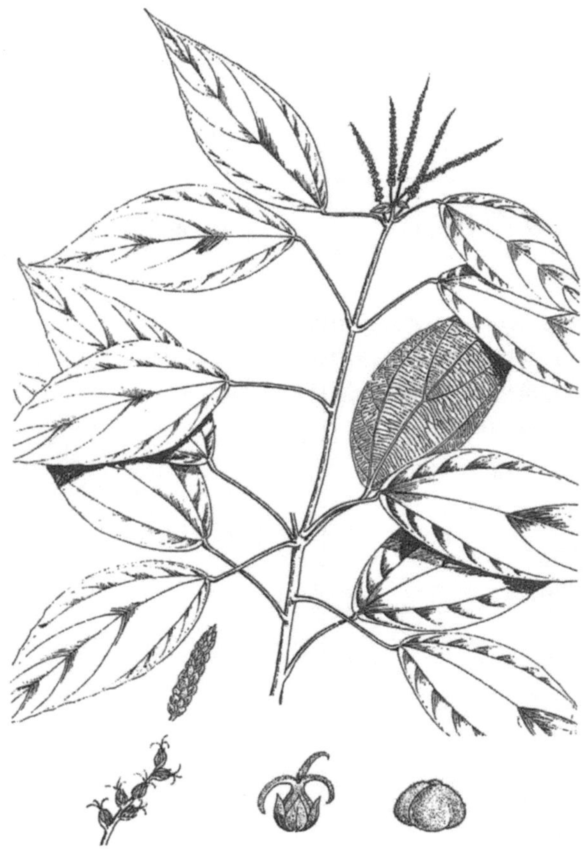

Figure 1 Mallotus philippensis [ADPS]

Habitat

Throughout tropical India, up to 1500 m in the outer Himalayas, and on hills.

Classical and common names

Ayurvedic: Kampillaka, Kampilla, Karkash, Raktaang, Rechan. Not to be confused with Kamala (Nelumbo nucifera).
Unani: Kamilaa.
Siddha: Kamela.
English: Kamala Tree.

Parts used

Gland and hair of the fruit.

Dose

500 mg–1 g.

Classical use

Charaka and Sushruta prescribed the fruit and flower-pollen internally in acute constipation, intestinal paralysis, abdominal swellings, and as a purgative. Sushruta gave oil of the seeds internally for intestinal parasites, polyurea and urinary diseases. Kampillaka, included in a medicinal oil, was prescribed for external application in chronic skin diseases, wounds and malignant ulcers.

5 g fruit-hairs and flowers were prescribed with jaggery for expelling intestinal parasites (Bhaavaprakaasha). For treating diseases of the spleen (hardening and swelling), Kampillaka, mixed with profuse honey, was prescribed (Vrindamaadhava).

In Unani medicine, Raughan-Kamilaa is prescribed externally for weeping eczema, herpes, scabies, ringworm and ulcers; diluted with hair oil, it is massaged over scalp in baldness.

Active principles and pharmacology

The heartwood gave betulin-3-acetate, lupeol, lupeol acetate, sitosterol and bergenin. Bark yielded acetylalcuritolic acid, sitosterol, its glucoside, bergenin and alpha-amyrin. Leaves gave amyrin.

The plant also gave the phloroglucinol derivatives rottlerin and a "red compound" as main constituents, along with small amounts or traces of isoallo-rotterin, the "yellow compound" methylene-bis-methylphloroacetophenone and two more compounds designated as kamalins 1 and 2. Other phlorogucinol compounds isolated from Kamila include methylene-bis-methylacetylphlorogucinol and rotteron, a red C-methylated cinnamoyl chromen and the other yellow chromeno-flavanone. The last two compounds are minor components.

Hair of capsule contains rottlerin, which was reported to show antifertility effect in female rats; isorottlerin was found almost inactive. Studies at the CDRI, Lucknow, with 50 % ethanolic extract of the fruits did not reveal any antifertility activity in mice. Petroleum-ether, alcohol, aqueous extracts of seed, administered from 1–4 day of post coitum, did not exhibit antifertility activity. (CCRAS.)

Kampilla in 10 and 20 mg dose exhibited 35.69 % and 78.21 % anthelmintic effect respectively on naturally infected (infected with tapeworm) albino rats. Kampilla exhibited non-irritant purgative activity in 50 mg/kg in rats. LD_{50} in chronic toxicity test was 200 mg.

The drug showed a significant anti-inflammatory effect in a dose of 5 mg i.m. against mice hind paw oedema by palutin s.c. It produced anti-inflammatory effect against turpentine-induced pleurisy in rats.

Kampilla showed considerable haemostatic activity, which was confirmed from prolongation of bleeding time as compared to adrenaline.

20 % Kampilla ointment in wax, applied locally against Taenia crusis and Taenia corposis, exhibited good antifungal activity.

20 % Kampilla ointment, applied on wounds, contracted the wound earlier. The drug played a significant role in stimulation of healing process and increasing the tensile strength as observed from experimentally produced tear wounds in rats.

The resin obtained from the plant exhibited antispasmodic effect against histamine on guinea-pig ileum. The saponin exhibited cardiac depressant activity blocked by atropine; exhibited inhibition of histamine-induced contraction in guinea-pig ileum and tracheal chain preparation.

The 50 % ethanolic extract of the fruits showed antibacterial activity against B. subtilis and Staph. aureus, which could not be confirmed in fractionated extract.

The extract also showed hypoglycaemic activity in rats and anticancer activity against human epidermoid carcinoma of nasopharynx in tissue culture and sarcoma 180 in mouse. (CCRAS.)

In a clinical study, Kampilla powder was administered to 52 children infested with hook worm (H. nana) in a single dose for two days. There was a satisfactory conversion of stools from positive to negative in 96 % cases. The drug perhaps acted not only on the worms in intestinal lumen, but also on cysticeroid stage lodging in the intestinal villi.

Malvaceae

MALVA

Malva sylvestris Linn.

Habitat

Temperate Himalayas from Kashmir to Kumaon up to 2400 m, Bihar and Peninsular India.

Classical and common names

Unani: Khubhaaji, Bhubbaaji Bustaani, Gul-Khair.
English: Dwarf Mallow.

Parts used

Fruits, seeds.

Dose

5–7 g.

Classical use

Fruits and seeds are imported into India from Persia and used in Unani medicine for their mucilaginous and demulcent properties in coughs, affections of mucous membranes of the pulmonary tract and urinary bladder.

The drug is employed as emollient, antiphlogistic, expectorant, antitussive, laxative and diuretic. It finds an important place in hooping cough compounds.

The drug is added to strong purgatives to modify their violent action.

Active principles and pharmacology

Leaves gave the flavonol glycosides gossypetin-8-O-beta-D-glucoside-3-sulphate and gossypetin-8-O-beta-D-glucuronide-3-sulphate; mucilage and tannins. Flowers contain malvin (an anthocyanin).

The flowers and leaves show emollient and soothing properties when applied to sensitive areas of the skin; applied as a poultice they reduce affected areas of the skin, reduce swelling, and draw out toxins.

Leaves given internally reduce gut irritation and exert a laxative effect.

Anacardiaceae

MANGIFERA

Mangifera indica Linn.

Figure 1 Mangifera indica [CCRAS]

Figure 2 Mangifera indica [CCRAS]

Habitat

Cultivated all over India.

Classical and common names

Ayurvedic: Aamra, Amb, Rasaal, Sahakaar, Pikavallabha, Madhudoot.
Unani: Ambaj.
English: Mango.

Parts used

Fruits, kernel, bark, tender leaves.

Dose

Powder of seed-kernel 1–3 g, decoction 50–100 ml, juice of tender leaves 10–20 ml.

Classical use

According to Ayurvedic classical texts (Bhaavaprakaasha, Maadhav-dravyaguna), Aamra is a cardiac tonic, promotes complexion, semen, and strength, increases digestive power, cures urinary diseases, and disorders caused by vitiated blood. It belongs to the astringent group of drugs. The seed kernel cures vomiting and diarrhoea; also cures burning sensation in the chest due to acidity.

Charaka gave fresh fruit in assimilation disorders. Sushruta gave unripe fruits in obesity, haemoptysis and vaginal discharges.

Juice of ripe mango, added with honey was prescribed in enlargement of spleen. (Vrindamaadhava, Bhaavarakaasha.)

Charaka gave seed-kernel of mango, in prescriptions, for diarrhoea and intrinsic haemorrhage. A decoction of the seed-kernel, mixed with honey and sugar was administered as a single drug in emesis and diarrhoea. (Vrindamaadhava, Bhaavaprakaasha.)

Seed-kernel of mango, pounded with Terminalia chebula (Haritaki), was applied to the scalp as a treatment for dandruff. A paste made of the seed-kernel, honey and camphor was recommended for treating relaxed vagina. (Shaarangadhara Samhitaa.)

Tender leaves of mango, also the tree bark, entered into many compound formulations for diarrhoea and dysentery during the 16th century.

The seed-kernel of mango is an ingredient in Vrihat Gangaadhar Churna (Shaarangadhara Samhitaa), prescribed in diarrhoea and dysentery. Pushyaanuga Churna (Bhaishajya Ratnaavali), a reputed Ayurvedic uterine tonic, also contains the seed-kernel. Both the compounds are available over-the-counter.

Active principles and pharmacology

Leaves, stem bark, heartwood and roots of Mangifera indica yield mangiferin. Fruits contain gallic acid, m-digallic acid, gallotanin, ellagic acid and mangiferin.

In ripe and mature stage m-trigallic acid has been found, while quercetin and iso-quercetin were reported only in mature and tender stages. The unripe fruit contained beta-glucogallin.

The seed pulp contained gallotanin. The ether extract of dry seed coat yielded beta-sitosterol, (+)-catechin, (±)-catechin, (−)-epicatechin and ethyl-gallate.

The steam distilled oil from Mangifera indica has been reported to contain mangiferin and indicine as major hydrocarbon constituents.

The roots yield, in addition to mangiferin, beta-sitosterol and friedelin. The rootbark on extraction with n-hexane gave friedelan-3-beta-ol, alpha-amyrin, cycloartenol, mangiferonic acid, mangiferolic acid, friedelin and beta-sitosterol. The acetone extract of stem bark yielded mangiferin, butin, fisetin, (−)-leucocyanidin, gallic acid, protocatechuic acid and quercetin.

The major components of mango (ripe Alphonso variety) pulp oil have been characterised as a triglyceride consisting of myristic acid (2.3 %), palmitic acid (23.8 %), palmitoleic acid (25.2 %), stearic acid (2.2 %), oleic acid (29.3 %), linoleic acid (3.8 %), and linolenic acid (13.4 %). (ICMR.)

The seed kernel, an important ingredient of Ayurvedic compounds, contains protein 9.5, fat 10.7, starch 72.8, sugar 1.7, tannin 0.11 and ash 3.66 %, silica 0.41, iron 0.03, calcium 0.23, magnesium 0.34, phosphorus 0.66, sodium 0.28, potassium 1.31, sulphur 0.23 and carbonate 0.09 %.

The amino acids of the kernel proteins are: cystine, aspartic acid, glutamic acid, glycine, threonine, alanine, tyrosine, histidine, arginine, lysine, proline, valine, leucines and phenylalanine.

The kernel fat contains stearic acid, oleic acids as major components, along with smaller proportions of linoleic and linolenic acids.

Blossoms of mango contained tannins in an amount of 15 % on dry weight basis.

Mangiferin imparts cardiotonic, diuretic and hepatoprotective properties to the mango. Mangiferin has been reported to show protection against CCl_4-induced liver injury in rats. Mangiferin exhibited definite signs of CNS-stimulation in mice and rats, cardiostimulant action in perfused frog heart, potentiated the effect of subanalgesic doses of morphine in rats. Mangiferin also produced hypocholeretic effect in dogs, as revealed by moderate increase in the bile flow. It did not show anticonvulsant and analgesic activity in rats.

The leaf of Mangifera indica revealed antibacterial action against B. subtilis, Staph. albus and Vib. cholerae.

Asclepiadaceae

MARSDENIA

Marsdenia tenacissima Wright & Arn.

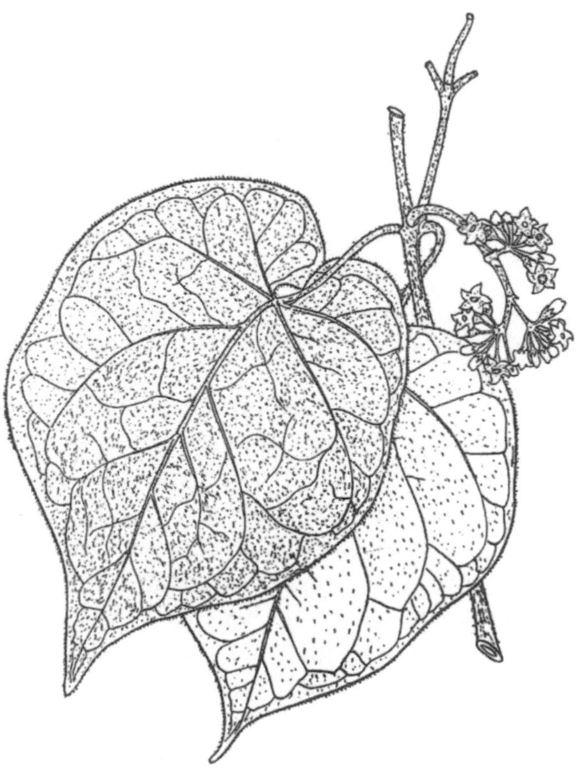

Figure 1 Marsdenia tenacissima—flowering branch [WOI]

Habitat

Himalaya, from Kumaon to Assam, up to 1500 m; Bihar, Madhya Pradesh, Orissa and South India.

Classical and common names

Muurvaa (CCRAS). In the South, Marsdenia volubilis T. Cooke is used as Muurvaa. Chonemorpha macrophylla G. Don is also used as a variety of Muurvaa. In West Bengal, Sansevieria roxburghiana Schult. is used as Muurvaa. INSA scientists also equated it with Muurvaa. Muurvaa has been wrongly equated with Safed Nishoth, the white variety of Turpeth (Operculina turpethum). The drug does not possess purgative properties attributed to black Turpeth.

Atirasaa, Gokarni, Morataa (Dhanvantri, Raaja Nighantu).

Parts used

Root.

Dose

Powder 3–5 g, decoction 50–100 ml.

Classical use

Charaka used the dried flowers and leaves of Muurvaa, in prescriptions, internally for cough, vomiting, fever; and as an appetizer.

Sushruta also used Muurvaa as an appetizer, in vomiting, indigestion, colic pain, intestinal catarrh, fevers. As an ingredient of an ointment, Muurvaa was applied externally over major burns. A medicinal oil, incorporating the herb, was prescribed internally, as well as externally, in skin diseases, fractures, convulsions, hemiplegia, facial paralysis, cephalalgia. The herb was considered to possess antitoxic properties.

Active principles and pharmacology

The root and seed are rich in pregnane glycosides of 2-deoxysugars, which on hydrolysis gave genins and sugars including cissogenin, tenasogenin, tenacissigenin, iso-drevogenin P, drevogenin Q, D-cymarose, asclepobiose, D-canarose, 3-O-methyl-6-deoxy-D-allose, 17-alpha-17-beta-marsdenin, beta-O-glucosyl-L-thevetose, cinnamic and acetic acids. Stems yielded five more glycosides named tenacissoides A-E.

Ethanolic extract (70 %) of the root showed mild CNS-depressant effect on mice, anthelmintic activity against earthworms, and antispasmodic activity against spasms induced in isolated guinea-pig ileum by histamine and acetylcholine. The extract was devoid of antibacterial activity.

Marsileaceae

MARSILEA

Marsilea minuta Linn.
Marsilea quadrifolia Linn.

Habitat
Widely distributed in tropical and temperate regions of world. Found throughout India, usually at the edges of ponds and canals; grows as a weed in marshy places.

Classical and common names
Ayurvedic: Sunishannaka (Charaka, Sushruta), Parnaka, Vastika parnika (Charaka); Chatushpatri; Susunishaak, Chaupaitra. Shitiwaara has now been equated with Celosia argentea L.

Parts used
Whole plant, leaves.

Dose
Juice 10–20 ml.

Classical use
Charaka gave sprouts cooked as vegetable in cough and spastic conditions of leg muscles. Sushruta prescribed the plant, cooked as a potherb, to harmonize internal body functions and for invigorating eyesight. The vegetable of Sunishannaka, fried in clarified butter, was administered in intrinsic haemorrhage. As a vegetable, it was also given in cases of poisoning.

In folk medicine, the herb is used as a vegetable for inducing sleep.

Active principles and pharmacology
Marsilin (1-triacontanol-cerotate), 3-hydroxy-triacontan-11-one, hentriacontan-6-ol, methylamine, beta-sitosterol, marsileagenin A, flavonol-O-mono- and -diglycoside, C-glucosylflavones and C-glucosylxanthones have been isolated from the plant.

The crude extract of Marsilea quadrifolia caused prompt hypotensive response in B. P. of anaesthetised dogs which was similar to ACh; caused cardiac depressant action on isolated rabbit heart which was blocked by atropine; showed ACh-like action in isolated smooth muscle (rabbit ileum); inhibited contractile response of noradrenaline; showed histamine-like contraction on isolated guinea-pig ileum which was blocked by antihistaminics; potentiated histamine on guinea-pig tracheal chain; potentiated bronchoconstrictor response elicited with acetylcholine. Thus the results suggested indirect cholinergic action of the extract, probably due to inhibition of cholinesterase. (CCRAS.)

Marsilin (400 mg/kg p.o.) was found to be active against electroconvulsions as well as metrazol-induced convulsions, while the leaf extract could not protect the rats against convulsions.

Both aqueous and alcoholic extracts of defatted leaves as well as marsilin (400 mg/kg) markedly potentiated thiopental sleep in mice, marsilin being most active.

Pre-treatment of rabbits with marsilin almost completely inhibited convulsions and altered EEG-pattern induced by metrazol.

Marsilin (400 mg/kg) significantly increased the mouse brain serotonin content, the activity being maximum after 60 minutes of its administration. At this dose level, it inhibited acetylcholinesterase in cerebral tissue in mice initially, followed by activation and inhibition alternately. Marsilin decreased the content of GABA and increased glutamic acid-level in mice. (ICMR.)

The alcoholic extract of Marsilea quadrifolia leaves did not reveal any narcotic activity even at a dose 100 mg/kg (oral) in mice, although it led to potentiation of pentobarbital-induced hypnosis at a dose of 5–10 mg/kg. Higher doses did not enhance this potentiation.

The 50 % ethanolic extract of Marsilea minuta (whole plant) did not exhibit anthelmintic, hypoglycaemic or diuretic effect experimentally.

Leguminosae

MEDICAGO

Medicago sativa Linn.

Figure 1 Medicago sativa—flowering and fruiting branch [WOI]

Habitat

Cultivated in Punjab, Uttar Pradesh, Maharashtra, Gujarat, Tamil Nadu.

Classical and common names

Vernacular: Wilaayati-gawuth, Lasunghaas, Lusan (The Wealth of India); wrongly identified as Vanmethika. Ahittha (Dhanvantari Nighantu), also Hispittha and Ashwabalaa are doubtful synonyms. *English:* Alfalfa.

Parts used

Sprouted seeds.

Classical use

Alfalfa is used in Homoeopathy as one of the most versatile herbs for favourably influencing nutrition and toning up appetite and digestion; to improve mental and physical vigour. Neurasthenia, nervousness, insomnia, nervous indigestion are within its therapeutic range. It corrects tissue waste. It is claimed to allay vescial irritability of prostatic hypertrophy. The rheumatic diathesis seems especially amenable to its action. It induces a general feeling of well-being; also quiet and refreshing sleep. (Boericke.)

One of the most common uses of Alfalfa in Chinese medicine is for the treatment of ulcers. Alfalfa is used to strengthen the digestive tract and to stimulate appetite.

Active principles and pharmacology

Leaves contain carotenoids (including among others, lutein); triterpene saponins (Sojasapogenols A-E, aglycones, medicagenic acid, hederagenins); isoflavonoids (including among others, formononetin glycosides, genistein, daidzein); coumarins (coumestrol, 3'-methoxy-coumestrol, lucernol, sativol, trifoliol); triterpenes (including, among others, stigmasterol, spinasterol; cyanogenic glucosides (corresponding to less than 80 mg HCN/ 100 g).

Seeds contain L-canavaine; betaine (stachydrine, homostachydrine); trigonelline and fatty oil. (PDR.)

Seeds gave following values: moisture 11.7, protein 33.2, fat 10.6, N-free extr. 32.0, fibre 8.1 and mineral matter 4.4 %.

Alfalfa is rich in some important nutrients, including calcium, phosphorus, iron, potassium, magnesium, essential enzymes, choline, sodium and silicon.

Alfalfa is a valuable source of vitamin A, B6, B12, D, K, and E. The vitamin P or rutin of Alfalfa builds capillary strength and reduces inflammation of the stomach lining. Vitamin A helps maintain the stomach's health and enzymes present in Alfalfa aid in food assimilation.

Alfalfa contains one and a half times more protein than grains like wheat and corn and its carbodydrate content is only half of that found in grains. Alfalfa proteins contain essential amino acids such as arginine, lysine, theronine and tryptophane.

The enzymes reported in Alfalfa lucerne are amylase, emulsin, coagulase, peroxidase, erepsin, lipase, invertase and pectinase.

A fibre-free concentrate, fit for human consumption, prepared by steeping tender leaves in cold water followed by grinding in a blender, extraction with water and drying under vacuum (yield 13 %) contains: protein 44.2, fibre 0.86, ether extr. 3.55, N-free extr. 30.69, ash 13.2, calcium 1.90, and phosphorus 0.52 %, carotene 110.1 mg, ascorbic acid 51.6 mg and thiamine 1.15 mg/100 g. (The Wealth of India.)

The extracts of Alfalfa leaf containing less than or equal to 1 ppm canavaine and also less than or equal to 9 ppm coumestrol act as anticholesterolemic and hypolipidemic drugs respectively.

In India, hypolipidemic effects of Alfalfa seed extracts (50 % EtOH) have been assessed under experimental conditions in rabbits.

Alfalfa meals prevented hypercholesterolemia, triglyceridemia and atherogenesis in cholesterol-fed rabbits and cynomologus monkey. Alfalfa seed extract feeding induced a significant reduction in various functions of lipid and lipoprotein cholesterol, while the HDL-cholesterol:total cholesterol ratio increased significantly.

Alfalfa-saponins reduced intestinal absorption of endogenous cholesterol in rats and increased bile acid excretion and prevented cholesterolemia in cholesterol-fed monkeys.

Increased serum-cholesterol and LDL-cholesterol were reduced by 38–41.7 % and 48–53.3 % respectively when fed with Alfalfa seed extract from the beginning or in established hyperlipidemic model. LDL-cholesterol lowering was maximum (64.4 %) in a model fed with Alfalfa meals without cholesterol. An increase in HDL-cholesterol:total cholesterol is suggestive of a beneficial role, since it is associated with reduced incidence of atherosclerosis.

Alfalfa, in the form of tea, provides vital alkalizing benefits for hyperacidity. It tends to control the flow of hydrochloric acid, aiding the action of gastric enzyme, pepsin. Alfalfa helps alkalize the food residues in the body.

Aqueous and ethereal extracts of the plant show anti-bacterial activity against Gram-negative bacteria.

Use in Western herbal

Alfalfa is used as a dietary supplement in diabetes, malfunctioning of thyroid gland, in high blood pressure, in certain arterial problems. It is given to convalescents who require easily assimilated nutrients. It has isolated use as a diuretic.

Alfalfa leaf tablets and capsules are available in herb outlets, and are used as a source of nutrients, including vitamins C, D, E, K, folic acid, biotin and several minerals.

While contemporary herbalists generally endorse the traditional view that Alfalfa can treat ulcers, there is no scientific proof of it, nor is there any scientific proof of diuretic property of Alfalfa. Some supplement manufacturers promote Alfalfa tablets for asthma and hay fever, but it has been proved in a study that Alfalfa contains neither bronchodilators nor antihistamines.

Caution

Alfalfa seeds should be consumed only in medicinal quantity. They contain toxic amino acid canavaine, which may cause blood disorders.

Meliaceae

MELIA

Melia azedarach Linn.

Habitat

Wild in sub-Himalayan tract at 700–1000 m, cultivated and naturalized throughout India.

Classical and common names

Ayurvedic: Mahaanimba, Ramyaka, Dreka.
Unani: Bakaayan.
English: Persian Lilac.

Parts used

Leaves, seed-kernel, bark.

Dose

Decoction 50–100 ml, seed powder 1–3 g (7–8 seeds prove toxic)

Figure 1 Melia azedarach—flowering branch [WOI]

Classical use

Sushruta prescribed Mahaanimba fruits internally in indigestion, colic and intestinal catarrh. (Seeds were taken with adjuvants like rice-water and clarified butter.) Ramayak Ghrita of Sushruta was a specific remedy for gout.

Shaarangadhara prescribed a paste of the root for alleviating sciatica; seeds, in prescriptions, for urinary disorders. Ashroghna Vati, a classical compound of the 16th century, was prescribed for piles. Juice of leaves (40 ml) was prescribed as an anthelmintic (Vaidyavallabha).

In folk medicine, a decoction of leaves is regarded as astringent and stomachic; a poultice of flowers is applied to eruptive skin diseases, and for killing lice; an infusion of bark is given in ascariasis; also as a blood-purifying aid.

Active principles and pharmacology

Leaves gave the two flavonoids quercitrin and rutin; and the tetranortriterpenoids salannin and vilasinin.

Fruit and heartwood yielded a bitter principle bakayanin and a lactone backalactone respectively.

The bitter constituents are present exclusively in the pericarp, not in the kernel, as in the case of Neem fruit. (Bakayanin has been isolated from the pericarp.)

Inner bark is bitter, but not astringent; exhibits anthelmintic activity. The outer bark is astringent, and is the seat of tannins. The bark also contains the alkaloid azaridine and paraisine.

In addition to bakalactone, a liquid with terpenic odour, a resinous material and tannins have been isolated from the heartwood. Aqueous extracts of the heartwood give good results in asthma.

Aqueous and alcoholic extracts of leaves showed significant anthelmintic activity against tapeworms and hookworms, the aqueous extract being more effective.

An unsaponifiable matter obtained from fixed oil of seeds showed antibacterial action against Staph. aureus, Xanth. citri, Proteus spp.; E. coli and B. subtilis in particular. It was inactive against Klebsiella spp., Sh. negater and Ps. aeruginosa. The stem bark extract (50 % ethanolic) did not show any antibacterial, antifungal and antiprotozoal activities.

50 % ethanolic extract of the stem bark was also devoid of hypoglycaemic, CNS and antifertility activities.

Papilionaceae

MELILOTUS

Melilotus officinalis (L.) Medik. ex Desr.

Habitat

Melilotus alba Linn., the white-flowered variety, is cultivated for fodder in Punjab and Uttar Pradesh; to a small extent in Madhya Pradesh and Coimbatore. Melilotus parviflora Desf. syn. Melilotus indica All., the yellow-flowered variety, is cultivated in Punjab and Uttar Pradesh. Melilotus officinalis; also a yellow-flowered variety, is found in Ladakh and Nubra at altitudes of 3000–4000 m.

Classical and common names

Unani: Iklil-ul-Malik. Asaab-ul-Malik; Naakhunaa.
English: Sweet Clover, Melilot.

Figure 1 Melilotus alba [WOI]

Parts used

Dried or fresh leaf and flowering branches, beans.

Dose

3–4 g.

Classical use

In Ayurvedic reference books, Melilotus parviflora is equated with Methikaa, Banamethikaa, Bahupatrikaa, Bahubijaa. Properties of Trigonella foenum-graecum have been attributed to M. parviflora. Methikaa is a different herb and is no way related to Melilotus spp.

In Unani medicine, Iklil-ul-Malik is a discutient, styptic, emollient, anodyne, astringent and diuretic. Its decoction is given in inflammatory conditions of liver, spleen, intestines, uterus, and rectum, also in hemiplegia; paste, poultice or plaster is applied externally. In Unani practice, it is considered closer to Matriacaria chamomilla than to Trigonella spp.

Active principles and pharmacology

Plant gave coumarins; hydroxycoumarins (including among others umbelliferone, scopoletin, herniarin, fraxidin); flavonoids (including among others kaempferol and quercetin glycosides); triterpene saponins. Seeds are narcotic. (PDR.)

The drug exhibits anti-oedematous properties, which explain its use for inflammatory and congestive oedema. It increases venous reflux and improves lymphatic kinetics.

In a study in Japan, intraperitoneal injection of Melilotus extract (ME) immediately after burn greatly reduced the amount of swelling and effectively inhibited the occurrence of necrosis and induration in the injured leg-skin of the rat, as compared with the saline controls in which a third degree of thermal injury was observed.

Furthermore, either intraperitoneal or subcutaneous local injection of ME four hours before burn was effective in reducing the oedema and thermal injury.

Use in Western herbal

Nicholas Culpeper (1616–1654) wrote of it: "Melilot, boiled in wine, and applied, mollifies all hard tumours and inflammations—swelling tumours in the spleen or other parts of the body. The head often washed with the distilled water of the herb and flower, or a lye made therewith, is effectual for those that suddenly lose their senses, as also to strengthen the memory, to comfort the head and brain, and to preserve them from pain, and the opoplexy."

The Irish herbalist K'Eogh reported in 1735:"A woman had a swelling for a year or more on her right side. It was cured by three or four times rubbing the grieved part with the oil made of Melilot."

German Commission E recognized the efficacy of Melilotus officinalis in:

- Blunt injuries
- Haemorrhoids
- Venous conditions.

The drug is used internally for problems arising from chronic venous insufficiency, such as pain and heaviness in legs, night cramps in legs, swelling, and for the supportive treatment of thrombophlebitis, post-thrombotic syndrome, haemorrhoids and lymphatic congestion. Externally the drug is used for contusions and superficial effusions of blood. (PDR).

Melilot helps reduce the risk of phlebitis and thrombosis.

The plant is also given for insomnia and anxiety for its mild sedative and antispasmodic properties.

Commuted drug for infusion; ointments, liniments, cataplasms and herbal sachets for external use; ointments and suppositories for rectal use are available in herb stores.

Caution

Contraindicated with anticoagulants. The potential bishydroxycoumarin-content may produce prolonged prothrombin and coagulation time.

Administration of the drug in higher doses lead to headache; transitory liver damage is possible.

Avoid using in combination with salicylates. (Francis Brinker.)

Lamiaceae

MELISSA

Melissa officinalis Linn.
Melissa parviflora Benth. (Indian spp.)
Melissa axillaris (Benth.) Bakh. f.

Habitat

Temperate Himalayas from Garhwal to Sikkim, Darjeeling and Khasi, Aka and Mishmi hills, at altitudes of 1200–3000 m.

Classical and common names

Unani: Baadranjbuuyaa, Billilotan. (According to the National Formulary of Unani Medicine.) Nepeta cataria Linn. and Nepeta hindostana Haines. (bearing blue flowers) are also known as Billilotan. *English:* Mountain Balm, Sweet or Lemon Balm. Nepeta cataria is known as Catmint or Catnip.

Parts used

Whole plant.

Figure 1 Nepeta hindostana [WOI]

Dose

5–7 g.

Classical use

In Unani medicine, Baadranjbuuyaa is used as a mood elevator; also as a sedative and nervine tonic in paralysis, hemiplegia, epilepsy. Its paste is applied externally on rheumatic joints. Its squash or liquid extract is prescribed as an expectorant and soothing agent in cough and bronchitis. Its decoction is used as a mild diaphoretic in fevers and for allaying nervousness and melancholia during protracted illness.

Baadranjbuuyaa is an ingredient in Khameera-e-Gaozaban (Qarabadeen-e-Sarkari), a reputed Unani tonic for melancholia, palpitation and debility. Many Unani tonics, which produce a sense of well-being and exhilarition, contain Baadranjbuuyaa. Muffareh-e-Azam and Muffareh-e-Dilkusha (Qarabadeen-e-Jadeed) are prescribed for nervous debility, neurasthenia and insomnia; Mufarreh Sosambari (Bayaz-e-Kabir) and Mufarreh Yaqooti (Qarabadeen-e-Azam-o-Akmal) are prescribed as tonics for brain, heart and liver in inflammatory conditions and asthenia.

Active principles and pharmacology

Essential oil contains geranial (citral a), neral (citral b), 6-methyl-5-hepten-2-one, citronellal, geranyl acetate, beta-caryophyllene and beta-caryophyllene oxide as chief constituents.

Aerial parts contain flavonoids (including, among others, cynarosides, cosmosiin, rhamnocitrin, isoquercitrin), glucosides, caffeic acid derivatives, triterpene acids, polyphenols, tannins.

Research has shown that the volatile oil and, in particular, citral and citronellal calm the central nervous system. The oil also exhibited strong antispasmodic activity. Polyphenols contribute to the antibacterial, antiviral activity of the drug and make it an effective agent for combating the Herpes simplex-virus, which produces cold sores. In a study, the average healing time of cold sores was halved to about 5 days and the time between outbreaks doubled.

The hydroalcoholic lemon balm extract exhibited CNS-sedative activity in animal studies.

Anti-thyroid activity of the herb has also been reported.

Use in Western herbal

John Evelyn (1620–1706) wrote of Lemon Balm: "Balm is sovereign for brain, strengthening the memory and powerfully chasing away melancholy."

German commission E recognized the efficacy of Melissa officinalis in:

- Nervous sleeping disorder
- Functional gastrointestinal complaints.

The decoction of the drug is utilized for nervous complains, lower abdominal disorders, nervous gastric complaints, hysteria, melancholia, chronic bronchial catarrh, nervous palpitation, nervous debility. Externally lemon balm creams are used on insect bites, sores, and slow-healing wounds. The essential oil is used in aromatherapy.

Lemon balm is also an ingredient in throat lozenges and cough drops.

Caution

Melissa officinalis should not be confused with Nepeta cataria (PDR). A major constituent of Nepeta cataria is neptalactone (80–95 %), which is quite similar in its chemical structure to the valepotriates, the sedative principle of valerian root. Mice given Nepeta cataria extract experienced a reduction of overall activity and an increase in their sleeping time. (John Heinerman.)

An active principle of the herb, *cis*-transneptalactone produces stimulatory responses in cats only when it is smelled, not when it is administered internally (Tyler). For this, Nepeta cataria is known as Catnip.

The Indian herbs Baadranjbuuyaa or Billilotan do not attract cats nor does their aroma stimulate them.

Melissa officinalis and Nepeta cataria should be treated as two different herbs.

Lamiaceae

MENTHA

Mentha spicata Linn.
Mentha arvensis Linn.

Habitat:

Cultivated in the plains of Punjab, Uttar Pradesh, Maharashtra and Delhi.

Mentha arvensis L., Japanese Mint, is cultivated in Jammu & Kashmir, Uttar Pradesh and Punjab at 270–1500 m.

Classical and common names

Ayurvedic: Pudinaa, Podinaka, Puutihaa, Rochini.
Unani: Naanaa; Pudinaa Kohi.
English: Spearmint.

Classical use

The drug is not mentioned in classic Ayurvedic texts.

Araq-e-Naanaa (Bayaz-e-Kabir) is a single drug preparation (distillation of Mentha spicata) of Unani medicine, prescribed in flatulence, stomachache, dyspepsia, nausea, vomiting, cholera. Araq-Ajeeb (Bayaz-e-Kabir) contains camphor, Pudinaa and Ajwaayin (Trachyspermum ammi), is prescribed in stomachache, colic, diarrhoea, cardalgia, catarrh, migraine.

Figure 1 Mentha arvensis [ZANDU]

The extract of Mentha spicata is an ingredient in many commercial compounds sold over-the-counter, for digestive disorders.

Pudinaa Barri is equated with Mentha sylvestris Linn., Pudinaa Kohi with Mentha spicata and Pudinaa Nahari with Mentha aquatica in Unani medicine. Pudinaa Kohi is an ingredient in Majoon-e-Fotnaji (Bayaz-e-Kabir), prescribed in flatulence, stomachache, dyspepsia, hepatalgia, chronic fevers.

Active principles and pharmacology

Major constituent of the essential oil is carvone; other constituents include L-limonene, dihydrocarvone, carvomenthone, iso-menthone and dihydrocarveylacetate. Leaves gave the flavonoids diometin-7-glucoside, diosmin, diosmetin-7-O-beta-D-glucuronide and luteolin-3-O-beta-D-glucuronide.

Leaves and flowering tops exhibit stimulant, carminative, antispasmodic, nervine, antiemetic, antihysteric properties.

Use in Western herbal

Spearmint (Mentha spicata) is considered much gentler than peppermint (Mentha piperita) and is used for minor problems, such as wind, indigestion, children's fever and stomach upsets.

Mentha arvensis var. piperascens is used in common cold, cough, bronchitis, inflammation of the mouth and pharynx and for liver and gallbladder complaints; externally for myalgia and neurologic ailments.

Over-the-counter drug Nervous Dyspepsia Tablets (Frank Roberts) contain spearmint.

Peppermint is used in a number of digestive remedies. Peppermint Oil capsules and tablets (50 mg, 200 mg, 250 mg) are readily available over-the-counter. The oil is included in a number of products for catarrh and colds. Rubs for rheumatism and creams for skin problems are also available.

Clusiaceae

MESUA

Mesua ferrea auct. non Linn.

Figure 1 Mesua ferrea [CCRAS]

Habitat

Himalayas from Nepal eastwards, in north-eastern India, Deccan Peninsula and the Andaman Islands.

Classical and common names

Ayurvedic: Naagakeshara, Naagapushpa (Sushruta); Chaampeya, Naaga, Naagakinjalka, Ahipushpa.
Unani: Naaremushk (flower, stamens).
English: Mesu.

Parts used

Stamens of flowers.

Dose

Powder 1–3 g.

Classical use

Charaka prescribed Naagakeshara, butter and sugar for bleeding haemorrhoids.

Sushruta prescribed stamens of flowers internally in haemoptysis, blood poisoning, chronic dysentery, diarrhoea with blood, toxic conditions, skin eruptions, and for purifying breast milk, for treating hiccough and spasmodic affections. As a tonic, the drug was included in a medicated liquor; as an antiseptic and healing agent in an ointment for ulcers, sinuses and swellings.

For leucorrhoea, Naagakeshara was prescribed with buttermilk; for conception, powder of Naagakeshara and Areca nut was administered (Bangasena). For conception, powdered Naagakeshara with cow's Ghee (clarified butter), during the season, was also prescribed (Gadanigraha).

Among over-the-counter compounds, Kanakaasava (Bhaishajya Ratnaavali), Elaadi Churna (Gadanigraha), Jaatiphalaadi Churna (Shaarangadhara Samhitaa), Karpuraadi Churna (Sahasrayoga) are prescribed in coughs, bronchitis and inflammations of throat and chest. Bhaaskar Lavana Churna (Shaarangadhara Samhitaa) is a reputed digestive, carminative and appetizer. Soubhaagya Shunthi (Ratnaakar Samhitaa) is an efficacious uterine tonic after delivery.

Mesua ferrea is an ingredient in Yelaathi Churanam of Siddha medicine, prescribed internally for chancres, ulcers, leprosy.

Unani physicians soak stamens of flowers (3–5 g) in water overnight and administer with purified sugar candy or honey for bleeding piles.

Jawarish-e-Naaremushk is prescribed in liver and intestinal diseases.

Ayurvedic and Unani prescriptions containing Mesua ferrea are administered in cases of uterine bleeding.

Active principles and pharmacology

Stamens gave alpha- and beta-amyrin, beta-sitosterol, biflavonoids-mesuaferrones A and B, and mesuanic acid.

Seed oil yielded several 4-phenylcoumarin analogues: mesuol, mesuone, mammeigin, mesuagin.

Mesuaxanthones A and B and a xanthone extracted from heartwood exhibited anti-inflammatory, CNS-depressant and antimicrobial properties.

Mesuol and mesuone were found to inhibit Staph. aureus, E. coli, Eberthella typhosa, Vib. cholerae, B. friendlanderi and M. phlei; were most active against Staph. aureus. The oil showed antibacterial activity against B. subtilis, Vib. cholerae, Staph. aureus, Staph. albus, Sh. dysenteriae, Sal. typhi, Sar. lutea, Strep. faecalis, B. pumilus, Strep. pyogenes and Ps. solanacearum.

The essential oil from fruits showed antifungal activity against Trichophyton terrestre, Trichophyton tonsurans, Trichophyton mentagrophytes, Histoplasma capsulatum (maximum inhibition). The essential oil from the stamens showed activity against C. tropicalis; also anthelmintic activity against tapeworms and hookworms.

The ethanolic extract (50 %) of the whole plant (excluding roots) showed hypotensive action in cat/dog, antispasmodic activity in isolated guinea-pig ileum and diuretic activity in rat.

Phenolic constituents of the seed oil revealed a potent antiasthmatic effect.

Methanol extract of flowers caused sedation in mice; partially blocked nicotine-induced convulsion in mice. Acetone extract of flowers produced relaxation of guinea-pig ileum. Acetone extract of stamens exhibited beta-blocking activity which was exemplified by potentiation of alpha-effects of adrenaline, markedly decreased duration of isoprenaline effects; exhibited antispasmodic activity on guinea-pig ileum. The seed oil was found to potentiate bronchodilator activity of isoprenaline both in vitro and in vivo. (CCRAS.)

Magnoliaceae

MICHELIA

Michelia champaca Linn.

Figure 1 Michelia champaca [CCRAS]

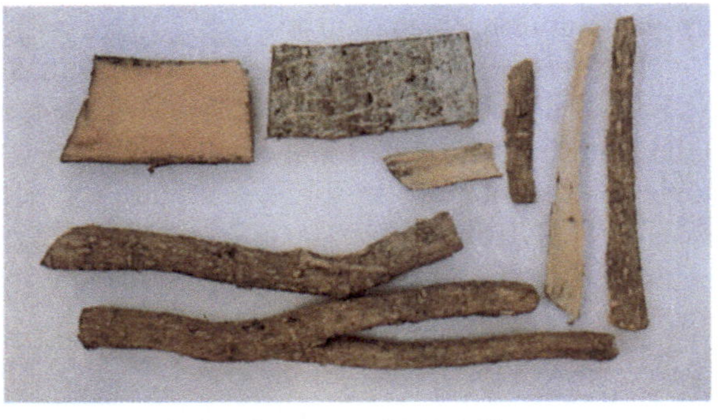

Figure 2 Michelia champaca [CCRAS]

Habitat

Eastern Himalayas, lower hills of Assam and hills of South India, up to 1000 m; cultivated in various parts of India.

Classical and common names

Ayurvedic: Champaka (Sushruta), Svarna Champaka, Hemapushpa, Chaampeya. (yellow-flowered variety). Three more varieties have been mentioned in Ayurvedic texts: Kshira Champaka, equated with Plumeria acuminata Ait., Apocynaceae; Panasagandhi Champaka, equated with Artabotrys hexapetalus (Linn. f.) Bhandari, Annonaceae; Bhuuchampaka, equated with Ochna pumila Buch-Ham ex D. Don., Ochnaceae.
Unani: Champaa.
Siddha: Sambangi.
English: Golden Champa.

Parts used

Flowers, leaves, bark.

Dose

Decoction 50–100 ml, powder 3–6 g.

Classical use

Sushruta prescribed flowers internally in haemoptysis and excessive bile secretion; also as an antidote to toxic manifestations.

Leaves of Champaka were included in a vaginal pessary recommended for treating vaginal foul smell and infections (Gadanigraha).

During the 16th century, Champaka was used for its astringent, disinfectant and cooling properties in poisoning, parasitic infections, dysuria and diseases due to vitiated blood. It was included in the Sarvasugandhi group of herbs, prescribed in psychoneurosis, fevers, nervousness, and skin diseases.

In folk medicine, an infusion or decoction of flowers is given in dyspepsia, nausea and fevers; also for preventing scalding in renal diseases. Flowers, macerated in oil, are applied externally in cephalalgia, opthalmia, nasal infections, sinus, rheumatism, gout and vertigo.

Juice of leaves is given with honey in colic. Leaves are also applied to indolent swellings.

A decoction of the bark is prescribed in mild gastritis. A paste of dried root or rootbark is applied to abscesses. (Bark is emetic and purgative and should be used internally with caution.)

Mahaalaakshaadi Tailams (Bhaavaprakaasha) is available over-the-counter and is used for massage in fevers, disturbed sleep and general debility.

Active principles and pharmacology

Essential oil from flowers gave a sesquiterpene hydrocarbon, champacene and the leaves gave linalool, linalylacetate, methylheptenone and geraniol.

Essential oil from the fruit rind gave phenylethyl-alcohol, pinocamphone, phenylethyl-acetate, 1,8-cineole, linalool, pinocampheol, alpha-phellandrene and ester of pinocampheol.

Leaves gave a polyisoprenoid and beta-sitosterol and its glucoside and liriodenine.

Stembark and roots yielded an alkaloid liriodenine. Rootbark yielded sesquiterpenes lactone—costunolide, parthenolide, dihydroparthenolide and micheliolide.

Seeds gave the fatty acids myristic, palmitic, hexadecadienoic, stearic, oleic, linoleic, arachidic; eicosenoic and hexadecenoic acids.

The ethanolic extract of the stembark showed hypoglycaemic action in rats and hypertensive effect in cat and dog. The extract did not show antibacterial and antifungal activity, Michelia nilagirica, a related spp., showed a diuretic action in rats and antispasmodic activity on isolated guinea-pig ileum.

Dried root and root bark exhibit purgative and emmenagogue activity.

Mimosaceae

MIMOSA

Mimosa pudica Linn.

Figure 1 Mimosa pudica [ZANDU]

Habitat

Native of tropical America. Found in gardens and fields throughout India as a hedge.

Classical and common names

Ayurvedic: Lajjaalu, Laajavanti, Namaskaari, Samangaa, Samkochini, Shamipatraa, Khadirkaa, Raktapaadi. Alambushaa is considered a related spp. of Lajjaalu, provisionally equated with Biophytum sensitivum (Linn.) DC., Geraniaceae, also with Mimosa netans Roxb., and Neptunia oleracea Lour.

Siddha: Thottal chinungi.

English: Sensitive Plant.

Parts used

Whole plant, leaves, root.

Dose

Juice 10–20 ml, decoction 50–100 ml.

Classical use

Charaka and Sushruta prescribed a decoction, with Samangaa as an important ingredient, in haemothermia, piles, diarrhoea, persistent dysentery. Included in an ointment, the herb was applied over piles, ulcers and wounds.

During the 16th century, Lajjaalu was a popular herb for treating piles and diseases of the female genital tract.

Samangaadi Churna (Ayurvedic Formulary of India) is available over-the-counter, prescribed internally in bleeding piles.

Active principles and pharmacology

Fresh tissues gave nor-epinephrine, D-pinitol (3-mono-Me-ether of inositol) and beta-sitosterol. Leaves contained alkaloids.

Decoction of leaves of Mimosa pudica in doses of 250, 500, 1000 and 2000 mg/kg in rats and dogs exhibited diuretic activity (considering urinary output, Na^+-, K^+- and Cl^--excretion). The activity in rats at 250 mg/kg dose was found to be 82 % of the standard diuretic (hydrochlorthiazide 2.5 mg/kg) treated group of rats. There was significant reduction (above 50 %) of Na^+- and Cl^--excretion without affecting K^+-excretion. The drug can be combined as a moderate diuretic with any modern synthetic diuretic causing K^+-loss. (CCRAS.)

Ethanol extract (50 %) of the whole plant exhibited spasmogenetic activity in isolated guinea-pig ileum.

An extract, administered in a dose of 1.6 mg/100 g parenterally every 4th day up to 120 days in rats having experimental injury of sciatic nerve, exhibited 30 to 40 % higher results in the process of regeneration of sciatic nerve as compared to hydrocortisone treated group.

Aqueous extract of root powder in pilot studies on patients with dysfunctional ulerine bleeding gave promising results (CCRAS.)

Sapotaceae

MIMUSOPS

Mimusops elengi Linn.

Figure 1 Mimusops elengi [CCRAS]

Habitat

Cultivated in North and Peninsular India and the Andaman Islands.

Classical and common names

Ayurvedic: Bakula, Keshara, Simhakeshara, Sthiraa, Vishaarada, Dhanvi, Madhupushpa (Charaka, Sushruta); Madhugandha, Chirpushpa; Maulsiri. Keshara is a misleading synonym of Bakula. In Ayurvedic medicine, it is equated with Crocus sativus, Nelumbo nucifera, Mesua ferrea and Osmanthus fragrans.

Figure 2 Mimusops elengi [CCRAS]

Siddha: Magilam.

English: West India Medlar, Spanish Cherry, Bullet Wood.

Classical use

Charaka prescribed extract of the bark and root with honey or in a medicated Ghee (clarified butter) in helminthiasis and fevers. Sushruta gave flowers internally in coughs and bilious derangement; as an ingredient of a medicinal liquor in diseases of the urinary tract.

A decoction of rootbark was prescribed as a gargle. Seeds were also used. (Vrindamaadhava, Raaja Maarttanda, Gadanigraha, Vaidyamanoramaa.)

The bark powder is an ingredient in a number of commercial tooth powders, prescribed in diseases of gums and teeth. An infusion of the bark, in folk medicine, is prescribed internally in diseases of the bladder and urethra.

Fruits and flowers, in the form of a lotion, are applied to wounds and ulcers. Powder of dried fruits, when sniffed, produces copious discharge from nose, is given in headaches. A paste of bark is applied externally to relieve headache.

In folk medicine, bark is considered to promote fertility in women, but this claim could not be validated scientifically. (The saponins, steroids and triterpenoids from the seed showed spermicidal activity.)

Active principles and pharmacology

The bark contained an alkaloid consisting largely of a tiglate ester of a base with a mass spectrum identical to those of laburinine and iso-retronecanol and a saponin which on hydrolysis gave beta-amyrin and bassic acid.

Beta-sitosterol and its glucoside, alpha-sipnasterol, quercitol, teraxerol and lupeol and its acetate are present in the aerial parts as well as in roots and seeds. The aerial parts also gave quercetin, dihydroquercetin, myricetin glucosides, hederagenin, ursolic acid, hentriacontane and beta-carotene.

Seed oil comprises capric, lauric, myristic, palmitic, stearic, arachidic, oleic and linoleic acids.

Benzene and methanol extracts of the plant exhibited significant antipyretic effect in febrile rats.

The saponins from seeds showed spasmolytic activity on isolated smooth muscle preparation. Pre-treatment of the drug effectively blocked ACh, histamine and barium chloride. The antihistaminic property was most significant. (CCRAS.)

Saponin from seeds has been reported to have spermicidal activity at a dilution of 0.06 per cent in human semen.

Ethanolic extract (90 %) of the fruits and leaves showed hypotensive effect in dogs; 50 % extract of the whole plant, excluding roots, showed diuretic action in rats.

The leaf extract showed antibacterial activity in vitro against B. anthracis, B. mycoides, B. pumilus, B. subtilis, Sal. paratyphi, Staph. albus, Vib. cholerae, Xanth. campestris and Xanth. malvacearum; the inhibition being significant against Xanth. campestris and B. anthracis.

Cucurbitaceae

MOMORDICA

Momordica charantia L.

Habitat

Met with as an escape up to 1500 m. Cultivated all over India.

Classical and common names

Ayurvedic: Kaaravellaka, Kaaravella, Kaathilla, Sushaavi.
Unani: Kerelaa.

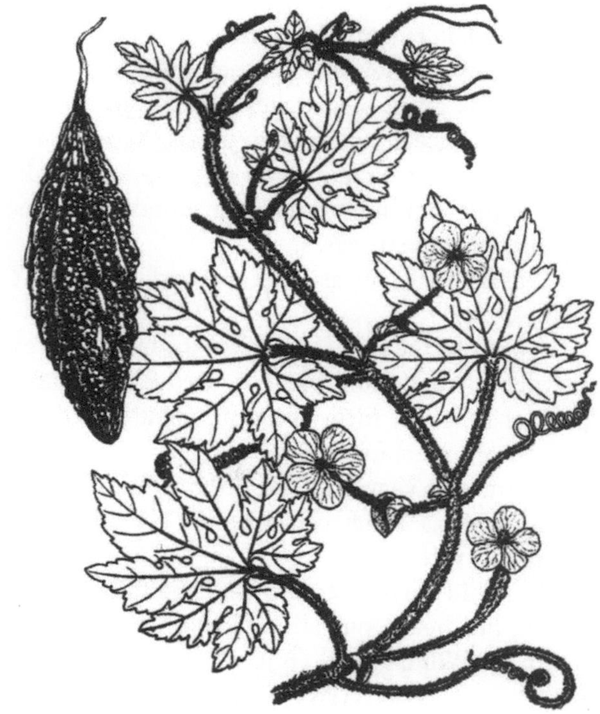

Figure 1 Momordica charantia—flowering branch and fruit [WOI]

Siddha: Pagal.
English: Bitter Gourd, Balsam Pear, Carilla.

Parts used

Fruits, leaves.

Dose

Juice 10–20 ml.

Classical use

Charaka prescribed a decoction of fruits and leaves, in prescriptions, internally in haemothermia and cough. Sushruta prescribed expressed juice of the fruits internally in toxicity, as an antiseptic and purgative; fruits as a potherb in the form of a diet in fevers, hiccup, cough, urinary diseases, skin diseases, gout.

Juice of the leaves, mixed with the powder of Curcuma longa (Haridraa) was prescribed in measles with fever, smallpox and other eruptions (Vrindamaadhava).

Fruits, after the upper skin, ridges and tubercles are removed, are kept in salt water and then boiled and squeezed, before they are cooked as a vegetable, which is considered useful in diabetes, diseases of liver and spleen, diseases due to vitiated blood, in rheumatism and gout.

As a popular household remedy, the juice of the raw fruit is given in diabetes. (See Active Principles.) The juice is supposed to possess blood-purifying properties. Practitioners of traditional medicine administer it internally in jaundice, piles, leprosy and skin diseases.

Leaf juice is given as an emetic and purgative in biliousness.

A paste of the root is applied externally to piles. Powder of the whole plant is used as a dusting powder or in the form of an ointment on leprous and other intractable ulcers.

Kaaravelli, Vaarivelli and Brihatvalli have been mentioned as other varieties of Kaaravella in Ayurvedic texts.

The wild variety, known as Karkota or Jangali Kareli, has been equated with Momordica dioica Roxb. ex Willd. (Small Bitter Gourd). The root, with milk, was prescribed for expelling calculi and gravels (Gadanigraha). It was also prescribed as an urinary tract disinfectant to women by Ayurvedic practitioners.

Active principles and pharmacology

The seed gave triterpene glycosides, named momordicosides A, B, C, D and E. Leaves and vines gave the tetracyclic triterpenes momordicines I, II, and III (bitter principles). Immature fruits gave several non-bitter and two bitter cucurbitacin glycosides.

Fruits, seeds, tissue culture gave a polypeptide which contained 17 types of amino acids. Fruits also gave 5-hydroxytryptamine and a neutral compound charantin (a steriodal glucoside), diosgenin, cholesterol, lanosterol and beta-sitosterol.

Momordica charantia fruit powder in a single dose of 100 g daily for two weeks was tried on normal controls and 25 patients of diabetes mellitus. The investigations did not observe any hypoglycaemic effect either in controls or in diabetic patients.

It was observed that Momordica charantia extract leads to a false negative test for sugar, thus misleading diabetic patients and physicians. (ICMR.)

Chronic administration of Momordica charantia fruit extract (1.75 g day for 60 days) to dogs led to testicular lesions with mass atrophy of spermatogenic elements, through the extract did not affect other biological parameters concerned with liver function and lipid metabolism. (ICMR.)

Only the seed powder exhibited hypoglycaemic activity due to charantin (momordicin). It exhibited pancreatic and extra pancreatic action.

Hypoglycaemic activity in powdered seeds of Momordica charantia was tested against Streptozotocin-induced diabetes mellitus in rabbits (1–3 g/day) and was found to be comparable to that of glybenclamide and positively associated with lipogenesis. Momordica charantia seeds, however, did not improve the renal function impairment caused by streptozotocin.

Charantin led to prolonged hypoglycaemia in varying doses in normal fasting rabbits, the fall being gradual and steady for 4 h and then recovering slowly. Charantin given in dose of 50 mg/kg orally lowered the blood sugar by 42 % at the 4th hour, the mean fall during 5 h being 28 %. The cumulative hypoglycaemic potency curve was not linear but tended to flatten with higher doses of charantin. In depancreatised cat the hypoglycaemic effect of charantin was considerably less than in non-depancreatised cats. The result suggested a pancreatic as well as extrapancreatic action of charantin.

The action of Momordica charantia seeds seems to be due to increased insulin release from the pancreas. There was associated hypolipaemic effect and a loss of body weight which could be attributed to the breakdown of stored protein to utilize more glucose.

Juice and extracts from Momordica charantia fruits showed no appreciable hypoglycaemic property on normal or alloxanised rabbits; on the contrary, 10 mg/kg p.o. of juice was found toxic to the experimental animals. But the fruits of smaller variety (Kareli) exhibited effectiveness on glucose tolerance test, without showing any toxicity. (CCRAS.)

All the rats which received M. charantia juice 15–40 mg/kg i.p. became sluggish within one hour followed by death in 6–18 hours. (CCRAS.)

Oral administration of alcoholic extracts of the plant did not produce any hypoglycaemic action (The Wealth of India).

The unsaponfiable matter obtained from the seed oil of Momordica charantia exhibited pronounced antibacterial activity against Gram-negative strains; it was completely devoid of any inhibitory action against Gram-positive organisms. The seed oil exhibited no antifungal activity.

Use in Western herbal

Momordica balsamina Linn. a related spp., is used for piles, burns, chapped skin in the form of a liniment prepared by adding the pulped fruit (without the seeds) to almond oil. The pulp is also used as a poultice. The fluid extract is used for dropsy.

Momordica balsamina is found in India in Gujarat, Rajasthan, Punjab, Uttar Pradesh, up to 300 m. It is known as Jangali Karelaa in Punjab, and possesses hypoglycaemic activity. (CIMAP.)

The seeds of Momordica cochinchinesis Spreng. are applied as a poultice to abscesses, haemorrhoids and scrofula. A paste of seeds gives encouraging result in psoriasis and ringworm.

Momordica charantia ripe fruit is given for inducing menstruation. Seeds were tested in China as a contraceptive in the 1980s. Research suggests that the plant may prove harmful to liver.

Moringaceae

MORINGA

Moringa oleifera Lam.

Habitat

Indigenous to the sub-Himalayan tracts from Chenab to Uttar Pradesh; cultivated throughout India.

Classical and common names

Ayurvedic: Shigru, Sigra, Shobhaanjana, Haritashaaka, Raktaka (Charaka); Murangi (Sushruta); Mochaka, Akshiva. The white variety is known as Shigru, the red one as Madhu-Shigru, equated with Momordica concanensis Nimmo. The blue variety, mentioned in Raaja Nighantu, still remains unidentified.
Unani: Sahajan.
Siddha: Murungi.
English: Drumstick, Horse-Radish.

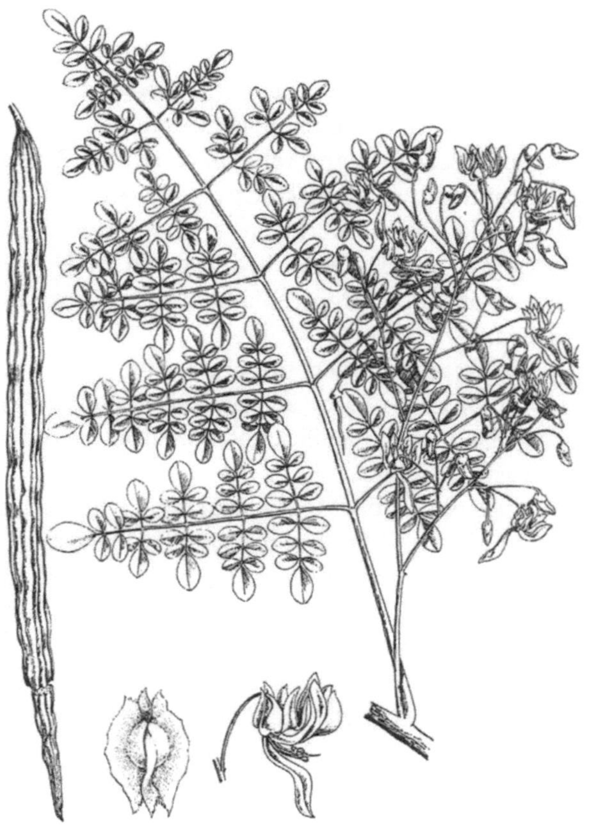

Figure 1 Moringa oleifera [ADPS]

Parts used

Leaves, flowers, seeds, fruits, bark.

Dose

Decoction 50–100 ml, powder 1–3 g.

Classical use

Charaka used the powdered bark, root and dried sap, in prescriptions or in steaming mixtures, in oedema, anosmia, asthma, fainting, chronic skin eruptions, painful piles.

Sushruta gave the drug internally in migraine, abdominal swelling, dropsy, for reducing obesity; for external application on nonsuppurating boils, scrofula and leprotic wounds.

The juice of Shigru root with honey was prescribed for treating internal abscesses (Vrindamaadhava).

During the 16th century, Shigru was prescribed in gout, obesity, headache, eye diseases, and as a digestive stimulant. Its principal use was in oedema.

In folk medicine, leaves are prescribed internally in catarrhal affections (emetic in high doses). A decoction of the root is used as a gargle.

Leaves, flowers, immature capsules are eaten as vegetables, said to give relief in cough that lingers after influenza. Young unripe pods (drumsticks) are cooked with curries as a preventive against intestinal worms. Seeds are given in case of an ascites resulting from enlargement of liver and spleen.

Unani practitioners prescribe soup of the pods in sub-acute cases of enlarged liver and spleen, articular pains, debility, neurological disorders and skin diseases.

Juice of the rootbark is prescribed with milk as a diuretic, antilithic and digestive; its decoction is given in enlarged spleen or liver, internal inflammations, calculous affections, hiccup, asthma, rheumatism, gout, lumbago.

Poultice of the leaves, also of the rootbark, is used for reducing glandular swellings (as a blistering agent).

Oil of the seeds is massaged in gout and acute inflammations.

Gum rubbed with milk and made into a paste is applied to temples in headache; also in rheumatism.

Active principles and pharmacology

Leaves gave a number of amino acids: aspartic acid, glutamic acid, serine, glycine, threonine, alpha-alanine, valine, leucine, iso-leucine, histidine, lysine, arginine, phenylalanine, tryptophan, cystine and methionine and alpha- and beta-carotene.

Flowers gave kaempferol, rhamnetin and their-3-glycosides, quercetin-3-glycoside.

Seeds and roots yielded 4-(alpha-L-rhamnosyloxy)-benzoyl-isothiocyanate, and the glycoside moringyne. Roots yielded benzoyl isothiocyanate.

Gum gave aldotriouronic acid.

Stem gave 4-hydroxymellein, vanillin, beta-sitosterone, beta-sitosterol and octacosanoic acid. Rootbark gave moringine, bark the triterpenenoid baurenol.

Alcoholic extract of the rootbark administered in doses of 500–1000 mg/kg p.o. in rats exhibited anti-inflammatory activity against formaldehyde and cotton pellet granuloma pouch. It also showed analgesic activity. The aqueous extract of the bark did not show effect against formaldehyde-induced arthritis in albino rats. The alcoholic extract of leaves did not show hepatoprotective property.

Moringa oleifera, in the form of a topical application in tuberculoid leprosy with uncomplicated trophic ulcers, did not show any significant improvement even after 8 months.

The juice from the leaves and stem bark inhibited Staph. aureus but not E. coli. Antitubercular activity was reported against M. phlei and M. 607. The ether extract of leaves showed bacteriostatic activity against Staph. aureus and Sal. typhosa. The fruit extract did not inhibit E. coli. The 50 % ethanolic extract of the fruit and root bark was devoid of any antibacterial activity.

The bark extract showed antifungal activity against Microsporum gypseum, Trichophyton mentagrophytes, Candida albicans and Helminthosporium sativum.

Papilionaceae

MUCUNA

Mucuna pruriens (Linn.) DC.

Mucuna prurita Hook.

Habitat

Throughout India including the Andaman and Nicobar Islands.

Classical and common names

Ayurvedic: Aatmaguptaa (Charaka, Sushruta); Rshabhi, Kapikacchuu, Adhigandhaa, Ajadaa, Kacchuraa, Laanguli, Rshyaproktaa, Svaguptaa, Shyaamguptaa, Markati, Kanduraa, Kevaanch, Shuukashimbi.
Unani: Konchh.
Siddha: Poonaikkali.
English: Cowhage.

Parts used

Seeds.

Dose

Powder 3–6 g.

Figure 1 Mucuna pruriens [ZANDU]

Classical use

Charaka gave an extract of seeds and root in muscular atrophy, facial paralysis and debility.

Sushruta prescribed seeds internally in respiratory trouble and wasting diseases; externally as an ointment for wounds. Agastya Rasaayan of Charak was specific for chronic bronchitis and asthma. Sushruta's confection of Aatmaguptaa seeds and fruits of Tribulus terrestris (Gokshura) was a reputed sex tonic. Vaanari Vatikaa (Bhaavaprakaasha), a confection of decorticated seeds of Aatmaguptaa, was prescribed in spermatorrhoea.

Ayurvedic sex tonics—Shataavaryaadi Churna (Yoga Ratnaakara), Mushali Paaka (Yoga Chintaajmani) and Dhaatu Paushtika Churna (Ayurveda Saar Sangraha) and Unani sex tonic—Majoon-e-Bandkushad (Qarabadeen-e-Azam-o-Akmal) contain Mucuna pruriens seeds.

In folk medicine, the seeds, as well as the root, are prescribed in diseases of the nervous system, facial paralysis, hemiplegia, and arthritis. Root is given for delirium in fevers. Root, made into an ointment, is used for elephantiasis.

Active principles and pharmacology

Seeds contain L-DOPA (yield 1.5 %) and some other aminoacids, glutathione, lecithin, gallic acid and beta-sitosterol. Mucunine, mucunadine, prurienine, prurieninine (unidentified bases) were reported from seeds. Serotonin was present only in pods. Other bases isolated from the pods, seeds, leaves and roots include indole-3-alkylamines-N, N-dimethyltryptamine; its N-oxide, bufotenine and 5-methoxy-N, N-dimethyltryptamine. Leaves gave 6-methoxyharman also.

The seeds of Mucuna pruriens in a dose of 3 g was tried in India as an anti-depressant herb in 25 cases of depressive illness. Results suggested that the herb increased the level of dopamine, serotonin and other catecholamines in the brain and body to the extent that it induced mood elevation and relieved depression. Withania somnifera (Ashwagandhaa powder) (3 g) daily was also prescribed to add sedative, tranquillizing and tonic effect to the treatment, as most of the patients were having symptoms of sleep disturbance, anxiety, and weight loss. Only the cases of endogenous and neurotic type depression were included in the study. After two months of treatment, 48 % were cured, 36 % improved, 8 % did not show any improvement and relapse was reported in 8 %.

In another study, 32 cases of depressive neurosis were treated with Mucuna pruriens (6 g once a day) along with Withania somnifera powder (10 g once a day). Results were quite encouraging: inital scores of nervousness, insomnia, irritability and depressive index were 2.62 ± 0.49, 2.39 ± 0.50, 2.52 ± 0.66 and 12.56 ± 3.17; after one month of treatment 0.87 ± 0.38, 0.71 ± 0.21, 0.65 ± 0.38 and 6.61 ± 2.13 respectively were recorded. Significant reduction in the degree of clinical anxiety and depression was observed.

Mucuna pruriens seeds show CNS-stimulant effect at low doses and CNS-depressant effect at high doses.

Seeds also act as a potent nervine tonic in cases of depressed libido.

The total alkaloids from the seeds were found to induce a noteworthy increase in the population of spermatozoa and in the weights of body testes, seminal vesicles and prostate of the treated rats. The exhibited activity was found to stimulate testosterone-enanthate-induced androgenic activity.

Mucuna pruriens seed powder, administered in a dose of 75 mg/kg body weight daily as an aqueous suspension, increased the sexual activity of male albino rats substantially. Though L-DOPA was found to arouse sexual activity, the low level content in Mucuna pruriens powder (0.60 per cent) could not account for the activity shown by the dose employed. This was, obviously, the synergistic action of the total herb.

Another study showed that in male albino rats the drug significantly augmented all parameters of mating behaviour.

When tested on frogs, prurieninine slows down the heart, dilates the blood vessels, depresses blood pressure and increases the peristaltic action of intestines.

Musaceae

MUSA

Musa paradisiaca Linn.

Habitat
Cultivated throughout India.

Classical and common names
Ayurvedic: Kadali, Rambhaa, Sakrtphala (Charaka, Sushruta).
Unani: Kelaa.
Siddha: Vazhai
English: Banana, Plantain.

Parts used
Unripe fruits, flowers, leaves, stem.

Dose
Juice 10–20 ml.

Classical use
Charaka gave the pith, bulb, root, fermented fruit and alkaline extract of ashes, in prescriptions, internally in piles, urinary diseases, abdominal diseases, ulcers, haematemesis, dermatosis, leucoderma. Sushruta prescribed fruits for their antitoxic, antiseptic, astringent and styptic properties.

In metrorrhagia, Kadali fruits, mixed with clarified butter, were prescribed. (Raaja Maarttanda, Gadanigraha.)

In chronic dysentery and diarrhoea, steamed unripe fruits were prescribed with curd (Siddhabheshaj-manimaalaa).

Active principles and pharmacology
The raw as well as ripe banana (skin and pulp) have been reported to contain a high amount of 5-hydroxytryptamine, ranging from 0.36 to 8.40 mcg of dry tissue as base, the content being highest in raw banana; and nor-epinephrine.

Serotonin (5-hydroxytryptamine) inhibits gastric secretion and stimulates smooth muscle in the intestine and elsewhere. Nor-epinephrine acts as a mediator of autonomic function as a vasoconstrictor agent. The efficacy of banana in coeliac disease, constipation, and peptic ulcer is attributed to these active principles.

Flowers contain saponins, tannins, reducing and non-reducing sugars, sterols, glycosides and fatty acids.

Administration of banana extract into the stomachs of guinea-pigs resulted in a marked reduction of histamine-induced gastric secretion. Banana feeding also protected guinea-pigs against chronic gastric ulceration and perforation, induced by repeated injections of histamine.

The effect of unripe banana pulp on phenylbutazone-induced gastric ulcers in guinea-pigs was studied. It protected the guinea-pigs against gastric ulcers. Unripe banana powder was also found to afford significant protection against stress ulcers produced in rats.

The unripe banana, dried in shade at room temperature of 26–32° C and powdered, has shown to have anti-ulcer activity in a variety of chemically-induced ulcers in guinea-pigs; gastric ulcers produced by intraperitoneal injections of histamine, aspirin-induced ulcers and indomethacin-induced ulcers.

During a clinical trial in India, 80 cases of clinically and endoscopically proved cases of uncomplicated deodenal ulcer were given two capsules 4 times a day for 12 weeks. Each capsule contained 500 mg of crude powder of dried raw banana. The ulcer healing was observed by repeat endoscopy (after 6 weeks and 12 weeks). Another trial in 25 cases revealed 68 % complete clinical relief, 20 % partial relief and no relief in 12 % of peptic ulcer patients.

Musa paradisiaca flower extract showed hypoglycaemic effect in rabbits. Unripe fruits and cooked flowers are found useful in diabetes.

Green banana tannin was found to contain high concentration of radio active iron. In an experimental study, this tannin reduced the normal iron content in liver and bone marrow with increase in total iron binding capacity and serum iron concentration in normal and anaemic albino rats.

Alkali prepared from Musa paradisiaca stem revealed a good diuretic activity in rats, due to its high potassium content. It showed no toxic effect.

According to M. Grieve, the use of plantain juice as an antidote for snake-bite in the East has been reported. In an experiment, a Kadali-based Siddha drug, advocated in Tamil Nadu as a remedy for snake-bite, was administered orally in four dose levels against LD_{50} of venom (0.82 mg/kg s.c.). It could not afford protection against envenomation, resulting in 100 % mortality in rats. This invalidated the claimed efficacy of Kadali juice in cobra venom poisoning. (CCRAS.)

Myristicaceae

MYRISTICA

Myristica fragrans Houtt.

Habitat
Native of eastern Moluccas; now cultivated in Tamil Nadu, Kerala, Karnataka, Andhra Pradesh, Assam, West Bengal.

Classical and common names
Ayurvedic: Jaatiphala, Jaatikosha (Charaka, Sushruta); Maalatiphala. (Seed kernel). Jaatipatri (Mace, arillus surrounding the seed).
Unani: Jauzbuwaa (Seed), Bisbaasaa (Mace).
Siddha: Masikkai.
English: Nutmeg, Mace.

Parts used
Seeds, arils.

Dose
1–2 g.

Classical use
Extract of fruit in a medicinal oil was prescribed by Charaka in hard, cutaneous eruptions and pruritus.

Sushruta used the extract of root in a medicinal Ghee (clarified butter) for cleansing and sterilizing interior of ulcers; prescribed fruits internally in halitosis, deranged digestion.

Jaatiphala was used, in prescriptions, internally for diarrhoea and dysentery with mucus. (Chakradatta, Bhaishajya Ratnaavali, Siddha-bhesajamanimaalaa.)

A paste of aril (Jaatipatri) was advised for local application on freckles. (Bhaavaprakaasha, Raaja Maarttanda.) A paste of Jaatiphala was applied over cracks in feet (Bangasena).

Jaatiphalaadi Churna (Shaarangadhara Samhitaa) is available over-the-counter, prescribed for diarrhoea and dysentery; with honey for cough, common cold and bronchitis; also in anorexia.

Both Jaatiphala and Jaatipatri are an ingredient in Karpuraadi Churna (Sahasrayoga), prescribed as expectorant and sedative in coughs, bronchitis, asthma, hiccup and common cold.

Jawarish-e-Bisbaasaa (Qarabadeen-e-Kabir) is a Unani tonic, prescribed for indigestion, dyspepsia and stomach-ache.

Roughan-e-Jauzbuwaa (Kitabul-Adviya) is a Unani oil for massage in rheumatic affections.

Nutmeg and Mace are ingredients in a number of Ayurvedic and Unani "sex tonics." Chandrodaya Makaradhwaja, Makaradhwaj, Naagavallyaadya Churnam, Arjakaadi Vatikaa (Bhaishajya Ratnaavali) are prescribed in sexual debility.

The Unani compounds Habb-e-Jadwar, Habb-e-Nishat, Majoon-e-Falaksair, and Majoon-e-Ispand Sokhtaani are prescribed in premature ejaculation; Majoon-e-Arad Khurma in spermatorrhoea; Habb-e-Ambar Momyaee, Majoon-e-Saleb, and Majoon-e-Bandkushad in sexual debility. (National Formularly of Unani Medicine.)

Active principles and pharmacology

Arils and seeds gave neolignans-dl-dehydrodiisoeugenol and dimeric phenylpropanoides. Seeds contain a triterpenic saponin and myristic acid.

Essential oil contains a number of constituents including terpinene, alpha- and beta-pinene, myrcene, terpinolene, camphene, limonene, sabinene, alpha- and beta-phellandrene, delta-carene, alpha-thujene, geraniol, alpha- and beta-terpineol, citronellol, linalool, caryophyllene, p-cymene, terpinic esters, methyleugenol, myristicin, elemicin, safrole and eugenol.

Kernels exhibit antidiarrhoeal, and anti-inflammatory properties. An excessive dose of seeds produces a narcotic effect and epileptic convulsions.

Oil from kernel or aril is a gentle purgative, externally stimulant in rheumatism and sprains. The presence of myristicin makes the oil toxic, if not used in medicinal doses in inflammations of the bladder and urinary tract.

When included in "sex tonics", myristicin and its companion constituents impart stimulant and hallucinogenic properties to them and prolong their action.

Valerianaceae

NARDOSTACHYS

Nardostachys jatamansi DC.

Figure 1 Nardostachys jatmansi—with rootstocks [WOI]

Nardostachys grandiflora DC.
Valeriana jatamansi auct. non Jones.

Habitat

Alpine Himalaya from Kumaon to Sikkim at 3200–5000 m.

Classical and common names

Ayurvedic: Maansi, Jataamaansi, Bhuutajataa, Tapaswini, Sulomashaa, Naladaa, Jatilaa. Tagara has been equated with Valeriana jatamansi.
Unani: Sumbul-e-Hindi, Naardeen-e-Hindi, Baalchhar. Sumbul-ut-Teeb is equated with Valeriana officinalis Linn.
Siddha: Sadamanjil. In the South, Valeriana wallichii is used as Tagara. (V. hardwickii Wall., V. dubia and V. jatamansi are considered as substitutes for the drug Valerian.)
English: Indian Spikenard, Musk-root.

Parts used

Rhizomes, root.

Dose

2–4 g.

Classical use

Mahaapaishaachika Ghrita of Charaka was prescribed in insanity (in addition to Jatilaa, it contained Bacopa monnieri, Centella asiatica, Mucuna prurita, Acorus calamus), and in epilepsy.

Smoking pillet containing Maansi was prescribed by Charaka and Sushruta in cough, hiccough and asthma. It was also used for fumigation for warding off fear-psychosis (Ashtaanga Hridaya).

Charaka and Sushruta incorporated Maansi in many medicinal oils, prescribed in oedema, piles, arthritis, gout, fractures, and obstinate skin diseases, mostly for external application.

Maansi was also used as a tonic for promoting growth of hair (Ashtaanga Hridaya).

In Unani medicine, Muffareh Yaqooti, Anoshdaru, Jawarish Jalinoos, Khamira-e-Abresham (National Formulary of Unani Medicine) are prescribed in nervousness and palpitation of heart; Barshasha in insomnia (also contains opium); Majoon-e-Lana in epilepsy, facial paralysis (also contains Strychnos nux-vomica).

Zimaad-e-Sumbul-ut-Teeb (ointment) is prescribed externally for inflammations.

Active principles and pharmacology

Rhizomes, rich in sesquiterpenoids, yielded jatamanshic acid, jatamansone, patchouli alcohol, nor-seychelanone, seychellen, alpha- and beta-patchoulene. Also obtained were valeranone, valeranal, nardol, calarenol, nardostachone, n-hexacosanyl arachidate and isovalerate, n-hexacosanol, calarene, n-hexacosane, beta-sitosterol, and the terpenic coumarins oroselol, jatamansin and sclinidin.

The oil yielded several hydrocarbons, a new oxide, beta-cudesmol, elemol, angelicin and jatamansinol.

An alkaloid, actinidin, was also reported from the plant.

Valeriana officinalis auct. non Linn.: Roots and rhizomes yielded the alkaloids dipyridylmethylketone, actinidine, iso-valeramide and valerianine; sesquiterpene ketone—valeranone; leaf and root oils contained alpha- and beta-pinene, caryophyllene and bornyl acetate. Bornyl acetate is the major constituent of root oil (31.5 %), whereas it is only 6.6 % in leaf oil. Leaf oil also contained borneol and D-limonene. Bornyl isovalerate is reported from root oil.

In Indian medicine, rhizomes and rootlets of all valarian spp. are used, in prescriptions, for their sedative, antidepressant, antiepileptic, antihysteric, hypotensive, antispasmodic, anti-inflammatory and cardiotonic properties.

Application of Nardostachys jatamansi as a single drug in epilepsy could not be validated experimentally. Nardostachys jatamansi in a dose of 100 mg/kg i.p. did not show antiepileptic effect in albino rats. With a higher dose of 200 mg/kg i.p. 60 % protection was obtained. With a still higher dose of 400 mg/kg i.p. better effect was not obtained; on the contrary, mortality of the animals increased. (CCRAS.)

Various extracts (light petroleum, benzene and ethanol) of Nardostachys jatamansi root showed sedative action in rats and the hypotensive activity in cats. The ethanolic extract was found most active.

The alcoholic extract reduced the rat brain serotonin content.

Further studies on the effect of jatamansone on the biosynthesis and metabolism of serotonin in rabbit brain revealed an impairment of biosynthesis of serotonin in the brain tissue, thus leading to a reduction in the brain levels of 5-hydroxytryptamine. The degradation of serotonin was unaffected. The mode of action of jatamansone was thus at variance with that of reserpine which has a direct action on the cell to liberate serotonin.

Antiarrhythmic activity of the volatile oil of Nardostachys jatamansi was first reported against acetylcholine-induced fibrillations in mongrel dogs. The volatile oil was compared with quinidine in another study and was found to be less active than quinidine as an antiarrhythmic agent against isolated rabbit auricular fibrillation, experimental auricular flutter in anaesthetized dogs, auricular fibrillation induced by aconitine and acetylcholine in dogs. The volatile oil had no effect on digitalis-induced ventricular arrhythmia in dogs. It showed an advantage over quinidine in causing a lesser degree of slowing of conduction, as tested by the ECG changes in the cat. The acute i.v.-toxicity of the oil was also lower than that of quinidine.

Bronchodilator effect of powdered Nardostachys jatamansi fumes and aerosols against histamine-induced bronchial asthma in guinea-pigs has been reported by Gupta S.S. et al. In two different sets of experiments, Nardostachys jatamansi fumes could not only revive guinea-pigs with acute dyspnoea induced by histamine aerosol, but also protected the animals (when pretreated with N. jatamansi alcoholic extract) against developing dyspnoea on exposure to histamine aerosol (Chemical abstracts (1964) 60:16386f). This validates the use of smoking pillets containing Maansi for asthma.

The alcoholic extract of Nardostachys jatamansi inhibited the constrictor response induced by histamine, serotonin and acetylcholine in isolated smooth muscles (i.e. trachea, colon, intestine and uterus). The extract also showed a direct papaverine-like anti-spasmodic effect on the intestine.

A compound Unani preparation (Khamira Abresham Arshadwala), having Nardostachys jatamansi as one of the ingredients, lowered the blood pressure in hypertensive rats and also exerted an antiarrhythmic action.

A compound Ayurvedic preparation, comprising Nardostachys jatamansi, Centella asiatica, Acorus calamus, Rauvolfia serpentina, Saussurea lappa, was administered to 20 patients of schizophrenia. The Ayurvedic preparation and chlorpromazine were found to be almost equal in reducing the various mental symptoms of the schizophrenic patients. The compound preparation was also significantly more effective in this respect than the single drug Tagara.

Jatamansone was tested in a double blind clinical trial covering 28 patients of moderate to severe hypertension. It was given in an oral dose of 10 mg/day for 16 days. Jatamansone lowered the blood pressure in patients in supine as well as standing position. The hypotensive effect manifested within 6–8 hours, the peak effect occurring between 12 to 14 hours, and was occasionally delayed up to 48 hours.

A double blind clinical trial was carried out with jatamansone and placebo in 28 hyperkinetic children (below the age of 14 years) for a period of 11 months, in comparison with two standard drugs viz. D-amphetamine and chlorpromazine. Both jatamansone and D-amphetamine showed significant improvement in the behaviour of the children. Jatamansone showed less improvement in restlessness and aggressiveness, as compared to D-amphetamine. Hyperactivity, however, responded equally favourably to jatamansone and D-amphetamine. Children with marked mental retardation showed a lesser response to all the three drugs.

Preliminary clinical studies with the powdered root as well as an oily emulsion of Nardostachys jatamansi in patients of neurocirculatory asthenia, functional cardio-neurosis, cardiac arrhythmias, menopausal syndromes, hypertensive states etc., showed encouraging results, particularly with respect to improvement in subjective parameters.

The essential oil of Nardostachys jatamansi showed weak antibacterial action against Staph. aureus, E. coli, Sal. typhosum, Vib. cholerae and Sh. flexneri and against Vib. cholerae, Sal. faecalis, Sal. typhi, C. diphtheriae and Strep. pyogenes.

The ethanolic extract (50 %) of the rhizomes showed antifungal activity against Candida albicans, Cryptococcus neoformans, Trichophyton mentagrophytes, Microsporum canis and Aspergillus niger, whereas the essential oil has been reported to reveal potent antifungal activity against Colletotrichum capsici, Helminthosporium turcicum and Alternaria helianthi. The essential oil also showed anthelmintic activity against Taenia solium.

In China, clinical studies on the treatment of heart disease with Valeriana officinalis var. latifolia demonstrated encouraging results. Preparation of a volatile oil fractionated from the root (VOL) has been used to treat 82 CHD-patients with angina pectoris, among whom ST-T ischemic changes appeared on ECG in 50 cases before treatment. Its total effective rate for the simple angina (without detectable ischemic findings) was 87.80 %; the angina with ischemic findings, 88.00 %. For comparisons, another 34 patients with the same conditions, 24 cases among them belonged to the angina with ischemic findings, were treated with a composite injection of Salvia miltiorrhiza (SMCo); the total effective rates for the simple angina and for the angina with ischemic findings were 41.18 % and 37.50 % respectively. The differences between VOL and SMCo were highly significant either in the simple angina or in the angina with ischemic findings. The results showed VOL was superior to SMCo on matter in the remission of symptoms, decreasing the attack frequency and shortening the duration of angina, or in restoring the blood supply to ischemic myocardium. In addition, it was discovered that VOL could lower plasma lipids as well. No toxic actions to liver, kidney, hemopoietic tissue, have been found. (http://www.ibiblio.org)

Use in Western herbal

Valerian entered the U.S. Pharmacopoeia as a tranquillizer in 1820 and remained there till 1942. It was listed in the National Formulary, the Pharmacists' Guide, until 1950.

The 19th century Eclectics prescribed Valerian as a "calmative...for epilepsy...mild spasmodic affections...hypochondria". David Hoffmann, a contemporary herbalist, calls it "one of the most useful relaxing herbs". According to Tyler, Valerian is perhaps best characterized as a minor tranquillizer.

In a double-blind study, valerian (450 mg or 900 mg of an aqueous root extract) significantly decreased sleep latency compared with a placebo. The higher dose of valerian did not further decrease sleep latency (WHO).

The effect of an aqueous extract of Valeriana officinalis root on subjectively rated sleep measures was studied on 128 people. Each person received 9 samples to test (3 containing placebo, 3 containing 400 mg valerian extract and 3 containing a proprietary over-the-counter valerian preparation). The samples, identified only by a code number, and presented in random order, were taken on non-consecutive nights. Valerian produced a significant decrease in subjectively evaluated sleep latency scores and a significant improvement in sleep quality, the latter was most notable among people who considered themselves poor or irregular sleepers, smokers, and people who thought they normally had long sleep latencies. Night awakenings, dream recall and somnolence the next morning were relatively unaffected by valerian. With the proprietary valerian-containing preparation, the only change was a significant increase in reports of feeling more sleepy than normal the next morning. (Leathwood et al. (1982) Pharmacol Biochem Behav jul.)

See Valeriana spp.

Nelumbonaceae/Nymphaeceae

NELUMBO/NYMPHAEA

Nelumbo nucifera Gaertn.

See figure 1.

Nelumbium nelumbo Druce
Nelumbium speciosum Willd.
Nymphaea spp.

Habitat
Throughout the warmer regions of India, in ponds, up to 1800 m. Cultivated often for its flowers.

Figure 1 Nelumbo nucifera [ZANDU]

Classical and common names
Ayurvedic: Nelumbo spp.—Kamala, Padma, Nalina, Arvind, Mahotpala, Raajeeva, Pushkara. Nymphaea spp.—Kumud, Utpala.
Unani: Nilofar (Nymphaea alba Linn., Nelumbo stellata Willd., Nelumbo lotus Linn.)
Siddha: Ambel.
English: Sacred Lotus. Water-lily.

Parts used
Flowers, seeds, filaments, leaves, roots.

Dose
Juice 10–20 ml, seed powder 3–6 g.

Classical use
Three varieties of water-lily, white, red and blue, are used in Indian medicine; their therapeutic efficacy and uses are similar. White water-lily is equated with Nymphaea alba Linn., red with Nelumbo rubra Roxb., blue with Nelumbo stellata Willd.

Nelumbo spp. are equated with Kamala, Nymphaea spp. with Kumud or Utpala. In Ayurvedic texts the white lotus was known as Pundarika, the red one as Kokanad. Indivara was the blue variety, which actually belonged to Nymphaea spp. (Nilotpala of Sushruta and Nilofar of Unani texts.)

Charaka and Sushruta gave lotus stalk in intrinsic haemorrhage and dysuria. Sushruta prescribed tendrils of water-lily internally in persistent dysentery, non-healing ulcers; also for promoting adhesion of fractured bones.

Both Kamala and Utpala were used in prescriptions for diarrhoea with blood, bleeding piles, prolapse of rectum. (Ashtaanga Hridaya, Bhaavaprakasha, Vrindamaadhava.) Stamens of white lotus, pounded with rice-water and mixed with sugar-candy were prescribed in dysentery (Bangasena).

Lotus seeds, mixed with honey, were prescribed in cough. Clarified butter processed with the paste of stalk, stem, stamens, leaves, and seeds of lotus was prescribed as a tonic for virility. (Ashtaanga Hridaya).

In folk medicine, decoction of flowers is given in cholera, haemorrhages, menorrhagia, fever, stranguary and as a cardiac tonic. Powdered rhizome is prescribed in burning sensation of the body, in nausea, chronic dyspepsia; externally in the form of a paste in scabies and ringworm. An arrowroot of rhizomes is given to children for dysentery and diarrhoea. Seeds are used for their antiemetic, diuretic and refrigerant properties.

Among over-the-counter compounds, Arvindaasava (Bhaishajya Ratnaavali) is prescribed to children as a restorative for improving their digestion and appetite.

Sharbat-e-Nilofar, a Unani squash, is prescribed as a cardiac tonic and refrigerant.

Active principles and pharmacology

Nelumbo nucifera: The plant contains a number of alkaloids—roemerine, nuciferine, nornuciferine, liensinine, isoliensinine, neferine, lotusine, armepavin, liriodenine, asimilobine. Beta-sitosterol, glucose, palmitic acid and 10-nonacosanol are also reported.

Kaempferol-3-galactorhamnoglucoside (robinin) is present in the flower, while the receptacle of the flower shows the presence of quercetin and luteolin. The petals and stamens contain isoquercitrin and glucoluteolin.

The seed contained 2.11 % oil, comprising myristic, palmitic, oleic, linoleic and linolenic acid.

Chinese arrowroot is said to be from the root of Nelumbium speciosum.

The 50 per cent ethanolic extract of Nelumbo nucifera showed CNS-depressant effect and potentiated barbiturate-induced hypnosis in mice. It showed diuretic activity in rats.

Nuciferine, the major alkaloid isolated from Nelumbo nucifera, and atherosperminine, the Hofmann degradation product of this alkaloid, showed divergent psychopharmacological effects. The behavioural effects of nuciferine indicated dopamine receptor blockade viz., catalepsy, inhibition of amphetamine toxicity, stereotypy and blockade of conditioned avoidance response (CAR). Atherosperminine, on the other hand, showed effects associated with dopamine receptor stimulation i.e. stereotypy, increase in spontaneous motor activity and amphetamine toxicity, reversal of haloperidol-induced catalepsy and inhibition of morphine analgesia and also inhibition of condition avoidance response. Potentiation of the anticonvulsant action of diphenylhydantoin was the only activity common to both the compounds. The study suggested that while nuciferine behaved as a dopamine-receptor antagonist like other neuroleptics which exhibit a chlorpromazine-like profile of activity, its degraded product atherosperminine acted as a dopamine-receptor agonist.

The 50 per cent ethanolic extract of Nelumbo nucifera rhizomes and seeds did not show any anti-inflammatory, hypoglycaemic and antipyretic activity.

Nymphaea alba: leaves yielded alpha-aminoadipic acid, luteolin-8-C-beta-D-glucopyranoside and gossypitrin; flowers cardiac glucoside nymphalin. Roots yield nympheine.

Nymphalin show sedative action in small doses. Alcoholic extract of rhizomes exhibits mild sedative and spasmolytic action and does not significantly depress the heart. A decoction of flowers works as a cardiac tonic in palpitation.

The decoction of Nymphaea alba flowers showed mild diuretic activity in rats.

The alcoholic extract of defatted fruits of Nelumbo stellata produced mild sedation and ataxia, potentiated hexobarbitone-induced hypnosis in mice and also produced a sharp and transient hypotension blocked by pre-treatment with atropine. In large doses, when given after atropinization, it produced a rise in blood pressure as also a stimulant effect on the guinea-pig ileum indicating the presence of some unstable cholinergic principle. The extract had a significant analgesic activity as revealed by aconitine-induced writhing in mice, antipyretic activity against Brewer's yeast induced pyrexia in rats, and anti-inflammatory activity against carrageenin-induced rat paw oedema. The anti-inflammatory activity compared well with that of hydrocortisone.

The petroleum ether extract of Nelumbo stellata seeds was tested against carbon-tetrachloride-induced hepatotoxicity in rats and mice. The extract at a dose of 300 mg/kg i.p. markedly reduced the prolongation of sleeping time and significantly prevented the increase in the weight and volume of the liver, and mortality. The extract also prevented the necrosis of the liver tissue and promoted to some extent liver regeneration. The alcoholic extract of the defatted seeds was not active in this respect.

Use in Western herbal

Nicholas Culpeper (1616–1654) wrote of water-lily: "The leaves do cool all inflammations, both outward and inward heat of agues, and so doth the flowers also, either by syrup or conserve; the syrup helps much to procure rest, and to settle the brain of frantic persons, by cooling the hot distemperature of the head.

"The seed as well as the root is effectual to stay fluxes of blood, either of wounds or of the belly, but the roots are most used, and more effectual to cool, bind, and restrain all fluxes in man or woman. The root is likewise very good for those whose urine is hot and sharp, to be boiled in wine and water, and the decoction drank."

"The distilled water of flowers is very effectual for all the diseases aforesaid, both inwardly taken and outwardly applied, and is much commended to take away freckles, spots, sunburn and morphew from the face or other parts of the body."

In addition to all this, the white water-lily is used as an anaphrodisiac for reducing the sexual drive.

Apocynaceae

NERIUM

Nerium indicum Mill.
Nerium odorum Soland
Nerium oleander Linn.

Figure 1 Nerium oleander [ADPS]

Thevetia peruviana (Pers.) Merill

Habitat

Native of mediterranean region. Distributed in the Himalayas from Nepal westwards to Kashmir up to 1950 m, extending to Baluchistan and Afghanistan. Also found throughout India in gardens. The white and red-flowered variety is equated with Nerium indicum, the yellow-flowered one with Thevetia peruviana/Thevetia neriifolia Juss.

Classical and common names

Ayurvedic: Karavira, Viraka, Ashvamaarka, Hayamaara, Gauripushpa, Siddhapushpa (white-flowered variety). Rakta pushpa, Raktaprasava, Ravipriya (red-flowered variety).

Unani: Kaner, Diflaa, Samm-ul-maar, Khar-zahar-ah.
Siddha: Alari, Arabivayr.
English: Indian Oleander.

Parts used

Leaves, root. (Extracts in oil, or in the form of a paste, only externally.)

Classical use

Charaka prescribed the leaves of the white-flowered variety externally in chronic and obstinate skin diseases of serious nature, including leprosy. Sushruta included Karavira in a medicinal paste for application in alopecia. For alleviating venereal diseases, the root pounded with water was applied (Bhaavaprakaasha). Karavira Taila of Chakradatta was specific for eczema, impeligo and other skin diseases.

The powder of the leaves was used as a snuff for treating epilepsy.

Karavira is an ingredient in Maaltyaadi Tailam, available over-the-counter for alopecia. Charaka's Shvetakarviraadya Taila and Shvetakarvira-pallavaadya Taila were prescribed for serious skin diseases.

All parts of the plant, especially roots, were known to be highly poisonous when taken internally. In Charaka Samhitaa, the drug was mentioned as Ashvamaaraka or the "horse-killer". The Persian and Arabic names Diflaa, Samm-ul-maar and Khar-zaharah also signify "Ass's Bane". Italian Ammazza Asino, applied to the allied European plant, has the same meaning.

Active principles and pharmacology

Root gave plumericin, alpha-amyrin, beta-sitosterol, kaempferol; cardioactive glycosides named odorosides A-H obtained from the rootbark.

Leaves contained the cardiac glycosides kaneroside, neriumoside, digitoxigenin-alpha-L-oleandroside, 5-alpha-adynerin and several oleasides.

Odorosides are cardioactive glycosides. In a search for cardenolides from leaves, gentiobiosyloleandrin, odoroside A and oleandrin were the main glycosides identified. Oleandrin, if hypodermically injected, causes heartbeats to fall from 75 or 100 to 10 or 12; if continued for some time, the heart ceases to beat and with it respiration.

The whole plant exhibited potent cardiotonic activity; digitalis-like effect on EKG and heart-lung preparation. Tincture from leaves was found two times more potent than tincture digitalis assayed on frogs and was half as tonic on myocardium. Plumieride, a glycoside, did not possess cardiotonic activity, but possessed antistress activity. (CCRAS.)

The alcoholic extract of Nerium indicum leaves produced a positive inotropic effect and bradycardia in isolated frog and rabbit hearts. No significant change in the coronary flow was observed in rabbits. The extract reversed experimentally induced heart failure in the isolated rabbit heart and caused a transient rise in blood pressure and a decrease in the heart rate of dogs. The electrocardiographic changes were similar to those produced by strophanthin on the heart-lung preparation of dog. The extract had a negative chronotropic effect in both the control and hypodynamic heart-lung preparation. A reversal of experimentally induced heart failure in the heart-lung preparation of the dog was produced by the extract. A potent cardiotonic activity and synergism with calcium indicated the presence of a potent cardiac stimulant compound in the extract.

The 50 per cent ethanolic extract of the roots of Nerium indicum showed spasmogenic activity on the isolated guinea-pig ileum and CNS-depressant effects in mice.

Plumieride, isolated from Nerium indicum roots, showed significant antipyretic effect against Brewer's yeast-induced pyrexia and anti-inflammatory activity against carrageenin-induced oedema in rat paw as well as analgesic activity against aconitine-induced writhing in mice.

A glycoside obtained from the roots of Nerium indicum increased adaptability of rats and mice against stressful conditions, as evidenced by better swimming performance and inhibition of aspirin-induced gastric ulcers, milk-induced leucocytosis and pentobarbitone-induced hypnosis.

In subacute toxicity studies in rats and rabbits and chronic toxicity studies in rats, Nerium indicum did not produce any macroscopic or microscopic changes in various organs. In the heart-lung preparation of dog, therapeutic doses of Nerium indicum and strophanthin were 13 % and 16.6 % of their toxic doses respectively. The study revealed Nerium indicum to be more potent than digitalis with a margin of safety similar to that of strophanthin.

Tincture of flowers exhibited cardiotonic, root CNS-active and spasmolytic activity. Externally, root exhibited healing properties for haemorrhoids and ulcers. Oil of the rootbark gave good results in leprosy.

In Homoeopathy, tincture of Nerium oleander (red laurel) leaves is used in diseases of the nervous system, hemiplegia and paralytic conditions under strict medical supervision.

Ranunculaceae

NIGELLA

Nigella sativa Linn.

Habitat
Cultivated in Punjab, Himachal Pradesh, Bihar and Assam.

Classical and common names
Ayurvedic: Upakunchikaa, Kaalaajaaji, Kalikaa, Prthvikaa.
Unani: Kalongi.
Siddha: Karum seeragm.
English: Small Fennel, Black Cumin.

Parts used
Seed.

Dose
1–3 g.

Classical use
Upakunchikaa was used as a carminative, diuretic, emmenagogue, galactagogue, anthelmintic, and in the treatment of mild puerperal fever; externally in skin eruptions.

Charaka included Upakunchikaa in Naraayan Churna for diseases of the digestive tract. Seeds were given as a corrective of purgatives. A decoction of seeds was given in amenorrhoea and dysmenorrhoea; also for expelling placenta and stimulating uterine contraction.

Figure 1 Nigella sativa—flowering branch and fruit [WOI]

The seeds were prescribed with jaggery for irregular fevers (Bhaavaprakaasha). In intrinsic haemorrhage 2.5 g seeds, mixed with double the quantity of sugar, were given (Vrindamaadhava).

In Ashtaanga Hridaya, Upakunchikaa entered into many prescriptions for internal tumours, piles, affections of female genitals.

According to Ayurvedic commentators, Kaarvi (Carum carvi Linn.) and Upakunchikaa possess similar properties.

Active principles and pharmacology
Essential oil from seeds contains nigellone and 2-methyl-4-isopropyl-p-quinone. Seeds contain fatty acids including palmitic, myristic, oleic, linoleic

and linolenic. Beta-sitosterol is also present in the seeds.

The seeds of Nigella sativa showed marked galactagogue action in rats, as revealed by an increase in the weight of the litters of rats fed with the seeds. Histological examination showed more proliferation of acini and secretory activity in breast tissues of the mother rats fed with N. sativa seeds, the action being more potent than that of Jeera (Cuminum cyminum) which is also considered to be a galactagogue.

In a study, the galactagogue principle was found to occur in the ether extract of the seed which, at a concentration of 1.8 %, showed a more powerful galactagogue effect than that induced by 0.5 mcg of estrogen injection given daily to lactating rats. The galactagogue action was assessed by an increase in weight of the litters. Histology of the lactating breast tissues confirmed the enhanced secretory activity in rats on a diet of ether extract of Nigella sativa and estrogen as compared to controls. The rats fed with the ether extract showed more distension in the breast tissues.

The alcoholic extract of Nigella sativa seeds did not show significant anti-inflammatory activity against carrageenin-induced rat paw edema.

The ethanolic extract (50 %) of Nigella sativa seeds showed hypotensive action in cat/dog, weak spasmolytic activity in isolated guinea-pig ileum and exerted a CNS-depressant effect in mice. It also showed antagonism to amphetamine hyperactivity in mice and anticancer activity against Lewis lung carcinoma. The extract did not show any antiviral and hypoglycaemic activity.

The ethanolic extract of Nigella sativa seeds showed diuretic activity, whereas the seed oil was found to be inactive.

The aqueous extract of the roasted seeds of Nigella sativa showed cholinomimetic action on rabbit duodenum, guinea-pig ileum and on dog blood pressure. It also exhibited a nicotinic effect after the blockade of its muscarinic activity by atropine on the rectus abdominis muscle of toad.

The essential oil from the seeds and the unsaponifiable matter showed a depressant action on frog heart. The oil had a relaxant effect on the isolated smooth muscle of rat. The unsaponifiable matter showed a CNS-depressant action as well as a spasmogenic effect on the isolated rat ileum. It has no effect on the isolated rat uterus.

The ethanolic extract of the seeds in a dose of 200 mg/kg could exert little (20 %) inhibition of copper acetate-induced ovulation in rabbits.

The aqueous, ethanolic (90 %) and the ether extracts of the seeds administered orally did not show any anti-implantation and resorptive activity in albino rats but revealed significant estrogenic activity in immature rats.

A preparation containing seeds of Nigella sativa as one of the ingredients, apart from Foeniculum vulgare (seeds), Mesua ferrea (flowers). Terminalia chebula (seeds), Curcuma zedoaria (tubers) and Terminalia arjuna (stem bark), exhibited, on oral administration, anti-implantation activity (38.9 %) in female albino rats.

The alcoholic extract of Nigella sativa seeds was inhibitory of E. coli and Staph. aureus, whilst the aqueous extract was inactive. The 50 % ethanolic extract of the seeds had no antibacterial activity. The petroleum extract was found active against Mic. pyogenes var. aureus, B. subtilis, Dip. pneumoniae and Strep. pyogenes.

The essential oil from the seeds showed antibacterial activity against B. subtilis, B. shigella dysenteriae, Sh. boydi, B. anthracis, B. pumilus, Sal. paratyphi, Sh. niger, Sh. sonnie and B. cereus, Staph. albus, Staph. aureus, Staph. lutea, E. coli and Sal. typhi. The oil also showed antibacterial activity against Sal. typhi, Ps. aeruginosa and Sh. shigae and was less active against E. coli, B. anthracis, B. subtilis, C. pyogenes, P. vulgaris, Staph. aureus and K. pneumoniae.

The essential oil from the seeds was active against A. albicans, A. flavus, A. niger, Curvularia lunata, M. gypseum and Trichoderma viride. The oil was inactive against C. oryzae, Penicillium chrysogenum and P. javanicum.

The essential and fixed oils and the unsaponifiable matter of the seeds of Nigella sativa showed antifungal activity against Fusarium solani, F. moniliforme, Helminthosporium oryzae, H. turcicum, Pythium vexans, Rhizoctonia bataticola, R. solani, Alternaria helianthi, Colletotrichum capsici and Pyricularia setariae.

Nyctanthaceae

NYCTANTHES

Nyctanthes arbor-tristis L.

Habitat
Outer Himalaya, Assam and West Bengal. Cultivated in many parts of India.

Classical and common names
Ayurvedic: Paarijaata, Shephaali, Shephaalikaa, Mandara, Raagapushpi. Paaribhadra has now been equated with Erythrina variegata Linn. var. orientalis (Linn.) Merill. (Indian Coral Tree).
Unani: Harsingaar.
Siddha: Pavazha mattigai.
English: Night Jasmine, Coral Jasmine.

Parts used
Leaves.

Dose
Juice 10–20 ml.

Classical use
There is no mention of Paarijaata, Paaribhadra and Shephaali in Charaka Samhitaa. Shephaalikaa of Sushruta Samhitaa and Ashtaanga Hridaya is now equated with Nyctanthes spp., which is known as Paarijaata and Harsingaar in Northern India, Madhya Pradesh and Maharashtra-Gujarat. According to NAA, Paarijaata has wrongly been equated with Adansonia digitata Linn. (Bombacaceae).

Paarijaata, in the form of a decoction, was prescribed by Sushruta as a specific remedy for urinary disorders including diabetes.

The drug was included in a number of compound preparations for polyuria (Vrindamaadhava, Bhaavaprakaasha). A decoction of the root was given in arthritis (Raaja Maarattanda), of leaves in sciatica (Chakradatta). Chewing of the root was advised in tonsillitis (Chakradatta, Raaja maarattanda). Leaf juice was also used as a cholagogue, anthelmintic and laxative; with honey in bilious fevers. Externally fresh juice of leaves or powdered seeds were used for cutaneous affections, alopecia and dandruff.

Active principles and pharmacology
Leaves afforded beta-amyrin, beta-sitosterol, hentriacontane, benzoic acid, astragalin (kaempferol-3-glucoside, nicotiflorin (kaempferol-3-rhamnoglucoside), nyctanthoside, nyctanthic acid, friedelin, lupeol and oleanolic acid.

The stem gave beta-sitosterol and a glycoside naringenin-4'-O-beta-glucopyranosyl-alpha-xylopyranoside.

Seeds gave the iridoid glucosides arbortristosides A and B, nyctanthoside, nyctanthic acid and beta-sitosterol.

Corolla tubes gave monogentiobioside ester of alpha-crocetin as a major constituent.

The ethanolic extract of the whole plant showed CNS-depressant effect and hypothermia in mice.

The alcoholic extract of the leaves showed significant tranquillizing as well as dose-dependent antipyretic effect and histamine-antagonistic and purgative activities.

The water-soluble portion of alcoholic extract of the leaves showed significant anti-inflammatory activity against acute, subacute and chronic models of inflammation in rats. It inhibited oedema induced by different phlogistic agents.

Experimentally, the extract of the whole plant was found efficacious in reducing inflammatory swellings of the knee joint and arthritis which was inhibited in acute as well as chronic phases.

The ethanolic extract (50 %) of the whole plant did not exhibit antibacterial, antifungal, antiviral activities. Diuretic activity was also absent. The bark extract showed inhibition against potato virus X.

Lamiaceae

OCIMUM

Ocimum basilicum Linn.

Figure 1 Ocimum basilicum [ADPS]

Ocimum caryophyllatum Roxb.
Ocimum pilosum Willd.

Habitat
Indigenous to lower hills of Punjab. Cultivated throughout India.

Classical and common names
Ayurvedic: Barbari, Tuvari, Tungi, Kharpushpa, Ajgandhikaa. Baabui Tulasi.
Unani: Faranjmishk, Raihaan (also equated with Ocimum sanctum).
Siddha: Karunthu.
English: Sweet Basil.

Parts used
Whole plant.

Dose
Decoction 50–100 ml, seed powder 1–3 g.

Classical use
In Bhaavaprakaasha, three species of Barbari have been mentioned: the white-flowered, the purple-flowered, and the one with leaves of bigger size. These have been equated with Ocimum var. basilicum, var. pilosum and var. majus.

According to Bhaavaprakaasha and Kaiyadeva Nighantu, Barbari is aromatic, stimulant, carminative, diuretic, anthelmintic, anti-inflammatory, antitoxic, and antiseptic. It was prescribed for asthma, catarrh, allergic conditions, skin eruptions, piles, diarrhoea, dysentery, nephritis, and cystitis.

Leaves were mostly used for flavouring purposes. Sumukha of Sushruta has been equated with Ocimum pilosum Linn. by INSA scientists. It was prescribed internally by Sushruta in dyspepsia, catarrh, cough and asthma.

The white (shveta) Barbari (syn. Kharpushpa) of Bhaavaprakaasha was prescribed in pediatric medicine for cough and helminthic infestations. Phanijjaka of the 16th century seems to be the Sanskritized name of Faranjmishk of Unani medicine.

In Unani medicine, a decoction of leaves is prescribed in amenorrhea and retention of urine. A mucilaginous jelly of seeds, prepared by steeping one teaspoonful of seeds in water, and mixed with sugar or its syrup, is prescribed as a tonic to heart, brain, liver and intestines.

Leaves are an ingredient in Khamira-e-Abresham Sada and seeds in Khamira-e-Gaozaban Sada, prescribed in melancholia, palpitation and debility.

Active principles and pharmacology

Essential oil contains 1, 8-cineole, eugenol, limonene, ocimene, geranial, *cis*-3-hexenol, citronellol, alpha-terpineol, camphor, methyleugenol, methyl cinnamate as minor and linalool, methyl chavicol (estragole) as major components. Methyl cinnamate may be a major constituent in some chemovars.

The essential oil obtained from Ocimum basilicum showed antibacterial activity against B. typhosa, B. pumilus, Sh. nigesta, Sar. lutea, P. vulgaris, K. pneumoniae, B. subtilis, B. anthracis, Sal. paratyphi and Xanth. campestris. The oil was also reported to be active against B. mycoides, E. coli and Sal. typhi. (A few contradictory findings have also been recorded).

The oil showed antifungal activity against Candida albicans, C. utilis, Penicillium notatum, Microsporum gypseum, Malbranchea pulchella, Chrysosporum tropicum, Penicillium liliacinum, Aspergillus fumigatus, and also against A. niger, Trichophyton mentagrophytes, Microsporum canis and Epidermophyton floccosum. Contradictory findings have been recorded about A. niger.

Ocimum basilicum var. crispum was found most effective against earthworms; 0.1 % suspension of the oil from the leaves induced paralysis in 9 minutes, as compared to similar response shown by a 0.1 % (w/v) solution of piperazine citrate in 30 minutes. The oil gave potent antiworm response.

Local application of Ocimum basilicum leaf-juice against acne vulgaris gave encouraging results.

Flowers exhibited carminative, diuretic, stimulant and demulcent; seeds antidysenteric properties.

Use in Western herbal

Views associated with Basil have been mixed down the ages. To the ancient Greeks and Romans, Basil was a symbol of hostility and insanity. It was associated with hatred and misfortune. Disocorides said that it should never be taken internally.

Other traditions associate the herb with love. When an Italian woman placed a potted Basil plant on her balcony, it was indicated that she is welcoming her lover. In northern Europe, lovers exchanged Basil sprigs as a sign of faithfulness. (Michael Castleman.)

Later, the Roman naturalist Pliny and Arab physicians defended Basil as a great healer. So did the Chinese, who used Basil to treat stomach, kidney and ailments due to vitiated blood. By the 17th century, Basil was widely used in Europe to treat cold, cough, bronchitis, nervous exhaustion. American herbalists suggest Basil as a digestive aid and appetite stimulant. Preparations of Basil are used as a supportive therapy for feeling of fullness and flatulence, for the stimulation of appetite and as a diuretic. Basil is not given to infants. (PDR.)

Ocimum sanctum L.
Ocimum tenuiflorum L.

Habitat

Found throughout India, up to 1800 m, in the Himalayas.

Classical and common names

Ayurvedic: Tulasi, Surasaa, Bhuutaghni, Suravalli, Sulabhaa, Manjarikaa (Charaka, Sushruta); Bahumanjari, Devadumdubhi, Apetaraakshasi, Shuulaghni.
Unani: Tulasi.
Siddha: Thulasi.
English: Holy Basil, Sacred Basil.

Parts used

Whole plant.

Dose

Decoction 50–100 ml, powder 1–3 g, fresh juice 5–10 ml.

Classical use

Charaka and Sushruta prescribed a paste or extract of the leaves internally, in prescriptions, for fever, splenic affections, toxicosis, and skin eruptions; also for catarrh, cough, asthma, and dyspepsia.

Surasaa, Shveta-Surasaa and Phanijjaka (Ocimum sanctum and Ocimum basilicum) belonged to the Surasaadi group of herbs, considered specific for alleviating anorexia, parasitic infections, rhinitis, cough, cold, asthma.

According to Bhaavaprakaasha and Ashtaanga Hridaya, Tulasi stimulates digestion, cures dysurea, vitiation of blood, urticaria, chronic skin diseases, piles, parasitic infections, vomiting, earache, conjunctivitis, post-parturition pain, chest diseases; pharynx, bronchial and lung infections.

For expelling intestinal worms, Surasaadi group of herbs were given with honey (Sushruta Samhitaa, Ashtaanga Hridaya). For cleansing wounds, Surasaadi group was prescribed (Vrindamaadhava). For conjunctivitis, the juice of Tulasi, mixed with honey, was used as collyrium. For cough, the juice of Krishna (black) Tulasi, mixed with honey, was a popular recipe (Charaka Samhitaa). For checking malarial fever juice of Tulasi, mixed with Piper nigrum (Maricha) powder was administered.

In Ashtaanga Hridaya, Tulasi is recommended for arresting greying of hair. It is an ingredient of Sahacharaadi Taila.

The Krishna (black) variety of Tulasi, bearing dark greenish-purplish leaves, is preferred in Indian medicine.

In folk medicine, in addition to the classical use, seeds are given in disorders of the genitourinary system. A decoction of fresh roots, stems and leaves is given as an antipyretic and diaphoretic in fevers. Mixed with a little ginger, leaf juice is given for colic in children, with black pepper (Piper nigrum) in catarrhal fever. Fresh juice is given for checking vomiting. Leaves, ground with water, are applied on bad boils, mosquito bites, and urticaria. A paste of fresh leaves, mixed with lemon juice, is applied on eczema, ringworm and other skin diseases. Dried leaves are used as snuff in myiosis and ozoena. Leaf juice is taken as a prophylactic during epidemics, as an adaptogen and internal antiseptic.

Active principles and pharmacology

The major components of the essential oil are eugenol, carvacrol, nerol and eugenolmethylether. Others include caryophyllene, terpinen-4-ol, decylaldehyde, si-selinene, alpha- and beta-pinene and camphor. The leaves have been reported to contain ursolic acid, apigenin, luteolin, apigenin-7-O-glucuronide, luteolin-7-O-glucuronide, orientin and molludistin.

The plant increased physical and mental endurance, protected animals from physical, chemical, biological and emotional stress. It exhibited a potent antistress property attributed to Eleutherococcus and Panax ginseng. (CCRAS.)

Ocimum sanctum plant's dried powder enhanced the physical endurance and survival time of swimming mice, prevented stress-induced gastric ulcers in rats and protected mice and rats against hepatotoxicity induced by carbon tetrachloride. It also prevented milk-induced leucocytosis in mice. Thus the plant manifested a non-specific type of protection against a variety of stress-induced biological changes (1995.)

The effect of Ocimum sanctum on the immune responses in experimental animals was studied during 1987–88. Ocimum sanctum enhanced anti-SRBC haemagglutination titre and IgE antibody titre, as measured by passive cutaneous anaphylaxis in rats. Antigen-induced histamine release from peritoneal mast cells was significantly inhibited by Ocimum sanctum. It also antagonised responses to various spasmogens on isolated guinea-pig ileum. Thus, it appears that Ocimum sanctum modulates the immune responses by acting at various levels in the immune mechanism, such as antibody production, release of mediators of hypersensitivity reactions and tissue responses to these mediators in the target organs.

The antiasthmatic activity of a 50 per cent aqueous ethanol extract of dried leaves, and the volatile and fixed oils of Ocimum sanctum was evaluated against histamine and acetylcholine-induced pre-convulsive dyspnoea (PCD) in guinea-pigs. The drug protected the guinea-pigs against histamine and PCD.

Ocimum sanctum leaves were tried in cases of bronchial asthma and stress-related hypertension. In both these conditions the drug was found to be quite efficacious.

Therapeutic effect of Ocimum sanctum leaves on hypertension was tested on rats and dogs with induced hypertension and then on human subjects suffering from essential hypertension. 60 ml of 75 per cent leaf extract administered orally in two halves resulted in a fall of diastolic as well as systolic pressure to the normal level with no adverse effects.

Fresh leaves of Ocimum sanctum mixed as 1 g and 2 g in 100 g of diet, given for four weeks, brought about significant changes in lipid profile of normal albino rats. This resulted in a significant lowering of serum total cholesterol, triglyceride, phospholipid and LDL-cholesterol and significant increase in HDL-cholesterol and total faecal sterol contents.

Ocimum sanctum was found to reduce gastrointestinal transit in albino rats. This new profile suggests that Ocimum sanctum can be used with benefit in emotional tension where intestinal mobility is usually increased.

The antiulcerogenic property of Ocimum sanctum was studied during 1993. The extract of Ocimum sanctum reduced the ulcer index in aspirin-treated rats. Seven days pre-treatment with the drug increased the mucous secretion also. The drug reduces acid secretion and increases mucous secretion.

Oral administration of alcoholic extract of Ocimum sanctum leaf led to marked lowering of blood sugar level in normal, glucose fed hyperglycaemic and streptozotocin-induced diabetic rats. The extract also potentiated the action of exogenous insulin in normal rats. The activity of the extract was 91.55 and 70.43 per cent to that of tolbutamide in normal and diabetic rats (1993).

The leaf-extract of Ocimum sanctum, given i.p. either as a single or multiple doses, exhibited radio-protective effect (1995).

The ether extract of Ocimum sanctum leaves showed antibacterial activity against E. coli, Staph. aureus, M. tuberculosis and Mic. pyogenes var. aureus.

The essential oil from the leaves exhibited antibacterial activity against E. coli, B. anthracis, B. subtilis, Sal. pullorum, Sal. richmond, Sal. newport, Sal. stanley, Sal. typhimurium, Staph. aureus, P. vulgaris and Ps. aeruginosa; being most active against Sal. stanlay. Eugenol was found effective against Arthrobactor globiformis, B. megatherium, E. coli and Pseudomonas spp.

The essential oil from leaves showed antifungal activity against Aspergillus niger, Rhizopus stolonifer and Penicillium digitatum. The oil, as also eugenol and methyl eugenol, showed activity against Alternaria solani, Candida guillermondii, Collectotrichum capsici. Curvularia spp., Fusarium solani and Helminthosporium oryzae; eugenol being most active.

The essential oil from the plant exhibited antifungal activity against the dermatophytes Epidermophyton floccosum, Trichophyton mentagrophytes and Microsporum canis.

Ocimum sanctum leaves were found active against most of the organisms tested, particularly Curvularia lunata, Rhizopus nigricans, Fusarium oxysporum, Kl. pneumoniae, Saccharomyces cerevisiae and Candida torulopsis.

Use in Western herbal

Western herbalists feel that virtues of Basil have been exaggerated in India. They are not prepared to believe that it is a better adaptogen than Panax ginseng, as claimed by CCRAS.

The leaves of Ocimum viride, a native of Western Africa, is used for its febrifugal properties. In Sierra Leone, it is known as "Fever Plant"; its decoction is prescribed as a remedy for fevers. The leaves of Ocimum canum, Ocimum gratissimum and Ocimum crispum (a Japanese spp.) are used as a medicine in bilious fever. Ocimum guineense is employed by negroes as a medicine in bilious fever. Ocimum americanum is used in cases of chest troubles and dysentery.

A tincture of Ocimum canum leaves is used in homoeopathy in the diseases of kidneys, bladder, urethra, uric acid diathesis, and for renal colic.

Boraginaceae

ONOSMA

Onosma bracteatum Wall.

Habitat

Distributed in north-western Himalayas from Kashmir to Kumaun at altitudes of 3500–4500 m.

Classical and common names

Ayurvedic: Gojihvaa, Kharpatra, Darvipatraa, Vrishjhvaa.
Unani: Gaozabaan (equated with Borago officinalis Linn. and other spp. of Boraginaceae by National Formulary of Unani Medicine)
Siddha: Ununjil.
English: Borage.

Parts used

Leaves, flowers.

Dose

Leaves 3–6 g, flowers 3–6 g.

Dried leaves and flowers of Onosma bracteatum, imported from Iran, constitute the drug Gaozabaan. Anchusa strigosa Labill. Caccinia glauca Savi and Macrotomia benthiamii DC. are sold as Gaozabaan. In Sindh and Punjab, Trichodesma indicum Br. and Trichodesma zeylanicum are known as Gaozabaan. Elephantopus scaber Linn. has also been equated with Gaozabaan.

Classical use

Sushruta used the decoction of Gojihvaa as a gargle in catarrh; as a potherb in anorexia, fever, cough; in haemoptysis, urinary troubles, skin diseases.

In Unani medicine, Araq-e-Gaozabaan (Bayaz-e-Kabir), prepared from the leaves of Borage, is prescribed as a tonic in palpitation, weakness of heart, brain and liver, depression, catarrh. Flowers are main ingredient in Khamira-e-Gaozabaan (National Formulary of Unani Medicine), a reputed brain and cardiac tonic.

Active principles and pharmacology

See Borago officinalis.

Ayurvedic Formulary of India equates Gojihvaa with Onosma bractaetum. All other identifications are controversial.

Many eminent Unani scholars, including Hakeem Daljeet Singh, have shown preference for Caccinia glauca Savi, while discussing various applications of Gaozabaan.

Caccinia glauca (syn. Caccinia crassifolia O. Kuntze), a native of Baluchistan, is imported into India from Iran, Afghanistan and Pakistan as Gaozabaan.

Flowers of Caccinia glauca contain pyrrolizidine the alkaloid retronccine-7, 9-dibenzoate.

Stems and leaves gave the saponins caccigenin, caccigenin lactone, and 23-deoxycaccigenin, rutin, and a saponin derived from caccigenin. The leaves also gave the glucoside caccinin and its aglucone caccinetin, which is the dimethylallyl ester of caffeic acid.

The aerial parts contain supinine and heliotridine or retronecine trachelanthate.

Caccinin and its aglucone caccinetin show diuretic, saponins anti-inflammatory activity.

According to the Wealth of India, the imported drug Gaozabaan, based on morphological studies, consists of the leaves and nutlets of Anchusa strigosa Labill (Boraginaceae) and the flowers of Echium amoenum Fisch. & Mey.

Anchusa strigosa exhibits stimulant, diuretic, demulcent and tonic properties and is found useful in cough, asthma, stones in the bladder and kidney and in bilious complaints. It is administered as a decoction or syrup. (The Wealth of India.)

The flowers contain hexatriacontane, 1, 14-heptadecanediol, 1, 2-octadecanediol, 10-nonadecanone, malvidin, pelargonidin, rhamnose, fructose, glucose, arabinose, galacturonic acid and glucuronic acid. Flowers are rich in amino acids.

The alcoholic extract of the flowers exhibited cardiac stimulant action on an isolated guinea-pig heart.

The ethanolic extract of the plant showed strong inhibitory activity on aryl-hydrocarbon-hydroxylase activity and 3H-benzo(a)pyrene binding to rat liver microsomal protein.

The ethereal fraction of the aqueous extract of the plant showed moderate antimicrobial activity against several microorganisms. (The Wealth of India Suppl.)

Use in Western herbal

Borage was known as Bugloss in the classical Western herbal. Bugloss belongs to Anchusa and Echium spp. The leaves make a good cordial on infusion, are diaphoretic, diuretic and pectorial, alleviate fevers, headaches and nervous complaints, provide relief in inflammatory pains.

Parkinson says of it: "The water distilled in glasses or the roote itself taken is good against the passions and tremblings of the heart as also against swoonings, sadness and melancholy." (M. Grieve.)

Convolvulaceae

OPERCULINA

Operculina turpethum (Linn.) S. Manso.

Figure 1 Operculina turpethum [ZANDU]

Ipomoea turpethum R. Br.
Merremia turpethum (L.) Shah & Bhat

Habitat

Throughout India up to an altitude of 900 m. Occasionally grown in gardens for ornament.

Classical and common names

Ayurvedic: Trivrta, Trivrtaa, Tribhandi, Triputaa, Saralaa, Suvahaa, Rechani, Nishotra.
Unani: Turbud, Nisoth.
Siddha: Sivathai.
English: Indian Jalap.

Parts used

Root.

Dose

Powder 3–6 g.

Classical use

Charaka and Sushruta prescribed Trivrta internally as a purgative in acute constipation, for intestinal paralysis and abdominal swellings; externally as a paste for malignant ulcers, as an ingredient of ointment for fistulas. A medicated oil, with Trivrta as an ingredient, was used for cleansing ulcers. Charaka and Sushruta also prescribed Trivrta powder, mixed with Zingiber officinale (Shunthi) in anaemia, jaundice and inflammations.

Among over-the-counter drugs, Trivrta Churna (Bhaavaprakaasha) is prescribed for habitual constipation and to patients suffering from piles. Trivrta is a major constituent in Avipattikara Churna (Bhaishajya Ratnaavali), prescribed as a laxative, digestive and carminative.

Trivrta is an ingredient in Chandraprabhaa Vati (Bhaishjya Ratnaavali), prescribed in urinary disorders, anaemia, jaundice, cough, bronchitis, and skin diseases. Ashwagandhaarishta, a reputed restorative tonic, also contains Trivrta.

Among over-the-counter Unani drugs, Jawarish-e-Shahreyaran, prescribed for constipation, colic, hepatitis, derangement of digestive functions; Habb-e-Iyarij, a purgative, prescribed as an adjuvant in neurological disorders. Turbud is a major ingredient in Majoon-e-Kalkalanaj, prescribed for its anti-inflammatory, diuretic and deobstruent properties.

Nishoth is used in Indian medicine as an adjuvant in the treatment of anaemia accompanied by splenomegaly also in periodic fevers; with Picrorhiza Kurroa (Katukaa) in jaundice. Traditionally administered against dropsy due to heart kidney or liver diseases.

Active principles and pharmacology

Operculina turpethum is regarded as an effective substitute for imported Ipomoea purga, as the percentage of resin present in the roots of Operculina turpethum (9–12 %) is comparable with 10–15 % of Ipomoea purga. The two varieties of white and black ones (Safed and Krishna Nishoth) have been considered to be varieties of Operculina turpethum. On the basis of pharmacognostical examination, Operculina turpethum has now been equated with black Nishoth. The white Nishoth of the market is actually Marsdenia tenacissima W & A., (Muurvaa), which is an adulterant, devoid of purgative property.

Operculina turpethum contains resin in varying amounts of 9–12 %. The resin contained the glucosides turpethin, alpha- and beta-turpethein. It also contained a coumarin, scopoletin, glucose, rhamnose and fucose.

Alpha- and beta-turpethein give laxative action to the drug.

The ethanolic, aqueous and ethereal extracts of the root showed anti-inflammatory activity against carrageenin induced rat paw oedema as well as cotton pellet-induced granuloma and formalin-induced arthritis in rats. The aqueous extract was found to be most potent fraction against all the three models of experimental inflammation. It was found superior to sodium salicylate. (CCRAS.)

Alcoholic extracts of the fresh root show antibacterial activity against Micrococcus pyogenes var. aureus and E. coli.

A decoction of the roots of Operculina turpethum and Piper nigrum was found effective against intestinal worms.

Orchidaceae

ORCHIS

Orchis latifolia Linn.

Habitat
Found in damp places in Himalayas from Kashmir to Nepal at altitudes of 2500–5000 m. O. latifolia yield Salep of commerce. Most of the Salep used in Unani medicine is imported from Iran and Afghanistan. The source of original Salep is Orchis mascula Linn., Orchis maculata, Orchis latifolia. The tuberous roots of Eulophia campestris spp. are also used as Salep.

Classical and common names
Ayurvedic: Munjaataka, Saalam-misri.
Unani: Saalab-misri, Khusyat-us-Saalab.
Siddha: Silamishri.
English: Salep.

Parts used
Tubers.

Dose
Powder 3–5 g.

Classical use
Charaka used the juice of Munjaataka tubers alone; also in prescriptions, as an aphrodisiac. It was prescribed as a tonic in leucorrhoea, morbid thirst and irregular fever. Mahaamaayura Ghrita, a nervine and brain tonic of Charaka, contained Munjaataka. Dvipanchmuulaadya (Jivaniya) Ghrita and Sukumaaraka Taila, prescribed for gout, also contained Munjaataka.

Sushruta included Munjaataka in an ointment for erysipelas and other extensive skin eruptions.

During the 16th century, Munjaataka was used extensively as a cardiac tonic, and entered into Unani compound preparations as an aphrodisiac.

In Unani medicine, Saalab is credited with anabolic and rejuvenating properties. For promoting spermatogenesis and for sexual stimulation the drug has been incorporated into a number of "sex tonics". Majoon-e-Salab, Habb-e-Ambar Momyaee, Habb-e-Jalinoos, Luboob-e-Barid, Laboob Kabir, Majoon-e-Bandkushad, Majoon-e-Piyaz, Majoon-e-Ra-hul Momineen, Majoon-e-Mughalliz, Halwa-e-Baiza-e-Murgh, and Halwa-e-Gazar, (all incorporated in the National Formulary of Unani Medicine), prescribed for "sexual debility", attenuated semen and spermatorrhoea, are available over-the-counter.

Salep, mixed with water, in the form of a mucilageous jelly, is administered in diarrhoea, dysentery and chronic fevers. A decoction of Salep, mixed with sugar and flavoured with spices, is used as a food supplement during convalescence.

Active principles and pharmacology

Orchis spp. contains mucilage (up to 50 %), glucans, glucomannans (partially acetylised)-starch (25 %) and proteins (5–15 %). The mucus is rich in mannose and starch. Also yields 2 % of ash, consisting chiefly of phosphates and chlorides of potassium and calcium.

All these constituents make Salep a wholesome and nutritious tonic food and a demulcent agent.

Use in Western herbal

Before World War 2, Salep used to be sold at stalls in the streets of London, and was held in great esteem in herbal medicine. It was a popular hot drink during the winter.

Sailing ships, on long voyages, used to stock Salep as an article of diet. One ounce (25 g) of Salep dissolved in 2 quarts (2.27 litres) of boiling water was considered sufficient subsistence for each man per day.

As the legend goes, "witches" were supposed to use the tubers of Salep in their philtres; the fresh tuber being given to promote true love, and the withered one to check wrong passions.

Nicholas Culpeper (1616–1654) wrote: "Orchis provoke lust exceedingly. The withered roots do restrain."

Folk wisdom maintained that a pregnant woman, eating the larger of the two tubers of Salep, would give birth to a boy. In Thessaly in Greece, women ate the tubers to increase their sexual urge.

Once believed to have "aphrodisiac" powers, Salep's current medicinal use in Western herbal is confined to the treatment of unspecified diarrhoea, particularly in children and to indigestion, heartburn and flatulence.

Bignoniaceae

OROXYLUM

Oroxylum indicum Vent.

Figure 1 Oroxylum indicum [CCRAS]

Figure 2 Oroxylum indicum [CCRAS]

Calosanthes indica (L.) Bl.

Habitat

Throughout the greater part of India up to an altitude of 1200 m, chiefly met with in ravines and moist places in the forests. A native of Indo-Malaysian region and China.

Classical and common names

Ayurvedic: Shyonaaka, Shoshana, Tuntuka, Kutunnata, Mandukparna, Patrorna, Bhalluka, Prthushimba. Sonaapaathaa. (Dirghavranta, Katvanga and Aralu have now been equated with Ailanthus excelsa Roxb. These were mentioned as synonyms of Shyonaaka in Bhaavaprakaasha.)

English: Indian Trumpet Flowers.

Parts used

Bark, root.

Dose

Decoction 50–100 ml.

Classical use

Charaka and Sushruta prescribed Shyonaaka internally for persistent dysentery, non-healing ulcers, gynaecological disorders, as an antiseptic, antitoxic, astringent and styptic. The drug was also administered for promoting adhesion of fractured bone. In the form of a medicated Ghee (clarified butter), Shyonaaka was specific for chronic dysentery. A paste of leaves was applied in the falling of hair and baldness.

Shyonaaka, in addition to diarrhoea, dysentery and abdominal diseases, was used in rheumatism, diseases of ear, nose and throat, high fever, partial paralysis, diabetes and dysuria. (Chakradatta, Ashtaanga Hridaya, Bangasena, Bhaavaprakaasha.)

In Ayurvedic medicine, roots are used fresh, as they lose their vitality after a few months. Tender fruits and seeds are used as stomachic and purgative; leaves, externally, for enlarged spleen, headache and ulcers.

Root and rootbark of Oroxylum indicum form one of the components of Ayurvedic preparation Dashamuula (the Group of Ten Roots). Market samples of this Group are found to consist of mainly dry pieces of stems and not of the root or rootbark. Dashmuulaarishta (Bhaishajya Ratnaavali), available over-the-counter, is prescribed as a bitter tonic and alterative. More than fifty drugs have been included in the preparation. The contribution of the Ten Roots is difficult to assess.

Active principles and pharmacology

Leaves contained flavones and their glycosides, baicalein and its 6- and 7-glucuronides, scutellarein and its 7-glucuronide, as well as anthraquinone, and aloe-emodin.

Stem-bark gave the flavonoids oroxylin A, baicalein, its 7-glucuronide, scutellarein and its 7-rutinoside, chrysin and p-coumaric acid.

Seeds gave another flavone, oroxindin.

Heartwood yielded beta-sitosterol and isoflavone, prunetin.

Bark of the root gave chrysin, baicalein and oroxylin. Bark also contained dihydrobaicalcein.

A decoction of bark showed good diuretic activity in rats, comparing favourably with potassium acetate, and being more potent than urea. 50 % ethanolic extract of the fruit showed spasmogenic action on isolated guinea-pig ileum, whereas the root, stem and stembark extracts had no such effects.

In an uncontrolled study on 21 confirmed patients of intestinal amoebiasis, the oral administration of concentrated aqueous extract powder of Oroxylum indicum led to symptomatic improvement as well as absence of Ent. histolytica cysts in the stool of 19 patients.

Poaceae

ORYZA

***Oryza sativa* Linn.**

Habitat

Cultivated for rice. An annual and perennial grass without rhizome. Grains are mostly white, but occasionally red, purple or brown. The red variety of rice is used in Indian medicine.

Classical and common names

Ayurvedic: Shaali, Vrihi-dhaanya.
Siddha: Neb.
English: Rice, Paddy.

Parts used

Grains, root.

Classical use

Rakta-Shaali, or the red variety of rice is used in Indian medicine as a nutritive diet which provides a vital support to various body functions, promotes vitality and acts as a cardiac tonic. Also used as lactagogue.

An alcoholic drink, prepared of the paste of Punarnavaa (Boerhavia diffusa) and Shaali, was known as Vaaruni. It shared the properties of wine and was prescribed in chronic cold, flatulence and colic (Bhaavaprakaasha).

Dhaanyaamla, prepared of the powder of Shaali and Kodrava (Paspalum scrobiculatum Linn.), was prescribed in anorexia and digestive disorders, especially for the persons residing on the sea coast (Bhaavaprakaasha).

Vrihi-dhaanya, a variety of Shaali, though sweet, produces acidity. Eatables prepared of the paste of this variety were considered heavy and were not recommended during illness as they produced burning sensation. The white variety of rice, known as Gaura Shashtika, was considered sweet, cooling and light, and was used when the red variety was not readily available.

Charaka prescribed a decoction of Shaali, in prescriptions, for dysuria; liquid gruel of Rakta-Shaali in chest pain; Shaali and Shashtika rice along with butter, goat's meat and fresh wine for bleeding piles. The root of Shaali was an ingredient in Charaka's rejuvenating tonic Brahma Rasaayana.

Sushruta gave Shaali rice and barley as a diet in anaemia; old Shaali rice with meat soups in diet during the treatment of wounds; the husk of Shaali rice, mixed with purified butter, for external application in severe burns.

A confection of Rakta-Shaali and milk, added with honey, was prescribed in meno-metrorrhagia; also as a galactagogue (Raaja Maarttanda).

A decoction of rice, called rice-water, is used in Indian medicine as a demulcent and refrigerant in febrile and inflammatory diseases. It is given as a post prandial drink with many compound preparations.

Active principles and pharmacology

Analysis of 14 types of husked rice from different parts of India gave the following range of values: moisture 10.90–13.78; ether extract 0.59–2.59; protein 5.50–9.32; carbohydrates 73.35–80.81; fibre 0.18–0.95 and mineral matter 0.79–2.00 %.

Coloured and coarse grained rice is generally richer in protein, phosphorus and calcium than the fine grained type.

Starch is the major constituent of rice. The amylase content of the starch varies according to grain type. Superior types contain up to 17.5 % of amylase.

The nutritive value of rice protein as determined both by animal experiments and human metabolism studies is of a high order, being superior to wheat and other cereal proteins. The essential amino acid make up of the proteins of 6 Indian types of husked rice was as follows: arginine 7.92–12.46; histidine 1.3–2.0; lysine 3.2–4.4; tryptophane 1.15–1.50; phenylalanine 4.3–5.0; methionine 2.1–2.6; threonine 3.25–3.62; leucine 7.38–9.41; iso-leucine 5.46–6.74; and valine 5.60–7.80 g/16 g N. (Moisture 8.9–11.4; and total N 1.09–1.58 %.) The proteins appear to be rich in arginine compared to those of other cereals. They are deficient in lysine and threonine. Mineral contents of rice resembles other cereals.

Whole rice is a good source of B-vitamins, particularly thiamine, pantothenic acid and pyridoxine. The riboflavin content is low and ascorbic acid is practically absent. The amount of fat-soluble vitamins A and D is negligible, but vitamin E content is considerable.

The darker type of rice contains higher amount of thiamine than the white one.

Rice contains sodium (20 mg/100 g); potassium (100 mg/100 g); magnesium, sulphur and chlorine. Being low in sodium content, rice is found useful as a staple cereal in low-sodium diets.

Trace elements reported in the grain include aluminium, manganese, copper, zinc, arsenic, boron, chromium, nickel, cobalt, iodine (more than 65 % of the iodine is found in the bran fraction), fluorine, selenium, titanium, molybdenum, vanadium lead, tin, strontium, barium, rubidium and lithium.

The enzymes present in rice include alpha-amylase, beta-amylase, amylosynthetase, catalase, protease, lipase, phenolase, oxidase, peroxidase; and an inhibitor which interferes with the oxidation of pyrogallol.

The pigments occurring in the coloured types of rice are a mixture of monoglycosides of cyanidin and delphinidin. The dark rice contains a diglycoside anthocyanin.

Among other constituents of rice are: citric, acetic, fumaric, succinic, oxalic, malic and several aromatic acids including ferulic, vanillic and p-coumaric acids.

The excess water, which is drained off after cooking, contains vitamins, minerals, amino acids, sugars and gelatinised starch.

Papilionaceae

OUGEINIA

Ougeinia oojeinensis (Roxb.) Hochr.

Figure 1 Ougeinia oojeinensis—flowering and fruiting branches [WOI]

Ougeinia dalbergioides Benth.

Habitat

Outer Himalayas and sub-Himalayan tracts from Jammu to Bhutan up to an altitude of 1500 m, and extending through the whole northern and central India into the greater part of Deccan Peninsula.

Classical and common names

Ayurvedic: Tinishaa, Tinisha (Charaka, Sushruta); Syandana (Bhaavaprakaasha); Nemi, Sarvasaara, Ashmagarbhaka, Vajjala, Chitrakrt.
Siddha: Narivengayam.
English: Chariot Tree, Punjab Kino, Sandan.

Parts used

Heartwood.

Dose

Decoction 50–100 ml.

Classical use

In Ayurvedic texts, both Tinishaa and Tinisha have been mentioned. Tinishaa was used for alleviating burning syndrome during illness, while Tinisha was administered for skin diseases, urinary disorders, anaemia, and for treating obesity. It was also known as Atimuktaka, the drug for reducing obesity, which has now been equated with Hiptage benghalensis.

Charaka prescribed fresh juice of the bark and stalks, in prescriptions, internally for fevers, debility and as a tonic for recuperation. Sushruta administered the drug internally in obesity, jaundice, urethral discharges, chronic skin diseases; oil of the seeds as a digestive in bilious affections. A decoction of Tinisha, in combination with other intestinal antiseptics and astringents, was administered in haemorrhagic diarrhoea and dysentery.

In folk medicine, the bark is used as a febrifuge. A kino-like exudation from the incised bark is used in diarrhoea and dysentery.

Active principles and pharmacology

Leaves and heartwood contained the iso-flavonoids dalbergioidin, homoferreirin and ougenin. Leaves, in addition, contained the flavonoids quercetin and kaempferol, and leucopelaragonidin. Stem-bark gave the triterpenes lupeol and betulin.

The bark contains 7 % tannins.

50 % ethanolic extract of Ougeinia oojeinensis stembark showed antispasmodic action on isolated guinea-pig ileum and weak CNS-depressant effect in mice. The extract of the whole plant showed anti-inflammatory effect against carrageenin-induced paw oedema and analgesic effect in rats. The extract was devoid of diuretic, hypoglycaemic, antipyretic and anticoagulant activities.

The ethanolic extract of the stembark and the whole plant showed hypotensive action in cat/dog.

Oxalidaceae

OXALIS

Oxalis corniculata Linn.

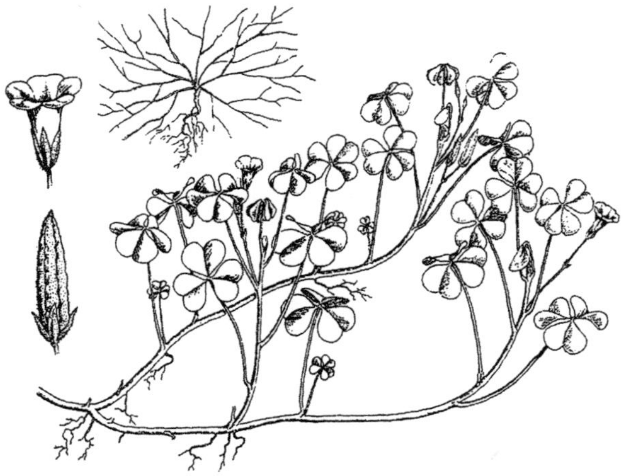

Figure 1 Oxalis corniculata [ADPS]

Habitat

A very common weed in cultivated and fallow lands, gardens and wastelands, particularly in moist and shady localities.

Classical an common names

Ayurvedic: Chaangeri, Amlapatrikaa, Amlikaa, Chukraa, Chukrikaa, Chhatraamlikaa. Chatuhcchaddaa is a wrong synonym of the herb, as it does not bear four leaflets.
Siddha: Pulai Kiri.
English: Indian Sorrel.

Parts used

Whole plant.

Dose

Juice 10–20 ml.

Classical use

Charaka and Sushruta use Chaangeri as a potherb for treating piles and mesenteric disorders.

Chaangeri Ghrita of Charaka, still available over-the-counter, was specific for diarrhoea and prolapse of rectum; Sunishannaka-Chaangeri Ghrita and Naagaraadi Ghrita was prescribed for piles.

Sushruta prescribed clarified butter, cooked with Chaangeri juice, paste of Piperaceae drugs and curd, for chronic diarrhoea.

Chaangeri juice, sour gruel and jaggery in equal quantity, churned together, was prescribed in insanity (Bangasena).

In folk medicine, fresh juice of the weed is given in dyspepsia and tympanitis; in fevers with biliousness. Leaf juice is also given to counteract the intoxication produced by the seeds of Dhattura (Datura metel).

Active principles and pharmacology

Aerial parts gave 2"-O-(beta-D-glucopyranosyl)-isovitexin.

Leaves are a good source of vitamin C (125 mg/100 g) and carotene (3.6 mg/100 g). They are rich in calcium (5.6 % dry material), but contain high contents of oxalates (12 % dry material). The leaves and stem contain tartaric and citric acids; stems contain also malic acid.

A crystalline principle, which produced fatal hypoglycaemic convulsions in rabbits, has been isolated from the plant. In Australia, the plant is suspected of causing sheep mortalities with symptoms of staggering and trembling. The plant, if eaten by dairy cows, affects the composition of milk and its consistency. Leaves are injurious if eaten in excess. (Wealth of India.)

An aqueous extract of the plant shows activity against Micrococcus pyogenes var. aureus. Expressed juice of the entire plant shows activity against Gram-positive bacteria.

Use in Western herbal

Oxalis acetosella (Wood Sorrel) is used for liver and digestive disorders. In the past, fresh leaves were also used to treat scurvy and wounds, and inflammation of gums.

The plant contains oxalic acid, a high vitamin C content, and clover acid, which in small amounts is diuretic.

Nicholas Culpeper (1616–1654) wrote of Wood Sorrel: "Wood Sorrel serves to all the purposes as other Sorrels do, and is more effectual in hindering putrefaction of blood, and ulcers in the mouth and body, and to quench thirst, to strengthen a weak stomach, to procure an appetite, to stay vomiting, and very excellent in any contagious sickness or pestilential fevers. The syrup, made of the juice, is effectual in all the aforesaid cases, and so is the distilled water of the herb."

According to PDR, the poisoning mentioned in older scientific literature seems dubious. Oxalic acid poisonings can occur only through the ingestion of very large quantities of the leaves, as, for example, in salad.

P

PAEDERIA

Paederia foetida auct. non L.

Figure 1 Paederia foetida [CCRAS]

Paederia scandens (Lour.) Merr.
Paederia tomentosa Bl.

Habitat

Himalayas, from Dehradun eastward up to an altitude of 1800 m, also in Bihar, Orissa, Bengal and Assam.

Classical and common names

Ayurvedic: Gandhaprasaarini. (The Wealth of India has equated Paederia foetida with Prasaarani, which is a synonym of Raaja Balaa.) Paederia foetida is a different drug. Raaja Balaa belongs to Malvaceae.
Siddha: Talanili, Pamarisangai.
English: Chinese Flower Plant, King's Tonic.

Parts used

Whole plant.

Dose

Juice 10–20 ml, decoction 50–100 ml.

Classical use

In Bhaavaprakaasha, Puutigandhaa is a synonym of Prasaarani. The drug used in Prasaarani Taila, Mahaamaashaadi Taila, Mahaanaaraayana Taila (all prescribed for massage in rheumatic affections) should be equated with Paederia foetida, as it possesses foetid smell, while Raaja Balaa (Sida veronicaefolia) is odourless. Kubjaa Prasaarini Taila of Chakradatta was also prescribed for massage in rheumatism with contraction and stiffness of joints.

In folk medicine, all parts of Gandhaprasaarini are employed for rheumatic affections.

A poultice of the leaves is applied to abdomen to relieve distension due to flatulence; also used in herpes. The juice of the root is prescribed in piles, inflammation of spleen, and pain in liver.

In the Philippines, boiled and mashed leaves are applied to the abdomen in cases of retention of urine; a decoction of the leaves is reported to possess diuretic properties and also to dissolve vesical calculi. (The Wealth of India.)

Active principles and pharmacology

The plant gave friedelan-3-one, beta-sitosterol and epifriedelinol.

Leaves are stems gave the iridoid glycosides asperuloside, paederoside and scandoside; sitosterol, stigmasterol, campesterol; ursolic acid, hentriacontane, hentriacontanol, ceryl alcohol, palmitic acid and methyl mercaptan.

Methyl mercaptan has been shown to be responsible for the foetid odour of the plant.

The decoction of Paederia foetida given orally in a daily dose of 1.5 ml (representing 0.75 g of the dry powder of the drug) for 10 days, showed significant anti-inflammatory action against formaldehyde-induced arthritis in non-adrenalectomized albino rats. Against carrageenin-induced rat paw oedema, decoction of whole plant showed a mild degree of anti-inflammatory activity.

The ethanolic extract (50 %) of Paederia foetida leaves exhibited antispasmodic activity on the isolated guinea-pig ileum.

The juice of leaves showed potent anthelmintic effect against bovine helminths, viz. Strongyloides spp., Trichostrongylus spp., Haemonchus spp. (100 %), Bunostomum spp. and Moniezia spp. (50–70 %) in young calves.

Paeoniaceae

PAEONIA

Paeonia emodi Wall. ex Royle

Habitat
North-western Himalayas, from Kashmir to Kumaon at 2000–3000 m.

Classical and common names
Unani: Ood Saleeb, Ood Gharqi, Phaavaania. (National Formulary of Unani Medicine also equated Ood Salib with Orchis latifolia, which is a different drug—Saalab misri, Khusyat-us-Saalab.)
English: Himalayan Peony.

Parts used
Root.

Dose
1–3 g (toxic in higher doses).

Classical use
Habb-e-Ood Saleeb is a reputed Unani medicine, prescribed in epilepsy, hysteria, paralysis, convulsions, insanity, nervous affections.

The drug is also employed in diseases of the liver, spleen and bladder.

The drug was imported into India from the Mediterranean region in the heyday of Unani medicine.

Active principles and pharmacology
Root oil gave a mixture of n-alkanes, beta-amyrin, butyrospermol, cycloartenol, lupeol, 24-methylenecycloartenol, cholesterol, campesterol, sitosterol; octanoic, decanoic, lauric, myristic, myristoleic, palmitic, palmitoleic, stearic, oleic, linoleic acids, and ethyl gallate.

Salicylaldehyde is the chief component of the essential oil.

Tubers and seeds of an allied European species, Paeonia officinalis Linn., contain a toxic alkaloid which produces contraction of the renal capillaries and increases the coagulability of blood.

Use in Western herbal
Since the time of Hippocrates (470–377 BC), Peony (Paeonia officinalis) was used for treating epilepsy. Ibn-el-Beitar, a medieval Arab physician, also recommended it for epilepsy.

Nicholas Culpeper (1616–1654) wrote of it: "The roots of male Peony, having been found by experience, cure the falling illness... The root is also effectual for women that are not sufficiently cleansed after child-birth."

Chinese Peony (Paeonia lactiflora) is used in prescriptions for headache, vertigo; spastic pains in the stomach and intestines, abdominal pain due to diarrhoea or dysentery abdominal mass, appendicitis; abnormal, painful or difficult menstruation.

Experimentally, Chinese Peony exhibits analgesic and spasmodic action in animals. The herb has been shown to prevent gastric secretion and may help to prevent gastric ulcers resulting from nervousness.

In experiments with mice, peony prolonged sleeping time, acted as anticonvulsive, lowered blood pressure and was anti-inflammatory. (Steven Foster and Yue chongxi.) The drug is used in the United States by traditional Chinese practitioners.

Caution
Contraindicated during pregnancy. Canadian regulations do not allow Paeonia officinalis as a non-medicinal ingredient for oral use products. (Michael McGuffin et al.)

Araliaceae

PANAX

Panax ginseng May
Panax schinseng Nees

Habitat
Eastern Asia. Cultivated in northern China, Korea and Japan. Imported into India.

Classical and common names
English: Asiatic Ginseng or Chinese Ginseng. American Ginseng (Panax quinquefolium).
Chinese: Ren-Shen, Jen-Shen (Asian Ginseng); Xi-yang-Shen, Hsi-yang-shen, (American Ginseng).

Use in Indian medicine
The most misused herb in India is Ginseng due to the misconception that it is an "infallible aphrodisiac", a "sexual rejuvenator" and a Chinese "panacea for longevity".

Ginseng with multivitamins and minerals was only the beginning of the long trail which brought all Indian adaptogens, antistress drugs and "sex tonics" into its fold. Over-the-counter, Ginseng is available in various combinations in which Ocimum sanctum, Withania somnifera, Asparagus racemosus, Mucuna pruriens, Dioscorea bulbifera, Argyreia speciosa, Tribulus terrestris have been indiscreetly incorporated. Even Shilaajeet (Asphaltum) has been added to Ginseng.

Active principles and pharmacology
Saponins are the primary active components of Panax species, termed ginsenosides or panaxosides; of the 13 or more ginsenosides in Ginseng, those that have been best studied as isolated compounds include ginsenosides Rg 1 and Rb 1. Rg 1 has been shown to slightly stimulate CNS-activity, to counter fatigue, and slightly increase motor activity; while Rb 1 works as a CNS-depressant, is anticonvulsant, analgesic, tranquillizing, hypotensive, and antipsychotic.

Ginsenosides, primarily responsible for the adaptogenic effect of Ginseng, are generally divided into two groups—"diols" and "triols"—based on two dammarine-type core triterpenes called protopanaxadiol and protopanaxatriol. Ginsenosides Rb and Rc are diols; Rg is a triol. The Rg "triols" are found to be arousing; the Rb and Rc "diols" sedative. ("Diols" predominate the American Ginseng.)

Other constituents of Panax Ginseng are: panax acid; glycosides (panaxin, panaquilin, ginsenin); alpha-sitosterol; stigmasterol, campesterol, a sesquiterpene (pancene); polyacetylenes (alpha-elemene, panaxynol); kaempferol; choline; an anti-oxidant ("maltol")' essential oil.

In India, the activity of Panax ginseng, Ocimum sanctum, Withania somnifera and other adaptogens was evaluated against stress by subjecting albino rats and mice to different stress situations (1981, 1982, 1984, 1985). In physical endurance test (swimming performance) and anoxia tolerance test, the score of Panax ginseng was shown lower than that of Ocimum sanctum, Withania somnifera and Altinga excelsa. (CCRAS 1999.)

There is no scientific evidence of enhanced sexual experience or potency resulting from Ginseng's use (Tyler). Of 37 clinical studies (between 1968–1990), not even one study could be cited in favour of Ginseng's use in impotence. There are question marks over Ginseng's use by women during their reproductive years. (Simon Y. Mills.)

Use in Western herbal
German Commission E recognized Ginseng as a tonic for invigoration and fortification in times of fatigue and debility, and during convalescence.

Pandanaceae

PANDANUS

Pandanus fascicularis Lam.

See figure 1 and 2.

Pandanus odoratissimus Linn. f.

Habitat
Sea coast of Indian Peninsula on both sides and the Andaman Islands. (Synonyms: Pandanus tec-

Figure 1 Pandanus fascicularis [CCRAS]

Figure 2 Pandanus fascicularis [CCRAS]

torius, Pandanus laevis, Pandanus variegatus, Pandanus latifolius, Pandanus amaryllifolius.)

Classical and common names

Ayurvedic: Ketaka, Ketaki, Suuchikaa pushpaa, (flowering season: August-September). Yellow-flowered variety: Svarna Ketaki, Laghu-pushpaa (flowering season: February-March).
Unani: Keoraa.
Siddha: Tali, Thaalay.
English: Screw-pine, Thatch Screw-pine.

Parts used

Flower, root.

Dose

Syrup 20–50 ml, infusion 20–50 ml, decoction 50–100 ml, paste 2.5–5 g.

Classical use

During the 16th century, Ketaka was an ingredient in compound preparations for osteo-arthritis, dysuria, abdominal mass, gynaecological disorders, and toxic conditions. Mahaanaarayan Taila of Bhaavaprakaasha was prescribed for rheumatic pain and gout. It is still available over-the-counter.

Root juice was administered in gynaecological disorders. Root, brayed in milk, was given in threatened abortion.

In epilepsy, a powder made of anthers and tops of bracts was given as a snuff.

An extract of the root was applied topically for arresting premature greying of hair (Bhaavaprakaasha).

In Unani medicine, Araq-e-Keoraa (Bayaz-e-Kabir) is administered as a cardiac tonic in palpitation. Sharbat-e-Keoraa (Qarabadeen-e-Jadeed) is given as a refrigerant and vitaliser. An infusion of fruit or leaves is prescribed in menorrhagia and amenorrhoea.

Active principles and pharmacology

Essential oil from flowers gave methyl ether of beta-phenylethyl-alcohol, benzylbenzoate, benzyl salicylate, benzyl acetate, geraniol, linalool, linalyl acetate, bromostyrene, guaiacol, beta-phenylethyl-alcohol, beta-phenylethyl-aldehyde, santalol. benzyl alcohol, dipentene, citral, phenylethyl-acetate, caproic acid, stearoptene and ester of phthalic acid.

Alcoholic extract of Ketaki exhibited depressant effect on perfused frog's heart and stoppage of heart in the diastole following higher doses. The effect remained unaffected by treatment of atropine. It did not show anti-inflammatory property against acute model of inflammation (carrageenin): on the contrary, an increase was observed in the paw oedema in such model of inflammation. It showed insignificant anticholinergic effect.

The extract in a dose of 0.5 mg/kg p.o. reduced urinary flow in rats. (CCRAS.)

Fruit infusion drunk twice a day showed an improvement in dysuria. The infusion also showed effectiveness against menorrhagia and secondary amenorrhoea; infusion of leaves exhibited styptic activity in vaginal bleeding.

Papaveraceae

PAPAVER

Papaver somniferum Linn.

Habitat

Native of Asia Minor and indigenous to Mediterranean costal region. Cultivated in Rajasthan, Uttar Pradesh and Madhya Pradesh under licenses issued by the direct opium officers of different areas. Cultivation is undertaken on behalf of the Government. For sale of opium and opium-based products, licences are required.

Classical and common names

Ayruvedic: Ahiphena, Aaphuuka. Dodaa (capsule). Poshta (outer skin of the fruit). Poshtadaanaa, Khashkhash (seeds). Afim (air-dried concrete milky latex or exudation obtained by incising the unripe seed-capsule of poppy. Opium).
Unani: Afyun (opium). Koknaar, Dodaa (capsule). Khashkhaash (seeds).
Siddha: Abini.
English: Opium Poppy, White Poppy, Opium. Red-flowered variety has been equated with Papaver somniferum var. glabrum Boiss., purple-flowered, known as black variety, is equated with Papaver somniferum var. nigrum.

Classical use

Aphiphena was not in use during the period of Charaka and Sushruta. Its first reference is found in Shaarangadhara Samhitaa and later on in Bhaavaprakaasha (16th century).

Aakaarkarbhaadi Churna of Shaarangadhara was prescribed as a sex-tonic; Yavaanyaadi Kwaatha (Bhaavaprakaasha) in obstinate cough and bronchitis.

Bhaishajya Ratnaavali incorporated many Ahiphena-based compound preparations for diarrhoea and dysentery—Ahiphenaasava, Ahiphena Vatikaa, Brihad Gangaadhara Churna, Karpura Rasa, to mention but a few. Dugdha Vati was prescribed in inflammatory conditions of liver, lungs, stomach, and intestines.

During the 16th century Ahiphena and Khashkhash entered into a number of compounds for rheumatic affections, neuralgias, haemorrhagic conditions, chronic bronchitis, insomnia, and sexual debility associated with psychosomatic disturbances.

In Unani medicine, Habb-e-Jadwaar is prescribed in premature ejaculation. For chronic bronchitis, restlessness, nervousness and insomnia, a number of Unani compounds have been incorporated in National Formulary of Unani Medicine—Barshasha, Dayaquzaa, Khamira-e-Khashkhaash, Habb-e-Khashkhaash, Habb-e-Shakeeqaa, Qurs Musallas, Habb-e-Sil. Sharbat-e-Khashkhaash is administered in diarrhoea and dysentery. For external application, Zimaad-e-Mubarrid is prescribed in cephalalgia as frigorific and sedative.

In practice, Unani physicians consider opium as an "infallible aphrodisiac, " which arouses sexual urge, helps in erectile consistency, prolongs the duration of coitus. This myth has now been eroded scientifically. Opium, after a temporary stimulation, diminishes sexual desire and its long-term use leads to drug-dependence and impotency.

Unani physicians use a camphorated opium liniment externally in rheumatism, lumbago, muscular and neuralgic pains, spasms, enlarged glands, mumps.

A mucilage of Khashkhaash and Tukhme-Kaahu (Lactuca sativa seeds), 2 and 1 part respectively, mixed with sugar, is given for insomnia.

In Indian medicine, Khashkhaash is considered nutritive (it is devoid of narcotic properties). Khashkhaash is used as a demulcent and emollient in obstinate constipation and catarrh of the bladder.

Seeds, pounded with milk, are applied over scalp for treating dandruff. Seed oil is used against scalds.

Active principles and pharmacology

Opium Poppy contains isoquinoline alkaloids (20–30 %): chief alkaloids are morphine (3–23 %), narcotine (2–10 %), codeine (0.2–3.5 %), papaverine (0.5–3 %), thebaine (0.2–1 %). The alkaloids are present as salts of meconic acid, lactic acid or fumaric acid. (Opium contains more than 40 alkaloids.) It also contains meconic acid, albumin, mucilage, sugars, resin and wax.

Morphine is an analgesic even in small doses, causes euphoria, then narcotic sleep. It slows down evacuation of the stomach, causing constipation and urine retention. Codeine is antitussive. Papaverine is spasmolytic and is used for gallbladder colic, bronchial and urogenital spasm. (PDR.)

Analysis of seeds of 5 types of Indian Poppy gave the following range of values: moisture 4.3–5.2, protein 22.3–24.4, ether extr. 46.5–49.1, N-free extr. 11.7–14.3, crude fibre 4.8–5.8, ash 5.6–6.0, calcium 1.03–1.45 and phosphorus 0.79–0.89 %; iron 8.5–11.1 mg/100 g; thiamine 740–1181, riboflavin 756–1203, and nicotinic acid 800–1280 mcg/100 g; carotene was absent. Minor minerals in the seeds include iodine, manganese, copper, magnesium, sodium, potassium and zinc. The seeds also contain lecithin, oxalic acid, pentosans, traces of narcotine and an amorphous alkaloid; the enzymes diatase, emulsin, lipase and nuclease.

The proteins are deficient in lysine and methionine.

A polysaccharide isolated from seeds showed antitumour activity in mice bearing yoshida sarcoma.

Caution

Overdoses and drug-abuse of opium leads to reduction of mental capacity, reactive euphoria, bradycardia, pulmonary and brain oedemas.

Opium is obsolete as a herbal drug. It is subject to legal restrictions in most of the countries.

Codeine 7–14 mg/kg in adults, 5 mg/kg in children is lethal. (Francis Brinker.)

Parmeliaceae

PARMELIA

Parmelia perlata Ach.

Habitat

A foliose lichen widely distributed in the plains of India and temperate Himalayas on tree bark, rocks, old walls.

Classical and common names

Ayurvedic: Shaileya, Shaileyaka, Shilaadaaru, Shailaka (Charaka); Shaila (Bhaavaprakaasha); Shilaapushpa, Kaalaanu-saarthaka, Bhuri-charilaa, Shilodbhava, Sthavira, Vrddha.
Unani: Chharelaa, Chhadelaa, Ushnah (equated with Usnea longissima Asch.)
Siddha: Karpasi.

Parts used

Lichen.

Dose

1–3 g.

Classical use

Charaka prescribed Shaileya for toxicosis and fevers. Mrtasanjivana Agada of Charaka, Taarkshya Agada and Mahaasugandhi Agada of Sushruta were specific compound preparations for counteracting poisons.

Shaileya was an ingredient in a number of medicinal oils for massage. Balaadya Taila, Shataavari Taila, Chandanaadi Taila of Shaarangadhara Samhitaa; Mahaabalaa Taila, Madhyam and Mahaa-naaraayana Taila, Mahaasugandhi Taila, Mahaa-chandandanaadi Taila, Gandha Taila of Bhaavaprakaasha were prescribed for oedema, arthritis, hemiplegia and wasting diseases.

Shaileya was an important ingredient in Shiroroga Yoga of Shaarangadhara Samhitaa and Jwarghni Vatikaa of Bhaavaprakaasha, prescribed for headache and fever. Sthaulya Yoga of Bhaavaprakaasha was for obesity.

Shaileya belonged to the Sarvaushadhi group of herbs, prescribed during the 16th century for fevers, neurological affections and psychoneurosis; also for eczema and oedema.

In folk medicine, Chharelaa is used in compound preparations for dyspepsia, diarrhoea, dysentery, spermatorrhoea, amenorrhoea. Its powder is applied to promote healing of wounds. In the form of a poultice Chharelaa is applied to renal and lumber regions in retention of urine. As a liniment, it is applied to the forehead during headache; also used as an incense.

In Unani medicine, Chharella is used in confections prescribed as cardiac tonics and refrigerants.

Chharelaa is an ingredient in Raughan-e-Surkh (Bayaz-e-Kabir), prescribed for massage in rheumatism, sciatica and gout.

Only fragrant species are used in Indian medicine.

Active principles and pharmacology

Lichens, in India, generally known as Chharelaa, constitute a class of perennial plants which is actually a combination of two organisms—a fungus and an alga—growing together in symbiotic association. Some 700 species are reported to occur in India.

Chharelaa has been equated with Parmelia perlata by Ayurvedic scholars, while Unani scholars treats it as Parmelia kamtschadalis Eschew. syn. Parmelia cirrhata Fr. Chharelaa is a synonym of Ushnah, equated with Usnea longissima Asch. by the National Formulary of Unani Medicine.

Chemical components of Parmelia perlata are lecanoric acid and atranorin. Usnea spp. contains usnic acid and salazinic acid. Kodaikanal spp. contains d-usnic acid, barbatic acid, stictic acid and caperatic acid.

An analysis of Parmelia kamtschadalis gave the following values: Crude protein 8.2, ether extract 2.6, iso-linchenin 22.5, crude fibre 11.0 and ash 12.3 %. Calcium 398 mg, phosphorus 156 mg, iron 52 mg, ascorbic acid 4.4 mg and riboflavin 20 mcg/100 g.

Parmelia cirrhata contains atranorin (0.6 %), salazinic acid (2.0 %) and d-protolichesterinic acid (0.6 %).

Some samples of lichen contain as high as 20–25 % lecanoric acid, though the bulk of Indian specimen contain around 5.0 %.

Usnea longissima is a pendulous lichen common in temperate and alpine Himalayas on trees. An analysis gave the following values: crude protein 5.4, ether extr. 2.2, iso-lichenin 46.3, crude fibre 3.8, and ash 3.5 %. Calcium 1028 mg, phosphorus 80 mg, iron 90 mg, ascorbic acid 3.4 mcg, and riboflavin 310 mcg/100 g.

Parmelia spp. exhibit astringent, resolvent, aperient and diuretic properties.

Usnea spp. contains a high percentage of usnic acid (3.4); they also contain barbatic acid and arabitol. The lichen is reported to be used in China as an expectorant and in the treatment of ulcers.

Active compounds of several lichens (Parmeliaceae and Usneaceae) are found active against Gram-positive bacteria. Usnic acid is found active against Streptococcus haemolyticus and Pneumococcus spp. and inhibits growth of tubercle bacillus. Barbatic acid and dibasic roccellic acid, in the form of its half esters or half amides, possess antitubercular activity. Barbatic acid also produce haemolysis.

Use in Western herbal

Usnea of Western herbal consists of the dried thallus of Usnea barbata, Usnea florida, Usnea hirta and Usnea plicata. It is used for its antimicrobial activity in inflammation of mouth and pharynx, in the form of lozenges equivalent to 100 mg herb (one lozenge 3–6 times daily). Usnea cough lozenges are sold over-the-counter. For external use, Liquorice and Propolis Spray, containing Usnea, is also available.

Malvaceae

PAVONIA

Pavonia odorata Willd.

Habitat

Found in open woods and waste places in the Deccan Peninsula, parts of West Bengal, Bihar, Orissa, Uttar Pradesh and Rajasthan. Also cultivated in gardens.

Classical and common names

Ayurvedic: Baalaka, Barhishtha, Hrivera (Sushruta); Baala, Ambu, Jala, Nira, Paya, Toya, Udichi, Udichya, Vaari, Murdhaja, Kacha (Bhaavaprakaasha). Sugandhabaalaa (not to be confused with Valeriana wallichii or Valeriana jatamansi, also known as Sugandhabaalaa). In the South, Coleus vetiveroides K.C. Jacob (Labiateae) is used as Baalaka.

Siddha: Permutti.

English: Fragrant Sticky Mallow.

Parts used

Leaves, roots.

Figure 1 Pavonia odorata Willd [ADPS]

Classical use

Sushruta gave the plant-preparations internally in skin eruptions and diseases due to vitiated blood. In the form of a medicated Ghrita (purified butter) it was prescribed in phthisis, asthma, emaciation, debility and neurological disorders. Externally, a paste of the herb was applied in erysipelas.

During the 16th century, the drug entered into a number of compound preparations for fevers of various origins; diarrhoea, dysentery; arthritis, gout; cough, and asthma.

Sudarshana Churna (Shaarangadhara Samhitaa) is still available over-the-counter. It is prescribed in intermittent fevers, as an antiperiodic, blood-purifier and digestive. Raasnaadi Churna is for fevers accompanied with cold and catarrhal symptoms. Chandan-Balaa-Lakshaadi Taila (Yoga Ratnaakara) is for external application in fevers.

In folk medicine, the roots are used for their refrigerant, antipyretic, stomachic and astringent properties. The root, with Aegle marmelos, is administered in dysentery.

Active principles and pharmacology

Roots gave an essential oil containing isovaleric acid, isovaleraldehyde, aromadendrene, pavonene, alpha-terpinene, azulene and pavonenol.

Biological activity of the plant was found to be spasmolytic, antiprotozoal and antipyretic.

Pedaliaceae/Zygophyllaceae

PEDALIUM

(a)Pedalium murex Linn.

Figure 1 Pedalium murex [ZANDU]

(b)Tribulus lenuginosus Linn.
Tribulus terrestris Linn.

Habitat

Western and Coromandal Coasts of South India. Tribulus terrestris (Zygophyllaceae) is distributed throughout India.

Classical and common names

Ayurvedic: (a) Brihatgokshura, Kshourak, Trikantaka. (b) Gokshura, Gokshuraka, Gokanta, Gokantaka, Trikantaka, Svaadukantaka, Shvadamshtraa (Bhaavaprakaasha). The fruit has four spines, hence the synonym Trikantaka (three thorns) is not applicable to any of these two.

Unani: (a) Khaar-e-Khasak Kalaan, Gokharu Kalaan. (b) Khaar-e-Khasak Khurd, Gokharu Khurd.

English: Caltrops (large variety, small variety).

Parts used

Whole plant, fruits seeds.

Dose

Powder 3–6 g, decoction 50–100 ml.

(A visid mucilage, resembling that of gum arabic, separates out by simple agitation of young twigs, leaves, fruits and seeds in water or milk. This mucilaginous infusion is administered in diseases of urinogenital system in Indian medicine.)

Classical use

A gruel prepared with Gokshura and Kantakaari (Solanum surattense) and added with jaggery (Charaka Samhitaa, Ashtaanga Sangraha); a decoction of Gokshura, Varuna (Crateva nurvala) and Shunthi (dry Zingiber officinale), mixed with honey (Bhaavaprakaasha); a decoction of Gokshura with Yavakshaara—impure carbonate of potash—(Vrindamaadhava); were specific preparations, recommended for dysuria, gravels and calculus.

In general practice, a simple decoction of Gokshura seeds or of the whole plant, mixed with honey, or a paste of Gokshura with coconut water was prescribed for dysuria. (Shaarangadhara Samhitaa, Bhaavaprakaasha, Ashtaanga Hridaya, Bangasena, Vaidya-manoramaa.)

Gokshuraadya Leha (Bhaavaprakaasha) was recommended for painful micturition, suppression of urine, haematuria, calculus affections. Gokshuraadi Churnam, Gokshuraadi Guggula, Dashmuul Kwaatha were prescribed for phosphaturia, spermatorrhoea, diseases of the genitourinary system, internal inflammations, and sexual debility.

As a "sex tonic" and for treating "impotence", Gokshura was administered alone, cooked in milk, as well as in the form of compound preparations with other nervine and rejuvenating tonics—sesamum seeds, Aamalaka (Emblica officinalis), Guduuchi (Tinospora cordifolia), Shataavari (Asparagus racemosus), Kapikacchuu (Mucuna prurita), Balaa (Sida cordifolia), Atibalaa (Abutilon indicum). (Ashtaanga Hridaya, Gadanigraha, Vrindamaadhava.)

Powder of Gokshura fruits and Ashwagandhaa (Withania somnifera), mixed with honey and taken with milk, was prescribed for alleviating consumption and cough (Raaja Maarttanda.)

A decoction of Gokshura and Shunthi (dry Zingiber officinale) was prescribed for rheumatism and lumbago.

Both the varieties of Gokshura were used in Indian medicine for similar properties. As the fruit is conical and has four spines, the synonym Trikantaka is still creating confusion. Ayurvedic scholars suggest that Acanthospermum hispidium DC. or Martynia annua Linn. is the source of the classical drug Trikantaka.

Gokshura was an ingredient of Dashmuula (the Ten Roots) of Ayurvedic texts for its diuretic properties. Dashmuulaarishta (Baishajya Ratnaavali) is a restorative tonic for women, suffering from urinogenital diseases.

Dashmuula Kwaatha Churna (Bhaishajya Ratnaavali) is prescribed in enteric fevers as a tonic. Dashmuula-Katutrayaadi Kwaatha Churna (Sahasrayoga) is a digestive, carminative and antiperiodic tonic. Chitraka Haritaki (Chakradatta); Chyavanapraasha Avaleha (Charaka), Dashmuula Haritaki (Ashtaanga Hridaya) also contain Dashmuula. (All available over-the-counter.)

Among other over-the-counter tonics, Gokshura is an ingredient in Shataavari Lehya (Sahasrayoga), prescribed as a diuretic and restorative tonic in menorrhagia, haemoptysis, jaundice, urinary disorders.

Sharbat Buzoori (Bayaz-e-Kabir) is a popular Unani compound for anuria, amenorrhoea and hepatic obstruction.

Sufoof-Gond Kateera, Sufoof-Beejband, Sufoof-Daarchini, Sufoof-e-Maghaz-e-Kamalgattaa (Bayaz-e-Kabir), are prescribed in nocturnal discharge, spermatorrhoea and "impotency".

Active principles and pharmacology

Pedalium murex: Leaves gave the flavonoids pedalitin and its glucoside (pedaliin), diometin, dinatin, dinatin-7-glucuronide and diosmetin-7-glucuronide.

Fruits and leaves yield phenolic acids which include caffeic acid, protocatechuic acid, p-coumaric acid and ferulic acid.

Fruits also yielded hepatriacontan-4-one, tetratriacontanyl octacosanoate, pentatricontane, sitosterol, hexatria-contanoic acid, hentriacontanoic acid, ursolic acid and vanillin.

Flowers gave dinatin, quercetin and quercetin-7-glucoside.

Glucoside obtained from the plant as well as decoction from it exhibited diuretic activity in rats (CCRAS.)

Pet.-ether extract in higher concentrations was found effective against Anopheles stephensi, Culex quinquefasciadus and Acdes aegypti larve.

Tribulus terrestris: Plant contains carboline alkaloids, steroidal sapogenins, saponins and flavonoids. Aerial parts with fruits contain the flavonoids rutin, quercetin, and kaempferol. Other steroidal glycosides reported from the plant are trillin, gracillin and dio-genin-D-glucoside.

Roots, also flowers, contain the phytosterols campesterol, beta-sitosterol, and stigmasterol, and amino acids. Plant contains diosgenin also. Seeds contain the carboline alkaloids harmane and harmine. Harmol is also reported.

The ether extract of fruits caused diuresis, increased the creatinine renal clearance in anaesthetised dogs. (CCRAS.)

The litholytic action of the fruit is attributed to the aspartic and glutamic acid content.

The furoctanol biglycosides are reported to be active in stimulating spermatogenesis and sertoli cell activity in rats. Oral application of saponin terrestrioside-F in male rats increased libido and sexual response. In female rats, this compound potentiated oestrus and increased fertility. (Chemexcil.)

Alcoholic extract of fruit showed antibiotic activity against E. coli and Stap. aureus; antifungal effect against T. montagrophytes, M. tonsurans, T. rubrum, M. gypsum, M. canis, E. flocossam and C. albicans.

Arecaceae

PHOENIX

Phoenix sylvestris Roxb.

Habitat

Found throughout India up to an altitude of 1500 m, in alluvial soil.

Classical and common names

Ayurvedic: Kharjura, Kharjuraka, Kharjurikaa. Chhuhaaraa (dry date).
Unani: Khurmaa.
Siddha: Periya itcham.
English: Date.

Classical use

Charaka prescribed clarified butter processed with Kharjura, Draakshaa (dried Vitis vinifera) and Pippali (Piper longum) as a tonic in emaciation, consumption, cough, asthma, fever; a fermented juice of Kharjura, Draakshaa and sugarcane for diarrhoea; a linctus of Kharjura pulp, coconut and Draakshaa in vomiting. Sushruta gave Kharjura, mixed with honey in intrinsic haemorrhage. The fruit was included in prescriptions for haematurea (Vrindamaadhava).

A linctus made of Kharjura, Draakshaa and Pippali was a household remedy for cough during the 16th century.

A potion prepared of Kharjura, Draakshaa and Daadima (pomegranate), mixed with sugar, powder of parched rice (red variety), and honey, was used as a refreshing drink for pacifying obstinate cough (Yoga Ratnaakar).

Pinda-kharjura, the bigger variety, has been equated with Phoenix dactylifera Linn. Raaja Nighantu mentioned five varieties: Kharjuri, Pinda Kharjuri, Raajkharjuri, Madhukharjuri, Bhuukharjuri. Bhuukharjuri is considered as Phoenix acaulis Roxb. or Phoenix humilis Royle by Ayurvedic scholars.

According to Bhaavaprakaasha, all the varieties of Kharjura and Kharjuri are cooling, sweet, unctous, appetiser, nourishing, refreshing, cardiac tonic and aphrodisiac, spermatogenic. They cure phthisis, consumption, bronchitis, asthma, haemothermia, fainting, alcoholism, intoxication.

Majoon-e-Aarad khurmaa (Bayaz-e-Kabir) of Unani medicine is a reputed tonic for spermatorrhoea, attenuated semen, oligospermia, sexual debility.

Milk, boiled with Khurmaa (Chhuhaaraa), is given as a general tonic.

Active principles and pharmacology

Charged C-glycosylflavones and caffeylshikimic acid, leucocyanidin are characteristically present in the plant. Flavonol glycosides are also common;

the 3-glucosides, 3-rutinosides and 3, 4'-digluco-sides of quercetin are variously present. Tricin-7-glucoside, luteolin-7-rutinoside and several uncharged C-glycosylflavones were detected.

Fruits contain sugar (50–85 %)—saccharose, inverted sugar; leucoanthocyanidine; piperidine derivatives—pipecolic acid, 5-hydroxy-pipecolic acid, and fatty oil (10 %). Baikiain is found in the seed only.

The main component of ripe dates are sugars, up to 85 % of total solids. Sucrose is the predominant sugar in dry varieties, sucrose and invert sugar in semi-dry types, and invert sugar predominates in the soft dates. Besides sucrose and invert sugars, rhamnose, xylose, arabinose, galactose and galacturonic acid have been identified in the fruit.

Analysis of the edible matter of freshly dried dates (edible matter 86 %, seed 14 %) obtained from Coonoor market (South India) gave the following values: moisture 15.3, protein 2.5, fat 0.4, carbohydrates 75.8, fibre 3.9, mineral matter 2.1, phosphorus 0.05, and calcium 0.12 %. Iron 7.3 mg, carotene as vitamin A 44 I.U., thiamine 0.011, riboflavin 0.023, nicotinic acid 0.9, and ascorbic acid 3 mg/100 g. Calorific value 317 cal/100 g.

Analysis of the edible portion (80 %) of the dried dates (Aarada, Chhuhaaraa) gave the following values: moisture 11.9, protein 2.9, fat 0.5, carbohydrates 82.9 and ash 1.8 %. Calcium 35.9, phosphorus 129.3 and iron 3.4 mg. Calorific value 347 cal/100 g.

Jaggery, made out of sap (sucrose content 10 %), contained moisture 9.6, carbohydrates 86.1, protein 1.5, fat 0.3; minerals 2.6, calcium 0.36 and phosphorus 0.06 %.

Dates exhibit demulcent, expectorant, laxative and tonic properties and are found efficacious in respiratory diseases, fevers, emaciation and low vitality. In Yeaman, they are administered in memory disturbances.

An estrogenic substance has been isolated from the non-saponifiable fraction of fatty oil from dried pollen. Pollen exhibited on gonadotropic activity in immature rats. The presence of cholesterol and stroke has been reported from the pollen.

Euphorbiaceae

PHYLLANTHUS

Phyllanthus fraternus Webster

See figure 1.

Phyllanthus amarus Schum & Thonn.

See figure 2.

Phyllanthus urinaria Linn.

Habitat

Bhuumyaamalaki is a small herb (up to 60 cm in height) while Aamlaki is a tree.

Phyllanthus fraternus/ Phyllanthus amarus is found occurring as a winter weed throughout the hotter parts of India, particularly on cultivated land up to 1000 m.

Phyllanthus urinaria is a diffusely branched herb, met with as a weed of cultivated areas throughout India, in the plains from Punjab to Assam and southward to Kerala up to 1000 m.

Most of the treatises on Ayurvedic herbs wrongly identify the drug Bhuumyaamalaki as Phyllanthus niruri Linn. Ayurvedic Formulary of India equates Taamlaki with Phyllanthus niruri. But taxonomic studies done on this genus by Prof Webster of California University reveal that Phyllanthus niruri is an American species. The Indian drug assigned to Phyllanthus niruri has been subsequently equated with Phyllanthus fraternus Webster. (ADPS.)

Classical and common names

Ayurvedic: Bhuumyaamalaki, Bahupatri, Bhuudhaatri, Taamalaki, Bahuphalaa.
Unani: Bhui Aamalaa.
Siddha: Kizhai-nelli, Shivappu-nelli.

Parts used

Whole plant, root.

Dose

Juice 10–20 ml, powder 3–6 g, infusion 14–28 ml.

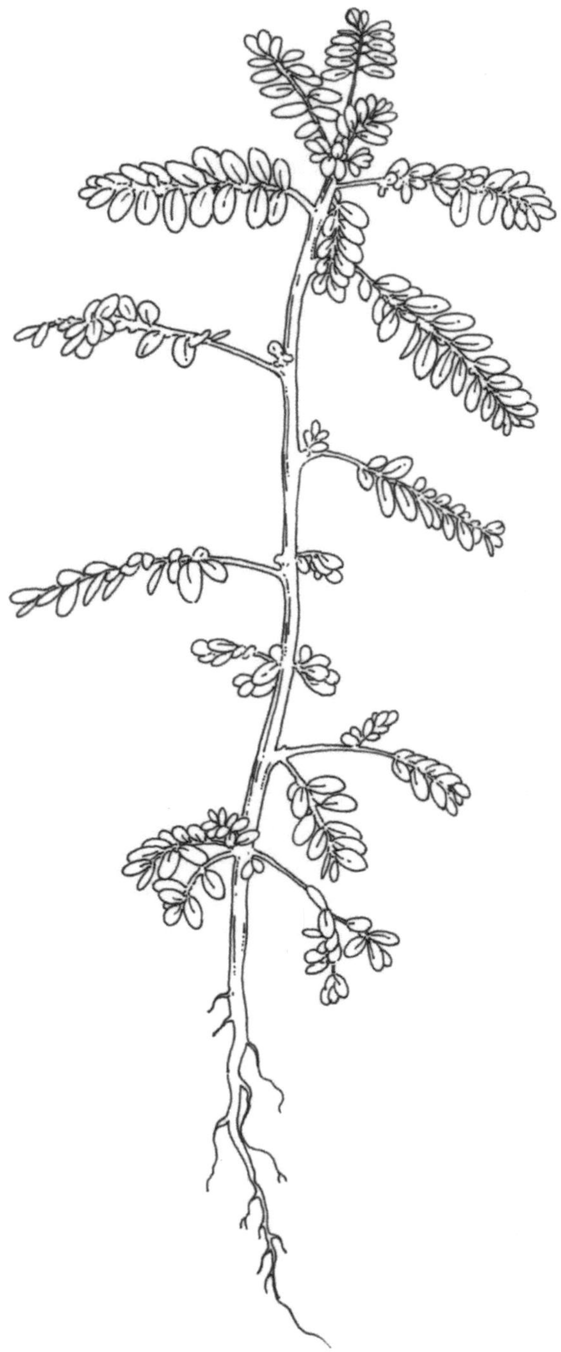

Figure 1 Phyllanthus fraternus [WOI]

Classical use

A paste of Bhuumyaamalaki made with butter milk was prescribed in jaundice. A decoction of the root or dry leaves was also administered (Vaidyamanoramaa).

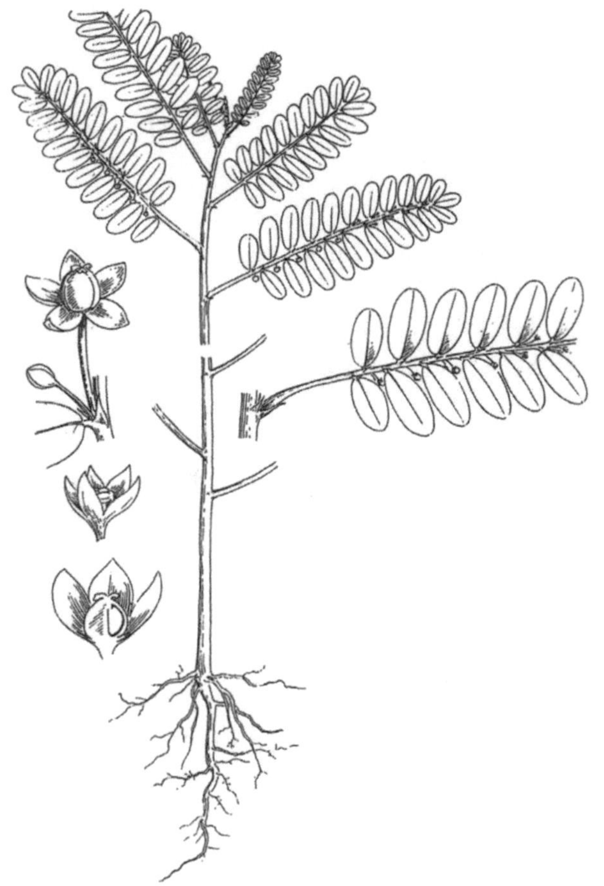

Figure 2 Phyllanthus amarus [ADPS]

Root, also seed, as powder or paste, with rice-water, was given in menometrorrhagia, haematuria, diarrhoea with blood (Bangasena).

Whole plant of Bhuumyaamalaki (20 g) and Maricha (Piper nigrum, 20 in number), pounded together, were prescribed for alleviating urinary disorders (Yogaratnaakara).

In folk medicine, an infusion of tender leaves is administered in chronic dysentery; a poultice of the root and whole plant, combined with rice water, is used for ulcers, sores and swellings; a poultice of leaves, mixed with salt, for skin affections.

In Unani medicine, the plant is used in jaundice as deobstruent, diuretic, cooling and astringent.

Active principles and pharmacology

Phyllanthus fraternus dry leaves contain 0.4 % of a toxic bitter principle phyllanthin, traces of a non-bitter substance hypophyllanthin, and about 5 %

of a colourless wax, made up mostly of esters (85 %) of long chain fatty acids and alcohols, free fatty acids and hydrocarbons. The leaves are rich in potassium (0.83 %), which is considered responsible for their diuretic activity.

Alcoholic extracts of leaves and roots show antibacterial activity against Micrococcus pyogenes var. aureus and E. coli.

In vitro tests were carried out to test the ability of various extracts of Phyllanthus fraternus leaves to inactivate the hepatitis B surface antigen (HBsAg). The crude extracts as well as "red pigment" obtained by TLC-purification showed potent in vitro immunoinactivating ability against HBsAg.

In a clinical trial involving 160 children (age group 1–12 year) at the Indian Institute of Research in Yoga and Allied Sciences, Tirupati, Phyllanthus fraternus (50 mg/kg in 3 divided daily dosage) has been claimed to be effective in the treatment of infective hepatitis. Out of these 160 children, 101 cases were cured and 59 cases dropped out. In a majority of cases, while disappearance of jaundice and hepatic tenderness was seen in the second week, it took five weeks for all the clinical features to disappear from all the cases and six weeks for bile salts and pigments. Improvement of appetite was quicker and more pronounced than other clinical features, as a majority of the cases (77.22 %) regained their normal appetite within one week. No side-effects were observed during the study (1982).

Phyllanthus fraternus has been shown to be effective as an adjunct, along with other drugs, in the treatment of jaundice due to infective hepatitis.

The aqueous extract of Phyllanthus fraternus leaves were reported to produce hypoglycaemic action in normal as well as alloxan-diabetic rabbits. The extract lowered the blood sugar level even when it was administered one hour after glucose administration. The hypoglycaemic activity of the leaf extract appeared to be higher than that of tolbutamide.

Phyllanthus amarus gave the lignans, a diarylbutane, phyllanthin and an aryltetrahydronapthalene, hypophyllanthin; the hydrolysable tannins phyllanthusiin D, amariin, amarulone and amarinic acid; the alkaloids ent-norsecurinine, sobubbialine, epibubbialine, a diarybutane, nyrphyllin and a neolignan, phyllnirurin.

Phyllanthus amarus exhibited antiviral actions on hepatitis B in human subjects. Protected the liver damage in rats induced by carbon tetrachloride. The lignans phyllanthin and hypophyllanthin exhibited antihepatotoxic activity in primary cultured rat hepatocytes.

The drug also showed diuretic and hypotensive effect in humans. (RRL, Jammu.)

Bitterness value of Phyllanthus amarus is 70 times more than Phyllanthus fraternus and Phyllanthus urinaria and 14 times more than Phyllanthus maderaspatensis. The bitterness imparts the therapeutic efficacy to the drug.

Leaves and stem of Phyllanthus urinaria gave the flavonoids quercetin, astragalin, quercitrin, iso-quercitrin and rutin.

Phyllanthus urinaria (whole plant) extract was reported to exert antibacterial activity against E. coli, Sal. typhosa, Vib. comma and Sh. dysenteriae.

Phyllanthus maderaspatensis Linn.

Habitat

A herb, or sometimes an under-shrub (30–90 cm high), occurs throughout the drier parts of India.

Classical and common names

Unani: Kanochaa, Isfahaan Marv.

Classical use

An infusion of leaves is prescribed for headache by Unani physicians. Seeds are used as a laxative, carminative and diuretic.

A mucilaginous preparation of the herb is administered in diarrhoea and dysentery.

Active principles and pharmacology

The dried seeds on extraction with light petroleum gave 8.1 % of a clear deep yellow oil. The fatty acid composition (saturated 10.2 and unsaturated 89.8 %) revealed the presence of myristic, palmitic, stearic, oleic, linoleic and linolenic acids. Beta-sitosterol has been identified in the unsaponifiable matter.

The defatted seeds yield 15.3 % of a fibrous mucilage which yields galactose, arabinose, rhamnose and aldobionic acid.

The presence of a reddish brown colouring matter, maderin, and an essential oil has also been reported.

Scrophulariaceae

PICRORHIZA

Picrorhiza kurroa Royle ex Benth.

Figure 1 Picrorhiza kurroa [WOI]

Habitat

Alpine Himalayas from Kashmir to Sikkim at altitudes of 2700–4500 m.

Classical and common names

Ayurvedic: Katukaa, Katurohini, Kattarohini, Katuki, Katukikaa, Shatparvaa, Shakuldaani, Arishta, Ashokarohinya (Charaka, Sushruta).
Unani: Kharbaq-e-Hindi, Kutki.
Siddha: Kaduguragini.
English: Picrorhiza.

Parts used

Roots, tubers.

Dose

Powder 1–3 g.

Classical use

Charaka and Sushruta gave a decoction of roots and tubers internally in jaundice, intestinal catarrh, colic, indigestion, dermatosis, vaginal discharges and other female diseases, including deficient lactation. Charaka's Katukaadya Ghrita was specific for anaemia and jaundice. A paste of Katukaa and Madhuka (Glycyrrhiza glabra), mixed with sugar, was prescribed as a cardiac tonic.

Charaka and Sushruta attributed antitoxic and febrifuge properties to the drug.

Katukaa mixed with sugar was prescribed for acid-dyspepsia (Vrindamaadhava), also for bronchial asthma and fevers due to derangement of digestive system (Ashtaanga Hridaya, Vrindamaadhava, Gadanigraha, Vaidyamanoramaa).

Katuki is an important ingredient in a number of over-the-counter compound preparations: Punarnavaa Mandoor (Bhaishajya Ratnaavali) for anaemia and jaundice; Sudarshana Churna (Ashtaanga Hridaya) for intermittent malarial fevers and skin diseases; Kutaja Phanita (Ashtaanga Hridaya) for diarrhoea, dysentery, enteric affections, peptic ulcers, haemorrhoids; Triphala Guggulu for piles, fistula and other inflammatory conditions; Mahaayogaraaja Guggula (Shaarangadhara Samhitaa) for arthritis, rheumatism; Panch-Tikta-Guggul Ghrita (Ashtaanga Hridaya), Manjishthhaadi Kwaatha Churna (Shaarangadhara Samhitaa), Panch-Tikta-Kwaatha Churna (IMPCOPS) for chronic skin diseases; Phalakalyaan Ghrita (Bhaishajya Ratnaavali) for uterine affections; Prabhanjan Vimardan Tailam (Sahasrayoga) for nervous and neurological disorders.

The most popular and widely prescribed compound preparation of Katuki is Aarogyavardhani Vati (Rasa Ratna Samuchchaya) for jaundice, anaemia, liver and spleen enlargements, chronic skin diseases and for fevers as an antiperiodic. The drug contains a number of minerals—calcined iron, mica, copper; purified mercury, sulphur, and asphaltum.

Active principles and pharmacology

Moideen Sheriff was the first Indian writer (1869) to equate Picrorhiza kurroa with Kutki. He also asserted that Kharbaq-e-asvad and kharbaq-e-siyaah of Unani medicine should be equated with Helleborus niger of Europe.

Roots of Picrorhiza kurroa yield a glycosidal bitter principle kutkin, found to be a mixture of two iridoid glycosides—picroside I and kutkoside (6'-O-*trans*-cinnamoylcatalpol and 10-O-vanilloylcatalpol). Also, D-mannitol, kutkiol, kutkisterol and a ketone subsequently found identical with apocynin.

Minor constituents include: picroside II, veronicoside, minecoside, the phenol glycosides picein and androsin, a number of cucurbitacin glycosides and 4-hydro-3-methoxyacetophenone.

The decoction of Picrorhiza kurroa, when administered to rats with carbon tetrachloride-induced liver injury, showed lesser structural damage in the hepatic tissue, as compared to the control rats. The alcoholic extract showed protective action against hepatotoxicity induced by carbon tetrachloride in rabbits, as evidenced by marked regression of SGOT and SGPT and serum alkaline phosphatase levels. The percentage retention of bromosulphathalein also showed comparatively lower values in the drug treated group.

The water soluble fraction of the 90 per cent alcoholic extract of Picrorhiza kurroa exhibited smooth muscle relaxant, laxative (in rats) and choleretic (in dogs) activities. It was devoid of analgesic and anticonvulsant actions and did not modify barbiturate-induced hypnosis.

The aqueous extract of the roots showed moderate antibacterial activity against Staph. aureus and Sal. typhi and marked inhibition against E. coli.

In an uncontrolled clinical study on 6 patients of infective hepatitis with jaundice, a decoction of Picrorhiza kurroa was reported to have led to a rapid fall in serum bilirubin levels towards normal range and a quicker clinical recovery of patients.

In the same study, two compound Ayurvedic preparations, viz. Aarogyavardhani and Phalatrikaadi Kwaatha (decoction)—both having Picrorhiza kurroa as the main ingredient—were also used in 6 patients each. The results were similar to those of Picrorhiza kurroa given alone.

Aarogyavardhani, a herbo-mineral preparation containing Picrorhiza kurroa as the major ingredient (along with Terminalia chebula, Terminalia belerica, Emblica officinalis, Commiphora mukul, Plumbago zeylanica and calcined minerals, was evaluated in a pilot clinical trial on 24 patients of viral hepatitis. Administered orally in a dose of 500 mg thrice a day for 3 weeks, the drug led to a significant degree of improvement in clinical symptoms and signs including appetite and liver size and significant lowering of the serum transaminases (SGOT, SGPT) and serum bilirubin in the first week of the treatment. There were no major side-effects and the extract was well tolerated.

These results were confirmed in a double blind clinical trial on 38 patients during an epidemic of acute viral hepatitis in Mumbai, when Aarogyavardhani was administered orally for 14 days (in a dose of 750 mg thrice a day). The drug markedly reduced tenderness and hepatomegaly as also the total serum bilirubin and the SGPT levels in the first week.

Another compound herbal preparation, which in addition to Picrorhiza kurroa contained Commiphora mukul, Myrobalans (Triphala), Boerhavia diffusa, and Strychnos nux-vomica, was evaluated in a clinical trial on 50 patients of infective hepatitis. This open trial indicated alleviation of the symptoms within 4 days, whereas jaundice was reported to have disappeared in 80 per cent of the patients within 7 days. The drug restored the liver functions to normal, as assessed by a fall of serum bilirubin as well as SGOT and SGPT levels. No side-effect was observed during therapy.

Punarnavaadi Kwaatha—yet another compound preparation—containing Picrorhiza kurroa along with Boerhavia diffusa, Azadirachta indica, Tinospora cordifolia, Terminalia chebula, Berberis aristata, Trichosanthes dioica and Zingiber officinale, was also evaluated for its efficacy in the management of viral hepatitis in 24 patients in a pilot trial. The compound showed encouraging results. (CCRAS, ICMR.)

Picrorhiza kurroa has also been claimed to have led to beneficial results in the management of bronchial asthma. In a study on 20 patients of bronchial asthma, a preparation containing the crude extract P. kurroa roots gave encouraging results in 10 patients. Patients with mild to moderate asthma responded with clinical relief, reduction in the need to use of bronchodilators and by better performance in pulmonary function tests.

In an open-ended clinical trial of Aarogyavardhani as a hypocholesterolemic agent on 101 patients with hypercholesterolemia, the drug was reported to produce marked reduction in serum cholesterol (with a mean fall of 29.1 %) and coagulation time. The hypocholesterolemic action of Aarogyavardhani was attributed to the presence of Picrorhiza kurroa, which is a proved cholagogue.

Picrorhiza kurroa (300 mg) on hypercholesterolemic patients produced significant reduction in cholesterol level in 80 % cases. (CCRAS.) Open trials in bronchial asthma gave encouraging prophylactic response with prolonged administration. (Chemexcil.)

Caution

Kutkin free extracts are not only devoid of any hepatoprotective activity but may aggravate galactosamine toxicity and should therefore be avoided in the treatment of liver disorders. Adulterants and substitutes should be checked for the presence of kutkin.

Herbs substituted in the name of Katuki are: Gentiana Kurroo Royal, Gentiana decumbens Linn. f., Gentiana tenella Fries, Helleborus niger Linn.

Commercial samples are found adulterated with Latotis cashmiriana (Royle ex Benth.) Bupr., Selaginaceae, which does not possess hepatoprotective properties.

Apiaceae

PIMPINELLA

Pimpinella anisum Linn.

Habitat

Native of the eastern Mediterranean. Grown in Madhya Pradesh in India.

Classical and common names

Unani: Anisoon, Baadiyaan-roomi
English: Anise, Aniseed, Spanish Aniseed.

Classical use

Not used in Ayurvedic medicine.

In Unani medicine, seeds are used as carminative, diuretic, diaphoretic, and a moderate expectorant; and in prescriptions for allaying griping of purgatives. Water-extract (araq) of seeds is used as an antispasmodic in flatulence and intestinal colic. Oil of seeds is applied locally to the forehead in headache.

In Unani composite drugs, tonics and confections, Anisoon is included for its carminative properties as an adjuvant. The root is used in fevers.

Active principles and pharmacology

Constituents of the volatile oil are *trans*-anethole (principal component), chavicol methyl ether (estragole), anisaldehyde. Fruits contain traces of furocoumarin-5-methoxypsoralen and 8-methoxypsoralen. Seeds contain 4-(beta-D-glucopyranosyloxy)-benzoic acid, caffeic acid containing protein and myristicin and flavone glycosides. The root oil contained geijerene, pregeijerene and beta-bisabolene as major components.

Anise oil of commerce in India is actually derived from Star Anise (Illicum verum). Both aniseed and Star Anise are recognized by B.P. and U.S.P. as the official sources of Anise Oil.

Anise Oil possesses carminative properties, relieves flatulence, griping in infants.

Fresh leaves of the plant contain vitamin C (8.7 mg/100 g).

Research findings suggest that Anise may also have a tonic effect on liver. It is also found mildly estrogenic, exhibiting galactagogue properties.

Use in Western herbal

German Commission E recognized the efficacy of Pimpinella anisum in the following areas:

- Loss of appetite
- Liver and gallbladder complaints
- Common cold
- Cough, bronchitis.

Over-the-counter, Culpeper Herbal Mixture for Acidity, Indigestion and Flatulence, Culpeper

Cough Relief Mixture, Aniseed and Liquorice Lozenges are popular commercial products.

Anise essential oil is used externally to treat lice and scabies.

Caution

Essential oil should be taken internally only under professional supervision.

Pinaceae

PINUS

Pinus roxburghii Sargent

Figure 1 Pinus roxburghii [WOI]

Pinus longifolia Roxb.

Habitat

Himalayas from Kashmir to Bhutan and in the Siwalik hills at altitudes of 450–2400 m. Occurs in the outer hills and valleys of the Himalayas which receive the bulk of the rainfall during monsoon.

Classical and common names

Ayurvedic: Sarala (Charaka, Sushruta); Oleo-resin— Shriveshtaka (Sushruta), Gandha-Birojaa.
Unani: Sanobar-ul-Hindi. Oleo-resin—Gandh-Bihrojaa, Qinn, Berzad.
Siddha: Seemaidevadaru.
English: Chir Pine.

Parts used

Wood, purified oleo-resin, turpentine oil.

Dose

Wood-powder 1–3 g, oil 1–3 drops, as anthelmintic 6–10 drops with caution.

Classical use

Charaka gave a decoction of leaves and bark, in prescriptions, internally for fever and loss of appetite. For facial paralysis and paralysis of lower extremities, a paste of Piper longum (fruits and roots) and Sarala and Devadaaru (Pinus deodara) was prescribed with honey. Sushruta administrated the drug internally in skin eruptions and blood poisoning; used it externally as an ingredient of a plaster for infantile paralysis and atrophy of muscles; as a paste for swellings and as an ingredient of an ointment for herpes, cutaneous, venereal sores. Dipikaa Taila, extracted from the wood, was prescribed for ear-ache (Sushruta Samhitaa).

The oils of Sarala and Devadaaru were used as anthelmintic, especially for Tapeworm (Ashtaanga Hridaya).

Wounds were fumigated with Shriveshtaka (Sushruta Samhitaa). It was an ingredient in Kushtaadya Taila, prescribed for massage in atrophy of muscles and paralysis (Shaarangadhara Samhitaa). Shriveshtaka was also applied over wounds and ulcers.

Active principles and pharmacology

Essential oil from oleo-resin contains chiefly alpha- and beta-pinene, delta-4-carene and longifoline.

The Indian turpentine is characterized by its comparatively low pinene (alpha-pinene 20–30 %, beta-pinene 5–10 %) and high carene (55–65 %) contents and is similar to Russian turpentine oil from Pinus sylvestris in some respects. Indian turpentine is considered inferior to American turpentine due to its low pinene content. American turpentine consists almost entirely of alpha- and beta-pinenes.

The turpentine oil is included in Indian Pharmaceutical Codex under the name oleum terebinthinae. The oil is locally irritant and feebly antiseptic. During its elimination through the mucous membrane of the lungs it acts as an expectorant in chronic bronchitis. It has also been found beneficial as a carminative in flatulent colic, and for arresting haemorrhages. Externally, it is used as a ruberfacient in various rheumatic affections, such as lumbago, arthritis, neuralgia. In the form of turpentine stupe it is used as a counter irritant in various deep-seated inflammations, especially of the abdomen.

On steam distillation, the needles of Pinus roxburghii, collected from hilly region of Jammu and Kashmir gave 0.26 % of a volatile oil containing alpha- and beta-pinenes, delta-3-carene, alpha-limonene, alpha-phellandrene, borneol, borneol acetate, longifolene and alpha-cadinene.

The oil showed antibacterial activity against E. coli, Staphylococcus aureus, Sal. typhosa and S. paratyphi.

The bark contains 11–14 % of tannins.

The Chir Pine often occurs mixed with other species which are also used in Indian medicine—Pinus gerardiana Wall., Pinus wallichiana A. B. Jackson, Pinus excelsa Wall. ex D. Don., Pinus griffithiiM'clelland.

Essential oil from the leaves of all these species also contains alpha- and beta-pinene as major constituents.

Use in Western herbal

Pinus sylvestris (Scotch Pine) is used in Western herbal.

Leaves contain a volatile oil consisting mainly of alpha- and beta-pinene, D-limonene, bornyl acetate, cadinene, Delta-3-carene, phellandrene.

Shoots and essential oil (under professional supervision) are used internally for catarrhal conditions of the upper and lower respiratory tract; externally for mild muscular pain and neuralgia, nasal congestion. Purified turpentine oil is used internally for chronic diseases of the bronchi with profuse secretion; externally for rheumatic and neuralgic affections.

The seeds are used for bronchitis, tuberculosis and bladder infection.

Pine oil, distilled from the needles, is included in a number of over-the-counter decongestants and expectorant mixtures for coughs and colds; in a variety of external rubs for muscular pains.

Caution

Contraindicated in bronchial asthma and whooping cough. Poisoning is possible with the intake of large doses.

Piperaceae

PIPER

Piper betle Linn.

Habitat

Cultivated in India for its leaves.

Classical and common names

Ayurvedic: Taambuula, Naagvallari, Naagini, Taambuulvalli, Bhujangalataa (Charaka, Sushruta).
Unani: Taambuula, Paan.
Siddha: Naagavalli.
English: Betel.

Parts used

Leaves.

Dose

Juice 5–10 ml.

Figure 1 Piper betle [ADPS]

Classical use

Charaka and Sushruta prescribed the leaf internally in halitosis, hoarseness, catarrhal inflammations of the throat, larynx and bronchi; also in cough and indigestion. A paste of betel leaves, mixed with salt and hot water, was administered for filaria (Bhaavaprakaash). For treating obesity one betel leaf, mixed with Maricha (Piper nigrum) was prescribed for two months (Vaidyamanoramaa).

Taambuula Leham (Sahasrayoga), specific for whopping cough with scanty and difficult expectoration, is available over-the-counter in the South.

In folk medicine, leaf juice, mixed with honey or with fresh ginger (Zingiber officinale), is used as a pectoral. Warm leaves, smeared with oil, are applied to the chest in cases of bronchitis, cough, chest congestion, affections of liver, enlarged glands. Fresh leaves, smeared with oil, are also applied as dressing for blistered surfaces or inflamed areas of wounds.

Unani physicians triturate "aphrodisiac" herbs in the juice of betel leaves for accentuating their properties.

Active principles and pharmacology

Leaves contain beta- and-gamma-sitosterol, hentriacontane, pentatriacontane, n-triacontanol, stearic acid and chavicol.

Essential oil from leaves, contained carvacrol, euganol, chavicol, allyl catechol, cineole, estragol, caryophyllene, cadinene, p-cymene and eugenol methyl ether.

Analysis of a sample of fresh leaves gave following values: moisture 85.4, protein 3.1, fat 0.8, carbohydrates 6.1, fibre 2.3 and mineral matter 2.3 %; calcium 230 mg, phosphorus 40 mg, iron 7 mg, ionisable iron 3.5 mg, carotene as vitamin A 9.600 I.U., thiamine 70 mcg, riboflavin 30 mcg, nicotinic acid 0.7 mg and vitamin C 5 mg/100 g. Betel leaves contain 3.4 mcg/100 g of iodine; also a high content of potassium nitrate (0.26–0.42 %). Betel leaves from Mumbai contained reducing sugars as glucose 1.4–3.2, non-reducing sugars as sucrose 0.6–2.5, total sugars 2.4–5.6, starch 1.0–1.2, essential oil 0.8–1.8, and tannin 1.9–1.3 %.

The leaf juice is acidic in nature; malic and oxalic acids have been reported.

Diastase and catalase are among the enzymes present in the leaves.

The leaves contain significant amounts of all the essential amino acids except lysine, histidine and arginine, which occur in traces.

The essential oil and extracts of leaves possess activity against Gram-positive and Gram-negative bacteria, probably due to the presence of chavicol.

An attempt was made to investigate the effect of Piper betle leaves on gastric secretion and mucosa. In 16 human volunteers to whom fresh aqueous extract of the leaves was administered, gastric secretion increased significantly, as tested by fractional test meal. When administered to guinea-pig, Piper betle aqueous extract (0.5 g/kg) led to gastric ulceration in a large proportion of the animals, which could only be prevented partially by pre-treatment with atropine.

A study was undertaken on chewing of betel quid (Paan, consisting of betel leaf, betel nut, trace amounts of lime and catechu) with and without tobacco in human volunteers, with the aim of determining the effect of betel quid on the salivary levels of nitrites and other factors in the formation of nitrosamines and the possible relevance of these findings to the development of oral cancer. A substantial rise in the salivary nitrite levels was observed following betel quid chewing by those volunteers who had never chewed tobacco, although no such increase was noted when the betel quid was chewed without tobacco. Salivary pH was also increased more with betel quid alone.

In a study on Swiss mice, the effects of betel leaves and betel nut extracts were studied for tumorigenicity. While betel nut polyphenol fraction led to a 100 per cent tumour incidence in the mice, aqueous extract of betel leaf administered alone as well as with betel nut extract did not produce any tumour.

Caution

Long-term use of betel quid, especially with tobacco, increases the incidence of cancer of the mouth and tongue.

Piper cubeba Linn. f.

Habitat

Native of Indonesia. Fruits are largely imported for use in Unani medicine. Cultivated in Assam and Karnataka.

Classical and common names

Ayurvedic: Kankola, Kokkola, Kankolaka, Takkola, Koraka (Charaka); Kolaka, Kashphala. Shitalchini.
Unani: Kabaabchini, Habb-ul-Uruus.
Siddha: Valmilagu.
English: Cubebs, Tailed Pepper.

Parts used

Fruits.

Dose

Powder 1–3 g.

Classical use

Charaka and Sushruta prescribed a paste of the fruits as a mouthwash, or dried fruits internally for oral and dental diseases, loss of voice, halitosis, fevers, cough.

According to Raaja Nighantu, Kankola is stimulant, carminative, diuretic, expectorant, deobstruent.

Unani physicians use a paste of the fruit externally on male and female genital organs to intensify the sexual pleasure during coitus. Due to this attributed property cubebs were called "Habb-ul-Uruus". They prescribe the drug for its stimulating and irritating properties which provide a psychological euphoria and a false feeling of confidence.

Cubebs are used traditionally for urethritis, leucorrhoea, cystitis, amenorrhoea, diseases of spleen and liver; in addition to its common use in digestive problems and chronic bronchitis.

Active principles and pharmacology

Essential oil gave sesquiterpene hydrocarbons. Seed oil contains palmitic, oleic, linoleic, linolenic, stearic, behenic and hexadecenoic acids.

The essential oil of Piper cubeba showed antibacterial activity against B. subtilis, Vib. cholerae, C. diphtheriae, Sal. typhi, Strep. faecalis, Strep. pyogenes, B. pumilus and Ps. solanacearum; being most active against Ps. solanacearum. The essential oil also showed antibacterial effect against B. subtilis, E. coli, Staph. aureus, Sal. typhi, Sal. paratyphi and Pestalotia spp., but not against Sh. niger.

The oil showed significant antifungal activity against Aspergillus flavus, A. fumigatus, Trichoderma viridi, Curvularia lunata, Alternaria tenuis, Penicillium funiclosum, P. javanicum, P. striatum and Fusarium solani; revealed anthelmintic activity against earth worms and tape worms in vitro.

Caution

Contraindicated in nephritis (Michael McGuffin et al); also in inflammatory conditions of the digestive tract.

Piper longum Linn.

Figure 2 Piper longum [ZANDU]

Habitat

Hotter parts of India, from the central Himalayas to Assam, Khasi and Mikir hills; lower hills of Bengal; evergreen forests of Western Ghats from Konkan to Travancore; also the Car Nicobar Islands. Imported from Malaysia and Singapore; a small quantity is exported to Sri Lanka, Pakistan, and Afghanistan. Imports are mostly of Java type Piper retrofractum; exports are of Piper longum and Piper peepuloides. In some hilly areas of the Vishakhapatnam district in Andhra Pradesh, long pepper is grown for its roots, used as Pippaliamuula in Ayurvedic and Unani medicine.

Classical and common names

Ayurvedic: Pippali, Maagadhi, Vaidehi (Charaka, Sushruta); Upakulyaa (Sushruta); Maagadha, Maagadhaa, Maagadhikaa, Magadhodbhavaa, Pippalikam, Chapalaa (Bhaavaprakaasha); Kanaa, Krishnaa (Shaarngadhara Samhitaa). Root: Pippalimuula, Pippali-Jataa, Granthi, Shadgranthi, Granthika, Maagadhi-muula, Kanaamuula, Krishnaamuula (Bhaavaprakaasha, Shaarangadhara Samhitaa.)
Unani: Filfil Daraaz, Daarfilfil. Root: Bekh-Filfil Daraaz, Bekh-Daarfilfil.
Siddha: Tippili.
English: Indian Long Pepper, Joborandi.

Parts used

Fruits, roots.

Dose

Powder 1–3 g.

Classical use

Charaka, Sushruta and Vagabhatta mentioned two varieties of Pippali; these have now been identified as Pippali and Hastipippali (syn. Gajapippali). Raaja Nighantu mentioned four varieties of Pippali: Pippali, Gajapippali, Sainghali and Vanapippali. Gajapippali has been equated with Piper chabaHunter, Sainghali with Piper retrofractum Vahl. and Vana-pippali (provisonally) with Piper peepuloides Roxb. or Piper sylvaticum Roxb.

In Indian medicine, Pippali grown in Magadha and Videha, belonging to Bihar, is used.

Charaka gave boiled milk added with Pippali powder; also Pippali with jaggery for cough, dyspnoea; Pippali with honey for diarrhoea; Sushruta gave Pippali with Shunthi (dry Zingiber officinale) for colic; Pippali and honey as a gargle for diseases of the mouth.

Pippali Vardhamaan Rasaayana was prescribed by Charaka and Sushruta for chronic and malarial fevers, arthritis, gout, asthma, chronic bronchitis and wasting diseases. An infusion of three Long peppers was given with honey or sugar on the first day, then for the successive days the dose was increased by three peppers every day; thus on tenth day the patient took 30 peppers as one dose. After this, the dose was reduced by three peppers daily, and finally the drug was withdrawn.

Pippali Vardhamaan Rasaayana was also prescribed for allergic conditions of the skin (Vrindamaadhava, Ashtaanga Hridaya, Vaidyamanoramaa); for muscular atrophy (Vrindamaadhava); for goitre (Gadanigraha).

Pippali, triturated in its own decoction for 192 hours, was known as Chatusshashti Prahari Pippali. It was prescribed as a potent tonic for cough, cold, and asthma (Ayurveda Saar-Sangraha).

Shunthi (dry Zingiber officinale), Pippali and Maricha (Piper nigrum), taken together are known as Tri-ushna or Trikatu (the Three Pungents) group of drugs in Ayurvedic medicine. This group is considered specific for cough, urinary disorders, skin diseases, derangement of digestive system, excess lipids. This group is incorporated in a number of tonic preparations, as it stimulates the assimilation of vital constituents of companion drugs into the body; increases bioavailability of companion drugs; increases the body's resistance to diseases and acts as an antiallergic and antitoxic agent.

The group of the Five Pungents, called the Panchakolaa group, consists of Pippali, Pippalimuula, Chavikaa (Piper chaba), Chitraka (Plumbago zeylanica) and Shunthi (Zingiber officinale). When Maricha (Piper nigrum) is added, it is called the group of Six Pungents. This group is specific for colic, anorexia, abdominal obstructions, tympanitis.

For contraception, Pippali, Vidanga (Embelia ribes) and Tankana (borax), in equal quantity, with milk, was prescribed during season (Bhaavaprakaasha).

Available over-the-counter are: Pippalaadyaasava (Shaarangadhara Samhitaa), Naaraayana Churna (Bhaavaprakaasha, Yoga Ratnaakara), Chitrakaadi Vati (Bhaishajya Ratnaavali), Panchkolaa Churnam (Sahasrayoga) for diarrhoea, dysentery; Guda Pippali (Bhaishajya Ratnaavali) for intestinal disorders, enlarged liver and spleen; Agastya Rasaayan (Charaka Samhitaa), Agastya Haritaki Avaleha (Shaarangadhara Samhitaa), Vasaavaleha (Bhaishajya Ratnaavali) for cough, chronic bronchitis, asthma; Pippalaadi Lauha (Bhaishajya Ratnaavali), containing calx of iron, for anaemia, enlarged spleen.

Jawarish-e-Falaafali, a reputed Unani medicine for dyspepsia, stomach-ache and constipation; Majoon-e-Falasifa, an adjuvant for polyuria, dysuria, are also available over-the-counter.

A common use of Piper longum in Indian medicine is for the prevention of recurrent attack of bronchial asthma, recurrent infection of throat, and for flatulence and dyspepsia.

In folk medicine, a paste of the drug is applied locally for muscular pains and inflammations.

Active principles and pharmacology

Piper longum root gave the alkaloids piperine, Piplartine (piperlongumine) and piperlonguminine, sesamin, methyl-3, 4, 5-trimethoxycinnamate.

Stem gave tricontane, 22, 23-dihydrostigmasterol.

Fruit gave N-isobutyldeca-*trans*-2-*trans*-4-dienamide.

Essential oil from fruits contained n-hexadecane, n-heptadecane, n-octadecane, n-nonadecane, n-eicosane, n-heneicosane, alpha-thujene, terpinolene, zingiberene, p-cymene, p-methoxyacetophenone, dihydrocarveol, phenylethylalcohol and caryophyllene. Fruits also gave L-tyrosine, L-cysteine hydrochloride, DL-serine and L-aspartic acid.

N-isobutyldeca-*trans*-2-*trans*-4-dienamide showed antitubercular; essential oil antibacterial, antimicrobial, anthelmintic; fruit decoction anti-inflammatory activity.

Milk extract of fruit effectively reduced passive cutaneous anaphylaxis in rats and protected guinea-pigs against antigen-induced bronchospasm. It did not show significant effect on total quantity of histamine in lungs, trachea, and intestines or on release of histamine on antigenic challenge.

Piperine exhibited hypotensive, antipyretic, CNS-stimulant, analeptic activity. The herb, especially piperine, improved drug availability in experimental animals.

Piperine was tested on Plasmodium berghei infected albino mice at doses of 5 and 10 mg/kg p.o. for five consecutive days. The drug showed 50 to 60 % antimalarial activity. (CCRAS.)

In a double-blind study carried out at the LTMG Hospital, Mumbai, 240 children (age ranging from neonatal period to 12 years) suffering from bronchial asthma were administered orally with Piper longum for two to three total courses. Each course constituted a gradual daily increase of the dose of Piper longum from a minimum of 1 g to a maximum of 30 g (depending on the age) and then daily reduction of the dose to the original dose. In 60 patients, significant effect in controlling the frequency and severity of the asthmatic attack was observed.

In another open trial at Wadia Hospital for Children and Institute of Child Health, Mumbai, on 20 paediatric patients of asthma (aged 1–2 years), Piper longum (powder in capsules) was administered with milk, in a gradually increasing dose for a period of 5 weeks. This regimen significantly decreased the severity and frequency of asthmatic attacks. The serum IgE-levels estimated in 6 patients did not yield any conclusive results. At the end of 1 year, 3 out of 20 patients failed to show satisfactory response. Three others who had a history of allergy to certain food items could consume these very food items and tolerate them well after this treatment. (See Classical use, Pippali Vardhamaan Rasaayana.)

A combination of Embelia ribes, Piper longum and borax was subjected to a detailed study at the Institute for Research in Reproduction, Mumbai. This preparation impaired the fertility of female mice and rats and produced sterility in male mice. The effect was reversible in the female animals, but not in the males. The antifertility effect in the male mice was significant, inhibiting spermatogenesis at a dose of 50–100 g/kg. The preparation, however, did not show any androgenic or antiandrogenic effect. When administered to virgin rats, it did not show any estrogenic or progestational activity. The results suggested that the plant preparation did not exert its antifertility action by interfering with the activity of ovarian hormones on the uterus. When administered to female rats (10 % of the diet for two weeks) and guinea-pigs (0.5 g for 6 days), slight prolongation of the estrus phase of the cycle was observed in both the species and there was an enhanced activity of alkaline phosphatase in the uteri.

In a clinical trial by CCRAS, 162 women were given one capsule (each capsule contained 500 mg of powdered Piper longum, Embelia ribes and borax in equal quantity) per day from day-5 for full circle, 540 were given two capsules from day-5 for full circle, 72 were given two capsules from day-1 for the whole month. 7.06 % and 1.20 % drug failure was recorded in group 1 and 2. There was no drug failure in group 3. (See Classical Use.)

The CNS-activities of three alkaloids, viz. piperine, piper-longumine, and piper-longuminine isolated from Piper longum fruits, were tested in frogs, mice, rats and dogs. Piperine showed a marked central stimulant activity, the main site of action being the medulla. Based on this study, the clinical use of piperine in the treatment of respiratory depression due to narcotic poisoning has been suggested.

A comparative study was undertaken between piperine and nalorphine for their effect against morphine-induced respiratory depression and analgesia. Both the compounds reversed the respiratory depression, but unlike nalorphine, piperine did not antagonise morphine-induced analgesia in rats.

To investigate the possible scientific basis of the extensive usage of Trikatu (Piper nigrum, Piper longum and Zingiber officinale) in a large number of Ayurvedic preparations, its effect on vasicine was studied. Trikatu as a whole influenced bioavailability to a great extent. Individually, Piper longum as well as Piper nigrum were found to be almost equi-effective in enhancing the bioavailability of vasicine in rats, while piperine (suspension) in 40 mg/kg dose, when co-administered orally with vasicine, enhanced the bioavailability of the latter by more than 300 %.

Research showed that piperine increases the bioavailbility of antitubercular drugs. (CSIR, Jammu.)

Piper nigrum Linn.

Figure 3 Piper nigrum [CCRAS]

Habitat

Cultivated in the hot and moist parts of India, Sri Lanka and other tropical countries.

Classical and common names

Ayurvedic: Maricha (Charaka, Sushruta); Vellaja, Uushana, Suvrrita, Krishna.
Unani: Filfil siyaah, Filfil safed (white variety); Al-philphil asvad.
Siddha: Milagu.
English: Black Pepper.

Parts used

Fruit.

Dose

Powder 500 mg–1 g.

Classical use

Dried seeds were used by Charaka, in prescriptions, for cough, rhinitis, consumption, anaemia, fainting. Sushruta gave seeds internally in indigestion, colic, intestinal catarrh; in persistent skin diseases, urinary discharges, hyperlipemia. Externally, Sushruta prescribed a paste of Maricha for baldness and alopecia.

As a single drug, Sushruta gave Maricha with sugar candy or honey for cough. Powder of Pippali or Maricha was administered for chronic dysentery (Ashtaanga Hridaya). Jaggery water, boiled and when cool, added with Maricha and honey, was a remedy for coryza (Vrindamaadhava, Raaja Maarttanda). For obesity, one betel leaf, pounded with 10 grains of Maricha, was given with cold water for at least two months (Vaidyamanoramaa).

Maricha is used commonly for cold, sore throat, flatulent dyspepsia. As an ingredient of Trikatu (see Piper longum), it is given as an adjuvant in viral hepatitis.

Marichaadi Churna (Charaka) is available over-the-counter for poor digestion, diarrhoea, dysentery; Ashta Churna (Shaarangadhara Samhitaa), Trikatu Churna (Bhaavaprakaasha) for hyperacidity and indigestion; Marichaadi Vati (Shaarangadhara Samhitaa) for cough and catarrh.

Unani compound, Jawarish Falaafali contains Piper longum and both black and white varieties of Piper nigrum. Jawarish Kamooni is prescribed for hyperacidity and flatulence.

Active principles and pharmacology

Fruits yielded piperine, piperetine and piperidine; amides—peperylin, piperoleins A and B and N-iso-butyl-cicosa-*trans*-2-*trans*-4-dienamide. Fruits also gave phenolic amides.

Essential oil from fruits contained alpha- and beta-pinene, sabinene, myrcene, limonene, terpinene, p-humulene, its oxides, selinene, camphene, linalool, terpineol and nerolidol in varying amounts.

Water extract of Piper nigrum exhibited antagonistic action on n-methyl-D-aspartate (NMDA)-receptor which might contribute to the drug's anticonvulsant property.

The major constituent piperine (2–5 %) showed CNS-depressant, antipyretic, analgesic, anti-inflammatory, anti-oxidant and hepatoprotective properties.

The drug stimulates the thermal receptors and increases secretion of saliva and gastric mucous. It influences liver and metabolic functions.

The extract of Piper nigrum was reported to be inhibitory of E. coli, Aerobacter aerogenes, L. casei, Staph. faecalis, Staph. aureus and Sh. sonnei. The essential oil was reported to inhibit Vib. cholerae, Staph. albus, C. diphtheriae, Sh. dysenteriae, Sar. lutea, Strep. faecalis, B. pumilus, Ps. pyogenes, Strep. pyogenes, Micrococcus spp. and Ps. solanacearum, B. subtilis, Staph. aureus, Sal. typhi, Sal. paratyphi and Pestalotia spp. Piper nigrum oil had no action against E. coli.

The essential oil of Piper nigrum showed antifungal activity particularly against Trichophyton terrestre, T. tonsurans, Candida albicans, Monosporium apiospermum, Aspergillus niger, A. fumigatus, A. nidulans, Sporotrichum schenckii and Fusarium oxysporum. The oil was also active against Trichophyton mentagrophytes, Histoplasma capsulatum, Beauveria spp., Cryptococcus neoformans, Aspergillus oryzae, Microsporum gypseum, Nocardia brasiliensis, Alternaria tenuis, Trichoderma viride, Penicillium javanicum, P. striatum and Fusarium solani.

Caution

The drug is contraindicated with alcohol.

Safrole occurs in Piper nigrum. Dietary safrole levels of 0.25, 0.5 and 1 % produced growth retardation. (Lloydia, 30, 1961.)

Anacardiaceae

PISTACIA

Pistacia chinensis Bunge subspp. integerrima (Stewart) Rech. f.
Pistacia integerrima Stew. ex Brandis.

Habitat

Himalayas from Indus to Kumaon from 350 to 2400 m, cultivated in Punjab plains.

Figure 1 Pistacia integerrima—fruiting branch [WOI]

Classical and common names

Ayurvedic: Karkatashringi, Shringi, Karkatashringikaa, Karkataakhya, Kulirashringaaya (Charaka, Sushruta.)
Unani: Kaakaraasingi.
Siddha: Karkatagasingi.

Parts used

Galls.

Dose

Powder 3–6.

Classical use

In Bhaavaprakaasha, Karkatashringi and Meshashringi have been equated with Ajashringi. Meshashringi and Ajashringi are different drugs, equated with Gymnema sylvestre and Dolichandrone falcata. Galls of Rhus succedanea Linn, syn. Rhus acuminata DC. are used as substitute for Karkatashringi.

Charaka gave the powder of Karkatashringi, mixed with oil, for cough; a gruel cooked with the drug for asthma.

Powdered galls of Karkatashringi and seeds of Raphanus sativus (Muulaka) were prescribed for whooping cough and asthma in children (Bangasena). Powdered galls and Mustaa (Cyperus rotundus) was given for checking vomiting (Charaka).

Sushruta prescribed Karkatashringi internally as a spermatogenetic and spermatopoietic agent.

For vigour and vitality, a paste of Karkatashringi was administered with milk (Ashtaanga Sangraha).

Karkatashringi of Charaka and Sushruta belonged to Kaakolyaadi group of herbs, which was considered vitalizing and nourishing. But now, in practice, the drug is not used as a tonic; in all Ayurvedic and Unani compound preparations it is specific for respiratory troubles. Other drugs of the Kaakolyaadi group are sweet, while powdered galls of Karkatashringi have a very astringent and slightly bitter taste and a terebinthine odour, exert depressant action on the central nervous system. Tonic properties of the classical Karkatashringi cannot be attributed to Pistacia integerrima galls, now in use in Indian medicine.

Over-the-counter, Shringyaadi Churna (Shaarangadhara Samhitaa) is available for cough, vomiting, and fever in children.

Active principles and pharmacology

The galls contain 20–75 % tannins, 1.3–1.8 % essential oil and 5 % resin, which is identical with gum mastic from Pistacia lentiscus L.

Essential oil from galls contain alpha-pinene, camphene, DL-limonene, 1, 8-cineole, alpha-terpineol, aromadendrene and carprylic acid.

Other constituents of the galls are beta-sitosterol and the triterpene acids pistacienoic acids A and B.

Heartwood gave the flavonoids (–)-eriodictyol, luteolin, and dihydroquercetin.

The oil, in moderate doses, exerts an antispasmodic action on involuntary muscles inhibiting excessive peristaltic movements of intestine. It showed a depressant action on the central nervous system of guinea-pigs and white rats when given in sub-lethal doses. The animals became deeply unconscious in about an hour. (Lethal dose: 0.1 cc/100 g body weight) The oil showed a slight irritant action on mucous membrane.

Essential oil from galls is antibacterial, antiprotozoal, anthelmintic, antiamoebic, antimicrobial; stem-bark spasmolytic. Galls possess antiasthmatic, astringent, and expectorant properties. (CIMAP.)

Pistacia lentiscus Linn.

Habitat

Chiefly in the Mediterranean region. Mastic resin is imported into India. Used in Unani medicine.

Classical and common names

Unani: Mastagi, Roomi Mastagi, Mastaki.
English: Mastic Gum.

Parts used

Gum-resin.

Dose

1–2 g.

Classical use

Jawarish-e-Mastagi (Qarabadeen-e-Azam-o-Akmal) is a single drug confection of Unani medicine, prescribed for diseases of the stomach and diarrhoea.

Raughan-e-Mastagi (Qarabadeen-e-Jadeed) is for topical application on weeping eczema, scabies and other skin eruptions.

Mastagi entered into Ayurvedic medicine after the 16th century, and was prescribed for cough, asthma, dysuria, dysmenorrhoea.

Active principles and pharmacology

The resin contains alpha- and beta-masticoresins, a volatile oil comprising mainly alpha-pinene, tannins, masticin and mastic acid. Pinenes exhibit strong antiseptic properties. The masticated resin releases active principles that tighten the gums.

Mastic is credited with carminative, stimulant, expectorant and diuretic properties.

Araceae

PISTIA

Pistia stratiotes Linn. var. cuneata Engl.

Habitat

A floating, stoloniferous herb found in ponds and streams almost throughout India up to a height of 1000 m.

Classical and common names

Ayurvedic: Jalakumbhi, Vaariparni, Vaarimuuli.
Siddha: Agasatamarai.
English: Water Lettuce, Tropical Duckweed.

Parts used

Whole plant.

Dose

Juice 10–20 ml, ash 1–3 g.

Classical use

Jalakumbhi was a specific drug for goitre. Internally, Jalakumbhi, Pippali (Piper longum) and rock-salt was administered in the morning; ash of Jalakumbhi, mixed with mustard oil, was applied externally. (Bhaavaprakaasha.)

For obstinate skin diseases, including leprosy, powder of the whole plant mixed with honey was prescribed. Fine powder of Jalakumbhi and Kaidarya (Myrica esculenta Buch-Ham) leaves was given internally for haemorrhoids (Vaidyamanoramaa).

Active principles and pharmacology

Plant gave 2-di-C-glycosylflavones of vicenin and lucenin type, anthocyanin-cyanidin-3-glucoside, luteolin-7-glycoside and mono-C-glycosylflavones—vitexin and orientin.

An analysis of leaves and stems gave: moisture 92.9, protein 1.4, fat 0.3, carbohydrates 2.6, fibre 0.9, ash 1.9, calcium 0.20, phosphorus 0.06 and dig. protein 1.2 %. Leaves are rich in vitamins A and C, and also contain B-vitamins. The ash is rich in potassium chloride and sulphate.

Plant exhibits antiseptic, antitubercular and anodyne; root laxative, diuretic; leaf antidysenteric, anthelmintic, antileprotic, antisyphilitic properties. (CIMAP.)

The ashes of the plant, applied to the ringworm of the scalp, give good results. Juice of leaves, boiled in coconut oil, is found efficacious externally in chronic skin diseases. The drug was not screened for goitre.

Plantaginaceae

PLANTAGO

Plantago ovata Forsk.
Plantago ispaghula Roxb.

Habitat
Indigenous to the Mediterranean region and West Asia; introduced into India and cultivated in North Gujarat, Rajasthan and Haryana.

Classical and common names
Ayurvedic: Isabgola; Ashvagola (non-classical). Persian Aspaghol has been translated into Sanskrit as Ashwakarna, while the same is equated with Shorea robusta.
Unani: Aspaghol.
Siddha: Isappa.
English: Ispaghula, Spogel Seeds.

Parts used
Seeds, husk.

Dose
Husk 5–10 g.

Classical use
Not mentioned in classical texts. During the 18th century, first reference was found in a treatise "Vaidyaamrita", written by Vaidya Moreshwara. The drug was recommended for fever with diarrhoea.

In Unani medicine, Aspaghol is prescribed with Naagarmothaa (Cyperus rotundus), also with rose-water, in chronic diarrhoea and dysentery.

Ayurvedic, Unani and Siddha practitioners use dried seeds and husk as emollient, demulcent and laxative in the treatment of chronic constipation, bacillary dysentery and diarrhoea due to irritative conditions of the gastro-intestinal tract. Seeds are prescribed for their cooling and diuretic properties in febrile conditions and affections of kidneys, bladder and urethra. A decoction of seeds is given in cough and cold. Crushed seeds, made into a poultice, are applied to rheumatic and glandular swellings.

In Unani medicine, seeds and leaves of Plantago lanceolata Linn. and Plantago major Linn., an allied species of Plantago ovata, known as Baartang, are used as an astringent, styptic and anodyne in internal haemorrhages, bleeding piles, diarrhoea and dysentery with blood (dose 5–7 g). The seeds of Plantago major, like those of other Plantago spp., contain mucilaginous matter mainly in the seed coat. They are used as an adulterant for Ispaghula.

Active principles and pharmacology
Mucilage polysaccharide (20–30 %) consists of highly branched weakly acidic arbinoxylans (up to 85 %) with xylan backbone and branches of arabinose, xylose and 2-O-(galacturonic acid)-rhamnose residues.

Fixed oil (approx 2.5 %) contains mainly linoleic, oleic and palmitic acids with aliphatic hydrocarbons and oxygenated fatty acids, iridoids and proteins.

Antidiarrhoeal effects of Plantago seeds have been extensively investigated in patients with acute and chronic diarrhoea. An increase in the viscosity of the intestinal contents due to the binding of fluid and an increased colonic transit time (decreased frequency of defecation) were observed in patients treated with the drug (WHO).

Plantago seeds, when used for constipation, increase the volume of the faeces by absorbing water in the gastrointestinal tract, which stimulates peristalsis. The intraluminal pressure is decreased, colon transit is increased and the frequency of defecation is increased.

When mixed with water, the therapeutic efficacy of the drug is due to swelling of mucilaginous seed coat which gives bulk and lubrication. Plantago seeds increase stool weight and water content owing to the water-bound fibre residue and an increased faecal bacterial mass (WHO).

The drug also reduces serum cholesterol levels due to increased excretion of bile acids which in turn stimulates the synthesis of bile acids from cholesterol, the sole precursor of bile acids.

Plantago seeds consumed with water prior to meal causes reduction in the consumption of fat and calories. (Int J Obesity (1995) 19:338.)

In a comparative study of the effects of codeine phosphate, diphenoxylate (as Lomotil) and Ispaghula (as Isogel), ingested between meals, on ileostomy output in 18 patients, Isogel produced most viscid output.

Ispaghula (as psyllium) has been compared with methylcellulose and sodium carboxymethyl cellulose as a hydrophilic laxative and was found to be superior.

The effect of bran, Ispaghula and lactulose were also observed in 31 patients of diverticular disease. Ispaghula had a significant effect on the stool weight and increased basal motility, though all the three test substances had a laxative effect. (Chemexcil.)

The efficacy of a new technique for sustained release form, viz. imbibed Ispaghula husk, has been reported. The effect of concentration on the absorption of atropine sulphate and ferrous sulphate by Ispaghula husk and also the release rates of these two drugs imbibed with Ispaghula husk were also investigated. The study showed that nearly half of the drugs was released in the first 25 minutes, after which the release was slow and gradual, indicating an uneven distribution of the drugs within the inter-carrier, the drug being concentrated mainly on the surface of the husk. This phenomenon may prove advantageous in therapeutics to ensure a high and rapid initial load followed by sustained concentration of the drugs subsequently. (ICMR.)

Plantago lanceolata (English plantain) and Plantago major (broad-leaf plantain), used in Unani medicine, contain 2–6.5 % mucilage composed of polysaccharides; 6.5 % tannins; iridoid glycosides; phenolic carboxylic acids; flavonoids; and minerals including zinc and potassium.

The drug showed bacteriostatic activity in vitro.

In a preliminary clinical trial on 53 patients of menorrhagia of varying aetiology, the seeds of Plantago major showed encouraging haemostatic results. The plant checked the bleeding in two to three days, when the seeds were given orally in doses of 3 g twice a day. Better results were obtained in inpatients than in outdoor patients. No toxic or untoward effects were observed. (ICMR.)

Use in Western herbal

Plantago ispaghula (psyllium) is used for habitual constipation where easy bowel movement is required, especially in patients with anal fissures and haemorrhoids; after rectal surgery and during pregnancy.

German Commission E recognized Plantago ispaghula's efficacy in the following areas:

- Constipation
- Diarrhoea
- Raised levels of cholesterol.

Plantago lanceolata and Plantago major leaves (known as Baartang in Unani medicine) are used internally for catarrh of the respiratory tract and inflammations of the oral and pharyngeal mucosa; externally for inflammatory conditions of the skin.

German Commission E recognized the efficacy of Plantago lanceolata in the following areas:

- Common cold
- Inflammation of the mouth and pharynx
- Cough, bronchitis
- Fevers and colds
- Inflammation of the skin.

A tea is given for phlegm congestion. Fresh or dried above ground parts are used.

In folk medicine, the pressed juice is used to treat wounds and as a styptic.

Caution

Plantago preparations may affect the absorption of other drugs taken simultaneously.

Bulking agents have been reported to diminish the absorption of some minerals (calcium, magnesium, copper and zinc), vitamin B12, cardiac glycosides and coumarin derivatives. Ingestion of lithium salts or carbamazepine and Plantago seeds should be separated in time as far as possible. (WHO.)

Plantago products should be taken with an adequate amount of fluid.

Asteraceae

PLUCHEA

***Pluchea lanceolata* Oliver & Hiern.**

Figure 1 Pluchea lanceolata—flowering branch [WOI]

Habitat
Sandy or saline soils of Punjab, upper Gangetic plains, Rajasthan and Gujarat.

Classical and common names
Ayurvedic: Raasanaa, Rasanaa (Charaka, Sushruta); Raasnikaa, Rasaa (Bhaavaprakaasha), Naakuli and Gandha-naakuli are doubtful synonyms.

English: Indian Groundsel.

Parts used
Leaves.

Dose
Decoction 50–100 ml.

Classical use
Plants of different genera and families are used in Indian medicine as Raasanaa: Pluchea lanceolata in Uttar Pradesh, Punjab and Gujarat (accepted as the classical drug), Vanda roxburghii R. Br. in Eastern Uttar Pradesh and West Bengal, Alpinia galanga Willd. in South India, and Dodonaea viscosa Linn. in Andhra Pradesh. All these drugs are used for arthritis.

According to Charaka, Naakuli and Gandhanaakuli were allied species of Raasanaa. Sushruta considered Naakuli and Sarpagandhaa as one drug. On the basis of clinical applications and classical references, Raasanaa is now being equated with Pluchea lanceolata, Naakuli with Rauvolfia serpentina Benth. ex Kurz. and Gandhanaakuli with Aristolochia indica Linn.

Ayurvedic Formulary of India suggests Pluchea lanceolata as the real Raasanaa and Alpinia galanga as its substitute.

Paste of the root and rhizomes of Raasanaa were used by Charaka, in prescriptions, internally, for rheumatic conditions, muscular pains, oedema, fever; externally as a massage oil; as a poultice for haemorrhoids. Sushruta prescribed Raasanaa internally for skin diseases and parasitic infestation; externally for swellings, cutaneous affections and venereal sores.

Raasanaadi Ghrita and Raasanaadi Taila were specific compound preparations for rheumatic affections, sciatica and swellings (Charaka Samhitaa, Vrindamaadhava, Bangasena, Bhaavaprakaasha.)

A paste of Raasanaa, Naakuli and Gandhanaakuli was applied on psoriasis (Kaashyapa Samhitaa).

In the North, Pluchea lanceolata is an ingredient in Mahaamaasha Taila (Bhaishajya Ratnaavali), Mahaanaaraayana Taila (ibid), Naaraayana Taila (Shaarangadhara Samhitaa) and Prasaarini Taila (ibid); available over-the-counter for rheumatic affections. Shadbindu Taila (Bhaishajya Ratnaavali) is for headache, cold, and sinusitis.

Mahaa-Raasanaadi Kwaatha (Shaarangadhara Samhitaa) is prescribed internally for rheumatoid arthritis.

In the South, Alpinia galanga is used in compound preparations as Raasanaa. Aama Vaata Taila, Karpaasaadya Taila, Mahaamaasha Taila, Naaraayana Taila, Sadbindu Taila are a few such products available over-the-counter. Mahaa-Raasanaadi Kwaatha and Raasanaadi Churna of the South also contain Alpinia galanga instead of Pluchea lanceolata.

Active principles and pharmacology

Plant gave choline, pluchine, taraxasterol, beta- and gamma-sitosterol. Leaves gave quercetin and iso-rhamnetin.

Pharmacological investigations indicated that the drug had two primary actions, viz. acetylcholine-like action and smooth muscle relaxant-spasmolytic action on different muscle preparations. The only central nervous system activity detected in the drug was that of potentiation of barbiturate hypnosis.

The water soluble fraction of the 90 per cent alcoholc extract Pluchea lanceolata stem significantly showed anti-inflammatory activity against formalin arthritis and granuloma pouch in albino rats comparing favourably with betamethasone. The extract significantly decreased the adrenal gland weight and increased the adrenal ascorbic acid and cholesterol contents. The plant extract appeared, however, to supress the delayed peri-articular changes more as compared to acute inflammatory reactions, whereas betamethasone suppressed both these reactions. Gastric ulcer and haemorrhage occurred in 5 out of 12 rats treated with Pluchea lanceolata extract, as against 10 out of 12 in betamethasone treated rats. Continuous treatment with 100 to 200 mg/day for 15 days did not produce any apparent toxic effect. The mortality was 1 out of 12 as compared to 12 out of 29 rats treated with betamethasone.

Pluchea lanceolata and five other plants used under the name Raasanaa were tested for comparative anti-inflammatory activity against formalin arthritis, carrageenin and for maldehyde-induced acute rat paw oedema. The water soluble fraction of the alcoholic extract of Pluchea lanceolata was found significantly effective, although less active than extract of Alpinia galanga and more active than Vanda roxburghii, Tylophora asthmatica, and Aristolochia indica extracts. (Tylophora asthmatica (L.f.) W. & A. is now equated with Antamuula.)

Plumbaginaceae

PLUMBAGO

Plumbago indica Linn.
Plumbago rosea Linn. (red variety)
Plumbago zeylanica Linn. (white variety)

Figure 1 Plumbago zeylanica [ZANDU]

Habitat

Red variety: wild or indigenous to Sikkim and Khasi hills. White variety more widespread and common than Plumbago indica, cultivated in gardens throughout India.

Classical and common names

Ayurvedic: Chitraka, Paathi, Daaruna, Ushna. Rakta-chitraka (Charaka); Hutaashna (Sushruta); Agni, Analnaamaa, Vyaala.
Unani: Sheetraj Hindi.
Siddha: Venkodiveli.
English: Ceylon Lead Wort, Lead Wort.

Parts used

Rootbark.

Dose

Powder 1–3 g.

Classical use

Charaka and Sushruta prescribed roots and leaves, in prescriptions, for intestinal catarrh, indigestion, colic, internal abscesses, jaundice, intestinal parasites, piles, urinary calculi, polyuria, spermaturia, vaginal discharges, deficient lactation, virulent skin diseases, poisoning.

Chitrakaadya Gudikaa and Chitraka Ghrita were prescribed for diarrhoea and dysentery; Chitrakaadi Leha for cough and diseases of the throat (Charaka Samhitaa). Chitraka Ghrita was also prescribed for oedema, swellings of stomach and intestines.

According to Sushruta, Chitraka acts like a fire during digestion (it is a strong digestive stimulant). Cures oedema, piles, parasitic infections and obstinate skin diseases, including leprosy.

Chitraka roots, with honey, were prescribed for hyperlipemia and obesity (Bangasena).

Among over-the-counter drugs, Chitraka Haritaki (Bhaishajya Ratnaavali) is for chronic cold; Chitrakaadi Gutikaa (ibid) for sprue, indigestion, colic, and loss of appetite. In the South, Chitraka Haritaki (Chakradatta) is prescribed as a digestive, carminative and expectorant.

In Unani medicine, Habb-Sheetraj is prescribed as a tonic to liver and digestive system.

Active principles and pharmacology

Roots of Plumbago zeylanica yielded the naphthaquinone derivatives plumbagin, 3-chloroplumbagin, 3, 3'-biplumbagin, elliptinone, chitranone, droserone, zeylanone, iso-zeylanone, 1, 2(3)-tetrahydro-3, 3'-biplumbagin and plumbazeylanone.

Root powder showed presence of protease enzyme; trace quantity of vitamins A, B1, B2 and C and was found to be a gastrointestinal flora normaliser. Root powder stimulated the proliferation of coliform bacteria in mice.

The anticoagulant activity of plumbagin was studied in rats. The compound significantly increased the prothrombin time, total proteins, GPT and alkaline phosphatase-levels in the liver tissue and decreased the GPT-levels in the serum. An anti-vitamin-K-activity on the part of plumbagin has been suggested along with an important role played by the hydroxyl group attached to the naphthoquinone ring of the compound.

Topical application of plumbagin isolated from Plumbago indica has been found to be useful in patients with common warts.

Root powder of Plumbago rosea produced 100 % antifertility activity when administered in a dose of 100 mg/100 g body weight orally from D1 to D7 of pregnancy in albino rats.

Plumbagin (2-methyl-5-hydroxy-1:4-naphthaquinone), isolated from the root of Plumbago rosea, in a dose of 20 mg/kg showed 100 % antifertility activity. But it was devoid of antioestrogenic, androgenic, antiandrogenic, progestational and anti-progestational activities. However, it exhibited oestrogenic, anti-gonadotrophic and anti-ovulatory activities.

The drug inhibited ovulation in rabbits at a dose of 200 mg/kg except in one animal, which may be due to biological variation. The drug showed satisfactory anti-implantation activity in the lowest dose. In rats, 58.33, 83.33 and 100 % activity was observed with doses of 50, 100 and 200 mg/kg respectively. (CCRAS.)

The chloroform extracts of Plumbago zeylanica roots showed antibacterial activity against B. mycoides, B. pumilus, B. subtilis, Sal. typhi, Sal. paratyphi, Sar. lutea, Staph. aureus, Xanth. citri and Xanth. malvacearum.

Plumbagin was reported to exhibit antibacterial activity against both Gram-positive and Gram-negative organisms. Plumbagin inhibited Staph. aureus, Staph. citreus, Staph. albus, Sal. paratyphi, Sal. dublin, and Kl. pneumoniae at a concentration of 20 mcg/ml. In another investigation, however, plumbagin did not show antibacterial activity against Proteus spp. and Staph. aureus.

The root extract of Plumbago zeylanica did not exhibit antiprotozoal, anthelmintic or antiviral activities.

Caution

Contraindicated during pregnancy.

Papilionaceae

PONGAMIA

***Pongamia pinnata* (L.) Merr.**

Figure 1 Pongamia pinnata [CCRAS]

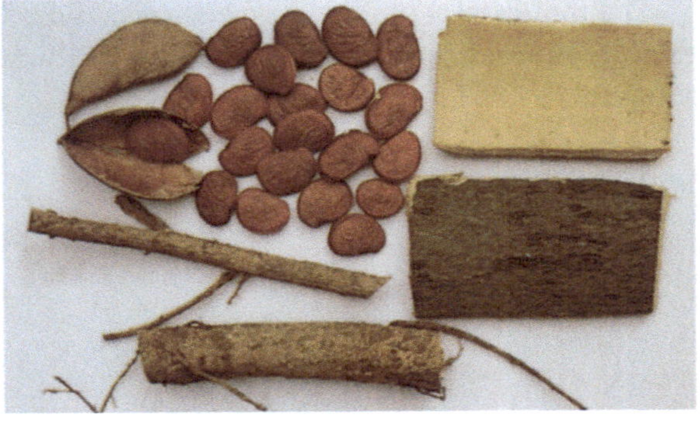

Figure 2 Pongamia pinnata [CCRAS]

***Pongamia glabra* Vent.**
***Derris indica* (Lam.) Bennett**

Habitat

Native of Western Ghats, chiefly found along the banks of streams and rivers or near sea coast in beach and tidal forests, up to 1200 m.

Classical and common names

Ayurvedic: Karanja (Charaka, Sushruta); Karanjaka, Ghrita-kaaranja, Naktaahva, Naktamaala, Prakiryaa, Udakiryaa. Puutika, Naktamaala are doubtful synonyms of Karanja; these are also equated with Holoptelea integrifolia Planch (Chirbilera). Kantaki-karanja, Lataa-Karanja, Karanji, Karanjwaa are equated with Caesalpinia crista Linn. (Fever Nut).
Unani: Karanj.
Siddha: Pungu.
English: Indian Beech, Karanj, Pongam Oil Tree.

Parts used

Bark, seeds, leaves.

Dose

Leaf-juice 10–20 ml, bark-decoction 50–100 ml.

Classical use

Charaka and Sushruta gave Karanja, in prescriptions, for eczema, scabies, leprosy, ulcers, piles, acute constipation, enlargement of spleen, jaundice, anorexia, biliousness, abdominal abscesses, intestinal paralysis, poisoning, fevers.

Sushruta prescribed Karanja seeds with honey for intrinsic haemorrhage; expressed oil from seeds, as laxative, for intestinal parasites, cutaneous affections; fruits for urinary and vaginal discharges. The oil was an ingredient of a hair oil, prescribed by Sushruta for baldness. The oil was also used externally in dermatosis, rheumatic affections, muscular atrophy.

Flowers were prescribed in glycosuria and other urinary disorders (Kaiyadeva Nighantu).

During the 16th century, fruits of Karanja, Kimsuka (Butea monosperma) and Arishta (Sapindus trifoliatus) were used for parasitic infections, obstinate urinary diseases, diabetes.

Karanjaadi Churna, Karanjaadi Ghrita, Karanjaadi Taila of Ayurvedic texts are no more available over-the-counter.

Karanja root is an ingredient in Dhanvantaram Ghritam, available in the South, prescribed for diabetes and rheumatic diseases.

Active principles and pharmacology

The tree is rich in flavonoids and related compounds. These include simple flavones, furanoflavonoids, chromenoflavones, chromenochalcones, coumarones, flavone glucosides, sterols, triterpenes and a modified phenylalanine dipeptide.

Seeds and seed oil gave karanjin, pongamol, pongapin and kanjone. Seeds, in addition, gave lanceolatin B, iso-pongaflavone and pongol. Seed oil also contained iso-pongachromene and pongaglabrone. Glabrachalcone was also isolated from the seed oil, which also gave a number of fatty acids, oleic and linolenic acids being major constituents.

Flowers yielded simple flavones, hydroxyfuranoflavones, furanoflavones, a chromenoflavanone, triterpenes, beta-sitosterol glucoside and aurantiamide acetate. (CIMAP.)

Karanjin is the active principle responsible for curative effect of the oil in skin diseases. Clinical experiments indicate that it is free from highly irritating and inflammatory effects of coumarin compounds. Its application in solution with vegetable oil or groundnut oil is reported to be better than when incorporated in a paraffin base. The oil gives good results in scabies, herpes, leucoderma and other cutaneous diseases. It can also be used internally with benefit as stomachic and cholagogue in cases of dyspepsia and sluggish liver.

Various parts of the tree exhibited following properties: leaf—carminative, antidiarrhoeal, bechic; stem-bark—styptic, anti-inflammatory; rind of pod and seed—febrifuge, antispasmodic, decongestant; leaf and seed—antileprotic; oil—antirheumatic. Root can be tried for scrofulous enlargements; stem-bark for bleeding piles, beriberi; rind of pod and seed for whooping cough, bronchitis; seed oil locally over chest for pneumonia, cold and chest pain; root and leaf for foul ulcers.

Pongamia pinnata exhibited significant hypoglycaemic effect in normal and alloxan diabetic rabbits.

The oil of Pongamia pinnata showed antibacterial activity against Mic. pyogenes var. aureus, Mic. pyogenes var. citreus, B. subtilis, C. diphtheriae, Sal. typhosa, Sal. paratyphi A, Sal. paratyphi B and E. coli; being most active against Sal. paratyphi A. The oil was found to be more active than neem oil. The oil was, however, inactive against P. vulgaris and Ps. pyocyana.

The essential oil from Pongamia pinnata showed mild antifungal activity against keratinophillic fungi, viz. Verticillium tenuipes, Malbranchea pulchella, Keratinophytontereum, and Chrysosporium tropicum.

In *in vitro* and *in vivo* evaluation of soaps prepared from some inedible oils, Pongamia pinnata oil soap did not reveal any antibacterial activity against Mic. pyogenes var. aureus, Mic. pyogenes var. albus, E. coli, Sal. typhi and Sh. dysenteriae, whereas it showed some inhibition against Aspergillus niger and not Epidermophyton floccosum and Trichophyton gypseum.

Portulacaceae

PORTULACA

Portulaca oleracea Linn.

See figure 1.

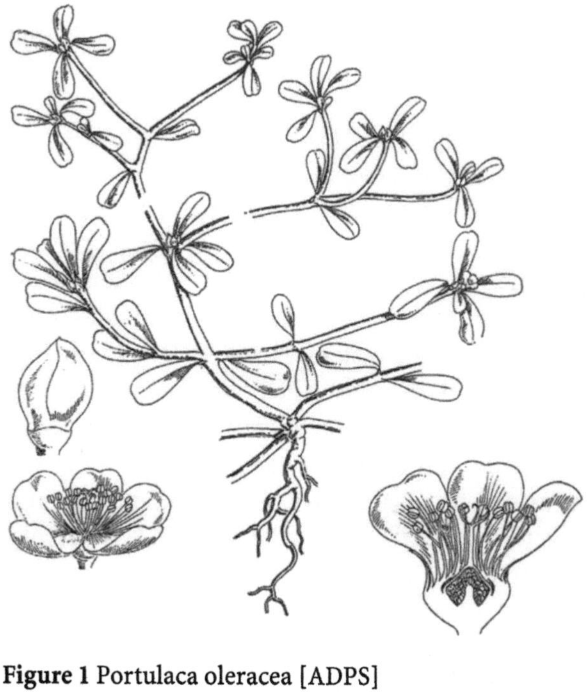

Figure 1 Portulaca oleracea [ADPS]

Portulaca quadrifida Linn.

Habitat
All over India up to 1700 m in the Himalayas; also cultivated as a vegetable.

Classical and common names
Ayurvedic: Lonikaa, Lonaa, Loni (Charaka, Sushruta); Gholika, Upodika.
Unani: Khurfaa, Kulfaa.
Siddha: Pulitari.
English: Common Indian Purslane.

Parts used
Whole plant, leaves, seeds.

Dose
Juice 1–20 ml, powder 125–400 mg.

Classical use
Charaka gave the plant cooked as a vegetable while treating piles. Sushruta used the plant as a potherb for its laxative properties.

In Unani medicine, leaves and seeds are used as a part of treatment for affections of kidneys and bladder. The drug is also prescribed as a diuretic and blood-purifier.

Powdered seeds are administered with Sharbat Buzoori (Bayaz-e-Kabir) in diarrhoea, hepatitis, heartburn, dysuria, painful micturition. (Cichorium intybus is the principal drug in Sharbat Buzoori.)

Active principles and pharmacology
Portulaca oleracea is the bigger variety of Purslane. Smaller variety is equated with Portulaca quadrifida.

The bigger variety yielded l-noradrenaline, dopamine and L-dopa. The stem yielded two red-violet pigments—oleracin I and II—which have been found to be acylated betacyanins. Ferulic acid and two pigments identified as 5-O-beta-cellobiosides of betanidin and iso-betanidin were obtained on alkaline hydrolysis of the betacyanin fraction. Aerial parts gave nicotinic acid; also tocopherol.

The diuretic and calmative effect of juice makes it an efficacious remedy for bladder affections and a soothing agent for diarrhoea and dysentery.

Research in China suggests that Purslane is a valuable herb for bacillary dysentery. In China, the herb is also used for appendicitis.

The juice of Purslane exhibited anthelmintic activity, especially against hookworms.

Research in India suggests that Purslane plant can be used in stomatitis, diseases of liver, spleen, kidney and bladder, cardiovascular system, on the basis of its depurative, antiscorbutic, aperient, blood-purifying and antibacterial properties.

Juice of the plant exhibits analgesic; flowering tops antihaemorrhagic; seed diuretic, antidysenteric and vermifugal activity.

The aqueous and ether extracts of the herb showed activity against Gram-negative bacteria.

The oral administration of homogenates of Portulaca oleracea reduced blood sugar level of alloxan-diabetic rabbits to normal.

The concentration of l-noradrenaline in the fresh plant (2.5 mg/g in one sample) is likely to be greater than that extractable from suprarenal glands of the mammals.

Roasted seeds (prescribed in Unani medicine), exhibited diuretic and antidysenteric property. A paste of seeds, when applied over burns and scalds, gave good results.

Verbenaceae

PREMNA

Premna serratifolia Linn.
Premna integrifolia Linn.
Premna obtusifolia R. Br.
Premna corymbosa auct. non Rottl. & Willd.

Habitat
Common along with the Indian and the Andaman coasts, plains of Assam and in the Khasi hills.

Classical and common names
Ayurvedic: Agnimantha, Shriparni, Jayee (Charaka, Sushruta). Vijayaa is a doubtful synonym. Clerodendrum phlomidis Linn. f. is equated with the smaller variety of Agnimantha, but it is a different drug known as Tarkaari in Ayurveda.
Siddha: Thamthalai.

Figure 1 Premna corymbosa [ADPS]

English: Headache Tree.

Parts used

Leaves, rootbark.

Dose

Decoction 50–100 ml.

Classical use

The roots of Agnimantha form a constituent of the Dashmuula group of roots, prescribed as a stomachic, laxative, cordial and tonic, as a cure for obstinate fever.

The entire plant was used by Charaka and Sushruta, internally, in prescriptions, for constipation, internal obstructions and abscesses, misperistalsis, piles, urinary diseases, calculi.

Shushruta gave roots internally as an appetizer for deranged digestion and dyspepsia; with calcined iron in virulent skin diseases, consumption, swellings; alkaline ashes for topical application in cutaneous diseases.

Charaka prescribed a decoction of Agnimantha for obesity.

In glandular enlargements and erysipelas, a paste of bamboo leaves and Agnimantha was given for topical application; a paste of bark, pounded with milk, for freckles; pounded rootbark with clarified butter was administered internally for urticaria and other allergic conditions of the skin. (Charaka Samhitaa, Chakradatta, Gadanigraha.)

Dashmuulaarishta (Bhaishajya Ratnaavali), available over-the-counter, is prescribed extensively as a restorative tonic for women. Dashmuula Kwaatha (Shaarangadhara Samhitaa) is a post-delivery tonic for women. Dashmuula Taila (Bhaishajya Ratnaavali) is for external application in sinusitis, headache.

Active principles and pharmacology

Leaves gave beta-sitosterol and polyisoprenoid; stembark contained spermine alkaloids, the major one was characterized as aphelandrine. Leaves contained a flavonoid, luteolin. Stems gave beta-sitosterol and betulin.

Decoction of Premna obtusifolia exhibited marked anti-inflammatory and antiarthritic activity against acute, subacute and chronic inflammation induced in both immunological and non-immunological experimental models. This activity did not appear to be mediated through the pituitary adrenal axis, as revealed by the failure of bilateral adrenalectomy to block the anti-inflammatory action of the extract. The extract did not produce any gastric ulceration in albino rats.

Premna obtusifolia stem-bark extract showed hypoglycaemic action in rats and hypotensive activity in cat/dog. The root extract also exhibited hypoglycaemic action in rats.

The root showed hepatoprotective; leaf carminative, galactogenic, bechic, febrifuge; rootbark antirheumatic and astringent properties. Root also exhibited antiarrhythmic activity (CIMAP).

Phenolic compound isolated from the fresh root-bark of Premna obtusifolia, showed antibacterial activity against Mic. aureus, B. subtilis and Strep. haemolyticus and was found inactive against E. coli, Sal. typhosa and B. dysenteriae.

Rosaceae

PRUNUS

Prunus amygdalus Batsch.
Prunus communis Fritsch
Amygdalus communis Linn.

Habitat

The species include three varieties, viz. var. amygdalus, var. amara (DC.) Focke and var. sativa (Ludw.) Focke. The first one includes wild types found in West Asia, Greece and North America; the second and third include a large number of cultivated types; var. amara comprising mostly the bitter almond and var. sativa the sweet almond.

In India, the almond is cultivated in Kashmir at elevations of 760–2400 m; also in Himachal Pradesh, Uttar Pradesh.

Sweet almond is used in Ayurvedic and Unani medicine.

Classical and common names

Ayurvedic: Vaataama (Charaka, Sushruta).
Unani: Baadaam Shireen.
English: Sweet Almond.

Classical use

Charaka and Sushruta used kernels of Vaataam alone or in prescriptions, internally, for debility, wasting diseases, emaciation, anaemia and as a revitalizing tonic.

Vaataam kernels were included in Amritapraasha Ghrita (for wasting diseases), Mahaa-Maayura Ghrita (for regulating menstrual cycle, also for alleviating spermic deficiency), and Jivaniya Ghrita (for gout).

In Ayurvedic texts Vaataam was known as Suphalam (a valuable fruit), as it was prescribed for promoting strength and semen.

Among over-the-counter Unani compound preparations, Majoon-e-Aarad Khurma is prescribed for spermatorrhoea, attentuated semen, oligospermia and debility; Laooq-e-Baadaam for bronchitis; Laooq-e-Shamoon for cold, cough, catarrh; Laooq-e-Zeeq-un-Nafas for asthma. Baadaam is also included in reputed Unani "sex-tonics"—Luboob Kabir, Luboob Sagheer, Majoon-e-Falaksair, Majoon-e-Mughalliz, Halva-e-Saalab. (National Formulary of Unani Medicine.)

Raughan-e-Baadaam Shireen (almond oil) is used in insomnia, and as a massage oil.

Almond flour, made from the residue left after expressing almond oil and almond butter, is used for the preparation of diabetic food.

Active principles and pharmacology

An analysis of kernels of Indian sweet almonds gives the following values: moisture 5.2, protein 20.8, fat 58.9, carbohydrates 10.5, fibre 1.7 and mineral matter 2.9 %. Calcium 230, oxalic acid 407, phosphorus 490, iron 4.5, thiamine 0.24, nicotinic acid 2.5 and riboflavin 0.15 mg/100 g. Ascorbic acid and vitamin A are absent. The kernel also contained folic acid (0.45 ppm), alpha-tocopherol (15 mg/100 g) and gamma-tocopherol (0.5 mg/100 g).

The carbohydrates present include sucrose (4.4–4.7 %), pentosans and hemicelluloses; starch is absent.

The chief protein of almond is a globulin, amandin; an albumin is also reported. The essential amino acid make-up of amandin is: arginine 11.9, histidine 1.6, lysine 0.7, phenylalanine 2.5, leucine 4.5 and valine 0.2 %. Tryptophan (1.4 %), methionine (0.7 %) and cystine (0.8 %) are also present. Amandin has a high arginine content.

At a 9 % level of intake, almond protein has a high digestibility co-efficient (94 %) but comparatively low biological value (50.8 %).

In Indian medicine, almonds are considered highly nutritions, demulcent and stimulant nervine tonic. They are also considered diuretic.

Bitter almonds (var. amara) are poisonous, contain cyanogenic glycosides—amygdalin 0.2–8.5 % (12–500 mg prussic acid per 100 g accordingly).

Use in Western herbal

John Gerard (1545–1612) wrote of almond oil: "The oil newly pressed out of sweet almonds is a mitigator of pain and all manner of aches, therefore it is good in pleurisy and colic. The oil of almonds makes smooth hands and face of delicate."

Culpepper (1616–1654) wrote of almond butter: "...very wholesome and commodius for students, for it rejoiceth the heart and comforteth the brain and qualifieth the heart and comforteth the brain and qualifieth the heat of the liver."

Almond oil is employed in bronchial diseases, in tickling coughs, hoarseness, costiveness, nephritic pains.

Almond flour is given to patients suffering from diabetes.

Blanched and beaten into an emulsion with barley water, sweet almonds are administered for the stone, gravel, stranguary and other disorders of the kidneys, bladder and biliary ducts.

Prunus cerasoides D. Don.

Habitat

Temperate Himalayas from Kashmir to Bhutan; hills of North-eastern India at 900–2300 m, introduced and cultivated in the Nilgiris.

Classical and common names

Ayurvedic: Padmaka, Padmagandhi (Charaka); Padmaadyaa.
English: Himalayan Wild Cherry, Bird Cherry.

Parts used

Bark.

Dose

Powder 3–6 g, decoction 50–100 ml.

Classical use

Rootstalks or fruits were used by Charaka, in prescriptions, internally as well as externally, for fever, malarial fever, high fever with delirium, gout, rheumatism, headache, haemothermia, neurological disorders. Padmaadi Leha of Charaka was for cough and chest diseases; Padmaka Taila for gout, arthritis, muscular atrophy.

Sushruta prescribed the bark for haemoptysis, bilious fever, persistent dysentery, peptic ulcers and for adhesion of fractured bones; also as a spermatogenetic and spermatopoietic tonic, as a galactagogue and uterine tonic for females.

Inhalation of Padmaka fumes was recommended during asthmatic attacks (Ashtaaga Hridaya).

Water extract of twigs was administered for treating threatened abortion (Dhanvantari Nighantu, Bhaavaprakaasha).

Active principles and pharmacology

Stembark contains flavones and iso-flavones—sakuranetin, genkwanin, prunetin, padmakastein and glycosides—sakuranin and padmakastin. Stem sapwood gave flavone glucoside—puddumin A. seeds contain flavonoid glycosides.

The leaves, twigs, bark and kernels contain a cyanogenetic substance. Kernels contain an oil similar to that of bitter almonds with a strong flavour of prussic acid.

The extract of Prunus cerasoides stem showed antispasmodic activity in isolated guinea-pig ileum.

Papilionaceae

PSORALEA

Psoralea corylifolia Linn.

Figure 1 Psoralea corylifolia [CCRAS]

Figure 2 Psoralea corylifolia [CCRAS]

Habitat

Rajasthan and eastern districts of Punjab, adjoining Uttar Pradesh.

Classical and common names

Ayurvedic: Somaraaji, Somavalli, Somavallikaa, Soma, Chaandri (Charaka); Vaakuchi, Baakuchi (Bhaavaprakaasha). Avalguja. Somaraaji and Avalguja have also been equated with Centratherum anthelminthicum Kuntze (Asteraceae).
Unani: Bakuchi, Baabchi.
Siddha: Karpoogaarisi.
English: Purple Flea-bane.

Parts used

Seeds, leaves, root. (Seeds are actually indehiscent pods of the plant, the pericarp closely adhering to the seed.)

Dose

1–3 g.

Classical use

Charaka prescribed seeds of Somaraaji, internally, for toxicosis, boils, tumours, skin eruptions. Shveta-avalgujaphal Rasaayana of Sushruta, Baakuchi Rasaayana of Astaanga Sangraha and Somaraaji Rasaayana of Ashtaanga Hridaya were reputed antitoxic and blood-purifying tonics. Panch-nimbaka Avaleha, Mahaabhallaataka Avaleha, Somaraaji Ghrita (Bhaavaprakaasha) were prescribed for leucoderma. Mahaa-Manjishthaadi Kwaatha (Shaarangadhara Samhitaa) also contained Baakuchi as an important ingredient.

Baakuchi was the drug of choice for treating obstinate skin diseases which included: itching red papules, itching eruptions, extensive eczema with thickened dermis, ringworm, rough and discoloured dermatosis, dermatosis with fissures, scabies, leprous leisons, leucoderma, vitiligo.

Baakuchi was given with hot water and the patient was exposed to the sun (Vrindamaadhava, Bangasena). Powder of Baakuchi, mixed with Shunthi (Zingiber officinale), was applied externally. Triturated powders of Baakuchi, Turvaraka seeds (Hydnocarpus laurifolia), Bhallaataka (Semecarpus anacardium), Chitraka (Plumbago zeylanica root) and Shilaajitu (Asphaltum mineral) were prescribed in obstinate cases (Ashtaanga Hridaya).

For vitiligo, a powder of Baakuchi seeds was administered with the decoction of Bibhitaka (Terminalia bellirica bark) and Kaakodumbara (Ficus hispida). A decoction of Aamalaka (Emblica officinalis) and heartwood of Khadira (Acacia catechu), added with Baakuchi powder, was also prescribed. (Shaarangadhara Samhitaa, Bangasena.)

For ringworm, one part of Tila (sesamum seeds) mixed with two parts of Baakuchi seeds was prescribed (Raaja Maarttanda).

In Unani medicine, Baabchi seeds are used, in prescriptions, in the form of qurs (pills) and zimads (ointments) for leucoderma and leprous leisons. The Central Council for Research in Unani Medicine runs a research centre at Hyderabad, Andhra Pradesh, where patients suffering from leucoderma, vitiligo and other obstinate skin diseases are treated.

Active principles and pharmacology

Seeds and roots contain chalcones, flavones, isoflavones, furanocoumarins and coumesterol group of compounds. These include bavachromanol, isoneobavachalcone, trilaurin, angelicin, psoralen, daidzein, coumesterol, coumestan, bakuchiol, bavachromene, bavachinin, iso-bavachin, bavachalcone, and iso-bavachalcone.

The essential oil from seeds contain beta-caryophyllene oxide, bakuchiol, beta-caryophyllene and linalool as chief constituents. Geranyl acetate, 4-terpineol and limonene are present in small amounts.

Preliminary clinical studies at the Calcutta School of Tropical Medicine were carried out with different preparations of Psoralea corylifolia applied locally in various skin affections. The oleo-resinous extract of the seeds (containing most of the essential oil) was found to be the most effective preparation, when applied locally on the patches of leucoderma (1956).

In view of the clinical use of Psoralea corylifolia seeds by Ayurvedic physicians through oral route, a preliminary clinical study was undertaken on the oral administration of the powdered seeds. Oral administration of the seeds was associated with severe undesirable side-effects like nausea, vomiting, malaise, headache and sometimes purging, indicating intense gastric irritation. To overcome these side-effects, the CDRI, Lucknow, used a mixture of psoralen-isopsoralen in all subsequent clinical studies. The pooled data of clinical trials carried out on patients of leucoderma with this mixture of psoralen-isopsoralen administered orally 10–30 mg/day at CDRI, Lucknow, as well as at the Christian Medical College, Vellore, (along with local application of the same mixture in liquid form simultaneously with oral administration, daily for 3 weeks) showed encouraging results. The recovery of pigmentation could be enhanced by supplementing this treatment with ultra-violet exposure (from a mercury vapour lamp) for 2–25 minutes. Oral administration of the furocoumarin mixture gave better results compared to external application of crude Psoralea oil alone (1956).

Local application of an oily mixture of psoralen-isopsoralen in 24 patients of leucoderma (6 male and 18 female) for periods varying from 1 to 12 months revealed erythema and pigmentation in 21 patients.

In 4 patients with leucoderma, psoralen used locally on the patches for periods varying from 7 days to 1 month, showed pigmentation which was maintained during a follow-up period of 10 months. Local irritation was a side-effect observed. On withdrawal of treatment, the patches reappeared. Application of an ointment of psoralen on patches of leucoderma in 16 patients daily for a period of 1–8 months also gave encouraging results, although the effect was slow.

8-Methoxy-psoralen (8-MOP) was given orally (40 mg b.d.) as part of photochemotherapy to 48 patients of psoriasis. The patients were exposed to sunlight for 5–30 minutes daily for 1–7 weeks. Of 38 patients of psoriasis vulgaris, 21 showed excellent or good response; while all the 6 patients of psoriasis erythroderma showed excellent or good response.

Clinical trial was also carried out with another chemical derivative of psoralen, viz. trioxalen (4, 5, 8-trimethylpsoralen) in 18 patients of psoriasis, in combination with sunlight. The results showed a significantly faster improvement with a combination of psoralen and trioxalen than with sunlight alone. (Research summary: ICMR.)

The seed extracts of Psoralea corylifolia inhibit the growth of Staphylococcus citreus, Staphylococcus aureus and Staphylococcus albus including strains resistant to penicillins. A highly potent staphylococcal fraction has been obtained from the seeds.

The seeds possess anthelmintic activity against earthworm, psoralen being the active principle.

The essential oil shows a selective activity again the skin streptococci. This probably accounts for its use in indigenous medicine for treatment of skin affections.

The oil has a marked antimicrobial action; in a dilution of 1 to 10, 000, streptococci and Paramecia are killed in ten minutes.

The oil exhibits a distinct stimulant action on voluntary muscles in high dilutions up to 1 to 100, 000. It increases the tone of the uterus and stimulates the intestinal smooth muscles of experimental animals.

Papilionaceae

PTEROCARPUS

Pterocarpus marsupium Roxb.

Figure 1 Pterocarpus marsupium—flowering and fruiting branch [WOI]

Habitat

Common in hilly regions throughout the Deccan Peninsula, and extending to Gujarat, Madhya Pradesh, Uttar Pradesh, Bihar and Orissa.

Classical and common names

Ayurvedic: Asana, Bijaka, Priyaka (Charaka, Sushruta).
Unani: Bijayasaar.
Siddha: Pirasaram.
English: Malabar Kino Tree.

Parts used

Heartwood, kino, flowers.

Dose

Decoction 50–100 ml, powder 3–6 g, gum-kino 1–3 g.

Classical use

Charaka gave a decoction of the bark, in prescriptions, internally for rheumatism, urinary disorders, fevers, hyperlipemia, obesity. Sushruta prescribed flowers in jaundice, chronic skin diseases, urethral discharges; included decoction of flowers in a hair oil for retarding greying of hair.

An alkali of Asana and Madhuuka (Madhuca indica) was prescribed for intrinsic haemorrhage (Charaka).

For venereal ulcers, a decoction of Asana and Khadira (Acacia catechu) was prescribed internally. Externally their paste, mixed with Guggulu (Commiphora mukul gum-resin), was applied.

Bijaka and Khadira were the drugs of choice for obstinate skin diseases during the classical period. In addition to the heartwood, flowers were also prescribed.

Asana belongs to the Saalasaaraadi group of herbs, considered specific for skin diseases, polyuria, anaemia, bronchial affections and obesity. Asana and Simshapaa (Dalbergia sissoo) group was used for allaying burning syndrome and oedema.

Asana was also used as an age-sustaining tonic for averting senility. A decoction-based linctus, mixed with Triphala (Chebulic myrobalan, Beleric myrobalan and Emblic myrobalan) and honey was administered (Charaka Samhitaa, Ashtaanga Hridaya). For improving vision, a decoction of Asana mixed with Bhringaraaja juice (Eclipta alba) was given (Ashtaanga Hridaya).

An alcoholic preparation, Bijakasaaraarishta, was prescribed for anaemia (Ashtaanga Sangraha).

As a paediatric medicine, pills of dried and powdered Asana flowers were given with rice-water for muscular spasms and nervine affections (Bangasena).

Asana was the drug of choice for "prameha" (polyuria including diabetes) during the 16th century.

Also used as a mouthwash and gargle in folk medicine.

In folk medicine, water stored in a tumbler made of Asana wood, is given to patients of diabetes mellitus. Kino and bark as powder or decoction is prescribed for diarrhoea, and pyrosis.

Active principles and pharmacology ;

Heartwood and roots contain isoflavonoids, terpenoids and tannins.

Heartwood gave (–)-epicatechin, 4-4'-dihydroxy-alpha-methylhydrobenzoin (marsupol), 5, 4'-dimethoxy-8-methyl-isoflavone, retusin-7-glucoside, irisolidone-7-rhamnoside, 5, 7-dihydroxy-6-methoxy-isoflavone-7-rhamnoside, carpusin, propterols A and B.

Roots gave liqui-ritigenin, garbanzol, isoliquiritigenin, p-hydroxybenzaldehyde, pterosupin, pseu-dobaptigenin and 5-deoxykaempferol.

Pterocarpus marsupium showed hypocholesterolemic effect in normal rabbits. A significant lowering of endogenous cholesterol level has been observed in rats fed with capsaicin. (See Classical use.)

In a series of studies carried out on rats, Chakravarthy and co-authors claimed a novel antidiabetic mechanism of Pterocarpus marsupium as revealed by pancreatic beta-cell regeneration by the flavonoid fraction and the pure flavonoid (–)-epicatechin isolated from the ethyl acetate soluble fraction of the ethanol extract of the bark. The authors further claimed significant antidiabetic effect in an authentic sample of (–)-epicatechin obtained from Australia, by the same mechanism. They also reported functional beta-cell regeneration by (–)-epicatechin in the islets of pancreas in alloxan diabetic rats, as assessed by immunoreactive insulin. These claims of Chakravarthy et al. have not been confirmed, however, by other workers. (See Lancet II (8292), 272. For lack of diabetic effect Lancet I (8284) 1303.)

Aqueous extract of Pterocarpus marsupium revealed hypoglycaemic effect in both acute and chronic conditions in normal rabbits. The aqueous extract was more potent than alcoholic fraction. In alloxan diabetic rabbits, the aqueous extract of Pterocarpus marsupium showed a more potent hypoglycaemic effect than Gymnema sylvestre and Coccinia indica. The hypoglycaemic activity of Pterocarpus marsupium extracts was confirmed in normal rabbits.

Tests on mice and rabbits with alcohol and aqueous extracts of the heartwood are said to have shown hypoglycaemic action, probably, by hindering the absorption of glucose in the intestines. (The Wealth of India.)

Pterocarpus marsupium stem was reported to be devoid of antibacterial activity in vitro against E. coli.

Propterol, propterol B, carpusin, marsupol, pterostilbine, oleanolic acid, liquiritigenin and isoliquiritigenin, isolated from Pterocarpus marsupium showed inhibition against Strep. faecalis R, and Staph. aureus R, streptococcus strain being more sensitive than the staphylococcus strain. Propterol, which was most active, was nearly 36–60 % active as compared to penicillin G in inhibiting the growth of these organisms. These compounds showed some inhibition against E. coli also.

The efficacy of Pterocarpus marsupium in the form of an ointment as an antifungal agent was evaluated in a clinical trial on 50 patients suffering from dermatophyte infection i.e. Tinea cruris (22 patients), Tinea corporis (14 patients) and mixed infections (14 patients). The ointment was locally applied for 7–10 days. The ointment made of the alcoholic extract of the wood was found more effective than that prepared from the aqueous extract.

Kino is included in IPC. It contains a non-glucosidal tannin kinotannic acid (25–80 %), kinoin and kino-red, in addition to small quantities of catechol (pyrocatechin), procatechuic acid, resin, pectin and gallic acid. Kino-red is the anhydride of kinoin, which is a phlobaphene produced from kino-tannic acid by the action of an oxidase enzyme present in the kino. The therapeutic value of kino is due to kinotannic acid which forms 70–85 % of the drug of good quality.

Kino is powerfully astringent and was formerly used widely in the treatment of diarrhoea and dysentery, and for relieving irritation caused by intestinal infection and colitis.

Pterocarpus santalinus Linn.

Figure 2 Pterocarpus santalinus [ZANDU]

Figure 3 Pterocarpus santalinus [ZANDU]

Habitat

Cuddapah district of Andhra Pradesh, neighbouring areas of Tamil Nadu and Karnataka.

Classical and common names
Ayurvedic: Raktachandana, Raktasaara.
Unani: Sandal Surkh, Sandal-e-Ahmar.
Siddha: Shivappu-Chandanam.
English: Red Sandal Wood, Red Sanders.

Parts used
Heartwood.

Dose
Powder 3–5 g, decoction 50–100 ml.

Classical use
Following are the botanical sources of different varieties of Chandana, used in Ayurvedic medicine: Santalum album (Santalaceae) Chandana, Bhadrashri; Pterocarpus santalinus (Papilionaceae) Raktachandana; Coscinium fenestratum (Menispermaceae) Pitachandana, Kaaliyaka, Kalambaka; Caesalpinia sappan (Caesalpiniaceae) Kuchandana, Pattanga. Harichandana (also called Bhadrashri) is equated with Syzygium aromaticum (Myrtaceae).

According to Bhaavaprakaasha, powder, linctus and alcoholic extract (Aasava) should be prepared with the white variety of Chandana, whereas the red one, Raktachandana, should be used in decoctions and pastes. In the absence of the white variety, camphor should be used; in the absence of both, red species should be used; in its absence, fresh Ushira (Vetiveria zizanioides) should be used.

Charaka and Sushruta prescribed a decoction of leaves or powdered bark of Raktachandana, internally, for fevers, burning syndrome, diarrhoea, bleeding piles, morbid menstrual discharges, dysuria.

Raktachandan was included in the Bhadrashriyaadi group of herbs, prescribed for intrinsic haemorrhage. Pushyaanuga Churna of Charaka (still available over-the-counter) was prescribed for leucorrhoea, metrorrhagia; Triphalaadya Ghrita of Sushruta for chronic fevers; Laakshaadi Ghrita of Sushruta for promoting complexion of the face. A paste of Raktachandana, Manjishthaa (Rubia cordifolia), Kushtha (Saussurea lappa) Lodhra (Symplocos racemosa), Priyangu (Prunus mahaleb), buds of Vata. (Ficus benghalensis) and roasted Masura (Lens culinaris) used for freckles (Ashtaanga Hridaya).

In folk medicine, a paste of the wood is used as a cooling external application for inflammations and headache.

Active principles and pharmacology
Heartwood contains the terpenoids eudesmol, isopterocarpolone, pterocarpol, cryptomeridiol, pterocarptriol and pterocarpdiolone. Pigments santalin permethylether, santalins A and B were isolated.

Sapwood gave acetyl oleanolic aldehyde, acetyl oleanolic acid and erythrodiol.

Bark contains the triterpenoids beta-amyrone, lupenone, epilupeol, lupeol and lup-(20)29-en-2-alpha-3-alpha-diol.

General pharmacological studies carried out on ethanol extract (90 %) of heartwood and bark of Pterocarpus santalinus showed a CNS-depressant activity including tranquillizing action, anticonvulsant action against electro-shock convulsions (but not metrazol-induced seizures). It also inhibited carrageenin-induced oedema in rat paw.

Pterocarpus santalinus (different parts of plant) could not show antibacterial, antifungal, anthelmintic, antiviral, antifertility and diuretic activities.

The extract of stem showed antispasmodic activity in isolated guinea-pig ileum, whereas the stem bark extract revealed semen coagulant activity, and the seed extract manifested hypoglycaemic activity in rats. The root extract was completely inactive.

Negative findings were obtained for hypothermic, diuretic, antipyretic and analgesic activities. (CCRAS.)

Sterculiaceae

PTEROSPERMUM

Pterospermum acerifolium (L.) Willd.

See figure 1.

Pterospermum aceroides Wall.

Habitat
The subtropical Himalayan tract and the outer valleys from Yamuna eastwards to West Bengal and in

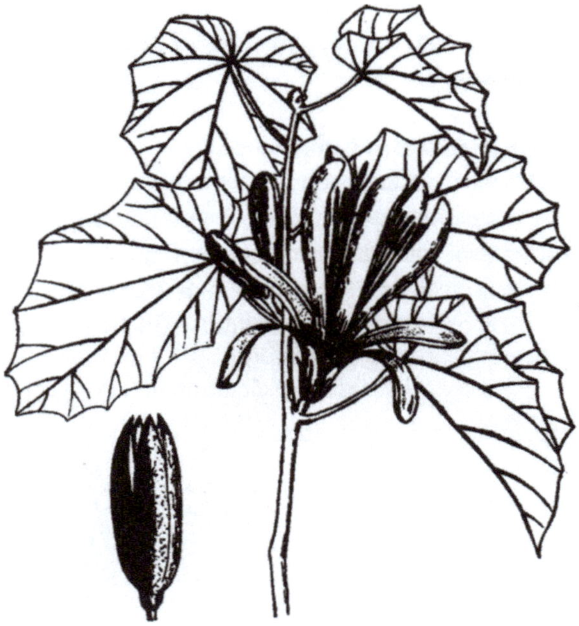

Figure 1 Pterospermum acerifolium—flowering branch and fruit [WOI]

North-eastern states, Bihar, Western Ghats and Andaman Island up to 1200 m.

Classical and common names

Ayurvedic: Muchukunda, Muchakunda, Kshatra-vriksha, Chivuka, Prativishnuka. Muchukunda Champaa. Pterospermum aceroides is also equated with Karnikaara, a synonym of Cassia fistula.
Unani: Gul-e-Muchkun.
Siddha: Vennangu.

Classical use

Muchukunda flowers, credited with blood-purifying, anti-inflammatory and antiseptic properties, were used in diseases due to vitiated blood and derangement of biliary functions—pruritus, skin eruptions, swellings, piles, ulcerations; headache; sore throat, cough, bronchitis. (Raaja Nighantu, Bhaavaprakaasha.)

A paste of flowers was applied over the forehead for providing relief during headache, hemicrania (Shaarangadhara Samhitaa, Vrindamaadhava). It was also used externally to small pox.

Muchukunda flowers were an ingredient in Himaanshu Taila. The oil was used externally for its antiseptic and cooling properties.

In folk medicine, the flowers and bark are charred and mixed with the powder of Mallotus philippinensis flowers for application to small pox.

Active principles and pharmacology

Flowers contain kaempferol-3-O-galactoside, luteolin-7-O-glucoside, luteolin-7-O-glucuronide, kaempferol, kaempferide-7-O-beta-D-glucopyranoside, 24-beta-ethylcholest-5-en-3 beta-O-alpha-cellobioside, 3, 7-dimethyl-7-methyl-1:5-pentacosanolide, n-hexacosane-1, 26-diol dilignocerate, friedelanol, beta-amyrin, beta-sitosterol and a mixture of acids and saturated hydrocarbons.

The acids identified were palmitic, stearic, arachidic, behenic, myristic, lignoceric, oleic, linoleic and linolenic acids. Seed oil gave malvalic acid. Trunk-bark and seeds gave the amino acids tryosine, cystine, glycine, alanine and leucine.

Flowers, leaves and bark exhibit anticephalagic, anti-inflammatory, haemostatic, antileprotic and antiseptic properties.

Punicaceae

PUNICA

Punica granatum Linn.

Figure 1 Punica granatum [ZANDU]

Habitat

Native of Iran, Afghanistan and Baluchistan. Found growing wild in the warm valleys and outer hills of the Himalayas between 900–1800 m. Cultivated throughout India.

Classical and common names

Ayurvedic: Daadima, Daadimba, Raktapushpa (Charaka, Sushruta), Dantabijaa, Raktakusumaa.
Unani: Anaar.
Siddha: Mathulai.
English: Pomegranate.

Parts used

Rind of the fruit, leaves, rootbark.

Dose

Decoction of bark 40–80 ml, powder of dried rootbark, seeds 325 mg–1.8 g.

Classical use

Charaka and Sushruta gave the juice of Daadima fruit alone or in prescriptions for diarrhoea, oedema, cough, emesis, and anorexia; fruit rind with Mustaka (Cyperus rotundus) for bleeding piles.

A decoction of the bark of Daadima and Kutaja (Holarrhena antidysenterica), mixed with honey, was administered for diarrhoea with blood (Ashtaanga Hridaya). Juice of the ripe fruit, a decoction of rind, paste of flower buds mixed with honey, juice of flowers mixed with honey, were prescribed for diarrhoea and intrinsic haemorrhage. (Ashtaanga Hridaya, Shaarangadhara Samhitaa, Siddha-bhesaja-manimaalaa.)

Fresh rind of Daadima, mixed with milk, was applied externally for freckles.

Among over-the-counter compound preparations, Daadimaashtaka Churna (Ashtaanga Hridaya) is for diarrhoea and dysentery; Lavan Bhaaskar Churna (Shaarangadhara Samhitaa) for indigestion and anorexia; Daadimaadi Ghrita (Charaka Samhitaa) for anaemia, jaundice, dyspepsia.

In Unani medicine, pomegranate fruit is known as Anaar Shireen or Rummaan Huluvv; Anaar Tursh or Rummaan Haamiz; and Anaar Maikhosh or Rummaan Muzz (sweet, sour and sweet-sour varieties respectively).

Sharbat-e-Anaar Shireen, Sharbat-e-Annar Tursh (Qarabadeen-e-Jadeed) are prescribed in chronic diarrhoea, biliousness and vomiting; Jawarish-e-Anarain (Qarabadeen-e-Azam) in hepatitis, anorexia, nausea, vomiting, bilious diarrhoea.

Qurs-e-Gulnaar (Qarabadeen-e-Jadeed) is given in haemorrhages, haematemesis, haemoptysis. (Gulnaar is the flower of male pomegranate tree which does not bear fruits.)

Active principles and pharmacology

Fruit rind gave an ellagitannin named granantin B, punicalagin, punicalin and ellagic acid.

Bark contained the alkaloids iso-pelletierine, pseudopelletierine, methylisopelletierine, methylpelletierine, pelletierine as well as isoquercetin, friedelin, D-mannitol and estrone.

Flowers gave pelargonidin-3, 5-diglucoside. Flowers also gave sitosterol, ursolic acid, maslinic acid, asiatic acid, sitosterol-beta-D-glucoside and gallic acid.

Rind gave pentose glycoside of malvidin and pentunidin. Seeds gave malvidin pentose glycoside. Fluoride, calcium, magnesium, vitamin C and phosphate are also reported from fruits.

Leaves gave elligatannins—granatins A and B and punicafolin.

Fresh rind of fruit contains: wax 0.8, resin 4.5, mannitol 1.8, non-crystallized sugars 2.7, gums 3.2, inulin 1.0, mucilage 0.6, tannin 10.4, gallic acid 4.0 and calcium oxalate 4.0 %. Pectin occurs to the extent of 2.4 %.

Tannins occur in all parts of the tree, particularly in fruit pulp (up to 26 % in dried rind), stembark (10–25 %), rootbark (28 %) and leaves (11 %).

Analysis of the edible portion (68 %) of pomegranates from Coonoor (South India) gave: moisture 78, protein 1.6, fat 0.1, fibre 5.1, other carbohydrates 14.5 and mineral matter 0.7 %. Calcium 10, magnesium 12, oxalic acid 14, phosphorus 70.0, iron 0.3, sodium 0.9, potassium 133.0, copper 0.2, sulphur 12.0, chlorine 2.0, carotene 0, thiamine 0.06, riboflavin 0.10, nicotinic acid 0.30 and vitamin C 14 mg/100 g. The fruit is a good source of sugars and vitamin C but is poor in calcium. The fruit contains 0.27 % of pectin (as calcium pectate); during ripening the insoluble pectin changes into soluble pectin.

The antidiarrhoeal activity of aqueous and alcoholic extracts of the fruit rind of P. granatum was investigated in three experimental models using albino rats. The extracts exhibited significant activity in rats when compared to loperamide hydrochloride, an antidiarrhoeal drug.

Antimicrobial evaluations of different solvent extracts from six plants (Berberis aristata, Bombax malabaricum, Gossypium herbaceum, Holarrhena antidysenterica, Myristica fragrans and Punica granatum) to being therapeutic agents in diarrhoea and dysentery showed P. granatum as most effective against four species of Shigella and E. coli.

The *in vitro* and *in vivo* studies, carried out on a virulent strain of Entamoeba histolytica, exhibited encouraging results. Good results were achieved with 40 mg/ml and 20 mg/ml rind and 5 mg/ml, 4 mg/ml, 2 mg/ml flowers. Mentronidazole served as a control drug. *In vivo* P. granatum-treated rats revealed normal cecum and intestines with no thickening or ulceration or any other lesions, and no motile trophozoites seen. No toxic effects were shown in the rats during the medication. In vivo studies showed complete inhibition of growth of E. histolytica after treating with rind 10 mg/10 g body weight and flowers 25 mg/10 g.

The fixed oil from the seeds of P. granatum showed inhibitory action against Sal. typhosa, Sal. paratyphi, Sh. flexneri and Kl. pneumoniae.

The extracts of the rind, flower, leaves and stembark exhibited antibacterial activity against B. anthracis, B. pumilus, B. subtilis, Sal. paratyphi, Vib. cholerae, Xanth. campestris and Xanth. malvacearum. Staph. albus, however, was resistant to all extracts of the plant.

The acetone extract of P. granatum was fungitoxic to Pyricularia oryzae and Colletotrichum falcatum. The extracts of the bark, fruit, pulp, flowers and leaves of P. granatum completely inhibited the spore germination of the fungi viz., Dreschlera rostrata and Curvularia lunata.

The alcoholic extract of P. granatum showed anthelmintic activity, as revealed by a dose-dependent inhibition of transformation of eggs into filariform larvae of Haemonchus contortus. The chloroform extract of the stem and root of P. granatum at a dose of 150 mg/kg twice a day for three consecutive days completely cleared Hymenolepis nana infestations in mice. However, at a lower dose of 100 mg/kg twice a day for 3 days, P. granatum was able to eliminate 40 % worms in mice. No effect could be seen against Syphacia obvelata and Nippostrongylus brasiliense.

Caution

Dried rootbark and seeds contain toxins. (Francis Brinker.) More than 80 g powder is fatal. Contraindicated during pregnancy due to its emmenagogue and possible abortifacient effects, confirmed by its uterine stimulant activity in animals.

Bark extracts are subject to legal restrictions in some countries.

Euphorbiaceae

PUTRANJIVA

Putranjiva roxburghii Wall.

Figure 1 Putranjiva roxburghii—flowers and fruit [WOI]

Drypetes roxburghii (Wall.) Hurusawa

Habitat

Found wild or cultivated in almost all parts of India, ascending up to an altitude of 750 m.

Classical and common names

Ayurvedic: Putranjiva, Putrajivaka, Putrajiva, Garbhadaa, Garbhakaraa; Yashtipushpa, Artha-saadhana. Jiyaapotaa.
Siddha: Karupali.

Parts used

Leaves, seed.

Dose

Juice of leaves 10–20 ml, seed powder 3–6 g.

Classical use

In Bhaavaprakaasha, Putrajiva is not a fertility-promoting herb. It was prescribed for "visphota", suppurative or eruptive diseases, possibly venereal diseases, which were to be treated before copulation. Paste of the drug was used only externally. Another herb, Putrakamanjari, is mentioned in Bhaavaprakaasha, which was credited with fertility-promoting properties. Infusion of the root was prescribed internally. It was used in "Punsavana Yoga" (Conception-promoting Compound). One more herb, Karkotikaa-Vandhyaa, is found in Bhaavaprakaasha as a synonym of Putrakamanjari (NAA). Root of the male plant was used as a paste for diseases of the breast (Mammalgia).

Rosaries of the hard stones of the fruits of Putrajiva were worn by children "(for warding off infections)" so that they remain in good health. This again suggests that the classical drug was antibacterial and antiviral.

In folk medicine, it is claimed that leaves and stones give good results in cases of leucorrhoea, menorrhagia and habitual abortion. A general physician of Gujarat claimed that he tried the herb on 25 women. 20 women conceived after 23 months. One woman, who was suffering from habitual abortion or stillborn baby, was also treated successfully. The physician also quoted a compound preparation, designated as "Putradaa Vati," which contained Putranjiva (commonly known as Jiyaapotaa) as the principal drug.

In folk medicine, the stones of the fruit are prescribed for spermatorrhoea, oligospermia and vaginal discharges. The drug has assumed importance in folk medicine due to its classical designation.

Active principles and pharmacology

Seed kernel on steam distillation yield 0.5 % of a sharp-smelling essential oil. The oil contains isopropyl and 2-butyl-isothiocyanates as the main constituents and 2-methyl-butyl-isothiocyanate as a minor component; the isothiocyanates are produced on enzymic hydrolysis of glucoputranjivin, glucocochlearin and glucojiaputin respectively.

An additional glucoside, glucocleomin has been identified in the seed kernels. On enzymic hydrolysis it affords a non-volatile mustard oil cleomin. A glucoside pattern similar to that in the seeds is reported in the shoots and roots.

The fruit pulp contains a large proportion of mannitol, and small quantities of saponin glucoside and an alkaloid. The alkaloid is also present in a small quantity in the stones of the fruit. (The Wealth of India.)

The drug should be examined for the following areas: venereal diseases, urinogenital diseases of women, threatened abortion, tetratogenic implications.

Q

Fagaceae

Quercus

Quercus infectoria Oliv.

Habitat

A small oak, indigenous to Asia Minor and Persia, is found in the Kumaun, Garhwal and Bijnor forests.

The oak galls, used in commerce and medicine, are excrescences on the Quercus infectoria, as a result of the puncture of the bark of young twigs by the female gall-wasp, Cynips Gallae-tinctoriae, who lays eggs inside. The young larva that hatches from the eggs feeds upon the tissues of the oak and secretes in its mouth a pecular fluid, which stimulates the cells of the tissues to a rapid division and abnormal development, resulting in the formation of a gall.

The larva thus becomes completely enclosed in a nearly spherical mass, which projects from the twig, furnishing it with a supply of starch and other nutritive material.

The growth of the gall continues only as long as the egg or larva lives or reaches maturity and passes into a chrysalis, from which the fully developed gall-wasp emerges and escapes into the air through a hole bored with its mandibles in the side of the gall.

For medicinal use, galls are collected before the insects escape.

Classical and common names

Ayurvedic: Maajuphalaka (Bhaavaprakaasha), Maayaaphala.
Unani: Maajuphal; Maazu (Persian), Maaphala.
Siddha: Mochakai, Mashikkai.
English: Oak galls, Aleppo galls, Mecca galls.

Parts used

Gall.

Dose

3–5 g.

Classical use

Although Maajuphalaka powder was included in an aphrodisiac sweet preparation—Kaameshwara Modaka (Bhaavaprakaasha)—aphrodisiac properties cannot be attributed to it. Many drugs of the compound preparation are heavy, difficult to digest, likely to upset the stomach and may precipitate diarrhoea; keeping all these aspects in mind, Maajuphalaka was included in the recipe as an astringent.

Maajuphal is considered as one of the most powerful vegetable astringents in Indian medicine. It is used in the form of powder, decoction, infusion and ointment. Powder is included in a number of dental powders, prescribed for strengthening gums and teeth. A decoction or infusion is used as a gargle for relaxed throat, tonsillitis, stomatitis. Ointment (1 part of finely powdered galls, 4–6 parts of vaseline or benzoated lard) is used for haemorrhoids, anal fissures, chapped nipples. Galls are no longer used internally for bowel complaints.

In Unani medicine, powdered galls mixed with vinegar are used for ringworm and alopecia. An ointment is applied for vaginal laxity, prolapse of rectum and diseases of the anus. A vaginal pessary is used for excessive vaginal discharges.

For local application on wound and ulcers, an ash of Maazu, or powdered Maazu and Punica granatum (Anaar) fruit skin is used as a dusting powder.

Active principles and pharmacology

Common oak (Quercus robur) dried bark contains: tannins (12–16 %), gallo-tannins, ellagic tannins, including among others, castalagin, pedunculagin, vesvalagin; flavano-ellagic tannins (acutissimine A and B, eugenigrandin, gujavacine B, stenophyllanin C); the catechin tannins oligomeric proanthocyanidins, monomeric, dimeric catechins and leucocyanidins. (PDR.)

Oak gall contains 50–70 % tannins (gallotannic acid). It also contains gallic acid (2–4 %), ellagic acid, nyctanthic acid and rubric acid; gum, starch, sugar and essential oil. Amentoflavone hexamethyl ether, iso-cryptomerin and beta-sitosterol have also been isolated.

The alcoholic extract of Quercus infectoria showed no significant anti-inflammatory effect against formaldehyde-induced paw oedema in rats. (CCRAS.)

Fraction-A (acetone-treated methanol extract in doses of 30, 100 and 200 mg/kg i.p.) exhibited analgesic effect in rats. Fraction-B (Chr-menthol extract) caused local anaesthetic effect, as revealed by complete blockade of sciatic gastroenemias of frog. Methanolic extract of galls of Quercus infectoria, having the composition of syringic acid, in 250 mg/kg i.p. produced barbiturate potentiation in mice; antitremorine activity in 500 mg/kg i.p. in mice; local anaesthetic effect in sciatic nerve stimulation in frog; hypnotic activity in mice. (CCRAS.)

Use in Western herbal

A decoction of common oak bark is used as a gargle for sore throat and tonsillitis. The bark, in the form of a lotion or ointment, is used for treating haemorrhoids, anal fissures, and weeping eczema. The drug is occasionally present in tea mixtures and gastro-intestinal remedies. Hot tea is recommended for swollen varicose veins, internally as well as externally for fomentation (alternate hot and cold packs are advised).

Galls are also used in the same way.

Bark extracts are available over the counter.

Caution

Internal use of Oak bark or galls may reduce or inhibit the absorption of alkaloids and other alkaline drugs.

Internal administration may lead to digestive complaints due to drug's secretion-inhibiting effect.

For treating diseases of gums and teeth, raw powder of galls should not be used. It must be diluted with other antiseptic and styptic herbs.

Contraindicated for external use with extensive skin surface damage.

R

Rubiaceae

RANDIA

Randia spinosa Poir.
Randia dumetorum (Retz.) Poir.

Figure 1 Randia dumetorum [CCRAS]

Figure 2 Randia dumetorum [CCRAS]

Catunaregam spinosa (Thunb.) Tiruv

Habitat
Throughout India, up to an elevation of 1350 m in the hills.

Classical and common names
Ayurvedic: Madana, Chhardana, Pindi, Shalayaka, Vishapushpaka.
Unani: Mainphal.
Siddha: Marukkallankay, Mad Karai.
English: Emetic Nut.

Parts used
Fruit.

Dose
For emesis 3–6 g powder; otherwise 1–2 g.

Classical use
Charaka prescribed Madana with Pippali (Piper longum) and Indrayava (Wrightia tinctoria seeds), with warm water, for inducing emesis during fevers. It was also used as a single drug for aborting accumulated phlegm, bile and internal toxic

substances. (Charaka Samhitaa, Ashtaanga Hridaya.)

A compound powder, called Madanaadivamana, composed of Madana, Arka (Calotropis gigantea) and Yashti-madhu (Glycyrrhiza glabra) was prescribed as an expectorant in doses of 300–900 mg; as emetic in 1.2–3.6 g doses.

For emesis, the rind and the pulp of the fruit was considered safe and free from toxic side-effects.

A paste of the nut was applied externally in rheumatism, and to disperse abscesses.

In Indian medicine, the pulp of the fruit, dried and powdered, is used as a substitute for Cephaelis ipecacuanha. The drug is considered to have anthelmintic and abortifacient properties.

Active principles and pharmacology

The activity of the drug is attributed to the presence of saponins which occur to the extract of 2–3 % in fresh fruits (about 10 % in dried whole fruit).

The saponins are concentrated mostly in the pulp. A mixture of two saponins, viz. randia or neutral saponin and randia acid or acid saponin, has been isolated from the pulp. The two saponins occur in the fruit at all stages of ripening. On complete hydrolysis, both the saponins yield oleanolic acid as the sapogenin. The acid saponin is believed to be natural prosapogenin of neutral saponin. It is formed by the combination of oleanolic acid with a molecule each of glucose, fructose and xylose and probably two molecules of glycuronic acid. Later, investigations gave a new saponin designated as ursosaponin, isolated from ethanolic extract of dried whole fruit. On hydrolysis it gave ursolic acid and glucose.

Besides saponins, the fruit contains a triterpene, an acid resin and an essential oil.

The ethanolic extract of the pulp showed a stimulant action on isolated guinea-pig uterus. In experimental animals, crude saponin produced salivation; on contact it caused a generalized irritation of the mucous membranes, producing sneezing, vomiting and bleeding from the urinary tract. The cornea was inflamed and the drug caused haemolysis both *in vitro* and *in vivo*. The perfused frog heart was arrested in a few minutes; with higher doses practically instantaneously. The drug was rapidly detoxicated by the liver. (The Wealth of India.)

The seeds are reported to be free from saponins. They contain fat (1.5 %), protein (14.2 %), mucilage, resin, organic acid (1.4 %), and a minute quantity of an unidentified alkaloid.

The bark contains scopoletin, D-mannitol and a mixture of saponins. The saponins on hydrolysis yield glucose, xylose, rhamnose and two triterpenic acid sapogenins—designated as randialic acid A and randialic acid B.

The roots contain scopoletin and D-mannitol.

Brassicacea

RAPHANUS

Raphanus sativus Linn.

Figure 1 Raphanus sativus [ZANDU]

Habitat

Cultivated throughout India; up to 3000 m in the Himalayas and other hilly regions. A cool season crop.

Classical and common names

Ayurvedic: Muulaka, Laghumuulaka, Muulakapotikaa.
Unani: Muuli, Turb.
Siddha: Mullangi
English: Radish.

Parts used

Root, seed, leaf.

Dose

Juice 10–20 ml, decoction 50–100 ml.

Classical use

Charaka gave soup of Muulaka root for diarrhoea; also for chronic coryza; soup of dried Muulaka for hiccough and asthma, with Bilva (Aegle marmelos fruit) for piles. Externally, a paste of Muulaka seeds and Baakuchi (Psoralea corylifolia) seeds, or Muulaka seeds pounded with Apaamaraga (Achyranthes aspera) seeds, was prescribed for skin diseases. For erysipelas hot paste of dried Muulaka was applied topically. Oedema was fomented with the hot decoction of Muulaka.

For jaundice and anaemia, juice of tender radish (40 ml, added with sugar) was administered (Siddha-bheshaja-manimaalaa).

In Unani medicine, seeds are used in the genitourinary tract infections and as an emmenagogue; roots for piles, gastrodynic pains, dysuria, stranguary and as cholagogue.

An ash-salt of leaves and roots is prescribed for diseases of the abdomen as a deobstruent.

In folk medicine, radish preparations are prescribed for liver and gall bladder affections; seeds as peptic, carminative, expectorant, diuretic.

In Indian homes, radish is served as a salad before meals for stimulating appetite and gastric juices. Juice of fresh leaves is used as diuretic and laxative. Pods, commonly known as Mungraa, are eaten raw or cooked as a vegetable.

Active principles and pharmacology

Radish is a good source of ascorbic acid (15–40 mg/100 g) and supplies a variety of mineral salts. Trace elements include aluminium, barium, lithium, manganese, silicon, titanium, fluorine and iodine (up to 18 mcg/100 g.)

White radish root contains vitamin A 5.0 I.U., thiamine 0.06, riboflavin 0.02, nicotinic acid 0.5, vitamin C 15.0 mg/100 g. Leafy tops are particularly rich in calcium, iron, ascorbic acid and are considered to be one of the richest sources of vitamin A among the leafy vegetables. Oxalic acid content of leaves is reported to be high (total oxalate 0.68 %, soluble oxalates 0.39 %) and as such may affect the calcium availability to the system. The vegetable is a fair source of calcium if consumed with rice. (The Wealth of India.)

Radish contains glucose as the major sugar and smaller quantities of fructose and sucrose. Pectin (0.3 %, as calcium pectate) and pentosans are also reported, while starch is absent.

Organic acids detected in radish include p-coumaric, caffeic, ferulic, phenyl pyruvic, gentistic and p-hydroxybenzoic acids. The total acid content is less in white radish than in red radish. Non-protein nitrogen accounts for about 40 per cent of total nitrogen; the free amino acids identified include most of the essential amino acids.

The pungent flavour of radish has been attributed to the presence of volatile oil iso-thiocyanate (mustard oils). A volatile oil distilled from white radish was reported to contain butyl-crotonyl iso-thiocyanate-sulphide as the chief constituent. Methyl mercaptan has been reported to be responsible for the disagreeable odour of radish oil.

Radish seeds contains glycosidically bound mustard oils of which allyl, methyl and isopropyl isothiocyanates and sulphoraphene have been identified in different varieties; 4-methylsulfinyl-3-butenyl-cyanide, the corresponding nitrile of sulphoraphene and cleavage product of its glucoside, has also been isolated.

Sulphoraphene exhibits antibacterial activity against Streptococcus, Pyococcus, Pneumococcus and E. coli.

Another sulphur-containing oil, raphanin, has been separated from aqueous extract of the seeds. It is active against several Gram-positive and Gram-negative bacteria. It has been suggested that raphanin is identical to sulphoraphene.

The seeds are also reported to contain a broad-spectrum antibiotic, machrolysin, specific against Mycobacterium tuberculosis.

The seeds have a high tocopherol content and possess strong antioxidant activity. The flowers possess cholagogue properties.

Use in Western herbal

Radish is considered secretagogue for the upper gastrointestinal tract, motility promoting, and antimicrobial.

German Commission E recognized the efficacy of Raphanus sativus roots in the following areas:

- Fevers and colds
- Loss of appetite
- Cough, bronchitis.

Radish is used for peptic disorders, especially those related to dyskinesia of the bile ducts, catarrh of the upper respiratory tract (PDR).

Apocynaceae

RAUVOLFIA

Rauvolfia serpentina Benth. ex Kurtz

Figure 1 Rauvolfia serpentina [ZANDU]

Habitat

Distributed in the sub-Himalayan tract from Punjab eastward to Nepal, Sikkim, in Assam, in the lower hills of the Gangetic plains, eastern and Western ghats, in some parts of Central India and in the Andamans.

Classical and common names

Ayurvedic: Sarpagandhaa, Sarpasugandhaa, Sarpachhatrikaa. Synonyms: Naakuli, Gandha-Naakuli.
English: Rauvolfia Root, Serpentine or Serpentina Root.

Classical use

Sarpagandhaa of Ayurvedic text was not the Sarpagandhaa of modern medicine (identified by the German scientist Rauwolf).

It is a wrong notion that Rauvolfia (as a herb) has been employed "for centuries" (in India) for the relief of various central nervous system disorders, both psychic and motor, including anxiety states, excitement, maniacal behaviour associated with psychosis, schizophrenia, insanity, insomnia and epilepsy.

There is no mention of Sarpagandhaa in Charaka Samhitaa. But later, Naakuli and Gandha-Naakuli of Charaka were considered as synonyms of Sarpagandhaa.

While commenting on the drugs of Sushruta, Dalhan equated Sarpagandhaa with Naakuli, Sarpachhatrikaa and Varshaasu Chhatraakaaraa (poisonous fungi and fungi of the rainy season). Another commentator, Arundatta, equated the two varieties of Naakuli with Sarpaakshi and Sarpasugandhaa. In Ashtaaga Hridaya, Sarpagandhaa and Sarpasugandhaa were used as synonyms of Naakuli. Sarpaakshi was also equated with the red-flowered variety of Shankhapushpi (Convolvulaceae) by Dalhan.

Charaka included Chhatraka (mushroom) species among edible vegetables, excluded Sarpachhatraka due to its poisonous character.

These classical references prove that Ayurvedic scholars have wrongly equated Naakuli with the Sarpagandhaa of modern medicine (Rauvolfia serpentina) and Gandha-Naakuli with Aristolochia indica (when its round fruit bursts, it assumes an umbrella or mushroom type shape).

Dalhan's equation of Sarpagandhaa with Sarpachhatrikaa and Varshaasu Chhatraakaaraa clearly indicates that the Sarpagandhaa of Ayurvedic texts was a fungi.

Unani scholars have equated Naakuli and Sarpagandhaa with Ishwarmuula, Ishwari, Israul, Zarawand-e-Hindi (Aristolochia indica, Indian birthwort).

In an insanity-specific classical compound, Mahaa-paishaachika Ghrita (Bhaishajya Ratnaavali), in different texts, Chhatra, Saati-chhatraa (smaller and bigger varieties of mushrooms); both the varieties of Naakuli (Naakuli and Gandha-Naakuli) and both the varieties of Sarpagandhaa (possibly Sarpagandhaa and Sarpasugandhaa) have been incorporated. In no way was the Sarpagandhaa of modern medicine used in the compound.

Active principles and pharmacology

Mushrooms: Agaricus campestris Linn; Cantharellus cibarius Fr. are edible mushrooms, excellent source of minerals (calcium, phosphorus, potassium, iron, copper and iodine), vitamins of B complex, vitamin C, K and D. Field mushroom contains amylase, maltase, glycogenase, protease, catalase, tyrosinase, two phosphomonoesterases, two polyphosphatases and polyphenoloxidase.

Amanita phalloides Secr. is toxic due to the presence of phalloidine, alpha-amanitine and beta-amanitine. A thermolabile hemolysin, phallin, is also present. Phalloidine is a peptide toxic to liver.

Amanita muscaria Pers. contains toxic principles—cholin, muscarine, mycetoatropine (muscaridine).

The albunoid phallin (found in Amanita phalloides) and muscarine (found in Amanita muscaria) appear to be related to serpent venom (cobra venom is used therapeutically in epilepsy). Muscarine, which prolongs the diastolic action of the heart and acts as a decided depressant upon the vaso-motor system and respiratory centre, has been employed as a remedy for epilepsy (M. Grieve).

Aristolochia indica: the root contains the nitrophenanthrene compounds aristolochic acid and aristolochic acid D. The root exhibits emmenagogue and (dangerous) abortifacient activity. Aristolochia serpentaria Linn. of the Central and Southern United States possesses stimulant, tonic, diaphoretic and properties resembling those of valerian (Valeriana officinalis) and Cascarilla (Croton). Aristolochia spp. are used as a counter-poison.

Rauvolfia serpentina: contains indole alkaloids (1–3 %), including resperine, rescinnamine, ajmaline and yohimbine. Rauvolfia serpentina, as a medicinal plant, was first recorded in Europe in 1785 but its efficacy was not screened until 1946. It was tried in modern medicine to lower high blood pressure and for controlling symptoms of mental illness.

The drug is now used only sparingly and is contraindicated in depression. Combined with alcohol, it severely increases the impairment of motor reactions. Severe bradycardia occurs in combination with digitalis glycosides. When combined with cough or flu remedies or with appetite depressors, an initial significant increase in blood pressure can occur.

Polygonaceae

RHEUM

***Rheum australe* D. Don**
***Rheum emodi* Wall. ex Meissn.**

Habitat

Himalayas from Kashmir to Sikkim, at altitudes of 3300–5200 m. Also cultivated in Assam for its leaves consumed as a vegetable.

Classical and common names

Ayurvedic: Amlaparni, Pitamuuli, Revatikaa (non-classical). Revandachini.
Unani: Revandachini.
Siddha: Nattirevaichini.
English: Indian Rhubarb.

Parts used

Rhizomes, dried roots.

Dose

Powder 120–360 mg for constipation, 1.5–2.0 g as purgative.

Classical use

Three types of rhubarbs are used in Unani medicine—the Chinese rhubarb, the Indian or Himalayan rhubarb and the Rhapontic rhubarb. Rheum palmatum, the Chinese rhubarb, is esteemed as the best variety. Rheum emodi is the chief source of Himalayan rhubarb. Rhapontic rhubarb consists of the rhizomes of Rheum rhaponticum (it is avoided as it causes gripe).

Unani physicians use rhubarb as a purgative and astringent tonic for constipation and atonic dyspepsia.

Cooked stalks are also given as a purgative. Powdered roots are used for cleaning teeth; also used as a dusting powder for ulcers.

Internally, Revandachini is administered in combination with Zingiber officinale or Foeniculum vulgare.

Qurs-e-Mulaiyin is available over-the-counter for constipation and colic.

Active principles and pharmacology

The drug contains a number of anthraquinone derivatives—rhein and emodin. Glycosidic active principles—sennosides A-F—are present with free anthraquinones. In addition to 0.05 % of an essential oil, roots contain an unidentified terpenic alcohol and a constituent believed to be methyl-n-heptyl ketone. Eugenol is present in the essential oil.

The astringent principle consists chiefly of gallic acid, present as gluogallin, together with small amount of tannin and possibly catechin. Glucogallin on hydrolysis yields gallic acid and glucose. The drug also contains cinnamic and rheinolic acids, volatile oil, starch and calcium oxalate.

The total oxalic acid content of the leaves and stems are reported to be 0.65 and 0.81 % respectively. Rutin (0.32 %) is present in the leaves and flowers.

Use in Western herbal

Rheum palmatum (dried underground parts) is used as rhubarb. The drug contains anthracene derivatives (3–12 %), flavonoids (2–3 %), naphthohydroquinone glycosides, tannins. Its main effect is laxative, primarily due to its influence on the motility of the colon, inhibiting stationary and stimulating propulsive contractions. This results in an accelerated intestinal passage and because of the active chloride secretion, increases water and electrolyte content of the stool.

Nicholas Culpeper (1616–1654) wrote of it: "A dram of the dried root of Monk's Rhubarb, with a scruple of Ginger made into powder, and taken into fasting in a draught or mess of warm broth, purges choler and phlegm downwards very gently and safely without danger.

The seed thereof contrary doth bind the belly, and helps to stay any sort of lasks or bloody-flux.

The distilled water thereof is very profitably used to heal scabs; also foul ulcerous sores... The juice of the leaves or roots or the decoction of them in vinegar, is used as the most effectual remedy to heal scabs and running sores.

The decoction of Bastard Rhubarb, being drank, heals of the jaundice.

The powder of the Rhubarb, taken with a little mummia and madder roots in some red wine, dissolves clotted blood in the body happening by any fall or bruise..."

Caution

Rhubarb is contraindicated in cases of acute inflammatory intestinal diseases, intestinal obstructions, appendicitis and abdominal pain of unknown origin. (PDR.)

Also contraindicated during pregnancy or lactation; in renal disorders, gout, rheumatism.

Potassium deficiency can cause an increase in the effect cardiac glucosides.

Acanthaceae

RHINACANTHUS

***Rhinacanthus communis* Nees**
***Rhinacanthus nasuta* Kurz.**

Figure 1 Rhinacanthus nasuta—flowering branch [WOI]

***Justicia nasuta* L.**

Habitat

Cultivated throughout India.

Classical and common names
Ayurvedic: Yuuthiparni, Yuuthikaparni, Paalaka-juhi.
Unani: Gul-baglaa.
Siddha: Nagamalli.
English: Snake Jasmine.

Parts used
Leaves, seeds, root.

Dose
Powder 3–5 g, Juice 5–10 ml.

Classical use
Unani practitioners apply a paste of Paalaka-juhi root and lime juice on ringworm, washerman's itch and other fungal infections; fresh juice of leaves on freckles and blemishes. Leaves are also applied for prickly heat.

The root, boiled in milk, is prescribed as an aphrodisiac.

Active principles and pharmacology
Roots contain a resinous active principle rhinacanthin (1.9 %). Exhibit antiseptic and antiparasitic properties.

The plant is rich in potassium salts; also contains oxymethyl anathraquinone.

Rhinacanthus calcaratus Nees, found in Assam and Bengal, is used as an anthelmintic. A preparation of root with citrus juice and pepper (Piper nigrum) is found efficacious in herpes.

Anacardiaceae

RHUS

Rhus parviflora Roxb.

Habitat
Dry, hot slopes of the Himalayas from Punjab to Nepal and in the hills of Madhya Pradesh and South India, at altitudes of 600–2000 m.

Classical and common names
Ayurvedic: Tintidi, Tintidika (Bhaavaprakaasha).
Unani: Sumaaq.
English: Sumach.

Parts used
Fruits.

Dose
Powder 3–6 g.

Classical use
Sushruta included ripe fruits of Tintidika, in prescriptions, as astringent, stomachic and appetiser. (Sushruta treated Amlikaa as a separate drug, which was also used as an appetizer and digestive stimulant. Chakrapaani created a confusion by equating Amlikaa with Tintidi. Amlikaa has now been equated with Tamarindus indica Linn.)

Fully matured fruits were prescribed as a cardiac tonic (Maadhava Dravya Guna); also in diarrhoea, dysentery, polyuria, fevers (Bhaavaprakaasha).

In folk medicine, fruit juice is used as a vermifuge. Paste, mixed with charcoal powder, is applied to unhealthy ulcers and suppurating piles.

Active principles and pharmacology
Leaves contain the flavonoids myricetin, quercetin and kaempferol and their 3-O-rhamnosides. Stems and leaves also gave hentriacontane, hentriacontanol, beta-sitosterol, lignoceric acid and iso-rhamnetin-3-alpha-L-arabinoside.

Fruits exhibit astringent and styptic activity and are found efficacious in bilious diarrhoea, dysentery, haemoptysis, leucorrhoea and diuresis. A gargle prepared of fruits gives good results in catarrhal affections of the pharynx.

Euphorbiaceae

RICINUS

Ricinus communis Linn.

Habitat
A native of India as well as of Africa; mostly under cultivation up to an elevation of 2000 m.

Figure 1 Ricinus communis [ZANDU]

Classical and common names

Ayurvedic: Eranda, Chitrabija, Triputi, Tribija, Chanchu, Manda, Uruvaka, Gandharva-hastaa (Charaka, Sushruta). (Not to be confused with Erandakarkati, equated with Carica papaya L.)
Unani: Bedanjeer, Arand.
Siddha: Ammanakku.
English: Castor, Castorseed.

Parts used

Seed oil, root.

Dose

Oil 5–10 ml, decoction 50–100 ml, seeds 1–5 in number.

Classical use

Eranda root, mixed with warm milk, was prescribed for misperistalsis, constipated bowels, diarrhoea with blood and mucus. (Charaka Samhitaa, Sushruta Samhitaa, Ashtaanga Hridaya.)

The expressed oil, with Haritaki (Terminalia chebula) or with Triphalaa (Terminalia chebula, Terminalia bellirica and Emblica officinalis), was prescribed, internally, in rheumatic affections; inflammatory and irritable conditions of intestines, diarrhoea, dysentery; and in inflammatory diseases of rectum and urinogenital tract. (Sushruta Samhitaa, Vrindamaadhava, Ashtaanga Hridaya.) Milky exudation or a decoction of Eranda fruits was also given (Sushruta Samhitaa).

A decoction of Eranda root, mixed with the paste of Shunthi (dry Zingiber officinale) or a paste of seeds boiled in milk and water, was given in rheumatic affections, lumbago, sciatica, plurodynia. (Shaarangadhara Samhitaa, Bhaavaprakaasha, Bangasena.)

For promoting conception, seeds of Eranda and Maatulunga (Citrus medica), pounded with clarified butter, were prescribed (Gadanigraha).

For improving lactation, leaves warmed over a fire, were applied to the breasts of nursing mother; a fluid extract or juice of leaves was also administered internally.

According to Ayurvedic classics, there are three kinds of Eranda—red, yellow and black; but only the red and white varieties have been identified. The white variety is used especially for fevers; the red one for swellings, pallor and mental diseases. The flowers of the red variety are prescribed for painful micturition and seeds for hepatitis and splenitis. Tender leaves are prescribed for allaying pain in the bladder.

A decoction of root and rootbark is used in rheumatic disorders, gastrointestinal affections like flatulence and constipation; juice of leaves in jaundice.

Active principles and pharmacology

The seeds yield from 46 to 53 % of fixed oil, which consists of the glycerides of ricinoleic, isoricinoleic, stearic and dihydroxystearic acids. The seeds have lipase and a crystalline alkaloid, ricinine. Leaves and stem also contain ricinine. The cake, left after expression of oil, has a poisonous lectin, ricin.

Rootbark shows the presence of steroid and considerable amount of gallotannins along with many inorganic ions. (CCRAS.)

The rootbark, extracted with different solvents, exhibited anti-inflammatory activity in various animal models. Crude alcohol extract showed 72.2 % inhibition of carrageenin-induced oedema in rats.

Ricinoleic acid is responsible for purgative action of the drug. Within small intestines pancreatic lipases hydrolyse the oil to glycerol and ricinoleic acid. Ricinoleate, like other anionic surfactants, reduces net absorption of fluid and electrolytes and stimulates peristalsis.

Ethanolic extract of the leaves showed significant protection against carbon tetrachloride and galactosamine-induced damage in rats. Leaves also exhibited significant growth of mammary glands in albino mice.

Caution
Long-term use of castor oil can cause a reduction of absorption of nutrients. It should not be taken during pregnancy, as it can cause uterine contraction.

Castor oil is contraindicated in intestinal obstruction, acute inflammatory intestinal diseases, appendicitis. Therapeutic doses should be maintained to avoid toxic symptoms of ricin poisoning.

Rubiaceae

RUBIA

Rubia cordifolia Linn.

Figure 1 Rubia cordifolia [ADPS]

Habitat
Common throughout India, ascending to an altitude of 3750 m.

Classical and common names
Ayurvedic: Manjishthaa, Vikasaa, Samangaa, Yojanavalli, Kaalameshika, Raktaangi, Raktayashti, Arunaa, Gandira, Jingi.
Unani: Manjeeth Iraani (Rubia tinctorum).
Siddha: Manjitti.
English: Indian Madder.

Parts used
Root.

Dose
Decoction 50–100 ml, powder 1–3 g.

Classical use
Charaka gave powdered dried roots and fruits, in internal prescriptions, for eczematous dermatosis and diseases of the spleen.

Manjishthaa, extracted in purified butter, was prescribed by Sushruta for bleeding piles; a decoction of Manjishthaa and Chandana (Santalum album) for polyuria. Externally, Manjishthaa was applied on major burns; mixed with honey on freckles and blemishes.

Manjishthaa, throughout, remained a potent drug for obstinate skin diseases, erysipelas and oedema, and an excellent aid for the promotion of complexion.

In Unani medicine, the drug is used as a deobstruent and tonic to liver and spleen, emmenagogue, diuretic and as an anti-inflammatory agent. It is prescribed with Sikanjbeen (lemon squash).

Externally, a paste of Manjeeth, smeared over betel leaf, is applied over rheumatic swellings. For freckles, it is mixed with vinegar and applied.

In folk medicine, Manjishthaa is administered in amenorrhoea, jaundice and venereal diseases.

Shaarangadhara Samhitaa's compound preparations, Mahaa-Manjishaadi Kwaatha and Mahaa-Manjishthaadyaarishta, are available over-the-counter for eczema, scabies and chronic skin diseases.

Over-the-counter drugs, with Manjishthaa as an important ingredient, include: Chandanaasava (Bhaishajya Ratnaavali), prescribed for urinogential disorders; Jaatyaadi Ghrita (Ashtaanga Hridaya) externally for chronic and septic ulcers; Phala Sarpi Ghrita (Ashtaanga Hridaya) and Phal Kalyaan Ghrita (Bhaishajya Ratnaavali) for amenorrhoea and uterine affections; Manjishthaadi

Taila (Sahasrayoga) for headaches; Marichaadi Taila (Bhaishaajya Ratnaavali) for ringworm and other fungal infections.

Active principles and pharmacology

Roots contain purpurin, xanthopurpurin, munjistin, free alizarin as well as its glucoside, pseudopurpurin, 1-hydroxy-2-methylanthraquinone, 1, 4-dihydroxy-6-methylanthraquinone, physcion, nordamnacanthal, 1, 4-dihydroxy-2-methylanthraquinone, 1, 5-diydroxy-2-methylanthraquinone, 3-prenylmethoxy-1, 4-naphthaquinone. Whole plant yielded pentacyclic triterpenic acids—rubicoumaric acid and rubifolic acid.

The colouring matter present in the root in a mixture of purpurin (trihydroxy anthraquinone) and munjistin (xanthopurpurin-2-carboxylic acid). Munjistin occurs in the form of a glycoside and is accompanied by an isomeride.

When ingested, the drug imparts a red colour to bones, milk and urine (earlier it was through that it induces haematuria).

10 and 20 ml/kg p.o. of water extract of the whole plant exhibited anti-inflammatory effect against carrageenin-induced inflammation in rats. The activity was comparable to phenylbutazone. (CCRAS.)

The drug was found effective in scabies and in infections with Tinea pedis (Chemexcil.).

Ethanolic extract of the aerial parts of the plant showed hypoglycaemic activity in albino rats.

Alcoholic and aqueous extracts of the plant exhibited antibacterial effect against Staph. aureus (Gram-positive). Aqueous extract was active against E. coli (Gram-negative).

Use in Western herbal

Madder fell largely out of use in the 19th century and is now rarely used to treat kidney and bladder stones; and menstrual and urinary disorders. There exists a strong indication that lucidin (an active principle of the drug) is mutagenic and carcinogenic.

After administration of the drug, decreased calcium oxalate crystallization in the kidney has been observed.

Nicholas Culpeper (1616–1654) wrote of Madder (Rubia tinctorum): "It hath an opening quality, and afterwards to bind and strengthen. It is a sure remedy for yellow jaundice, by opening the obstructions of liver and gall, and cleansing those parts; it opens also the obstructions of the spleen and diminishes the melancholy humour. It is available for the palsy and sciatica, and effectual for bruises inward and outward... The root for all those aforesaid purposes, is to be boiled in wine or water, as the cause requires.

The seed thereof taken in vinegar and honey, helps the swelling and hardness of spleen.

The leaves and roots beaten and applied to any part that is discoloured with freckles, morphew, the white scruf, or any such deformity of the skin, cleanses thoroughly, and takes them away."

Polygonaceae

RUMEX

Rumex acetosa Linn.

Classical and common names

Ayurvedic: Chukrikaa (Charaka). Chuko. Chukra is a confusing synonym. In Ayurvedic texts, Chukrikaa is described as a vegetable, while Chukra as the dried juice of sour fruits.
Unani: Hummaaz, Shaaka-turshak, Jangali Paalak. Seeds—Tukhm-e-Tursh, Beejaband (seeds of Sida cordifolia are also used as Beejaband).
English: Golden Sorrel.

Parts used

Leaves, seeds, root.

Dose

Juice 60–75 ml, root or seed powder 3–5 g.

Classical use

Charaka prescribed Chukrikaa, Balaa (Sida cordifolia and Prishnparni (Uraria picta) as a potherb to patients suffering from bleeding piles. Chukrikaa was also cooked with Naagakeshara (Mesua ferrea) and Utpala (Nelumbo nucifera). The potherb was prescribed with curd and pomegranate seeds for diarrhoea.

Tuberous roots of Rumex nepalensis variety, which grows abundantly in India, are used as a substitute for rhubarb and are sold as Revanadchini. In folk medicine, 120–360 mg of the drug is given in constipation.

In Unani medicine, Beejaband surkh and siyaah, an ingredient of spermatic tonic Majoon-e-Bandkushad, are the seeds of Sida cordifolia. Seeds of Jangali Paalak are given as refrigerant, diaphoretic and antiscorbutic during fevers, bronchial affections and skin diseases. An infusion of seeds and aerial parts is used for dysentery, dyspepsia, nausea, hepatic disorders, loss of appetite. The drug is used like rhubarb and sarsaparilla.

Active principles and pharmacology

Roots of Rumex acetosa gave the hydroxyanthraquinones chrysophanol, aloe-emodin, emodin, physcion, 8-O-beta-D-glucosylchrysophanol, 8-O-beta-glucosyl-emodin, sitosterol and omega-acetoxyaloe-emodin. Aerial parts gave rutin, hyperin and quercetin glucoside. Leaves gave vitexin.

The plant exhibits antiscorbutic and leaf diuretic properties. The sour taste of Sorrel is due to the acid oxalate of potash. Tartaric and tannic acids are also present.

Use in Western herbal

Aerial parts of Rumex acetosa are used for acute and chronic inflammation of nasal passages and respiratory tract. Also used as an additional measure in antibacterial therapy and for its detoxifying effect.

Nicholas Culpeper (1616–1654) wrote of Sorrel: "Sorrel is prevalent in all hot diseases, to cool any inflammation and heat of blood in agues... and fainting arising from heat... and to refresh the overspent spirits... to quench thirst and procure an appetite in decaying stomachs. It resists the putrefaction of the blood, kills worms, and is cordial to the heart, which the seed doth more effectually, being more drying and binding, and thereby stays the hot fluxes of women's courses, or of humours in the bloody flux, or flux of the stomach.

The root also in decoction, or in powder, is effectual of all the said purposes... The decoction of the root is taken to help the jaundice and to expel the gravel and the stone in the reins or kidneys."

Caution

Calcium oxalate crystals in the leaves mechanically penetrate tissue, thus creating many small wounds. Anthrones in the fresh root irritate the gastrointestinal mucosa. (Francis Brinker.)

Rutaceae

RUTA

Ruta graveolens Linn.

Habitat

Distributed in the Mediterranean region and temperate Asia. Cultivated in gardens.

Classical and common names

Unani: Sudaab, Suddaab. Somalataa (CIMAP) is a confusing synonym.
English: Garden Rue.

Parts used

Leaves, whole plant, oil.

Dose

Juice 5–10 ml, powder 1–3 g, infusion 10–20 ml, oil 1–5 drops.

Classical use

In Unani medicine, Sudaab is used as a stimulant to the nervous system and uterus. The drug is prescribed for hysteria, nervous disorders, amenorrhoea, dysmenorrhoea; also as stomachic, antispasmodic, anthelmintic, resolvent and diuretic.

Leaves are an ingredient in over-the-counter drug, Javarish-e-Kamooni Saadaa (Qarabadeen-e-Jadeed), and Javarish-e-Kamooni Kabir, for dyspepsia, diarrhoea, gastro-intestinal disorders, sluggish liver; Raughan-e-Sudaab, for muscular pain, also for otalgia, coxalgia; Zimad-e-Tehal, for massage in splenitis. An infusion of leaves is prescribed for amenorrhoea and dysmenorrhoea; during fevers as diuretic and diaphoretic.

Somalataa is a confusing synonym of Ruta graveolens in Indian medicine. According to Sushruta, it is a creeper which develops gradually 15 leaves during first 15 days of the moon-month;

leaves fall gradually during the declining moon-month. Its tuber was supposed to be like a nectar which prevented old age and death. It is difficult to believe that Ruta graveolens or 8 other drugs, which have been suggested as Soma by different scholars, were used as Soma during the period of Sushruta. (Other drugs are: Amanita muscaria Linn., Cannabis sativa Linn., Ceropegia spp., Ephedra gerardiania Wall., Peganum harmala Linn., Periploca aphylla Decne., Rheum emodi Wall., Sarcostemma brevistigma W. & A.)

Active principles and pharmacology

Rue contains about 0.5 % volatile oil (including 50–90 % 2-undecanone), flavonoids (including rutin), furanocoumarins (including fagarine, arborinine, skimmianine, and others).

Plant is spasmolytic due to the presence of bergapten, xanthotoxin.

Rue oil consists mainly of methyl nonyl ketone (80–90 %) with a small amount of methyl heptyl ketone. Other constituents present are: L-alpha-pinene, L-limonene, cineole, methyl-n-heptyl-carbinol and methyl-n-nonyl-carbinol; ethyl valerate, methyl salicylate and methyl anthranilate.

The oil exhibits anthelmintic, antispasmodic, antiepileptic, rubefacient and emmengagogue properties. It is used extensively in veterinary medicine. In large doses, the oil acts as acronarcotic poison.

Furanocoumarins are reported to have phototoxic and mutagenic action. (German Commission E.)

An alcoholic extract of the herb shows antibacterial activity against Micrococcus pyogenes var. aureus and E. coli.

Use in Western herbal

Rue is chiefly used to encourage the onset of menstruation. It stimulates the muscles of the uterus. In European herbal medicine, Rue has also been used to treat hysteria, epilepsy, vertigo, multiple sclerosis, Bell's palsy. An infusion is used as an eyewash for strained and tired eyes.

Nicholas Culpeper (1616–1654) wrote of Garden Rue: "It provokes women's urine and women's courses, being taken either in meat or drink... A decoction (of leaves) with dried dill leaves and flowers, eases all pains and torments, inwardly to be drank, and outwardly to be applied warm to the place grieved... It kills and drives forth the worms of the belly... It takes away wheals and pimples, it being bruised with a few myrtle leaves, it be made up with wax, and applied... The juice of it and fennel, with a little honey, helps the dimness of the eye-sight."

Caution

Phototoxic reactions causing dermatosis have been observed.

Even therapeutic doses can cause melancholic moods, sleep disorders, spasm.

Fresh juice of leaves can lead to painful irritation of stomach and intestines. Contraindicated during pregnancy. (German Commission E.)

Deaths of pregnant women have been reported upon usage of Rue.

Poaceae

SACCHARUM

Saccharum officinarum Linn.

Habitat

Extensively cultivated throughout India in several varieties.

Classical and common names

Ayurvedic: Ikshu, Dirgha-chhada, Bhuurirasa, Asipatra, Madhutrna, Gudamuula.
Unani: Gannaa, Naishaakar.
English: Sugarcane, Noble Cane.

Parts used

Cane, root.

Dose

Juice 20–40 ml, root decoction 50–100 ml.

Classical use

Properties of Ikshu and its derivatives are documented in Maadhava Dravyaguna. Sugarcane juice alleviates diseases due to derangement of liver and digestive functions. Sugarcane, when chewed, mixed with saliva was considered a nourishing tonic, while the juice extracted mechanically was considered constipative. Boiled juice was not prescribed.

Phanita (sugarcane juice boiled down to one-fourth) was prescribed for exhaustion and for cleansing urine and urinary bladder. Guda (jaggery or crude sugar) was also prescribed for cleansing urine. Jaggery was administered in biliary derangement and was considered superior to other derivatives of sugarcane. It was also used as a cardiac tonic.

Sitopalaa (double purified sugar candy, known as Misri) was used as demulcent and pectoral.

Charaka and Sushruta used Ikshu-rasa (sugarcane-juice) as a drink, or ingredient of mixed drinks, as a tonic; with milk or honey for haemoptysis, anaemia, urinary diseases; and as a spermatopoietic. Roots were included in prescriptions as a urinary tract disinfectant.

Fermented juice of Ikshu, Draakshaa (Vitis vinifera) and Kharjura (Phoenix dactylifera fruits) was prescribed for diarrhoea and dysentery (Charaka Samhitaa.)

The root of sugarcane has been included in Panch-trinamuula (Roots of Five Grasses) of Ayurvedic medicine. (Others being roots of Saccharum spontaneum, Desmostachya bipinnata, Imperata cylindrica, Oryza sativa.) These are reputed diuretic drugs, singly or in combination. The root of Ikshu was used in prescriptions for dysuria, anuria, retention of urine, for urinary affections during pregnancy (Bhaavaprakaasha).

For proper assimilation of food and digestion, sugarcane derivatives were prescribed with the juice of Aadraka (fresh Zingiber officinale).

Sitopalaadi Churna (Shaarangadhara Samhitaa) contains 16 parts of double-purified sugar candy (Sitopalaa), 8 parts of Bamboo-manna (Vansalochana), 4 parts of Long Pepper (Piper longum), 2 parts of Cardamoms and one part of Cinnamom bark. Available over-the-counter, this compound preparation is a popular remedy for cold, cough and bronchitis.

Active principles and pharmacology

Indian sugarcanes have generally the following range of values: water 70–75 %; sugars (sucrose, glucose, fructose) 12–15 %; fibre 12–20 %; nitrogenous substances 0.3–04 %; fats and waxes 0.15 %–0.25 %; gums and pectins 0.15–0.25 %; free acids 0.08–0.12 %; combined acids 0.10–0.12 % and ash 0.3–0.5 %. The internode of the cane is richer in sucrose than the node; maximum concentration of sucrose has been reported in the central internodes of mature canes. The amount of reducing sugars, especially fructose, decreases as the cane ripens.

Non-sugar constituents present in the cane juice are: carbohydrates other than sugars, viz. hemicelluloses, pentosans, starch, myo-inositol etc. 8.5 %; pectins 1.5 %; proteins 7.0 %; albuminoses and peptoses 2.0 %; amino acids 9.5 %, acid amides (asparagine and glutamine) 15.5 %; carboxylic acids 13.0 %; colouring matter 17.0 %; wax 17.0 %; inorganic salts 7.0 % and silica 2 %. Nitrogenous substances identified in the cane include protein and peptones, amides, amino acids, nitrates and ammonium salts. Asparagine and glutamine are predominant amino acids in the juice.

Aconitic acid constitutes about three-fourth of the total carboxylic acids present in the juice.

Indian cane juices contain 0.4–0.7 %, minerals. Vitamins present are: thiamine 53, riboflavin 31, niacin 49, pantothenic acid 2180; biotin 22, vitamin D 176 mcg/100 g.

Enzymes present in the cane are: invertase, oxidases, reducing and catalysing enzymes, diastase, peptonizing enzymes, lactase, peroxidase, tyrosinase.

Phenols in the cane juice are mainly polyphenols from the tannin and anthocyanin of the rind. The amount of phenols range from 28 to 102 mg/100 g. Other organic constituents present in sugarcane juice are pectin, phosphatides, hemicelluloses and nitrogenous substances.

A steroid compound, found to alleviate joint-stiffness induced in guinea-pigs, has been isolated from cane juice.

Glycans A, B, C, D, E, and F, yielded from the non-sucrose portion of Saccharum officinarum, exert remarkable hypoglycaemic action in normal and alloxan-produced hyperglyacaemic mice.

An aromatic ester glycoside isolated from roots of Saccharum officinarum has been characterised as vanilloyl-1-O-beta-D-glucoside-acetate.

Food yeast and baker's yeast is prepared by the fermentation of molasses.

In homoeopathic medicine, Saccharum officinarum is an antiseptic which combats infection and putrefaction; is considered a nutrient and tonic in a variety of cardio-vascular affections.

Hippocrateaceae

SALACIA

***Salacia chinensis* Linn.**
***Salacia latifolia* Wall. ex M. Laws**
***Salacia prinoides* DC.**

Habitat

Throughout India, including the Andaman Islands, along seashore and river banks, as well as in forests at altitudes up to 750 m.

Classical and common names

Ayurvedic: Saptachakraa, Swarnmuula. Saptarangi (trade).

Parts used

Root.

Dose

Decoction 50–100 ml.

Traditional use

Roots are credited with antidiabetic properties. A decoction of root is used in amenorrhoea and dysmenorrhoea and venereal diseases.

Active principles and pharmacology

Stem yielded gutta (a linear isomer of natural rubber), dulcitol and proanthocyanidin consisting of a dimer of leucopelargonidin.

Leaves yielded proanthocyanidins and also gutta.

Roots gave proanthocyanidins consisting of monomeric leucopelargonidin, its monomer, dimer and tetramer.

Rootbark contains triterpenoids, mangiferine, phlobatannin, glycosidal tannins and a triterpene ether.

On the experimental level, both infusion and decoction of the rootbark exhibited hypoglycaemic activity. In 1 g/kg p.o. dose level on fasting as well as glucose-fed rabbits, the effect was comparable to that of tolbutamide (250 mg/kg). Significant reduction in blood sugar level in 0.5 and 1 g/kg of the infusion-treated, as well as 0.5 g/kg decoction-treated groups at the end of the third hour was observed. The plant showed encouraging results in clinical studies at the Clinical Research unit (Ayurveda), Safdarjung Hospital and the All India Institute of Medical Sciences, New Delhi. Extracts prepared from leaves also showed considerable hypoglycaemia.

The active principle responsible for producing hypoglycaemia deteriorated on storage of roots for six months. (CCRAS.)

Leucoanthocyanidin, obtained from the roots, produced indifferent changes in blood sugar.

All samples of roots, leaves, leucoanthocyanidin caused significant purging in dogs. (CCRAS.)

Both root and leaves of Salacia fruticosa Heyne (Saptarangaa) were studied on fasting blood sugar in rabbits. The results suggest that both the leaves and root of Salacia fruticosa contain active principle for producing hypoglycaemic effect, but in the roots the response was variable. No hypoglycaemic effect was noted with the infusion of rootbark (1 g/kg) in normal fasting and glucose-fed rabbits. (CCRAS.)

A comparative study with Salacia prinoides, Salacia fruticosa and Salacia macrosperma was performed to find out the efficacy of each species in diabetes. In this study, only Salacia prinoides exhibited significant hypoglycaemic property. However, Salacia prinoides in a dose of 25 g twice daily did not give encouraging results in patients of diabetes mellitus. (CCRAS.)

Salicaceae

SALIX

Salix caprea Linn.

Figure 1 Salix caprea—male and female flowering branch [WOI]

Salix alba Linn.

Habitat

Planted for ornament in Kashmir, Punjab, Himachal Pradesh and Uttar Pradesh, known as Goat Willow. Salix alba Linn. (white Willow) is frequently cultivated in the Western Himalayas.

Classical and common names

Ayurvedic: Vetas, Vaanira, Vidula. Jalavetas (Salix alba).
Unani: Bedmushk. Bedsaadaa (Salix alba).
English: Common Willow.

Parts used

Bark, root, leaves.

Dose

Decoction 50–100 ml.

Classical use

Vetas and Jalavetas were used as substitutes for each other in Ayurvedic medicine. (Vetas has been equated with Salix spp., Amlavetas with Garcinia spp. and Vetra with Calamus rotang Linn. Both Vetas and Vetra were included by Charaka in the same group of herbs.)

Charaka and Sushruta, included Vetas in prescriptions for fevers with rigor, for internal bleeding, menstrual and vaginal disorders, obesity. Sushruta gave a decoction of Vetas and Nala (Arundo donax) for chronic fever. A confection prepared with Vetas leaves was administered for diarrhoea.

A decoction of Vetas was used as a douche for making relaxed vagina firm.

Sushruta included Vetas in an ointment for venereal sores.

In Unani medicine, a decoction of the bark is employed in fevers, acute rheumatism and gout, headaches, palpitation. Leaves, honey of flowers, sugars (known as Bed-angbeen), and a distillate of flowers (known as Araq-e-Bedmushk) are also used.

Araq-e-Bedmushk and Araq-e-Bedsaadaa are available over-the-counter. Unani physicians prescribe them in nervousness and palpitation, as a supporting medicine in psychological stress and cardiac affections.

Active principles and pharmacology

Salix caprea: All parts of the plant gave glycosides.

Leaves contained the flavonoid glycosides chryseoeryol-7-(2-alpha-1-arabinofuranoside)-beta-D-galactopyranoside (salicaprene), chryseoeryol-7-beta-D-galactopyranoside (salicapriine), anthocyanidins—delphidinin and cyanidin; piperidine imino acid—pipecolic acid.

Bark gave the phenolic glycosides fragilin, picein, salicin, salicortin, salireproside, triandrin and vimaline.

Staminate inflorescence gave the flavonoids iso-rhamnetin and diosmentin, and diosmetin glucosides capreoside and salicapreoside.

Active principles of male racemes consisting of alkaloids, glycosides and saponins increase the amplitude and slow the beat and act more rapidly than digitalis on isolated frog heart. Fatal doses stop the heart in systole. (CIMAP.)

Salix alba: Leaves contain the flavonoid glycosides alboside characterised as rhamnazine-3-beta-D-glucoside, apigenin-7-O-(4-p-coumaryl)-glucoside, terniflorin (an artefact), quercetin-3-O-glucoside, quercetin-3-O-rutinoside, iso-rhamnetin-3-O-glucoside, iso-rhamnetin-3-O-rutinoside, quercetin-7, 3'-dimethylether-3-O-glucoside and luteolin-7-O-beta-D-glucopyranoside; leucoanthocyanidins; phenolic compounds—salicin, salicortin, salidroside, fragitin, grandidentatin and triandrin; piperidine-imino-acid-5-hydroxypipecolic-acid.

Bark contains salicin, triandrin, +-gallocatechol, +-catechol, (−)-epi-gallocatechol, gallate, epi-catechol gallate and polysaccharides.

The bark exhibits astringent, antidiarrhoeal and sedative properties; found efficacious in haemoptysis and febrile diseases of rheumatism. When the powdered drug is used, salicin is not released quantitatively. 500 mg of aspirin, moreover, is equal to 794 mg of salicin, contained in no less than 88 g of Willow bark.

Use in Western herbal

Salicylic acid, the main constituent of Willow was first isolated in 1838. It was the forerunner of aspirin, a chemical drug first produced in 1899. Salicylic acid inhibits prostaglandin production, relieves pains and affords relief during fevers. Unlike aspirin, it does not thin the blood, nor does it irritate the stomach lining.

Diocorides, a Greek physician of the 1st century, suggested Willow leaves, mashed with a little pepper (Piper nigrum) and drunk with wine, for lower back pain.

Nicholas Culpeper (1616–1654) wrote of it: "Both the leaves and bark, and the seed, are used to stanch bleeding of wounds, and at mouth and nose, spitting of blood, and other fluxes of blood in man or woman, and to stay vomiting and provocation thereunto, if the decoction of them in wine be drunk.

The leaves bruised with some pepper, and drank in wine helps much in the wind colic. The boughs of the tree are very convenient to be placed in the chamber of one sick of a fever.

The decoction of the leaves or bark in wine, takes away scruff and dandruff by washing the place with it. Burnt ashes of the bark, being mixed with vinegar, takes away warts, corns, being applied to places."

German Commission E recognized the efficacy of Salix alba and other comparable Salix spp. in diseases accompanied by fever, rheumatic ailments, headaches.

The herb is a common ingredient in diaphoretic and antirheumatic teas.

Willow is used as a first aid for medication in poisoning. Activated charcoal powder of the wood (1 tablespoonful), powdered ginger root (one teaspoonful) and white sugar (1/2 teaspoonful) in one large glass of distilled water is prescribed as an antidote; also for diarrhoea, gas, heartburn and other general gastrointestinal distress.

John Heinerman recommends a decoction of Willow bark (5 tablespoons of Willow bark in 5 cups of water) for external application for sore gums, festering sores, open wounds, serious burns, sweaty feet and tonsillitis (as a gargle).

Bombacaceae

SALMALIA

Salmalia malabarica (DC.) Schott & Endl.

See figure 1.

Figure 1 Salmalia malabarica [ZANDU]

Bombax ceiba Linn.
Bombax malabaricum DC.

Habitat
Widely distributed throughout India up to 1500 m or even more.

Classical and common names
Ayurvedic: Shaalmali, Mochaa (Charaka); Pichhila, Raktapushpa, Sthiraayu, Kankataadhya, Tuulini.
Unani: Semal.
Siddha: Purani.
English: Silk Cotton, Red Cotton.

Parts used
Flowers, root, bark, gum (Mocharasa).

Classical use
Charaka prescribed fresh flowers or resin, internally and externally, for haemothermia, dysentery and rectal prolapse. Sushruta used paste of the bark as styptic.

Petioles or exudation of Shaalmali boiled in milk, added with purified butter, or cold infusion of Shaalmali petioles, added with Glycyrrhiza glabra, was administered for diarrhoea. (Charaka Samhitaa, Sushruta Samhitaa.)

For leucorrhoea and metrorrhagia, Shaalmali Ghrita (Bangasena) was the compound preparation of choice during the 16th century. Vegetables of Shaalmali flowers (Bhaavaprakaasha) or powder of flowers with honey (Chakradatta) were also prescribed.

Flowers of Shaalmali, well-steamed and mixed with Raajikaa (Brassica juncea) were prescribed for splenomegaly.

Shaalmali tuber, mixed with cow-milk, was used as an aphrodisiac (Haarita Samhitaa).

Paste of Shaalmali thorns, mixed with milk, was a popular facial cosmetic for freckles and for promoting complexion (Vrindamaadhava).

Mocharasa (exudate of Shaalmali) was used by Charaka in combination with Chandana (Santalum album) for intrinsic haemorrhage. Milk boiled with Mocharasa was also administered.

Charaka and Sushruta incorporated Mocharasa in prescriptions for bleeding piles, diarrhoea, dysentery, prolapse of anus, dysuria, menorrhagia, leucorrhoea.

Over-the-counter, Shaalmali Ghritam (Arya Vaidya Sala) is available for urethral and vaginal discharges.

Majoon-e-Muqawwi-e-Rahem (National Formulary of Unani Medicine) is prescribed for atony of uterus, polymenorrhagia and leucorrhoea.

Active principles and pharmacology

Rootbark contains hemigossypolone, hemigossypolone-7-methylether, iso-hemigossypol-1-methylether (hemigossypol-6-methylether), iso-hemigossypol-1, 2-dimethylether, 8-formyl-7-hydroxy-5-isopropyl-2-methoxy-3-methyl-1, 4-napthaquinone, 7-hydroxycadalene. Rootbark also contain lupeol and beta-sitosterol which are also present in the stem-bark. Stem-bark, in addition, contains beta-sitosterol-beta-D-glucoside.

Flowers contain beta-sitosterol-beta-D-glucoside, beta-sitosterol, hentriacontane, hentriacontanol, kaempferol and quercetin.

Purified gum contains 8.9 % mineral matter and a considerable proportion of catechol tannin; on complete hydrolysis it yields a mixture of L-arabinose, D-galactose, D-galacturonic acid and possibly rhamnose (traces). The gum also contains tannic and gallic acids.

The aqueous extract of the seeds showed moderate oxytocic activity on rat uteri and on guinea-pig and rabbit uterine strips. It exerted a negligible blood pressure elevating action in anaesthetized dogs.

The gum, due to tannic and gallic acids, gave good results in cases requiring astringents.

Salvadoraceae

SALVADORA

Salvadora persica L.

Habitat

Dry and acrid regions, saline lands and coastal tracts.

Classical and common names

Ayurvedic: Pilu, Piluu, Pilukaa (Charaka).
Unani: Miswaak.
English: Mustard Tree, Salt Bush Tree, Toothbrush Tree. (The bigger variety has been equated with Salvadora oleoides Decne.)

Classical use

Charaka and Sushruta prescribed fruits, also oil of seeds, as laxative, in rhinitis, hemicrania, intestinal parasites, urinary disorders and suppurating skin diseases.

Fresh fruits of Pilu were prescribed with buttermilk for piles (Ashtaanga Samgraha, Ashtaanga Hridaya).

In Unani medicine, fruits are used for diuretic and leaves for purgative properties. Rootbark is used for amenorrhoea.

Active principles and pharmacology

Roots gave elemental gamma-monoclinic-sulphur, benzyl-glucosinolate, a methoxylbenzyl derivative of urea named salvadourea, m-anisic acid and sitosterol.

Rootbark and stem-bark contain trimethylamine.

Seed fats contain myristic, lauric and palmitic acids.

Air-dried rootbark yields 27.1 % of ash, containing considerable amounts of salts, mostly as chlorides, and a fairly large amount of alkaloidal constituents, small amounts of resin, colouring matter and traces of tannins and saponin.

Stem-bark is spasmolytic, emmenagogue, ascarifuge and febrifuge. Stem and leaf are antidote to poisons. Fruits are deobstruent, stomachic, antibilious, lithontriptic; seeds diuretic and purgative. Leaves posses antiscorbutic and astringent properties. A decoction gives good results in asthma and cough; a poultice can be applied to painful tumours, piles, rheumatic swellings. Dried leaves in small doses afford relief in flatulent dyspepsia.

Santalaceae

SANTALUM

Santalum album Linn.

Figure 1 Santalum album [CCRAS]

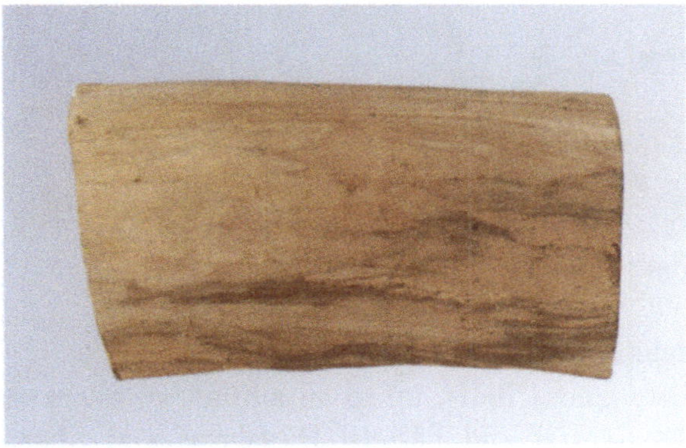

Figure 2 Santalum album [CCRAS]

Habitat
Distributed in the peninsular India, especially common in Karnataka and Tamil Nadu at 1200 m.

Classical and common names
Ayurvedic: Chandana, Shrikhanda, Bhadra-Shree, Gandhsaara, Malayaj.
Unani: Sandal-e-Abyaz, Sandal safed.
Siddha: Chandanam.
English: Sandalwood.

Parts used
Heartwood.

Dose
Powder 3–5 g, decoction 50–100 ml.

Classical use
Charaka and Sushruta gave powdered wood or paste in internal and external prescriptions for toxicosis, diarrhoea, dysentery, vomiting, jaundice, oedema of vagina, polyuria, bleeding piles, haemorrhages, urethral discharges, chronic skin diseases, obesity. Sushruta incorporated paste of the wood in an ointment for major burns, and bark for inflammations and traumatic swellings.

Chandanaadya Ghrita of Charaka was specific for diarrhoea and dysentery; Chandanaadi Kwaatha for gout.

In case of menstrual blood being foetid and purulent, a decoction of both white and red sandalwood (Pterocarpus santalinus) was administered. (Sushruta Samhitaa.)

Chandana, mixed with sugar and rice-water, was prescribed for haematuria (Vrindamaadhavaa). For inflammations, it was mixed with the powdered barks of Kshiri (Ficus) spp. (Bhaavaprakaasha); for urticaria and pruritus, with Guduuchi (Tinospora cordifolia).

In Ayurvedic texts, four variants of Chandana have been mentioned. Chandana has been equated with Santalum album Linn.; Rakta Chandana or Sandal Surkh of Unani medicine with Pterocarpus santalinus Linn. f.; Kuchandana syn. Patanga with Caesalpinia sappan Linn., Kaaliyaka syn. Pita Chandana with Coscinium fenestratum (Graetn) Colebr., and Harichandana syn. Bhadrashriya with Syzygium aromaticum (Linn.) Merr. & L. M. Perry.

Among over-the-counter products, Chandanaasava (Bhaisajya Ratnaavali) and Chandanaadi Vati (IMPCOPS) are prescribed for urinary and urinogenital affections; Chandanaadi Taila (Bhaisajya Ratnaavali) for external application in pruritus and skin eruptions. A decoction of Shadanga-Kwaath Churna (Ashtaanga Hridaya) is prescribed to quench excessive thirst and alleviate burning sensation during high fevers.

Unani practitioners prescribe Khamira-e-Sandal (Qarabadeen-e-Azam) as a cardiac tonic in palpitation; Sharbat-e-Feryad (Qarabadeen-e-Jadeed) in cold, cough and inflammations of respiratory tract; Raughan-e-Sandal (Kitabul Adviya) internally in burning micturation, cystitis, gleet and venereal diseases (dose: 5–10 drops on sugar).

Active principles and pharmacology

Bark contains a triterpene, urs-12-en-3-beta-yl-palmitate.

Chief constituents of essential oil from heartwood are alpha- and beta-santalol. Other constituents include the sesquiterpene hydrocarbons alpha- and beta-epi-beta-santalene, alpha- and beta-curcumene and beta-farnesene. Dihydroagarofuran is also present in the essential oil.

Leaves contain a peptide, a polyamine, amino acids, glycoprotein.

The essential oil of sandalwood possesses urinary disinfecting effect. If used in higher doses and for longer periods it can be toxic to the kidneys. (PDR.)

Use in Western herbal

German Commission E recognized the efficacy of Santalum album in the following areas:

- Cold, cough, bronchitis
- Fevers and colds
- Infections of the urinary tract
- Inflammations of the mouth and pharynx
- Liver and gallbladder complaints.

Caution

Sandalwood is contraindicated in diseases of the kidney parenchyma. (Michael McGuffin.) Sandalwood oil should be taken in an enteric-coated form. (PDR.)

Caesalpiniaceae

SARACA

Saraca asoca (Roxb.) De Wilde
Saraca indica auct. non L.

Figure 1 Saraca indica [ZANDU]

Habitat

Throughout India, up to an altitude of 750 m in the Central and Eastern Himalayas and Khasi, Garo and Lushai hills; also in the Andaman Islands. Saraca indica does not occur in India, but is recorded as grown in the Botanical Garden, Kolkata.

Classical and common names

Ayurvedic: Ashoka, Hempushpa, Taamrapallava, Pindapushpa, Gandhapushpa.
Siddha: Asoku.

Parts used
Bark, seed, flowers.

Dose
Decoction 50–100 ml, powder 3–6 g.

Classical use
Charaka and Sushruta incorporated flowering buds, seeds and bark in internal prescriptions for haemothermia, haemorrhages, gynaecological disorders; also for arthritis, sciatica, neuralgia and neurological affections.

Milk, boiled with the decoction of Ashoka bark, was prescribed for meno-metrorrhagia (Vrindamaadhava); Ashoka seeds for calculus and retention or suppression of urine (Chakradatta).

Sedative, antitoxic, antiseptic, astringent and styptic properties were attributed to the drug in Indian medicine. It is used extensively for the treatment of excessive uterine bleeding, dysmenorrhoea, leucorrhoea, and depression in women.

Practitioners of Indian medicine also use the bark and flowers for biliousness, dyspepsia, colic, dysentery, haemorrhagic diarrhoea, piles, ulcers and pimples; dried flowers in diabetes.

Ashokaarishta and Ashoka Ghrita (Bhaishajya Ratnaavali) are reputed over-the-counter drugs in India for menstrual disorders. Ashoka Vati (IMPCOPS) contains Ashoka bark and clax of Green Vitreol. It is administered with Ashokaarishta while treating cases of dysmenorrhoea and amenorrhoea.

Active principles and pharmacology
Flowers gave beta-sitosterol, flavonoids and the flavone glycosides quercetin, kaempferol-3-O-beta-D-glucoside, quercetin-3-O-beta-D-glucoside and apigenin-7-O-beta-D-glucoside; the anthocyanins pelargonidin-3, 5-diglucoside and cyanidin-3, 5-diglucoside.

Bark yielded catechol and sterols, a wax containing n-alkalines, esters and free primary alcohols.

Phenolic glycoside and non-phenolic glycoside, obtained from the water soluble fraction and ethanolic extract of bark, were studied for their biological activities.

Non-phenolic glucoside was observed as cardiac stimulant on frog heart and dog's ventricle.

Spasmodic activity of phenolic glycoside on smooth muscle was due to direct myotropic action. Spasmodic activity in rabbit ileum remained unaffected by pentolinium and slightly affected by atropine; whereas spasmodic effect of non-phenolic glycoside on rabbit ileum was modified by pentolinium and blocked by atropine, suggesting thereby that its spasmodic activity on smooth muscle was cholinergic in nature. Spasmodic action of phenolic glycoside on uterus also was not blocked by pentolinium and atropine.

Pure phenolic glycoside exhibited potent oxytocic activity on uterus of different mammals and was similar in nature to pitocin. The sensitivity increased by oestrogen and progesterone. Pure phenolic glycoside was devoid of any toxicity and side-effects up to a concentration of 250 mg/kg in animal studies.

The bark powder was used directly to screen its anti-implantation activity in proven fertile female rats. The drug showed 40 % anti-implantation activity at a dose of 200 mg/kg administered from Day 1 to Day 5 of pregnancy. (CCRAS.)

Aqueous extract (glycoside) of the bark increased the life span of mice in case of Eh. ascites carcinoma by 24 %. In case of sarcoma S-180, it decreased the weight of tumour by 24 %, suggesting antitumour activity of the glucoside. (CCRAS.)

Water soluble fraction of methanolic fraction of flower and flower bud exhibited significant inhibitory effect against Sh. boydis, S. viballerup, Sh. dysenteriae, E. coli, Vibro cholerae, of which flower bud was more potent. Chloroform extract screened against Nunnizia fulva (–), N. fulva (+), Trichophyton equinum and C. crassitunicatum exhibited good inhibitory zones in all test fungi.

Asteraceae

SAUSSUREA

Saussurea lappa C. B. Clarke

Habitat

Valley of Kashmir at altitude of 2500–3000 m. Also cultivated in Kashmir and neighbouring Himalayan regions.

Classical and common names

Ayurvedic: Kushtha (Charaka, Sushruta), Vaapya, Kaashmir. (Kushtha has no connection with the botanical genus costus.)
Unani: Qust. (Saussurea hypoleuca Spreng)
Siddha: Kottam.

Parts used

Root.

Dose

Powder 0.25–1.25 g.

Classical use

Charaka and Sushruta incorporated Kushtha in internal prescriptions for toxicosis, intestinal colic, indigestion, anaemia, deficient lactation, uterine and vaginal affections, chronic skin diseases. The root was an ingredient in pastes and ointments for ringworm and other cutaneous eruptions. Fruits were prescribed for their astringent and tonic properties.

For mental diseases, purified butter processed with the roots of Kushtha and Vachaa (Acorus calamus), and fresh juice of Braahmi (Bacopa monnieri) and Shankhapushpi (Convolvulus pluricaulis) was prescribed. A linctus of Kushtha, Vachaa and Swarna Bhasma (calcined gold) was administered as a brain tonic. (Charaka Samhitaa, Sushruta Samhitaa.)

By the 16th century, Kushta entered into a number of rejuvenating and intellect-promoting tonics.

Kushtha was a member of Elaadi Gana (the Cardamomum group) which was considered specific for curing toxicosis, urticaria, skin eruptions and for promoting complexion. It was also included in the group of herbs called Sarv-aushadhi, specific for psychosomatic and neurological disorders, fevers, skin diseases.

Charaka incorporated Kushtha in his reputed compound, Agastya Rasaayana, still available over-the-counter, for asthma and chronic bronchitis. More than twenty compound preparations, with Kushtha as an important ingredient, are available in herbal stores.

Unani physicians prescribe Anqaruya-e-Kabir (Qarabadeen-e-Qadri) internally and Raughan-e-Qust externally in hemiplegia, paralysis and other neurological disorders.

In Unani medicine, Qust-e-Arabi (Sweet Kushtha) and Qust-e-Hindi (bitter Kushtha) are used. Sweet Kushtha of Unani medicine is actually Orris root of Iris spp.

Active principles and pharmacology

Roots yielded 22-dihydrostigmasterol.

Roots of the Punjab variety gave costunolide, dehydrocostuslactone, costic acid, palmitic and linoleic acids, beta-sitosterol and alpha-cyclocostunolide.

The Kashmir variety gave, in addition to costunolide, dehydrocostuslactone and alpha-cyclocostunolide, alantolctone, beta-cyclocostunolide and iso-alantolactone.

A novel terpenoid C14 ketone, characterized as (E)-9-isopropyl-6-methyldeca-5, 9-dien-2-one, has also been reported from root oil. The acidic fraction of essential oil from roots contained acetic acid, 3-methylbutyric acid, hexanoic acid, heptanoic acid, octanoic acid, 7-octenoic acid, 3-isopropylpenthanaoic acid, 4-ethyloctanoic acid and alpha-amorphenic acid.

Leaves contain taraxasterol and its acetate, alpha-amyrin stearate, beta-amyrin and lupeol palmitates and an alkaloid, saussurine.

The alkaloid saussurine exhibits a smooth muscle relaxant activity. On rabbit ileum *in vitro* saussurine antagonized the contractions produced by histamine and acetylcholine. On peripheral vessels (perfused ear vessels of rabbit), the alkaloid produced initial dilatation followed by subsequent constriction. The alkaloid did not block the effects of adrenaline and histamine on the vessels. On guinea-pig tracheal chain, it did relax the bronchoconstriction produced by histamine in this preparation. On systemic administration, under experimental conditions, saussurine displays a positive inotropic action on myocardium. (Chemexcil.)

The essential oil exhibits marked carminative properties. It inhibits peristaltic movements of the gut and produces relaxation. Injection of the essential oil produces vasodilatation in splanchnic area and produces a definite stimulation on circulation. It is absorbed from gastro-intestinal tract and partly excreted by lungs producing an expectorant action and partly by the kidneys, producing a diuretic action.

Delactonized oil and some lactone fractions of the oil exhibit hypotensive, spasmolytic and bronchiodilatory effects.

The essential oil showed antiprotozoal effect in vitro (1 to 10, 000 dilution). It also showed antibacterial effect against streptococci and staphylococci.

Cyperaceae

SCIRPUS

Scirpus kysoor Roxb.
Scirpus grossus Linn. f.

Habitat

Distributed throughout India, especially in swamps, up to an altitude of 700 m.

Classical and common names

Ayurvedic: Kaseru, Raaja Kasheruka.
Unani: Kaseru.
Siddha: Karundan.

Parts used

Tuber.

Dose

Dried and ground tuber 5–10 g.

Classical use

Paste of Kaseru, Shrngaataka (Trapa natans), and lotus root, boiled with milk, was prescribed for treating threatened abortion (Sushruta Samhitaa). Tonic properties of Kaseru were used for promoting spermatogenesis and lactation. Flowers were prescribed for jaundice; the tuber, pounded with rose-water, for vomiting, diarrhoea, burning sensation and eruptions. (Kaiyadeva Nighantu, Bhaavaprakaasha.)

Unani physicians prescribe a linctus of Kaseru during fevers, after recovery from gastroenteritis, and as a cardiac tonic in debility.

Active principles and pharmacology

Tubers showed presence of steroids including small amounts of progesterone. Also present are sugars, tannins, starches and saponins.

Methanol extract of tubers produced biphasic hypotensive effect on anaesthetised cats. The pet.-ether and chloroform extracts exhibited no effect on blood pressure.

Both pet.-ether and methanol extracts caused non-specific antispasmodic effect against acetylcholine, histamine and barium chloride-induced spasm on smooth muscle preparation of guinea-pig ileum.

Methanol extract potentiated the effect of isoprenaline; also of adrenaline at higher doses (400 mcg/ml).

The tuber exhibited antidiarrhoeal, antiemetic and digestive properties. Amylase is present in the tuber.

The tuber contains 62–67 % of digestible carbohydrates and 7.5–11.8 % of proteins.

Anacardiaceae

SEMECARPUS

Semecarpus anacardium Linn. f.

Habitat

Outer Himalayas from Sutlej to Sikkim, and fairly common throughout the hotter parts of India, as far east as Assam.

Classical and common names

Ayurvedic: Bhallaataka, Arushkar (Charaka, Sushruta), Agnik, Agnimukha, Shophkrit, Viravrksha.
Unani: Bhilaavaa, Balaadur, Bilaadur.
Siddha: Sorankottai.

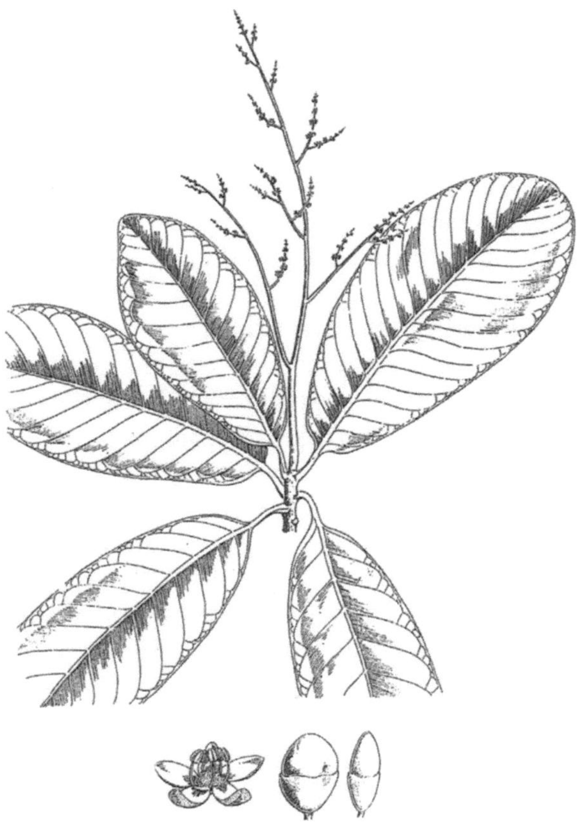

Figure 1 Semecarpus anacardium [ADPS]

English: Marking Nut.

Parts used

Detoxified nut, oil.

Dose

As advised by the physician.

Classical use

Detoxified nuts of Bhallaataka were incorporated in prescriptions for toxic conditions, obstinate skin diseases, tumours, malignant growths, fevers, haemoptysis, excessive menstruation, vaginal discharges, deficient lactation, constipation, intestinal parasites (Charaka, Sushruta).

Bhallataka, Haritaki (Terminalia chebula) and Jiraka (Cuminum cyminum), mixed with jaggery and made into a sweet bolus, were administered in splenomegaly (Vrindamaadhava).

A decoction of bruised fruits (1 to 8) in a dose of 25 ml was given for asthma.

A decoction with milk and purified butter, in gradually increasing doses, was given in peripheral neuritis, sciatica, facial paralysis, hemiplegia. Bhallaataka-vardhamaana (Vrindamaadhava) was used as a nervine tonic.

Bhallaataka nuts (4), boiled with 1.25 l milk, was given as an aphrodiasic (Ashtaanga Sangraha). Bhallaataka Rasaayana of Charaka was prescribed as a potent rejuvenating tonic.

Purified butter, cooked with the paste and decoction of Bhallaataka, mixed with sugar, was administered for treating tumours (Chakradatta.)

Bhallataka nut (1) pounded and mixed with double quantity of jaggery was prescribed for eliminating guinea worm. Bhallataka oil, mixed with Vidanga (Embelia ribes), was also used. (Ashtaanga Hridaya, Gadanigraha, Siddhabheshaja-manimaalaa.)

Juice of the pericarp and the oil of Bhallaataka are powerful escharotics; used in Indian medicine in small doses (0.03–0.06 cc diluted with clarified butter, cream or honey ten times its volume). Even external application of the oil may cause painful micturition and bloody stools. The oil of Bhallaataka, mitigated with butter, or mustard oil in which the fruits of Bhallaataka are fried, is used externally. Ripe fruits, boiled with a solution of cow-dung, are used internally. Nuts are used only after curing under medical supervision.

Among over-the-counter Ayurvedic compound preparations, with Bhallaataka as one of the ingredients, Narasimha Ghrita (Ashtaanga Hridaya), Bhallataka Vati (Bhaishajya Ratnaavali) are used as blood-purifiers and haematinic tonics; Kalyaanka Kshaara (Ashtaanga Hridaya) for gastrointestinal disorders; and Sanjivani Vati (Shaarangadhara Samhitaa) for fevers.

Bhallataka enters into many important formulations of the South—Guggulu-tikta Ghritam, Nimbaamritaasavam, Naarsimha-Rasaayanam, Varnaadi Kashayam, Mahaaraaja-prasaarani Tailam.

In Unani medicine, Majoon-e-Asal-e-Balaadur is prescribed for neurasthenia; Majoon-e-Balaadur for dementia, amensia; Raughan-e-Balaadur externally in paralysis, hemiplegia, Bells Palsy. Anqaruya-e-Kabir is also prescribed for neurological affections.

Active principles and pharmacology

Nut shells contain the biflavonoids: biflavones A, C, A1, A2, tetrahydrorobustaflavone, B (tetrahydroamentoflavone), jeediflavone, semecarpuflavone and gulluflavone.

Oil from nuts, Bhilavinol, contains a mixture of phenolic compounds mainly of 1, 2-dihydoxy-3-(pentadecadienyl-8, 11)-benzene and 1, 2-dihydoxy-3-(pentadecadienyl-8', 11')-benzene.

Leaves contain amentoflavone as the sole biflavonoid.

The plant was screened for its antiarthritic, antitumour, hypotensive, antispasmodic and antiallergic properties. (CCRAS.)

Milk extract of the nut showed anti-inflammatory property against carrageenin, 5-HT and formaldehyde-induced rat paw oedema (in acute anti-inflammatory studies). The extract was inactive on sub-acute models of inflammation. The extract suppressed acute primary inflammation of adjuvant arthritis in albino rats. It had no effect on secondary lesion. It effectively suppressed the tuberculin reaction in sensitised rats. Most of the animals treated with the milk extract of the nut gained in weight, but about 20 % of animals developed alopecia as well as gangrene of limbs, tails and ears. (CCRAS.)

A Siddha drug, Raasanaadhi Mezhugu (containing castor oil in which Marking Nut was fried), showed 34.5 % reduction of carrageenin-induced acute inflammation in rats in 100 mg/kg dose. The oil was also found to possess significant analgesic effect in a dose of 300 mg/kg against acetic acid-induced writhing response in mice.

Raasanaadi Guggulu tablet, containing Semecarpus anacardium, administered to patients, 2 tablets thrice daily, produced significant anti-inflammatory effect. (CCRAS.)

The nut extract exhibited negative inotropic effect on isolated frog heart. It antagonised carbachol, histamine and pitocin on rat, guinea-pig and rabbit ileum and rat uterus. It produced delayed hypotensive effect on anaesthetised dogs which was found unaltered by prior administration of atropine but blocked by mepyramine.

The nut exhibited antitumour activity in the nasopharynx. (CCRAS.)

An experiment was conduced to elucidate the ultrastructural changes in degenerating hippocampal neuron cell bodies of stressed rates and protective effect of Marking Nut extract. After the treatment, the number of cells demonstrating degenerating characteristics was significantly reduced (80 %).

Pedaliaceae

SESAMUM

Sesamum indicum Linn.
Sesamum orientale Linn.

Habitat

Cultivated throughout the plains of India, up to an altitude of 1200 m.

Classical and common names

Ayurvedic: Tila, Snehphala (Charaka). Not to be confused with Tilaparni, equated with Cleome gynandra Linn.
Unani: Kunjad.
Siddha: Ely, Ellu.
English: Sesame, Gingelly.

Parts used

Seed, oil, alkaline ash.

Classical use

Black sesamum seeds (5 parts) mixed with sugar (1 part) was prescribed with milk for diarrhoea with blood (Charaka); 80 or 40 g black sesamum seeds with cold water in the morning for haemorrhoids (Sushruta).

Tila, Bhallaataka (cured Semecarpus anacardium nuts), Haritaki (Terminalia chebula) and jaggery, in equal quantity, was prescribed for alleviating piles, anaemia, splenomegaly, cough, asthma (Vrindamaadhava).

A decoction of Tila, mixed with purified butter, jaggery, Bhaargi (Clerodendrum spp.) and Trikatu (dry ginger, Long pepper, Black pepper), was administered for amenorrhoea. (Ashtaanga Hridaya, Vrindamaadhava, Raaja Maarttanda.)

Black Til seeds (40 g), mixed with Aamalaka (Emblica officinalis) or Bibhitaka (Terminalia bellirica) or Haritaki (Terminalia chebula), were prescribed as a rejuvenating tonic (Ashtaanga Hridaya.)

Sushruta included alkali of the plant in an ointment for burns.

In Unani medicine, Tila seeds are considered emollient, diuretic, lactagogue and a nourishing tonic. Tila seeds and kernels of Akharot (Juglans regia kernels) are prescribed for bleeding piles. Black Tila seeds are an important ingredient in Majoon-e-Balaadur, prescribed for neurasthenia.

In folk medicine, equal parts of sesame oil and lime water is applied as a dressing for burns and scalds.

Two distinct types of seeds of Tila, white and black, are used in Indian medicine. The black variety is used internally, while the oil of white variety is used in medicinal oils. The white variety yields more oil than the black one.

Active principles and pharmacology

Leaves gave a flavonoid, pedalin.

Non-saponifiable fraction of seed oil gave sterols, a lignan-sesamin and a nitrolactone, sesamolin. Sesamol, alpha- and beta-tocopherol are also present in seed oil.

Basic aroma compounds of roasted seeds consisted of 2, 4 (or 5)-dimethylthiazole and substituted pyrazines.

Pinoresinol has also been reported from the plant.

Analysis of dry seeds, grown in various parts of the world gave values in the following ranges (in g/100 g): moisture 4.1–6.5; ether extr. 43.0–56.8; protein 17.6–26.4; crude fibre 2.9–8.6; carbohydrates 9.1–25.3; and mineral matter 4.1–7.4; Ca 1.06–1.45; and P 0.47–0.62.

Nearly one-half of two-thirds of the calcium in the seed is present as oxalate and a major part of it is located in the outer epidermis. Trace elements present in the seed include iron, iodine, zinc, cobalt, molybdenum and nickel.

Sesame seeds are fairly rich in thiamine and niacin. Ascorbic acid is present in traces. Vitamin A potency ranges from 15 to 100 I.U./100 g seed. Refined sesame oil contains about 0.05 % tocopherols; the combined alpha- and beta-tocopherol content of the unsaponifiable matter of the oil is reported to be 0.005 %.

The principal protein of sesame seed is a globulin. There is no appreciable difference in amino acid composition of proteins from black and white varieties of the seed. Sesame protein is rich in sulphur amino acids, particularly methionine. It is also rich in tryptophan.

In experiments undertaken using laboratory animals, sesame seeds have been shown to lower blood sugar levels and also to raise the levels of stored carbohydrates (glycogen). (Andrew Chevallier.)

Papilionaceae

SESBANIA

Sesbania grandiflora (L.) Poir.

Figure 1 Sesbania grandiflora [CCRAS]

Agati grandiflora Desv.

Habitat

A native of Malaysia. Grown in many parts of India—Punjab, Delhi, Bengal, Assam, the Andamans.

Classical and common names

Ayurvedic: Agastya (Sushruta); Munidrum, Bangasen, Vakrapushpa.
Siddha: Agathi.
English: Agati Sesban, Swamp Pea.

Parts used

Whole plant.

Dose

Juice 10–20 ml, decoction 50–100 ml.

Classical use

Sushruta gave flowers for cough and emaciation; processed with clarified butter as a tonic in night-blindness.

Butter, extracted from the milk curdled with the powder of flowers, was prescribed for gout (Vrindamaadhava). Juice of leaves, in the form of nasal drops, was administered during intermittent malarial fevers; mixed with Maricha (Piper nigrum) for alleviating epileptic fits. (Vrindamaadhava, Bhaavaprakaasha, Haarita Samhitaa.)

In traditional medicine, the juice of flowers is prescribed for nasal catarrh and headache; the juice of root with honey as an expectorant; a decoction of bark in small doses for diarrhoea (emetic and aperient in large doses); also for stomatitis, ulceration of the alimentary canal, for thrush and infantile disorders of the stomach; pods for anaemia, tumours and glandular enlargements and for dispelling toxins.

An infusion of the bark is given during the first stage of smallpox and eruptive fevers.

Seeds are used as an emmenagogue.

Externally, a paste of root is applied to painful swellings; also to scabies and skin eruptions. A poultice made from the leaf juice is applied to bruises.

Leaves are chewed for disinfecting mouth and throat.

An allied species, Sesbania sesban Merrill, has been equated with Balaamotaa of Dhanvantari Nighantu and Kaiyadeva Nighantu; also with Jayanti or Jaint of folk medicine. (Jayanti is also a synonym of Agnimantha (Premna serratifolia Linn.) in Ayurvedic texts.)

Active principles and pharmacology

Sesbania grandiflora: Flowers gave methyl ester of oleanolic acid, nonacosan-6-one and kaempferol-3-rutinoside.

Seeds gave kaempferol-3, 7-diglucoside, (+)-leucocyanidin and cyanidin-3-glucoside. Seeds also contain galactomannan.

The aqueous extract of flowers have been found to produce haemolysis and sheet erythrocytes even at low concentration. The active principle responsible for the haemolytic effect is characterised as the methyl ester of oleanic acid.

Aerial parts exhibit spasmolytic, hypothermic and diuretic; bark astringent properties.

Sesbania sesban: Pollen and pollen tubes contain alpha-ketoglutaric, oxaloacetic and pyruvic acids. Flowers contain cyanidin and delphinidin glycosides.

Pods and leaves contain cholesterol, campesterol and beta-sitosterol.

Unsaponfiable matter of fixed oil from seeds showed antibacterial, cardiac depressant and hypoglycaemic activity. (CIMAP.)

Dipterocarpaceae

SHOREA

Shorea robusta Gaertn. f.

Figure 1 Shorea robusta—flowering branch and fruit [WOI]

Habitat

Found extensively in parts of North-East and Central India, up to 1700 m.

Classical and common names

Ayurvedic: Shaala (Charaka, Sushruta); Saalasaara, Dhuupavraksha.
Unani: Raal (resin).
Siddha: Kungilyam.

English: Sal.

Parts used
Heartwood, flowers, gum-resin.

Dose
Powder 3–5 g, decoction 50–100 ml, gum-resin 1–3 g.

Classical use
Powdered flowers or heartwood was used alone or in prescriptions, for polyuria, emaciation and anaemia due to loss of blood, goitre and diseases due to vitiated blood. (Charaka Samhitaa, Sushruta Samhitaa, Ashtaanga Hridayaa.) Fumes of gum-resin were inhaled during asthmatic attacks and bouts of cough. For leucorrhoea and metrorrhagia, heartwood powder, as well as the gum-resin, was administered internally.

Externally the gum-resin was applied over foul ulcers, infected wounds and skin eruptions.

Shaalsaaraadi Leha (Bhaishajya Raatnaavali) was a specific compound preparation for polyuria.

Unani physicians prescribe gum-resin internally for diarrhoea with blood, bleeding piles, urethral discharges.

The Marham-e-Raal of Unani medicine is available over-the-counter for syphilitic ulcers, fistula, nasal ulcers. It stimulates healthy granulation and cleanses the wound of pus. The ointment is also used for freckles, blemishes and cracks of hands and feet.

Active principles and pharmacology
Heartwood contains a chalcone, glycoside-4'-hydroxychalcone-4'-O-beta-D-glucopyranoside, a polyphenol-hopeaphenol, leucoanthocyanidin.

Essential oil contains p-cymene, tetrahydro-gamma-cadinene, cadalene, 1, 6-dimethylnaphthalene, 1, 2, 6, 8-tetramethylnaphthalene and a few oxygenated benzene derivatives.

The bark of Shorea robusta is rich in tannins. Tannin content of various parts is as follows: bark 7–12 %, young leaves 22 % and powder dust 12 %. The spray-dried aqueous extract of the bark contains 39.6 % of tannins with trans/non-trans ratio of 0.73.

Resin exhibits stomachic, astringent, antidysenteric, styptic, antiseptic, and antigonorrhoeic properties.

Malvaceae

SIDA

Sida cordifolia Linn.

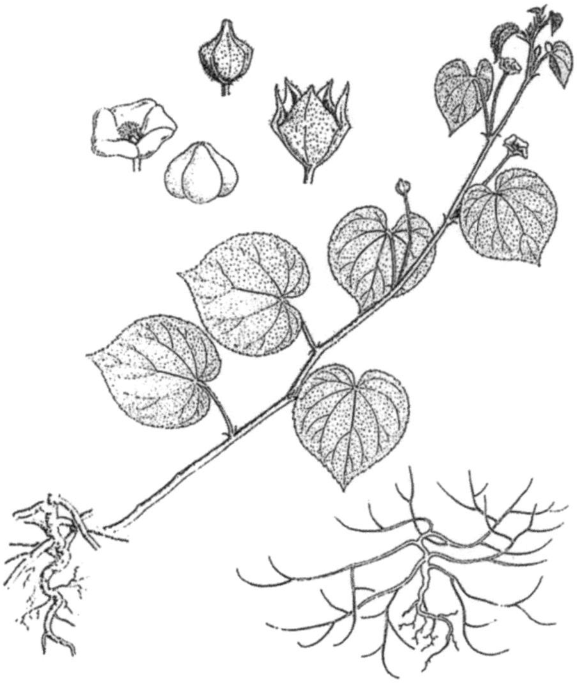

Figure 1 Sida cordifolia [ADPS]

Habitat
Distributed in moist places throughout tropical and sub-tropical India and Nepal to an altitude of 1050 m.

Classical and common names
Ayurvedic: Balaa, Sumanganaa, Khara-yashtikaa, Balini, Bhadrabalaa, Bhadraudani (Charaka); Vaatyaalikaa.
Unani: Bariyaara, Khirhati, Khireti.
Siddha: Nilathuththi, Kuruntotti, Balaa.
English: Country Mallow.

Parts used
Root, seeds.

Classical use
Charaka and Sushruta prescribed a decoction of leaves in urinary diseases, splenic disorders, for rejuvenation and promoting fertility.

A medicinal oil of Balaa, also soup of the plant, was prescribed for rheumatism, gout, hemiplegia, neurological affections, muscular atrophy, torticollis. (Charaka Samhitaa, Sushruta Samhitaa, Vrindamaadhava, Bangasena.) Sahasrapaaka, Shatapaaka, Dashpaaka (processed one thousand times, hundred times and ten times respectively) Balaa oil extracts were used extensively in Indian medicine (Charaka Samhitaa, Ashtaanga Hridaya, Vrindamaadhava).

For goitre, an extract of Balaa, Atibalaa (Abutilon indicum) and Devadaaru wood (Cedrus deodara) in oil was administered internally (Sushruta Samhitaa).

For bleeding piles, a liquid gruel of parched paddy, processed with Balaa and Prshniparni (Uraria picta); for intrinsic haemorrhage, cow's milk cooked with Balaa; for meno-metrorrhagia, the root of Balaa with honey; was prescribed (Charaka Samhitaa, Sushruta Samhitaa, Vrindamaadhava).

Balaa Rasaayana (Sushruta Samhitaa) was prescribed as a rejuvenating and nervine tonic.

For mental diseases, the white-flowered Balaa was prescribed along with the powdered root of Apaamaarga (Achyranthes aspera) and milk (Bangasena). The root of white-flowered Balaa, pounded with cow's milk, was prescribed for promoting fertility (Chakradatta). Entire plant of Atibalaa (syns: vaatyaayani, Vaatapushpi), by way of various extracts, was prescribed by Charaka for loss of vitality.

The two varieties of Balaa (Balaa-dwya) included Balaa and Atibalaa (used together in most of the cases during the period of Charaka and Sushruta); the three-varieties additionally included Naagabalaa (Grewia hirsuta Vahl); the four-varieties included Mahaabalaa (Sida rhombifolia Linn. Mast.) and the five-varieties included Raajabalaa (Sida veronicaefolia Lam.).

Balaa, Atibalaa, Nagabalaa and Mahaabalaa constitute the group of four Balaa species of Bhaavaprakaasha.

During the 16th century the three-varieties of Balaa are mentioned as one group (Ayurveda Saukhyam): Balaa, Mahaabalaa and Naagabalaa. All the three were used as rejuvenating and age-sustaining tonics for promoting vitality and longevity. Sida rhombifolia was used for alleviating dysuria and Grewia populifolia (Gaangeruki) for consumption, emaciation and debility.

Kerala physicians use Sida rhombifolia subspp. retusa as Balaa.

Over-the-counter, Balaarishta (Bhaishajya Ratnaavali) is available for nervous debility and rheumatic affections; Balaashwagandhaa-lakshaadi Taila (Sahasrayoga) for internal as well as external use in neuritis and emaciation; Balaa-guduchyaadi Taila (Sahasrayoga) for inflammatory and rheumatic swelling; Bhuunaaga Taila (IMPCOPS), Dhanvantara Taila (Sarvaroga Chikitsaa-Ratna), Kaarpaasaathyaadi Taila (Sahasrayoga) for hemiplegia, paralysis, aphasia.

Ksheerbalaa Tailam, the reputed medicinal oil from Kerala and Tamil Nadu, is available in various strengths (medicated and processed 7 times, 28 times. 101 times, based on Sahasrayoga). The oil is prescribed internally as well as externally for acute and chronic rheumatism, general paralysis, paraplegia, neuralgia, hemicrania, cephalalgia, diseases of the brain and spinal cord, epilepsy, hysteria, and for a host of diseases connected directly or indirectly with the nervous system.

In Unani medicine, earlier, Bariyaara and Khirhati (Khireti) were treated as separate drugs (Muhit Aazam), now both are used as Bariyaaraa, equated with Sida cordifolia and Sida rhombifolia.

Unani physicians prescribe a paste of the white-flowered Bariyaaraa for spermaturea, fresh juice of the whole plant for spermatogensis.

Powdered root, mixed with sweetened milk, is prescribed for leucorrhoea and dysuria.

An extract of the rootbark in Tila oil externally, and powdered drug with milk internally, is administered for facial paralysis and sciatica.

A poultice of yellow-flowered Bariyaara (leaves) is applied to inflammations. A paste of Khireti is prescribed internally and externally for improving complexion and for alleviating erysipelas.

For diarrhoea, 6 g of Khireti pounded with 2–4 numbers of Piper nigrum is administered. During fevers, a decoction of root, with ginger powder, is prescribed.

Seeds, known as Beejaband, are used in aphrodisiac and spermatic tonics of Unani medicine.

See Abutilon indicum.

Active principles and pharmacology

Roots contain the alkaloids ephedrine and si-ephedrine, beta-phenethylamine, carboxylated tryptamines, the quinazoline alkaloids vasicinone, vasicine, vasicinol. In addition, choline and betaine have also been obtained.

Alkaloids present in the roots are also present in stems and leaves in different amounts.

Ephedrine and si-ephedrine are major alkaloids in the aerial parts.

Roots of all other Sida spp. (Sida acuta, Sida rhombifolia, Sida spinosa) also contain the alkaloids beta-phenethylamine, ephedrine, si-ephedrine, vasicinone, vasicine, vasicinol, choline, betaine.

The aqueous extract of (Sida rhombifolia) root at 5–10 mg/kg p.o. exhibited antiarthritic effects in animal studies. It was non-toxic up to 15 g/kg p.o.

The alcoholic extract of the root in 10 mg/kg p.o. suppressed carrageenin-induced rat hind paw oedema. It produced hypertension in anaesthetised dogs. Exhibited significant bronchodilatory activity. The drug also showed cardiac stimulant and antispasmodic properties.

Anticonvulsant and antipyretic activities of Sida cordifolia have been observed.

Sida cordifolia exhibited antibacterial, antifungal and antiviral activity. Antiprotozoal activity against E. histolytica and anthelmintic activity against H. nana and A. galli has also been observed.

Asteraceae

SILYBUM

Silybum marianum (L.) Gaertn.

Habitat

Kashmir at altitudes of 1800–2400 m. Cultivated for ornament in gardens (florets—rose-purple).

Classical and common names

English: Holy Thistle, Milk Thistle.

Traditional use

Not used in Ayurvedic and Unani medicine. In folk medicine, flowering heads are consumed by diabetics. Young leaves are used as salad. Roots are eaten boiled as a potherb.

As a diet or in infusion, the herb is prescribed as a galactagogue. The herb is also used for intermittent fevers, dropsy, uterine affections.

Seeds are used for their demulcent, antispasmodic and anti-haemorrhagic properties. Though used for the treatment of jaundice and calculi of liver and gall-bladder, the drug could not yet enter into compound herbal preparations for hepatic affections. Its principal use is still as a galactagogue.

Active principles and pharmacology

Milk Thistle seed contain 1.5–3 % flavone lignans, collectively referred to as silymarin; 20–30 % fixed oil, of which (approximately) 60 % is linoleic acid, 30 % oleic acid and 90 % palmitic acid; 25–30 % protein; 0.038 % tocopherol; 0.63 % sterols, including cholesterol, campesterol, stigmasterol and sitosterol; and some mucilage.

The therapeutic activity of silymarin is based on following mechanisms of action: It alters the structure of outer cell membrane of hepatocytes in such a way as to prevent penetration of the liver toxin into the interior of the cell. It stimulates the action of nucleolar polymerase A, resulting in an increase in ribosomal protein synthesis, and thus stimulates the regenerative ability of the liver and the formation of new hepatocytes.

Milk Thistle extract provides hepatocellular protection by stabilizing hepatic cell membranes.

Other actions include interruption of enterohepatic recirculation of toxins, stimulation of protein synthesis and regeneration of damaged hepatocytes, as well as antioxidant activity.

Silibinin stimulates the activity of the DNA-dependent RNA-polymerase I, causing an increase in rRNA-synthesis and an accelerated formation of intact ribosomes. This results in general increase in the rate of synthesis of all cellular proteins. (Expanded German Commission E Monographs, 2000 edition.)

Use in Western herbal

Preparations of Milk Thistle are used as a tonic, for functional disorders of liver and gallbladder, as well as for jaundice, gall-bladder colic and diseases of the spleen. (PDR.) It is also used as an antidote to Death-Cap poisoning.

In the past, the plant was used as a galactagogue.

Over-the counter, Milk Thistle Vegicaps (520 mg), Milk Thistle capsules (100 mg), Milk Thistle tablets, Silymarin Extract are available.

Cruciferae

SISYMBRIUM

Sisymbrium irio Linn.

Habitat

Kashmir, Punjab, Haryana and from Rajasthan to Uttar Pradesh, especially on moist soils.

Classical and common names

Unani: Khaaksi, Khoobkalaan.
English: London Rocket.

Classical use

Seeds are prescribed by Unani physicians for treating throat affections, fevers, cough, asthma, loss of voice. A decoction of seeds is administered for aborting smallpox and measles. During cholera the drug is administered with rose-water; for diarrhoea with the fresh juice of Cichorium intybus (Kaasani) leaves.

Seeds are employed externally as a stimulating poultice.

Active principles and pharmacology

Seeds contain a flavonoid, iso-rhamnetin; fatty oil from seeds contains linolenic and oleic acids as chief constituents; other constituents are erucic acid, linolenic acid, palmitic acid and stearic acid; arachidic acid is present in traces. Sitosterol is reported from unsaponifiable matter. 24-methylenecholesterol is the principal sterol present in the pollens.

The plant is highly variable. Besides the diploids, triploids, tetraploids and hexaploids also exist.

Leaves are rich in protein, vitamin C, beta-carotene (vitamin A potency) and minerals. On analysis, the tender leaves of the wild plants gave: moisture 81.5, protein 7.0, ether extr. 1.2, fibre 0.4, carbohydrates 7.3; and mineral matter 2.6 %; calcium 485.0, phosphorus 125.0, iron 2.5; and vitamin C 176 mg/100 g; and vitamin A potency 10, 000 I.U./100 g.

An infusion of leaves gives relief in the affections of the throat and chest. Seeds exhibit rubefacient, febrifugal and expectorant properties.

The fatty acid composition of the seed oil showed: palmitic 8.2, stearic 3.1, oleic 27.5, linoleic 35.3, linolenic 8.3, and erucic 17.6 %. Seeds contain isorhamnetin and possibly a mustard oil.

Use in Western herbal

Sisymbrium officinale (St. Barbara's Hedge Mustard) is named by the French as the "Singer Plant", as it was considered an infallible remedy for loss of voice.

A strong infusion of the whole plant was taken in former days for all diseases of the throat.

Juice of Sisymbrium sophia Linn. (Flixweed or Fluxweed), mixed with an equal quantity of honey or vinegar, was recommended for chronic coughs, hoarseness and ulcerated sore throats. A strong infusion of the herb was administered for asthma. Seeds were used as a special remedy for sciatica. (M. Grieve.)

Alliaceae

SMILAX

Smilax china Linn.

Habitat

Indigenous to China and Japan. Imported into India and commonly found in the Indian market. Smilax species also grow in the eastern Himalayas and Burma.

Classical and common names

Ayurvedic: Chobachini, Dweepaantar-Vachaa; Madusnuhi.
Unani: Chobchini.

Figure 1 Smilax china [ZANDU]

English: China Root.

Parts used

Root.

Dose

Powder 50–100 mg.

Classical use

Chobachini entered into Indian medicine during the 16th century. Bhaavamishra of Bhaavaprakaasha used the common local name (Chobachini) and coined a new Sanskrit name Dweepaantar-Vachaa for the herb. He prescribed Chobachini powder, in a dose of approximately 4 g with honey, for syphilis. This was the first prescription containing Chobachini as a single herb for syphilis in Ayurvedic medicine.

Among over-the-counter drugs, Madhusnuhi Churna (Bhaavaprakaasha) and Madhusnuhi Rasaayana (Sahasrayoga) are prescribed as blood-purifying tonics in syphilitic or leprotic affections of the skin.

In Unani medicine, Araq-e-Chobchini (Qarabadeen-e-Azam) is prescribed for psychoneurosis and melancholia; Majoon-e-Chobchini (Qarabadeen-e-Azam-o-Akmal) and Majoon-e-Ushbaa (Bayaz-e-Kabir) for syphilis and related skin diseases, neurological affections and rheumatism.

Smilax china root is the main ingredient in Parangi Pattai Choornam of Siddha medicine, prescribed for venereal diseases, leprosy and leucoderma. Smilax china is also an important ingredient in Mahaa Vallaathi Legiyam and Parangi Rasaayanam of Siddha medicine, prescribed for obstinate skin diseases, venereal diseases, chancres, ulcers, and fistula.

Chobachini is imported into India from China and Malaysia, though part of the supplies of the drug sold in India is derived from the tubers of indigenous species—Smilax glabra Roxb., Smilax lanceifolia Roxb. and Smilax macrophylla Roxb. non Willd. (known as bigger Chobachini, Hindi Chobachini and Jangali or Desi Chobachini). All these species are used as substitutes for sarsaparilla.

Active principles and pharmacology

Smilax china: Tannin, resin and cinchonin, along with smilacin, and steroidal saponins (0.5–3 %) yielding on hydrolysis sarsasapogenin, have been identified in the tuberous roots. The presence of flavonoid glycosides, and three saponins one of which yields diosgenin, has also been reported.

Leaves contain rutin.

Smilax glabra: Beta-sitosterol and stigmasterol have been isolated from rhizome. Alcoholic extract of the plant is reported to contain a glucoside and a colouring matter. Starch content of the tubers is 69.8 %.

A decoction of the fresh root is used by some hill tribes in Assam for the treatment of venereal sores. (Found in Garo and Khasi hills.)

Smilax lanceifolia: The juice expressed from the root is taken in rheumatic pains and the refuse after extraction of juice is used as a poultice over the affected part. (Found in the Sikkim Himalayas and in the hills of Assam and Manipur.

Smilax macrophylla, syn. Smilax ovalifolia: Locally, the plant is used in the treatment of venereal diseases, also applied in rheumatism. The roots are used in urinary complaints and dysentery. (Found almost all over tropical parts of India.)

Smilax perfoliata Lour, syn. Smilax prolifera Roxb. is used like Smilax ovalifolia in indigenous medicine. (Found with Smilax ovalifolia.)

Smilax aspera Linn.: A high tannin content is reported in the root. Roots and stems contain a heteroside parillin, a complex phenolic acid and potassium nitrate. The diuretic action of the roots is said to be due to potassium nitrate. The roots contain 31-norcycloartanol and beta-sitosterol. The crude saponin fraction from roots gave, on hydrolysis, three crystalline sapogenins—sarsasapogenin, a sapogenin named asperagenin, and a steroid sapogenin. Asperagenin belongs to the rare type of 25-hydroxysteroidal sapogenins.

Sarsaparilla of herbal medicine contains a mixture of saponins derived mainly from sarsasapogenin and smilagenin. The saponins display a strong diuretic action as well as some diaphoretic, expectorant and laxative properties.

Neither the whole drug nor its contained saponins is effective in the treatment of syphilis or as a "blood-purifier" (Tyler). It is also claimed that sarsaparilla contains testosterone and can be substituted for steroidal drugs. Testosterone has never been detected in any plant, including sarsaparilla (Tyler).

According to M. Blumenthal (American Botanical Council), the most deceptive practice with respect to sarsaparilla is the substitution for it, in some commercial herb products, for Hemidesmus indicus, the Indian sarsaparilla. Hemidesmus indicus belongs to an entirely different plant family (Asclepiadaceae) and it does not contain the same saponins or other plant constituents found in sarsaparilla (Tyler). According to CCRAS, Smilax china belongs to Alliaceae.

Use in Western herbal

Western use of Smilax china is largely confined to skin preparations. Sarsaparilla Root Vegicaps (520 mg) and Jamaican Sarsaparilla capsules are available over-the-counter. Sarsaparilla is also an ingredient in a number of remedies and supplements—Sarsaparilla Complex, Echinacea Complex, Blue Flag Root, Seaweed and Sarsaparilla Tablets, Burdock and Nettle Formula, Clear Complexion Tablet, Skin Eruption Mixture.

A number of Smilax spp., originating in Central and South America, are used medicinally—Smilax regelii, Smilax febrifuga, as well as the Mexican variety; as cleansing, antiseptic and anti-inflammatory drugs, helpful for irritant and chronic skin problems, including psoriasis. (Efficacy in psoriasis has not been documented.)

Caution

According to German Commission E, taking sarsaparilla preparations (may) lead to gastric irritation and temporary impairment (diuresis). The absorption of simultaneously administered drugs, for example, digitalis glycosides or bismuth, is increased. The elimination of other substances (e.g. hypnotics) is accelerated.

Solanaceae

SOLANUM

Solanum ferox Linn.
Solanum indicum Linn.

Habitat

Throughout warmer parts of India, up to an elevation of 1500 m.

Classical and common names

Ayurvedic: Brhati, Brhatikaa, Mahati, Hinguli, Prasaha, Vartaki, Kaantaa (Charaka), Kshudrabhantaki, Sinhi; Bana-bhantaa. Kateri (bigger variety).
Unani: Katai Kalaan.
Siddha: Papparanulli.
English: Poison-Berry.

Parts used

Root, fruit.

Dose

Powder 3–6 g, decoction 50–100 ml.

Classical use

Charaka used the leaves and roots of Brhati, in internal and external prescriptions, for toxicosis, pain and oedema of vagina.

Figure 1 Solanum indicum—flowering and fruiting branch [WOI]

Brhati was prescribed with buttermilk for diarrhoea; the fruit pounded and cooked with purified butter for cough; the fruit boiled in alkaline water of Koshaataki (Luffa acutangula) for piles; the root of the white-flowered variety for promoting conception. (Charaka Samhitaa, Shaarangadhara Samhitaa, Bangasena, Vrindamaadhava.)

Externally the juice of Brhati fruit, mixed with honey, was applied for alopecia. (Shaarangadhara Samhitaa, Gadanigraha, Vrindamaadhava.)

Brhati fruit, though used for treating piles and other inflammatory conditions, is contraindicated in haemothermia (Bhaavaprakaasha).

In traditional medicine, the fruit is prescribed in asthma, dry cough, bronchitis, chronic febrile conditions, also in dysuria. It is specific for bronchospasm. A decoction of the root is used in difficult parturition.

Two varieties of Brhati are mentioned as one group in Ayurvedic texts, these are equated with Solanum indicum and Solanum surattense(Kantakaari).

Other species, Solanum melongena Lam. var. incanum Kuntze and Solanum torvum Swartz are also used as Brhati.

Among over-the-counter drugs, Brhati is an ingredient in Bhaarangyaadi-Kwaatha Churna (Sahasrayoga), prescribed for cough, bronchitis, and fever; in all Dashamuula (the Group of Ten Roots) preparations—Dashamuula Kwaath Churna (Sahasrayoga), Dashamuulaarishta (Bhaishajya Ratnaavali), Dashamuula-Haritaki (Ashtaanga Hridaya). Both Brhati and Kantakaari are included in the group of Ten Roots. This group is specific for cough, bronchitis, asthma, oedema, headache, fever, anorexia, tympanitis, convulsions.

Active principles and pharmacology

Plant contains solasonine, diosgenin, beta-sitosterol, lanosterol, solamargine, solasodine and tomatidenol. Seed oil contains carpesterol.

The presence of solanine has been reported in roots, leaves and fruits.

Air dried leaves contain 0.32 % of total alkaloids. The total alkaloid content of the fruit varies from 0.2 to 1.8 % (dry weight basis). Plants growing in Jammu and Kashmir bear fruits with high alkaloid content (crude glycoalkaloids 4.8 % and total alkaloids 1.8 %).

Fruits contain carbohydrases, maltase, saccharase and melibiase; a proteolytic enzyme, similar to pancreatic trypsin is also present.

Prostaglandin content of the endometrium from 10 women was examined before and after the administration of a few drops of water extract of the whole plant through nasal route. A significant increase in prostaglandin E and F was observed. 17-oxysteriods in the urine of women patients also increased on exposure to nasal spray of the drug. (Chemexcil.)

The plant has been investigated for its cytotoxic activity. Both chloroform-soluble and insoluble fractions of ethanolic extract showed cytotoxicity to colo-205 (colon), KB (nasopharynx), HeLa (uterine cervix), HA (hepatoma), Hep-2 (laryngeal epidermoid), GBM 8401/TSGH (glioma) and H1477 (melanoma) cells.

Solanum nigrum Linn.

Figure 2 Solanum nigrum [CCRAS]

Habitat
Common to all parts of India; up to 3000 m in the Western Himalayas.

Classical and common names
Ayurvedic: Kaakamaachi, Kaakaahya, Kaakamaataa (Charaka).
Unani: Mako.
Siddha: Manattakkali.
English: Black Nightshade

Parts used
Whole plant.

Dose
Juice 10–20 ml, extract 20–50 ml, decoction 50–100 ml.

Classical use
Charaka prescribed the entire plant, cooked as a vegetable, for skin lesions, oedema, cough, and muscular atrophy.

A decoction of Kaakamaachi, mixed with Pippali (Piper longum) or Maricha (Piper nigrum) and jaggery; Kaakamaachi juice, cooked with purified butter, was prescribed as a revitalizing tonic. (Charaka Samhitaa, Vaidyamanoramaa.)

Kaakamaachi whole plant is included in the Four Pastes of Charaka Samhitaa; others being those of Shirish (Albizia lebbeck bark), Aaragavadha (Cassia fistula leaves), Kaarpaasa (Gossypium herbaceum flowers). This group of Four Pastes was specific for cutaneous diseases.

Shunthi (Zingiber officinale) pounded with the juice of Kaakamaachi was administered for urticaria.

Kaakamaachi belonged to Sushruta's Surasaadi group of herbs, specific for cough, asthma, rhinitis, anorexia and parasitic infections.

In Bhaavaprakaasha, the root has been suggested in two prescriptions for insomnia; as a vegetable for gout. The root was used as a single drug.

For conception, the juice of Kaakamaachi was prescribed (Vaidyamanoramaa).

In Unani medicine, Araq-e-Mako (Bayaz-e-Kabir) is prescribed for visceritis and hepatitis. Freshly prepared extract of the plant is administered for the treatment of cirrhosis of liver; viral hepatitis, as an adjuvant to hepatotoxic drugs, and for inflammation of joints.

Active principles and pharmacology
Berries contain the steroidal alkaloid glycosides solasonine, alpha- and beta-solamargine, alpha- and beta-solanigrine; steroidal sapogenins: diosgenin and tigogenin; solasodine and solasodiene. Steroidal glycosides: desgalactotigonin and 26-O-(alpha-D-glucopyranosyl)-22-methoxy-25-D-5-furost-3-alpha-26-diol-3-O-alpha-lycotetraoside have also been obtained from berries.

Stems and root yielded a spirostanol glycoside-uttronin A, furostanol glycosides: uttrosides A and B.

Solanine has also been reported from the plant and consists of six components—alpha, beta- and gamma-solanine and alpha-, beta- and gamma-chaconine.

The plant was observed to have smooth muscle relaxant activity, and enhanced microcirculation.

Infusions or decoctions of the plant, after transient stimulation, depress the central nervous system and reflexes of the spinal cord. Small doses increase, and large doses decrease cardiac activity. In the isolated rabbit ear, vasodilatation has been observed. Extracts of the plant affect the rate and amplitude of respiration; they also have an effect on the isolated ileum of guinea-pig. A decoction of the plant was used for the treatment of ascites in dogs.

The alkaloidal fraction of the plant is teratogenic. Alcoholic extract of Solanum nigrum protected against chlorpromazine-induced hepatic damage in adult albino rats.

Crude extract from fruits of the plant was found inhibitory to TMV (Tobacco Mosaic Virus) and SRV (Sunnhemp Rosette Virus). Solasonine and solamargine were also effective.

Caution

Leaves, when used in overdose, in the treatment of scrofulous affections, produce diaphoresis, nausea, purging and nervous disturbances. The juice of fresh leaves is reported to produce dilatation of the pupil.

Solanine is toxic in doses of 200–400 mg.

Solanum virginianum Linn.
Solanum xanthocarpum Schard & Wendl.
Solanum surattense Burm. f.

Habitat

Distributed throughout India, particularly in Deccan, Kerala and Punjab.

Classical and common names

Ayurvedic: Kantakaari, Kantakaarikaa, Vyaaghri (Charaka). Kateri.
Unani: Katai Khurd.
Siddha: Kandamkathiri.
English: Wild Egg-Plant, Yellow-Berried Nightshade.

Parts used

Whole plant.

Dose

Decoction 50–100 ml.

Figure 3 Solanum surattense [ZANDU]

Classical use

Charaka and Sushruta used the extract of the entire plant and fruits in internal prescriptions for bronchial asthma, tympanitis, misperistalsis, piles, dysuria, and for rejuvenation. Kantakaari Ghrita of Charaka was specific for cough and asthma.

A decoction of Kantakaari, or its extract in purified butter processed with Aahdraka (fresh Zingiber officinale), or Haritaki (Terminalia chebula) or Kaasamarda (Cassia occidentalis), Vaasaa (Adhatoda vasica) and Bhringaraaja (Eclipta alba), or Pippali (Piper longum) was prescribed for cough.

A decoction of Kantakaari, Guduuchi (Tinospora cordifolia), Shunthi (dry Zingiber officinale), added with the powder of Pippali (Piper longum) was the drug of choice for cough, asthma, chronic coryza, hoarseness of voice, indigestion, anorexia, abdominal pain, and fever. (Shaarangadhara Samhitaa.)

A linctus prepared with stamens of Kantakaari flowers was prescribed for chronic cough in children (Bangasena).

A decoction of Kantakaari was administered for treating suppression and retention of urine (Sushruta Samhitaa, Bhaavaprakaasha).

For promoting fertility and impregnation, the fruit of Kantakaari was prescribed with a uterine tonic—Phala-Sarpi (Ghrita) of Ashtaanga Hridaya.

Kantakaari has been suggested as a substitute for fertility-promoting classical drug Lakshmanaa (not yet properly identified).

Mustard oil, cooked with Kantakaari, was applied externally for lichen. (Sushruta Samhitaa.)

Among over-the-counter drugs, Kantakaari Vati (Ashtaanga Hridaya), Kantakaari Ghrita (Sahasrayoga), Kantakaari Avaleha (Shaarangadhara Samhitaa), Vaasaa-Kantakaari Lehya (IMPCOPS), Kanakaasava (Bhaisajya Ratnaavali) are prescribed for cough, bronchitis, asthma and chest diseases. Vidaaryaadi Ghrita (Ashtaanga Hriday) is a restorative and cardiac tonic. Panch-Tikta Ghrita (Ashtaanga Hridaya) is a blood-purifier, antiperiodic, antiseptic, anti-inflammatory. Punarnavaasava (Bhaishajya Ratnaavali) is a diuretic, haematinic, cholagogue.

Active principles and pharmacology

Plant gave solasonine, solamargine, beta-solamargine and solasodine. Petals yielded apigenin. Stamens gave quercetin diglycoside and sitosterol.

A glucoalkaloid named solanocarpine is found in the fruits. Solanocarpidine and a sterol known as carpesterol are also present. Potassium nitrate, a fatty acid, a resinous and phenolic substance, diosgenin and sitosterol are present.

The glucoalkaloid content of fruits collected from the plants growing in Jammu & Kashmir in reported to be 3.5 % (total alkaloid 1.1 %).

The leaves, roots and fruits yielded coumarins, scopolin, esculin and esculetin.

Seed and fruit extract of Xanthocarpum exhibited non-specific mild relaxation against histamine, acetylcholine and barium chloride in guinea-pig tracheal chain preparation. Only a mild antispasmodic action against acetylcholine-induced asthma was observed.

The extract produced relaxant effect against barium chloride-induced spasm in rabbit, rat and guinea-pig ileum.

Glucoalkaloidal saponin from Solanum xanthocarpum caused cardiotonic effect in hypodynamic heart and produced significant expectorant effect, caused release of histamine, fall in blood pressure due to peripheral vasodilatation which was considered for its diaphoretic effects. (CCRAS.) The beneficial effect of the drug on bronchial asthma may be attributed to the depletion of histamine from bronchial and lung tissue.

Cold alcoholic extract of dry leaf and stem in a dose of 2 ml to each rat exhibited protection against MES in 54 % animals. It potentiated dilantin sodium (55.55 %).

Solanine exhibited anti-inflammatory effect against cotton pellet granuloma and carrageenin-induced oedema in albino rats; also analgesic property by clip method and acetic acid writhing on albino rats.

Powdered drug of the whole plant in a dose of 1 g two to three times a day for a month on 305 patients suffering from bronchial asthma exhibited complete relief in 50 % patients. (CCRAS.)

Extracts of the whole plant show antiviral activity against Ranikhet disease virus and also against Sarcoma 180 in mice. Extracts of shoot and fruit show antibacterial activity against Staphylococcus aureus and E. coli.

Asteraceae

SPHAERANTHUS

Sphaeranthus indicus Linn.

See figure 1.

Sphaeranthus hirtus Willd.

Habitat

Found abundantly in damp places in the plains all over India, ascending to an altitude of 1500 m in the hills.

Classical and common names

Unani: Mundi.
Siddha: Kottaikaranthtai.

Figure 1 Sphaeranthus indicus—flowering branch [WOI]

English: East Indian Globe Thistle.

Parts used
Whole plant.

Dose
Juice 10–50 ml, infusion 50–100 ml.

Classical use
Mundi and Mahaa-mundi entered into Indian medicine through Unani medicine. Mahaa-mundi has been equated with Sphaeranthus africanus Linn.

Mundatikaa, Munditikaa, Mundi, Shraavani of Ayurvedic classics are confusing and misleading synonyms of Mundi of Unani medicine. Mundi of Ayurvedic classics was used for haemothermia, gout, rheumatism, polyuria; as a nourishing and rejuvenating tonic; while Mundi of Unani medicine is specific for putrefaction of blood.

It is wrong to presume that Shraavani and Mahaa-shraavani, the "divine" plants of Sushruta, were treated as Mundi and Mahaa-mundi during the later period. Shraavani was a latex-bearing plant, sweet, refrigerant, age-sustaining and rejuvenating; while Mundi does not bear latex, it is hot, bitter, pungent, blood-purifying. (Chakrapaani equated Munditikaa with Shraavani and Brihat Munditikaa with Mahaa-Shraavani.)

The whole plant of Mundi is collected, dried in the shade and used as a dry herb in the form of a powder, infusion, syrup or distillate.

Among preparations available over-the-counter, Araq-e-Mundi (Bayaz-e-Kabir) is prescribed for putrefaction of blood, asthenopia; Araq-e-Musaffi-e-Khoon Qawi is for acne, pimples, boils, scabies, pruritus, white or black scales of skin; Majoon-e-Mundi (Qarabadeen-e-Azam-o-Akmal) is a blood-purifying confection.

In Unani medicine, Mundi is considered to possess the properties of Tricholepis glaberrima DC. and Tephrosia purpurea (L.) Pers (Brahmadandi and Sarphonkaa). These plants can be used as a substitute for Mundi. (CCRUM.)

Active principles and pharmacology
Essential oil contains methylchavicol, alpha-ionone, D-cadinene and p-methoxycinnamaldehyde as major constituents.

Herb also yields the alkaloid sphacranthine. Beta-sitosterol, stigmasterol, beta-sitosterol-beta-D-glucoside and hentriacontane were obtained from the capitulum comprising the inflorescence.

The drug (consisting capitula (inflorescence) of the plant) contained alkaloids 0.91 %, glycosides 1.97 %, tannins 1.01 %, reducing sugars 19.65 %, volatile oil 0.25 %.

The juice of the plant is styptic. The essential oil is active against Vibrio cholera and Micrococcus pyogenes var. aureus.

The Wealth of India included Hapushaa (Havushaa) among synonyms of Mundi (Sphaeranthus indicus). Arya Vaidya Sala, Kottakkal (Kerala), and IMPCOPS, Chennai (Tamil Nadu), also equate Rakta Hapushaa with the red-flowered Sphaeranthus indicus and the white-flowered variety with Solanum africanus.

Like Mundi, the source of the drug Hapushaa has not been identified properly. Many authorities (PV Sharma, KM Nadkarni, NAA, INSA) equate the drug with Juniperus communis Linn. of Cupressaceae, a dense shrub found in the Himalayas at altitudes of 1600–4200 m. The larger shrub is identified as Juniperus macropoda Boiss. Bhaavaprakaasha mentioned smaller and larger varieties of Hapushaa, but these referred to the size of fruits—Ashwatha phala and Pratham phala. Ashwatha phala possessed an aroma of Ficus religiosa fruit, while Pratham phala emitted a fish-odour.

The berry of Juniper is used in medicine for cutaneous diseases and is credited with antigonorrhoeic, antileucorrhoeic, antidropsical, diuretic and carminative properties. (CIMAP.)

Juniper berry contains a volatile oil (chief components—monoterpene hydrocarbons), diterpenes, catechin tannins, flavonoids, inverted sugar. It is strongly antiseptic within the urinary tract and helps relieve fluid retention (contraindicated in kidney diseases). In Western herbal, Oil of Cade, produced from Juniperus oxycedrus is applied to treat skin rashes.

In Bhaavaprakaasha, Hapushaa was incorporated in prescriptions for dysuria, polyuria and calculus affections of the bladder.

This analysis should prove that Hapushaa and Mundi are two different drugs and should not be treated as synonyms.

Loganiaceae

STRYCHNOS

Strychnos nux-vomica Linn.

Habitat

Throughout tropical India up to an altitude of 360 m. The tree occurs to a considerable extent in Uttar Pradesh, Bihar, Orissa, the Coromandel coast, Andhra Pradesh, Mysore, and is common in the monsoonal forests along with Western Coast.

Classical and common names

Ayurvedic: Kuchalaa (Trade name). Classical synonyms quoted in Ayurvedic reference books: Vishamushtikaa, Vishamushti, Vishatinduka, Kapilu, Kuchilaa.
Unani: Azaaraaqi, Kuchlaa.
Siddha: Etti.
English: Nux-vomica.

Figure 1 Strychnos nux-vomica [ZANDU]

Parts used

Detoxified seeds.

Dose

60–250 mg.

Classical use

Vishamushtikaa (Sushruta Samhitaa), Vishamushti (Shaarangadhara Samhitaa), Vishatinduka (Bhaavaprakaasha), Kapilu, Kuchalla, Kaaraskara (Raaja Nighantu) were synonyms of a poisonous fruit of Tinduka species (Diospyros montana Roxb.) during the classical period. During the medieval period, these were equated with Mahaanimba or Bakaayana (Melia azedarach).

Sushruta prescribed Vishmushtikaa internally for dyspepsia, catarrh, cough and asthma. Vishamushti of Shaarangadhara Samhitaa was Kuchal or Kuchelaa, incorporated in Agnitunda Vati for indigestion and abdominal colic.

Vishamushtikaa of Bhaavaprakaasha was used as a drug belonging to the Surasaadi Group (Ocimum spp., Lamiaceae) of Sushruta for cough, bronchitis, asthma, parasitic infestations; Vishatinduka or Kuchilaa for chronic skin diseases.

It is a wrong presumption that Nux-vomica was exported to Europe from South India and Ceylon since "ancient times".

In Europe, Strychnos was considered a synonym of Atropa for a long time. A German scientist brought to light the properties of Nux-vomica during the year 1540. The drug appeared in drug stores in London only during 1940. The drug was used as a poison for killing cats, dogs and crows.

John Henry Clarke wrote in 1901 in (Homoeopathic) Practical Materia Medica: "The wood of the tree (of Strychnos nux-vomica) is used in India in intermittent fever. A decoction of leaves is used externally in rheumatism. Under Brucea antidysenterica, the bark of the tree was imported into Europe by mistake."

Strychnos colubrina Linn., known as Kuchalaa creeper or Kaajarbel, is equated with Anjanki of Sushruta Samhitaa in Ayurvedic reference books. Anjanki was prescribed for biliary obstructions as a purgative and anthelmintic.

Strychnos ignatii Bergius (seeds contain strychnine) was used in Unani medicine. It was known by its Arabic, Persian, and Spanish name Papitaa. Habb-e-Papitaa Vilaayati of Unani medicine was prescribed for cholera; Habb-e-Papitaa Desi for indigestion. Strychnos ignatii and Carica papaya are totally different drugs, though both were known as Papitaa; now suffixed as Vilaayati and Desi by CCRUM.

Among over-the-counter drugs, containing Strychnos nux-vomica, Agnitundi Vati is for dysentery, diarrhoea and deranged digestion; Vaatagajaankusha Rasa for paraplegia and hemiplegia; Vishamushti Taila (Bhaishajya Ratnaavali) for external application in inflammations; Krimimudgara Rasa (Vaidya Chintaamani) for round worms and thread worms; Vishatinduka Vati (Zandu) for sexual debility.

Among-over-counter Unani drugs, Habb-e-Azaaraaqi, Davaa-e-Azaaraaqi, Majoon-e-Azaaraaqi (Bayaz-e-Kabir) are for hemiplegia, Bells Palsy and tremor; Raughan-e-Azaaraaqi (Qarabadeen-e-Kabir) for massage in rheumatic affection.

Strychnos nux-vomica, Allium sativum and Holoptelea integrifolia bark extracts in oil, known as Vida Muttith Thhailam of Siddha medicine, is prescribed for massage in rheumatic diseases.

For curing, Strychnos nux-vomica seeds are cut into four pieces and boiled into the decoction of Amaranthus roots. After drying in the sun, the seeds are rubbed with purified butter. Seeds are also boiled in milk or in a mixture of milk and water. Cotyledons are scraped apart and the embryo is removed. They are boiled again. The process of boiling exerts a mitigating effect on Nux-vomica.

Nux-vomica is always prescribed with sufficient quantity of milk, butter or purified butter.

All preparations of Nux-vomica are prescribed under medical supervision.

Active principles and pharmacology

Plant is rich in alkaloids. Rootbark and seeds contain the alkaloids brucine, strychnine, pseudobrucine (3-hydroxy-brucine), pseudostrychnine (3-hydroxystrychnine), beta-colubrine, 3-hydroxy-beta-colubrine (vomicine) and iso-strychnine (novacine).

Seeds also contain alpha-colubrine, 3-hydroxy-alpha-colubrine (icajine) and 15-hydroxystrychnine. Cycloartenylpalmitate, 16-hydroxy-alpha-colubrine, 16-hydroxy-beta-colubrine and 3-methoxyicajine have also been reported from seeds.

Brucine and strychnine are also present in stem-bark and leaves.

Strychnos potatorum Linn.

Habitat

In the deciduous forests of West Bengal, Central and South India, up to 1200 m.

Classical and common names

Ayurvedic: Kataka, Payah-prasaadi, Chakshushya; Nirmali.
Siddha: Thettran, Thetrankottai.
English: Clearing Nut.

Parts used

Seeds, root.

Dose

Seed powder 1–3 g, root decoction 50–100 ml.

Classical use

Kataka seeds, pounded with buttermilk, were prescribed with honey for alleviating polyuria (Yoga Ratnaakara). Powder of the seed, with milk, was given internally in diseases of genitourinary tract.

Seeds were used to clarify foul and muddy water. They were sliced and rubbed around the sides of unglazed earthern vessels in which water was stored.

Half to one full seed, rubbed up into a fine paste with some buttermilk was given internally for chronic diarrhoea.

Powdered seed, mixed with honey, was applied to boils to hasten suppuration. Pulp of the fruit was used as a substitute for ipecac in the treatment of dysentery and bronchitis. The fruit was regarded as an emetic.

Active principles and pharmacology

Seeds, leaves and trunk-bark gave diaboline and acetyldiaboline. Seeds also gave brucine, strychnine, novacine, icajine, oleanolic acid, 3-beta-acetoxyloleanolic-acid and oleanolic acid glycoside. Leaves and bark gave isomotiol, stigmasterol, campesterol and sitosterol.

Albumin, present in the seed, acts as a mechanical precipitant of suspended matter present in the water. Allum aided by ripe seeds has been found to be very effective in removing suspended impurities from coal-washery wastes. The clarification is due to the combined action of colloids and alkaloids in the seed.

The seeds resemble those of nux-vomica and, though non-poisonous, are used as adulterants of the latter. (The Wealth of India.)

Gentianaceae

SWERTIA

Swertia chirata (Wall.) C. B. Clarke

See figure 1.

Figure 1 Swertia chirata—flowering branch [WOI]

Swertia chirayita (Roxb. ex Flem.) Karst.

Habitat

Found in the temperate Himalayas at altitudes of 1200–3000 m from Kashmir to Bhutan, and in the Khasi hills; in Meghalaya at 1200–1500 m.

Classical and common names

Ayurvedic: Kiraatatikta, Kiraatatiktaka, Katutikta, Trnanimba (Charaka); Kiraata, Bhuunimba.
Unani: Chiraitaa.
Siddha: Nilavembu.
English: Chiretta, Chirata.

Parts used

Whole plant.

Dose

Powder 1–3 g, infusion 50–100 ml.

Classical use

Charaka and Sushruta prescribed the entire plant as a paste or decoction, or cooked as a potherb, internally, for purification of vitiated blood, chronic skin diseases, poisoning, oedema, fevers, cough, intrinsic haemorrhage and affections of the urinary tract.

Charaka gave a decoction of Kiraatatikta for purifying breast milk. Kiraatyaadi Churna and Bhuunimbaadi Churna of Charaka were specific compounds for sprue; Bhuunimbaadi Kashaaya of Gadanigraha for eruptive boils.

Hot infusion of Kiraatatikta, mixed with Dhaanyaka (Coriandrum sativum) leaves was administered for alleviating fever; a paste of Kiraatatikta and Shunthi (Zingiber officinale) followed by a decoction of Punarnavaa (Boerhavia diffusa) for generalised oedema. (Siddha-bheshaja-manimaalaa, Vrindamaadhava.)

In folk medicine, Chiraitaa is a household drug for chronic fevers, anaemia, gouty affections and boils. Its decoction is administered for liver disorders, gastrointestinal disorders, dyspepsia, anorexia. It is used as a tonic to prevent malaria.

Chiraitaa is the principal ingredient in Sudarshana Churna (Shaarangadhara Samhitaa), available over-the-counter for intermittent and malarial fevers. Other over-the-counter classical drugs include: Chinnodbhavaadi Kwaatha Churna (Sahasrayoga), digestive, carminative, antiperiodic, antiseptic and expectorant, prescribed as an adjuvant in intermittent fevers; Manjishthaadi Kwaath Churna (Shaarangadhara Samhitaa), prescribed as a blood purifier in chronic skin diseases.

Chiraitaa talkh (bitter) is an ingredient in Majoon Musaffi-e-Khoon (Bayaz-e-Kabir), prescribed in putrefaction of blood.

Chiraitaa shireen (sweet, equated with Swertia angustifolia Ham.) is an ingredient in Majoon-e-Masik-ul-Baul, prescribed in polyuria, bed-wetting in children and spermatorrhoea. Chiraitaa shireen is also an ingredient in Jawarish-e-Jalinoos, a reputed Unani tonic for liver.

Figure 2 Swertia angustifolia—flowering branch [WOI]

Active principles and pharmacology

Whole plant gave the alkaloids gentianine, gentiocrucine and enicoflavone. Aerial parts gave nine tetraoxygenated xanthones, including swerchirin and mangiferin. Roots also contained xanthones but the content of mangiferin was found to be lower.

Swertia angustifolia, used as a substitute for Swertia chirata or true Chiretta, gave ursolic acid, xanthones and beta-sitosterol.

Chloroform extract of the root of Swertia chirata exhibited neuromuscular blocking activity on rat phrenic nerve diaphragm.

The experiments carried out showed definite antispasmodic activity.

Total xanthones (extract), emulsified in gum tragacanth, exhibited 42.85 % and 54.80 % inhibition on sub-acute models of inflammations—cotton pellet implantation and granuloma pouch, respectively (in 50 mg/kg p.o. dose).

Swerchirin, a xanthone from Swertia chirata, exhibited potent antimalarial property against P. berghii infected rats, on oral as well as subcutaneous administration compared to primaquine-treated and untreated control (infected) rats. Swerchirin at both oral and subcutaneous dose levels was found effective even at one-fifth standard dose of primaquine. Mangiferin, however, was found to have no antimalarial property in 15 mg/kg/oral/ subcutaneous dose.

Extracts of Chirata also exhibited antihepatotoxic, anticholinergic, antiulcerogenic and hypoglycaemic activities in animal models. (RRL Jammu.)

Pet.-ether extract of Chirata showed anthelmintic activity (50 to 60 %), but chloroform, ethanol and water extracts turned out to be ineffective.

Chirata possesses the property of a bitter tonic, but, unlike most other bitters, it does not constipate the bowels. Instead, it tends to produce a regular action and causes a free discharge of bile.

Symplocaceae

SYMPLOCOS

Symplocos racemosa Roxb.

Habitat

Abundant in the plains and lower hills throughout North and East India, ascending in the Himalayas up to an elevation of 1400 m.

Classical and common names

Ayurvedic: Lodhra, Rodhra, Shaavara.
Unani: Lodh Pathaani.
Siddha: Velli-lethi.
English: Sweetleaf, Sapphire Berry.

Parts used

Bark.

Dose

Powder 3–5 g, decoction 50–100 ml.

Classical use

Charaka and Sushruta gave dried rootbark, in internal and external prescriptions, for haemothermia, menstrual and urinary disorders, obstinate skin diseases, persistent dysentery, acute constipation, piles, abdominal inflammation, intestinal paralysis, dropsy.

For leucorrhoea, Charaka prescribed paste of Lodhra bark with the decoction of Nyagrodha (Ficus benghalensis) bark.

Sushruta used alkali of the bark as styptic; paste of the bark for swellings; powder of the bark for quick healing of wounds.

For pimples, a paste of Lodhra, Dhaanyaka (Coriandrum sativum) and Vachaa (Acorus calamus) was prescribed (Vrindamaadhava).

Lodhra bark and Alaabu (Lagenaria siceraria) leaves, in equal parts, were pounded and the paste was applied to the vagina for keeping it firm and normal (Bhaavaprakaasha).

Lodhraasava (Ashtaanga Hridaya), available over-the-counter, is a widely prescribed tonic for gynaecological disorders, especially leucorrhoea, menorrhagia.

Lodhra is an ingredient in Pushyaanuga Churna (Bhaishajya Ratnaavali), extensively used in leucorrhoea, menorrhagia, dysmenorrhoea and other uterine disorders.

Active principles and pharmacology

Betulinic acid, oleanolic acid, acetyl oleanolic acid and ellagic acid are reported from the plant. Stem-bark gave proanthocyanidin-3-monoglucofuranoside of 7-O-methyl-leucopelargonidin. A new flavan glycoside, symposide, has been isolated from stem bark.

Alpha-spinasterol, isolated from the plant, was found to have highly significant anti-inflammatory property in carrageenin-induced acute rat paw oedema in doses of 10 and 25 mg/kg i.p. The effect was observed 57.9 % at 25 mg/kg dose. The drug was found more potent than phenylbutazone, but less potent than betamethasone and metyrapone.

Glycosides, isolated from the plant, produced marked relaxant effect on dog's ileum in situ equipotent with the relaxation effect of adrenaline (CCRAS.)

In a pharmacological study, a crystalline fraction from the bark was found to reduce the frequency and intensity of contractions *in vitro* of both pregnant and non-pregnant uteri of animals; another fraction from the bark, besides showing action on uteri, was spasmogenic on the various parts of the gastrointestinal tract and could be antagonized by atropine.

Glycoside 3-monoglucofuranoside of 7-O-methyl-leucopelargonidin is highly astringent and is reported to be responsible for the medicinal properties of the bark.

Symposide and (–)-epiafzelechin showed anti-fibrinolytic activity.

A crystaline fraction from the bark was found to inhibit the growth of Micrococcus pyogenes var. aureus, E. coli and enteric and dysenteric group of organisms.

Lodhra (Shaavara Lodhra of classical texts) is also equated with Symplocos paniculata (Thumb.) Miq. syn. Symplocos crataegoides Buch.-Ham. ex D. Don. The water-soluble fraction from the bark has been reported to exhibit anti-oxytocic activity. Ethanolic extract of the leaves, also of stems, affected blood pressure in dogs and cats and showed activity on the ileum of guinea-pigs.

Ethanolic extract of the leaves showed activity against Entamoeba histolytica strain STA, Ascaridia galli and Ranikhet-disease virus.

T

Asteraceae

TAGETES

Tagetes erecta Linn.

Habitat
Native to Mexico. Extensively cultivated in gardens all over India for its flowers.

Classical and common names
Ayurvedic: Jhandu; Gendaa.
Unani: Sadbarg, Gul-hazaaraa, Gul-jaafari.
English: Aztec or African Marigold.

Parts used
Leaves, seeds, florets.

Dose
Juice 10–20 ml, paste 5–10 g.

Classical use
A paste of the leaves of Jhandu, Nimbaa (Azadirachta indica) and Mahaa-Nimbaa (Melia azedarach), taken internally, alleviates bleeding piles. Tender leaves of Jhandu plant pounded with Maricha (Piper nigrum) are also administered internally for bleeding piles, venereal sores and discharges. (Siddha-bheshaja-manimaalaa.)

In Unani medicine, a confection of tender leaves and purified sugar is prescribed in anuria, retention of urine and kidney troubles.

An aqueous extract drawn from soaked leaves of Tagetes erecta and the root of Kelaa (Musa paradisiaca) is given for bleeding piles.

Leaves, pounded with black pepper, are prescribed, internally as well as externally, for insect bites.

In folk medicine, an infusion of the plant is used against cold, bronchitis and rheumatism. The leaves and florets are used as emmenagogue.

Dried and powdered seeds of Tagetes erecta are prescribed for attenuated of semen.

Active principles and pharmacology
Flowers gave lutein esters of dipalmitate, dimyristate and monomyristate. Fresh petals gave hydroxyflavonols, quercetagetin and tagetiin.

Plant yields an essential oil with strong, sweet and lasting odour. The major components of the oil are d-limonene, ocimene, L-linalyl acetate, L-linalool, tagetone and n-nonylaldehyde.

Quercetagetin and quercetagetrin have been isolated from the Indian variety and kaempferitrin and helenien from Rumanian varieties. Several carotenoids have been identified in the flowers from Surrey (UK) and from Florida (USA).

Seeds gave positive tests for alkaloids. The roots contain a number of bithienyls and a little terthienyl.

The aqueous extract of flowers showed activity against Gram-positive bacteria.

Tender leaves, as well as flowers, exhibit styptic, anti-inflammatory, blood purifying, antiseptic and diuretic properties.

Use in Western herbal
Calendula officinalis Linn., a related species of Tagetes erecta, is used in Western herbal.

Nicholas Culpeper (1616–1654) wrote of Marigolds: "The juice of Marigold leaves mixed with vinegar, and any hot swelling bathed with it, intensely gives ease and assuage it. The flowers, either green or dried, are much used in possets, broths and drink, as a comforter of the heart and spirits, and to expel any malignant or pestilential quality which might annoy them."

In the West, Marigold is a popular remedy for minor skin problems, cuts, grazes and wounds; minor burns, sunburn; acne, thrush, nappy rash, inflamed nipples.

Marigold is also used externally for varicosis, phlebitis, thrombophlebitis; dry dermatosis, eczema and ringworm.

An infusion or tincture of Marigold is used internally for gastritis, colitis, peptic ulcers, regional ileitis. It helps treat toxicity responsible for many fevers and infections and acts as a cleansing agent for the liver and gallbladder.

Marigold is also used as a mild estrogenic drug for reducing period pain and for regulating menstrual bleeding. An infusion of Marigold is used as a douche for vaginal thrush.

German Commission E recognized the efficacy of Calendula flower in the following areas:

- Inflammation of the oral and pharyngeal mucosa
- (Externally for) Poor healing wounds
- Ulcus cruris.

Over-the-counter, Marigold largely features in external applications.

Caesalpiniaceae

TAMARINDUS

Tamarindus indica Linn.

Habitat
Found naturalised or cultivated almost throughout the plains and sub-Himalayan tracts of India; particularly in the South.

Classical and common names
Ayurvedic: Amlikaa, Suktaa (Charaka, Sushruta); Amli, Chukraa, Chukrikaa, Chinchaa, Chandikaa.
Unani: Tamar Hindi.
Siddha: Puli, Amilan.
English: Tamarind Tree.

Figure 1 Tamarindus indica—flowering branch and fruit [WOI]

Parts used
Fruit, seed, leaf, flower, ash.

Classical use
According to Ayurved Saukhyam, unripe fruit is highly acidic; ripe fruit is a laxative, appetiser, digestive stimulant, cleanses urinary bladder; dried fruit pulp is a cardiac tonic, cures exhaustion, giddiness, mental fatigue and morbid thirst.

Amlikaa, Muurvaa (Marsdenia tenacissima), Chitraka (Plumbago zeylanica), Soorana (Amorphophallus) and Saindhava (sea-salt)—these drugs taken together were called Panch Hutaashnaa (Panchaagni). This group was prescribed for treating piles, diseases of the liver and deranged digestion.

Pulp of ripe fruit was prescribed as adjunct to other purgatives. It was used with dates, figs, cloves, cardamoms in atony of the liver, stomach and intestines. Charaka and Sushruta used the fruit as stomachic, astringent, appetizer, digestive; in prescriptions for diarrhoea, dysentery, and oedema.

A decoction of Amlikaa root was prescribed for alleviating sprue syndrome and piles; also alcoholism (Ayurveda Saukhyam).

A paste of the rootbark was applied topically for removing freckles (Ashtaanga Sangraha).

Ash of the bark, with common salt, was prescribed for internal obstructions, colic, indigestion (Bhaavaprakaasha).

Ash was also used as a gargle for sore throat and aphthous ulcers.

Fire of bark was used for fomentation of stiff joints and inflammations due to gout and arthritis. (Vaidyamanoramaa).

Seed kernel, pounded with milk, was prescribed for polyuria and urethral discharge (Vaidaymanoramaa).

Paste of fried seeds was applied on anus, after setting the tract in position, for treating rectal prolapse (Siddha-bheshaja-manimaalaa).

Powder of outer shell of seeds, Shunthi (dry Zingiber officinale) and rock-salt was prescribed with buttermilk for diarrhoea and dysentery.

Boiled seeds were used as a poultice to boils. Ash of burnt skin of the ripe fruit was used as an alkaline, in prescriptions, for enlargement of spleen.

Amlikaa leaves, cooked as vegetables, were prescribed for bleeding piles (Charaka Samhitaa). Leaves, crushed with water and expressed, were used in bilious fever and scalding of urine. Juice of leaves, warmed by dipping a red hot iron, was given in dysentery. A decoction of leaves was used as a gargle in throat infections; also as a wash for indolent ulcers.

Poultice of leaves was applied over rheumatic affections. Hot decoction was used for fomentation. Paste of leaves and Haridraa (Curcuma longa) was prescribed as a prophylactic for smallpox (Chakradatta). Poultice of flowers was used in conjunctivitis.

Juice of leaves was applied topically for treating ringworm (Vaidyamanoramaa). Juice expressed from flowers was prescribed for bleeding piles.

In Unani medicine, seeds are roasted and the kernel thus recovered is used in spermatorrhoea, nocturnal emissions and seminal debility. For treating atony of vagina, a pessary prepared from kernel of seeds is prescribed.

Active principles and pharmacology

Leaves gave the flavone C-glycosides orientin, vitexin, iso-orientin and iso-vitexin. Leaves and fruits gave tartaric acid and malic acid.

Fruit pulp yielded the amino acids serine, beta-alanine, proline, pipecolinic acid, phenylalanine and leucine.

Seed oil consisted of lauric, palmitic, myristic, stearic, arachidic, behenic, oleic, linoleic and lignoceric acids.

A new bitter principle, tamarindienal, has been isolated from fruit pulp. It showed fungicidal activity against Aspergillus niger and Candida albicans, and strong bactericidal activity against Bacillus subtilis, Staphylococcus aureus, E. coli and Pseudomonas aeruginosa.

A polysaccharide isolated from fruit pulp showed immunomodulatory activities, such as phagocytic enhancement, leukocyte migration inhibition and inhibition of lymphocyte proliferation.

From seeds, 2-hydroxy-3', 4'-dihydroxyacetophenone (TAO), methyl-3,4-dihydroxybenzoate (TA1), 3,4-dihydoxyphenylacetate (TA2) and (-)-epicatechin have been isolated. These active principles exhibited strong antioxidant activity in linoleic acid auto-oxidation system.

Ethanolic extract of seed coat also exhibited potent antioxidant activity.

The chemical utilization of the pulp has been aimed at the isolation of tartaric acid. Fruit contains 3–10 % tartaric acid.

The fruit acts as an aperient by increasing the volume of liquid in the large intestines.

Vitamin C was formerly believed to be among the constituents of tamarind, but this is now disputed.

Tamaricaceae

TAMARIX

Tamarix aphylla (Linn.) Karst.
Tamarix dioica Roxb.

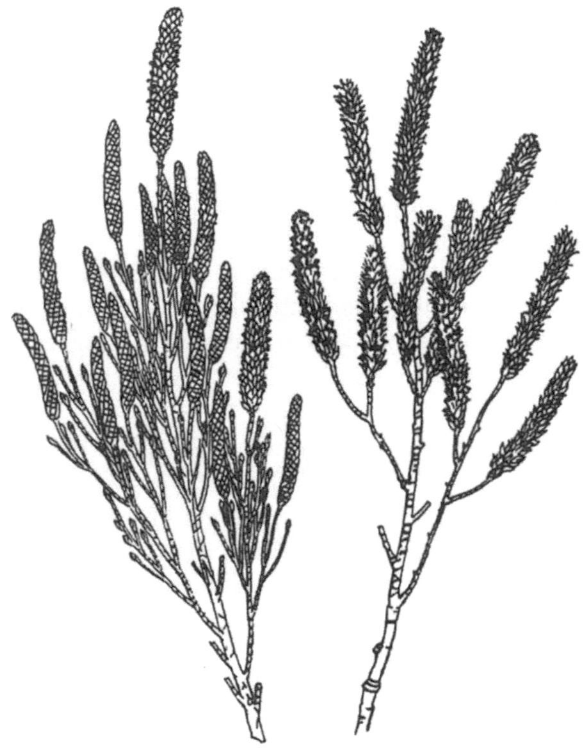

Figure 1 Tamarix dioica—flowering branches [WOI]

Tamarix gallica Linn.
Tamarix indica Willd.

Habitat

Found in North India in sandy or gravelly areas and low-lying saline soils and on the bank of rivers.

Classical and common names

Ayurvedic: Jhavuka, Shaavaka, Bahugranthikaa.
Unani: Maayeen Kalaan (large-galls), Maayeen khurd (small galls).
English: Takut Galls.

Parts used

Root, leaves, galls.

Dose

Decoction 50–100 ml, powder 1–3 g.

Classical use

Maayeen Kalaan and Mayeen khurd are used in Unani medicine like Oak galls (Maajuphala). A powder or decoction is administered in watery fluxes or chronic discharges.

A strong infusion is used as a gargle in sore-throat and stomatitis. Powdered galls, in the form of an ointment, are applied to ulcerating piles and fissures; also used in the form of pessaries for atony of the vagina.

An infusion of the bark or galls is usually combined with Chiretta for diarrhoea and dysentery.

Tamarix Manna is used as a mild expectorant and laxative.

Maayeen Khurd is an ingredient in Habb-e-Pechis, prescribed for diarrhoea, dysentery and haemorrhages. Sunoon-e-Muluk is to be applied in toothache and oral sepsis.

Maayeen Kalaan is an ingredient in Saffof-e-Habis-ud-Dam and Safoof-e-Habb-ur-Rumman, prescribed in haemorrhagic diarrhoea, polymenorrhagia and haemorrhages.

Active principles and pharmacology

Galls of Tamarix aphylla, syn. Tamarix orientalis Forsk. contain the polyphenols gallic acid, ellagic acid, dehydrodigallic acid, dihydrojuglone-5-glucoside, isoferulic acid and optically active corilagin isomer, juglanin; the flavonoids quercetin, its glucoside, isoquercitrin, its methyl derivative, tamarixetin, and tamarixin.

Leaves contain the polyphenols tamarixin, isoquercitrin- and isoferulic acid and its sodium salt.

The glucoside tamarixin, present in the leaves of Tamarix gallica, has been characterised as 4'-methylether of quercetin.

Fresh leaves of Tamarix dioica gave trans-2-hydroxy-4-methoxycinnamic-acid and iso-rhamnetin.

The bigger variety of galls contain 45 % of tannin; the smaller variety 50 %.

Several species of Tamarix produce Manna; these include Tamarix aphylla, Tamarix dioica, Tamarix gallica.

Asterceae

TARAXACUM

Taraxacum officinale Weber ex Wiggers

Habitat

Temperate Himalayas and the Khasi Hills of Meghalaya, Mishmi Hills of Arunachal Pradesh and hills of South India at 3000–5000 m. Also occurs in Gujarat.

Classical and common names

Unani: Kanaful, Kaasani Dashti, Kaasani Sahraayi, Hindbaa-al-Barri.
English: Dandelion.

Parts used

Root.

Dose

Powder 500 mg–2 g.

Classical use

Kaasani Dashti entered into Unani medicine during the 16th century, as a hepatic stimulant. It was prescribed in chronic disorders of liver, and visceral diseases. It was also used as a mild cholagogue, diaphoretic, diuretic and in chronic skin diseases.

Powdered root in doses of 600–900 mg is considered efficacious as a hepatic stimulant.

Active principles and pharmacology

The roots afforded triterpenes and steroids; 4-alpha-15,11-beta-13-tetrahydroridentin B,1,3-epimer of 4-alpha-15-dihydroerivanin, si-taraxasterolacetate, beta-sitosterol and its beta-D-glucopyranoside, taraxacolide-1'-O-beta-D-glucopyranoside, 11,13-dihydrotaraxinacid-1'-O-beta-D-glucopyranoside and taraxin acid-1'-O-beta-D-glucopyranoside.

Rhizomes are used as a substitute for chicory (Cichorium intybus).

Dandelion leaves are an extremely rich source of potassium. When the whole plant is used as a diuretic, body's natural balance is maintained. The leaves are credited with antiscorbutic properties.

An analysis of leaves and edible flowers from Pakistan gave the following values: moisture 88.8, protein 3.6, ether extr. 1.6, total carbohydrates 3.7, fibre 0.44 and ash 2.3 %; phosphorus 59.1, calcium 473.5, iron 3.3, and vitamin C 73 mg, vitamin A value 6700 I.U./100 g. Other vitamins reported in the leaves are: thiamine 0.19, riboflavin 0.14 and niacin 0.8 mg/100 g. Pectic substances (7.81 % as calcium pectate, dry basis), starch and alpha-amylase are also present.

Use in Western herbal

Dandelion entered into Western herbal after the 15th century.

German Commission E recognized the efficacy of dried leaves and roots of Taraxacum officinale in the following areas:

- Dyspeptic complaints
- Liver and gallbladder complaints
- Disturbances of bile flow.

Dandelion is mainly included in over-the-counter remedies for kidney and liver problems. It is also used as a cleansing remedy in rheumatism and skin disorders. A diabetic infusion is made from the root and leaves.

Dandelion capsules (250 mg and 520 mg) are sold over-the-counter.

Caution

Contraindicated in obstruction of bile duct. In case of gallstones, should be used only after consultation with a physician. The drug is included in many over-the-counter slimming preparations as a diuretic and mild laxative. These products should not be used for long in an attempt to lose body fluids for losing weight.

Taxaceae

TAXUS

Taxus baccata Linn.

Figure 1 Taxus baccata [WOI]

Habitat

Found in the Himalayas at altitudes of 1800–2300 m, and in the hills of Meghalaya and Manipur at an altitude of 1500 m. In the Sikkim Himalayas, it accompanies Silver Fir, Hemlock-spruce and Rhododendron spp.

Classical and common names

Ayurvedic: Thunera. Sthauneya and Sthauneyaka (Charaka, Sushruta, Bhaavaprakaasha) are equated with Taxus baccata by CCRAS. INSA scientists have equated Sthauneyaka of Charaka and Sushruta with Clerodendrum infortunatum Linn. (Glory Tree, equated with Bhandira by CCRAS). Taalisapatra, Brahmi, Birmi, Jarnab are misleading synonyms of Sthauneyaka.
English: Yew.

Classical use

Charaka prescribed a decoction of Sthauneyaka bark during fever; Sushruta in skin diseases and diseases due to vitiated blood.

Agurvaadya Taila of Charaka was prescribed in fever with cold. Mrtasanjivana Agada, Taarkshya Agada and Mahaasugandhi Agada were prescribed for rheumatic affections.

During the 16th century, Gada, a synonym of Kushtha (Saussurea lappa), was used as a substitute for Sthauneyaka. Kushtha was used as a substitute for Pushkara Muula (Inula racemosa); also for Tagara (Valeriana spp.)

On the basis of classical usage, it is difficult to equate Sthauneya or Sthauneyaka of Ayurvedic medicine with Taxus baccata. Shauneyaka was used mostly in massage oils (Mahaanaaraayana Taila and Mahaasugandhi Taila of Bhaavaprakaasha) and was not included in internal prescriptions. (In Shaarangadhara Samhitaa and Bhaavaprakaasha, Taalisa or Taalisapatra was incorporated in Vyosaadi Vati, Jaatiphalaadi Churna, Sudarshana Churna, Lavanbhaaskara Churna.)

Inclusion of Sthauneyaka in Elaadi (Elettaria) group of aromatic herbs by Sushruta raises further doubts about its equation with Taxus baccata.

According to the Wealth of India, all parts of Taxus baccata are poisonous due to Taxin, except fleshy aril. Non-poisonous and fleshy aril is eaten by the tribes and is used for its carminative, stomachic and expectorant properties.

It is possible that instead of needles, dried aril of a particular species of Taxus was used in medicinal oils during the period of Charaka, Sushruta and Bhaavaprakaasha. It is also possible that during collection dried leaves of Taxus baccata and Rhododendron spp. get mixed up (Taxus baccata, Silver Fir, and Rhododendron spp. grow together.)

In Europe, poisonous properties of yew have been recognized and disputed from ancient times. Gerarde claimed that he ate berries of yew without any ill effect, while a patient in an asylum died after eating yew leaves (John Henry Clarke).

In any case, the use of Taxus baccata in Indian medicine is still disputed.

Active principles and pharmacology

Yew leaves contain diterpene esters of taxane type (taxine 0.6–2.0 %), including, among others, taxine A, taxine B, taxol; flavonoids, including, among others, sciadopytisin, ginkgetin, sequoria flavone. (PDR.)

The pigments rhodoxanthin, lycopene, betacarotone and zeaxanthin are present in the fruit.

The presence of eschescholtzxanthone has been reported in the fresh, ripe, red fruit.

Taxol inhibits cell division and is being researched as a potential anticancer drug.

Yew is no longer used in the West as a herbal medicine. In homoeopathy a tincture of berries and fresh young shoots are used (up to third potency).

C. infortunatum: Flowers contain fumaric acid, beta-sitosterol and methyl esters of caffeic acid, a diterpenoid and a sterol. Leaves contain the flavonoids scutellarein-7-O-glucuronide and hispidulin-7-O-glucuronide. Leaf and root exhibit antitumour, vermifuge and febrifuge properties. (CIMAP.)

Rhododendron anthopogon D. Don (Ericaceae), known as Taalisfar in Kashmir, is found in the Himalayas at altitudes of 3000–4000 m. According to the Wealth of India, leaves are known as Talisri. Leaves gave ursolic acid and quercetin. Quercetin is also present in the bark.

Papilionaceae

TEPHROSIA

Figure 1 Tephrosia purpurea—fruiting branch [WOI]

Tephrosia purpurea (Linn.) Pers.

See figure 1.

Tephrosia hamiltonii Drumm.

Habitat

Grows abundantly in upper Gangetic plains, and Western Himalayas. Commonly grown as a green manure in paddy fields in India and tobacco and rubber plantations in other countries.

Classical and common names

Ayurvedic: Sharapunkhaa, Vishikha-punkhaa (Bhaavaprakaasha); Sarphokaa.
Unani: Sarphukaa.
Siddha: Mullukaivelai.
English: Purple Tephrosia.

Parts used

Root, seeds, ash.

Dose

Powder 3–5 g, ash 1–3 g.

Classical use

Sharapunkhaa root, chewed and swallowed, or a paste of the root with buttermilk was prescribed in splenomegaly. (Bhaavaprakaasha, Raaja Maarttanda, Vrindamaadhava, Chakradatta.)

Ash of Sharapunkhaa and Haritaki (Terminalia chebula), in equal quantity, in a dose of 2.5 g, was prescribed in hardening and swelling of the spleen.

Root of the white variety was prescribed as a fertility drug to barren women. It was supposed to check premature ejaculation in men (Bhaaavaprakaasha, Vaidyamanoramaa).

Kantapunkhaa was prescribed in intestinal colic and helminthiasis (Raaja Nighantu).

The white variety of Sharpunkhaa is equated with Tephrosia procumbens Buch-Ham., and Tephrosia candida DC; Kantapunkhaa of Raaja

Nighantu with Tephrosia spinosa Pers. Unani scholars equate the white variety with Tephrosia villosa Pers. (though the plant bears light pink or blue flowers.)

In Unani medicine, Sarphukaa is used as a blood-purifying herb, in diseases of the skin, vitiation of blood, ailments of liver and spleen.

Ash is also used as an internal blood-purifier.

A paste of the root is applied externally on enlargement of cervical gland, and piles.

Sarphukaa is an ingredient in Araq-e-Musaffi-e-Khoon Qawi (Bayaz-e-Kabir), a reputed blood-purifying tonic of Unani medicine.

Active principles and pharmacology

Roots contain the closely related flavonoids apollinine, semiglabrin, semiglabrinol, tephroglabrin, tepurindiol, pongamol, iso-lonchocarpin, O-methylpongamol, lanceolatins A and B.

Leaves contain a flavonoid, rutin; a triterpenoid, lupeol; and a sterol, beta-sitosterol.

Seeds contain a diketonepongamol; a dimethylchromene flavone, iso-lonchocarpin; the furanoflavones karanjin and kanjone; a flavanone, purpurin; and a sterol, si-stosterol. A flavonoid, lanceolatin-B, is also present in seeds.

Benzene extract of sodium hydroxide soluble fraction of the drug was found to possess analgesic effect both orally and intraperitoneally. Sodium carbonate soluble fraction showed encouraging anti-inflammatory effect on rat paw oedema. (Toxic by oral route at 20 mg/kg.)

The alcoholic extract showed poor antipyretic property only, without having analgesic and anti-inflammatory property in rats and mice.

Pet.-ether and acetone extracts exhibited anti-inflammatory property.

Aqueous extract of seeds (60 mg/kg) on administration in normal and alloxanised diabetic rats exhibited 60–70 % hypoglycaemic effect compared to tolbutamide (250 mg/kg). The extract improved liver function without any adverse effect. (CCRAS.)

Pet.-ether, ether and benzene extracts showed significant sleep potentiation in mice. (CCRAS.) Benzene extract of the plant in a dose of 100 mg/kg potentiated pentobarbitone sleeping time in mice.

Rutin (2.5 %), found in leaves and seeds, is indicative of the drug's use in reversing capillary fragility; also its application in hypertension.

Hepatoprotective effect of aerial parts of Tephrosia purpurea was evaluated against (+)-galactosamine-induced and carbon tetrachloride-induced hepatotoxicity in rats. An oral dose of powdered aerial parts (0.5 g/kg) to rats prevented elevation of SGOP, SGPT and bilirubin levels.

Combretaceae

TERMINALIA

Terminalia arjuna (Roxb. ex DC.) W. & A.

Figure 1 Terminalia arjuna [CCRAS]

Figure 2 Terminalia arjuna [CCRAS]

Habitat

Commonly found in Madhya Pradesh, Bihar and Peninsular India; grows as an avenue tree.

Classical and common names

Ayurvedic: Arjuna, Dhananjaya, Kaakubha, Kakubha, Aartagala (Charaka, Sushruta); Indravriksha, Paartha, Virataru, Viravriksha (Bhaavaprakaasha).
Siddha: Maruthu.

Parts used

Bark.

Dose

For decoction with milk 5–10 g, decoction 50–100 ml, powder 3–6 g.

Classical use

Charaka used bark and leaves in internal prescriptions for cardiac disorders, diarrhoea, intrinsic haemorrhage, piles, ulcers; Sushruta in vaginal discharges, migraine, internal abscesses, obesity, haemoptysis, jaundice, urinary calculi, chronic skin diseases.

Sushruta used powdered bark externally as styptic; also as an ingredient of a hair tonic for treating greying of hair.

According to Vrindamaadhava, those who take the powder of Arjuna bark, processed with milk, purified butter or jaggery water, are relieved of cardiac disorders, chronic fevers, intrinsic haemorrhage and attain longevity.

Powdered bark of Arjuna, impregnated with the juice of Vaasaa (Adhatoda vasica); also in combination with powdered Naagabalaa (Grewia hirsuta) and Kapikacchuu (Mucuna prurita seeds), was prescribed for alleviating consumption, cough, internal haemorrhaage. (Vrindamaadhava, Bangasena, Chakradatta, Bhaavaprakaasha.)

A decoction of Arjuna, with Chandana (Santalum album) or with Dhava (Anogeissus latifolia), was administered for spermaturia (Sushruta Samhitaa, Haarita Samhitaa). In diarrhoea associated with blood, Arjuna bark, mixed with honey, was prescribed (Vrindamaadhava). In freckles, a paste of Arjuna bark, mixed with honey, was applied topically (Ashtaanga Hridaya, Vrindamaadhava, Shaarangadhara Samhitaa).

Arjunaarishta (Bhaishajya Ratnaavali), prescribed as a cardiac and respiratory tract tonic, is available over-the-counter.

Active principles and pharmacology

Bark contains a triterpene, arjungenine; triterpene glucosides I, II, III and IV. Stembark gave the flavones baicalein and arjunolone characterised as 6, 4'-dihydroxy-7-methoxyflavone. A new triterpene, glucoside-arjunolitin, has also been isolated and elucidated.

Dry bark from the stem contains 20–24 % tannin and dry bark from lower branches 15–18 %. Tannin is of a mixed type, containing both pyrogallol and catechol tannin, similar to the English Oak bark.

Fruits contain 7–20 % tannin; gave an ester together with beta-sitosterol; friedelin. methyl-oleanolate, gallic acid, ellagic acid, arjunic acid; flavanones include arjunone.

Alcoholic extract of the bark of Arjuna exhibited hypotention and bradycardia in anaesthetised dogs.

Intra-ventricular and intra-arterial administration of alcoholic extract produced long-lasting hypotension and bradycardia, suggesting thereby its activity through CNS-neurons.

The peripheral component of hypotention and bradycardial effect can be used as an antianginal agent in patients with tachyarrhythmia and hypertension. (CCRAS.)

Both alcoholic and aqueous extracts of bark produce fall in blood pressure. These extracts significantly increased rate and amplitude of hypodynamic heart in lower doses. But higher doses exhibited marked depressant effect followed by diastolic arrest.

Alcoholic extract produced spasmogenic effect on rabbit and guinea-pig ileum, which was abolished by atropine pretreatment.

The aqueous extract of the bark exhibited no significant diuretic effect, but increased sodium excretion in experimental conditions.

Pharmacological studies with alcoholic extract exhibited its ineffectiveness on pentobarbitone hypnosis, pyrexia, analgesia, uterine contractions. (information based on inputs from CCRAS.)

Terminalia bellirica Roxb.

Habitat

Found in deciduous forests throughout the greater part of India, but not in arid regions.

Figure 3 Terminalia bellirica [CCRAS]

Figure 4 Terminalia bellirica [CCRAS]

Classical and common names

Ayurvedic: Bibhitaka, Vibhitaka, Vibhitaki, Vibhita, Aksha, Kaamaghna (Charaka, Sushruta); Kalidruma (Bhaavaprakaasha).
Unani: Balelaa, Baheraa.
Siddha: Thandri.
English: Belliric Myrobalan.

Parts used

Fruit (rind).

Dose

Powder 3–6 g.

Classical use

Charaka and Sushruta used the dried fruits (after removing the seed), alone or in prescriptions, for cough, irregular fever, cardiac affections, gradual loss of vision, deficient lactation, urinary diseases, chronic skin diseases and as a purgative. The oil of the kernel of seed was used as an ingredient of Sushruta's Mahaa-Nila Taila, prescribed for preventing premature balding and greying of hair. The oil was also administered in sinus affections.

Bibhitaka powder (10 g), with honey, was prescribed for alleviating cough, dyspnoea and asthma (Ashtaanga Hridaya, Raaja Maarttanda, Chakradatta). A paste of Bibhitaka, with rice-water was administered for burning sensation and pain due to oedema (Ashtaanga Sangraha).

In Ayurvedic medicine, unripe fruits are used as purgative, dried ripe fruits as astringent in diarrhoea, piles, fevers, sore throat, bronchitis. Kernel, or the oil expressed from kernel, is used externally in rheumatism.

Bibhitaka is a constituent of Triphala (See Terminalia chebula.)

Active principles and pharmacology

Fruits contain beta-sitosterol, gallic, ellagic acid, ethyl gallate, galloyl glucose, chebulagic acid and a cardiac glycoside, bellaricanin.

The fruit contains all the components of Terminalia chebula, except corilagin and chebulic acid. Fleshy fruit pulp contains 21.4 % tannin.

A new triterpene, belleric acid, along with its glucoside, bellericoside; arjungenin and its glucoside have been isolated.

Alcoholic extract of the fruit possesses bile-stimulating activity. Alcoholic extract 30 mg/kg does not affect blood pressure, but in higher doses of 60 mg/kg produces a fall in blood pressure.

The extract of Terminalia bellirica was found to possess antihistaminic effect on experimental asthma in guinea-pigs. The antiasthmatic effect was confirmed by pollen-induced asthma in animals. The extract was found to be CNS-stimulant in dogs. It also antagonised morphine depression of respiratory tract. (CCRAS.)

Terminalia bellirica fruit powder was administered to 137 cases of chronic cough and asthma. The drug exhibited bronchodilatory, antispasmodic, antiphlegmatic, expectorant and sedative activity (CCRAS, Chemexcil).

The cold extract of Terminalia bellirica was studied for its effect against M. pyogenes, Staph. aureus, E. coli, B. dysenteriae, B. typhosus, Vib. choleriae. It showed potent activity against S. aureus, E. coli and E. officinalis.

Terminalia chebula Retz.

Figure 5 Terminalia chebula [ZANDU]

Habitat
Throughout India. In Sub-Himalayan tracts from the Ravi eastward to West Bengal and Assam, ascending up to an altitude of 1500 m.

Classical and common names
Ayruvedic: Haritaki, (Charaka, Sushruta); Kaayasthaa, Shreyasi (Bhaavaprakaasha); Jivanti, Puutanaa, Vijayaa, Abhayaa, Rohini, Chetaki, Amritaa.
Unani: Harad; Halelaa siyaah, Halelaa zard, Halelaa kaabuli (varieties).
Siddha: Kadukkai.
English: Chebulic Myrobalan.

Parts used
Fruit.

Dose
Powder 3–5 g.

Classical use
Charaka and Sushruta used dried fruits, alone or in prescriptions, for indigestion, oedema, dermatosis, urinary diseases, uterine and vaginal disorders, irregular fevers; as an over-all tonic, blood-purifier, galactagogue, purgative. Sushruta prescribed a decoction of fruits with clarified butter for hysteric convulsions and epileptic fits; pasty mass of fruits, with honey or treacle, in blood poisoning. The fruits were included in tonics for improving eyesight.

In Ayurvedic texts, seven varieties (now treated as synonyms) have been mentioned: Jivanti is golden in colour; Puutanaa bears bigger stones (specific for external application); Vijayaa is longish; Abhayaa has five pieces (specific for diseases of bones); Rohini is round in shape (specific for ulcers); Chetaki has three pieces (used in the form of powder); Amritaa bears three fruits in a bunch (specific for cleansing the body).

Haritaki was prescribed during summer with equal quantity of jaggery; during rainy season with adequate quantity of rock-salt; during autumn with equal quantity of purified sugar; in the beginning of winter with Shunthi (dry Zingiber officinale); in the later part of winter with Pippali (Piper longum); in the spring season with honey.

Haritaki was contraindicated during pregnancy.

When chewed, Haritaki promotes digestive powder; when taken in paste form, it cleanses the bowels; when used after frying, it harmonizes body functions; when used by steam-boiling, it is constipative.

All the seven varieties are now equated with Terminalia chebula. In Kerala, two varieties of the drug, known as Kattukka and Karuvilla kattukaa, are recognized as Haritaki. The later being either tender or pathogenetic seedless fruits of the same species.

In Ayurvedic and Unani medicine, three varieties of Harad are used: seedless, dried raw fruits known as Javaa Harad, Baala Harad, Halelaa siyaah; yellow variety, known as Halelaa zard; and bigger variety, known as Halelaa kaabuli. In many compound preparations of Unani medicine all the three varieties are used together.

Important grades of Harad, recognized in the trade, are Bhimlies (from Chennai), Jubbulpores (from Jabalpur, Madhya Pradesh), Rajpores (from Kolhapur, Maharashtra), Vingorlas (from the Mumbai forests) and myrobalans from the Madras coast (Tamil Nadu). Myrobalans from the Salem district of Tamil Nadu are considered the best (due to their high tannin content).

In folk medicine, Harad is used in constipation, tympanitis, vomiting, colic, sprue syndrome, jaundice, splenic disorders; for treating cough, asthma, hiccup, throat affections, and impaired voice.

One fruit of Abhayaa, two fruits of Bihitaki (Terminalia bellirica) and four fruits of Aamalaki (Emblica officinalis syn. Emblic myrobalan), taken together, were called Triphalaa (the Three Myrobalans). It was prescribed as a laxative, digestive, promoter of eyesight, intellect and longevity. It is credited with the properties which enhance body resistance against disease and induce immunity; and is included as an adjunct in a number of compound preparations.

Triphalaa is used in folk medicine, as a dentifrice for bleeding gums. Mixed with oil, it is applied to cuts, wounds, burns and scalds. Water, in which Triphalaa has been steeped overnight, is used as a cooling wash for eyes. Also for affording relief in conjunctivitis.

Among over-the-counter drugs, Haritaki is the main ingredient in Chitraka-Haritaki (Chakradatta), Dashmuula-Haritaki (Ashtaanga Hridaya); Haritaki Khanda (Bhaaishajya Ratnaavali), Haaridraa Khanda (ibid), Abhayaarishta (ibid); Agnimukha Churna (ibid).

Triphalaa Churna (Shaarangadhdara Samhitaa) is one of the largest-selling herbal products over-the-counter; used as a household remedy for regulating digestion.

Triphalaa is an important ingredient in Avipattikar Churna (Bhaishajya Ratnaavali), Sanjivani Vati (Shaarangadhara Samhitaa), Triphalaadi Taila (Sahasrayoga), Chinchaadikaa Leha (ibid).

Triphalaa Ghrita (Shaarangadhara Samhitaa) and Mahaa-Triphallaadya Ghrita (Bhaishajya Ratnaavali) are prescribed internally in diseases of the eye, and for promoting eyesight.

Naarikelaanjana (Aarogya Kalpadruma, IMPCOPS) a herbal eye-drop, containing Triphalaa, is prescribed in conjunctivitis, both acute and chronic.

Jaatyaadi Ghrita (Ashtaanga Hridaya) is for external application on chronic and septic ulcers.

Danta-Dhaavan Churna (IMPCOPS) is prescribed as a toothpowder for pyorrhoea, gingivitis and diseases of the teeth and gum.

In Unani medicine, Triphalaa is included in a number of confections, as a digestive, stomachic and laxative drug. Majoons, containing Triphalaa (Halelaa, Balelaa and Aamalaa), are called Itrifal. Halelaa siyah, Halelaa zard and Halelaa kaabuli (all the three varieties) are included in Itrifal-e-Deedan, Itrifal Kishneezi, Itrifal-e-Muqil Mulaiyin, Itrifal-e-Ustukhuddus, Itrifal Zamuni.

A number of pills of Unani medicine contain all the three varieties of Halelaa.

Active principles and pharmacology

Fruits contain chebulinic acid, tannic acid and chebulin.

Oil from kernels yielded palmitic, stearic, oleic, linoleic, arachidic and behenic acids.

Antioxidant constituents of the plant, phloroglucinol and pyrogallol, have been isolated along with ferulic, vanillic, p-coumaric and caffeic acids.

Ether extract showed higher antioxidant activity than BHA and BHT. Acid esters present in phenolic fraction of extract were found most effective.

A new ellagitannin—terchebulin—has been isolated from fruits along with punicalagin and terflavin A and its structure has been elucidated. Terflavins B, C and D, punicalagin and punicalin have been isolated from leaves.

Shikimic, gallic, triacontanoic and palmitic acids, beta-sitotsterol, daucosterol, triethyl ester of chebulic acid and ethyl ester of gallic acid from fruits have been isolated. A new triterpene—chebupentol—has been isolated from fruits; arjungenin, terminoic acid and arjunolic acid have also been isolated.

The oil in the kernel increased the motility of the gastrointestinal tract of the mouse. The action was comparable with castor oil. The oil itself is non-irritant, but releases an irritant principle when incubated with lipase.

The laxative activity is also exhibited by the fruit pulp.

The main purgative ingredient in Triphalaa is Terminalia chebula, possibly by rendering the irregular peristaltic movements uniformly progressive. The purgative principle in the pericarp of the fruit has been found to be a glycoside which may be similar to sennoside.

A comparative study of purgative activity of different commercial samples showed that the potency of 1 g of Survari Harad was found to be equal to that of 1.47 g of Baala Harad or 1.7 g of Javaa Harada.

Chebulin, found in the constituents of Triphalaa, exhibited antispasmodic activity similar to papaverine.

The anthelmintic activity of Triphalaa was found to be more than that of any of its three constituents (in diluted aqueous/extract), possibly due to their synergistic effect.

A fraction, obtained by treating the alcoholic (80 %) extract of the fruits (of Triphalaa) with hydrochloric acid and extracting with ether, showed reasonably high activity against a number of bacteria and fungi.

Menispermaceae

TINOSPORA

Tinospora cordifolia (Willd.) Miers. ex Hk. f. Thoms.

Figure 1 Tinospora cordifolia [CCRAS]

Figure 2 Tinospora cordifolia [CCRAS]

Habitat

Found throughout tropical India, ascending to an altitude of 300 m.

Classical and common names

Ayurvedic: Guduuchi, Amritaa (Sushruta); Amritaka, Amritalataa, Amritavalli, Chhinnaa, Madhuparni, Vatsaadani, Tantrikaa, Kundalini. Guduuchi Sattva (starch).

Unani: Gilo, Gulanchaa, Sat-e-Gilo (sedimented starch).

Siddha: Seenthil. Seenthil Sarkarai (sedimented starch).

English: Gulancha Tinospora.

Parts used
Stem, root, flowers.

Dose
Juice 5–10 ml, decoction 5–10 ml.

Classical use
A cooled decoction of Guduuchi mixed with honey, or a paste of Guduuchi leaves mixed with buttermilk, was administered for jaundice. (Charaka Samhitaa, Ashtaanga Hridaya, Bangasena, Bhaavaprakaasha.)

The root of Guduuchi, pounded with rice-water, was prescribed for splenomegaly.

Juice of Guduuchi and Shataavari (Asparagus racemosus), in equal quantity, was given for treating fever (Sushruta); a decoction of Guduuchi, Nimba (Azadirachta indica) and Aamalaki (Emblica officinalis), mixed with honey, for irregular fever (Sushruta); a simple decoction of Guduuchi or its juice added with Pippali (Piper longum) powder for chronic fever (Vrindamaadhava, Bhaavaprakaasha).

Juice of Guduuchi was prescribed for polyuria, associated with diabetes (Ashtaanga Hridaya, Shaarangadhara Samhitaa).

A decoction of Guduuchi, Nimba (Azadirachta indica) and Patola (Trichosanthes dioica) leaves, mixed with honey, was prescribed in acid gastritis.

A decoction prepared with 30 g Guduuchi, Shunthi (dry Zingiber officinale) and Dhaanyaka (Coriandrum sativum) was administered for alleviating gout, rheumatic affections and obstinate skin diseases (Vrindamaadhava, Bhaavaprakaasha). Regular use of juice, paste, powder or decoction of Guduuchi was recommended for gout. Amritaa Ghrita (Shaarangadhara), Guduuchi Ghrita (Vrindamaadhava), Amrityaadya Taila and Madhuparnyaadi Taila (Charaka) were specific compound preparations for gout.

Guduuchyaadi Ghrita of Charaka was specific for cough.

Juice of Guduuchi (including that of root and flowers), mixed with the juice of Manduukaparni (Centella asiatica), powder of Yashtimadhu (Glycyrrhiza glabra) and paste of Shankhapushpi (Convolvulus pluricaulis), was prescribed with milk as a rejuvenating tonic. (Charaka Samhitaa.)

Guduuchi was included in more than ninety compound formulations of Bhaavaprakaasha.

Among over-the-counter drugs, Guduuchi Sattva (Bhaavaprakaasha) is prescribed for chronic fevers, diabetes and wasting diseases; Amritaarishta (Bhaishajya Ratnaavali) for intermittent fever.

Guduuchi is incorporated in a number of over-the-counter compound preparations for its antiperiodic, antipyretic and anti-inflammatory properties—Saarivaadyaasava (Bhaishajya Ratnaavali), Sudarshana Churna (Shaarangadhara Samhitaa), Sanjivani Vati (ibid) Manjishthaadi Kwaatha Churna (ibid), Raasanaadi Kwaath Churna (Sahasrayoga), Guggulu Ghrita (Ashtaanga Hridaya), Chitraka-Haritaki (Chakradatta), to name but a few.

In Unani medicine, Sat-e-Gilo is incorporated in Habb-e-Tabaasheer, prescribed during fevers. It is also an ingredient in Qurs-e-Ziabetus Khaas, prescribed for diabetes mellitus.

Seenthil Sarkarai (Guuduchi Sattva) is the principal ingredient of Siddha compound Sandana Podi, prescribed for diabetes mellitus.

Active principles and pharmacology
The creeper contains tinosporon, tinosporic acid, tinosporol, alpha-sitosterol and cordifolide; a furanoid bitter principle, tinosporine, and quaternary alkaloids magnoflorine and tembetarine.

Leaves contain cordifolone and heptacosanol.

Stems contain cordifol, tinosporidine, a diterpenoid, tinosporide, and perberilin. Other constituents of stem are heptacosanol and beta-sitosterol. A new furanoditerpene glucoside—tinocrisposid—has been isolated from stems.

Most of the classical properties of Tinospora cordifolia—antipyretic, anti-inflammatory, antiarthritic, antiallergic, hepatoprotective, hypoglycaemic and CNS-depressant—have been confirmed after experimental trials. (CCRAS.)

The bitter principle of the drug produced hypoglycaemic effect in rats and increased the glucose uptake by tissues.

The drug produced antipyretic and analgesic effect in rats.

Significant anti-inflammatory activity was observed against carrageenin, 5-HT, formaldehyde, granuloma pouch and adjuvant arthritis in experimental animals. The anti-inflammatory activity of the plant resembles that of nonsteroidal anti-inflammatory agents.

Water extract of the drug increased the urine output. It decreased blood urea and a steady level was obtained on fourth and fifth day in animals. (CCRAS.)

The aqueous extract of stem antagonised the effects of agonists such as 5-hydroxytryptamine, histamine, bradykinin and prostaglandins E1 and E2 on the rabbit smooth muscle.

The hepatoprotective effect of Tinospora cordifolia extract has been studied in carbon tetrachloride-induced liver damage in rats. It proved effective in prevention of fibrosis, and in stimulating regeneration of hepatic tissue.

The drug is found to be nontoxic in acute toxicity studies.

Apiaceae

TRACHYSPERMUM

Trachyspermum ammi (L.) Sprague
Trachyspermum copticum Link.
Carum copticum Hiern

Habitat

Cultivated throughout India, mainly in the plains—Madhya Pradesh, Uttar Pradesh, Gujarat, Maharashtra, Rajasthan, Bihar and West Bengal.

Classical and common names

Ayurvedic: Yavaani, Yamaani (Charaka, Sushruta); Yavaanikaa, Yamaanikaa (Bhaavaprakaasha). (Yavaani is not a synonym of Ajamodaa.)
Unani: Naankhwaah, Desi Ajawaayin.
Siddha: Omam.
English: Ajowan. (Bishop's Weed is a confusing nomenclature. It is equated with Ammi visnaga, used for angina pectoris.)

Classical use

Charaka and Sushruta prescribed Yavaani for treating indigestion, colic pain, internal obstructions, anaemia, diarrhoea.

For piles, buttermilk added with Chitraka (Plumbago zeylanica) and Yavaani powder was prescribed. Yavaani, Shunthi (Zingiber officinale), Paathaa (Cissampelos pareira), pomegranate juice, jaggery, mixed with salt and buttermilk, formed the carminative mixture of Charaka.

For treating alcoholism, a decoction of Yavaani was prescribed for its pungent yet pleasant flavour and stimulating warmth for neutralising the urge for alcohol.

Yavaani and jaggery was prescribed for urticaria (Vrindamaadhava).

For diseases of the teeth, Yavaani and Vachaa (Acorus calamus) root was chewed. (Haarita Samhitaa.)

Naaraayana Churna (Shaarangadhara Samhitaa), prescribed for chronic constipation, contains both Yavaani and Ajamodaa (see Trachyspermum roxburghianum).

Among over-the-counter drugs of Unani medicine, Majoon-e-Naankhwaah (Bayaz-e-kabir) is prescribed in flatulence and anorexia; Araq-e-Ajawaayin as a carminative. Araq Ajeeb contains sublimated Ajawaayin and Pudinaa (Mentha arvensis). It is prescribed for dyspepsia, nausea, vomiting, diarrhoea, cholera.

In folk medicine, Ajawaayin is used in conjunction with asafoetida, myrobalans and rock salt or black salt. Taken with buttermilk Ajawaayin is a common remedy for relieving difficult expectoration due to dried-up phlegm.

After parturition, water fortified with the extract of Ajawaayin, is given for averting puerperal disorders.

Ajawaayin-kaa-Phool (Flower-of-Ajowan—crystallised thymol from the oil) is administered as an antiseptic and aromatic carminative.

The oil of Ajawaayin is used as an expectorant in emphysema, bronchial pneumonia and allied respiratory diseases.

The root of the plant is used for its diuretic and carminative property in febrile conditions and stomach disorders.

Ajawaayin is also used in antiseptic lotions and ointments. A paste of crushed fruits is applied externally for relieving colic pain. The oil is recommended as an external application in cases of rheumatism.

Active principles and pharmacology

Seeds contain the phenolic glucoside 2-methyl-3-glucosyloxy-5-isopropyphenol; sugars; and tannins.

Seeds contain: moisture 7.4; protein 17.1; fat 21.8; fibre 21.2; carbohydrates 24.6 %; vitamins and mineral matter—calcium, phosphorus, iron, sodium, potassium, thiamine, riboflavin, nicotinic acid and carotene.

Thymol is the major component of the essential oil; others include alpha- and beta-pinene, camphene, mycrene, delta-3-carene, limonene, gamma-terpinene, p-cymene and carvacrol.

Pharmacological studies of the oil indicated that it had a parasympathomimetic effect and produced contraction of isolated ileum, tracheal chain and bronchial musculature of guinea-pig. It depressed the cardiac musculature of guinea-pig. It also depressed the cardiac musculature in frogs and caused a marked fall in blood pressure in cats.

Thymol exhibits intestinal antiseptic and anthelmintic property, but is found toxic in high doses, which leads to fatal poisoning.

The action of Ajowan oil is similar to that of thymol. Extract of Ajowan and its oil is used with benefit in deodorant mouthwashes, gargles, toothpastes, antifungal creams and lotions.

Ajowan extract showed antibiotic activity against Salmonella typhosa, Micrococcus pyogenes var. aureus and E. coli.

Thymol exhibits a powerful activity against a variety of fungal infections of the skin.

Trachyspermum roxburghianum (DC.) Craib
Trachyspermum involucratum Wolff. non Marie
Carum roxburghianum Benth (DC.) ex Kurz

Habitat

Cultivated throughout India, mainly in Punjab and Uttar Pradesh.

Classical and common names

Ayurvedic: Ajamodaa (Charaka) (Fruits of Ajamodaa are bigger than those of Ajawaayin.)
Unani: Karafs-e-Hindi. (Karafs is equated with Apium graveolens Linn.)
Siddha: Ashamtagam.
English: Ajmud.

Classical use

Charaka and Sushruta used Ajamodaa internally as an appetizer and absorbent of intestinal mucus and unassimilative lymph-chyle; curative of catarrh, internal obstructions, jaundice, oedema, colic gastralgia.

A paste of Ajamodaa was applied externally for promoting suppuration and bursting of non-suppurating boils and carbuncles. A decoction was used as a gargle for catarrh due to bad throat. (Sushruta Samhitaa.)

A formulation containing Ajamodaa, Paathaa (Cissampelos pareira), Kutaja (Holarrhena antidysenterica), Utpala (Nymphaea stellata), Shunthi (Zingiber officinale) and Pippali (Piper longum), all in equal quantity, was prescribed for diarrhoea and dysentery (Sushruta Samhitaa).

Hingwaashtaka Churna (Bhaishajya Ratnnavli), with Ajamodaa as an important ingredient, is available over-the-counter for tympanitis and abdominal discomfort.

In folk medicine, Ajamodaa is also used in bronchitis and asthma as an expectorant. A decoction is administered in palpitation and cardiac affections. The drug is also used in prescriptions for amenorrhoea.

Active principles and pharmacology

Fruits yield an essential oil (up to 2.5 %), a fixed oil (4.5 %) and a crystalline ketonic compound (0.1 %).

Major constituents of the essential oil from seeds are: D-limolene 35.1; alpha-terpinene 19.4; D-linallol 4.7; DL-terpineol 5.7; and DL-piperitone 13.6 %. Other constituents are thymol, thymoquinol and cuminol.

Seeds yield the coumarins bergaptene, 7-methoxy-6-methylcoumarin, and umbelliferone. Beta-sitosterol is also reported from seeds.

Ajmud induced hyperactivity of the central nervous system in mice.

The crystalline ketonic substance exhibited powerful antispasmodic activity; the action was particularly marked on the smooth muscle in rabbit's gut.

The essential oil and the crystalline substance were found to lower blood pressure in dogs and cats; the effect was due to direct action on blood vessels.

The oil produced marked diuretic effect in rabbits.

The fruits, left after the extraction of the essential oil, showed pronounced cardiotonic activity.

Trapaceae

TRAPA

Trapa bispinosa Roxb.
Trapa natans Linn. var. bispinosa (Roxb.) Makino.

Habitat
An aquatic plant cultivated for its fruits in tanks, lakes, ponds throughout the greater part of India.

Classical and common names
Ayurvedic: Shrngaataka, Shrngamuula, Trikota (Charaka); Jalphala, Trikonaphala, Paaniyaphala, Jalkanda, trikona, Trika.
Unani: Singhaaraa.
Siddha: Singara.
English: Water Chestnut.

Parts used
Shelled fruit; dried kernel flour.

Dose
Powder 5–10 g.

Classical use
Charaka gave shelled fruits alone or in prescriptions, internally, for impotency; dysuria, polyuria and other urinary disorders; cough, consumption; after debilating illness; intrinsic haemorrhage.

To stabilize foetus and to prevent abortion, pregnant women were given to eat water chestnut, lotus seed and kashru tube. Water chestnut, lotus stem and Vidaari tuber (Pueraria tubercosa) was prescribed as a galactagogue. (Charaka Samhitaa, Sushruta Samhitaa.)

Amritapraasha Ghrita (Charaka, Sahasrayoga), available over-the-counter, is prescribed as a tonic for cough, debility and emaciation.

Singhaaraa flour is an important ingredient in Majoon-e-Aarad-e-Khurma (Bayaz-e-Kabir), prescribed by Unani physicians for treating spermatorrhoea, attenuated semen, oligospermia, sexual debility.

A confection of water chestnut flour and milk is prescribed in diarrhoea, dysentery, bilious affections, piles, leucorrhoea, menorrhagia. Due to its cooling, antibilous and nourishing properties, preparations of water chestnut flour are served after fasts.

Active principles and pharmacology
Analysis of kernels gave the following values: moisture 70; protein 4.7; fat 0.3; fibre 0.6; other carbohydrates 23.3; and mineral matter 1.1 %; calcium 20, phosphorus 150 and iron 0.8 mg/100 g. Other minerals reported are: copper 1.27 %; manganese 5.7; magnesium 38; sodium 49; and potassium 650 mg/100 g. Iodine (50.6 mcg/100 g) is also present.

The vitamin contents of kernel are: thiamine 0.05; riboflavin 0.07 %; nicotinic acid 0.6; and vitamin C 9 mg/100 g; vitamin A 20 I.U./100 g.

The kernels contain 15.8 mg/100 g oxalate on dry weight.

The presence of beta-amylase and a considerable amount of phosphorylase has been reported in the kernels.

The nutritive value of flour, prepared from dried kernels, is as follows: moisture 10.6; protein 8.0; fat 0.6; and minerals 2.6 %; calcium 69; phosphorus 343; iron 2.8; and thiamine 0.44 mg/100 g.

The biological value of the proteins of water chestnut was found to be higher than that of proteins in wheat (protein efficiency ratio at 7 % level: water chestnut 1.8, wheat 1.1).

The starch isolated from water chestnut flour consists of 15 % amylose and rest amylopectin.

Cucurbitaceae

TRICHOSANTHES

Trichosanthes dioica Roxb.

Habitat
Wild in the plains of North India from Punjab to Assam. Extensively cultivated all over warmer regions of India for its fruits, used as a vegetable. The bitter variety is equated with Trichosanthes cucumerina.

Classical and common names
Ayurvedic: Patola (Charaka, Sushruta).
Unani: Parwal.

Figure 1 Trichosanthes cucumerina [ADPS]

Siddha: Kombupudalai.
English: Pointed Gourd.

Parts used

Leaves, fruits.

Dose

Juice 10–20 ml, decoction 50–100 ml.

Classical use

Charaka and Sushruta used decoction of the leaves and gourds cooked as vegetable for fevers, diarrhoea, pruritus, erysipelas, urinary diseases, oedema, intrinsic haemorrhage, as an antitoxic, appetizer, febrifuge.

A decoction of Patola, Indrayava (Wrightia tinctoria) and Dhaanyaka (Coriandrum sativum), cooled and added with sugar and honey was prescribed for checking vomiting, diarrhoea, acid gastritis (Vrindamaadhava). A decoction of Patola and Dhaanyaka was prescribed as an appetizer, carminative, digestive, laxative, antipyretic (Chakradatta). Patola leaves and fruits were prescribed in the form of a soured soup for treating alcoholism (Charaka Samhitaa).

Juice of bitter Patola (known as Patolikaa) leaves was applied topically in alopecia (Shaarangadhara Samhitaa).

In Bhaavaprakaasha, Patola is an ingredient in more than eighty compound preparations.

Ayurvedic and Unani physicians give root of Parwal for drastic purgation; juice of leaves and fruits as cholagogue in sub-acute cases of enlarged liver and spleen; tender tops in the form of a potherb as a vermifuge; decoction of stalks as expectorant.

Over-the-counter, Patola is an ingredient in Sudarshana Churna (Shaarangadhara Samhitaa) and Nimbaadi Kwaatha Churna (Sahasrayoga), prescribed in intermittent fevers; in Jaatyaadi Ghrita (Ashtaanga Hridaya), for external application on chronic and septic ulcers.

Active principles and pharmacology

The root contains an amorphous saponin (1.5 %), a non-nitrogenous bitter amorphous glucoside called trichosanthin, resembling colocynthin, a phytosterol, hentriacontane, a small amount of an essential oil containing terpenes, a little fixed oil, starch and reducing sugars.

Extract of the seeds possess haemagglutinating activity.

Fatty acids from seeds comprise: trichosanic or eleostearic 27.8, linoleic 33.4, oleic 27.1 and saturated 11.7 %.

The fruit (edible matter 95 %) has the following composition: moisture 92.0; protein 2.0; fat 0.3; fibre 3.0; other carbohydrates 2.2 and mineral matter 0.5 %; vitamin C 29.0 mg/100 g; carotene 153 mcg/100 g of edible matter. Iodine and fluorine contents of the gourd are 0.66 and 2.1 ppm on dry edible matter. A trace of 5-hydroxytryptamine has also been detected.

Trichosanthes dioica seeds fed at 1 g per cent level for eight weeks significantly lowered fasting blood sugar, total cholesterol and triglycerides and increased the level of phospholipids and HDL-cholesterol in male albino rats.

Not only the seeds, Trichosanthes dioica whole fruit and pulp were also found to exert the same activity.

Cucurbitacin B has been isolated from the fruits of Trichosanthes cucumerina (Patolikaa). Meso-inositol was found to be present in the plant.

Trichosanthes cucumerina plant possessed two main active principles of opposite character, one having acetylcholine-like spasmodic action, while the other produced spasmolytic and antiacetylcholine-like action on smooth muscle preparations. (CCRAS.)

Both alcoholic and aqueous extracts produced hypotensive action on anaesthetised dogs, which was not antagonised by prior administration of atropine on dog's heart in situ, decrease in auricular contractions but increase in ventricular contractions.

Aqueous and alcoholic extracts potentiated pentobarbital hypnosis in rats, but failed to protect them for electroshock convulsions.

50 % ethanolic extract of the plant without root produced diuretic effect. (CCRAS.)

LD_{50} in mice was found to be approximately 3 g/kg.

Papilionaceae

TRIGONELLA

Trigonella foenum-graecum Linn.

Habitat

Found wild in Kashmir, Punjab and Upper Gangetic plains. Widely cultivated in many parts of India.

Classical and common names

Ayurvedic: Methikaa (Bhaavaprakaasha); Vastikaa, Selu (Ayurveda Saukhyam); Methini, Dipani, Bahupatrikaa, Bodhaini, Gandhaphala.
Unani: Hulbaa, Methi.
Siddha: Vendhayam.
English: Fenugreek.

Parts used

Seed.

Dose

Powder 3–5 g.

Classical use

Methikaa was not used during the period of Charaka and Sushruta. During the Vedic period, Methi (or Meth) was the pole with which sacrificial horses were tied. The wild variety of Methikaa, during that period, was known as Ahittha and was fed to horses. Methikaa entered into Indian medicine during the period of Bhaavamisra.

According to Maadhava Dravyaguna, Methikaa harmonizes all body functions, alleviates rheumatic affections and possesses digestive and laxative properties. In Bhaavaprakaasha, only two compound preparations have been mentioned—Panchjiraka-Paaka and Pugaa-Paaka. Panchjiraka Paaka was a uterine tonic for puerperal disorders; Pugaa-Paaka was a spermatogenic and aphrodisiac confection.

Antirheumatic, anticatarrhal, and febrifugal properties were attributed to Methika in Ayurvedic classics (Dhanvantari Nighantu, Bhaavaprakaasha).

Saubhaagya Shunthi Paaka (Bhaishajya Ratnaavali), with Methi as one of the ingredients, is a reputed over-the-counter uterine tonic, prescribed during post-delivery period.

Methi is included in a number of Ayurvedic compound preparations for its carminative, galactagogue and antidiabetic properties.

Active principles and pharmacology

The plant contains a number of steroidal sapogenins, especially diosgenin found in the oily embryo. Two furastanol glycosides, F-ring opened precursors of diosgenin have been reported, as also hederagin glycosides. The alkaloid trigonelline, trigocoumarin, trimethyl coumarin and nicotinic acid are also present.

Mucilage (25–40 %) is a prominent constituent of the seeds, contains mannogalactans.

Seed contains (45–60 %) carbohydrates, mainly mucilaginous fiber (galactomannans); 20–30 % proteins high in lysine and tryptophane; 5–10 % fixed oils (lipids); pyridine-type alkaloids, mainly trigonelline, choline, gentianine and carpaine; the flavonoids apigenin, luteolin, orientin, quercetin, vitexin and isovitexin; free amino acids; calcium and iron; saponins (0.6–1.7 %); glycosides yielding steroidal sapogenins on hydrolysis (diosgenin, yamogenin, tigogenin, neotigenin; cholesterol and sitosterol; vitamin A, B1, C and nicotinic acid; and 0.015 % volatile oil (n-alkanes and sesquiterpenes).

Seeds from India, Afghanistan and Pakistan contained 28.7–30.8 % protein and 5.5–6 % oil (dry weight.). Seeds yielded 14.1 % saponin.

Chronic administration of seed extract enhanced food consumption and motivation to eat in rats and also induced hyperinsulinemia and hypocholesterolaemia (Rastogi, Compendium 1998).

In rats, trigonelline counteracts the hyperglycaemic effect of cortisone given two hours before or simultaneously. Antiulcer activity has been studied in rats.

Hypocholesterolaemic and anti-inflammatory effects have been observed experimentally. The either extract of the seeds had an effect on inflammation induced in rats by cotton-pellet insertion, or formalin or carrageenin exposure, comparable to that of salicylates. (Chemexcil.)

Hypoglycaemic and hypocholesterolaemic activity in healthy individuals has been reported for whole seed extracts.

In two diabetic insulin-dependent subjects, daily administration of 25 g Fenugreek seed powder reduced fasting plasma-glucose profile, glycosuria and daily insulin requirements (56–20 units) after eight weeks. Significant reductions in serum cholesterol concentrations were also reported. (Expanded Commission E Monographs.)

A subsequent study investigated the lipid lowering activity of Fenugreek seeds in 60 non-insulin dependent diabetic subjects. Ingestion of an experimental diet containing 25 g Fenugreek seed powder daily resulted in significant reduction of total cholesterol, low density and very low density lipoprotein cholesterol and triglyceride levels. The effect on lipids also affected glucose and insulin levels. (Sharma RD, Phytotherapy Res (1996) 10:332)

Fenugreek seed powder is used in India by Ayurvedic and Unani physicians for rheumatic disorders and spondylosis, chronic bronchitis, hepato- and splenomegaly and obesity. Encouraging results have been achieved.

Use in Western herbal

German Commission E monograph reported secretolytic, hyperemic and mild antiseptic activity of Fenugreek seed. It recognized its internal use for loss of appetite and external use as a poultice for local inflammation.

The British Herbal Pharmacopoeia reported its actions as demulcent and hypoglycaemic.

Asclepiadaceae

TYLOPHORA

Tylophora indica (Burm. f.) Merril
Tylophora asthmatica (L. f.) W. & A.

Habitat

Found in Assam, West Bengal, Orissa and Peninsular India, ascending to an altitude of 900 m. One of the most common plants in the fields and low sandy jungles of South India.

Classical and common names

Ayurvedic: Antamuula, Muulini (non-classical). Muula and Muulaka have been equated with Raphanus sativus. Not to be confused with Anantamuula (Hemidesmus indicus.)
Siddha: Nay Palai, Nangilaippiratti.
English: Emetic Swallow-Wort; Indian Ipecacuanha or Country Ipecacuanha.

Parts used

Root, leaves.

Traditional use

Roots are used for the treatment of asthma, bronchitis, whooping cough, dysentery and diarrhoea; also in rheumatic and gouty pains.

The leaves and root are often employed as a substitute for ipecacuanha.

Active principles and pharmacology

Roots yielded the alkaloids tylophorinidine and gamma-fararine.

Whole plant yielded the alkaloids tylophorine, tylophorinine, desmethyltylophorine and demethyltylophorinine, and a flavonoid, kaempferol.

Leaves gave the alkaloids tylophorinidine, and D-septicine and D-iso-tylocrebrine; the triterpenoids alpha- and beta-amyrin; the sterols beta-sitosterol, stigmasterol and campesterol; phenylalanine; and a flavonoid, quercetin.

Aerial parts gave an alkaloid, skimmianine.

Ceryl alcohol has also been reported from the plant. Wax, resin, caoutchouc, tannin, glucose, calcium salts and potassium salts are found in the plant.

The dried plant contains 0.2–0.3 % of alkaloidal bases.

Dried leaves are reported to be more uniform and certain in their action than the roots.

The alkaloidal fraction of Tylophora indica caused bronchodilator effect and in smaller doses potentiated beta-adrenergic response of adrenaline. This potentiating effect has been attributed to the inhibition of aminoxidase enzyme or to the sensitisation of beta-adrenergic receptors. The beneficial effects of Tylophora indica have been attributed to membrane stabilising and immunosuppressive effects.

The plant antagonizes the contractile effects of histamine, acetylcholine and posterior pituitary extracts. The plant pretreatment prevented bronchospasm induced by Freund's adjuvant and bovine albumin in rats.

Tylophorine and tylophorinine also exhibited potential amoebicidal activity. Tylophorine inhibited growth of both axenic and polyaxenic strains of E. histolytica. It is five times less toxic than emetin. Tylophorine was found effective against hepatic and intestinal amoebiasis at non-toxic doses.

Alcoholic extract of Tylophora indica leaves showed promising immunosuppressive activity. It decreased the sheep red blood cell induced delayed type hypersensitivity and increased skin allograft rejection time in mice. It also significantly inhibited the humoral antibody production and phagocytic function of reticulo-endothelial system.

Tylophorinidine, isolated from the roots of three-year old plant (yield 0.009 %), and two other bases identified as (+)-septicine and (+)-isotylocrebrine exhibited antitumour activity. Extracts of leaves and stem showed promising results against two standard transplantable tumours, lymphoid leukemia L 1210 and lymphocytic leukemia P 388, in mice.

Plant extracts produced significant anti-inflammatory effect in rats as tested in adjuvant arthritis, hind-paw oedema, granuloma-pouch and cotton pellet implantation techniques.

Clinical trials at the Vallabhbhai Patel Chest Institute, Delhi, have indicated marked relief in symptoms of bronchial asthma and allergic rhinitis for a few weeks in 40–50 % of the cases, after a small dose of 3–6 leaves only. A high incidence of side-effects (sore mouth, loss of taste, vomiting) was observed in 75 per cent of the cases (ICMR). In another double-blind trial, using powder of dried leaves of Tylophora indica, significant effect was observed only in the group with perennial type of asthma (Gupta SS, CCRAS).

Caution

In doses commonly used, patients may have nausea and vomiting. Alkaloids are reported to have caused dermatitis. Tylophorine is toxic to Paramecium caudatum in a concentration of 1 in 50,000 or more. The MLD for frogs is 0.4 mg/kg body weight. Its toxicity for mice is very low.

In large doses, dry leaves may cause fatal poisoning.

Fabaceae

URARIA

Uraria picta Desv.

Habitat
Found in dry grass lands, waste places and open forests in sub-Himalayan tract from Kashmir to West Bengal and Assam up to an altitude of 1800 m. The Wealth of India has also equated Prishnparni with Uraria lagopodioides Desv. syn. Uraria lagopoides DC.

Classical and common names
Ayurvedic: Prishnparni, Prithakparni, Sinhapushpi (Charaka, Sushruta).
Siddha: Sittira Paladai.

Parts used
Root.

Dose
Decoction 50–100 ml.

Classical use
Charaka and Sushruta prescribed the entire plant, internally in prescriptions, in misperistalsis, diarrhoea, dysentery, cough, consumption, respiratory diseases, abdominal glands, fever; as an ingredient of a potion for increasing breast milk; for alleviating asthmatic attacks.

Charaka gave a liquid gruel cooked with Prishnparni for diarrhoea with blood; cooked with Prishnparni, parched paddy and processed with Balaa (Sida cordifolia) for bleeding piles and hemorrhage.

Sushruta gave milk, cooked with Prishnparni added with sugar and honey, for gout; powdered root of Prishnparni with meat soup for promoting adhesion of fractured bone.

The herb is an ingredient of Dashmuula (the Ten Roots) of Ayurvedic medicine, employed as an alcoholic or aqueous extract in remittent fever and for the treatment of inflammations.

In folk medicine, the root is used as febrifuge, bechic; root and pod in prolapse of anus in infants; fruit in stomatitis. Leaves are considered antiseptic and are used in gonorrhoea.

In Indian medicine, Prishnparni and Shaalaparni are used together due to common synonyms. Shaalaparni, equated with Desmodium gangeticumDC., was used by Charaka as a cardiac tonic; by Sushruta with Prishnparni and Brahati (Solanum indicum) for gout.

Prishnparni and Shaalaparni, both were included in Sudarshana Churna and Dashmuula Kwaatha of Shaarangadhara Samhitaa and Bhaavaprakaasha.

Active principles and pharmacology
Hot aqueous extract of the shoots showed oxytocic activity on both gravid and non-gravid uteri of experimental animals. The aqueous extract of the plant also showed anti-implantation activity on rats, and spasmogenic effect on the gut of rabbits and uteri of rats.

Total extract of Uraria picta has been found to effect better and quicker healing of fractures in experimental animals due to early accumulation of phosphorus and more deposition of calcium.

A decoction of leaves is prescribed in Malaysia for diarrhoea. In Ghana, the plant is employed for treating cardiac affections.

Liliaceae

URGINEA

Urginea indica (Roxb.) Kunth.
Scilla indica Roxb. non (Wt.) Baker

Habitat

Frequently cultivated in sandy soils near the seashore in Deccan Peninsula for its bulbs. Found in dry hills of the lower Himalayas up to an altitude of 1500 m.

Classical and common names

Ayurvedic: Vana-palaandu, Kolakanda, Vajrakanda.
Unani: Unsul-e-Hindi, Isqueel-e-Hindi, Piyaaz-Dasti, Piyaaz-Saharaayi. Jangli Piyaaz.
Siddha: Marovemgayam.
English: Indian Squill.

Parts used

Bulb.

Dose

Powder 120–200 mg.

Classical use

Vajrakanda of Sushruta and Kolakanda of Raaja Nighantu have been equated with Urginea indica. Vana-palaandu is a non-classical and commonly used synonym. Vajrakanda was earlier equated with Euphorbia neriifolia (Snuhi). The source of Vajjurkanda, found in Madhya Pradesh, is Urginea indica. It is used as a cardiac tonic and for inflammations and ulcers. Sushruta used the oil extract of the tuber in diseases of the sinus.

Kolakanda of Raaja Nighantu was used as an antitoxic and anthelmintic drug. (Kola kanda is not to be confused with the root of Zizyphus jujuba Lam.)

In Unani medicine, the bulb is used as an expectorant, cardiac stimulant, diuretic, deobstruent and emmenagogue. Young tunicated bulbs are preferred; they lose their medicinal properties as they get older. Bulbs are not used in food preparations like onion.

Mixed with Anjir (Ficus carica), anise, Draakshaa (Vitis vinifera) and honey, syrup of Urginea indica is administered in bronchitis with emphysema and in spasmodic croup.

A powder of the bulb is applied topically to warts; roasted and crushed bulbs to corns on the soles of feet.

Sikanjabeen Ansali and Sirka-e-Unsul are single drug preparations of Isqueel in Unani medicine; used as diuretic, expectorant and cardiac tonics.

Active principles and pharmacology

Bulbs, roots and leaves of Urginea indica contain stigmasterol, sitosterol and campesterol. Bulbs also contain hentriacontanol, octacosanoic acid and cardiac glycosides—scillarens A and B.

Scillaren A is a crystalline substance sparingly soluble in water, whereas B is an amorphous powder freely soluble in water. Scillaren A yields on hydrolysis proscillaredin A, which in turn yields scillaredin A; whereas scillaren B yields scillaredin B. A possesses a six-numbered, doubly unsaturated ring in its structure resembling that of toad venom.

The bulbs contain, in minor quantities, another water-insoluble glucoside yielding an aglucone.

The powdered bulbs of Urginea indica contain approx. 51 % of an alcohol-insoluble mucilage. This distinguishes them from the European squill which contains sinistrin and related carbohydrates which are readily soluble in alcohol. Mucilage consists of mannose, glucose and xylose. Both the powder and the mucilage have a very high power of swelling in water.

The squill available in the Indian market consists chiefly of Urginea indica, admixed with a small portion of Scilla hyacinthiana.

The drug in small doses possesses cardio-tonic, stimulant, expectorant and diuretic properties; in large doses it is emetic, cathartic and may cause cardiac depression.

The action of squill as a cardiac tonic resembles that of digitalis. Though the squill acts more rapidly and is less cumulative in its effect, its use as a substitute for digitalis can not be advocated due to the irritating effect and poor absorption, entailing larger doses.

The tincture prepared from the Indian squill is considered to be as effective as that from European squill in the treatment of cough, chronic bronchitis and asthma.

The alcoholic extract of the bulb showed anticancer activity against human epidermoid carcinoma of the nasopharynx in tissue culture and against Friend-Virus leukaemia in the mice.

The extract also showed hypoglycaemic activity. The alcoholic extract is found active against Entamoeba histolytica strain STA.

Use in Western herbal

Urginea maritima, syn. Drimia maritima, is used in Western herbal. The drug contains cardioactive steroid glycosides (bufadienolides 1–3 %): chief components—glucoscillarene A, proscillaridin A, scillarene A; including among others, scillicyanoside, scilliglaucoside; and mucilage. (PDR.)

The drug is positively inotropic on myocardial work capacity and negatively chronotropic. There is a lowering of increased, left ventricular diastolic pressure and pathologically elevated venous pressure. (PDR.)

German Commission E recognized the efficacy of Urginea maritima in following areas:

- Milder cases of heart insufficiency
- Reduced kidney capacity.

The drug is contraindicated during therapy with digitalis glycosides, and in potassium deficiency.

Squill is used in a number of over-the-counter cough mixtures, chest-mixtures and cough pastilles.

Urticaceae

URTICA

Urtica dioica Linn.
Urtica pilulifera Linn.

Habitat

North-western Himalaya from Kashmir to Shimla at 2400–3600 m. Urtica pilulifera is found near houses in Shimla and other hill stations.

Classical and common names

Vernacular: Bichhuu Buuti, Bichhuu, Bichhuaa, Chichru.
Unani: Anjuraa.
English: Common Stinging Nettle; Roman Nettle (Urtica pilulifera)

Classical use

In Unani medicine, ash of the plant, mixed with vinegar is prescribed for cervical adenitis, enlarged thyroid and tumours; paste of seeds with honey for asthma; seed kernels with rose-water as purgative and antilithic. Nettles are also used in nephritis, haematuria, leucorrhoea and menorrhagia. An ash of dried leaves or powdered dried leaves are inhaled for alleviating asthmatic attacks.

The irritant property of the nettle is used externally to excite activity in paralysed limbs by slapping or pricking the patient with a bundle of fresh twigs.

Active principles and pharmacology

The urticating properties of the hair are attributed to the presence of acetylcholine, histamine and 5-hydroxytryptamine which are highly concentrated in stings. A histamine-liberating enzyme is also present in the hair (not present in underground parts).

Betaine and choline are reported from leaves.

Plant is rich in vitamins and carotenes. An Italian analysis of the whole young plant at the preflowering stage gave the following values (on dry basis): crude protein 30.4; fat 3.4; cellulose 10.3; N-free extr. 39.6; ash 16.3; calcium 2.97; phosphorus 0.68; iron 0.0322; magnesium 0.65; potassium 3.45; sodium 0.14; manganese 0.0043; sulphur 0.54; silicon 0.68; and chlorine 0.27 %; and beta-carotene 20.2 mg/100 g. Vitamin B1 80 mcg/100 g (of fresh plant, in another sample).

Plant contains flavonoids (0.7–1.8 %); glucoquinones; tannins; silicic acid (1–4 %); volatile oil (ketones including 2-methylhept-2-en-6-on); and a high content of potassium ions.

Research has indicated that nettles increase the excretion of uric acid from the body. Nettle gives encouraging results in the treatment of arthritic conditions and gout. The tannins account for astringent effect. The glucoquinone exhibited hypoglycaemic action in animals.

Extract of nettle has been found to slow the heart in experimental animals and dilate and constrict the blood vessels alternately in different conditions.

Intravenous administration of 96 % ethanolic extract of nettle to dogs under narcosis resulted in higher tones in smooth muscles of the uterus and small intestines. A similar effect was produced on the uterus by betaine compounds extracted from the nettle leaves. Both the preparations caused an acceleration of the coagulation of blood and increased plasma-tolerance to heparin, but the acceleration was more significant in case of betaine preparation. Compared to other common liquid extracts of nettle, ethanolic extract was faster in action.

Clinical experiments have confirmed the utility of the herb as a haemostatic in vomiting of blood, uterine haemorrhage and bleeding from nose. An infusion of nettle is administered, lukewarm, to control excessive menstrual flow. The seeds are prescribed for the treatment of severe diarrhoea and intestinal worms; an infusion of leaves as an expectorant and blood-purifier.

Externally, an infusion of leaves gives good results in dandruff.

Nettle extract inhibited Shigella paradysenteriae, Shigella ambigua, Shigella sonnel, Pasteurella aviseptica and particularly the antibiotic-resistant strains of Micrococcus pyogenes var. aureus; but was ineffective against E. coli, Proteus vulgaris, Klebsiella pneumoniae and Pseudomonas aeruginosa.

Use in Western herbal

German Commission E recognized the efficacy of nettle leaf and root in the following areas:

- Rheumatism
- Diseases of lower urinary tract
- Kidney stones
- Prostate complaints
- Irritable bladder.

In over-the-counter drugs, nettle is an ingredient in Alfavena, Tormentavena, Celery complex, Sarsparilla Complex, Kelp and Nettle compound, Burdock and Nettle Formula. Nettle leaf capsules (200 mg) are also available.

For external application, Hypericum and Urtica Ointment and Stinging Nettle Oil are sold. Plant juice is also available.

Caution

The drug is contraindicated when there is fluid retention due to reduced cardiac or renal activity.

The stings of the fresh plant produce weals and blisters, especially when arthritic and other inflammations and paralysed parts are flayed with fresh nettles for more than one minute.

V

Valerianaceae

VALERIANA

Valeriana wallichii DC.

Figure 1 Valeriana wallichii [ZANDU]

Valeriana jatamansi Jones
Nardostachys jatamansi (Jones) DC.

Habitat

Found in temperate Himalayas at an altitude of 3000 m; and in the Khasi and Jaintia hills between 1500 m and 1800 m.

Valeriana jatamansi is abundant in western Himalayas, whereas Valeriana hardwickii and Valeriana officinalis are rather scarce.

Classical and common names

Ayurvedic: Tagara. Synonyms (Nat, Vakra, Kutil, Nahush) are confusing and misleading. Tagara is sold as Tagara Ganthodaa (wrongly spelled as Taggar in the Wealth of India) in the market.
Unani: Jatamansi is recognized as Sumbul-ut-Teeb (National Formulary of Unani Medicine); Valeriana wallichii as Asaaroon Hindi (Asaaroon is a different drug equated with Asarum europeaum, syn. Wild Nard, Hazel Wort); Nardostachys jatamansi as Sumbul-e-Hindi, Sumbul-ut-Teeb-e-Hindi, Naarad-e-Hindi, Naardeen-e-Hindi. Unani scholars equate Valeriana officinalis with Naadeen or Sumbul-e-Roomi.
Siddha: Tagarai.
English: Indian Valerian.

Classical use

Dried rhizomes and roots of Valeriana jatamansi Jones, syn. Valeriana wallichii DC., as well as those of Valeriana officinalis were official as Valerian in the pre-1966 editions of Pharmocopoeia of India, but in 1966 edition only Valeriana jatamansi was retained, Valeriana officinalis was deleted, as it is mainly a European plant.

The Ayurvedic Pharmocopoeia of India, 1989, equates Tagara with Valeriana wallichii and recognizes Tagar Ganthodaa, Baala, Mushkbaalaa, Tagarai as its synonyms; and Maamsi, Jataa, Jatilaa, Baalchhara as synonyms of Nardostachys jatamansi.

National Academy of Ayurveda equated Baala, Baalaka and Hrivera with Valeriana wallichii, syn. jatamansi.

In classical medicine, Tagara was used in prescriptions for fevers of various aetiology, especially with delirium; aberrations; convulsions; irritating chronic cough and asthma; diarrhoea and sprue syndrome with spasm and colic; and for preventing "evil dreams" (fearpsychosis, psychoneurosis).

Baala, Baalaka, Hrivera, Sugandha-baalaa, Udichya of Ayurvedic texts were used as febrifuge, carminative and intestinal antiseptics. These synonyms do not carry properties of Tagara which was used as a sedative, antispasmodic, tranquillizer, and as a specific drug for CNS.

In Shaarangadhara Samhitaa and Bhaavaprakaasha, Baala, Baalaka, Hrivera, Udichya and Tagara have been included in prescriptions by their specific names. In Bhaavaprakaasha, Kushtha (Saussurea lappa) was identified as a substitute for Tagara (Kushtha is considered a herbal neuroleptic to some extent). Venu (Bambusa bambos) was identified as a substitute for Nat (a synonym of Tagara). Substitutes for other synonyms were not mentioned.

Valeriana jatamansi can be used as a substitute for Tagara. Baalaka, Hrivera and Udichya should be equated with Pavonia odorata Willd. (Malvaceae), according to the Wealth of India.

Though Kushtha was a substitute for Tagara, both Kushtha and Tagara were included in many compound preparations as independent herbs—Mahaa Chandanaadi Taila (Bhaishajya Ratnaavali); Mahaa-Naaraayana Taila (ibid); Dhanvantara Taila (Serva-Roga-Chikitsaa), to name but a few.

Tagara, Kushtha and Jataamansi, all the three, were included in Naaraayan Taila (Shaarangadhara Samhitaa); Manjishthaadi Taila (Sahasrayoga), Prabhanjana Vimardana Taila (ibid), Triphalaadi Taila (ibid); Pippalaadyaasava (Shaarangadhara Samhitaa).

Tagar, Kushtha and Hrivera were included in Sudarshana Churna (Shaarangadhara Samhitaa.)

Tagar and Jatamaansi were included in Himsaagara Taila (Bhaishajya Ratnaavali), Chandanaadi Taila (ibid).

In Unani medicine, Sumbul-ut-Teeb and Qust are used together in tonics and massage oils—Javaarish-e-Jalinoos (Qarabadeen-e-Jadeed); Safoof-e-Qaranful (Bayaz-e-Kabir); Raughan-e-Baladur (Qarabadeen-e-Ehsani).

Active principles and pharmacology

Rhizomes and roots of Valeriana wallichii contain cyclopentapyrans, acacetin-7-O-rutinosides, valtrate, didrovaltrate, linarin, iso-valerinate, valepotriates and an iridoid ester glycoside named valcrosidatum.

Essential oil from roots without rootlets contain calarene, beta-bergamotene, valeranone, arcurcumene, maalioxide, and maaliol. Main acids present in the oil are isovaleric acid and (+)-beta-methyl-valeric-acid. Other acidic constituents are formic, propionic, butyric, palmitic and stearic acids and isovaleryl ester of D-(–)-alpha-hydroxyvaleric-acid.

Essential oil from roots and rootlets contain beta-sitosterol, substantial amounts of alpha-, beta-and gamma-patchoulene and maaliol in traces; maalioxide is absent in the oil.

Valepotriates (yield 2.0 %) are found useful as tranquillizers and sedatives in formulations, similar to meprobamate.

Valtrate, didrovaltrate, deoxido-didrovaltrate were tried on rat hepatoma cells (HTC strain) and for their effect on synthesis of DNA and proteins. These compounds rapidly and essentially inhibited the synthesis of both DNA and proteins.

The leaf extract was found to have analgesic effect in rats.

Benzene and alcoholic extracts of rhizomes produced marked and mild hypothermia respectively. Both benzene and alcoholic extracts produced antipyretic effect. Benzene and alcoholic extracts produced mild CNS-depressant activity.

The extracts exhibited antibiotic activity against Micrococcus pyogenes and Entamoeba histolytica.

Use in western herbal

Constituents of Valeriana officinalis, used in western herbal, include valtrates, didrovaltrates and isovaltrates. Other constituents include monoterpenes and sesquiterpenes, caffeic, gamma-aminobutyric and chlorogenic acids, beta-sitosterol, methyl-l, 2-pyrrolketone, choline, tannins, gums, alkaloids and resin.

The co-occurrence of three cyclopentane-sesquiterpenoids (valerenic acid, acetoxyvalerenic acid and valerenal) is confined to Valeriana officinalis and permits its distinction from Valeriana edulis and Valeriana wallichii. (WHO.)

German Commission E recognizes the internal use of valerian for restlessness and sleeping disorders based on nervousness.

Valerian is a main component in an approved Commission E "Sedative Tea", composed of 40 % valerian, 30 % Passion Flower herb and 30 % Lemon Balm leaf.

Over-the-counter, Valerian tablets (150 mg), Root vegicaps (520 mg) and tincture of valerian is available. It ranks as one of the most widely used over-the-counter herbal sedatives and tranquillizers. It is also included in remedies for menopausal and premenstrual syndrome, and period pains.

In Germany alone, valerian is an active ingredient in more than 100 over-the-counter tranquillisers and sleep aids, some of which are especially formulated for children.

See Nardostachys jatamansi.

Dipterocarpaceae

VATERIA

Vateria indica Linn.

Figure 1 Vateria indica—flowers and fruit [WOI]

Vateria malabarica Bl.

Habitat

Peninsular India, from Kanara to Trivandrum, ascending to 360 m and in Coorg up to 1050 m.

Classical and common names

Ayurvedic: Sarja, Sarjaka, Marich-patrak, Ajakarna, Shaala (also a synonym of Shorea robusta). Sarjarasa (oleo-resin).
Unani: Raal (oleo-resin).
Siddha: Kundurukam.
English: White Dammer of South India, Indian Copal, Piney Resin.

Parts used

Resin, oil.

Classical use

Charaka used dried and pulverized plant of Ajakarna, in prescriptions, internally and externally, for obstinate skin diseases, piles, fistula, and as a sedative.

Sushruta prescribed Ajakarna in obesity, urethral discharges, jaundice, cutaneous affections, anomalies of urinary secretion; externally for ringworm, immature boils, lymphatic venereal disease.

Powder of Sarjarasa, mixed with jaggery, was prescribed for diarrhoea and sprue syndrome (Bhaavaprakaasha).

A paste of Sarjarasa, rock-salt, honey and mustard oil was prescribed topically for cracks in feet; powder of Sarjarasa was dusted over witlow (Vrindamaadhava).

Oleo-resin of Shaala (Shorea robusta) is used by Ayurvedic practitioners, and that of Vateria indica by Unani practitioners as Sarjarasa and Raal.

Among over-the-counter drugs, Habb-e-Raal (Bayaz-e-Kabir) of Unani medicine is prescribed internally for diarrhoea, cramps, gastric and duodenal ulcers; Marham-e-Raal (Qarabadeen-e-Qadri) externally for syphilitic ulcers, fistula, nasal ulcers; Marham-e-Khanazeer (Bayaz-e-Kabir) for cervical adenitis.

In folk medicine, the oleo-resin of Vateria indica is used for the treatment of several diseases—diarrhoea, tonsillitis, chronic bronchitis, tubercular glands, rheumatism, and piles. Mixed with sesamum oil, it is given in gonorrhoea, and with purified butter and long-pepper (Piper longum) for the treatment of syphilis and ulcers.

The juice of leaves is applied to cure burns. It is orally administered to prevent vomiting.

Active principles and pharmacology

The resin is a complex mixture of several triterpene hydrocarbons, ketones, alcohols and acids, along with small amounts of sesquiterpenes. On distillation, the oleo-resin gave an essential oil (76 %) with a stray balsamic odour. The oil is found to consist of phenolic constituents and azulenes, with the latter predominating. It contains D-camphaane, alpha- and beta-pinene, limonene, chamazulene, beta-caryophyllene, D-camphor, alpha-terpineol, D-borneol and thymol.

Bark contains polyphenols—DL-epi-catechin, laevorotatory isomers of fisetinidol, afzelechin and a coumarin—bergenin. Bergenin is also present in seeds.

Chief constituents of fatty acids from seeds are stearic and oleic. Other acids reported are myristic, palmitic, arachidic and linoleic. Non-saponifiable matter comprises mostly of sterol.

The essential oil shows marked antibacterial activity against Gram-positive and Gram-negative microorganism. Pronounced inhibition of growth has been recorded in vitro against Bacillus subtilis, Bacillus pumilis, Vibrio cholera, Micrococcus pyogenes var. aureus, Pseudonomas solanacearum, Salmonella typhi, Sarcina lutea, Shigella dysenteriae, Streptococcus faecalis and S. pyogenes, whereas lesser response was shown towards Staphylococcus albus and Corynebacterium diptheriae.

Poaceae

VETIVERIA

Vetiveria zizanioides (Linn.) Nash
Andropogon muricatus Retz.
Andropogon squarrosus auct. non L.
Anatherum zizanioides (L.) Hitchcock & Chase

Habitat

Found throughout the plains and lower hills of India, particularly on the riverbanks and in rich marshy soil, ascending to an altitude of about 1200 m. Cultivated in Kerala, Tamil Nadu, Karnataka, Andhra Pradesh.

Classical and common names

Ayurvedic: Ushira, Bahu-muulaka, Sugandhi-muula, Jataamedaa, Indragupta, Nalada (Charaka, Sushruta); Laamajjaka (Bhaavaprakaasha), Sevya (Shaarangadhara Samhitaa).
Unani: Khas.
Siddha: Vattiver.
English: Vetiver, Cuscus.

Parts used

Dried root, grass, oil.

Dose

Juice 25–50 ml, infusion 50–100 ml, powder 3–6 g.

Classical use

Charaka and Sushruta used powdered grass or its infusion in bilious fever, haemoptysis, toxicosis, suppurated conditions, chronic skin diseases, piles, debility, senility.

Charaka prescribed Ushira, with Aamalaka (Emblica officinalis) juice, sandalwood powder, sugar and rice-water, for treating burning sensation, thirst, loss of consciousness, vomiting, and haemorrhage.

Fresh Ushira was a substitute for Rakta Chandana (Pterocarpus santalinus) in Ayurvedic medicine. It belongs to the Saarivaadi Group of herbs, specific for morbid thirst, bilious fever, burning sensation, and haemorrhagic diseases.

Ushira was used externally for prickly heat and boils caused by excessive perspiration (Vrindamaadhava), and for alleviating oppressive heat.

Among drugs available over-the-counter Shadanga Kwaath Churna (Ashtaanga Hridaya) is prescribed during fevers; Ushiraasava (Bhaishajya Ratnaavali) in haemorrhages from respiratory and alimentary tract. Dashmuulaarishta is a restorative and nervine tonic; Arvindaasava a carminative; Balaarishta is for rheumatic affections. Ushira is also an ingredient in Chandanaadi Taila, a cooling massage oil. (All based on Bhaishajya Ratnaavali.)

In Unani medicine, Khas is used in fevers, headache and palpitation as a cooling tonic to heart and brain, and as a blood-purifier.

Essence of Khas or oil is administered in therapeutic doses for checking vomiting during cholera. Sharbat Khas is a cooling and refreshing drink of Unani medicine.

Active principles and pharmacology

Major constituents of essential oil are vetiselinenol and khusimol. Others include khusilal, elemol, 10-epi-eudesmol, beta-eudesmol, vetiverol, cyclocopacampherol, alpha- and beta-vetivone, laevojujenol and several other sesquiterpenoids including vetidiol.

The oil is the most complex of the essential oils. The laevorotatory North Indian Khas oil is unique in containing both cadinane and eudesmane sesquiterpene of the unusual antipodal configuration. The constituents of typical dextrorotatory oils (Haiti, Java, South India) are largely nootkatanes, vetispiranes and substances of tricyclic zizaane structure.

Roots from North India have been found to yield fructose, glucose, sucrose and glycerol, besides the oil.

Analysis of the grass gave the following values: crude protein 6.1–6.7; ether extract 1.1–2.1; crude fibre 34.7–42.7; N-free extr. 45.0–47.4; total ash 5.3–9.0; calcium 0.28–0.31 and phosphorus 0.05–0.60 %.

The oil, biologically, acts as a carminative in flatulence, colic and obstinate vomiting. It also exhibits a stimulant, diaphoretic and refrigerant properties. When locally applied in rheumatism, lumbago and sprain, it is a good embrocation and is found to afford relief.

Apocynaceae

VINCA

***Vinca rosea* Linn.**
***Lochnera rosea* (L.) Reichb.**
***Catharanthus rosus* (Linn.) G. Don**

Habitat

Commonly grown in Indian gardens. Native to Madagascar.

Classical and common names

Ayurvedic: Sadaapushpaa, Sadampushpa (non-classical), Nityakalyaani. Sadaabahaar.
Siddha: Sudukadu mallikai.
English: Madagascar Periwinkle.

Parts used

Whole plant.

Traditional use

The plant has been used as a folk remedy for diabetes in India (also in Sri Lanka and South Africa). An infusion of leaves is given for menorrhagia.

It is now an obsolete drug, not considered safe for diabetes. The plant was never used for its antileukemic activity in Indian medicine.

Active principles and pharmacology

Plant contains about 0.86 % of alkaloids in roots; 0.67 % in leaves and 0.31 % in stems.

More than 100 alkaloids and related compounds have so far been isolated and most of these characterized. These include a number of monomeric indole alkaloids, a few 2-acylindoles and oxindole, several alpha-methylene-indolines, several dihydroindoles, bisindole and more than a dozen alkaloids isolated during biosynthesis experiments.

Vinblastine and vincristine, known as vincaleucoblastine (VLB), (vin)leurocristine (VCR), vincoside and iso-vincoside (strictosidine) are most important, the first two from clinical and the rest from the biogenetic points of view.(CIMAP.)

Rose-purple flowers contain an anthocyanin.

L. rosea alkaloids inhibit the growth of Vibrio cholerae and Micrococcus pyogenes var. aureus, but possess no antibacterial action against enteric group of organism. Vindoline and other alkaloidal fractions from the leaves are active against Micrococcus pyogenes var. aureus and var. albus, Streptococcus haemolyticus, Cornybacterium diptheriae. Leaf extracts form a useful antibacterial agent for the treatment of streptococcal and staphylococcal infections.

VLB is used in the treatment of Hodgkin's disease; in testicular cancer, breast cancer, bladder cancer and non-small cell lung cancer. VCR, though highly active, is rarely used alone due to neurotoxicity as side-effect. It is used as a critical component of all major combinations in treatment of childhood leukemia, Wilm's tumour, embryonal rhabdomyosarcoma, non-Hodgkin's lymphomas; melanoma, small cell cancer.

VLB, as also VCR, are used in non-malignant diseases as well as in thrombotic thrombocytopenic purpura and drug-induced microangiopathic haemolytic anaemia.

VCR is of greater clinical importance than VLB.

A high degree of diuretic activity with catharanthine and vindolinine, and lesser activity with lochrovicine and vincolidine has been established in rats. Ajmalicine is antidiuretic and antihypertensive.

Hypoglycaemic activity, negative with the crude extractives, was positive in various degrees with a number of pure alkaloids—catharanthine, leurosine, lochnerine, tetrahydroalstonine, vindoline and vindolinine. A correlation was observed that several alkaloids and derivatives which exert hypoglycaemic response also produce diuresis. (CIMAP.)

Ajamalicin, also known as raubasin, extracted from root, is a vasodilator and hypotensive agent. Its sedative and tranquillizing properties are more marked than those of Rauvolfia serpentina.

Pericalline, perivine, periformyline, leurosivine, VCR, VLB, perividine, vindolinine and carosine showed in vitro antiviral activity.

Leaf extract 0.05–0.40 g/kg showed dose-dependent anti-inflammatory activity against carrageenin-induced rat hind paw oedema.

Vincristine sulphate showed absence of sperm in seminiferous tubules, depletion of germinal epithelial elements and formation of hypertypic giant cells in mature male rats. Continuous administration (10.0–20.0 mcg doses) of vincristine per day for 15 days to rats resulted in drastic regression of the entire male reproductive system.

Use in Western herbal

Periwinkle (Vinca minor) is included in the list of unapproved herbs of German Commission E.

In folk medicine, Periwinkle is used for circulatory disorders, cerebral circulatory impairment and support for the metabolism of the brain. (PDR.)

Vincamine as a pure substance is available for therapeutic use.

Tincture of the whole plant of Vinca minor is used in homoeopathy for "tremulousness in all blood-vessels". (John Henry Clarke.)

Caution

In animal studies, administration of Periwinkle (Vinca minor) caused destruction of blood components, manifested as leukocytopenia, lymphocytopenia and suppression of the immune system.

Indole alkaloids of Vinca rosea produce granulocytopenia and bone marrow depression through mitotic arrest. (Francis Brinker.)

Violaceae

VIOLA

Viola odorata Linn.

Habitat

Found in Kashmir and other parts of the western Himalayan regions at altitudes of 1500–1800 m; frequently cultivated in gardens.

Classical and common names

Unani: Banafshaa, Banafsaj, Kakosh, Fareer.
English: Sweet Violet.

Parts used

Flowers.

Dose

5–6 g.

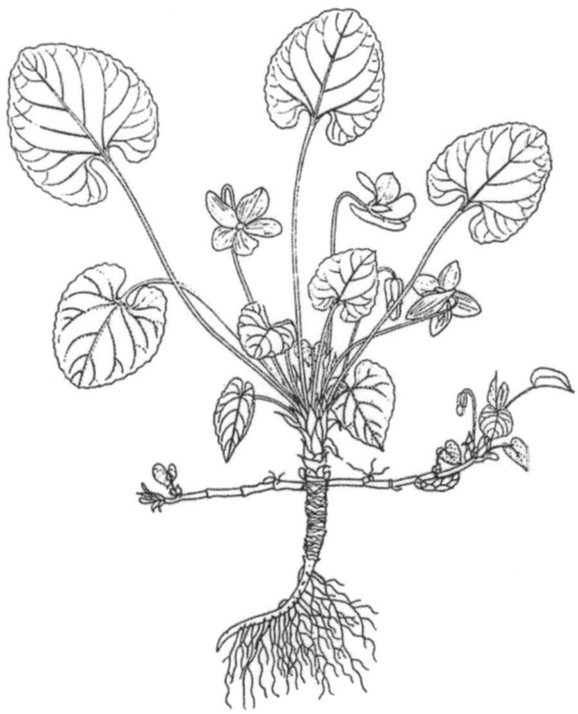

Figure 1 Viola odorata—flowering plant [WOI]

Classical use

In the olden days, Banafshaa was imported from Iran, now it is collected in Kashmir, Kangra and Chamba during its flowering season (April-July).

The herb (mature flowers) are used in Unani medicine as expectorant, diaphoretic, antipyretic and diuretic; alone or in prescriptions, for catarrhal and pulmonary affections. The drug is generally prescribed in the form of a decoction, infusion, jam (Gulkand) or syrup (Sharbat).

Khamira-e-Banafshaa and Sharbat-e-Banafshaa (Bayaz-e-Kabir) are prescribed for catarrh of the respiratory tract, bronchitis, fever. Gulkand Banfshaa is prescribed as laxative. Raughan-e-Banafshaa (Qarabadeen-e-Qadri) is prescribed for external massage in cephalalgia, insomnia, induration of joints.

Unani physicians attribute anti-inflammatory properties to Banafshaa. Its decoction or infusion is administered in pleurisy, diseases of liver and intestines.

Active principles and pharmacology

Flowers contain an emetic principle called violin (present in all parts of the plant), which is acrid and bitter, a volatile oil, rutin (2 %), cyanin (5.3 %), a colourless chromogen, a glycoside of methyl salicylate and sugar.

The volatile oil contains alpha- and beta-irones, alpha- and beta-ionones, violet-leaf aldehyde (3–5 %) and a number of alcohols.

The ketones are responsible for the characteristic odour of the flowers.

High content of tocopherol is also reported from flowers.

The leaves contain an essential oil, an alkaloid, colouring matter, friedelin (0.016 %), beta-sitosterol (0.033 %) and a straight chain alcohol.

The root stock contains saponins (0.1–2.5 %), which explains its usefulness as an expectorant, a glycoside of methyl salicylate, an essential oil (approx. 0.04 %) closely resembling that of flowers, 2-nitropropionic acid (0.001–0.002 %) and an alkaloid (1.41 %) with marked hypotensive activity.

The roots contain an alkaloid-odoratine, which was identified as triacetonamine.

In experiments conducted on rats, an extract of the herb (containing an emetine-like alkaloid) was found to be effective against induced inflammation.

The herb shows antimycotic and antibacterial activity, gives good results in the treatment of eczema. Poultice of leaves is found to give relief in cancerous growths.

The tincture of fresh flowering herb is used in homoeopathy for the treatment of dyspnoea with violent cough, whooping cough, sore throat, swelling of cervical glands.

Use in Western herbal

Pliny prescribed a liniment of violet root and vinegar for gout and disorders of spleen.

John Gerard (1545–1612) wrote of Violets: "The flowers are good for all inflammations, especially the sides and lungs, they take away hoarseness of the chest, the ruggednesse of the winde-pipe and jawes and takes away thirst... There is likewise made of Violets and sugar... especially it comforteth the heart and other inward parts."

Nicholas Culpeper (1616–1654) wrote: "All the violets are cold and moist while they are fresh and green, and are used to cool any heat or distemperature of the body, either inwardly or outwardly, as inflammations in the eyes, in the matrix or fundament, in imposthumes also, and hot swellings, to drink the decoction of the leaves and flowers made with water or wine, or to apply them poultice-wise to the grieved places... Flowers (fresh or dry) are effectual in pleurisy and all diseases of the lungs... good also for the liver and jaundice."

An infusion of leaves in boiling water (1 to 5) was recommended for cancer of the tongue; also for fomentation of the affected part. (M. Grieve.) The herb has long been used in cancer treatment notably for breast and skin cancers. (Penelope Ody.)

Violet Salve, a skin cream, is sold over-the-counter in Germany and Austria.

Verbenaceae

VITEX

Vitex negundo Linn.

Figure 1 Vitex negundo—flowering branch [WOI]

Habitat

Throughout the greater part of India, ascending to an altitude of approx. 1500 m in the outer Himalayas.

Classical and common names

Ayurvedic: Nirgundi, Nilanirgundi, Sindhuvaara (Charaka).
Unani: Sambhaalu, Fanjankisht.
Siddha: Noohi.
English: Five-leaved Chaste-Tree.

Parts used

Leaves, seeds.

Dose

2–3 g.

Classical use

Charaka used paste of the tubers and leaves, in prescriptions, externally, for neuralgic pain in the supraclavical region, sinus, fistula, scabies.

Sushruta prescribed flowers in dyspepsia, catarrh, cough and asthma.

In Bhaavaprakaasha two varieties were mentioned—the white-flowered was known as Sindhuvaara, the blue-flowered as Nirgundi or Shephaali. Nirgundi belongs to the Surasaadi group of herbs of Ayurvedic medicine, considered specific for cough, rhinitis, asthma. This group helps in the cleansing of ulcers.

Leaves were prescribed for dispelling inflammatory swellings of joints from acute rheumatism and of testes from gonorrhoeal epididynitis and orchitis. Eclipta alba, Ocimum sanctum, Trachyspermum ammi seeds were added to leaves.

Hot poultice of leaves was applied to affected parts. Paste of leaves was applied to temples during headache.

A decoction of leaves with long pepper (Piper longum) was prescribed in catarrhal fever (Bhaavaprakaasha). Purified butter, cooked with the juice of Nirgundi leaves, was a remedy for cough (Ashtaanga Hridaya); purified butter, cooked with the root, fruit and leaves, was prescribed for consumption (Chakradatta).

The root, pounded with water, was prescribed as a snuff (Vrindamaadhava) and Nirgundi Taila (Shaarangadhara Samhitaa, Vrindamaadhava) as a massage oil, for cervical adenitis.

In puerperal diseases, a warm decoction of Nirgundi, Lashuna (Allium sativum) and Shunthi (Zingiber officinale) added with Pippali (Piper longum), was administered (Yogaratnaakara).

Extract of Nirgundi in purified butter, mixed with black pepper (Piper nigrum) was prescribed for splenic enlargement. Leaves were applied as plaster externally.

Juice of leaves was applied topically for removing foetid discharge from wounds and ulcers.

A compound preparation of oil, Nirgundi Taila, available over-the-counter, is used as a dressing for venereal sores, chronic ulcers and skin diseases (Sarva-Roga-Chikitsaa-Ratnam).

In folk medicine, an oil prepared with the juice of leaves is applied to gangrenous wounds. Juice of leaves is applied for diseases of the scalp and as a hair tonic for promoting growth of hair.

Roots are used for their febrifugal, expectorant and diuretic properties. They are used in dyspepsia and rheumatism, and also for boils. The powdered root is consumed as an anthelmintic; and as a demulcent in dysentery; also for piles.

Flowers are used in fever, diarrhoea and liver complaints; fruits in headache, catarrh and coryza.

Ayurvedic physicians prescribe Nirgundi Kalpa (a rejuvenating tonic of Bhaishajya Ratnaavali) for retarding old age and for retaining and promoting virility. 30 g powdered root of the herb is mixed with 60 g honey and kept sealed in a non-absorbant earthen pot, preferably pretreated with clarified butter, for one month under moderate temperature. This tonic is prescribed 10 g to 20 g twice daily for one month.

In Unani medicine, seeds of Vitex agnus-castus Linn. are used. These are imported from Iran. (This plant is also five-leaved, sometimes seven-leaved.) Seeds are known as Tukhm-e-Sambhaalu. Leaves are used only externally.

Sambhaalu seeds are administered internally with Sikanjbeen (Sugarcane vinegar) for dispersing hard swellings, especially that of spleen; and as diuretic and deobstruent.

Sufoof Fanjankisht is prescribed in spermaturia. Powdered seeds (2 g–3 g/day) are prescribed for spermatorrhoea; also for spermatochesis.

For promoting sperm count, a compound of Sambhaalu seeds and Kushta-e-Jast (zinc) is prescribed.

For increasing the libido, Unani physicians give powdered seeds of Sambhaalu and dry ginger (Zingiber officinale) with milk (10 g/day).

Active principles and pharmacology

Seeds of Vitex negundo gave the hydrocarbons n-tritriacontane, n-hentriacontane, n-pentatriacontane, and n-nonacosane. Other constituents of seeds are beta-sitosterol, p-hydroxybenzoic acid and 5-oxyisophthalic acid.

Leaves yielded an alkaloid, nishindine; the flavonoids 5-dydroxy-3,6,7,3',4'-pentamethoxyflavone and casticin; the irridoid glycosides angusid, aanguusid and 2-p-hydroxy-benzoylmussaenosidic acid; and an essential oil. Other constituents of leaves are vitamin C, carotene, gluco-nonitol, p-hydroxybenzoic acid, 5-oxyisophthalic acid, 3,4-dioxybenzoic acid and sitosterol.

Stembark contains two leucoanthocyanidins. Bark contains a fatty acid, beta-sitosterol, vanillic acid, p-hydroxy benzoic acid and luteolin. Presence of flavonoid C-glycoside is also indicated. (CIMAP.)

The extract of leaves and twigs showed antibacterial activity against Micrococcus pyogenes var. aureus and E. coli; antifungal activity against Trichoderma viridiae, Fusarium helminthosporium.

Ethyl acetate extract of leaves of Vitex negundo in 50 mg/kg p.o. produced significant anti-inflammatory effect against carrageenin, bradykinin and 5-HT-induced hind paw oedema. The extract of leaves showed protection against turpentine pleurisy in rats.

Seeds of Vitex negundo exhibited significant anti-inflammatory effects as compared to betamethasone and phenylbutasone; produced marked reduction in exudates (67 %) and in control studies with betamethasone and phenylbutasone 70 % and 50 % respectively, markedly suppressed early and delayed inflammatory changes.

Butanol extract of the root showed anti-inflammatory and analgesic effect. Cold aqueous infusion and chloroform extracts of the root markedly inhibited oxytremorine-induced tremors in mice.

70 % ethanolic extract of the root in 400 mg/kg p.o. exhibited no antispermatogenic, antiovulatory activities in experimental animals. (CCRAS.)

Vitex negundo was clinically tried on patients of oligospermia by scientists of University College of Medicine, Kolkata. Dry powder of negundo plant leaf (250 mg) was given with cold water for 120 days. After 75 days, optimum increase of nearly 3.5 times in sperm count was seen. The ratio of live and dead sperm was also increased to optimum level from 1:1 to 4:1. Withdrawal of the drug affected the sperm count and it sloped down to base level within another 120 days (MAPA, 9602–0732.)

Use in Western herbal

Modern medical work with vitex began with the introduction of concentrated extract of the vitex fruit in the 1950s. From 1943 to 1997, approximately 32 clinical trials were conducted on a proprietary Vitex agnus-castus berry product (Angolyt, Madaus, Germany) for evaluating its effectiveness in treating PMS, mastitis and fibrocystic disease, menopausal symptoms, poor lactation, hyperprolactinemia, uterine bleeding disorders and menstrual irregularities. (Expanded Commission E Monographs.)

German Commission E approved the use of Vitex agnus-castus fruit for irregularities of the menstrual complaints and mastodynia.

T. Michael Murray and Joseph Pizzorno have quoted one study which shows that Vitex agnus-castus extract lowers prolactin levels in men at a dose of 480 mg per day. (Males with hyperprolactinemia frequently face impotence.)

In homoeopathic medicine, Vitex agnus-castus is used for various sexual debilities—marked depression of vital power, premature old age with apathy, self-contempt for sexual abuse, nervous debility in unmarried persons, feeble erection without sexual desire, suppressed sexual desire, emission of prostatic fluid when straining at stool, cold, hard, swollen, painful testicles (John Henry Clarke). In general practice, the drug is prescribed to females for leucorrhoea staining yellow, suppressed menses, scanty or suppressed breast milk, inflammation of the uterus.

Vitaceae

VITIS

Vitis vinifera Linn.

Habitat

Cultivated in many parts of India.

Classical and common names

Ayurvedic: Draakshaa (Charaka, Sushruta), Gostani.
Unani: Angoor. Dehydrated fruit—Daakh, Kismis, Maweez, Munaqqaa.
English: Grapevine.

Parts used

Leaves, stems, berries.

Classical use

Charaka and Sushruta used the shoots of the plant in prescriptions, the fruits alone or in drink, for cough, asthma, consumption, cardiac disorders, haemothermia, fever, digestive derangement, urinary affections.

Charaka prescribed juice of Draakshaa and Aamalaka (Emblica officinalis) in jaundice and anaemia. Draakshaa Ghrita of Charaka was specific for anaemia.

Sushruta prescribed a decoction of Draakshaa and Aaragvadha (Cassia fistula) for allaying burning sensation and excessive thirst during fevers; a linctus of Draakshaa, Pippali (Piper longum) and honey for consumption.

Raisins (Kismis), with powdered Terminalia chebula and honey, were prescribed for acid dyspepsia (Chakradatta); raisins, with Emblica officinalis, dates, long pepper, black pepper, were prescribed for cough, catarrh and respiratory diseases (Shaarangadhara Samhitaa).

During the 16th century, grapes were prescribed for dysuria, stranguary, jaundice, severe cold and fevers, thrush in children; raisins for jaundice, anaemia, sub-acute cases of enlarged liver and spleen; in addition to their known applications.

Grapes and raisins were incorporated in 25 compound preparations of Shaarangadhara Samhitaa and 60 of Bhaavaprakaasha.

Over-the-counter, Draakshaarishta of Gadanigraha, as well as that of Shaarangadhara Samhitaa, is available as a cardiac tonic, mild laxative, digestive and appetizer, alterative in convalescence. Draakshaasava (Yoga Ratnaakara) Draakshaavaleha (Vaidya Saar Sangraha) and Draakshaadi Churna (Vaidya Chintaamani) are also available.

Draakshaa is incorporated in a number of compound preparations of Sahasrayoga—Amritapraasha Ghrita, Shataavari Ghrita, Sukumar Ghrita, to name but a few.

Ashoka Ghrita, a uterine tonic of Bhaishajya Ratnaavali, also contains Draakshaa.

In folk medicine, a bigger variety of raisins with seeds, boiled with water and equal quantity of milk, are prescribed for diabetes. Juice of unripe grapes and leaves is used as astringent. Fixed oil of seeds or powdered seeds and leaves are prescribed for chronic diarrhoea.

Ash of the wood is administered as a preventive of stone in bladder and for treating cold swelling of the testes. The ash is also used internally and externally for piles.

In Unani medicine, tender leaves are prescribed internally for dull and numb headache, heaviness and dizziness. Their paste is applied topically for headache due to heat, inflammations of the eye and burning sensation. Juice of leaves is administered for intestinal affections, diarrhoea, haemoptysis. A syrup, prepared from the juice of leaves, is prescribed in palpitation, hangover and nausea.

Roasted Munaqquaa, with black-salt, is prescribed as a laxative during convalescence.

Pounded flowers are prescribed for piles; also vinegar of grapes.

Sap of young branches is applied externally to ringworm and other cutaneous affections.

Ash of the wood is prescribed as antilithic. Also for headaches of various aetiology.

Over-the-counter, Sharbat Maveez is a mild laxative, digestive and appetizer. Maweez Munaqqaa is the main drug of Majoon-e-Zabeeb (Bayaz-e-Kabir), specific for epilepsy.

Active principles and pharmacology

Composition of ripe berries of some types of grapes grown in India is as follows: moisture 72.9–85; total sugars 13.4–20.6; acidity and tartaric acid 0.41–1.49; tannins 0.10–0.60; crude protein 0.7–1.1; fat 0.0–0.3; ash 0.46–0.61 %.

Grapes are rich in reducing sugars, but poor in protein. They are a fair source of iron and other minerals, traces of bromide, iodide and fluoride are normally found; also copper in traces.

Grapes are a poor source of vitamins. Carotene 2 mcg; thiamine 0.07; riboflavin 0.19; niacin 0.7 and vitamin C 1.0 mg/100 g edible portion, are reported to be present in raisin-type grape. The types and environmental factor greatly influence the B vitamins content.

Vitamin content of grapes increases during maturation, except for biotin which decreases.

Leaves are rich in carotenoids. Their content reaches the maximum value when vine is at full bloom. Leaves contain thiamin, niacin, biotin and tocopherol, and are richer in ascorbic acid than berries.

Grapes are a good source of bioflavonoids (vitamin P), which are known to be useful in purpurea, capillary bleeding in diabetes, inflammation from injury, radiation damage and arteriosclerosis. Catechins and anthocyanogenic tannins present in grapes possess bioflavonoid activity. Vine-sprouts and leaves are rich in bioflavonoids.

The total level of phenols in the berry is subject to environmental influence. Vitis vinifera produces monoglucosides, whereas other species mostly produce diglucoside. Some species contain both mono- and diglucosides. The pigments of red and black types of Vitis vinifera consist of monoglucosides, delphinidin, petunidin, malvidin and/or cyanidin and peonidin. Caffeoyl and p-coumaroyl derivatives of anthocyanins also occur. All the monoglucosides are 3-glucosides and all diglucosides are 3, 5-diglucosides.

Flavonols are present in comparatively larger amounts in leaves than in berries. Skins of grapes contain quercitin, kaempferol and myricetin in the form of glucosides. Flavones generally appear to be absent in berries.

Catechin, epicatechin and epicatechin gallate are the major flavan-3-ols occurring in seeds of many types of grapes. The extract of grape seed has been classified as a chemoprotective agent and one of the botanical antioxidants.

Catechin is present in the leaves, roots and all parts of vine of several types. Stems are relatively rich in phenolic content.

Analysis of a market sample of raisins (dehydrated grapes) gave following values: moisture 15.6; crude protein 2.2; reducing sugars 72.8; acidic as tartaric 1.8; tannins 0.3 and ash 2.0 %. Ca 64.7 mg; P 76.9 mg; and Fe 4.1 mg/100 g.

Grapes, both fresh and dried, exhibit demulcent, expectorant, cooling, laxative, stomachic and diuretic properties. Leaves and sap of the young branches are astringent.

W

Solanaceae

WITHANIA

Withania coagulans Dunal
Puneeria coagulans Stocks.

Habitat
Drier parts of Punjab, also around Shimla, Garhwal and Kumaun.

Classical and common names
Unani: Desi Asgandh, Kaaknaj-e-Hindi, Paneer.
English: Vegetable Rennet.

Traditional use
Fruits are used as sedative, alterative, diuretic and emetic. In dried form they are employed in dyspepsia, flatulence, colic, intestinal affections, chronic complaints of liver, asthma and stranguary.

Sushruta's Ashwagandhaa was used externally for quick healing of wounds and as an emetic, and matches well with Withania coagulans. (Root of Withania somnifera is not used as an emetic.)

Active principles and pharmacology
Roots contain the withanolides 3-beta-hydroxy-2, 3-dihydro-withanolide,5,27-dihydroxy-6-alpha-7-alpha-epoxy-1-oxo-(5-alpha)-witha-2,24-dienolide,5,20-alpha-(R)-dihydoxy-6-alpha-7-alpha-epoxy-1-oxo-(5-alpha)-witha-2,24-dienolide and withaferin A. Later two compounds are also present in leaves.

Leaves, in addition, contain a withanolide, withanone.

Berries contain triacontane, dihydrostigmasterol and the free amino acids proline, hydroxyproline, valine, tryosine, aspartic acid, glycine, aspargine, cysteine and glutamic acid; and a milk-coagulating enzyme.

Total alkaloids from fruits exhibit anti-inflammatory; fruit and plant proteolytic; 3-beta-hydroxy-2, 3-dihydrowithanolide E anti-inflammatory activities.

Alcoholic extract of fruit was very active against Micrococcus pyogenes var. aureus and E. coli. Hexene, ethanol and aqueous extracts also showed antibacterial activity.

The essential oil is active against Micrococcus pyogenes var. aureus and Vibro cholerae; also showed anthelmintic activity.

No distinction is made between the berries obtained from Withiana coagulans and those from Withiana somnifera. Both species have identical macroscopical features. (Withiana coagulans and Withiana somnifera are known by the same name in most of the regional languages.)

Withania ashwagandha Kaul
Withania somnifera (Linn.) Dunal

Figure 1 Withania somnifera [CCRAS]

Figure 2 Withania somnifera [CCRAS]

Habitat

Throughout drier parts of India in waste places and on bunds. Extensively cultivated in Manaasaa (district Manasor) and Ganj Basodaa (district Vidisha) of Madhya Pradesh. Except for a limited collection of the roots from wild plants growing in Bikaner and Pilani areas of Rajasthan, most of the roots available in the market are obtained from the cultivated plants.

Indian plants differ from the known chemotypes of Israel in the composition of withanolides.

The cultivated plants are reported to differ from the wild ones not only in their morphological characters but also in their therapeutic action, though the alkaloids present are the same in both (synergistic action of all constituents is to be taken into account). In view of these differences some botanists consider the cultivated plant distinct from the wild one and have given it a new specific name, Withania ashwagandha Kaul (KN).

Classical and common names

Ayurvedic: Ashwagandhaa, Ashwakanda, Gandharvagandhaa, Turaga, Turagagandhaa, Turangagandhaa, Vaajigandhaa (Bhaavaprakaasha); Gokarnaa, Vrshaa, Varaahakarni, Varadaa, Balyaa, Vaajikari (Ayurveda Saukhyam.)
Unani: Asgandh.
Siddha: Amukkura.
English: Winter Cherry. (Not to be confused with Winter Cherry of Western herbal, equated with Physalis alkekengi.)

Parts used

Root, ash.

Dose

Root powder 3–6 g, ash 1–2 g.

Classical use

According to the Wealth of India, it is difficult to accept Withania somnifera as the source of classical Ashwagandhaa (of Charaka and Sushruta).

Charaka divided plant substances in fifty groups according to physiological actions of their decoctions. Ashwagandhaa of his period was included in Vaajikarana (aphrodisiac) group, but was not mentioned in Jivaniya (promoter of longevity), Vayasthaapana (age-sustainer) and Rasaayana (rejuvenator) groups. Sushruta also did not include Ashwagandhaa in Rasaayana (rejuvenating) group.

Ashwagandhaa and Kapikacchuu (Mucuna pruriens) were included by Charaka in the same group of (Vaajikaran) herbs, specific "sex drugs" for promoting virility. Ashwaka of Sushruta (Dalhan) is now being equated with Mucuna monosperma DC. It is quite possible that Ashwagandhaa of Charaka and Sushruta period was a related species of Mucuna.

During the 7th century, Withania somnifera was included as Ashwagandhaa among twenty age-sustaining compound preparations of Ashtaanga Hridaya.

By the 16th century, Withania somnifera entered into more than fifteen formulations of Shaarangadhara Samhitaa. It was considered spermatogenic and libido-stimulant. Externally it was applied to stimulate the growth of the penis; also to relaxed vagina for restoring its original shape. Ashwagandhaadi Tailam, also known as Linga-vardhana Tailam, is still available over-the-counter. (It contains extracts of Withania somnifera and Mucuna prurita.)

In Bhaavaprakaasha, Ashwagandhaa was incorporated in more than twenty compound preparations and ten medicinal oils, for cough, asthma, consumption, emaciation, haemothermia, hyperacidity, cardiac affections, anxiety neurosis, mental and neurological disorders, muscular atrophy, oedema, gout, arthritis, spermaturia, diseases of the female genital tract, menstrual disorders, and sterility.

The powder of Ashwagandhaa root, with milk or jaggery or purified butter, was prescribed as a nourishing, rejuvenating and restorative tonic, for promoting virility in men and fertility in women. (Ashtaanga Hridaya, Raaja Maarttanda.)

To promote intellect, life span and virility, Manduukaparni (Centella asiatica), Shankhapushpi (Convolvulus pluricaulis) and Shataavari (Asparagus racemosus) were added to Ashwagandhaa (Ashtaanga Hridaya). For cough and consumption, Pippali (Piper longum) was added (Vrindamaadhava). When prescribed as a cardiac tonic, Bibhitaka (Terminalia bellirica) was added.

Single drug, mixed with sugar and purified butter, was prescribed for insomnia (Bangasena).

For arthritis, purified butter cooked with the decoction and paste of Ashwagandhaa along with milk, was prescribed. Ashwagandhaa Ghrita of Vrindaamaadhava was specific for arthritis and debilated conditions.

Among over-the-counter drugs, Ashwagandhaarishta (Bhaishajya Ratnaavali) is prescribed for anxiety neurosis; Ashwagandhaa Churna (Bhaavaprakaasha) for stress, fatigue and rundown conditions. Balaa-Ashwaganadhaa-Lakshaadi Taila (Sahasrayoga) is prescribed internally as well as externally as a nervine tonic and restorative.

Amukkuraa (Ashwagandhaa) Churnam of Siddha medicine is prescribed in rheumatic diseases, sexual insufficiency and spermatorrhoea.

In Unani medicine, Habb-e-Asgandh is prescribed as a supporting aphrodisiac and nervine tonic.

Paste of the root is applied to penis for toning up eractile tissues; mixed with mother's milk a paste of the root is applied to sagging breasts.

Majoon-e-Saalab is for sexual debility and attenuated semen, Majoon-e-Baladur for neurasthenia, dementia, amnesia.

Active principles and pharmacology

Roots contain the alkaloids nicotine, somnine, somniferine, somniferinine, withanine, withananine, withananinine, pseudo-withanine, tropine, pseudotropine, 3-alpha-tigloloxytropane, choline cuscuolohygrine, anaferine and its DL-form, isopelletrierine, anahygrine, withasomine and visamine. Roots also contain withanolides.

Leaves contain 12 withanolides including withaferin A, 27-deoxywithaferin A and 27-deoxy-14-alpha-hydroxywithaferin A.

Withaferin A at 10.0 mg/kg dose showed significant protective effect against CCl_4-induced hepatotoxicity in rats. It was as effective as hydrocortisone dose for dose.

Effect of concomitant administration of Withania somnifera and Piper longum produced marked histopathological changes in induced hepatotoxity. Results indicate that the drugs can be used as important adjuvants with antitubercular drug therapy.

A double blind clinical trial, to study the effect of Ashwagandhaa on the prevention of the process of ageing in 101 healthy male adults in the age group of 50–59 years, indicated that the increase in haemoglobin, RBC, hair melanin and seated stature in the treated group was statistically significant in comparison with the placebo. The decrease in ESR was much higher in the treated group than in the placebo. Ashwagandhaa was found to be an excellent haematinic in this study.

Ashwagandhaa was also tried as a nervine tonic. 71.4% people reported improvement in their sexual performance. Though it is a subjective clinical improvement, still the statement of the majority of volunteers is worth noting.

The total extract of Ashwagandhaa exhibits marked sedative action on mice, rats, rabbits, dogs and monkeys and a relaxant and antispasmodic action on ileal and tracheal muscle. An extract of Ashwagandhaa has prolonged hypotensive, bradycardiac and respiratory actions.

The methanolic extract of roots has been shown to potentiate pentobarbitone-induced hypnosis and pentylenetetrazol (PTZ)-induced seizure latency.

Research on drugs acting primarily on CNS, establishes Withania somnifera as an antianxiety and adaptogenic drug. Withania somnifera extract induced a significant decrease in the arterial and diastolic blood pressure in "normotensive" pentobarbital anaesthetized dogs.

Withania somnifera was administered 500 mg twice daily to a group of mountaineers throughout a 29 day tenure which included 17, 000 ft altitude gain through trekking and 6 days training in that height including climb to 21, 000 ft and then descending. At different altitudes both psychological and physiological parameters were repeated for assessment. Altered behavioural pattern was improved by Withania somnifera, also responsiveness, alertness, state of awareness along with physical capabilities.

A study was conducted to evaluate the protective/inhibitory effect of Withania somnifera and Ocimum sanctum on stress-induced gastric ulcer in albino rats. Macroscopical and histopathological findings such as congestion, erosion, discrete and multiple haemorrhages, ulcers and perforation in the control group of rats served for scoring the intensity of the lesions. A significant decrease of scoring of pretreated rats indicated the protective action of these herbal extractions.

Antitumour and radiosensitizing effects of alcoholic root extracts of Ashwagandhaa and their modification by heat were studied in vivo on sarcoma-180 grown on sorsum of adult BALB/c mouse. Ashwagandhaa was injected (i.p.) in a dose of 500 mg/kg body weight for 10 consecutive days. Ashwagandhaa increased the effect of radiation on tumour regression as well as growth delay. The combination of Ashwagandhaa for 10 days with one local exposure to gamma radiation followed by hyperthermia significantly increased the tumour cure, growth delay of partially responding tumours and animal survival.

Withania somnifera 1 mg/kg oral drug produced significant anti-inflammatory activity of 60.09 per cent in the second hour of its administration against carrageenin-induced paw oedema in rats.

It is suggested that drugs which show profound influence on acute phase proteins (a 2M) not only provide symptomatic relief during inflammatory conditions but also possible beneficial effects on the disease process itself. Administration of Withania somnifera caused inhibition of a 2M-synthesis as well as twofold enhancement to total protein synthesis. Reduction of albumin synthesis has been reported during inflammatory conditions.

It has been observed in some studies that treatment of Withania somnifera caused effective suppression of a 2M and significant/insignificant increase of albumin levels, hence the wide spectrum of pharmacological activities by Withania somnifera can be attributed to its effects on serum proteins.

Root powder of Withania somnifera was studied in 46 patients of rheumatoid arthritis with a dose of 4, 6 or 9 g/day for a period of 3–4 weeks. Pain and swelling disappeared in 14 patients, considerable improvement was observed in 10 patients, 11 patients showed mild improvement. There was no improvement in 4 patients and 7 patients discontinued the treatment.

In a study, a combination of Withania somnifera 1 g twice daily and Cyperus rotundus 1 g thrice daily was administered to patients of rheumatoid arthritis for a period of more than one year. Results were better in maiden cases (excellent 28%, good 12%, fair 45%, poor 17% and no response 2%). The response was poorer in long-term cortisone-treated cases (excellent 23%, good 19%, fair 21%, poor 26% and no response 13%). The efficacy of these herbs in cases of osteoarthritis was relatively lesser (excellent 12%, good 23%, fair 34%, poor 15%, no response 16%).

The effect of Withania somnifera and Smilax china have been assessed on 25 confirmed cases of cervical spondylosis. Treatment was given for 30 days (4 g powder of Withania somnifera along with 100 ml decoction of Smilax china thrice daily). The herbal drug treatment was free from any side effect.

Fifty cases of keratitis and rough skin (due to malnutrition, polluted environment, chemical cosmetics, lack of hygiene) in the age group of 30 to 50 years were treated with Ashwagandhaa (3 g twice daily with water) for eight weeks. The result has been found to be quite encouraging. The drug also stopped the falling of hair in females after delivery.

Subacute toxicity studies with repeated injections of Withania somnifera extract at a dose of 100 mg/kg body wt for 30 days in Wistar rats of either sex did not result in any mortality or changes in peripheral blood constituents.

Allied species:

Chemotype I (Israeli): contains withaferin A as major constituent. Minor constituents are withanolides N and O.

Chemotype II (Israeli): contains withanolide D as a major constituent. Withanolide G, 27-hydroxywithanolide D, 14-alpha-hydoxywithanolide D and 17-alpha-hydroxywithanolide D are present in trace amounts.

Chemotype III (Israeli): contains withanolides E-M. Leaves of offspring F3 (from crossing of chemotype I and III), contain withanolides Q and R. Withanolides P and S are also present in leaves.

The plants raised in Israel from Jamnagar (Gujarat) seeds contained only three withanolides: withaferin A, 5-alpha-17-alpha-dihydroxy-1-oxo-6-alpha-7alpha-epoxy-witha-2, 24-dienolide and a minor withanolide.

Lythraceae

WOODFORDIA

Woodfordia fruticosa (L.) Kurz

See figure 1.

Woodfordia floribunda Salisb.

Habitat

Commonly occuring throughout North India, ascending to an altitude of about 1500 m in the Himalayas. Rather scarce in South India.

Classical and common names

Ayurvedic: Dhaataki, Dhaatri, Kunjaraa, Taamrapushpi (Charaka).
Siddha: Velakkai.
English: Fire-Flame Bush, Shiranjitea.

Classical use

Charaka and Sushruta used sweetened decoction of flowers for fever, haemothermia, persistent dysentery; included Dhaataki in conception-promoting group of herbs.

Figure 1 Woodfordia fruticosa [ZANDU]

Powder of Dhaataki flowers, mixed with honey, was prescribed for leucorrhoea (Vrindamaadhava).

A paste of Dhaataki and Nilotpala (Nymphaea stellata) flowers, with honey was prescribed to women during season for conception (Gadanigraha).

Liquid gruel, processed with the decoction of Dhaataki flowers, Shunthi (Zingiber officinale) and pomegranate seeds, was prescribed in fever, diarrhoea and abdominal pain (Bangasena).

A sweet bolus, prepared of Dhaataki flowers, parched paddy and sugar, was given to children suffering from diarrhoea (Ashtaanga Hridaya).

Dhaataki was used as a substitute for Yashtimadhu (Glycyrrhiza glabra) in Ayurvedic medicine.

Flowers have been incorporated in most of the alcoholic medicinal compound preparations of Ayurvedic medicine for fermentation.

Among over-the-counter drugs, Dhaataki is an important ingredient in Gangaadhara Churna (Shaarangadhara Samhitaa), prescribed for diarrhoea, dysentery; Samangaadi Churna (Bhaavaprakaasha) for bleeding piles.

In folk medicine, a decoction of flowers is prescribed in menorrhagia and seminal weakness.

A paste flowers is used for the treatment of coughs.

Flowers also enter into an ointment used on the pustules of smallpox.

Dried flowers are powdered and dusted over ulcers and wounds to eliminate discharge and promote granulation.

Active principles and pharmacology

Flowers and leaves gave the polyphenols ellagic acid, polystachoside and myricetin-3-galactoside. Flowers also gave the anthocyanins pelargonidin-3,5-diglucoside and cyanidin-3,5-diglucoside, octacosanol, chrysophanol-8-O-beta-D-glucopyranoside and beta-sitosterol. Hecogenin, mesoinositol and the flavone glycosides quercetin-3-rhamnoside, naringenin-7-glucoside, and kaempferol, are also reported from flowers.

Bark contains the C-glucoside, bergenin. Stems contain octacosanol, beta-sitosterol, nor-bergenin (C-glucoside of gallic acid), bergenin and gallic acid.

Benzoic acid and cinnamic acids are also reported from the plant (CIMAP).

Three dimeric hydrolysable tannins—woodfordins A, B and C—and oenothein B have been isolated from dried flowers.

Oneothein B exhibited remarkable host-mediated antitumour activity. Woodfordin C also showed antitumour activity and inhibitory activity towards DNA-topoisomerase II.

An extract of the plant was found to stimulate the contraction of small intestines, and investigations have corroborated the clinical use of the drug in bowel complaints.

The drug shows antipyretic action which compares favourably with that of acetylsalicylic acid.

Rutaceae

ZANTHOXYLUM

Zanthoxylum armatum DC.

Figure 1 Zanthoxylum armatum—flowering branch and fruits [WOI]

Zanthoxylum alatum Roxb. and var. planispinum Zanthoxylum planispinum Sieb. & Zucc.

Habitat

Found in the hot valleys of the Himalayas from Jammu to Bhutan at altitudes of 1000–2100 m, in Khasi hills at 600–1800 m, and in the Eastern ghats in Orissa and Andhra Pradesh at 1200 m. Planted for hedges in Assam.

Classical and common names

Ayurvedic: Tumburu (Charaka, Sushruta); Tejabala, Tejaswani. Tejovati (bark). Nepaali Dhaniyaa.
Unani: Faaghir, Kabaab-Khandaa.
English: Toothache Plant.

Parts used

Fruit.

Dose

Powder 1–3 g.

Classical use

Charaka used Tumburu, in prescriptions, as errhine for heaviness and pain in head, hemicrania, chronic coryza, loss of smell, maggots and loss of consciousness.

Sushruta used Tumburu as a pain-relieving drug for discomfort and congestion in chest, cardiac region also for piles and tumours.

Tumburu was an ingredient of a potion of Sushruta prescribed for hysterical convulsions.

Charaka included Tumburu in medicinal oils and ointments for skin diseases. Kanakaakshi Taila of Charaka was one such product. Tumburu was also an ingredient in a paste applied locally to muscles for treating numbness and atrophy.

Tejovati (bark) was used as toothpowder for diseases of the mouth and throat (Bhaavaprakaasha).

In Unani medicine, the fruits, seeds and bark are used as a carminative, stomachic and anthelmintic. An extract of the fruits is prescribed for round worms.

A decoction of fruits is prescribed as a gargle in stomatitis.

Fruits are chewed as a mouth-freshner. Twigs are chewed and used as toothbrush for cleaning teeth.

Active principles and pharmacology

Seeds contain the flavonoids tambulin and tambulol (tambaletin).

Essential oil from fruits contain chiefly linalool and linalyl acetate. Other constituents are limonene, sabinene, citral, geraniol, methylcinnamate, 1, 8-cineole, p-cymene, *cis*-ocimene, gamma-terpinene, camphor, alpha-fenchol, carvone, tagetinol, allo-aromadendrene, alpha-thujene, alpha-terpineol, beta-caryophyllene and a rare monoterpene triol-3, 7-dimethyl-1-octen-3, 6, 7-triol.

Essential oil from leaves contain methyl-n-nonylketone, linalyl acetate, sesquiterpenehydrocarbons and tricosane.

Dried bark and branches contain the lignans sesamin, fargesin, eudesmin and epi-eudesmin; a neutral lactone-pulviatide. Several alkaloids have been obtained from the plant.

The stem has exhibited hypoglycaemic activity.

The essential oil possesses deodorant and antiseptic properties. In a dilution of 1 to 40, 000, the oil killed 50 per cent of ciliates. It showed good knock-down effect against one-day-old nymphs of Locusta migratoria Linn., but gave only 40 per cent mortality after 24 hours.

The freshly distilled essential oil from seeds exhibited strong antibacterial activity against E. coli, Vibrio cholerae, Micrococcus pyogenes var. aureus, Shigella dysenteriae and Salmonella typhi.

The powdered bark also showed 100 per cent mortality against one-day-old nymphs of Locusta migratoria.

The oil obtained by steam-distillation of the fresh plant showed antifungal activity against a number of fungi.

Fruits give good results in dental diseases and their lotion in scabies.

Use in Western herbal

Zanthoxylum americanum (Prickly Ash, Toothache Tree) was used in the US during the 19th century as a circulatory stimulant and to treat arthritis. It stimulates blood flow to painful and stiff joints, conditions where the arteries of the limbs have narrowed, and prevent sufficient blood reaching the hand or leg muscle.

Prickly Ash is applied topically to treat leg ulcers and chronic pelvic inflammatory affections. (Andrew Chevallier.)

The powdered bark forms an excellent application to indolent ulcers and for cleansing, stimulating, drying up and healing old wounds.

The pulverized bark is also used for paralytic affections and nervous headaches. The bark is chewed, as a popular remedy for toothache.

Berries are used as carminative and antispasmodic.

Labiatae

ZATARIA

Zataria multiflora Boiss.

Habitat

Distributed in Iran, Afghanistan and Pakistan. Leaves and flowers, mixed with slender woody portions of the stem, are imported into India for use in Unani medicine.

Classical and common names

Unani: Saatar Faarsi, Al-Saatar. Origanon.
English: Wild Thyme.

Parts used

Dried plant.

Dose

Infusion 5–7 g of herb.

Classical use

Three varieties of Saatar are used in Unani medicine: Sahraayi (wild), Kohi (found on hills) and Bustaani (cultivated in gardens). Infusion of the drug possesses a minty odour.

Saatar is used in Unani medicine as a stimulant, diaphoretic, anodyne, diuretic, emmenagogue, anthelmintic, antilithic and anti-inflammatory drug. An infusion is prescribed for detoxifying liver, stomach and intestines.

In inflammatory conditions of the liver, Saatar is administered internally with vinegar. Its paste, prepared with vinegar, is applied externally. For treating cough, asthma and respiratory catarrh, Saatar is used as an expectorant with figs.

An infusion of Saatar is prescribed for the treatment of premature labour pains.

Mentha spicata Linn. or Mentha viridis Linn. is a substitute for Saatar.

Araq-e-Saatar is prescribed in stomachache; also in diseases of liver and stomach.

Active principles and pharmacology

The essential oil, recovered from the drug on steam-distillation, contained phenols 69 % and non-phenols 31 %. Phenols consist mostly of carvacrol. Traces of thymol and 0.5 % zatarol were the other phenols isolated. Non-phenols included p-cymene 17; cineol 1; borneol 3; zatarinol 2 % and an unidentified acid 0.6 %.

Besides essential oil, the drug contains five alkanes—n-nonacosane, n-hentriacontane, n-dotriacontane, n-tritriacontane and n-pentatriacontane; beta-sitosterol; betulin; four fatty acids—behenic, lignoceric, cerotic and montanic; oleanolic acid and two unidentified triterpenic ketones.

Use in Western herbal

Therapeutic applications of Saatar of Unani medicine and Wild Thyme of Western herbal are found to be similar. Wild Thyme is equated with Thymus serpyllum Linn. It contains 30 to 70 % of phenols.

Nicholas Culpeper (1616–1654) wrote of it: "It provokes urine and the terms, and eases the griping pain of the belly, cramps, ruptures, inflammation of the liver. If you make a vinegar of the herb, as vinegar of roses is made, and anoint the head with it, it stops the pain thereof. It is excellently good to be given in phrenzy or lethargy, although they are two contrary diseases. It helps spitting and voiding of blood, coughing and vomiting; it comforts and strengthens the head, stomach, reins and womb; expels wind and breaks the stone."

Culpeper's description of the drug, well matches the medicinal application of Saatar in Unani medicine.

Gramineae

ZEA

Zea mays Linn.

Habitat

Extensively cultivated in India both in the plains and in the hills, especially in the areas with hot summer.

Classical and common names

Unani: Makkaa, Zurraa Makkaa.
English: Maize, Indian Corn.

Parts used

Cornsilk (styles and stigmas), cob.

Dose

Infusion 2 teaspoonfuls of dry herb.

Classical use

Unani physicians prescribe an infusion made from two teaspoonfuls of dry cornsilk two to three times a day for anuria, dysuria and for cystitis.

Burnt and powdered cob is prescribed internally in metrorrhagia and bleeding piles; mixed with salt for treating cough.

Active principles and pharmacology

Key constituents of cornsilk are flavonoids (including among others maysin, maysin-3'-methylether); saponins (2–3 %), alkaloids (0.05 %); 6-methoxybenzoxazolinone; sterols (including among others beta-sitosterol, ergosterol); tannins; bitter substances; fatty oil; volatile oil (0.2 %) containing, among others, carvacrol, alpha-terpineol, menthol, thymol. (PDR.)

Fatty oil from stigmas obtained by petroleum ether extraction (2.0–2.5 %) contains a saturated hydrocarbon, stigmasterol, sitosterol and a trace of ergosterol.

Cornsilk also contains vitamin C, K, potassium.

Maize cob contains sizable amounts of B complex vitamins. Analysis of cobs of 25 varieties grown in the USA showed that cobs contained: riboflavin 0.8–2.6; biotin 0.02–0.06; niacin 4.3–

16.0; pantothenic acid 3.4–7.0; and pyridoxine 1.0–3.4 (mcg/g). Cobs contain less biotin, niacin and pyridoxine than the grain, but slightly more riboflavin; cobs are approximately equal to grain in pantothenic acid content.

Maize cob contains a number of polysaccharides constituting almost 80 % of the cob; the pentosan content varies from 28.1 to 34.9 %. Two polysaccharides have been obtained in pure state from the alkali extract of cob: holocellulose, an arabinoxylan containing 89.4 % xylose and a highly branched polysaccharide containing xylose 59.1, arabinose 21.9, monomethyl hexuronic acid 11.3 and galactose 7.6 %.

Alkali extracts of maize cob gave hemicellulose A and hemicellulose B. A-fraction yields a series of oligosaccharides composed only of xylose; B-fraction produces a mixture of monosaccharides and oligosaccharides, the latter containing xylose and uronic acid units. An aldotriouronic acid and some aldobiouronic acids have been identified. Aldotriouronic acid gave xylose, glucuronic acid and aldobiouronic acid on complete hydrolysis.

Cornsilk possesses chloretic, diuretic and astringent properties. It stimulates bile secretion mildly, acts as a urinary demulcent, gently lowers blood pressure and reduces blood-clotting time. It soothes and relaxes the lining of the urinary tubules and bladder, relieves irritation and improves urine flow and elimination. It is found efficacious in frequent urination and for difficulty in passing urine in prostate disorders.

Cornsilk is found useful as an adjunct to other therapies for acute cystitis.

Xylose, arabinose and galactose, obtained from stems, leaves and cobs of maize, exhibit antitumour activity.

Use in Western herbal

In 1994, German Commission E recommended a preparation for the treatment of BPH. It contains a complex extract of 92 % rye (Secale cereale) pollen, 5 % timothy (Phleum pratense) pollen and 3 % Zea mays pollen. One *in vitro* study showed a dose-dependent inhibition of inflammatory mediators; another demonstrated growth-inhibiting effects in cultured prostatic epithelial cells and fibroblasts. (Rational Phytotherapy.)

Corn silk is being used in Homoeopathy. (Tincture: one part, by weight of the silk, to two of alcohol or fluid extract). Its principal use is indicated in irritable conditions of the urinary tract, ureters, bladder and urethra, as a diuretic in renal disorders. In a case study, patient's urine contained albumin, mucus, some blood with specific gravity 1010. Pus cells and granular epithelial cells were found in abundance. Corn silk fluid extract (one teaspoonful every four hours) was given. In less than a week there was visible improvement. The constant calls to urinate (every forty-five minutes) ceased and the patient could spend the night without being disturbed more than once or twice. The specific gravity of the urine increased and pus, blood and albumin disappeared. (John Henry Clarke.)

Zingiberaceae

ZINGIBER

Zingiber officinale Rosc.

Figure 1 Zingiber officinale [ZANDU]

Habitat

Cultivated throughout tropical India. Indian Ginger is considered only second to Jamaican ginger in quality. There are two main types of Indian Ginger: Cochin Ginger, which comes from central Kerala; Calicut Ginger from Malabar. Kolkatta Ginger is the same as Calicut Ginger. Indian Ginger is more starchy, but is almost as pungent as Jamaican Ginger.

Classical and common names

Ayurvedic: Fresh—Aardraka, Aadrikaa, Shringvera, Katubhadra; dried—Shunthi, Naagara, Mahaushadha, Vishva, Vishvabheshaja.
Unani: Fresh—Zanjabeel-e-Ratab; Al-Zanjabeel; dried—Zanjabeel, Zanjabeel-e-Yaabis.
Siddha: Fresh—Ingi; dried—Chukku.
English: Ginger.

Parts used

Rhizome.

Dose

Fresh rhizome: Juice 5–10 ml, dried rhizome powder 1–3 g.

Classical use

Charaka and Sushruta prescribed Aardraka with equal quantity of jaggery for oedema; with Kulattha (Dolichos biflorus) for piles; milk processed with Aardraka for diseases of the abdomen; milk processed with Shunthi for coryza, bronchial asthma; Shunthi with Pippali (Piper longum) or Draakshaa (Vitis vinifera), mixed with jaggery, for cough; Shunthi mixed with jaggery for jaundice; Shunthi processed with Mushta (Cyperus rotundus) and Ativishaa (Aconitum heterophyllum) for diarrhoea; Shunthi and and Dhaanyaka (Coriandrum sativum) as a post-meal drink or Shunthi and Chitraka (Plumbago zeylanica) mixed with vinegar for piles; milk cooked with Shunthi, Draakshaa (Vitis vinifera) and Kharjura (Phoenix dactylifera) for influenza and fever.

Charaka prescribed Aardraka, mixed with equal quantity of jaggery (dose: 20 g), increasing 20 g per day up to 200 g. This regime was meant for alleviating oedema, abdominal diseases, jaundice, internal obstructions and congestions, polyuria, cough, bronchial asthma, consumption.

This therapy was also recommended in a lower dose of 10 g, increasing the dose 10 g per day up to 120 g (Vrindamaadhava). (In current practice, increasing doses of ginger juice, used as a strong diuretic, was not found efficacious in dropsy and chronic heart diseases.)

Water processed with Balaa (Sida cordifolia) or Prishnparni (Uraria picta) or Kantakaari (Solanum surattense) and added with Shunthi, was prescribed by Charaka as an antinarcotic drink in cases of alcoholism.

Aardraka, with equal quantity of vinegar, was prescribed before meals for deficient digestion (Ashtaanga Hridaya); salted ginger for stimulating appetite (Vrindamaadhava).

Powdered Shunthi was used as a snuff for headache, congestion of nasal passage and sinus. (Chakradatta, Vrindamaadhava.)

Shunthi with a decoction of Punarnavaa (Boerhavia diffusa) or with that of Guduuchi (Tinospora cordifolia) and Pippali (Piper longum) was a popular remedy for oedema, rheumatic affections, sciatica, lumbago during the 16th century. (Ashtaanga Sangraha, Vrindamaadhava, Bhaavaprakaasha.)

Hot decoction of Shunthi was prescribed as a stimulating and warming cardiac tonic (Vrindamaadhava).

Shunthi entered into more prescriptions than Aardraka, though both possess the same virtues.

In Shaarangadhara Samhitaa, Shunthi is incorporated in 68 prescriptions, Aardraka only in 13; in Bhaavaprakaasha, Shunthi features in 95 prescriptions, while Aardraka only in 35. Shunthi entered into a number of nervine tonics, massage oils and medicinally activated purified butter preparations.

A combination of Shunthi, Maricha (Piper nigrum) and Pippali (Piper longum), known as Trikatu, enters into a number of Indian compound formulations for its carminative, anti-inflammatory and antihistaminic properties and for enhancing the bioavailability of companion drugs.

Unani practitioners prescribe juice of ginger and onion (10 g of each), mixed together, for nausea, vomiting, retching and dyspepsia.

Dry ginger is prescribed as a corrective adjunct to purgatives for preventing nausea and griping.

A paste of dried ginger, cinnamon and clove is applied to forehead for neuralgic headache. Powdered ginger is also administered internally as a decongestant during headache.

A warm infusion of dried ginger (1 to 24) is prescribed in chronic rheumatism.

In viral hepatitis, a decoction of dried ginger, Piper nigrum and Piper longum is administered.

Among over-the-counter Ayurvedic drugs, Brihat Gangaadhara Churna (Shaarangadhara Samhitaa) is for diarrhoea and dysentery, Panchsakaara Churna (Siddha-bheshaja-manimaalaa) for constipation; Shunthi, Maricha and Pippali—all the three are included in Lavana Bhaaskara Churna (Shaarangadhara Samhitaa), Naaraayana Churna (ibid), Hingwaashtaka Churna (Bhaishajya Ratnaavali), Avipattikara Churna (ibid), for chronic constipation, abdominal distress, flatulence and hyperacidity. Taalisaadi Churna (Shaarangadhara Samhitaa) is for dry cough and fever; Saubhaagya Shunthi for post-delivery recuperation.

Among over-the-counter Unani drugs, Jawarish-e-Zanjabeel (Qarabadeen-e-Azam-o-Akmal) is for diarrhoea; Murabba-e-Zanjabeel (Bayaz-e-Kabir) and Jawarish-e-Fanjnosh for flatulence, stomach-ache; also for cough, coryza, bronchitis.

Inji Rasaayanam of Siddha medicine contains fresh ginger and is prescribed for biliousness and indigestion. Inji Vadagam also contains fresh ginger and is prescribed in diseases of liver and spleen.

Active principles and pharmacology

Chemical composition of ginger varies according to type and agro-climatic conditions under which it is grown.

Analysis of a market sample of green ginger gave the following values: moisture 80.9; protein 2.3; fat 0.9; fibre 2.4; carbohydrates 12.3; and minerals 1.2 %, Ca 20; P 60; and Fe 2.6 mg/100 g. Ginger also contains traces of iodine and fluorine.

The vitamins present in green ginger are: thiamine 0.06; riboflavin 0.03; niacin 0.60 and vitamin C 6.0 mg/100 g. The value reported for carotene in the fresh rhizome is 40 mcg/100 g.

Ginger contains small quantities of glucose, fructose and sucrose; raffinose is probably present in traces. The principal carbohydrate of the rhizome is starch. Besides starch, the rhizomes are reported to contain pentosans.

The yield of oil from dry ginger varies from 1.5 to 3.0 % (average 2 %). The oil contains sesquiterpene hydrocarbons (50 % or more), sesquiterpene alcohols, monoterpenoids and associated compounds. The occurrence of esters of acetic and caprylic acids and a trace of phenol has been reported.

The predominent sesquiterpene hydrocarbon is zingiberene, other sesquiterpenes present in the oil are ar-curcumene (17.7 %); farnesene (9.8 %) and relatively smaller amounts of beta-bisabolene, gamma-selinene, beta-elemene and beta-sesquiphellandrene. The sesquiterpene alcohol, zingiberol, a mixture of beta-eudesmol stereoisomers, occurs in the oil. Two other isomeric alcohols and a crystalline diol have been reported.

The monoterpene hydrocarbons present in the oil include camphane, alpha- and beta-pinene, cumene, myrcene, limonene, p-cymene and beta-phellandrene. The oxygenated monoterpenes and associated compounds present are 2-heptanol, 1, 8-cineole, borneol, bornyl acetate, linalool. geranial and neral.

The oil from Cochin ginger contains more citral (geranial and neral) than the oils from Jamaican or African ginger.

An oleoresin, known as gingerin, extracted from ginger, represent the pungent principles of ginger. Besides the pungent principles, the oleoresin contains essential oil (7–28 %) and also non-pungent substances (about 30 %). The non-pungent substances include carbohydrates, palmitic acid and other fatty acids. The pungent principles of ginger are oxymethyl phenols. Commercial preparations of oleoresin contain gingerol, shogaol and zingerone. Small quantities of paradol are also reported to be present.

The rhizome has been found to be a new source of proteolytic enzyme.

Clinical studies have demonstrated that oral administration of powdered ginger rhizome (940 mg) was more effective than dimenhydrinate (100 mg) in preventing gastrointestinal symptoms of motion sickness.

One study compared the effects of powdered ginger rhizome and seven over-the-counter and prescription antiemetic drugs on prevention of sea-sickness on 1489 subjects. It concluded that ginger was as effective as other antiemetic drugs tested.

In a double-blind randomized cross-over trial, oral administration of powdered ginger (250 mg 4 times daily) effectively treated pernicious vomiting in pregnancy.

One study in China reported that 113 patients with rheumatic pain and chronic lower painful points or reaction nodules, experienced full or partial relief of pain, decrease in joint swelling, and improvement on recovery in joint function (WHO).

In a clinical trial, a 42-year old migraine patient was prescribed 500–600 mg ginger powder with water at onset of the attack. The powder aborted the effect of migraine headache within half an hour. Use of 1.5–2 g powder per day after the attack of migraine decreased the frequency of attacks markedly. (J Ethnopharmaco V 29, 267–273, 1990, Mustafa and Srivastava.)

Ginger has demonstrated analgesic effect in experimental animals. The effect is thought to be a result of shogaol, inhibiting the release of substance P. Pungent principles of ginger are potent inhibitors of prostaglandin and leukotriene synthesis.

Alcoholic extracts of ginger have been found to stimulate the vasomotor and respiratory centres in anaesthetized cats. They also stimulate the heart. Gingerol has shown potent cardiotonic activity (positive ionotropic and chronotropic effects) on isolated guinea-pig left atria. These effects are a result of acceleration of calcium uptake by heart muscle.

Ginger inhibits platelet aggregation.

Ginger has been shown to reduce significantly serum and hepatic cholesterol levels in cholesterol-fed rats by impairing cholesterol absorption as well as by stimulating the conversion of cholesterol to bile acids.

Ginger has been shown to inhibit diarrhoea. It improves gastric motility, at the same time exerts antispasmodic effect.

Ginger demonstrates significant antiulcer effects in a variety of animal models. In one study roasted ginger demonstrated inhibition of ulcer formation in three ulcer models, while dry ginger had no such effect.

Ginger is reported to contain an antihistaminic factor.

Shogaol and zingerone have been shown to inhibit the growth of Salmonella typhi, Vibrio cholerae and Tricophyton violaceum. Aqueous extracts of ginger are found effective against Trichomonas vaginalis.

In China, fresh and dried ginger are treated as different remedies. Fresh ginger is given for chills, fever, headache and muscular pain. Dried ginger is used for "internal cold, " with symptoms such as cold hands, weak pulse, pale complexion. (Andrew Chevallier.)

Use in Western herbal

German Commission E recognized ginger's efficacy in the following areas:

- Dyspepsia
- Motion sickness.

The British Herbal Compendium indicates ginger for atonic dyspepsia, colic, prophylaxis of travel sickness and vomiting of pregnancy. Other uses include anorexia, bronchitis, and rheumatic complaints.

Ginger tablets, capsules, biscuits and Ginger Beer are available over-the-counter.

Tincture of dry rhizome is used in homoeopathic medicine in digestive disorders, colic, diarrhoea, flatulence, dry cough, asthma, stiffness of neck with headache and nausea; rheumatic pain, frontal headache, hemicrania.

Rhamnaceae

ZIZIPHUS

Ziziphus mauritiana Lam.

See figure 1.

Ziziphus jujuba (Lam.) Gaertn. non Mill.
Ziziphus nummularia (Burm f.) Wight & Arn.

Habitat

Indian jujube is found both wild and cultivated throughout the greater part of India, ascending to an altitude of about 1500 m in the Himalayas.

Classical and common names

Ayurvedic: Badara, Kola (smaller variety), Karkandhu.
Unani: Ber.
Siddha: Ilandi.
English: Indian Jujube.

Parts used

Fruit, leaf, root, bark.

Dose

Powder 3–5 g, for infusion 5–10 g.

Figure 1 Ziziphus mauritana—flowering and fruiting branch [WOI]

Classical use

Three varieties of jujuba have been mentioned by Charaka—Badara, Kola (Kuvala) and Karkandhu (bearing big, medium and small fruits respectively). All the three were mentioned separately. Karkandhu was also known as Shragaala-Badari. Sinchitikaa-phala, another variety, was also mentioned. Sushruta mentioned five varieties—Badara, Kola, Karkandhu, Sauvira and Sinchitikaa-phala. Later on, during the 16th century, only three varieties, Badara, Kola and Karkandhu, were used in Ayurvedic medicine.

Badara and kola have been equated with Ziziphus mauritiana and Karkandhu with Ziziphus nummularia.

Charaka gave a sweetened decoction of Badara leaves, in internal prescriptions, for loss of voice, giddiness, and for piles; a decoction of Karkandhu plant as purgative, also as warm enema.

Sushruta prescribed fruits of Badara internally in haemoptysis, menstrual and other vaginal disorders, for bilious and rheumatic affections; ripe fruits of Kola as demulcent and purgative; oil of Kola seeds internally for anuria.

Green fruits of Karkandhu were prescribed by Sushruta as diuretic, stomachic and laxative.

Kola belongs to Panchaamlaka (Five Sours) group of Charaka, others being Punica granatum, Garcinia indica, Oxalis corniculata and Rumex vesicarius. Sour fruits of Badara, Kola and Karkandhu were used for treating obesity. Paste of seed-kernel of Kola, with water, was prescribed for treating Bhasmaka (gluttonous appetite.)

A fine powder of Karkandhu, mixed with equal quantity of purified butter and jaggery, was prescribed for leucorrhoea (Raaja Maarttanda, Gadanigraha). In folk medicine, kernels are prescribed to stop vomiting, and for relief from abdominal pain; also for the treatment of diarrhroea and dysentery.

A decoction of bark is used as an astringent in gingivitis.

Active principles and pharmacology

Fruits gave cyclic AMP and cyclic GMP, ziziphus saponins I, II, III and jujuboside B and p-coumarylates of alphitolic acid.

Bark contains betulinic acid leucocyanidin and the cyclopeptide alkaloids mauritines A-F, amphibines A-F, and frangufoline.

Wood contains leucopelargonidin, betulinic acid and ceanothic acid.

Seeds gave the saponins jujubosides A and B. Acid hydrolysis of the saponin afforded ebelin lactone which yields the sapogenin jujubogenin.

The fruits are a good source of vitamin C and sugars, and contain appreciable amounts of minerals. One analysis of the fruit pulp is as follows: moisture 81.6; protein 0.8; fat 0.3; carbohydrates 17.0; and minerals 0.3 %; calcium 4; phosphorus 9; iron 1.8; carotene 0.021; thiamine 0.02; riboflavin 0.02; niacin 0.7; and vitamin C 76 mg/100 g. Presence of fluoride (0.1–0.2 ppm) is reported. Pectin content (as calcium pectate) is 2.2–3.4 % on moisture-free basis.

Fruit contains total sugars 5.4–10.5, reducing sugars 1.4–6.2, non-reducing sugars 3.2–8.0; and acidity (as citric acid) 0.2–1.1 %. Citric acid is the major acid in the fruit; malic and oxalic acids being present in small amounts.

The oil, extracted from seeds, contains oleic acid 71.7; linoleic acid 15.4; and saturated acid about 13.0 %.

Leaves contain crude protein 12.9–16.9; ether extr. (fat) 1.5–2.7; fibre 13.5–17.1; N-free extr. 55.3–56.7; total ash 10.2–11.7; calcium 1.42–3.59; and phosphorus 0.21–0.33 %.

Ceryl alcohol and two alkaloids—protopine and berberine, have been isolated from leaves.

Ziziphus nummularia (wild jujube): leaves contain ascorbic acid; rootbark contains tannins and the cyclopeptide alkaloids nummuiarines A, B, and C, mucronine D and amphibine H.

Leaves, on dry matter basis, contain: crude protein 14.24–14.70; ether extr. 3.16–5.91; crude fibre 8.20–20.60; N-free extr. 51.78–60.70; calcium 1.70–2.17; phosphorus 0.16–9.19 %.

The twig bark contains 12 % tannin.

The fruit is cooling and astringent, and is found efficacious in bilious affections.

Leaves are used in scabies and other skin diseases. Smoke from dried leaves is inhaled for the treatment of colds and coughs.

Ziziphus sativa Gaertn.
Ziziphus vulgaris Lam.
Ziziphus jujuba Mill.

Habitat

A native of North China, from where it was introduced into West Asia. It failed to survive in India. Fruits are imported into India from China, Afghanistan and Persia. Indian jujube is equated with Ziziphus jujuba (Lam.) Gaertn. non Mill.

Classical and common names

Unani: Unnaab.
English: Chinese Tsao (Ta-ts'ao).

Classical use

Unnaab is used in Unani medicine for infusing vital energies in the body. The fruit enters into prescriptions and confections for fatigue, exhaustion, malnutrition, hypertension, insomnia.

For treating depression, Unnaab and Prunus amygdalus kernels (Baadaam) are soaked in rosewater overnight, a paste is prepared with Borago officinalis (Gaozabaan) and Lavandula stoechas (Ustukhuddus) and the recipe is administered with honey, first thing in the morning. This is also prescribed to patients of migraine.

For treating dry cough, lozenges containing Unnaab, gum tragacanth, rose extract and sugar, are prescribed.

Powdered leaves (3 g/twice daily) are prescribed for polyuria due to diabetes.

Sharbat-e-Unnaab (Bayaz-e-Kabir) is prescribed for bronchitis as a cough syrup.

Unnaab is also an ingredient in other cough syrups—Sharbat-e-Ejaz and Sharbat-e-Sadar, prescribed for coryza, chronic catarrh, bronchitis. Dayaqooza, a confection for coryza and catarrh, contains Unnaab as one of the ingredients.

Active principles and pharmacology

The pulp of fresh and dried jujubes (Ziziphus jujuba Mill.), grown at the US Plant Introduction Garden, Chico, California, gave the following values respectively: moisture 65.4, 16.2; protein 1.2, 4.4; fat 0.3, nil; fibre 1.3, 3.2; reducing sugars 11.9, 39.2; sucrose 12.2, 20.8; acids as citric 0.4, 1.2; pectin 0.4, nil; and ash 0.9, 2.3 %. High contents of carotene (70 mg/100 g) and a fair amount of vitamin C are present in the pulp. The fruit contains 10 % tannin. The fruit contains very high levels of both cyclic AMP and cyclic GMP and shows antiallergic activity.

In China, jujubes are eaten after crushing the seed inside the fruit. A sample of seeds contained: moisture 21.3; protein 2.3; fat 0.4; fibre 2.2; N-free extr. 72.6; and ash 1.3 %.

Kernels cause a sedative effect and are found efficacious in insomnia.

Three cyclopeptide alkaloids—sanjoinine I, II and III—have been isolated from seeds.

The bark contains 7 % tannin. A new triterpenoid, zizyberanalic acid, has been isolated from the bark.

Leaves and stem contain terephthalic acid and its methyl esters.

Use in Western herbal

Jujuba is used as a nutrient to improve muscular strength and in the prophylaxis of liver diseases and stress ulcers. It is considered anti-allergenic and sedative.

Practitioners of Chinese medicine recommend fruits of Ziziphus vulgaris in nausea and vomiting, abdominal pain in pregnancy, and as a poultice in cancrum oris. The 3-year-old kernels of stones are considered especially efficacious in abdominal pain and as an application to wounds. The leaves are regarded as diaphoretic, and are prescribed in the typhoid fever of children. The heartwood of the tree is recommended in marasmus (progressive wasting, especially in children), and a decoction of it given as a haematinic tonic. The root is used in the eruptive fevers of children. The bark is used in decoction as a wash to inflammations of the eyes.

Kernels are given for insomnia, palpitation and night sweats.

A compound preparation, Equillence (Ragina), a combination of royal jelly and eight Chinese herbs including Chinese jujuba, is available over-the-counter for menstrual irregularities.

Appendix I

Classical Ayurvedic References

Charaka, Sushruta and Vagabhatta composed three Samhitaas, collectively known as Brihatrayee. These treatises belonged to the period till the 7th century. Thereafter, in the medieval period from the 8th to 16th century, a number of Nighantus were composed. During the medieval period, Maadhava Nidaana, Shaarangadhara Samhitaa and Bhaavaprakaasha were collectively known as Laghutrayee.

Ayurveda Saukhyam	16th century AD
Ashtaanga Hridaya	7th century AD
Ashtaanga Nighantu	8th century AD
Ashtaanga Sangraha	6th century AD
Bhaavaprakaasha	16th century AD
Bangasena	18th century AD
Bhaishajya Ratnaavali	17th century AD
Bhela Samhitaa	Prior to 7th century BC
Chakrapaani/Chakradatta	11th century AD
Charaka Samhitaa	1000 BC
Dalhan	12th century AD
Dhanvantari Nighantu	Prior to 13th century AD
Gadanigraha	12th century AD
Haarita Samhitaa	Prior to 7th century BC
Kaashyapa Samhitaa	Prior to 7th century BC
Kaiyadeva Nighanu	1450 AD
Maadhava Dravyaguna (Nighantu)	Prioir to 12th century AD
Madanpaal Nighantu	1374 AD
Nighantu Ratnaakara	1837 AD
Nighantu Sangraha	1748 AD
Raaja Maarttanda	11th century AD
Raaja Nighantu	14th century AD
Rajavallabha Nighantu	17th century AD
Shaarangadhara Samhitaa	13th century AD
Shaaligaraama Nighantu	1896 AD
Sahasrayoga	4th century AD
Shodhala Nighantu	12th century AD
Siddha-bheshaja-manimaalaa	18th century AD
Sushruta Samhitaa	1000 BC
Vagabhatta I, III	6th, 7th century AD
Vaidya Manoramaa	13th century AD
Vrindamaadhava	8th century AD
Yoga Ratnaakara	16th century AD

Appendix II

Acknowledgements

This encyclopedic work on Indian medicinal plants and herbal remedies is the outcome of literary research based on the material drawn from the following sources, and references cited therein:

American Botanical Council, P.O.Box 144345, Austin TX 78714-4345. www.herbalgram.org
Complete German Commission E Monograph. 1998. Ed. Mark Blumenthal et al.
Herbal Medicine—Expanded Commission E Monographs. 2000. Eds. Blumenthal, Goldberg, Brinckmann.

Basic Chemicals, Pharmaceuticals and Cosmetics Export Promotion Council, Mumbai. (CHEMEXCIL).
Selected Medicinal Plants of India. 1992.

Central Council for Research in Ayurveda & Siddha (CCRAS), 61-65, Institutional Area, Janak Puri, New Delhi-110058.
Pharmacognosy of Indigenous Drugs, Vol I, II, III. 1999.
Pharmacological Investigations of Certain Medicinal Plants and Compound Formulations Used in Ayurveda and Siddha. 1996.
Pharmacopoeial Standards of Ayurvedic Formulations. 1987.
Phytochemical Investigations of Certain Medicinal Plants Used in Ayurveda. 1990.
Sahasrayoga (Sanskrit-Hindi translation of Malayalam text). 1990.
Effect of Varuna (Crataeva nurvala) in Enclarged Prostrate and Associated Urinary Disorders. 1987.
A Check-list of Sanskrit Medical Manuscripts in India. 1972.
Research in Ayurveda and Siddha—Bibliography of CCRAS Contribution (1969–1997). 1999.
Researach papers presented during a seminar on Research Achievements of Ayurveda, Oct. 1999.
Annual Reports, survey reports, and unpublished research papers. Personal communication with Research Officers.
Journal of Research in Ayurveda & Siddha, bi-annual journal. 1990–2002 issues.

Central Council for Research in Unani Medicine, 61-65, Institutional Area, Janak Puri, New Delhi-110058.
Standardization of Single Drugs in Unani Medicine, Part 1, 2, 3. 1997.
Report of the workshop on Standardization of Unani Drugs. 1995. Annual Reports, survey reports, and unpublished research papers. Personal communication with Research Officers.

Central Drug Research Institute, Lucknow and NISCOM , Dr KS Krishnan Marg, New Delhi-110012.
Compendium of Indian Medicinal Plants, Vol 1–5. 1960–1994. Ed. Ram P Rastogi.

Central Institute of Medicinal & Aromatic Plants (CIMAP), Ram Sagar Misra Nagar, Lucknow-226016.
Dictionary of Indian Medicinal Plants. 1992. Eds. Akhtar Husain et al.

Council of Scientific & Industrial Research (CSIR), National Institute of Science Communication (NISCOM), Dr KS Krishnan Marg, New Delhi-110012.

Glossary of Indian Medicinal Plants, RN Chopra et al. 1956.
Supplement to Glossary of Indian Medicinal Plants, RN Chopra et al. 1968.
Second Supplement to Glossary of Indian Medicinal Plants with Active Principles, Part 1 (A–K) LV Asolkar et al. 1992.
The Wealth of India, Vol I–XI. Revised editions 1, 2, 3. First supplement series 1–4. 2000–2003.
Medicinal & Aromatic Plant Abstracts (MAPA). Bi-monthly journal. 1990–2002 issues.

Indian Council of Medical Research (ICMR), Ring Road, Ansari Nagar, New Delhi-110029.

Medicinal Plants of India, Vol 1, 2. 1976, 1987. Ed. GV Satyavati et al. (Revised editions in Press.)

Indian Drug Manufacturers' Association, Mumbai 400018, and Regional Research Laboratory, CSIR, Jammu-Tawi.

Indian Herbal Pharmacopoeia, Vol 1, 2. 1998, 1999. Eds. SS Handa et al.

Indian Medical Practitioners Cooperative Pharmacy & Stores Ltd (IMPCOPS), Lattice Bridge Road, Thiruvanmiyur, Chennai-600041.

Formulary of Ayurvedic Medicine.
Formulary of Unani Medicine.
Formulary of Siddha Medicine.

Indian National Science Academy (INSA), Indian National Commission for History of Science, Bahadurshah Zafar Marg, New Delhi-110002.

Charaka Samhita, A Scientific Synopsis. 1980. P Ray and HN Gupta.
Current Research in Pharmacology in India. 1984. Ed. BN Dhawan.
Current Research in Pharmacology in India. 1984. Eds. PK Das and BN Dhawan.
Sushruta Samhita, A Scientific Synopsis. 1993. P Ray, BN Gupta and M Roy.

Institute of History of Medicine & Medical Research, Hamdard Nagar, New Delhi 110062. Hamdard National Foundation, Hamdard Building, Asaf Ali Road, New Delhi-110002.

Research and Development of Indigenous Drugs. 1989. Eds. PC Dandiya and SB Vohora. Out of Print.
A Survey of Drugs, with Particular Reference to Arab (Unani) Medicine and Ayurveda. Second rev. ed. 1961. (Out of Print.) (Critical appraisal in Oct–Dec 1992 issue of *The Journal of Research and Education in Indian Medicine,* Varanasi 221005, by M Ali, Jamia Hamdard University, New Delhi-110062.)

Ministry of Health and Family Welfare, Deptt of Health, Government of India, New Delhi-110001.

The Ayurvedic Formulary of India, Part 1.
The Ayurvedic Pharmacopoeia of India, Part 1, Vol I, II.
National Formulary of Unani Medicine.

National Academy of Ayurveda (NAA)—Rashtriya Ayurveda Vidyapith. (An autonomus organization under Ministry of Health and Family Welfare, Government of India), Dhanvantari Bhavan, Road 66, Punjabi Bagh, New Delhi-110026.

Medicinal Plants Used in Ayurveda. 1998.
Plants of Bhava Prakasha. 1999.
Plants of Sharangdhara Samhita. 1999.

Society for New Age Herbals, B-1/211, Janak Puri, New Delhi-110058

Indian Herbal Therapies—Applications of Research Findings. 2003, second revised edn. CP Khare. Vishv Vijay Pvt. Ltd, M-12, Connaught Circus, New Delhi-110001.

World Health Organization.
Monographs on Selected Medicinal Plants, Vol 1,2. 1999, 2003.
Adverse Effects of Herbal Drugs 1, 1991. Eds. De Smet et al. Springer-Verlag.

Chandraraj Bhandari: *Vanaushadhi Chandrodaya* (Hindi), Vol. 1, 2. 1986. Chaukhambha Sanskrit Pratishthan, Gopal Mandir lane, Varanasi-221001.

William Boericke: *Homeopathic Materia Medica.* B Jain Publishers Pvt. Ltd, New Delhi-110052.

Francis Brinker: *The Toxicology of Botanical Medicine. 2000. Herb Contraindications and Drug Interactions.* 2001. Eclectic Medical Publications, Sandy, Oregon 97055.

Donald J Brown: *Herbal Prescriptions for Better Health.* 1996. Prima Publishing, P.O. Box 1260 BK, Rocklin, CA-95677. www.primahealth.com

Kitty Campion: *The Family Herbal.* 1988. Leopard Books, Random House, London SW1V 2SA.

Michael Castleman: *The Healing Herbs.* 1995. Bantam Books, 1540 Broadway, New York, NY-11036.

Andrew Chevallier: *The Encyclopedia of Medicinal Plants.* 1996. Dorling Kindersley, 9 Henrietta Street, London WC 2E 8PS.

RN Chopra et al: *Indigenous Drugs of India.* 1958. Academic Publishers, 12/1A Bankim Chatterjee Street, Kolkata-700073.

JH Clarke: *A Dictionary of Practical Materia Medica,* Vol 1–3. 1900. Reprint 1996. B Jain Publishers Pvt. Ltd, New Delhi-110052.

Nicholas Culpeper: *Complete Herbal.* 1653. Reprint 1995. Wordsworth Editions Ltd, Cumberland House, Cribb Street, Ware, Hertfordshire, SG12 9ET.

Bhagwan Dash and Lalitesh Kashyap: *Materia Medica of Ayurveda,* based on *Ayurveda Saukhyam* of *Todaraananda* (16th Cent. AD). 1987. Concept Publishing Company, H-13, Bali Nagar, New Delhi-110015.

Steven Foster and VE Tyler: *Tyler's Honest Herbal.* 2000. Haworth Herbal Press, 10 Alice Street, Binghamton, NY-13904-1580.

Steven Foster and Yue Chongxi: *Herbal Emissaries, Bringing Chinese Herbs to the West.* 1992. Healing Arts Press, Rochester, Vermont. USA.

John Gerard: *Historie of Plants.* 1927. Reprint 1998. Tiger Books International PLC, 26A York Street, Twickenham, Middlesex, TWI 3LJ, UK.

Das Govind: *Bhaishajya Ratnawali* (Hindi). Ed. Vaidya Lalchandra. 1988. Motilal Banarasi Das, Delhi-110007.

M Grieve: *A Modern Herbal.* 1973. Reprint 1994. Ed. Hilda Leyel. Tiger Books International, London.

Barbara Griggs: *New Green Pharmacy.* 1997. Vermillion, Random House UK Ltd, London SW1V 2 SA.

John Heinerman: *Encyclopedia of Healing Herbs and Spices.* 1996. Parker Publishing Company, West Nyack, NY-10995.

David Hoffmann: *Complete Herbal.* 1996. Element Books Inc., 160 North Washington Street, Boston MA-02114. *An Elder's Herbal.* 1993 *The Herbal Handbook.* 1998. Healing Arts Press, Rochester, Vermont. USA.

SG Joshi: *Medicinal Plants.* 2000. Oxford & IBH Publishing Company Pvt. Ltd, 66 Janpath, New Delhi-110001.

CP Khare: *Indian Herbal Therapies—Application of Research Findings.* Second revised edn. 2003. Vishv Vijay Pvt. Ltd, M-12, Connaught Circus, New Delhi-110001.

Robyn Landis, Khalsa Karta Purkh Singh: *Herbal Defence.* 1997. Warner Books, 1271 Avenue of Americas, New York, NY-10020.

Michael McGuffin et al: *Botanical Safety Handbook.* 1997. CRC Press, Baca Raton, Florida 3343.

Gale Maleskey and the Editors of Prevention Health Books: *Nature's Medicines, The Guide to Health Supplements.* 1999. Rodale Press Inc., Emmaus, Pennsylvania. www.healthyideas.com

Medical Economics Company: *PDR* for Herbal Medicines (Physicians' Desk Reference). 1999–2000. Montvale, NJ 07645-1742. (Phytopharm Consultancy Institute for Phytopharmaceuticals, Olafstr. 6, D–13467, Berlin, Germany, can be contacted over www.phytonet.com)

Lucinda G Miller et al: *Herbal Medicinals: A Clinician's Guide.* 1998. Pharmaceutical Products Press, The Haworth Press Inc. New York.

Simon Y Mills: *The Complete Guide to Modern Herbalism.* 1994. Thorsons, Harper Collins, 77–78 Fulham Palace Road, Hammersmith, London W6 8JB.

The Essential Book of Herbal Medicine. 1993. (First published under the title *"Out of Earth"* by Viking Arkana.) Penguin Books Ltd, London W8 5TZ.

Sheriff Moodeen: *A Catalogue of Indian Synonymes of the Medicinal Plants.* 1869. Reprint 1988. Periodical Expert Book Agency, New Delhi-110032.

Daniel B Mowrey: *Herbal Tonic Therapies:* 1993. Wing Books, Random House, Value Publishing Inc., New Jersey-07001.

Michael T Murrey: *The Healing Power of Herbs.* 1995. Prima Publishing, P.O. Box 1260 BK, Rocklin, CA 95677. www.primahealth.com

Michael T Murrey and Joseph Pizzorno: *Encyclopedia of Natural Medicine.* Rev. edition. 1998. Prima Publishing, CA 95677.

KM Nadkarni: *Indian Materia Medica,* Vol 1, 2. Rev. edition. 1954. Popular Prakashan, 35 C, Tardeo Road, Mumbai-400034.

Penelope Ody: *The Complete Medicinal Herbal.* 1997. Dorling Kindersley, 9 Henrietta Street, London WC 2E 8PS.

Handbook of Over-the-Counter Herbal Medicines. 1996. Kyle Cathie Ltd, 20 Vauxhall Bride Road, London SW1V 2SA.

Steven G Ottariano: *Medicinal Herbal Therapy, A Pharmacist's Viewpoint.* 1999. Nicolin Fields Publishing Inc., 2456 Lafayette Road, Portsmouth, NH 03801.

Miriam Polunin and Christopher Robbins: *The Natural Pharmacy.* 1992. Collier Books, Macmillan Publishing Company, 866 Third Avenue, New York, NY-10022.

Daniel Reid: *A Handbook of Chinese Healing Herbs.* 1995. Simon & Schuster, West Garden Place, Kendal St, London W2 2AQ.

James E Robbers and VE Tyler: *Tyler's Herbs of Choice.* 2000. The Haworth Herbal Press, 10 Alice Street, Binghamton, NY 13904-1580.

V Schulz, R Hansel, VE Tyler: *Rational Phytotherapy, A Physicians' Guide to Herbal Medicine.* Fourth edition. 2001. Springer-Verlag, Berlin/Heidelberg.

PV Sharma: *Dravya Guna Vijnaana* (Hindi), Vol 1, 2. 1986. Vol 5, 1998. Chaukhambha Sanskrit Sansthan, Gopal Mandir Lane, Varanasi-221001.

Classical Uses of Medicinal Plants. 1996. Chaukhambha Visvabharati, Gopal Mandir Lane, Varanasi-221001.

Dalhan and his Comments on Drugs. 1982. Munshiram Manoharlal, Rani Jhansi Road, New Delhi-110055.

Daljit Singh: *Unani Dravyagunaadarsh* (Hindi), Vol 2. 1974. Ayurvedic-Tibbi Academy, Uttar Pradesh, Lucknow-226001.

Unani Siddha Yoga Samgraha (Hindi). 1998. Shri Baidyanath Ayurveda Bhawan Ltd, Naini, Allahabad. UP.

Ramsushil Singh: *Vanoshadhi Nidarshikaa* (Hindi). 1983. Uttar Pradesh Hindi Sansthan, M.G. Marg, Lucknow-226001.

VV Sivarajan and Balachandran: *Ayurvedic Drugs and their Plant Sources.* 1994. Oxford & IBH Publishing Co Pvt. Ltd, 66 Janpath, New Delhi-110001.

GA Stuart: *Chinese Materia Medica.* 1911. Probsthian & Co, 41 Gt. Russell St., London WC1.

Maya Ram Uniyal: *Prayogatmak Abhinava Dravyaguna Vigyanam* (Hindi). 1999. Shri Baidyanath Ayurveda Bhavan Ltd, Patna-800001.

RC Wren: *Potters' Cyclopaedia of Botanical Drugs and Preparations.* 1941. Reprint 1988. Potter & Clarke Ltd, 60–64 Artillery Lane, London E; CW Daniel, Saffron Walden, Essex, UK.

Appendix III

Dosage

Dosage of plant juice, powder of plant parts, decoction and paste of the herb has been quoted as recommended by the Central Council for Research in Ayurveda & Siddha, New Delhi, Dr Priya Vrata Sharma in Dravya Guna Vijnaana, Vol II, and Hakeem Daljit Singh in Unani Dravyagunaadarsha.

For (suggested) dosage of dry extract, soft extract, liquid extract, decoction, infusion, powder and syrup, *Herbal Drug Industry,* Chief Editor: RD Chaudhary, Eastern Publishers, Greater Kailash II (Masjid Moth), New Delhi 110048, 1998, should be consulted. 195 plants have been covered.

In *Indian Herbal Pharmacopoeia,* jointly published by Regional Research Laboratory, CSIR, Jammu-Tawi and Indian Drug Manufacturers' Association, Mumbai, suggested dosage is based on the following sources:

Ayurvedic Pharmacopoeia of India. 1989. Ministry of Health, Government of India, New Delhi-110001.
British Herbal Compendium. 1993. British Herbal Medicine Association.
Handbook of Ayurvedic Medicinal Plants. 1990. Kapoor LD, CRC Press Inc., Boca Raton, Florida.
Herbal Medicines. 1996. CA Nowall et al, The Pharmaceutical Press, London.
Pharmaceutical Basis of Therapeutics. 1996. Goodman and Gilman. McGraw Hill, New York.
Pharmacopoeia of India. 1996. Ministry of Health, Government of India, New Delhi-110001.
Physicians Desk Reference (PDR). 1991.
Standard of Asian Herbal Medicine, Vol 1. 1993. Jakarta.

Information regarding the active life span of the herb after collection, called *veeryakaalaavadhi,* has been incorporated in *Vanoshadhi Nidarshikaa,* Ramsushil Singh, but the same has not been quoted. It is not that easy to predict shelf-life of a herb. Apart from protection against physical and chemical changes, the preservation against the insects or mould attack is also important. Most of the herbs need to be preserved at temperature between 1–5 degrees C. to prevent microbial growth. Details are available in Quality Control Methods for Medicinal Plant Materials, published by the World Health Organization, Geneva.

Appendix IV

Medicinal Plants Gardens and Herbariums

The fountainhead of information on medicinal plants is the Jawahar Lal Nehru Ayurvedic Plants Garden and Herbarium, Pune. The garden was taken over by the Central Council for Research in Ayurveda and Siddha (CCRAS) in 1978, and was restructured with an aim of collecting and cultivating medicinal plants, to carry out research on plant drugs, and to facilitate their proper identification and standardization.

Following medicinal plants are represented in the garden:

Abrus precatorius
Abutilon indicum
Acacia catechu
Acacia leucophloea
Acacia nilotica
Acacia sinuata
Acalypha indica
Achras zapota
Achyranthes aspera
Acorus calamus
Adansonia digitata
Adenanthera pavonia
Adhatoda zeylanica
Aegle marmelos
Ailanthus excelsa
Albizia lebbeck
Albizia procera
Aloe barbadensis
Alstonia scholaris
Anagalis arvensis
Andrographis paniculata
Annona squamosa
Aphanamixis polystachya
Argemone mexicana
Argyreia nervosa
Arisotolochia indica
Artabotrys hexapetalus
Artemisia nilagarica
Artocarpus lakucha
Artrocarpus heterophyllus
Asparagus adscendens
Asparagus racemosus
Azadirachta indica
Bacopa monnieri
Balanites aegyptiaca
Basella alba
Bauhinia purpurea
Bauhinia racemosa
Bauhinia tomentosa
Berberis aristata
Berberis lycium
Biophytum sensitivum
Bixa orellana
Blumea lacera
Boerhavia diffusa
Bombax ceiba
Boswellia serrata
Bryonopsis laciniosa
Buchanania lanzan
Bursera penicillata
Butea monosperma
Caesalpinia bonduc
Caesalpinia pulcherrima
Calotropis gigantea
Calotropis procera
Canna orientalis
Capparis zeylanica
Cardiospermum halicacabum
Carica papaya
Carissa congesta
Carvia callosa
Caryota urens

Cassia auriculata
Cassia fistula
Cassia sophera
Cassia surattensis
Cassia tora
Casuarina equisetifolia
Catharanthus roseus
Catunaregam spinosa
Catunaregam uliginosa
Cayretia carnosa
Celastrus paniculatus
Centella asiatica
Chenopodium ambrosioides
Cicca acida
Cinnamomum tamala
Cissampelos var. hirsuta
Cissus quadrangularis
Citrus aurantifolia
Clematis gouriana
Clematis hedysarifolia
Clematis triloba
Clerodendrum multiflorum
Clerodendrum serratum
Clitoria ternatea
Coccinia grandis
Cocculus hirsutus
Cocos nucifera
Coleus amboinicus
Commelina benghalensis
Commiphora agallocha
Commiphora wightii
Corallocarpus epigaeus
Corchorus trilocularis
Cordia dichotoma
Costus speciosus
Crateva magna
Crotalaria juncea
Cryptolepis buchanani
Curcuma aromatica
Curcuma longa
Cymbopogon citratus
Cymbopogon martinii
Cynodon dactylon
Cyperus rotundus
Dalbergia lanceolaria
Dalbergia latifolia
Dalbergia sissoo
Datura metel
Dendrophthoe falcata
Desmodium gangeticum
Digera muricata
Dioscorea bulbifera
Dioscorea esculenta
Dioscorea hispida
Dioscorea oppositifolia
Diospyros montana
Diospyros peregrina
Dodonaea viscosa
Dregea volubilis
Echinops echinatus
Eclipta prostata
Emblica officinalis
Enicostemma axillare
Ervatamia coronaria
Erythrina suberosa
Erythrina variegata
Eucalyptus
Euphorbia hirta
Euphorbia thymifolia
Euphorbia tirucallii
Evolvulus alsinoides
Ficus benghalensis
Ficus carica
Ficus hispida
Ficus racemosa
Ficus religiosa
Flacourtia indica
Garcinia indica
Gardenia resinifera
Gloriosa superba
Glossocardia bosvallia
Glycyrrhiza glabra
Gmelina arborea
Gossypium spp.
Grewia asiatica
Grewia tenax
Gymnema sylvestre
Haematoxylon campechianum
Helicteres isora
Hemidesmus indicus
Heterophragma quadriloculare
Hibiscus abelmoschus
Hibicus cannabinus
Hibiscus rosa-sinensis
Hiptage benghalensis
Holarrhena antidysenterica
Holoptelea intergrifolia
Homonoia riparia
Hyptis suaveolens
Indigofera glandulosa
Indigofera linifolia
Indigofera triata
Ipomoea indica
Ipomoea obscura

Ipomoea pescaprae
Jasminum humile
Jasminum officinale
Jatropha curcas
Jatropha nana
Kaempferia rotunda
Lantana camara
Lawsonia inermis
Leptadenia reticulata
Madhuca longifolia
Mangifera indica
Martynia annua
Mesua ferrea
Minosa pudica
Mirabilis jalapa
Morus alba
Mukia maderaspatana
Murraya koenigii
Nerium indicum
Nyctanthes arbor-tristis
Ocimum gratissimum
Oroxylum indicum
Paederia foetida
Passiflora edulis
Phyla nodiflora
Piper betel
Piper nigrum
Plumbago zeylanica
Pogostemon parviflorus
Pongamia pinnata
Psoralea corylifolia
Pterocarpus santalinus
Punica granatum
Rauvolfia serpentina
Ricinus communis
Ruta graveolens
Santalum album
Saraca asoca
Scirpus kysoor
Semecarpus anacardium
Sida acuta
Solanum surattense
Sonchus asper
Spilanthes acmella
Stenolobium stans
Stereospermum suaveolens
Syzygium cumini
Tectona grandis
Terminalia arjuna
Terminalia chebula
Tinospora cordifolia
Trichodesma indicum
Ixora coccinea
Jasminum multiflorum
Jasminum sambac
Jatropha gossypifolia
Kaempferia galanga
Kalanchoe pinnata
Lavandula bipinnata
Lepidagathis crisata
Limonia acidissima
Madhuca longifolia var. latifolia
Manihot esculenta
Melia azedarach
Michelia champaca
Mimusops elengi
Moringa oleifera
Mucuna pruriens
Muntinga calabura
Musa paradisiaca
Notonia grandiflora
Ocimum sanctum
Operculina turpethum
Oxalis corniculata
Pandanus tectorius
Phoenix sylvestris
Phyllanthus fraternus
Piper longum
Pithecellobium dulce
Plumeria rubra
Polyalthia longifolia
Psidium guajava
Pterocarpus marsupium
Pterospermum acerifolium
Quamoclit pennata
Rhynchosia minima
Rosa centifolia
Saccharum spontaneum
Sapindus trifoliatus
Sarcostemma acidum
Securinega leucopyrus
Sesbania sesban
Solanum nigrum
Solanum violaceum
Sopubia delphinifolia
Stachytarpheta jamaicensis var. indica
Sterculia foetida
Striga gesneroides
Tamarindus indicus
Tephrosia purpurea
Terminalia bellirica
Thevetia peruviana
Tribulus terrestris
Tylophora indica

Typha australis
Vernonia cinerea
Vigna trilobata
Vitex negundo
Woodfordia fruticosa
Valeriana wallichi
Vetiveria zizanioides
Viola serpens
Withania somnifera
Wrightia tinctoria

Plants, grown on a mass-scale in the Regional Research Centre, Jhansi, U.P., managed by CCRAS, include Abroma augusta, Aloe barbadensis, Asparagus racemosus, Glycyrrhiza glabra, Psoralia corylifolia, Rauvolfia serpentina, Uraria picta. Ledgerised medicinal plants, mentioned in Ayurvedic Formulary of India Part I, are also displayed in the Centre.

Rajasthan and Gujarat are two main Guggul (Commiphora wightii gum) producing states in India. Guggul herbal farm, Mangaliwas, near Ajmer in Rajasthan, is devoted to Guggul cultivation and propagation on a large scale.

In the Medicinal Plants Garden at Ranikhet, about 200 species are being maintained. Saffron Experimental Research Farm is situated in a western slope of Ranikhet.

Surveys on a state or regional basis and compilations on various plant sources of herbal drugs are available from CCRAS and CCRUM, Jawahar Lal Nehru Anusandhan Bhavan, Institutional Area, opp. D-Block, Janak Puri, New Delhi-110058.

A Medico-ethno-botanical survey programme is being carried out by the Medicinal Plants Units of CCRAS, functioning at five centres, or zones, located at Bangalore, Guwahati, Jhansi, Tarikhet and Pune. The survey party, a botanist, an Ayurvedist, a research assistant and other staff conducts survey tours almost all over the country, stretching from the Himalayas to the Kanyakumari in the South and from Gujarat to Kolkata in the East, besides covering the Andman and Nicobar and Lakshdeep Islands. It has covered 22 states and over 300 forest divisions of the country and collected over one lakh plant specimens which are housed in 18 Regional Herbariums and in the Central Herbarium and Museum. Some special survey tour programmes have also been conducted in the rich forest areas and tribal pockets to unearth the valuable data on the usage of medicinal plants by the tribes in various inaccessible forest areas. These areas include the Leh-Ladakh, Sikkim, Arunachal Pradesh, Khasia and Garo Hills, South Mirzapur, Panchmarhi, Amarkantak, Chitrakoot, Bastar, Nilgiris, Silent Valley, and the Andaman and Nicobar Islands.

Botanical Survey of India at P-8 Brabourne Road, Kolkata-700001, maintains the Central National Herbarium and Indian Botanic Garden at Kolkata. A directory of Botanical Gardens and Parks of India was published by BSI (1990).

National Botanical Research Institute (NBRI), Lucknow, U.P., is maintaining a well laid out botanical garden and a rich herbarium of plant specimens (totalling 2000 taxa), properly identified and classified.

The Botany section of the Central Drug Research Institute (CDRI), Lucknow, U.P., contains specimens of about 5000 species of medicinal plants.

Regional Research Laboratory (CSIR), Bhubaneshwar-751013, Orissa, is maintaining a herbarium which houses 7000 specimens from the Orissa region.

Many medicinal plants used in South India are being cultivated in the Herbal Garden of Arya Vaidyasala, Kottakkal-676503, Kerala.

Tamil Nadu Farms and Herbal Medicine Corporation Ltd., Arumbakkam, Chennai-600106, takes up cultivation of herbal plants commercially required for the production of herbal drugs.

Appendix V

Indian Systems of Medicine Research and Training Centres in India

Institute of Post-Graduate Training and Research in Ayurveda

The Institute was established by the Government of India at Gujarat Ayurveda University, Jamnagar, in 1956. Being an old and reputed institute, it holds a prestigious status of excellence in Ayurveda. It has produced a number of scholars who served and are serving the country through their expertise in Ayurveda. Institutes from countries like Japan, Italy, the Netherlands, Australia, Argentina, USA and Germany have entered into a memorandum of understanding with this Institute. The Institute has established an International Centre for Ayurvedic Studies which conducts a three-year introductory course in Ayurveda for foreign nationals. The Institute is well known for collaborative research projects with international organisations. The World Health Organisation established a collaborative unit for Indian systems of medicine in the Institute in 1997. The Institute conducts post-graduate courses in thirteen specialities of Ayurvedic pharmacy. It has state-of-the-art facilities in various specialities of Ayurveda.

Central Council for Research in Ayurveda and Siddha

The Central Council for Research in Ayurveda and Siddha (CCRAS) was established in 1969. It has forty-one units functioning all over the country. It is an apex body in India, fully devoted to research with regard to functions, coordination, development and promotion of Ayurveda and Siddha systems of medicine.

The research activities of the Council include clinical, drug, literary and family welfare research. As a result of these, the Council is successful in the development of various drugs and formulations such as Ayush-64 for malaria, Ayush-56 for epilepsy, Ayush-82 for diabetes mellitus, 777 oil for psoriasis, Poonimiali Chendooram for leucoderma and Pippalyaadi Yoga as an oral contraceptive. The Council has undertaken a medico-botanical survey and collected more than 120,000 plant specimens and published several monographs and books. It has been successful in obtaining fifteen process patents.

Pharmacopoeial Laboratory for Indian Medicine

The Pharmacopoeial Laboratory for Indian Medicine (PLIM) was established in the year 1970 as a standard-setting cum drug testing laboratory at the national level for drugs of Indian systems of medicine, i.e. the Ayurveda, Unani and Siddha systems. About 350 single drugs and 83 compound formulations of Ayurveda have been worked out, and 158 pharmacopoeial monographs have been published in Ayurvedic Pharmacopoeia of India part 1 (Vols I and II). PLIM also offers technical aid, advice and services in respect of standardization, testing and analyzing ISM drugs and medicinal plants.

National Institute of Ayurveda

The National Institute of Ayurveda was established in 1976 in Jaipur, Rajasthan. It is a Government of India establishment being developed as a state-of-the-art institute in the field of Ayurveda. The Institute is conducting under-graduate as well as post-graduate courses. The post-graduate courses are run in nine branches. Apart from conducting academic courses, the Institute is involved in various research activities. The areas of research include management of diabetes mellitus, vitiligo and other skin diseases, hypercholasteraemia and mental health. Many short-term research projects have been undertaken by the Institute. The Institute has a computerized document archive system for the preservation of rare documents and manuscripts. It provides speciality services in orthopaedics, cancer, Yoga and Panchkarma. Besides this, it also has TMT and audiometry units. The Institute is a regular training centre for reorientation programmes for teachers and physicians. It has entered into a memorandum of understanding with various drug-manufacturing firms.

Central Council for Research in Unani Medicine

The Central Council for Research in Unani Medicine (CCRUM) was established in March 1978. It has multifaceted research activities in the fields of clinical and drug research, survey, cultivation of medicinal plants, and literary research. Different phases of the clinical trials have been completed on 30 drugs so far. Seven drugs have been filed for grant of patent rights, while twenty-three drugs are in the process of patentization. Besides developing research drugs, the Council has also undertaken validation trials of twenty-six kit medicines for common ailments. Standardization work on forty-seven compound drugs and thirty-eight single drugs has been completed. About 5100 plant specimens have been collected, identified, processed and maintained on the herbarium sheets. The Council has recorded more than 1700 folklore claims from various tribal and other rural peoples in the course of field studies.

Rashtriya Ayurveda Vidyapeeth

Rashtriya Ayurveda Vidyapeeth (the National Academy of Ayurveda) (RAV) is an institution under the Ministry of Health and Family Welfare, Government of India, established in 1988 primarily to promote the knowledge of Ayurveda. The main objective of Vidyapeeth is to preserve the knowledge and skills of patient care possessed by traditional Vaidyas and eminent Ayurvedic scholars and arrange the transfer of this knowledge to the younger generation through the Indian traditional system of education.

To meet this primary objective, the Vidyapeeth runs two types of courses, where the selected students go to appointed Gurus (tradititonal scholars) and study the preferred texts of Ayurveda and acquired clinical acumen. The Institute also holds conferences and seminars every year, accords recognition and felicitation to eminent scholars of Ayurveda as Fellow of RAV for their significant contribution to the promotion of Ayurveda.

The Department of Indian Systems of Medicine and Homeopathy

A special department under the Ministry of Health and Family Welfare, Government of India, was established in March 1995. The department is functioning for the sole cause of promotion and development of all the systems of Indian medicine through various schemes and projects. Address: Indian Red Cross Building, 1 Red Cross Road, New Delhi-110 001.

Appendix VI

German Commission E Recommendations

Herb/Drug	Recommended for condition
Abies	Common cold
	Coughs and bronchitis
	Fevers and cold
	Inflammation of the pharynx
	Muscular and nerve pains
Achillea millefolium	Loss of appetite
	Dyspeptic complaints
	Liver complaints
	Gallbladder complaints
Aesculus hippocastanum	Venous conditions
	Chronic venous insufficiency
	Nocturnal cramps in the calves
	Pruritus and swellings of the legs
Allium cepa	Loss of appetite
	Dyspeptic complaints
	Common cold
	Coughs and bronchitis
	Fevers and colds
	Inflammation of the mouth and pharynx
	Hypertension
	Atherosclerosis
Allium sativum	Atherosclerosis
	Elevated levels of lipids in blood
	Common cold
	Coughs and bronchitis
Aloe vera	Constipation
Alpinia galanga	Loss of appetite
	Dyspeptic complaints
	Liver and gallbladder complaints
	Common cold
	Fevers and colds
	Cough and bronchitis
	Inflammation of the mouth and pharynx
Althaea officinalis	Coughs and bronchitis

Herb/Drug	Recommended for condition
Anethum	Loss of appetite Liver and gallbladder complaints Inflammation of pharynx Common cold, cough Bronchitis Fevers and colds
Angelica (European)	Loss of appetite Dyspeptic complaints Mild spasm of gastrointestinal tract
Artemisia	Loss of appetite Dyspeptic complaints Liver and gallbladder complaints
Asparagus officinalis	Infections of the urinary tract Kidney and bladder stones
Atropa	Arrhythmia Cardiac insufficiency (NYHA I and II) Nervous heart complaints Liver and gallbladder complaints
Barringtonia acutangula	Common cold Cough Bronchitis Inflammation of the mouth and pharynx Fever Diarrhoea
Betula	Infections of the urinary tract Kidney and bladder stones
Calendula officinalis	Inflammation of the mouth and pharynx Wounds and burns
Calendula officinalis	Inflammation of the mouth and pharynx Wounds and burns
Capsicum annuum	Muscle tensions Rheumatism
Carum carvi	Common cold Coughs and bronchitis Fevers and colds Inflammation of the mouth and pharynx Liver and gallbladder complaints Loss of appetite
Cichorium intybus	Loss of appetite Dyspeptic complaints Liver and gallbladder complaints
Cinnamomum camphora	Coughs and bronchitis Rheumatism Hypertension Hypotension Nervous heart complaints Arrhythmia Cardiac insufficiency NYHA I and II

Herb/Drug	Recommended for condition
Cinnamomum verum	Loss of appetite Dyspeptic complaints Coughs and bronchitis Fevers and cold Inflammation of the mouth and pharynx
Colchicum autumnale	Gout Mediterranean fever
Coriandrum sativum	Loss of appetite Dyspeptic complaints
Crataegus	Mild cardiac insufficiency (stage II NYHA)
Cucurbita pepo	Irritable bladder Prostate complaints
Curcuma domestica	Loss of appetite Liver and gallbladder complaints
Elettaria cardamomum	Loss of appetite Liver and gallbladder complaints Common cold Inflammation of the mouth and pharynx Coughs and bronchitis Fevers and colds
Eucalyptus globulus	Coughs and bronchitis Rheumatism
Foeniculum	Cough Bronchitis
Fumaria	Liver and gallbladder complaints
Gentiana lutea	Dyspeptic complaints Loss of appetite
Glycyrrhiza glabra	Coughs and bronchitis Gastritis
Hypericum perforatum	Anxiety Depressed moods Inflammation of the skin Blunt injuries Wounds and burns (first degree)
Hyssopus officinalis	Fevers and colds Liver and gallbladder complaints
Juniperus	Loss of appetite Infection of urinary tract Kidney and bladder stones
Lavandula angustifolia	Loss of appetite Nervousness and insomnia
Melilotus officinalis	Blunt injuries Haemorrhoids Venous conditions
Melissa officinalis	Nervous sleeping disorder Functional gastrointestinal complaints
Panax ginseng	Fatigue and debility During convalescence

Herb/Drug	Recommended for condition
Pimpinella anisum	Loss of appetite Liver and gallbladder complaints Common cold Cough, bronchitis
Plantago ispaghula	Constipation Diarrhoea Raised levels of cholesterol
Plantago lanceolata	Common cold Inflammation of the mouth and pharynx Cough, bronchitis Fevers and colds Inflammation of the skin
Raphanus sativus	Loss of appetite Fevers and colds Cough, bronchitis
Salix alba	Diseases accompanied by fever, rheumatic ailments, headaches
Santalum album	Cold, cough, bronchitis Fevers and colds Infections of the urinary tract Inflammations of the mouth and pharynx Liver and gallbladder complaints
Senna	Constipation
Tagetes erecta	Inflammation of the oral and pharyngeal mucosa Poor healing wounds Ulcus cruris
Taraxacum officinale	Dyspeptic complaints Liver and gallbladder complaints Disturbances of bile flow
Urginea maritima	Milder cases of heart insufficiency Reduced kidney capacity
Urtica	Rheumatism Diseases of lower urinary tract Kidney stones Prostate complaints Irritable bladder
Valeriana	Restlessness and sleeping disorders based on nervousness
Vitex agnus-castus	Menstrual complaints Mastodynia
Zea mays	BPH
Zingiber officinale	Dyspepsia Motion sickness

Credits

Picture Credits

ADPS: Drawings by Prof Vasudevan Nair, Palghat, Kerala, reproduced from Ayurvedic Drugs and their Plant Sources by V.V. Sivarajan and Indira Balachandran. Permission was given by Indira Balachandran.

CCRAS: Reproduced with permission from The Database series (5 volumes), published by the Central Council for Research in Ayurveda & Siddha, Department of Indian Systems of Medicine, Government of India, New Delhi.

WOI: Reproduced with permission from The Wealth of India (13 volumes), published by the National Institute of Science Communication, Council for Scientific & Industrial Research, Dr KS Krishnan Marg, New Delhi-110012.

ZANDU: Permission was granted by the Late Dr KM Parikh, the then Managing Director, Zandu Pharmaceuticals, Gokhale Road South, Mumbai-400 025. Reproduced in concurrence with Dr Girish G Parikh and Shri Bhaskar G Parikh.

Roman Spellings

Roman spellings of Ayurvedic synonyms, introduced for the first time by Rashtriya Ayurveda Vidyapeeth (National Academy of Ayurveda), New Delhi-110026, have been followed. Asiatic Society's markings are now obsolete.

Roman spellings of Siddha synonyms, adopted by Central Council for Research in Ayurveda and Siddha, New Delhi, have been followed instead of those incorporated in Siddha Formulary of India.

Index

B

C

D

M

N

O

T

About the Author

C. P. Khare

C.P. Khare was born on December 18 1932 and has been a herbalist for the last fifty years.

He was born into a family of herbal physicians. He studied the original Ayurvedic texts and acquired first-hand knowledge of the cultivation and processing of herbs and their usage even before going to college.

After graduation he took up journalism in 1952. Since then he has edited and contributed many features on health and personal problems in various journals.

He has been the director of one of the biggest publishing houses in India for the last twenty-five years—publishing twenty one magazines in seven languages (English, Hindi, Marathi, Gujarati, Kannada, Tamil and Telugu), commanding a total circulation of more than 4 million.

As a Founder President of the Society for New Age Herbals in Delhi, which started functioning in 1990, he sought the cooperation and participation of physicians of modern medicine, traditional medicine, pharmacologists and scientists to modify, restructure and re-examine the age-old herbal formulations in the light of pharmacological research and clinical trials being undertaken in various research institutes in India.

He has also been included among the directors of Dabur Ayurvedic Specialities Ltd., a herbal major in India.

This work is the result of the last ten years' efforts.

MIX
Papier aus verantwortungsvollen Quellen
Paper from responsible sources
FSC® C105338

If you have any concerns about our products,
you can contact us on
ProductSafety@springernature.com

In case Publisher is established outside the EU,
the EU authorized representative is:
Springer Nature Customer Service Center GmbH
Europaplatz 3, 69115 Heidelberg, Germany

Printed by Libri Plureos GmbH
in Hamburg, Germany